DC/AC Circuits and Electronics:

Principles & Applications

Robert J. Herrick

Purdue University
West Lafayette, Indiana

THOMSON

DELMAR LEARNING   Australia Canada Mexico Singapore Spain United Kingdom United States

DC/AC Circuits and Electronics: Principles & Applications

By Robert J. Herrick

Executive Director:
Alar Elken

Executive Editor:
Sandy Clark

Senior Acquisitions Editor:
Gregory L. Clayton

Senior Development Editor:
Michelle Ruelos Cannistraci

Executive Marketing Manager:
Maura Theriault

Channel Manager:
Fair Huntoon

Marketing Coordinator:
Brian McGrath

Executive Production Manager:
Mary Ellen Black

Production Manager:
Larry Main

Senior Project Editor:
Christopher Chien

Art/Design Coordinator:
David Arsenault

Technology Project Manager:
David Porush

Technology Project Specialist:
Kevin Smith

Senior Editorial Assistant:
Dawn Daugherty

Library of Congress Cataloging-in-Publication Data:

ISBN: 0-7668-2083-1

NOTICE TO THE READER

Publisher does not warrant or guarantee any of the products described herein or perform any independent analysis in connection with any of the product information contained herein. Publisher does not assume, and expressly disclaims, any obligation to obtain and include information other than that provided to it by the manufacturer.

The reader is expressly warned to consider and adopt all safety precautions that might be indicated by the activities herein and to avoid all potential hazards. By following the instructions contained herein, the reader willingly assumes all risks in connection with such instructions.

The publisher makes no representation or warranties of any kind, including but not limited to, the warranties of fitness for particular purpose or merchantability, nor are any such representations implied with respect to the material set forth herein, and the publisher takes no responsibility with respect to such material. The publisher shall not be liable for any special, consequential, or exemplary damages resulting, in whole or part, from the readers' use of, or reliance upon, this material.

Contents

5
KCL–Kirchhoff's Current Law

11
Series-Parallel Circuits

13
Capacitance and Reactance

14
RC Switching Circuits

15
Wave Shaping and Generation

19
Dependent Sources

Preface

Introduction to the Herrick & Jacob Series

The traditional approach to teaching fundamentals in Electronics Engineering Technology begins with a course for freshmen in DC Circuit Theory and Analysis. The underlying laws of the discipline are introduced and a host of tools are presented and applied to very simple resistive circuits. This is usually followed by an AC Circuit Theory and Analysis course. All of the topics, rules, and tools of the dc course are revisited, but this time using trigonometry and complex (i.e., real and imaginary) math. Again, applications are limited to passive (simple) resistor, capacitor, and inductor circuits. It is during this second semester, but often not until the third semester, that students are finally introduced to the world of *electronics* in a separate course or two. At this point they find what they have been looking for: amplifiers, power supplies, waveform generation, feedback to make everything behave, power amps, and radio frequency with communications examples. This approach has been in place for *decades*, and is the national model.

So what is wrong with this approach? Obviously, it has been made to work by many people. Currently, during the dc and ac courses students are told not to worry about why they are learning the material, only that they will have to remember and apply it later (provided they survive). These two courses (dc and ac) have become tools courses. A whole host of techniques are taught, one after the other, with the expectation (often in vain) that eventually when (or if) they are ever needed, the student will simply *remember*. Even for the most gifted teachers and the most dedicated students, these two courses have become "weed out" classes, where the message seems to be one of "if you show enough perseverance, talent, and faith, we will eventually (later) show you the *good* stuff (i.e., the electronics)." Conversely, the electronics courses are taught separately from the circuit analysis classes. It is expected that the student will quickly recall the needed circuit analysis tool (learned in the dc and ac courses taken several semesters before) when it is needed to understand how an amplifier works or a regulator

xvii

is designed. This leads to several results, all of which have a negative effect. First, the student sees no connection between dc and ac circuit theory and electronics. Each is treated as a separate body of knowledge, to be memorized. If the students hang on long enough, they will eventually get to the electronics courses where the circuits do something useful. Second, the teachers are frustrated because students are bored and uninterested in the first courses, where they are supposed to learn all of the fundamentals they will need. But when the students need the information in the later electronics courses, they do not remember. The result is a situation with which we have struggled for decades. This new series has been developed to address and solve this problem.

The Herrick & Jacob series offers a different approach. It integrates circuit theory tools with electronics, interleaving the topics as needed. Circuit analysis tools are taught on a *just-in-time* basis to support the development of the electronics circuits. Electronics are taught as applications of fundamental circuit analysis techniques, not as unique magical things with their own rules and incantations. Topics are visited and revisited in a helical fashion throughout the series, first on a simple, first approximation level. Later, as the students develop more sophistication and stronger mathematical underpinnings, the complex ac response and then the nonsinusoidal response of these same electronics circuits are investigated. Next, at the end of their two years of study, students probe these central electronic blocks more deeply, looking into their nonideal behavior, nonlinearity, responses to temperature, high power, and performance at radio frequencies, with many of the parasitic effects now understood. Finally, in the Advanced Analog Signal Processing book, Laplace transforms are applied to amplifiers, multistage filters, and other closed-loop processes. Their steady state and transient responses, as well as their stability, are investigated.

The pervading attitude is "Let's do interesting and useful things right from the start. We will develop and use the circuit theory as we need it. Electronics is *not* magic; it only requires the rigorous use of a few fundamental laws. As you (the student) learn more, we will enlarge the envelope of performance for these electronic circuits as you become ready." Learn and learn again. Teach and teach again. Around and around we go, spiraling ever upwards.

Introduction to This Text

The *DC/AC Circuits and Electronics: Principles and Applications* text is the first book in an integrated series of electronics engineering technology texts. This first text applies a new and innovative approach of incorporating basic laws and circuit analysis techniques to electronic circuits as well as elementary passive circuits. Instead of studying endless not-real-world circuits, the student is introduced to and begins to gain command of useful dc and ac electronic circuits that include components with diodes, LEDs, BJTs, MOSFETs, and op amps. Initially, the ideal and near-ideal characteristics of electronic devices are used to develop basic circuit analysis techniques and a global perspective of electronic circuit operations. As this textbook and the textbook series progress, additional device characteristics and details are introduced as needed.

Circuit laws and analysis techniques are integrated through a specific sequence and with a variety of applications. A greater distinction of the laws is made with entire chapters dedicated to Kirchhoff's current law, Kirchhoff's voltage law, and then Ohm's law for resistance. Only circuit theorems and analysis techniques that are practical and typically applied in real-world circuit analysis are incorporated. The philosophy of this textbook and this series is a just-in-time-to-apply approach to optimize the students' learning experience and sustain a high level of interest. Electronic circuits serve as examples to develop analysis skills that apply to current and future technology.

This pedagogical approach has been used very successfully for several years by the author and several other faculty at Purdue University. Students are engaged and able to enjoy electronics much sooner than in traditional approaches. With this new pedagogical approach of engaging students early in their field of interest, Professor Herrick and his colleagues have experienced an *exceptional retention rate* of students in their program.

Organization of the Text

The organization of this textbook and this series is unique and has proven very successful in engaging and retaining students. The emphasis is on problem solving through foundational principles, laws, and solid circuit analysis techniques. The *thought process* is the key to understanding current and especially future technology.

Features

Pedagogy

Each chapter begins with *performance-based learning objectives*. These are rigorously implemented throughout that chapter.

There are hundreds of *examples,* each providing fully worked-out and explained numerical illustration of the immediately preceding information. Every example is followed by a practice exercise that requires the reader to extend the techniques just illustrated with a different set of numbers, perhaps in a slightly different direction. Numerical results are given to allow students to check the accuracy of their own work.

This text uses ideal and nearly ideal components. Later texts in the series explore component specifications in greater detail; however, the ability to find *specifications*, parts, and vendors is a skill central to electronics technicians. For that reason, full data sheets are not typically provided in this textbook series. The reader should explore the web sites of semiconductor manufacturers and vendors, developing a set of bookmarks unique to his or her own needs.

The wide margins have been reserved to serve as visual keys for important information contained in the adjacent text. Significant effort has been made to keep these margins uncluttered. The occasional marginal figure or table supports the text beside it. Icons, notes, equations, and key specifications are placed so that the reader can quickly glance along the margins to find the appropriate details explained in the adjacent text. This is very much the way a good student annotates a book while studying. Plenty of room has been left for precisely that purpose.

Computer Simulation

Electronics Workbench's *MultiSIM*® and *Cadence PSpice*® are used to illustrate simulations of circuit and electronic applications. There are considerable differences between the two packages. Each has its own strengths and weaknesses. To rely only on one is to mislead the reader. This first textbook utilizes both simulators but primarily, because of its ease of use, MultiSIM is the primary simulator used. Later textbooks in the series use both simulators, depending upon which is the most effective for the application. Attempting to teach electronics without a simulation package seriously handicaps the reader, especially in nonlinear operations. Appropriate simulation is essential.

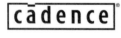

The simulations are *integrated* throughout the text, at the points in the instruction where they are needed to illustrate a point. They are *not* just included at the end of each chapter as add-ons, as if an afterthought. The appropriate icon is placed in the margin beside each example, as a key. Most of the simulations are MultiSIM, but several Cadence PSpice simulations are provided to show some feature differences. Files for these simulations are available on the accompanying CD.

Many of the end-of-chapter problems can be solved or the manual calculations verified by using one of the simulation software packages.

For faculty adopting this text, these files are also available.

The education version of MultiSIM 2001 and the evaluation version of Cadence PSpice 9.1 are used to illustrate the simulations in this text-book.

MultiSIM is a product of Electronics Workbench.
http://www.electronicsworkbench.com

Cadence information is available from
http://www.orcad.com

Problems

End-of-chapter problems are organized in sections that correspond to the sections within the chapter. Problems are frequently presented in pairs. Answers to the odd-numbered problems are given at the end of the text. Readers must solve the even-numbered problems without a target answer.

CD

The textbook edition of MultiSIM is provided on the accompanying CD. Also the MultiSIM and Cadence PSpice files used to create the examples and figures throughout the text are included there.

Supplements

Online Companion

Visit the textbook's companion Web site at

http://www.electronictech.com

There you will find:

MultiSIM and Cadence PSpice files for the examples and problem solutions.

A link to automated homework: a unique set with random values is generated each time you access the problems. Your answers are checked as soon as you enter the result of each step.

Over thirty downloadable PowerPoint presentations.

Link to the author's Purdue University course Web site.

Textbook updates.

Instructor's Guide

The instructor's guide contains solutions to end-of-chapter problems.
ISBN: 140185253X

Laboratory Exercises

A separate combination laboratory manual and workbook that follows the pedagogy of this textbook is available through Delmar Thomson Learning. ISBN: 1401880401

About the Author

Robert J. Herrick is the Robert A. Hoffer Distinguished Professor of Electrical Engineering Technology for teaching at Purdue University and is currently serving as the department head of that department. Teaching is his passion, education his profession. He has been awarded national, regional, university, school, and department awards for outstanding teaching. His awards include the HP Outstanding Laboratory Instruction Award, the ASEE Illinois-Indiana Outstanding Teaching Award, the Purdue University Murphy Award for Outstanding Undergraduate Teaching, the School of Technology Dwyer Award for Outstanding Undergraduate Teaching, and the EET department CTS Electronics Award for Outstanding Undergraduate Teaching. Professor Herrick has also taught and served as department head at the University of Toledo, and taught at Western Michigan University. His full-time engineering experience includes 13 years of design and development of electronic switching systems at Bell Telephone Laboratories and ITT Advanced International Technology Center. He has also served as an engineering consultant for several major industries in the areas of controls, instrumentation, technical publications, and computer software design and development, and served as an electronics technician of ground-based navigational aid systems in the United States Air Force.

His 23 years of accumulated engineering and engineering technology education experience have been dedicated to student learning. He is a charter member of the Purdue University Teaching Academy and served for several years on its first executive board. The Purdue Teaching Academy is dedicated to the advancement of outstanding teaching and education at Purdue University. Professor Herrick has been a national leader with IEEE and ASEE, including leadership and development of the international Frontiers in Education Conference, which is dedicated to the advancement of engineering and engineering technology education in the United States and throughout the world.

Professor Jacob, a primary author in this textbook series, introduced Professor Herrick to the world of cycling in 1990. Since then Professor Herrick has logged well over 50,000 miles, many of those with Professor Jacob. Many of the teaching concepts and techniques, as well as the idea for this text series, were germinated during their long rides in the countryside while occasionally bouncing creative ideas off each other.

Acknowledgements

The author and Delmar Learning wish to express their gratitude to the reviewers and production team that made this textbook possible. These include:

Venkata Anandu	Southwest Texas State, San Marcos, TX
Sohail Anwar	Penn State, Altoona, PA
G. Thomas Bellarmine	Florida A & M University, Tallahassee, FL
John Blakenship	DeVry University, Atlanta, GA
Joseph Booker	DeVry University, Addison, IL
Harold Broberg	Indiana University/Purdue University Indianapolis (IUPUI), Indianapolis, IN
Robert Fladby	University of Nebraska, Omaha, NE
Yolanda Guran	Oregon Institute of Technology, Portland, OR
Robert Hofinger	Purdue University School of Technology, Columbus, IN
Sang Lee	DeVry University, Addison, IL
John Meese	DeVry University, Columbus, OH
Jim Pannell	DeVry University, Irving, TX
Parker Sproul	DeVry University, Phoenix, AZ
Don Zinger	Northern Illinois University, De Kalb, IL

The author would like to give special thanks to Professor J. Michael Jacob for authoring the last chapter of this textbook, his copyedit work in polishing the textbook, his plethora of ideas he shared with the author in the creation of this textbook, and his dedication to our shared idea of this textbook series. The author also gives special thanks to Kenneth Eichenberger, a recent graduate of Purdue University, for working out and verifying all of the homework questions and solutions and Harold Meaux, a microelectronics manufacturing master technician, for verifying Ken Eichenberger's work and computerizing the solutions.

The author would also like to thank Delmar Learning for their belief in and commitment to this project, particularly Greg Clayton, acquisition editor, Michelle Ruelos Cannistraci, development editor, and the rest of their team, including Larry Main, Christopher Chien, and David Arsenault. Their willingness to venture down untraditional paths in pedagogy and text production has provided the essential elements necessary to convert this unique idea into a viable contribution in the spread of learning.

Avenue for feedback

No system ever performs exactly as intended. In fact the quality of the results is often a function of the quality of the negative feedback. So please contact either the author or the publisher with your suggestions for improvements:

Robert J. Herrick
1415 Knoy Hall of Technology
Purdue University
W. Lafayette, IN 47907-1415
rherrick@purdue.edu

Michelle Ruelos Cannistraci
Delmar Learning–Executive Woods
5 Maxwell Drive
Clifton Park, NY 12065-2919
michelle.cannistraci@delmar.com

Dedication

This book is personally dedicated to the many people who have made a difference: to my wonderful wife Becky, who has served lots of sloppy joes to my students at home help sessions; to my daughter Jennifer Herrick Tucker (an author of children's books), who along with my wife has endured many years of the university teaching lifestyle to let me follow my life's mission of teaching; to my good friend, cycling buddy, fellow author, and my catalyst for pushing my envelope, Professor J. Michael Jacob; to my lifelong friends and soul mates who have made such a great difference in my family's life, Harold and Janice; to my brother Bud, who lent me his awesome headphones to get through many a long night writing; to Danny and Kathy, who dare to dream to change the world; and to my students and all students who are investing in themselves and their future by working diligently today for tomorrow's success. And ultimately I wish to thank God for giving me the opportunity to serve so many, who are now or soon will be making a difference in our world.

Robert J. Herrick (Bob) June, 2002

1

Units and Number Notation

Introduction

Electricity and electronic technologies coupled with semiconductor material development are the foundation for today's high tech world. To understand the electrical and electronic technologies, you must have command of their fundamental language of units and numerical representation. You must be able to accurately manipulate mathematical expressions. You must express numeric values in an appropriate format with appropriate units.

This chapter establishes the fundamentals of units, number notation, numerical precision and accuracy, powers of 10, and their mathematical operations, unit conversion, and unit prefixes.

Objectives

Upon completion of this chapter you will be able to:

- Explain the need for units.

- Distinguish between and mathematically manipulate exact and measured numbers.

- State and interpret the meaning of significant figures.

- Apply unit identities to convert between units of measure.

- Use powers of 10 to express numeric values and perform mathematics.

- Express numbers in scientific and engineering notation.

- Define and effectively use unit prefixes.

1.1 Units of Measurement

What? Always identify *what* you are working with, investigating, or solving for. For example, if someone offers you 1000 for this book you had better ask, "1000 what?" Is that 1000 cents, 1000 pesos, 1000 francs, 1000 yen, or 1000 bananas; just what are you being offered?

Units of measure describe characteristics such as distance, volume, force, time, and temperature. In the United States, people are most familiar with the **English system** of units that includes the units of feet, gallons, pounds, degrees Fahrenheit, and seconds. Others throughout the world use the metric system of units that includes units of meters, liters, kilograms, degrees centigrade, and seconds. The **SI (International System)** system of units is now a world standard mostly based upon the metric system of units including the units of meters, liters, kilograms, degrees Kelvin, and seconds. The SI system is the standard of the Institute of Electrical and Electronics Engineers, Inc. (**IEEE**) and the American National Standards Institute (**ANSI**). For your convenience and reference, Appendix B provides tables of the systems of units. This text utilizes the SI system of units and introduces these units appropriately throughout the text as needed.

Be aware that, as a professional, you need to work with a variety of units. As economical cultures continue to cross-pollinate, the mixture of unit usage is a real issue. Units are critical; this text clearly establishes units and carries units throughout calculations. You must establish this habit as well. Numbers without units are meaningless. Blindly assuming units or carelessly handling units in mathematical calculations leads to errors.

English and SI systems of units

IEEE and ANSI standards

1.2 Significant Figures

Values are either **exact** or **measured** (approximate). For example, there are exactly 100 centimeters (abbreviated "cm") in a meter (abbreviated "m"). If a value is measured, then two key characteristics become important, namely, accuracy and precision.

Accuracy refers to the closeness of the measurement to the actual value being measured. The closer the measured value is to the actual value, the more accurate the measured value is. Accuracy depends upon the quality of the measuring instrument, the procedure, and the operator. You can express how accurate a measurement is by calculating its **error**. You can calculate the actual error, the difference between measured value

Exact or measured?

and actual value; however, it is typically more meaningful to express the error relatively as a **percent error**, comparing the measured value with the actual or expected value.

$$\% \ error = \frac{\text{measured value} - \text{expected value}}{\text{expected value}} \times 100 \%$$

Measured % error

The lower the percent error, the more accurate the measurement is. If the measured value is the same as the actual value, the percent error is 0% and there is no error in the reading. A positive percent error indicates that the measured value is greater than the actual value; while a negative value indicates the measured value is less than the actual value.

Example 1-1

I measured the width of the cover of this textbook with a metric ruler to be 20.7 cm. If the textbook publisher states that it is actually 20.5 cm, what is my percent error?

Solution

$$\% \ error \ = \ \frac{20.7 \ cm - 20.5 \ cm}{20.5 \ cm} \times 100\% = 0.98\%$$

The error is about one percent. It is also positive, indicating that the measured value is greater than the expected value. The cause of the error could be a faulty metric ruler or operator error (my measurement technique or my eyesight).

Practice

a. Use an English ruler and measure the length of this textbook page. Write down your answer.

b. Assuming the textbook page length is actually 9.25″, find your measured percent error.

c. Is your percent error positive or negative? What does that indicate?

Precision indicates how many **significant figures** (or digits) are assigned to a measured value. For example, 20.7 cm has three significant figures and indicates a more precise value than 21 cm. A measured value of 20.7 cm indicates a range of possible values:

Precision (significant figures)— number of meaningful digits

$$20.65 \ cm \ \leq \ \text{measured value} \ < \ 20.75 \ cm$$

A value of 20.65 cm or greater would **round up** to 20.7 cm, while a value less than 20.75 cm would **round down** to 20.7 cm. More sophisticated techniques of rounding off numbers are used to statistically analyze volumes of data but this book utilizes the simple technique just described.

Digits 1 through 9 are always considered **significant digits** or significant figures, while the significance of the digit 0 depends upon its use. **Leading zeros**, that is, zeros that occur before the first occurrence of a 1-9 digit, are not significant (for example, 0.0007 has one significant digit). Zeros that occur between digits 1-9 are significant (for example, 207 has three significant digits). Zeros that occur *after* a decimal point and after a 1-9 significant digit are significant (for example, 0.7000 has four significant digits). Table 1-1 demonstrates the proper use of significant figures (or digits).

Table 1-1 Significant figure examples

Number	Significant Figures	Rationale
7	1	one significant digit
74	2	two significant digits
704	3	zero between significant digits is significant
7000	1	trailing 0's are not significant without "."
70.000	5	trailing 0's are significant due to "."
0.007	1	leading 0's are not significant
0.7300	4	trailing 0's significant due to "."

Note: always express a decimal number with a **leading zero**; for example, "0.7", not ".7". Otherwise it is too easy for the decimal, without a leading "0", to be overlooked or lost in photocopying.

Example 1-2

Find the number of significant figures for the following numbers: 25, 2500, 25010, 0.00730, and 0.007301.

Solution

25	2 significant figures
2500	2 significant figures
25010	4 significant figures
0.00730	3 significant figures
0.007301	4 significant figures

Practice

Find the number of significant figures for the following numbers: 0.0425, 0.042050, 2000, and 2010.

Answer: 3, 5, 1, and 3 significant figures

When making measurements, *record your answers to the level of precision actually measured*. For example, I measured the textbook cover width to three significant figures, 20.7 cm. Three significant figures is reasonable based upon my measuring instrument, a standard metric ruler. A stated measured value of 20.6999 cm would be inappropriate. That implies that I measured to a precision of 6 significant figures. That is not possible with a standard metric ruler and my eyesight. Express your measured values to the appropriate precision. Do not understate or overstate the precision. Expressing my measurement as 21 cm is understating the precision; 20.6999 is overstating the precision; and 20.7 cm is the appropriate precision.

Typically, an analog meter (a continuous meter with a pointer) can be read to a precision of 2 or 3 figures. A digital meter displays the number of significant figures based upon the range selected. Always select the appropriate meter range to provide the most precise measurement, that is, the most sensitive range possible that does not produce an out-of-range measurement. When using a digital meter, only include the digits that are steady; do not include the least significant digits if they are changing.

Typically, when performing theoretical calculations, carry out your work to four significant figures. After several calculations in a series of calculations, round-off errors will affect the resulting least significant digit.

When adding and subtracting measured values, the number with the least resolution determines the answer's resolution. For example, the hundred's place has less resolution than the ten's place; the ten's place has less resolution than the unit's place, the unit's place has less resolution than the tenth's place, and so on.

Example 1-3

Measured values of 123 cm and 1.1 cm are added. What is the resulting sum?

Solution

123	cm	unit's place resolution
+ 1.1	cm	tenth's place resolution
124	cm	rounded to unit resolution

Unit's place resolution is the best we can do on this measured sum. Thus, the resulting sum of 124.1 cm is rounded to the unit's place.

Practice

Measured values of 9.2 cm, 4.82 cm, and 1.001 cm are added. What is the sum of the measured values?

Answer: 15.0 cm   (resolution to the tenth's place)

When multiplying and dividing numbers, the answer has the same number of significant figures as the multiplicand number with the lowest number of significant figures. Thus, the precision of a product of measured values is limited to the least precise measurement.

Example 1-4

Calculate the area of a rectangle based upon its measured sides of 123 cm and 1.1 cm.

Solution

123	cm	3 significant figures
× 1.1	cm	2 significant figures
140	cm^2	rounded to 2 significant figures

The product of 135.3 cm^2 is rounded up to 140 cm^2 since its precision is limited to 2 significant figures.

Practice

A bicyclist is measured to travel 34.65 meters in 3.25 seconds. What is his velocity based upon the measured values?

Answer: 10.7 meters per second   (three significant figures)

1.3 Powers of 10

In electronics, values can be very large or very small. Powers of 10 are a convenient way to express such values. For example, expressing the value 0.000001 as a decimal number can be very confusing, especially if you are attempting to convey it verbally. You must learn to use the powers of 10 effectively to perform the basic mathematical operations used in electronics.

Powers of 10 Greater Than 1

The powers of 10 for values of 1 or greater are shown in Table 1-2 for the values through 1,000,000. Note that n in 10^n represents the number of times 10 is multiplied times itself. Also, its resulting value is the value 1 with n trailing 0's.

Table 1-2 Powers of 10 greater than 1

Ten's Number	Ten's Product	Value	Number of Trailing 0's
10^0	1	1	0
10^1	10	10	1
10^2	10 x 10	100	2
10^3	10 x 10 x 10	1,000	3
10^4	10 x 10 x 10 x 10	10,000	4
10^5	10 x 10 x 10 x 10 x 10	100,000	5
10^6	10 x 10 x 10 x 10 x 10 x 10	1,000,000	6

A very handy technique is to start with the value 1.0 and move the decimal point n places to the right. For example, if $n = 2$, start with 1.0 and move the decimal point 2 places to the right increasing the value to 100.

Powers of 10 Less Than 1

The powers of 10 for values of one or less (decimals) are shown in Table 1-3 for values through 0.000001. Note that n in 10^{-n} represents the number of times 1 is divided by 10. Its value is 1.0 with the decimal point moved n places to the left. For example, if $n = -3$, then start with 1.0 and move the decimal point 3 places to the left, decreasing the value to 0.001.

Table 1-3   Powers of 10 less than 1

Ten's Number	Ten's Divided	Value	Number of Decimal Moves Left
10^{-0}	1	1	0
10^{-1}	$1 \div 10$	0.1	1
10^{-2}	$1 \div 10 \div 10$	0.01	2
10^{-3}	$1 \div 10 \div 10 \div 10$	0.001	3
10^{-4}	$1 \div 10 \div 10 \div 10 \div 10$	0.0001	4
10^{-5}	$1 \div 10 \div 10 \div 10 \div 10 \div 10$	0.00001	5
10^{-6}	$1 \div 10 \div 10 \div 10 \div 10 \div 10 \div 10$	0.000001	6

Expressing Numbers as Powers of 10

You can express numbers as powers of 10, for example, 120,000 can be expressed as 12×10^4. This represents the number 12 with four trailing 0's. It also represents the number 12 with the decimal point moved four places to the right.

$$120,000 = 12 \times 10,000 = 12 \times 10^4$$

The number 0.005 can be expressed as 5×10^{-3}. This represents the number 5 with the decimal place moved three places to the left.

$$0.005 = 5 \times 0.001 = 5 \times 10^{-3}$$

A number may be expressed in many ways as a function of 10. Table 1-4 demonstrates how 120,000 may be expressed as powers of 10.

Table 1-4   Expressing 120,000 as powers of 10

1.2	x 10^5
12	x 10^4
120	x 10^3
1200	x 10^2
12000	x 10^1
120000	x 10^0

Your scientific calculator has an <**EE**> or an <**EEX**> key that allows you to input numbers as powers of 10. The EE key provides the function of $\times 10^n$, where n can be a positive or a negative exponent.

The calculator representation of "5E3" is the number 5×10^3.

The calculator representation of "5E–3" is the number 5×10^{-3}.

When keying in a negative exponent, be sure to use the **unary negative key**, not the subtraction key.

Example 1-5

Key 5000 into the calculator using the <EE > key.

Solution

To key this number into the calculator, key in

5   <EE>   3   <Enter>

The results displayed is 5000 if the calculator is in the standard display mode.

Practice

With your scientific calculator in the standard display mode, enter the value 0.004 using your <EE> key, that is, 4E–3.

Answer: 0.004

1.4 Powers of 10 Algebra

Multiplication of Powers of 10

The product of powers of 10 is a power of 10 with the original exponents added.

Exponents add

$$10^m \times 10^n = 10^{m+n}$$

Note: m and n can be positive or negative numbers.

Example 1-6

The following are examples of powers of 10 multiplication.

$$10^2 \times 10^3 = 10^{2+3} = 10^5$$

$$10^6 \times 10^{-2} = 10^{6-2} = 10^4$$

Practice

Find $10^9 \times 10^{-2} \times 10^{-3}$

Answer: 10^4

Division of Powers of 10

The division of powers of 10 is a power of 10 with the original exponents subtracted. The denominator's exponent is subtracted from the numerator's exponent.

Exponents subtract

$$\frac{10^m}{10^n} = 10^{m-n}$$

Note: m and n can be positive or negative numbers.

Example 1-7

The following are examples of powers of 10 division.

$$\frac{10^5}{10^2} = 10^{5-2} = 10^3$$

$$\frac{10^5}{10^{-2}} = 10^{5-(-2)} = 10^7$$

Practice

Find $\dfrac{10^7}{10^4}$, $\dfrac{10^{-7}}{10^{-4}}$

Answer: 10^3, 10^{-3}

Reciprocals of Powers of 10

The reciprocal of a power of 10 is a power of 10 with the original exponent negated.

$$\frac{1}{10^n} = 10^{-n}$$

Negative of exponent

Note: n can be a positive or negative number. Also, note that you can express 1 as 10^0 and apply the division rule to obtain the reciprocal rule.

$$\frac{1}{10^n} = \frac{10^0}{10^n} = 10^{0-n} = 10^{-n}$$

Example 1-8

The following are powers of 10 reciprocal examples:

$$\frac{1}{10^2} = 10^{-2}$$

$$\frac{1}{10^{-4}} = 10^4$$

$$10^{-6} = \frac{1}{10^6}$$

Powers of Powers of 10

A power of 10 raised to a power is a power of 10 raised to the product of the original exponents.

Exponents multiply

$$(10^m)^n = 10^{mn}$$

Note: m and n can be positive or negative numbers.

Example 1-9
The following are examples of powers of 10 raised to a power:

$$(10^2)^3 = 10^{(2)(3)} = 10^6$$

$$(10^2)^{-3} = 10^{(2)(-3)} = 10^{-6}$$

$$(10^{-8})^{-1/2} = 10^{(-8)(-1/2)} = 10^4$$

Practice
Find $(10^6)^{-1/3}$, $(10^{-1})^{-1}$

Answer: 10^{-2}, 10^1

Addition and Subtraction of Powers of 10

Adding or subtracting numbers represented as powers of 10 requires first that each have the same exponent. Then the numbers are added and subtracted. The exponent of the power of 10 is the same as the original exponent.

Numbers add or subtract
Exponents must be the same

$$(A \times 10^n) \pm (B \times 10^n) = (A \pm B) \times 10^n$$

Note: n can be a positive or negative number but must be the same value for all numbers being added or subtracted.

Example 1-10

The following are examples of adding and subtracting numbers expressed as powers of 10:

$$(4\times10^{-2})+(5\times10^{-2}) = (4+5)\times10^{-2} = 9\times10^{-2}$$

In the following example, the first number must be modified to match exponents before subtraction can be performed.

$$(120\times10^{1})-(5\times10^{2})$$

The 10^{1} term must be converted to 10^{2} before subtraction.

$$120\times10^{1} = 12\times10^{2}$$

$$(12\times10^{2})-(5\times10^{2}) = (12-5)\times10^{2} = 7\times10^{2}$$

Practice

Find $(4\times10^{3})+(5\times10^{2})$

Answer: 45×10^{2}

Multiplication of Powers of 10

The product of numbers expressed in powers of 10 is a number expressed as a power of 10. Its number is the product of the original numbers. Its 10's exponent is the sum of the original exponents.

$$(A\times10^{m})\times(B\times10^{n}) = (A\times B)\times10^{m+n}$$

Note: m and n can be positive or negative numbers.

Numbers multiply
Exponents add

Example 1-11

The following are examples of multiplication of numbers expressed as powers of 10:

$$(7\times10^{2})(5\times10^{4}) = (7\times5)\times10^{(2+4)} = 35\times10^{6}$$

$$(7\times10^{-2})(5\times10^{3}) = (7\times5)\times10^{(-2+3)} = 35\times10^{1}$$

Practice

Find $(-4\times10^{5})\times(5\times10^{-2})$

Answer: -20×10^3

Division of Powers of 10

The division of numbers expressed in powers of 10 is a number expressed as a power of 10. Its number is the original numerator divided by the original denominator. Its 10's exponent is the original denominator exponent subtracted from the original numerator exponent.

Numbers divide
Exponents subtract

$$\frac{A \times 10^m}{B \times 10^n} = \frac{A}{B} \times 10^{m-n}$$

Example 1-12

The following are examples of division of numbers expressed as powers of 10:

$$\frac{18 \times 10^7}{3 \times 10^2} = \frac{18}{3} \times 10^{7-2} = 6 \times 10^5$$

$$\frac{6 \times 10^4}{2 \times 10^{-2}} = \frac{6}{2} \times 10^{4-(-2)} = 3 \times 10^6$$

Practice

Find $\dfrac{-28 \times 10^2}{7 \times 10^7}$

Answer: -4×10^{-5}

Application of the Rules of Powers of 10

The following example demonstrates powers of 10 application rules.

Example 1-13

Resolve the following expression by use of the rules of the powers of 10.

$$\left[\frac{(2 \times 10^{-3} + 4 \times 10^{-3})^2}{2 \times 10^2} \right]^{-2} \qquad \text{Original problem}$$

Solution

$$\left[\frac{(6\times10^{-3})^2}{2\times10^2} \right]^{-2}$$
Numerator summed

$$\left[\frac{6^2\times10^{-6}}{2\times10^2} \right]^{-2}$$
Numerator ×10 raised to power

$$\left[\frac{36\times10^{-6}}{2\times10^2} \right]^{-2}$$
6 squared

$$\left[18\times10^{-8} \right]^{-2}$$
Divided

$$18^{-2}\times10^{16}$$
Raised to power

Use the calculator to solve this as shown below. Note that (−) represents the unary negative key not the subtraction key. The ^ key represents *raising a number to the power of*. The keystrokes are:

18 ^ (−) 2 <EE> 16 <Enter>

The result is

$$3.086\text{E}13 \quad \text{or} \quad 3.086\times10^{13}$$

Practice

Find $$\left[\frac{(8\times10^2 - 6\times10^2)^3}{(2\times10^2)^{-8}} \right]^{-2}$$

Answer: 2384×10^{-54}

1.5 Scientific and Engineering Notation

As the names indicate, scientific notation is typically used by science and engineering notation is utilized primarily by engineering and engineering technology. The usefulness of engineering notation will become apparent in the next section in determining unit prefixes.

Scientific Notation

Scientific notation expresses a value as a power of 10 such that the numeric part is greater than or equal to 1 but less than 10. That is, the numeric part always has **one and only one digit left of the decimal point**. The form of scientific notation is

$1 \leq N < 10$

$$N \times 10^n$$

where

- N is a number between 1 and 10, that is $1 \leq N < 10$.

- N must have one and only one digit left of the decimal point.

- The number of significant figures of N corresponds to its original value's number of significant figures.

- The exponent n is an appropriate number to force the number N to be a value between 1 and 10.

- The exponent n can be a positive or a negative number.

Example 1-14

Express the number 123 in scientific notation.

Solution

1 2 3.

You need to move the decimal point 2 places to the left to get only one digit left of the decimal point, that is, from 123 to 1.23. To compensate, add 2 to the 10's exponent, that is, from 10^0 to 10^2.

$$1.23 \times 10^2 \qquad \text{Scientific notation}$$

Practice

Express 12,300,000 in scientific notation.

Answer: 1.23×10^7

Example 1-15

Express the number 0.00012 in scientific notation.

Solution

You need to move the decimal point 4 places to the right to get only one digit left of the decimal point, that is, from 0.00012 to 1.2. To compensate, subtract 4 from the 10's exponent, that is, from 10^0 to 10^{-4}.

$$1.2 \times 10^{-4} \qquad \text{Scientific notation}$$

0.00012

Practice

Express 0.00000012 in scientific notation.

Answer: 1.2×10^{-7}

Engineering Notation

Engineering notation expresses a number as a power of 10 such that the numeric value is greater than or equal to 1 but less than 1000, and the 10's exponent is a multiple of three. The form of engineering notation is

$$N \times 10^n$$

where

1≤N<1000
n is a multiple of 3

- N is a number between 1 and 1000, that is $1 \leq N < 1000$.

- The number of significant figures of N corresponds to its original value's number of significant figures.

- The exponent n must be a multiple of 3 and is an appropriate value to force the number N to be a value between 1 and 1000.

- The exponent n can be a positive or negative number.

Example 1-16

Express the number 12,300 in engineering notation.

Solution

You need to move the decimal point 3 (a multiple of 3) places to the left to get a number between 1 and 1000, that is, from 12,300 to 12.300.

To compensate, add 3 to the 10's exponent, that is, from 10^0 to 10^3.

$$12.3 \times 10^3 \qquad \text{Engineering notation}$$

Practice

Express 12,300,000 in engineering notation.

Answer: 12.3×10^6

Example 1-17

Express the number 0.00012 in engineering notation.

Solution

You need to move the decimal point 6 (a multiple of 3) places to the right to get a number between 1 and 1000, that is, from 0.00012 to 120. Note: *you had to pad with a trailing 0 to create the appropriate number*.

To compensate, subtract 6 from the 10's exponent, that is, from 10^0 to 10^{-6}.

$$120 \times 10^{-6} \qquad \text{Engineering notation}$$

Practice

Express 0.00000012 in engineering notation.

Answer: 120×10^{-9}

1.6 Unit Prefixes

Unit prefixes are a simple way to communicate very large and very small unit values. As humans we like to communicate in words. Expressing simple values like 10 seconds is very comfortable and meaningful for us; but, how about 0.000001 seconds? We realize this is a very small amount of time. But how small is it? It is difficult to visualize.

Expressing extreme numbers in scientific notation or engineering notation is more helpful. For example, the above value would be expressed as 1×10^{-6} seconds. It is also very helpful to establish names that represent large and small unit values. These names help us more effectively visualize and communicate values. The name "micro" stands for the value 10^{-6}. Thus 1×10^{-6} second can be written as 1 microsecond. The term 10^{-6} is replaced directly with the word "micro".

The unit prefix naming convention adopted for the SI system of units is based upon engineering notation. Look at Table 1-5. Use uppercase symbols for mega and above; use lowercase letters for kilo and below. The symbol μ for the prefix micro is a Greek letter. Be careful to clearly print the prefix symbols μ (*long leading line*) for micro (10^{-6}), lowercase m for *milli* (10^{-3}), and uppercase M for *mega* (10^{6}). Otherwise you will convey incredibly inaccurate information.

Table 1-5 SI unit prefixes

Prefix Name	Prefix Symbol	Value
tera	T	10^{12}
giga	G	10^{9}
mega	M	10^{6}
kilo	k	10^{3}
-	-	10^{0}
milli	m	10^{-3}
micro	μ	10^{-6}
nano	n	10^{-9}
pico	p	10^{-12}
femto	f	10^{-15}

Example 1-18

Express 1 kilogram in the basic unit of grams.

Solution

- 1 kilogram = 1 kg

- Directly replace "k" with its equivalent value of 10^3.

- $1 \text{ kg} = 1 \, (10^3) \text{ g} = 1 \times 10^3 \text{ g} = 1000 \text{ g}$

- Note: "k" is written in the lowercase. Do not use uppercase.

Practice

Express 2 megagrams in the basic unit of grams.

Answer: 2×10^6 g

Example 1-19

Express 2 milliseconds in the basic unit of seconds.

Solution

- 2 milliseconds = 2 ms

- Directly replace "m" with its equivalent value of 10^{-3}.

- $2 \text{ ms} = 2 \, (10^{-3}) \text{ s} = 2 \times 10^{-3} \text{ s} = 0.002 \text{ s}$

Practice

Express 2 nanoseconds in the basic unit of seconds.

Answer: 2×10^{-9} s

Example 1-20

Express 120,000,000 seconds with the proper prefix.

Solution

- Express in engineering notation: 120×10^6 s

- Replace 10^6 with the mega "M" prefix: 120 Ms

Example 1-21

     Express 0.00034 seconds with proper prefix.

Solution

- Express in engineering notation: 340×10^{-6} s

- Replace 10^{-6} with the micro "μ" prefix: 340 μs

1.7 Unit Prefix Conversion

Sometimes it is necessary to convert from one unit prefix to another. Remember a prefix is just a name for an engineering notation power of 10 whose exponent is a multiple of 3.

Example 1-22

     Convert 0.00003 kilograms to milligrams.

Solution

- Move decimal point to the right in groups of 3 digits and convert to the next prefix until the number is between 1 and 1000.

- Start with original value.   0.00003 kg

- Move decimal point 3 places right.   0.03 g

- Move decimal point 3 places right.   30 mg

- Note as the number gets larger, the prefix gets smaller.

Practice

     Convert 0.00000003 kilograms into micrograms.

Answer: 30 μg

Example 1-23

     Convert 4,000,000 picoseconds to milliseconds.

Solution

- Move decimal point to the left in groups of 3 digits and convert to the next prefix until the number is between 1 and 1000.

- Start with original value.   4,000,000 ps

- Move decimal point 3 places left: 4,000 ns

- Move decimal point 3 places left: 4 μs

- Move decimal point 3 places left: 0.004 ms

- Note as number gets smaller, the prefix gets larger.

Practice

Convert 50,000 milligrams into kilograms.

Answer: 0.05 kg

1.8 Unit Consistency and Conversions

You must take units very seriously as noted earlier in this chapter. It is sometimes necessary to convert to other units within the same system of units or to convert a value to a different system of units. You must carry units throughout your calculations to ensure accurate units in the resulting answer. Calculations without units are not meaningful.

To convert to a new unit, you must multiply by an equivalent value of 1 or a **unity multiplier**, for example, 60 seconds per minute. The following converts 5 minutes (min) to seconds (s):

$$time = 5 \text{ min} \times \left(\frac{60 \text{ s}}{\text{min}} \right) = 300 \text{ s}$$

Notice the unity multiplier of 60 seconds per minute does not change the value of the answer. It only changes the value's units from "minutes" to "seconds". In the unity term, "minutes" is strategically placed in the denominator to cancel the original "minutes" in the numerator. And the equivalent number of "seconds" per minute is placed in the numerator. The "minutes" cancel, and the answer is left in units of "seconds".

Example 1-24

You are offered a major contract at 2 cents per millisecond for 2 hours. The actual job cost is $100,000. Will you make money? Prove your answer. Use unit conversion, multiplying by a conversion factor that is an identity. Minutes are abbreviated "min" and hour is abbreviated "hr".

Solution

Note: units are set up in the conversion factor so as to cancel out the unit you want to eliminate. For example, if you want to cancel out a unit in the numerator then you must multiply by a conversion factor with that unit in the denominator.

$$\frac{2 \text{ cents}}{1 \text{ ms}} \times \left(\frac{1\$}{100 \text{ cents}}\right) \times \left(\frac{1000 \text{ ms}}{\text{s}}\right) \times \left(\frac{60 \text{ s}}{\text{min}}\right) \times \left(\frac{60 \text{ min}}{\text{hr}}\right) \times 2 \text{ hr} = \$144,000$$

Looks like an excellent deal, a profit of $44,000 in 2 hours.

Practice

Convert 3 cents per µs to dollars per minute.

Answer: 1.8×10^6 dollars per minute

Example 1-25

Convert 230,000 microseconds to seconds.

Solution

Use a unity multiplier to convert from µs to s.

$$230,000 \text{ µs} \times \left(\frac{10^{-6} \text{ s}}{\text{µs}}\right) = 0.23 \text{ s}$$

Practice

Convert 30,000,000 nanoseconds to seconds.

Answer: 0.03 s

Example 1-26
Convert 0.00005 grams to micrograms.

Solution

Use a unity multiplier to convert from g to µg.

$$0.00005 \text{ g} \times \left(\frac{10^6 \text{ µg}}{\text{g}} \right) = 50 \text{ µg}$$

Practice
Convert 0.0000007 seconds to nanoseconds.

Answer: 700 ns

Example 1-27
Convert 670,000,000 milligrams to kilograms.

Solution

Use a unity multiplier to convert from mg to g, then another unity multiplier to convert from g to kg.

$$670,000,000 \text{ mg} \times \left(\frac{10^{-3} \text{ g}}{\text{mg}} \right) \times \left(\frac{\text{kg}}{10^3 \text{ g}} \right) = 670 \text{ kg}$$

Practice
Convert 70,000 nanoseconds to microseconds using two unity multipliers: (from ns to s) and (from s to µs).

Answer: 70 µs

The next example demonstrates converting between systems of units, from metric to English.

Example 1-28
The Arecibo National Astronomy and Ionosphere Center in Puerto Rico has the most sensitive astronomical telescope in the world. The diameter of its spherical shaped reflector is 305 meters, the largest antenna dish in the world. Convert the diame-

ter expressed in the SI units of meters to the English system units of feet.

$$305 \text{ m} \times \left(\frac{100 \text{ cm}}{1 \text{ m}} \right) \times \left(\frac{1 \text{ inch}}{2.54 \text{ cm}} \right) \times \left(\frac{1 \text{ foot}}{12 \text{ inch}} \right) = 1000 \text{ feet}$$

Note that each fraction is a unity and its denominator is set up to eliminate (convert) a unit.

1.9 Order of Mathematical Operations

Introductory circuit analysis requires a fundamental knowledge of algebra and the manipulation of mathematical expressions. You must follow the rules (hierarchy) of mathematical operations. These are also the rules followed by a computer or a calculator when evaluating numerical expressions.

The following is an ordered list of mathematical operators with the highest priority operators first.

1. Unary negative -5 *not subtraction*

2. Powers and roots 10^2 $\sqrt{2}$

3. Multiplication and division 5×7 $12/3$

4. Addition and subtraction $5 + 2$ $5 - 3$

Note: Parentheses (...) are resolved before applying its resulting value.

The following example is a step-by-step reduction of a complicated formula. Each of the above rules is applied in hierarchical order until the problem is resolved to an answer. Intermediate calculations are shown to the right to help clarify intermediate steps. Simplified expressions are shown to the left after intermediate calculations are performed, eventually resulting in simple division and result.

Example 1-29

Solve $\dfrac{(-4)^2 + (2\sqrt{100} - 8)}{-2^2}$

$(-4)^2 = 16$

$\sqrt{100} = 10$

$2^2 = 4$

Now: $\dfrac{16 + (2 \times 10 - 8)}{-4}$

$(2 \times 10 - 8) = 12$

Now: $\dfrac{16 + 12}{-4}$

$16 + 12 = 28$

Simplifying: $\dfrac{28}{-4} = -7$

Summary

This chapter established the fundamentals of units, number notation, numerical precision and accuracy, powers of 10 and their mathematical operations, unit conversion, and prefix symbols.

The need for knowing and using units effectively was emphasized. Exact values (e.g., 12 per dozen) were compared to measured values (approximations). Measured values introduce error. Accuracy is the term that describes how close the measured value is to the actual or expected value. Percent error indicates what percent the measured value is relative to the expected value. The precision of a measured value is characterized by its number of significant figures. More significant figures indicate greater precision.

If measured values are added or subtracted, the result's resolution (least significant place) is limited to the measurement with the worst resolution. If measured values are multiplied or divided, the result's precision (number of significant figures) is limited to the measured value with the least precision (the measurement with fewest significant figures).

Powers of 10 express very large and very small numbers. Numbers expressed as powers of 10 can be added, subtracted, multiplied, divided, raised to a power, and the results expressed as a power of ten.

- **Add**: Match the exponents, then add the numbers. The resulting exponent is the matched exponent.

- **Subtract**: Match the exponents, then subtract the numbers. The resulting exponent is the matched exponent.

- **Multiply**: Multiply the numbers and add the exponents.

- **Divide**: Divide the numbers and subtract the exponents (numerator exponent – denominator exponent)

- **Power**: Raise the number to the power with multiplied exponents.

Power of 10 numbers are usually expressed in scientific notation form or in engineering notation form. Scientific calculators can display numbers in both of these formats. Unit prefixes provide names for the commonly used engineering notation powers of 10 (e.g., milli for 10^{-3} and micro for 10^{-6}). Unity multipliers are used to convert between units and prefixes. It is critical to carry units throughout your calculations and properly express values with the correct units and proper prefix.

The next chapter introduces the foundational electric phenomenon of current and voltage. These are the primary parameters used to describe electrical and electronic circuit behavior.

Problems

Units of Measurement

1-1 Find and list the units of length in the English, Metric MKS, Metric CGS, and SI systems of units. (See Appendix B)

1-2 Find and list the units of temperature in the English, Metric MKS, Metric CGS, and SI systems of units. (See Appendix B)

Significant Figures

1-3 A known length of 15 meters is measured to be 14.8 meters. Find the percent error of the measurement.

1-4 An expected time of 15 seconds is measured to be 15.3 seconds. Find the percent difference of the measurement.

1-5   Find the number of significant figures for each of the following.
   a. 100
   b. 101
   c. 0.1
   d. 0.001000

1-6   Find the number of significant figures for each of the following.
   a. 1000
   b. 1001
   c. 0.0001
   d. 0.00010.

1-7   What is the precision of a typical metric ruler?

1-8   What is the precision of a typical thermometer?

1-9   A measured value is 11 m. What are its possible minimum and maximum values?

1-10   A measured temperature is 25°C. What are its possible minimum and maximum values?

1-11   Find the sum of two measured values: $11.1 + 9$.

1-12   Find the difference of two measured values: $11.1 - 9$.

1-13   Find the product of two measured values: 80×12.

1-14   Find the quotient of two measured values: 155/5.

Powers of 10

1-15   Express each as a power of 10:
   a. 1
   b. 100
   c. 10,000
   d. 0.1
   e. 0.001
   f. 0.00001

1-16   Express each as a power of 10:
   a. 10
   b. 1,000
   c. 100,000
   d. 0.01
   e. 0.0001
   f. 0.0000000001

1-17   Express each as a power of 10 (a whole number without trailing zeros times a power of 10):

     a.   50

     b.   120,000

     c.   0.043

     d.   0.00007

1-18   Express each as a power of 10 (a whole number without trailing zeros times a power of 10):

     a.   430

     b.   80,000

     c.   0.93

     d.   0.000101

Powers of 10 Algebra

1-19   Perform the following as powers of 10:

     a.   (100)(1000)

     b.   (0.001)(10,000)

     c.   100,000/100

     d.   100/0.0001

     e. $(100)^3$

     f. $(0.001)^2$

     g. $-(10^2)^2$

     h. $(-0.01)^2$

1-20   Perform the following as powers of 10:

     a. $100,000 \times 100$

     b. 100×0.00001

     c. $10,000 \div 1,000$

     d. $0.001 \div 0.00001$

     e. $(1000)^2$

     f. $(0.01)^3$

     g. -100^{-2}

     h. $(-0.001)^{-2}$

1-21   Perform the following as powers of 10:

     a. $(2 \times 10^2) + (4 \times 10^2)$

     b. $(3 \times 10^3) + (3 \times 10^2)$

     c. $(4 \times 10^4) \times (2 \times 10^2)$

     d. $(4 \times 10^5) \times (3 \times 10^{-1})$

     e. $(12 \times 10^5) \div (2 \times 10^2)$

     f. $(12 \times 10^4) \div (2 \times 10^{-2})$

g. $(4\times10^3)^2$

h. $(2\times10^{-2})^3$

1-22   Perform the following as powers of 10:

a. $(2\times10^{-2}) + (4\times10^{-2})$

b. $(3\times10^{-3}) + (3\times10^{-2})$

c. $(4\times10^4) \times (2\times10^{-2})$

d. $(4\times10^{-5}) \times (3\times10^{-1})$

e. $(12\times10^{-5}) \div (2\times10^2)$

f. $(12\times10^{-4}) \div (2\times10^2)$

g. $(4\times10^3)^{-2}$

h. $(2\times10^{-2})^{-3}$

1-23   Perform the following as powers of 10:

a. $\dfrac{800^3 (0.01)^2}{20^2}$

b. $\dfrac{-(2)^{-2}\, 30^2}{20^{-5}}$

1-24   Perform the following as powers of 10:

a. $\dfrac{40^3 (0.001)^2}{20^2}$

b. $\dfrac{-(2)^{-3}\, 50^2}{40^{-5}}$

Scientific and Engineering Notation

1-25   Express in scientific notation:

a.   3400

b.   0.00560

1-26   Express in scientific notation:

a.   340,000

b.   0.0000560

1-27   Express in engineering notation:

a.   1200

b.   340,000

c.   0.0560

d.   0.00078

1-28   Express in engineering notation:
- **a.** 12,000
- **b.** 3,400,000
- **c.** 0.0050
- **d.** 0.00007

Unit Prefixes

1-29   Express with the most appropriate prefix:
- **a.** 1,200 s
- **b.** 3,400,000 s
- **c.** 0.0560 s
- **d.** 0.00078 s

1-30   Express with the most appropriate prefix:
- **a.** 12,000 s
- **b.** 340,000,000 s
- **c.** 0.00050 s
- **d.** 0.00000007 s

Unit Prefix Conversion

1-31   Convert to the indicated prefix:
- **a.** 120,000 s to ks
- **b.** 120,000 s to Ms
- **c.** 120,000 s to Gs
- **d.** 0.00003 s to ms
- **e.** 0.00003 s to μs
- **f.** 0.00003 s to ns

1-32   Convert to the indicated prefix:
- **a.** 45,000 s to ks
- **b.** 45,000 s to Ms
- **c.** 45,000 s to Gs
- **d.** 0.06 s to ms
- **e.** 0.06 s to μs
- **f.** 0.06 s to ns

1-33   Convert to the indicated prefix:
- **a.** 120,000 ks to Ms
- **b.** 120,000 ks to Gs
- **c.** 0.03 ms to μs
- **d.** 0.03 ms to ns

1-34   Convert to the indicated prefix:
    a.   1,200,000 ks to Ms
    b.   1,200,000 ks to Gs
    c.   0.0003 ms to μs
    d.   0.0003 ms to ns

Unit Consistency and Conversions

1-35   A corporation earns 3 cents per microsecond per 8 hour work day. How many dollars per work day is that?

1-36   A corporation earns $100,000,000 per 40 hour work week. How many cents per millisecond is that?

1-37   How long is a 100 yard football field in centimeters?

1-38   How many inches is a 100 meter swimming pool?

1-39   Convert to the indicated prefix using a *unity multiplier*:
    a.   120,000 s to ks
    b.   120,000 s to Ms
    c.   0.00003 s to μs
    d.   0.00003 s to ns

1-40   Convert to the indicated prefix using a *unity multiplier*:
    a.   4,500,000 s to Ms
    b.   4,500,000 s to Gs
    c.   0.06 s to ms
    d.   0.06 s to μs

1-41   Convert to the indicated prefix using a *unity multiplier*:
    a.   120,000 ks to Ms
    b.   0.03 ms to μs

1-42   Convert to the indicated prefix using a *unity multiplier*:
    a.   12,000,000 ks to Gs
    b.   0.0003 ms to ns

Order of Mathematical Operations

1-43   Evaluate and express in engineering notation

$$\frac{(2 \times \sqrt{25+75} + 5)^6}{(-5)^2 \, (2+3)^2}$$

1-44   Evaluate and express in engineering notation

$$\frac{(5+6 \times 5) \, (\sqrt{256}\,)^{1/2}}{-(-10)^4}$$

2

Current, Voltage, and Common

Introduction

The atom is the source of electric charge. Electric charge is the source of current and voltage, the two basic parameters of electricity. You must gain a fundamental understanding of current and voltage and how to properly measure them.

Objectives

Upon completion of this chapter you will be able to:

- Explain basic atomic structure and charged particle behavior.

- Describe the three types of basic electrical material: conductor, insulator, and semiconductor.

- Describe how current is created and calculate current based upon charge flow.

- Describe how voltage is created and calculate voltage based upon charge difference.

- Draw and describe how current meters and voltmeters are connected to a circuit.

- Draw the schematic symbols of and describe the three types of commons: earth, analog, and chassis return commons.

- Draw the schematic symbol of and apply ideal voltage supplies in circuits.

- Convert from a voltage supply schematic symbol to voltage supply bubble notation, and vice versa.

- Calculate node voltages and voltage drops in circuits.

2.1 Atomic Structure and Charge

To understand the source of electricity, electronic component behavior, and electronic circuit action, you need an elementary understanding of the atom and its structure. The Bohr model of the atom provides the simplest useful model. An atom is the smallest particle of an element (like copper or carbon) that still retains its properties.

The center or **nucleus** of the atom consists of electrically neutral particles called **neutrons** and positively charged particles called **protons**. Negatively charged particles called **electrons** orbit around the nucleus and are restricted to certain distances from the nucleus, forming bands of orbiting electrons as seen in Figure 2-1, a representation of the helium atom with 2 protons, 2 neutrons, and 2 electrons. The helium atom is assigned the **atomic number** of two since it has two protons.

The orbiting electrons are very light compared to the heavier protons and neutrons. Protons and neutrons are about 1836 times heavier than the electron. The heavy nucleus in an electrically conducting material like copper tends to stay and vibrate in place while the lighter, orbiting electrons can easily be pulled away from the nucleus and are then free to travel through the electrical material. The nucleus of an atom occupies less than 1% of its volume. The electrons orbit at relatively great distances from the nucleus, much like orbiting planets around a sun. The atom is mostly space.

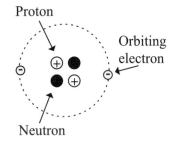

Proton

Orbiting electron

Neutron

Figure 2-1
Helium atom (not to scale)

Electrostatic Charge

When a glass rod is rubbed with a silk cloth, **electrostatic** (stationary) charge is developed on the silk cloth and on the glass rod. The silk cloth is deemed to have accumulated electrons from the glass rod, leaving a net **negative charge** on the silk cloth and a net **positive charge** on the glass rod. The following observations have been made after silk cloths have been rubbed on glass rods:

- The silk cloth is now attracted to the glass rod

- Two silk cloths repel each other

- Two glass rods repel each other

This observation of electrostatic charge was the beginning of the understanding of electricity. If two oppositely charged objects are placed too close to each other, the electrons from one will fly to the other object causing an arc of electricity. This is an electrostatic discharge, something we have all experienced, for example arcing from your hand to a door-knob on a very dry day. The air breaks down and the electrons move in the same direction (**electrodynamic** or moving charge). Lightning is a form of electrostatic discharge.

Charge

Charge is the most fundamental electrical property of matter. Again, the source of electrical charge is the electron and the proton of the atom. The electron and proton are of *equal* but *opposite* charge. The electron is considered to have a **negative charge**; the proton, a **positive charge**. A **balanced atom** has an equal number of electrons and protons, thus the atom does not have a net charge. The atom with all of its neutrons, protons, and electrons is charge neutral.

Figure 2-2
Opposite charges
attract
(pull together)

Opposite charges attract each other and *like charges repel each other* as demonstrated in Figures 2-2 and 2-3. Charge is assigned the symbol Q. The unit of charge is the **coulomb**, abbreviated **C**. A coulomb is equivalent to a charge created by 6.242×10^{18} electrons (or protons). Thus, the number of electrons per coulomb is expressed as

$$6.242 \times 10^{18} \ \frac{e}{C}$$

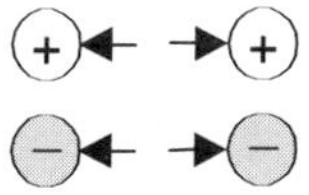

where

    e = number of electrons or protons
    C = coulomb, the unit of charge

Figure 2-3
Like charges *repel*
(push apart)

The reciprocal of the number of electrons per coulomb yields the charge per electron or proton.

$$1.602 \times 10^{-19} \ \frac{C}{e}$$

Example 2-1
    Find the charge in coulombs of 200 electrons created by rubbing a glass rod with silk.

Solution

$$Q = (200\,e) \times \left(1.602 \times 10^{-19} \ \frac{C}{e} \right) = 3.204 \times 10^{-17} \ C$$

Practice
    Find the number of electrons generating 0.3 coulombs of charge.

Answer: 1.873×10^{18} electrons

Atomic Balancing Forces

Since opposites attract, why do the lightweight, orbiting electrons not just fly into the nucleus, attaching themselves to the protons and destroying the atom? This is not a pleasant thought if these are your atoms.

The answer is that electrons also have mass as well as electrical charge. The electrical force of attraction between opposite charges (the electrons and protons) is balanced by an equal and opposite mechanical centrifugal force acting on the mass of the electrons to pull them out of orbit.

Both of these balancing forces are needed. If the electrical force were not present to pull the electrons toward the nucleus, the centrifugal force would cause all the orbiting electrons to fly off into space. That would also destroy the atom.

Life is a balance of forces. In the atom, the electrical force between the nucleus and orbiting electrons is balanced by the centrifugal force of the mass of the orbiting electrons.

Coulomb's Law

Coulomb's law states the **electrical force** of attraction or repulsion between two charges Q_1 and Q_2 is:

$$F = \frac{kQ_1Q_2}{r^2} \tag{2-1}$$

where

F = force expressed in newtons (N)

k = constant value of 9.0×10^9 Nm^2 / C^2

Q_1 = first charge expressed in coulombs (C)

Q_2 = second charge expressed in coulombs (C)

r = distance between the charges expressed in meters (m)

Example 2-2

For Figure 2-4, find the force of electrical repulsion between a positive charge of 2 coulombs and a positive charge of 3 coulombs that are 2 millimeters apart.

Solution

First, compare the units expressed in the problem with Equation 2-1 that is used to solve this problem. Do the units match Equation 2-1? The constant k has predetermined units. The two charges of Q_1 and Q_2 are specified in coulombs (the proper units of charge). The distance r is expressed in millimeters, which must be converted to the proper units of meters.

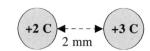

Figure 2-4
Repelling charges of Example 2-2

$$r = 2 \text{ mm} = 2\ (10^{-3})\ \text{m}$$

Now the units are consistent with Equation 2-1 and we can directly substitute into Equation 2-1 to solve for the electrical force of repulsion.

$$F = \frac{\left(9.0 \times 10^9\ \dfrac{\text{Nm}^2}{\text{C}^2}\right)(2\ \text{C})(3\ \text{C})}{(2 \times 10^{-3}\ \text{m})^2}$$

Simplify the expression by multiplying (2C) (3C) to get $6C^2$. Remember you must also multiply the units.

$$F = \frac{\left(9.0 \times 10^9\ \dfrac{\text{Nm}^2}{\text{C}^2}\right)(6\ \text{C}^2)}{(2 \times 10^{-3}\ \text{m})^2}$$

Multiply and divide the numeric values and simplify to the following expression.

$$F = 1.35 \times 10^{16} \frac{\left(\dfrac{\text{Nm}^2}{\text{C}^2}\right)(\text{C}^2)}{\text{m}^2}$$

Cancel the C^2 terms and simplify to the following expression.

$$F = 1.35 \times 10^{16} \frac{\text{Nm}^2}{\text{m}^2}$$

Cancel the m^2 terms and simplify to the following expression.

$$F = 1.35 \times 10^{16} \text{ N}$$

All the units *properly cancelled* to produce the correct resulting unit of force, the newton.

Practice

Find the electrical force of attraction between a positive charge of 4 microcoulombs and a negative charge of 6 microcoulombs that are 3 centimeters apart.

Answer: 240 N

Electron Shells and Subshells

Orbiting electrons are restricted to specific distances from the nucleus, forming **shells** around the nucleus. **Subshells** of electrons are formed within the shells. The regions between shells are called **forbidden regions** in which electrons cannot orbit. Figure 2-5 illustrates these principles for the nickel atom.

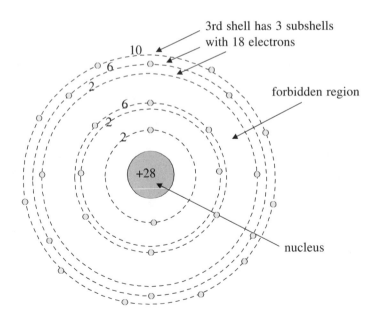

Figure 2-5   Nickel atom with atomic number of 28 (not to scale)

The *maximum number of electrons in a shell* is calculated by the formula $2N^2$.

- 1st shell with $N=1$ has a maximum of 2 electrons

- 2nd shell with $N=2$ has a maximum of 8 electrons

- 3rd shell with $N=3$ has a maximum of 18 electrons

Also, notice from Figure 2-5 that this creates an orderly pattern of the maximum electrons in the subshells (*4 more than the preceding subshell*):

- 1st subshell has a maximum of 2 electrons

- 2nd subshell has a maximum of 6 electrons

- 3rd subshell has a maximum of 10 electrons

Valence Band

The outermost shell or band of orbiting electrons is called the **valence band**. The electrons in this band are called **valence band electrons**. It is this outer shell that interacts with the outer shells of other atoms to form molecules. Since it is the valence band that characterizes the interaction of atoms to form molecules, we find it convenient to represent the nucleus and the inner shells as the **core** of the atom. Then the core and the valence band are used to represent a simplified view of the atom when studying its interaction with other atoms, as seen in Figure 2-6.

2.2 Conductor, Insulator, Semiconductor

Electrical material is categorized as a conductor, an insulator, or a semiconductor based upon how easily the material's electrons can be extracted from their parent atoms by external energy and how easily free electrons flow through a material.

Conductor

Metals such as copper, silver, and gold are classified as **conductors** because very little energy (often heat) is needed for their outer electrons to leave their parent atoms and move freely through the material. These conductive elements have a common characteristic, a single electron in

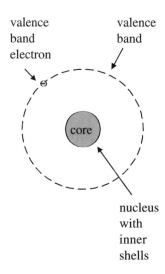

Figure 2-6
Valence band

their outer shell or outer subshell. Figure 2-7 represents such an atom, copper, which is an excellent conductor.

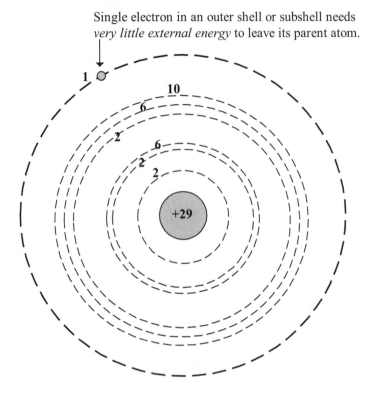

Figure 2-7 Copper, a good conductor with one valence band electron

Free Electrons

The electrical force on the outer electron of the copper atom by its nucleus is relatively weak. Recall that the electrical force on two charges is inversely proportional to the square of their distance. Also, there are several orbiting electrons between the outer shell and the nucleus, weakening the hold of the nucleus on the outer electron. Very little thermal energy (heat) is needed for this outer electron to leave its parent atom and become a **free electron**. The electron is then free to move easily through the conducting material. Without an external charge field, the motion of the free electrons is random. There is no net flow of electrons in any given direction. This concept is pictured in the conductor of Figure 2-8.

Figure 2-8 Random motion of free electrons

If an external charge difference is provided to a conductor, a positive charge on one end and a negative charge on the other end, the free electrons are repulsed by the external negative charge and are attracted toward the external positive charge. Figure 2-9 represents a conductor (for example, a copper wire) with an externally applied charge difference.

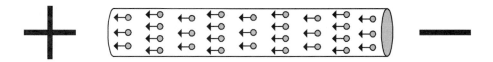

Figure 2-9 Drift of free electrons toward external positive charge

As electrons are collected from the material into the external positive charge, an equal number of electrons can be supplied to the material from the external negative charge. The net effect is a continuous drift of electrons toward the positive charge. This creates an electron flow that is called a **current**. Current is a fundamental electrical parameter that we shall cover in more detail later.

Ion

The net charge of a balanced atom is neutral since its total number of electrons equals its total number of protons. However, a **parent atom** that has lost an electron is electrically unbalanced. It has a net positive charge since it lacks an electron, and is called a **positive ion**. Free electrons are continuously recombining with positive ions while new free electrons are being formed. Thus the term **electron drift** is very appropriate since a free electron may not make the entire journey but rather there is an over-all net effect of free electrons drifting toward the external positive charge. An atom that has gained an additional electron is called a **negative ion**.

Insulator

Insulators or insulating material such as glass, mica, and rubber do not easily create free electrons nor allow the flow of free electrons. Chemical-ly, it is the atom's valence band electrons that bond with other atoms' valence band electrons to create molecules and compounds. The maxi-mum number of valence band electrons that can be shared between atoms is 8. Materials in the periodic table of elements with more than 4 valence band electrons are considered natural insulators. These atoms do not easi-ly give up their electrons to create free electrons.

Electrically, insulators do not allow free electrons to be created with-in the insulating material; moreover, insulators do not allow external free electrons to enter and flow through the insulating material. Insulating materials play a significant role in electricity and electronics. Insulating material is needed to separate conductors, for example, insulation sur-rounding a copper wire. Also, insulation may be an integral ingredient of an electronic component.

Semiconductor

The **semiconductor** falls between the conductor and the insulator in terms of its conducting ability. Semiconductors allow electron flow but not as readily or easily as a good conductor. Good semiconductor ele-ments, namely germanium and silicon, are characterized by a *valence band with four electrons*. These atoms share their four valence band elec-trons with four neighboring atoms to provide the parent atom with the equivalent of eight outer band electrons.

The term semiconductor is generally reserved and used for a very special category of semiconductor crystals. The semiconductor elements germanium and silicon are used to create many of the modern electronic components, namely, transistors and integrated circuits (or chips). They are the material building blocks of today's electronic age.

2.3 Current

As noted previously, a flow of electrons in the same direction creates a current; that is, a current of electrons. Current is a fundamental variable in electricity and electronics. It is represented by the symbol I. The letter I is based on the French word for current, **intensité**.

The electron flow must have a directed drift caused by an external charge difference. Figure 2-10 shows a diagram of a battery (the source of charge) connected to a lamp using copper wire to create a current path to light the lamp. There must be a *continuous path* for the electrons to flow from the negative post to the positive post of the battery. The lamp provides a continuous path with filament wire (a wire that heats and illuminates).

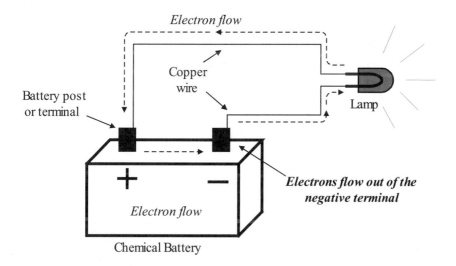

Figure 2-10 Electron flow current

The battery chemically creates a difference of charge, one post a concentrated positive charge and the other post a concentrated negative charge. Free electrons (1) flow out of the negative battery post, (2) flow through the conducting wire, (3) flow through the lamp (heating it and lighting it), (4) flow through the wire, and (5) flow into the positive battery post. Internally, the battery chemically attracts electrons from the positive post and deposits extra electrons at the negative post. An analogy is that electrons are *chemically pumped* from the positive terminal to the negative terminal inside the battery. This continues until the battery can no longer chemically create charges at the battery posts. At this point the battery must be recharged or properly disposed of and replaced. Based upon this electron flow view, the negative charge flow is from the external negative terminal through the circuit to the external positive terminal of the supply. Typically, the trades and the military use electron flow current.

Engineering and engineering technology use what is termed **conventional current**. In the early 1800s, electricity was considered to be a fluid. The terms "flow" and "current" are derived from this misconception (e.g., the flow or current of a river). Another misconception is that the positive charge is flowing, not the negative electron charge. By "convention," the charge was assumed by the scientific community to flow *from a battery's positive terminal through the circuit to the battery's negative terminal*, and thus it is labeled *conventional current*. Science has since revised its understanding but conventional current is still the standard for engineering and engineering technology. And, in many ways, conventional current provides a mathematical advantage since we are working with "positive" charge flow instead of "negative" charge flow.

Figure 2-11 shows the same circuit diagram but from the conventional current point of view. The current (positive charge) flows *from the positive battery terminal, through the external circuit, and returns to the negative battery terminal*. In the conventional current pump analogy, internally the battery chemically pumps positive charge from the negative post to the positive post to produce continuous current.

From the conventional current viewpoint, current (1) flows *out of* the battery positive post, (2) flows through the conducting wire, (3) flows through the lamp, (4) flows through the wire, and (5) flows *into* the battery's negative post. Realize that conventional current is just a mathematical concept. The electrons are actually flowing to create the current, but conventional current does work accurately and effectively as a conceptual mathematical tool.

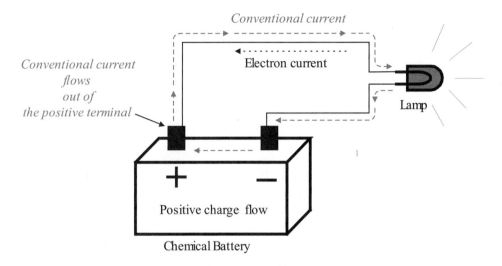

Figure 2-11 Conventional current (positive charge flow)

Ampere

The unit of current is the **ampere**, commonly abbreviated **amp**. The unit symbol for the ampere is the letter **A**. One ampere is equivalent to one coulomb of charge passing through a perpendicular cross-section of material in one second, as seen in Figure 2-12.

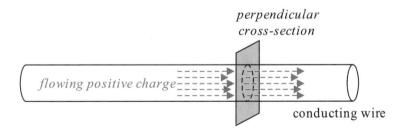

Figure 2-12 Flowing positive charge creates conventional current

$$1 \text{ ampere} = \frac{1 \text{ coulomb}}{1 \text{ second}}$$

or

$$1 \text{ A} = \frac{1 \text{ C}}{1 \text{ s}}$$

Recall that one coulomb of charge is equivalent to 6.242×10^{18} charged particles (electrons or protons). Thus a significant number of charged particles must be flowing to create any meaningful current. Electrical systems (like hair dryers, refrigerators, and light bulbs) operate in the ampere range. Electronic circuits (like power amplifiers, computers, and electronic watches) operate from amperes down to very small currents, for example, microamperes.

Example 2-3

A bicycle lamp draws 2 A of current. Find the number of coulombs of charge per second required to create this current. Then find the number of electrons per second required to create this current.

Solution

First, convert from amperes (A) to coulombs per second.

$$I = 2\text{A} = 2\,\frac{C}{s}$$

Next, convert from coulombs (C) to the number of charged particles (e); and cancel the coulomb (C) units.

$$I = \left(2\,\frac{C}{s}\right) \times \left(6.242 \times 10^{18}\,\frac{e}{C}\right) = 12.484 \times 10^{18}\,\frac{e}{s}$$

Practice

A 6 V flashlight bulb draws 9.363×10^{18} electrons per second from its battery. Find its current in amperes (A).

Answer: 1.5 A

Fixed Direct Current (dc)

This chapter covers fixed **direct current** (dc). That is, the current direction and values are fixed and do not vary with time. Current that varies with time is called alternating current (ac), which will be studied in a later chapter.

Current Definition

The most fundamental definition of current is based upon charge and is given by Equation 2-2.

$$I = \frac{Q}{t} \tag{2-2}$$

where
   I = current in amperes (A)
   Q = charge in coulombs (C)
   t = time in seconds (s)

Example 2-4
   A computer monitor draws a charge of 15 coulombs per second. Find its current.

Solution
   Apply Equation 2-2, the fundamental definition of current in terms of charge. Note: the units are already appropriate to substitute into the expression, namely, charge is in units of coulombs (C) and time is in units of seconds (s).

$$I = \frac{Q}{t} = \frac{15 \text{ C}}{1 \text{ s}} = 15 \text{ A}$$

Practice
   A bicycle lamp draws 50 C of charge in 10 seconds from its battery. Find the lamp's current.

Answer: 5 A

Example 2-5

240 C of charge flows through a 200 W light bulb in 2 minutes. Find the current through the bulb.

Solution

First convert to the proper units as needed. The charge is in coulombs and is appropriate. But, time is in units of minutes and must first be converted to seconds before applying the expression for current.

$$t = 2 \text{ min} \times \frac{60 \text{ s}}{\text{min}} = 120 \text{ s}$$

Watch your units. Always carry units throughout your calculations. The units are now appropriate to substitute into the expression for I, namely, charge is in units of coulombs (C) and time is in units of seconds (s).

$$I = \frac{Q}{t} = \frac{240 \text{ C}}{120 \text{ s}} = 2 \text{ A}$$

Practice

A light emitting diode (LED) emits red light for an hour. If 72 C of charge flows through the LED in that hour, find the current through the LED. Watch your units.

Answers: 20 mA

Example 2-6

A computer printer is rated at 400 mA. Find the amount of charge in coulombs needed to operate the printer for 2 hours. Then find the number of electrons that must flow to create this current.

Solution

First, examine the units. Time is in the units of hours (hr). It must be converted to the proper unit of seconds.

$$t = 2 \times \frac{3600 \text{ s}}{\text{hr}} = 7200 \text{ s}$$

Current is already in the proper unit of amperes but with a *milli* prefix. Second, algebraically rewrite Equation 2-2 and solve for charge Q.

$$Q = I\,t = 400 \text{ mA} \times 7200 \text{ s} = 2{,}880{,}000 \text{ mA} \cdot \text{s}$$

where the $\cdot$ represents multiplication of the units. Substitute the *milli* prefix equivalence of $m = 10^{-3}$.

$$Q = 2{,}880{,}000\left(10^{-3}\right) \text{ A} \cdot \text{s}$$

Substitute the ampere unit equivalence of $A = \dfrac{C}{s}$

$$Q = 2880\left(\frac{C}{s}\right)s = 2880 \text{ C}$$

The number of charged particles needed to create a charge of 2880 C is:

$$\text{\# of charges} = 2880 \text{ C} \times \left(6.242 \times 10^{18} \frac{e}{C}\right) = 1.80 \times 10^{22} \, e$$

Practice

Current flows through the wire of a toaster to create heat, which in turn toasts the bread. A toaster operates at 7.5 A for one and a half minutes. Find the amount of flowing charge in coulombs required to heat the wire that toasts the bread. How many electrons flowed during that time to create the 7.5 A?

Answer: 675 C; 4.213×10^{21} e.

Electricity—Safe or Dangerous? Your Choice!

The more you know about electricity and the more you use safe practices, the safer you are. Most of the people injured by electricity are injured at home and know very little about electrical safety.

Current is the thing that actually injures or kills. Electrons can move through your body just like they do through a wire or a lamp. The good news is that the body's skin resists electron flow more so than a wire or lamp. The human body's resistance to current (electron flow) depends upon the individual and environmental conditions. *Moisture significantly reduces the body's resistance to electron flow* and therefore causes current through the body to be increased. Thus, using an electrical hairdryer in the shower is a very bad idea. If your body is wet, your body conducts current more readily. A voltage shock would produce a higher current through your body than normal. Beware: perspiration (moisture) is basically mineral water and increases the body's conduction of current. A common saying, "Water and electricity do not mix," is a truism. Also, perspiration and electricity do not mix.

Some humans can feel electricity with only a few milliamperes of current. Up to 10 milliamperes is usually considered safe but should still be treated with caution. Anything over 10 mA must be considered dangerous. A current of 50 mA may cause shock, enough to put a heart into an irregular heartbeat, which can be fatal. The following are some common sense practices to help you get started in the laboratory. This list is in no way comprehensive for all situations. You must always be aware of your environment and keep safety in mind.

- Be aware of your input voltage supply and any quick-release safety switches.

- Do not use frayed electrical power cords; repair or replace them.

- Remove all metallic jewelry that can make contact with the circuit: rings, watches, and hanging jewelry.

- Do not handle a live circuit. That is, in a live circuit do not replace parts, break the circuit, touch hot leads, or generally handle the circuit. Always turn off the circuit's power supply when you must break the circuit or insert parts.

- Use common sense about wearing safety glasses when working on a live circuit, cutting leads, or soldering. In live circuits, parts can and do explode occasionally, especially when in the experimental mode.

- Be sure you and your environment are dry. Normally safe situations can be made extremely dangerous from moisture including your perspiration.

- When possible and reasonable, make hands-off measurements. That is, connect your test leads to the dead circuit, remove your hands from the test leads, and then power up the circuit. If that is not possible or reasonable, then with the circuit live, connect one test lead at a time using one hand only. The worst situation is connecting both hands simultaneously to the circuit creating a current path through your heart. Your heart is controlled by electrical signals in your body. Very little current is required to affect its rhythm, which can be fatal. Connect one test lead at a time, using only one hand.

Never touch the probe tips when making measurements. You are becoming part of the circuit and will draw some current. Minimally you affect the reading. Worst case, you will be shocked.

2.4 Current Meter Connection

Current is called a **through** parameter. The current flows within and through the wire. A current meter, also called an **ammeter,** is used to measure current. There are several steps you must follow to properly measure current. Figure 2-13 represents the state of the circuit before the ammeter is connected.

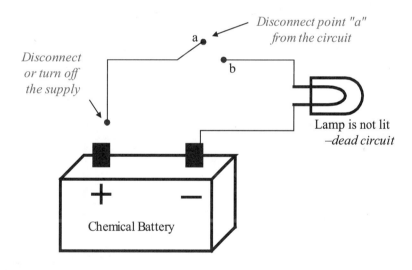

Disconnect point "a"
from the circuit

a

b

Disconnect
or turn off
the supply

Lamp is not lit
–dead circuit

Chemical Battery

Figure 2-13 Disconnect the circuit where the ammeter measurement is made

1. Power down the circuit (disconnect the battery or turn off the circuit's energy supply).

2. Identify on the circuit the point at which you want to measure the current. Hopefully you have a relatively convenient point like a lead wire or a component wire you can pull out.

3. Pull out that one lead. Lead point *a* has been pulled out of the circuit, leaving point *b* on the board.

4. Figure 2-14 demonstrates the circuit with the ammeter properly connected. Connect the ammeter + lead (red, mA) to point *a*. Connect the ammeter − lead (black, COM) to point *b*. This connection creates a positive conventional current reading with the current flowing into the ammeter red lead, through the ammeter, and out of the ammeter black lead.

5. Be sure the ammeter is in the proper range. If this is a multipurpose meter, be sure the meter is in the *current mode*.

6. Power up the circuit and take a hands-free reading.

7. When finished, power down the circuit, remove the ammeter, and reconnect the original circuit.

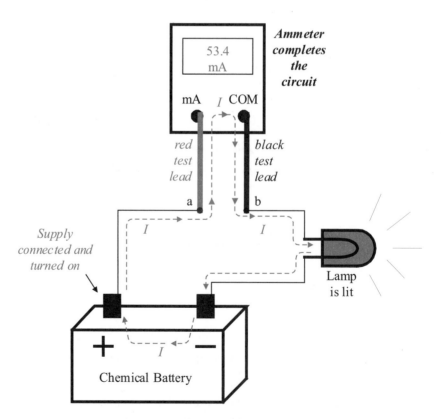

Figure 2-14   Ammeter completes the circuit for a conventional current reading

8. If you are finished with current measurements and the ammeter is a multimeter, place the meter in the voltage position. A voltmeter is an open while a current meter is essentially a short. Leaving the multimeter in the voltmeter position is the safest position in which to leave the meter.

Warning: To a first approximation, the *ammeter* is equivalent to an ideal *conductor* or a piece of *wire*, as demonstrated in Figure 2-15. Mentally visualize an ammeter as a piece of wire that measures current and also reconnects the circuit (re-establishing the circuit current you are measuring). For circuit analysis purposes, you can *model* the ideal ammeter as a piece of wire; that is, replace the ammeter with a piece of wire and then analyze the circuit.

Figure 2-15 Ammeter mode—ideally a conductor that measures current

2.5 Voltage

Voltage is the second fundamental parameter. Voltage is the electromotive force (electrical pressure) that causes electrons to drift in the same direction. The electrons become electrodynamic (moving) instead of electrostatic (stationary). These moving electrons do *work* just like moving water drops do work (e.g., moving a waterwheel). To understand the foundational principles of voltage, you must first understand the principle of doing work.

Mechanical Work

Work is performed if a *force* is applied over a *distance*. The general formula for work is:

$$W = f\,d \qquad\qquad \textbf{(2-3)}$$

where

W = work in joules (J)
f = force in newtons (N)
d = distance in meters (m)

Thus, the **joule (J)** unit is equivalent to a **newton-meter** (N·m). In the English system, force is in units of pounds (lb) and distance is in units of feet (ft), thus force is measured in the units of **foot-pounds** (ft·lb).

Example 2-7

Becky uses a 14 pound (lb) bowling ball. Find the work done for her to lift it 3 feet (ft) in English units and in SI units. Refer to Figure 2-16.

Solution

In English units, using Equation 2-3 the work performed is

$$W = 14 \text{ lb} \times 3 \text{ ft} = 42 \text{ ft} \cdot \text{lb}$$

To express this problem in SI units, you must convert from English to SI units. First convert the force of a 14-pound bowling ball from units of pounds (lb) to units of newtons (N).

$$f = 14 \text{ lb} \left(\frac{4.45 \text{ N}}{\text{lb}} \right) = 62.3 \text{ N}$$

Next, convert distance from units of feet (ft) to units of meters (m).

$$d = 3 \text{ ft} \left(\frac{0.3048 \text{ m}}{\text{ft}} \right) = 0.914 \text{ m}$$

In SI units, the work performed in units of joules is

$$W = f \, d = (62.3 \text{ N})(0.914 \text{ m}) = 56.9 \text{ J}$$

Recall that the unit of newton-meter (N·m) is equivalent to a joule (J). Note, it does not matter whether you lift the ball in one second or one hour, the work performed is the same. The same work done faster requires more **power**, not more work. We shall discuss power next.

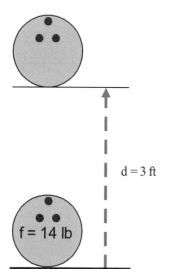

Figure 2-16
Work of lifting a bowling ball in Example 2-7

$d = 3$ ft

$f = 14$ lb

Practice

> Jenny, who weighs 100 pounds, climbs up a rope 20 feet long. Find the work done in English units and SI units. If that person climbs the rope 3 times as fast, find the work expended in SI units.

Answer: 2000 ft·lb, 2713 J, 2713 J (no change, time is not a factor in doing work)

Power

Power is the *rate of creating energy or expending energy (doing work)*. The general formula for power is:

$$P = \frac{W}{t}$$ (2-4)

where

> P = power in watts (W)
> W = work (or energy) in joules (J)
> t = time in seconds (s)

Be careful not to confuse the use of W as a variable for energy or work with W as the unit symbol for the watt. You must understand the context of the symbol's use to know whether the symbol is representing a variable or a measured unit.

Example 2-8

> Refer to Example 2-7. Find the power required for Becky to lift her bowling ball in 2 seconds.

Solution

> From the previous example, the work expended by Becky in lifting her bowling ball 3 feet is 56.9 J. Substitute the known values of energy and time into Equation 2-4.
>
> $$P = \frac{W}{t} = \frac{56.9 \text{ J}}{2 \text{ s}} = 28.5 \text{ W}$$

Practice

> Becky lifts her bowling ball 3 feet in ½ second. Find the power required for her to do this.

Answer: 114 W (Wow! This is about the power required to light a 100 W light bulb. Imagine that Becky must continuously lift her bowling ball 3 feet per half second to generate that same power needed to light a 100 W lamp. Impressive!)

Mechanical Potential Energy

Energy is the capacity to do work. For example, in lifting a pail of water, you apply external energy and perform work (force times distance). This concept can be seen in Figure 2-17.

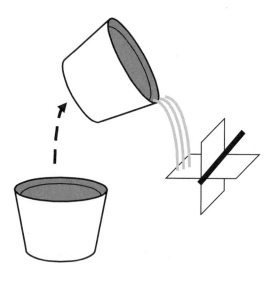

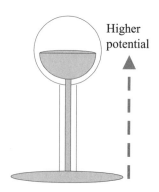

Higher potential

Figure 2-17 Work raising water creates potential energy, which in turn is used to do work to run the paddle wheel

Figure 2-18
Work is required to raise water up into the tank

Now that you have lifted the pail of water, it has potential energy (the capacity to do work) by virtue of its position. If you pour the water and it turns a paddle wheel while flowing downward, its potential energy is expended. If you had a water pump that would raise the water from a reservoir and into a tank, the water in the tank has potential energy. Figure 2-18 demonstrates this concept. The higher the water is raised, the more potential energy produced by virtue of its higher position. The water in the tank is at a higher potential than the water at the bottom in the reservoir by virtue of its position. The difference between the potential energy levels is called the potential difference. When the water is released, gravity pulls the water downward and it can do work, for example, turning a paddle wheel. The paddle wheel axle could be connected to the shaft of a generator, which produces electricity. Thus falling water is converted into mechanical energy to rotate the paddle wheel, and the mechanical energy is converted into electrical energy by a generator.

Voltage Definition

Electrical work is performed if an electrical force (electromotive force) moves a charge over a distance (similar to the mechanical definition). In this case, the electromotive force is called **voltage**. This is the electrical force that creates pressure to move electrons in an electrical circuit. An equivalent expression for electrical work is

$$W = VQ$$

where

W = work or energy in joules (J)
V = voltage in volts (V)
Q = charge in coulombs (C)

From this we derive our basic definition of voltage as

$$V = \frac{W}{Q} \qquad \textbf{(2-5)}$$

Voltage is also called **potential difference** because voltage is always the *difference in potential between two points*.

An analogy is shown in Figure 2-19. It shows a comparable electrical battery raising charge just like the pump raised water. The water pump raises water drops from the water reservoir to a higher potential in the water tank. Likewise, the chemical battery *raises positive electrical charge* from the negative battery post to the higher potential positive battery post using chemical energy. A potential difference of charge is formed, creating a voltage difference between the positive and negative terminals. Figure 2-19 assumes conventionally that the positive charge is moving inside the supply from − to +. *The more energy imparted or expended per charge, the higher the voltage or potential difference created.*

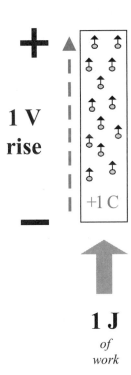

Figure 2-19
Work is required to create a voltage potential difference.

Example 2-9

A AA battery has a voltage of 1.5 V (the potential difference between its positive and negative posts). It is capable of generating 9000 C of charge over its life. Find the electrical work that can be produced over the life of this battery.

Solution

Rewrite Equation 2-5 to solve for electrical work and solve for it.

$$W = V \ Q = 1.5 \ \text{V} \times 9000 \ \text{C} = 13,500 \ \text{J}$$

The electrical potential energy of this battery created by external chemical energy is 13.5 kJ. The electrical charge in this battery is capable of doing work by flowing through a circuit just like water drops in the tank have potential of doing work flowing down a trough and turning a paddle wheel.

Note: it requires 19 J of work to lift a 14 pound bowling ball 1 foot. The equivalent mechanical work of 13.5 kJ would lift the 14 pound bowling ball 710 feet.

Practice

A battery can generate 1800 C of charge and create a potential energy (capacity to do work) of 16,200 J. Find the voltage (potential difference) created between the battery terminals.

Answer: 9 V

A more formal definition of voltage (potential difference) is: *a potential difference of 1 volt (V) exists between two points if 1 joule (J) of energy is used to move 1 coulomb (C) of charge from one point to the other point.* Refer to Figure 2-20(a) to help you visualize this key concept.

In this case, external energy is used to move the positive charge against a resisting force (the positive end of the device). That is, the *potential energy is increased* (for example, a battery). Note: the *potential is raised* as indicated by the (− to +) polarity notation. *As the charge moves from the bottom to the top (− to +) of the material, the voltage rises by 1 V (a gain in potential).*

If electrical energy is given off as the charge moves, energy is dissipated and the potential energy is reduced. For example, a lamp dissipates its energy in the form of light and heat. Voltage is dropped across the lamp. Figure 2-20(b) demonstrates a component dropping voltage (losing potential) and dissipating energy as charge moves through the component. *As the charge moves from the top to the bottom (+ to −), the voltage drops by 1 V.*

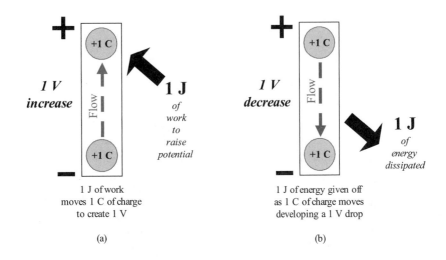

Figure 2-20 (a) Voltage potential *increase* (b) Voltage potential *decrease*

Example 2-10

Find the amount of charge moved through a 2 V lamp to dissipate 8 J of energy.

Solution

Rewrite Equation 2-5 and solve for charge:

$$Q = \frac{W}{V} = \frac{8 \text{ J}}{2 \text{ V}} = 4 \text{ C}$$

Practice

A battery uses 9 J of chemical energy to move 6 C of charge. Find the voltage created.

Answer: 1.5 V

Another way to look at a voltage supply is that a charge difference exists between the positive and negative posts of a battery. The charge difference creates a potential difference (or a voltage difference). When this charge source (voltage supply) is connected to a completed circuit, electrons are caused to move by the electrical pressure (potential difference) of the battery.

Fixed Voltage (DC)

Fixed voltage is also labeled dc voltage; it does not vary with time. Figure 2-21 is a sketch of a 6 V dc voltage versus time; it is a constant. The interpretation of dc is strictly translated as **direct current** but the term dc is applied to fixed voltage as well as fixed current. For example, the car battery is a 12.6 V dc battery. Ideally, it remains at a fixed 12.6 V during its operation. Naturally, a battery eventually discharges to lower voltages over an extended period of time, but it is intended to be a fixed supply of voltage. Under heavy load, for example cranking a car engine, the voltage may drop somewhat.

Warning: never use a conductor to connect the positive (+) and negative (−) terminals of a battery. Excessive current is created and could damage or even explode the battery.

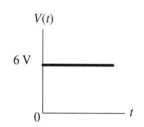

Figure 2-21
Sketch of dc versus time

2.6 Voltmeter Connection

Voltage is called an **across** parameter. The voltage is measured across components or between two external points in the circuit. Remember current is a much different measurement, measuring at a single point inside the circuit. Measuring voltage is much simpler than measuring current.

A voltmeter is used to measure voltage. Figure 2-22 shows a simple voltage measurement of a car battery. The **COM** or **common lead** of the voltmeter is connected to the negative (−) post, the lower potential voltage point of the battery. The common lead is typically colored *black* as a standard and is referred to as the **reference lead**.

The **V** or **voltage lead** of the voltmeter measures a voltage at a point relative to its reference lead. It is connected to the positive (+) post of the battery. Since the *red* lead is connected to the higher potential terminal on the battery, the voltmeter displays a *positive* reading.

Warning: What if the voltmeter test leads were reversed? If it were a *digital* voltmeter, the meter reading would be −12.60 and there is no harm done. The reversed leads are indicated by the negative (−) sign of the measurement. If it were an **analog** meter, the needle (meter movement) would attempt to go off scale in the reverse direction. This could cause meter damage, or an excessive voltage protection switch to trip to protect the meter.

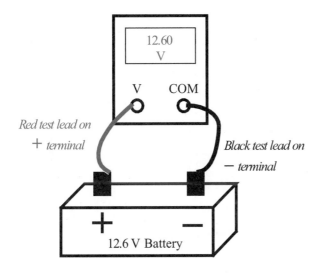

Figure 2-22 Voltmeter measurement of a car battery

Figure 2-23 demonstrates measuring voltage across a *load* rather than a supply. The lamp is *drawing current from the battery* and is therefore considered a load on the battery. As current flows through the load, the load drops voltage. The top of the lamp is at a 12.6 V higher potential than the bottom of the lamp.

The voltage plus (+) and minus (−) sign notation at the terminals of a component is called the **voltage polarity** of the component. Note the *polarity* of the voltage drop across the lamp, V_{lamp}. If properly labeled, this indicates that the top terminal is at a higher voltage potential than the bottom terminal. In this case, the top terminal is measured to be 12.6 V higher relative to the bottom terminal.

For a load, it is always true that the *entering* conventional current creates the higher potential terminal (labeled +) relative to its current exiting terminal (labeled −). As current flows through a load device, potential is lost. Thus, the top of the lamp is labeled with a positive (+) sign since current enters at that terminal. And the bottom of the lamp is labeled with a negative (–) sign since current exits at that terminal. *Essentially, for a load, conventional current always enters the relatively higher potential terminal* (+) *and exits the relatively lower potential* (–) *terminal.*

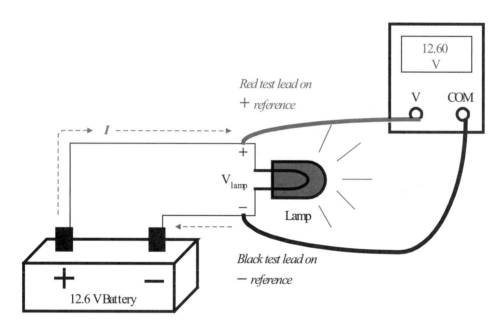

Figure 2-23   Voltmeter measurement of a car lamp

For a voltage supply, its voltage polarity is established by its construction of a positive terminal and a negative terminal. For a load, its voltage polarity is established by the current direction in the load. There are a few steps you must follow to properly measure voltage.

- The circuit must be powered up to take circuit voltage readings.

- You may connect the test leads to a live circuit, but connect them one at a time. In an electrically live circuit, connect the common lead of the meter first. It is best to use test leads that allow hands-free measurements. Do not use ragged edged grabbers like alligator clips to connect to the circuit; they damage component leads.

Warning: When jumping car batteries, the very last connection must be away from the battery to the metal chassis of one of the cars. Cars typically have the negative terminal of the battery connected to the car's chassis (a connecting cable runs from the negative battery post to the car chassis). This last connection typically arcs, which can be very dangerous near a battery if it is leaking any chemical gases.

Ideally, to a first approximation, the voltmeter does not draw current from the circuit. In Figure 2-23, the circuit current is shown flowing only through the lamp and not through the voltmeter. Essentially, the voltmeter is an insulator that measures voltage.

2.7 Common

Typically a voltage reference or **common** is established in electrical systems or electronic circuits. This common also serves as a *zero (0) volt reference point* (or node). There are three types of commons. Figure 2-24 shows the IEEE and ANSI standard symbol representations of these commons: earth, analog, and chassis return common.

Analog	Earth	Chassis
Common	Common	Return
		Common

Figure 2-24 Common symbols

Note, the term *ground* instead of common is in widespread use in the profession. Ground is a confusing term. We shall consistently use the term *common* to help you visualize its purpose. Also, be aware that the IEEE standards are not universally followed when using common (or ground) symbols, especially in manufacturer specifications (or spec sheets).

Analog Common

In an electronic circuit, it is important to reference a point common to the circuit. This point is called **analog common** and may also serve as a *0 volt reference*. Other voltages in the circuit can be measured relative to the analog common for comparison.

Systems that include digital or radio frequency (RF) circuits may have their own respective commons. Thus, a circuit might have an analog common, a digital common, and an RF common, which are physically different common points in the circuit.

Earth Common

Metal conducting rods are driven deep into the earth until they reach mineral enriched water. The minerals in the water conduct electrically and form a *common conducting point* (or **node**). This is often referred to as **electrical ground** but we shall refer to it as **earth common**. Note: it is neither the dirt nor the ground that is conductive; it is the minerals in a water solution *in the earth* that is common. Thus, the terms ground and grounding can be confusing. The term **grounding** does specifically refer to tying a point to earth common.

Chassis Return Common

As the name **chassis return common** indicates, the chassis of the equipment is considered a common point. An excellent practical example of this is the automobile chassis. Electrical wire is used to connect to components such as taillights, but the metallic chassis is used as a conductive return path. This saves wire, weight, and expense. Also, the large, thick conducting frame is a better conductor than wire.

2.8 Ideal Fixed (DC) Voltage Supply

An ideal fixed (dc) voltage supply delivers a constant voltage even though the current that it produces may vary. For example, an ideal D-cell battery would deliver a fixed, constant 1.5 V no matter how cold it is or how much current is being drawn from it. *The exception is that a conductor (wire) must not be connected across a voltage supply from its positive to negative terminal. This could damage the supply or cause a battery to explode.*

The four popular sources of dc voltage are:

- Batteries (electrochemical)

- Electronic power supplies (electrical/electronic)

- Generators (electromechanical)

- Photocells (photovoltaic)

Refer to Figure 2-25(a) for the schematic symbol of an ideal voltage cell (such as D cell battery). An electronic voltage supply or a supply which consists of several cells (such as a car battery) has the voltage supply symbol shown in Figure 2-25(b). These symbols are used in schematic drawings to replace and model the actual voltage cell or supply. The *long line* of the voltage supply E always represents the *positive* (+) terminal of the supply. The *short line* of the supply represents the *negative* (−) terminal of the supply.

The letter *E* represents a **voltage supply**. For example, E = 12 V for a battery, as demonstrated in Figure 2-26. The letter *V* represents a *device or circuit voltage drop*, for example V_L = 12 V for the voltage drop across a lamp. Each component has its unique symbol: the dc voltage supply, the switch, the lamp, and the analog common. A switch completes or breaks the circuit. We shall study switches in a later chapter. Typically the schematic components are *not* fully labeled. The circuit of Figure 2-26(a) can be labeled like that of Figure 2-26(b) without full labeling: E for voltage source, SW for switch, and L for lamp.

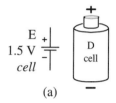

(a)

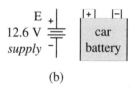

(b)

Figure 2-25
Ideal dc voltage supply
(symbol in left figure)

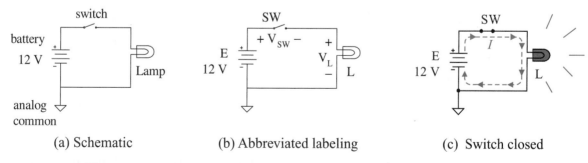

(a) Schematic   (b) Abbreviated labeling   (c) Switch closed

Figure 2-26 Schematic of a DC lamp circuit

In Figure 2-26(c) the switch is closed. The current flows from the positive (+) terminal of the dc voltage supply, through the closed switch, through the lamp (lighting it), and back to the negative (−) terminal of the dc voltage supply.

Computer Simulations

Computer simulation programs simulate a circuit and display simulated measurement values. The program operator creates the circuit from prepackaged or user-generated component icons in the program. Typically it is very easy to create a circuit by dragging and dropping the desired component icons and connecting components using the computer mouse. MultiSIM® is such a simulator software package. MultiSIM was previously named Electronics Workbench.

Figure 2-27(a) is a MultiSIM simulator schematic of the circuit of Figure 2-26(a) with the switch open. The schematic was created in Multi-SIM by dragging and dropping the schematic symbol icons onto a blank circuit sheet. The voltage supply, the analog common, the switch, and the lamp are components provided in the MultiSIM simulator package. A *common* is required in the circuit for the simulator to execute properly. The component labels can be edited and have been for this example.

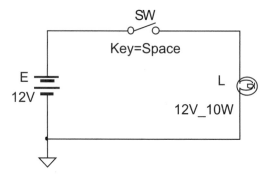

Figure 2-27(a) MultiSIM schematic of the circuit of Figure 2-26(a)

Notice that the switch SW has a notation below it that the computer keyboard *space bar* closes and opens the switch for the simulation. Also, the lamp has been specified as a 12 V lamp at 10 W of power just as a home light bulb would be specified as a 120 V lamp at 100 W of power.

To be useful, we must place meters in the circuit to take measurements. Figure 2-27(b) shows the circuit of Figure 2-26(a) with voltmeters placed across each circuit device to measure the simulated voltage of each device. With the switch open the device voltages are simulated to be:

$E = 12.000$ V

$V_{SW} = 12.000$ V

$V_{lamp} = 0.173$ mV (approximately 0 V)

Ideally the lamp voltage should be 0 V. However, instruments are electronic devices and when placed in the circuit become part of that circuit. Thus, instruments can impact the circuit and affect circuit voltage values. Also, be aware that this is a simulation and not the real physical circuit. You must use judgment perhaps supported by other means such as manual calculations to corroborate your simulated findings.

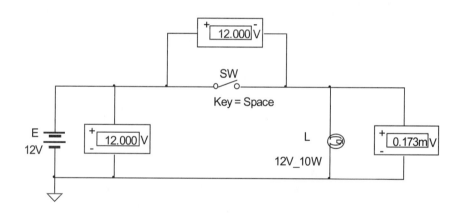

Figure 2-27(b) Simulation of the circuit of Figure 2-26(a) with the switch open

Figure 2-27(c) shows the simulated results with the switch closed. In the simulation, tap the space bar to close the switch and then run the simulation again to generate the new simulated measurements with the switch closed. With the switch closed, the simulated results are:

$E = 12.000$ V

$V_{SW} = 8.333$ nV (approximately 0 V)

$V_L = 12.000$ V

Notice that the switch is now shown as closed (that is, making contact with both of its terminals) and completing the circuit. With the switch closed, the supply voltage of 12 V is applied to the lamp to produce light. The simulation demonstrates that the lamp is lit with the lines radiating from the lamp symbol. Ideally the switch voltage should be 0 V; however, again, the instrumentation and the devices are simulated circuit devices. Instrument characteristics and component characteristics (not ideal characteristics) are incorporated into the simulation.

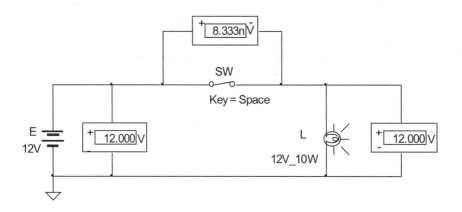

Figure 2-27(c) Simulation of the circuit of Figure 2-26(a) with the switch closed

multiSIM

Some of the major features of MultiSIM include:

a. Relatively user friendly and easy to edit
b. An on-off-pause button for simulation runs
c. Easy pick and place of components
d. Easy drawing of wire connections
e. The voltmeter must be placed into the circuit schematic just as it would be placed in the circuit
f. Analog and earth common symbols are available as shown in Figure 2-27(d)

Figure 2-27(d)
MultiSIM common symbols

We shall use MultiSIM in the textbook where appropriate to verify or demonstrate differences with manual calculations. We shall introduce other features of MultiSIM later in the textbook as needed. MultiSIM is not taught in this textbook. It is somewhat intuitive, and on-line tutorials and help aids are available elsewhere. This book places the MultiSIM icon in the margin next to MultiSIM simulations to help identify them.

how big can you dream?™

Cadence PSpice® is another widely used simulator. It is also used in this textbook where appropriate. These simulations are identified with the Cadence icon shown in the margin. The Cadence PSpice symbols for common are shown in Figure 2-27(e). Notice that Cadence PSpice and MultiSIM have common symbols for *analog common* and *earth common*. Cadence PSpice has a symbol for *chassis common* but MultiSIM does not.

Figure 2-27(e)
Cadence PSpice common symbols

Batteries

Batteries are the most commonly known dc voltage supply. Batteries *produce a charge difference (a voltage difference)* by chemical activity. The charge is then delivered to a continuous circuit to produce current. Batteries use an electrochemical action to deliver electrons to the negative post of a battery and pull electrons away from the positive post of the battery. Once the chemical activity is exhausted by the circuit, the battery must be replaced with a new battery or a recharged battery if it is a rechargeable battery. Many circuits incorporate a charging circuit to keep the battery charged, for example, the automobile battery is recharged while the engine is running.

The most popular *nonrechargeable* battery is the *alkaline* battery. They are used in flashlights, smoke detectors, and remote controls. The AAA, AA, C, and D cell batteries are rated at 1.5 V at increasing sizes and charge capacities. Typically, larger batteries can deliver more current longer. See Table 2-1 for a list of popular alkaline batteries.

Note the D cell generates about 57,000 coulombs of charge over its lifetime while the much smaller AAA cell generates about 4,000 coulombs of charge over its lifetime. There is also a 9V transistor battery and a 6 V lantern battery available.

Table 2-1 Alkaline batteries, their voltages, and capacities

Alkaline Battery	Rated Voltage	Current Capacity	Approximate Charge Capacity
AAA	1.5 V	300 mA for up to 4 hours	4,000 C
AA	1.5 V	500 mA for up to 5 hours	9,000 C
C	1.5 V	1 A for up to 7 hours	25,000 C
D	1.5 V	1 A for up to 16 hours	57,000 C
6 V Lantern	6 V	1.5 A for up to 15 hours	79,000 C
9 V Transistor	9 V	250 mA for up to 2 hours	1,800 C

Amp-hour Ratings

A more popular and useful way to express the capacity of a battery is in **amp-hours (Ah)** or **milliamp-hours (mAh).** In other words, the capacity of a battery is stated in how much current can be delivered for how long. The charge in *coulombs* is simply converted to its equivalent units in terms of *current* × *time*. Recall the fundamental definition of current is

$$I = \frac{Q}{t}$$

which can be written as

$$Q = I\,t$$

Thus 1 C (coulomb) is equivalent to 1 A·s (amp-second) or

$$1\,C = 1\,A \times 1\,s$$

Since there are 3600 seconds in one hour, this could be written as amp-hours.

$$3600\,C = 1\,A\,h$$

A popular term for battery capacity is **battery rating**. The battery rating is expressed in amp-hours or milliamp-hours instead of coulombs.

$$\textit{Battery Rating} = I\,t$$

Example 2-11

Find the *battery rating* of a C cell battery that can deliver a charge of 25,000 C.

Solution

$$Battery\ Rating = Q_{capacity} = 25,000\ C$$

Convert the coulombs (C) to amp-seconds (A·s). Then convert seconds (s) to hours (h).

$$Battery\ Rating = (25,000\ A\cdot s)\times\frac{1\ h}{3,600\ s} = 7\ Ah$$

Practice

A battery is rated for a capacity of 16 Ah. Find the total charge that can be delivered by this battery. Which battery in Table 2-1 is this battery?

Answer: 57,600 C. It is a D cell.

The amp-hour rating is very useful in forecasting how long a battery will last if the current drawn by the circuit is known. For example, a D cell can draw 1 A for 16 hours, 2 A for 8 hours, or 4 A for 4 hours. *Warning:* Batteries have a maximum current rating, which must not be exceeded.

Example 2-12

A 1.2 V watch battery is rated for 500 mAh. The watch draws 50 µA of current. How long does the battery last?

Solution

First, rewrite the battery capacity expression to solve for time t. Notice that the more current that is drawn, the shorter time the battery lasts.

$$t = \frac{Battery\ Capacity}{I}$$

Substitute the known values in the above expression, replace prefixes with powers of 10, cancel the ampere (A) units, and resolve the expression to a value.

$$t = \frac{500\ mAh}{50\ \mu A} = \frac{500\times10^{-3}\ Ah}{50\ \times\ 10^{-6}\ A} = 10^{4}\ h$$

Notice the simple cancellation of the units to produce units of time. 10^4 hours is equivalent to 416 days and 16 hours.

Practice

You just bought a new car battery rated at 40 amp-hours. Unfortunately, you leave your headlights and taillights on. Your headlights and taillights combined draw 20A of current. How long before the battery is fully discharged (dead)?

Answer: 2 hours

The true purpose of a battery is to create charge. How much charge can a battery discharge before it is exhausted? Rather than rating a battery in terms of how many coulombs of charge it can generate, the equivalent ampere-hour (Ah) or milliamp-hour (mAh) rating is used.

Table 2-2 expresses the capacity of alkaline batteries in amp-hours.

Table 2-2 Alkaline batteries' amp-hour ratings

Alkaline Battery	Approximate Amp-hour Capacity
AAA	1.1 Ah
AA	2.5 Ah
C	7.2 Ah
D	16.0 Ah
6 V Lantern	22.0 Ah
9 V Transistor	250 mAh

Amp-hours is a more popular and convenient way to express battery charge capacity. Alkaline batteries are relatively inexpensive but they are not rechargeable. So the alkaline battery is good for infrequent use such as flashlights. If used frequently, alkaline batteries are a very expensive energy solution.

A popular *rechargeable* battery is the *nickel-cadmium* battery (abbreviated NiCad) used in such things as flashlights, portable televisions, and bicycle lighting systems. Battery technology continues to evolve to produce more efficient, cost-effective energy.

A very popular *rechargeable* battery is the *lead acid* battery used in automobiles as shown in Figure 2-28. Actually, the lead acid battery consists of 6 individual 2.1 V cells connected to produce 12.6 V. Figure 2-28 provides a conceptual picture of a 12 V battery and its equivalent schematic symbol. Notice the difference of the schematic symbol for a voltage cell versus that for a voltage supply.

Think! *Start* at the (−) reference terminal on the right and proceed to the (+) terminal on the left. Each cell *raises* the voltage by 2.1 V. Thus, there is a total rise in voltage of 12.6 V. By connecting each cell end-to-end, the voltages add to create a total voltage. For example, a flashlight that requires 2 D cell batteries creates a voltage supply of 3 V, which in turn applies voltage to a 3 V lamp. What lamp should you use in a flashlight that requires 4 D cell batteries? The voltages add, so you need a 6 V lamp. If you were to put a 3 V lamp in such a flashlight, it would not last long, perhaps only seconds.

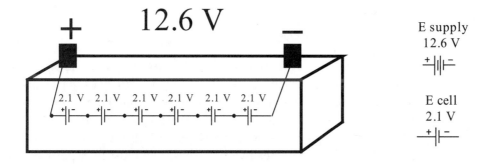

Figure 2-28 Car battery with six internal 2.1 V cells

Electronic Power Supplies

The electronic power supply takes commercial ac voltage and converts it to a dc voltage as demonstrated in Figure 2-29. Internally the power supply consists of several electronic devices. A full treatment of the dc power supply is presented later in this textbook.

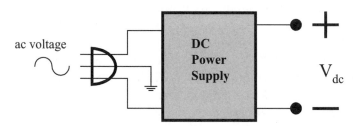

Figure 2-29 Block diagram of a DC power supply

Power supplies are built into equipment where commercial ac is available and dc voltage is required. This is a very convenient and inexpensive way to provide dc voltage for continuous use. Batteries are much more expensive and need replacing or recharging on a regular basis. A power supply produces dc voltage as long as the circuit has ac voltage into it. Also, different dc voltages can be provided by the power supply to feed different circuits in the equipment that require different dc voltages.

Positive DC Power Supply

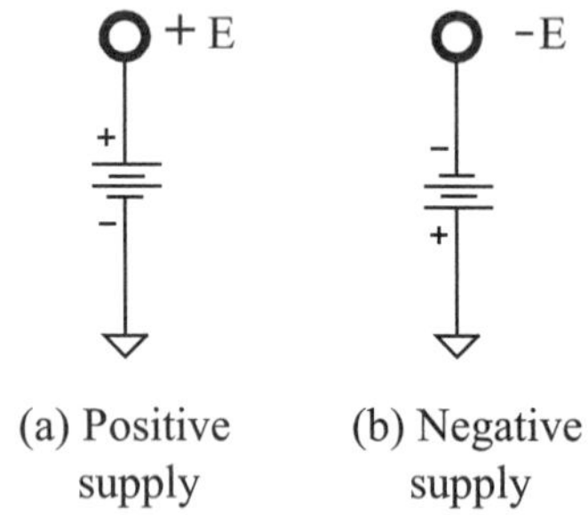

Figure 2-30(a) is the schematic symbol of a *positive power supply*. Note the orientation of this supply's *polarity*. The (–) terminal of the supply is tied to COM (0V). Going from its negative (–) terminal to its positive (+) terminal is a *rise* of E volts. Therefore, this is a *positive supply*. A quick look shows the negative (–) terminal is tied to COM, therefore the other supply terminal must be positive (+). Thus, this must be a positive (+) power supply.

Negative DC Power Supply

Figure 2-30(b) is the schematic symbol of a *negative power supply*. Note the orientation of this supply's *polarity*. The (+) terminal of the supply is tied to COM (0V). Going from its positive (+) terminal to its negative (–) terminal is a *fall* of E volts. Therefore, this must be a *negative supply*. A

(a) Positive supply

(b) Negative supply

Figure 2-30
Positive and negative power supply symbols

quick look shows the positive (+) terminal is tied to COM, therefore the supply terminal is negative (−). Thus, this is a negative (−) supply.

Triple Power Supplies

DC power supplies are also used as stand-alone equipment in the laboratory to produce various dc voltages for experimentation. Typically, a laboratory power supply provides several outputs that can vary from 0 V to maximum dc voltage. Figure 2-31 shows conceptually a triple power supply.

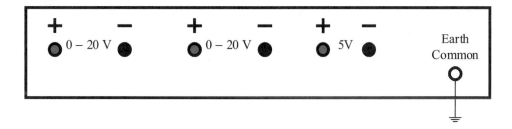

Figure 2-31 Triple independent dc power supply

In this dc power supply there are three separate power supplies. Two vary from 0 to 20 V and the third is a fixed 5 V supply (useful for 5 V digital electronic circuits). The three power supplies are independent of each other and have no common connection to each other. They are also not connected to earth common (not grounded).

If needed, the outputs could be physically connected to earth common as shown in Figure 2-32. *Common is typically considered to be the 0 V reference for the power supply, but earth common is an absolute 0 V reference.*

All of the negative (−) terminals are tied together and then tied to earth common. These terminals are now at 0 V potential. The left two power supplies vary from 0 to +20 V and the third terminal is now fixed at +5 V. These three positive supplies are available to supply an electronic circuit(s), but the common must also be connected to the circuit(s) to complete the conducting path.

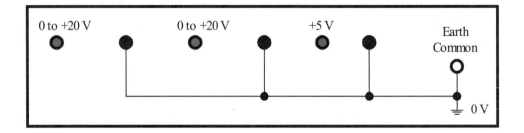

Figure 2-32 Positive triple dc power supply with earth common

Another popular triple power supply arrangement is shown in Figure 2-33. All three power supplies have a terminal *internally* connected to analog common (COM). The left power supply is a *positive supply* that can vary from 0 to +18 V. The middle power supply is a *positive supply* that can vary from 0 to +20 V. The right power supply is a *negative supply* that can vary from 0 to −20 V. *All terminal voltages are relative to COM.*

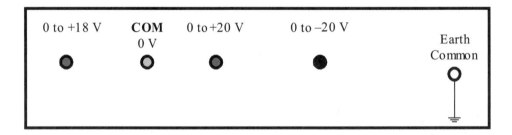

Figure 2-33 Triple dc power supply with internal analog common

Figure 2-34 shows the schematic representation of this dc power supply. The arrows through the power supply schematic symbols indicate that these are **variable power supplies**. The voltage being applied is a voltage difference between the power supply terminal and common. Both terminals must be connected to the circuit. Note the polarity of each supply with respect to COM; there are two positive supplies and one negative supply.

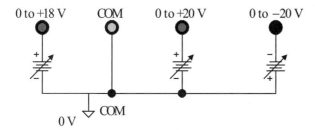

Figure 2-34 Triple dc power supply with analog common

Bubble Circuit Notation

A popular practice is to simplify the drawing of a complicated schematic by using *bubble notation* to represent a dc voltage supply. See Figure 2-35, which represents the above triple power supply in bubble notation.

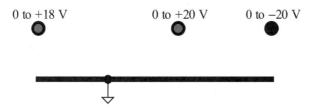

Figure 2-35 Bubble notation of triple power supply

A bubble is used to represent the terminal voltage provided by the power supply. *The schematic symbol is removed, but it is implied that the other terminal is connected to common.*

The *bubble notation* actually identifies the *power supply terminal* to connect to the circuit. Practically, the bubble notation is the best diagram to use when wiring your circuits. The bubbles represent the physical terminals of the supply.

Figure 2-36 demonstrates the conversion of a schematic notation of a dc voltage supply to its bubble notation.

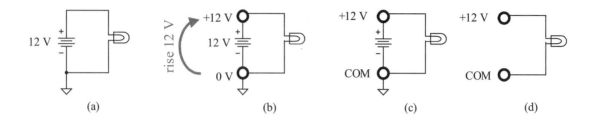

Figure 2-36(a,b,c,d) Converting to bubble notation.

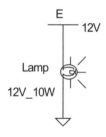

Figure 2-36(e)
MultiSIM bubble
notation

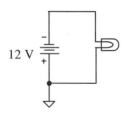

Figure 2-37
Example circuit

The schematic of Figure 2-36(a) is converted to the *bubble notation* of Figure 2-36(d) by the following steps:

- Figure 2-36(a): circuit with a 12 V dc supply

- Figure 2-36(b): the COM voltage is replaced by its 0 V equivalent and then the other battery terminal is identified as 12 V higher in potential (a **rise** from 0 V to +12 V).

- Figure 2-36(c): the top supply terminal is labeled with its terminal voltage value of +12 V and the bottom terminal is labeled COM for analog common.

- Figure 2-36(d): the schematic symbol is removed, leaving the bubble notation of +12 V.

Figure 2-36(d) is a much better wiring diagram to connect the circuit than is Figure 2-36(a). Figure 2-36(e) shows the MultiSIM bubble notation for this circuit with the line at the top representing voltage supply E with a voltage of 12 V.

Example 2-13
Using Figure 2-37, convert the circuit's dc supply to its bubble notation. Create MultiSIM schematics for both the circuit of Figure 2-37 and for its bubble notation.

Solution
- First, the *analog common* is identified at 0 V potential, refer to Figure 2-38(a). Next, in going from common (0 V) up to the other supply's terminal, there is a 12 V *fall* in potential. Thus, the other terminal must be at a −12 V potential *relative to common*.

- Figure 2-38(b) represents a *bubble notation* form often used in schematics to simplify schematic drawings.

- Figure 2-38(c) represents the *complete bubble notation* for this supply. The voltage supply symbol is removed and its –12 V terminal is drawn as a –12 V bubble and the analog common is replaced with its bubble notation.

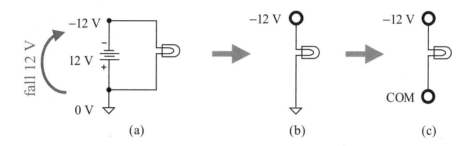

(a)   (b)   (c)

Figure 2-38(a,b,c) Example solution, converting to bubble notation.

The two equivalent schematics that can be created for this circuit are shown in Figure 2-38(d).

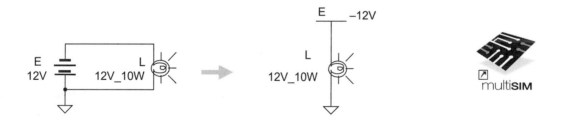

Figure 2-38(d) MultiSIM schematic and its equivalent bubble notation

---

Practice

A 6 V supply is replaced with a split supply of +3 V and –3 V. Using Figure 2-39, convert the circuit's dc supplies to their bubble notation. How much voltage is dropped across the lamp?

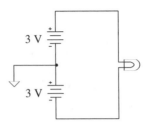

Figure 2-39 Practice circuit

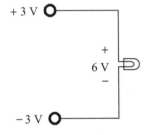

Figure 2-40
Practice circuit solution

Answer: See Figure 2-40. Note that this bubble notation does not show the analog common. *Analog common is assumed to be the other terminal* of the 3 V supply and the other terminal of the −3 V supply. Also, this is a much better wiring diagram for connecting your circuit. It shows that the +3 V terminal is connected to the top of the lamp, the −3 V supply is connected to the bottom of the lamp, and that common is not connected to the lamp. The lamp still drops 6 V.

You must also be able to convert bubble notation to dc supply schematic symbol notation to perform circuit analysis. The schematic symbol notation is often a more practical view of the circuit for circuit analysis. The schematic symbol is drawn between its bubble and common. If it is a *positive supply*, then the + terminal of the supply symbol is connected to the bubble's node and the − tied to common. If it is a *negative supply*, then the − terminal of the supply symbol is connected to the bubble's node and the + tied to common.

Example 2-14

Using Figure 2-41, convert the dc supply bubble notation to dc supply symbols and redraw the circuit.

Perform each of the individual steps to convert this bubble notation to its schematic symbol notation.

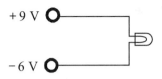

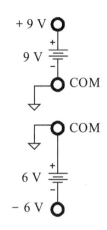

Figure 2-41 Example circuit

Figure 2-42
Example intermediate
solution

Solution

First convert each supply bubble to its equivalent schematic supply symbol as shown in Figure 2-42. For bubble notation, the other terminal of the power supply is always assumed to be common. The top circuit of Figure 2-42 shows the positive supply; the bottom circuit, the negative supply. Now substitute these voltage supply symbols into the original schematic in place of the bubble notation, placing the common conveniently in one place central to both supplies. See Figure 2-43 to visualize the complete conversion.

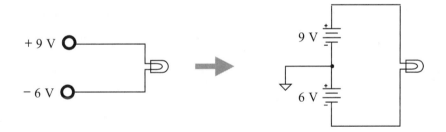

Figure 2-43 Example solution—converting to schematic symbols

2.9 Node Voltage

A **simple node** is a single point on a conductor of a circuit as seen in Figure 2-44. In this case, each of these nodes is on the same conducting wire. They are all at the same +3 V potential. We could use just *one of these* *nodes* to represent all of the node points on this wire; the rest are not needed. Thus, one of these node points would uniquely identify this node. This is also called a **voltage node** because it has the same voltage no matter which simple node point you pick.

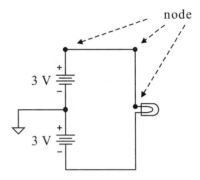

Figure 2-44 Simple node points

Figure 2-45 displays the 5 *unique nodes* of its circuit: nodes *a, b, c, d,* and *e.* From our previous work, we know that the *node voltage* of *e* is assumed to be 0 V (analog common); the node voltage at *a* is +3 V, and the node voltage at *d* is at −3 V. The node voltages of nodes *b* and *c* are unknown since we do not know the voltage drops of the lamps.

To better understand node voltage values, you must grasp the difference between node voltage values relative to common and physical component voltage drops.

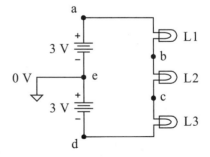

Figure 2-45 Unique nodes

Node Voltage Scale

The concept of a node voltage scale can be compared to that of temperature. For example, 2 V is a higher node voltage than −4 V just as 2 degrees is warmer than −4 degrees. The **node voltage scale** of Figure 2-46 is a helpful visual aid.

    Warning: be careful to distinguish between this relative node voltage scale and the actual physical voltage. A −100 V supply (magnitude of 100 V) is physically a much greater physical voltage than a 15 V supply (magnitude of 15 V). This is true even though, on the node voltage scale, 15 V is considered a relatively higher node voltage potential. This distinction is critical and will become clearer with practice.

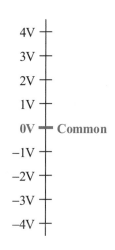

Example 2-15
> Which is the higher node voltage potential and which is the greater physical voltage?
>
> **a.** 3 V and 1 V
>
> **b.** 3 V and −1 V
>
> **c.** 3V and −4 V
>
> **d.** 0 V and −4 V
>
> **e.** −3 V and −4 V

Figure 2-46
Node voltage scale

Solution
> Use the node voltage scale to find the relationship between node voltages. Compare the magnitudes of the numbers to determine the actual physical relationship.
>
> **a.** Node voltage   3 V > 1 V,   *physically* |3 V| > |1 V|.
>
> **b.** Node voltage   3 V > −1 V,   *physically* |3 V| > |−1 V|.
>
> **c.** Node voltage   3 V > −4 V,   *physically* |−4 V| > |3 V|.
>
> **d.** Node voltage   0 V > −4 V,   *physically* |−4 V| > |0 V|.
>
> **e.** Node voltage   −3 V > −4 V,   *physically* |−4 V| > |−3 V|.

2.10 Circuit Voltage Notation

Voltage is always measured between 2 points (that is, 2 nodes). In Figure 2-47(a), node *a* and node *b* represent any two voltage nodes in a circuit. The notation V_{ab} notes that the voltage at node *a* is being stated with respect to node *b*. That is, if you were standing at node *b*, what is the voltage at node *a*? Is it greater than (+) or less than (–) your node's voltage? Greater or lesser by how much?

(a)   (b)

Figure 2-47 Node voltage difference

If a voltage is listed with only *one node subscript*, as shown in Figure 2-47(b), then the second node subscript is *assumed to be common*. V_a represents the voltage at node *a* relative to common (that is, the *reference node is common*).

The *voltage difference* of V_{ab} is defined to be the difference of the two nodes (node *a* voltage less the reference node *b* voltage)

$$V_{ab} = V_a - V_b \tag{2-6}$$

where V_a and V_b are the voltages with respect to common at nodes *a* and *b*, respectively. The popular name for a *voltage difference* is a **voltage drop**.

Example 2-16

V_a is 15 V and V_b is 3 V. Find V_{ab}.

a   b
●   + V_{ab} –   ●
15 V   3 V

Solution

By definition,

$$V_{ab} = V_a - V_b = 15 \text{ V} - (3 \text{ V}) = 12 \text{ V}$$

Practical approach: Start at node *b* (the reference node is at 3 V) and walk to node *a* (at 15 V). You stepped up 12 V.

Practice

V_a is 15 V and V_b is −3 V. Find V_{ab}.

Answer: 18 V; node *a* is 18 V higher than node *b*.

Equation 2-6 can also be used to find a node voltage if the other node voltage and the voltage difference (rise or fall) are known.

Example 2-17

V_a is 15 V and V_{ab} is 18 V. Find V_b.

Solution

Rewrite $V_{ab} = V_a - V_b$ to solve for the node *b* voltage V_b.

$$V_b = V_a - V_{ab} = 15 \text{ V} - 18 \text{ V} = -3 \text{ V}$$

Practical approach: Node *a* is at 15 V. Fall 18 V to get node *b*. The result is −3 V.

Practice

V_b is 3 V and V_{ab} is 12 V. Find V_a.

Answer: 15 V (do mathematically and using the *practical approach*)

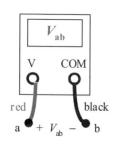

Figure 2-48
Measuring V_{ab}

Measuring Voltage Properly

A voltmeter must be connected properly to accurately measure voltage value and polarity between two nodes. Figure 2-48 demonstrates measuring the voltage V_{ab}. To measure the voltage V_{ab}, the voltmeter *must be connected* such that

- V (red lead) is connected to node a

- COM (black lead) is connected to node b

Thus the V (red lead) is connected to the first node subscript (node a) and the COM (black lead) is connected to the second subscript node (node b). It is critical that you understand and apply this voltage measuring technique to properly measure the correct polarity of a voltage difference. If you reverse the voltmeter test leads, you will incorrectly measure the voltage polarity.

Example 2-18

The voltmeter is connected such that the red lead is connected to a 15 V node and the black lead is connected to a −3 V node. Find the measured voltage drop.

Solution

$$V_{\text{red lead}} - V_{\text{black lead}} = (15 \text{ V}) - (-3 \text{ V}) = 18 \text{ V}$$

Practice

The voltmeter is connected such that the red lead is connected to a +9 V node and the black lead is connected to a +3 V node. Find the measured voltage drop. If the leads are reversed, find the measured voltage drop.

Answer: 6 V, −6 V

Circuit Voltage Notation

Figure 2-49 demonstrates node voltages, current direction impact on voltage drop polarities, and voltage drops.

Based upon the circuit of Figure 2-49(a), the net voltage being applied to the circuit is the difference between nodes a and d. Node a is 3 V *above* common, thus $V_a = +3$ V. Node d is 3 V *below* common, thus

$V_d = -3$ V. The difference between these two nodes, a and d, is the net voltage being applied to the circuit, namely 6 V.

$$E = V_{ad} = V_a - V_d = 3\text{ V} - (-3\text{ V}) = 6\text{ V}$$

See Figure 2-49(b), which shows the net voltage being applied to the load (the three lamps). The lamps are identical (ideally) and drop the same voltage, dividing up the 6 V equally. Thus each lamp drops 2 V. The current I exits the +E terminal, flows down and through the lamps, and returns to the −E terminal.

Note the *polarity* of the voltage drop of each lamp. As the current flows through each lamp, voltage potential is lost. This is indicated by the relative (+) and (−) polarity of the lamp's voltage drop. The (+) *polarity* indicates that this is the *higher potential terminal* of the device and the (−) indicates that this is the lower potential terminal of the device.

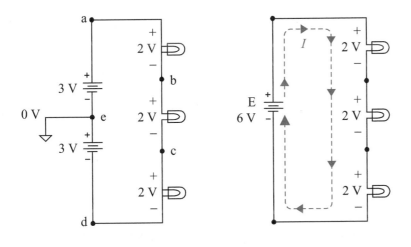

(a) Complete schematic   (b) Schematic showing net supply voltage

Figure 2-49 Sample circuit to study voltage notation

When dropping voltage across energy dissipating components, like these lamps, conventional current always creates a voltage drop such that the current *enters* the higher potential terminal (indicated by + sign) and *exits* the lower potential terminal (indicated by the − sign). *Note this does not apply to voltage supplies that are manufactured to have a positive terminal and a negative terminal regardless of circuit current direction.*

Example 2-19

Find all the node voltages of Figure 2-49(a).

Solution

Since *common* is assumed to be 0 V, it must be our *starting node*. It is the only known node voltage. Proceeding *clockwise* around the circuit, add voltage if you encounter a *rise in voltage* (− to +) and subtract voltage if you encounter a *fall in voltage* (+ to −). This rule is *not* device dependent; it is only polarity dependent.

- *Node e*
  common, thus $V_e = 0$ V.

- *Node a*
  node e is 0 V, top supply *raises* 3 V, thus $V_a = 3$ V

- *Node b*
  node a is 3 V, top lamp *falls* 2 V, thus $V_b = 1$ V

- *Node c*
  node b is 1 V, middle lamp *falls* 2 V, thus $V_c = -1$ V

- *Node d*
  node c is −1 V, bottom lamp *falls* 2 V, thus $V_d = -3$ V

- *Node e*
  node d is −3 V, bottom supply *raises* 3 V, thus $V_e = 0$ V

See Figure 2-50(a); it shows all of the resulting node voltages in the ovals. Starting at common at $V_e = 0$ V, rise 3 V to $V_a = +3$ V, fall 2 V to $V_b = 1$ V, fall 2 V to $V_c = -1$ V, fall 2 V to $V_d = -3$ V, and rise 3 V back to common and $V_e = 0$ V. Note: each dark line represents a single node. For example, *node a includes the + terminal of the top supply, the top terminal of the top lamp, and their connecting wire.*

Also note: we started at node *e* at 0 V, went clockwise around the circuit adding and subtracting voltage rises and falls, and ended back at node e at 0 V. As we proceeded clockwise around the circuit, voltages were added (with a device raise in potential) and subtracted (with a device fall in potential). This rule is not device dependent.

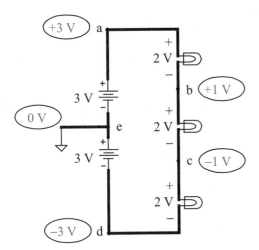

Figure 2-50(a) Example solution – *node voltages shown in ovals*

Practice

Using Figure 2-50(a), start at common and go *counterclockwise* around the circuit and find the node voltages. Note: follow the same rule as in the example. If you encounter a *voltage rise* (− to +) then add the voltage. If you encounter a *voltage fall* (+ to −) subtract the voltage. This rule is *not* device dependent.

Answer: You must get the same node voltage values whether you go clockwise or counterclockwise around a circuit. The answers are the same as previously found.

Let's find other useful voltage drops in the circuit of Figure 2-50(a).

Example 2-20

Find the following voltage differences (drops) of Figure 2-50(a): V_{ac}, V_{ca}, and V_{be}.

Solution

$$V_{ac} = V_a - V_c = (3\ \text{V}) - (-1\ \text{V}) = 4\ \text{V}$$

$$V_{ca} = V_c - V_a = (-1\ \text{V}) - (3\ \text{V}) = -4\ \text{V}$$

$V_{ca} = -V_{ac}$ This is equivalent to reversing the voltmeter leads.

$V_{be} = V_b = 1\ \text{V}$ Node e is common.

Practice

Find the following voltage differences (drops) of Figure 2-50(a): V_{bd}, V_{db}, and V_{de}.

Answer: 4 V, –4 V, –3 V

Exploration

Use MultiSIM and verify the example and practice results. Figure 2-50(b) demonstrates the verification of three measurements.

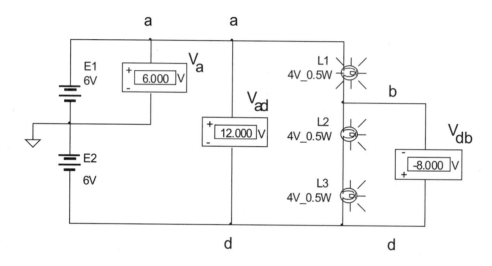

Figure 2-50(b) MultiSIM measurements of V_a, V_{ad}, and V_{db}

Notice that the + terminal of the voltmeter corresponds to the *first voltage subscript* and the – terminal corresponds to the *second voltage subscript* or common if the second subscript is missing. Notice that the V_{db} meter measures a –8 V.

Looking from another perspective, you can use the component voltage drops to calculate other voltages in a circuit. The following example demonstrates this approach.

Example 2-21

From the circuit of Figure 2-51, find: V_{ab}, V_a, V_{ac}, E, V_{ca}. Note: the lamps are different and drop different voltages.

Solution

In this case, use the component drops to find the desired voltages. For V_{ab}, start at the reference b node and raise 2 V to get to the target a node. Thus,

$$V_{ab} = 2\,V$$

V_a and V_{ac} are the same value; they both have common as their reference node. For V_{ac}, *start at reference node c* (common), raise 5 V, and then raise 2V to get to target node a.

$$V_a = V_{ac} = 5\,V + 2\,V = 7\,V$$

The supply E is the same as V_a and V_{ac}, therefore the supply voltage must also be 7 V. That is, E = 7 V.

V_{ca} is $-V_{ac}$, thus, $V_{ca} = -7$ V. Or, you can start at the reference a node and go to the target node c and get the same value (falling 2 V and falling 5 V is a total fall of 7 V).

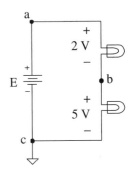

Figure 2-51
Example circuit

Practice

For the circuit of Figure 2-51, find: V_{bc}, V_b, V_{cb}.

Answer: 5 V, 5 V, −5 V

Summary

Electricity is created by a difference of electrical charge and the resulting flow of charge. The electron and the proton are the charged particles of the atom that produce electricity. Electrons are considered negatively charged particles; protons, positively charged particles. You must know the basics of atomic structure and how protons and electrons interact to understand the basics of electricity, electronics, and component composition.

How easily electrons flow depends on the nature of the material. Electrical material is categorized as conductors (electrons flow easily), insulators (electrons do not flow easily), and semiconductors (somewhere between a conductor and an insulator).

Current and voltage are the primary variables of electricity. Current is created by electrons moving in the same direction, like drops of water moving in a pipe. A voltage supply is created by separating charges of opposite polarity. That is, a cluster of positive charge separated from a cluster of negative charge creates a voltage difference between the two charged clusters. If a continuous circuit is connected to them, charge flows through that circuit.

An ideal dc voltage supply produces a fixed voltage under various load conditions. An ideal supply is the simplest model of a real supply. The ideal voltage supply model is very useful in analyzing circuits. Many real voltage supplies approach the ideal model.

You need to understand how to properly connect current and voltage meters to the circuit; and, if improperly connected, what are the possible consequences to the circuit and the meter.

Voltage is always measured between two points. Frequently a common return is used in a circuit. You must understand and properly use the three standard types of common returns that are used in electricity and electronics: earth, analog, and chassis return commons. Node voltage and bubble voltage supply notation are voltages measured relative to common; they are used extensively in electronics. You must clearly understand and properly apply voltage notation (node voltage notation and component voltage drop notation).

Problems

Atomic Structure and Charge

2-1   Name the three basic elements of an atom and the associated charge of each (positive, negative, or neutral).

2-2   Like charges attract, repel, or have no effect on each other?

2-3   Opposite charges attract, repel, or have no effect on each other?

2-4   Charge:
   a. What is the symbol for charge?
   b. What is the basic unit of charge?
   c. What is the symbol for the unit of charge?

2-5   Number of charged particles in a coulomb:
   a. How many electrons are needed to create 1 C of negative charge?
   b. How many protons are needed to create 1 C of positive charge?

2-6   Find the number of electrons needed to create 30 coulombs of charge.

2-7   Write Coulomb's law.

2-8   Find the electrical force of attraction of a 6 coulomb positive charge and an 18 coulomb negative charge separated by a distance of 12 centimeters.

2-9   Find the electrical force of repulsion of a 100 millicoulomb negative charge and a 20 millicoulomb negative charge separated by 50 millimeters.

2-10   Aluminum has an atomic number of 13.
  a. Draw its shell configuration.
  b. Prove that it is a conductor.

2-11   Silver has an atomic number of 47. Prove that it is a conductor. Based on its atomic structure, should silver be a better or worse conductor than copper?
  a. Draw its shell configuration.
  b. Prove that it is a conductor.

2-12   What is the valence band?

Conductor, Insulator, Semiconductor
2-13   Briefly describe the main characteristics of:
  a. Conductors
  b. Insulators
  c. Semiconductors

2-14   What is a free electron?

2-15   What is a positive ion?

2-16   What is a negative ion?

Current
2-17   Is a continuous circuit path required for current to exist?

2-18   Current fundamentals:
  a. What is the symbol for current?
  b. What is the basic unit of current?
  c. What is the symbol for the unit of current?

2-19   What is the difference between electron and conventional current flow?

2-20   A 100 watt light bulb draws 1 A of current.
    a. How many coulombs of charge must flow per second to create this current?
    b. How many coulombs of charge must flow per hour to create this current?

2-21   An electric furnace requires 1800 coulombs to flow per minute to operate. Find its current.

2-22   A hair dryer requires 240 coulombs of charge to flow in 2 minutes to operate. Find its current.

2-23   Find the period of time that 15 mC of charge must flow to create a current of 30 μA.

Current Meter Connection

2-24   Is current a *through* or an *across* parameter? Which is proper terminology, (a) the current *through* the lamp or (b) the current *across* the lamp?

2-25   Do you need to break the circuit to measure current?

2-26   If you misconnect the current meter, could it cause excessive current?

2-27   Draw the schematic of Figure 2-52 with the current meter properly connected to measure the current flowing out the voltage supply. Indicate the ammeter *red lead* and *black lead* to produce a positive current reading for conventional current. Should the voltage supply be *on* or *off* to measure current?

2-28   Draw the schematic of Figure 2-52 with the current meter properly connected to measure the current flowing into the lamp. Indicate the ammeter *red lead* and *black lead* to produce a positive current reading for conventional current. Should the voltage supply be *on* or *off* to measure current?

2-29   A digital current meter indicates a negative value. What does the negative indicate?

Voltage

2-30   Find the work performed to lift 150 pounds a distance of 100 yards in English units and in SI units. If the load is lifted the same distance twice as fast, is the work performed changed?

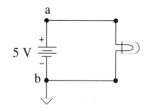

Figure 2-52

2-31 Find the work performed to lift 2000 pounds a distance of 100 yards in English units and in SI units. If the load is lifted the same distance twice as fast, is the work performed changed?

2-32 If the load of Problem 2-30 is lifted in 1 minute, find the power required.

2-33 If the load of Problem 2-30 is lifted in 20 seconds, find the power required.

2-34 Is a continuous circuit path required for a voltage to exist?

2-35 Voltage fundamentals:
a. What is the symbol for voltage?
b. What is the basic unit of voltage?
c. What is the symbol for the unit of voltage?

2-36 Find the work performed for a 1.5 V supply to raise 5 coulombs of charge.

2-37 Find the work performed for a 9 V supply to raise 10 coulombs of charge.

2-38 Find the charge required for a 12 V lamp to dissipate 6 joules of energy in heat and light.

2-39 Find the voltage dropped across a hearing aid that dissipates 600 millijoules of energy with 0.5 coulombs of charge.

2-40 What charge is required for a 2 V lamp to dissipate 100 J of energy in light and heat energy?

2-41 A 100 watt light bulb dissipates 100 joules of energy per coulomb of charge. Find its voltage drop.

2-42 A battery of 5 V can generate 57,000 C of charge over its lifetime. How much work is performed over its lifetime?

2-43 As current flows through a voltage supply, is there a rise or a fall in voltage potential?

2-44 As current flows through a lamp, is there a rise or a fall in voltage potential?

Voltmeter Connection

2-45   Is voltage a *through* or an *across* parameter? Which is proper terminology, (a) the voltage *through* the lamp or (b) the voltage *across* the lamp?

2-46   Do you need to break the circuit to measure voltage?

2-47   If you misconnect the voltmeter, could excessive current flow through it?

2-48   Draw the schematic of Figure 2-52 with the voltmeter properly connected to measure the voltage of the voltage supply. Indicate the voltmeter *red lead* and *black lead* to produce a positive voltage reading.

2-49   Draw the schematic of Figure 2-52 with the voltmeter properly connected to measure the voltage of the lamp. Indicate the voltmeter *red lead* and *black lead* to produce a positive voltage. Should the voltage supply be *on* or *off* to measure the voltage applied to the lamp?

2-50   Draw the schematic of Figure 2-52 with the voltmeter properly connected to measure the voltage V_a. Indicate connection of the voltmeter *red lead* and *black lead*.

2-51   Draw the schematic of Figure 2-52 with the voltmeter properly connected to measure the voltage V_{ab}. Indicate connection of the voltmeter *red lead* and *black lead*.

2-52   A voltmeter indicates a negative value. What does the negative indicate?

Common

2-53   Draw the schematic symbol for earth common and briefly describe the meaning of earth common.

2-54   Draw the schematic symbol for analog common and briefly describe the meaning of analog common.

2-55   Draw the schematic symbol for chassis return common and briefly describe the meaning of chassis common.

Ideal Fixed (DC) Voltage Supply

2-56   Draw the schematic symbol of an ideal voltage supply. Properly label its terminals as positive (+) and negative (−).

2-57   A 12.6 V car battery will operate for 3 hours while drawing a current of 15 amps.
  a. What is the battery rating in amp-hours?
  b. How much charge does it generate in 3 hours?
  c. How much energy (work) does the battery produce?

2-58   The same 12.6 V battery of the preceding problem draws a current of 30 A.
  a. How long until the battery is discharged?
  b. How much charge is discharged?
  c. The battery is recharged with a trickle current of 250 mA. How long does it take to recharge the battery?

Node Voltage

2-59   Use the schematic of Figure 2-53.
  a. The current flows *out of* or *into* a supply's positive terminal?
  b. The current flows *clockwise* or *counter*clockwise?
  c. Redraw the schematic in *bubble notation*.
  d. Is this a positive (+) or negative (−) voltage supply?
  e. Find V_b, V_a, and V_{ab}.
  f. If this supply is a battery that can supply a 2 A current for 2.5 hours, what is its battery capacity rating?
  g. The voltmeter's red lead is connected to node a and its black lead to node b. What voltage is displayed?

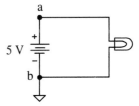

Figure 2-53

Circuit Voltage Notation

2-60   Use the schematic of Figure 2-54.
  a. The current flows *out of* or *into* the supply's positive terminal?
  b. The current flows *clockwise* or *counter*clockwise?
  c. Redraw the schematic in *bubble notation*.
  d. Is this a positive (+) or negative (−) voltage supply?
  e. Find V_b, V_a, and V_{ab}.
  f. The voltmeter's red lead is connected to node *a* and its black lead to node b. What voltage is displayed?
  g. The voltmeter's leads are then reversed. What voltage is displayed?
  h. If this supply is a battery rated at 15 Ah, how long can it supply a current of 3 A?

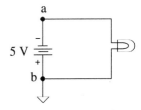

Figure 2-54

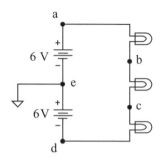

Figure 2 -55

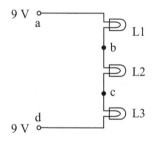

Figure 2-56

2-61   Use the schematic of Figure 2-55.
   a. Draw the schematic and show the direction that current flows.
   b. Find node voltages V_e, V_a, and V_d and show them on the schematic next to their nodes.
   c. To measure node voltage V_a, where would you place the red lead and the black lead of the voltmeter?
   d. Find V_{ad}. What is the total net voltage being applied to the lamps by the voltage supplies?
   e. If the lamps drop equal voltages, show the lamp voltage drops with polarity on your schematic.
   f. Find V_{ac} and V_{ca}.
   g. To measure V_{ac}, where would you place the red lead and the black lead of the voltmeter?
   h. Find V_{bd} and V_{db}.
   i. Redraw the schematic in *bubble notation*. Are any of the lamps connected to common?

2-62   Use the schematic of Figure 2-56.
   a. Draw the schematic, converting the bubble notation to the corresponding schematic symbols. Note the polarity and value of each voltage supply on your schematic.
   b. Find node voltages V_a and V_d and show them on your schematic next to their nodes.
   c. Find V_{ad}. What is the total net voltage being applied to the lamps by the voltage supplies?
   d. If the lamps drop equal voltages, find the lamps' voltage drop. Show the lamp voltage drops with polarity on your schematic.
   e. Find V_b and V_c.
   f. To measure node voltage V_b, where would you place the red lead and the black lead of the voltmeter?
   g. Find V_{ac} and V_{ca}.
   h. To measure V_{ca}, where would you place the red lead and the black lead of the voltmeter?
   i. Find V_{bd} and V_{db}.
   j. Are any of the lamps connected to common?

2-63   Assume the bottom supply's polarity of Figure 2-56 is accidentally reversed.
   a. Find the node voltages V_a and V_d and show them on the schematic next to their nodes.
   b. Find V_{ad}. What is the total net voltage being applied to the lamps by the voltage supplies?
   c. Will the lamps light?

3

Resistance

Introduction

In this chapter you will learn about a basic characteristic of electrical and electronic materials called resistance. All materials resist flowing charge, some more or less than others. You will learn to calculate resistance based upon the material's shape and inherent resistance to current. You will learn to characterize the resistive behavior of a component with a characteristic curve, which relates its current and voltage. You will learn about resistance extremes of the short and the open, which you will apply to the characteristics of a switch and a diode. You will learn about the carbon composite resistor and how to read its color bands to determine its value. You will learn to properly measure resistance and test diodes with a test instrument called an ohmmeter.

Objectives

Upon completion of this chapter you will be able to:

- Define and calculate resistance in terms of material resistivity and component geometry.

- Relate resistivity and resistance to conductor, insulator, and semiconductor materials.

- Define and calculate conductance and relate it to resistance.

- Interpret and apply component characteristic curves.

- Apply the concept of a short and an open.

- Apply switches in an electrical circuit.

- Model the ideal diode, the practical diode, and the LED.

- State the rating of and calculate the resistance range of a color-coded carbon-composite resistor.

- Properly use an ohmmeter to measure resistance and check diodes.

3.1 Resistance Fundamentals

All materials resist flowing charge. Electrons bounce off of atoms, recombine with ions, and regenerate electrons that create current.

The parameter symbol for electrical **resistance** is the letter **R**. The unit of electrical resistance is the **ohm**. The ohm's unit symbol is the Greek letter **omega (Ω)**. For example, a resistance with a value of 220 ohms is written as $R = 220\ \Omega$.

A more formal definition of electrical resistance, based upon a component's electrical material and geometry, is stated in Equation 3-1.

$$R = \rho\,\frac{l}{A} \tag{3-1}$$

where
R = resistance in ohms (Ω)
ρ = resistivity of the electrical material in ohm-meters (Ω-m)
l = length of material in meters (m)
A = cross-sectional area of material in square meters (m^2)

Resistivity

Resistivity is an inherent characteristic of the material. The resistivity of:

- A conductor is very low since it offers very little resistance to charge flow (i.e., current).

- An insulator is very high, since it significantly resists charge flow.

- A semiconductor is between that of a conductor and an insulator.

The symbol for resistivity is the Greek letter **rho (ρ)**. Table 3-1 lists the resistivity values for some typical materials.

Figure 3-1 shows two basic shapes (cylindrical and rectangular) commonly utilized in constructing conductors and resistors.

Resistance – General Geometric Observations

The following general observations can be made of the effect of the material and geometry of a resistance R:

- Double the resistivity R doubles

- Double the length R doubles

- Double the area R halves

The first two observations should be somewhat obvious. If the resistivity of the material is twice as much, then its resistance is twice as much. If the length of the material is twice as long, its resistance is twice as much.

The last observation is mathematically obvious but is also intuitive. A larger cross-sectional area allows more flowing charges to fit into the cross-sectional area. An analogy is water flowing in a pipe. A larger pipe allows more water to flow. The larger cross-sectional area offers less resistance to water flow. Likewise a larger cross-sectional area of wire offers less resistance to electrons flowing.

Table 3-1 Resistivity

Material	Resistivity ($\Omega \cdot$m)	Type of Material
Silver	16.4×10^{-9}	conductor
Copper	17.2×10^{-9}	conductor
Gold	24.4×10^{-9}	conductor
Aluminum	28.2×10^{-9}	conductor
Tungsten	54.7×10^{-9}	conductor
Mercury	954×10^{-9}	conductor
Carbon	0.00004	semiconductor
Germanium	0.47	semiconductor
Silicon	640	semiconductor
Paper	10×10^{9}	insulator
Mica	500×10^{9}	insulator
Glass	1000×10^{9}	insulator
Teflon	3000×10^{9}	insulator

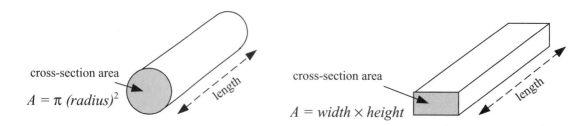

Figure 3-1 Basic geometric shapes used for resistors and conductors

Conductors

Conductors have a *very low resistivity* value. A conductor has a plentiful number of free electrons and allows them to travel easily through its material. Silver is listed in Table 3-1 as the best conductor, but its cost is high. Copper is a close second in lowest resistivity. Copper is a very malleable (easily shaped) metal that is not too brittle. Its cost is more reasonable than silver or gold, and therefore, it has become the conductor of choice in electrical and electronics. Gold is a good malleable conductor but is very expensive. Aluminum is relatively inexpensive but has a higher resistivity.

Example 3-1

Find the resistance of 100 feet of copper wire with a diameter of 50.8 mils.

Solution

Convert the diameter to a radius r and express its units in meters.

First, divide the diameter by 2 to obtain the radius. Second, convert from the unit of mils to inches (note that 1 mil = 10^{-3} inch). Third, convert inches to the SI unit of centimeters. Fourth, convert centimeters to the unit of meters. The following expression for the radius incorporates all four of these steps into one equation. Note that units cancel to produce the final desired result of radius r in units of meters.

$$r = \frac{50.8 \text{ mils}}{2} \times \frac{10^{-3} \text{ in}}{\text{mil}} \times \frac{2.54 \text{ cm}}{\text{in}} \times \frac{1 \text{ m}}{100 \text{ cm}} = 645.2 \times 10^{-6} \text{ m}$$

The cross-sectional area A is

$$A = \pi r^2 = \pi \, (645.2 \times 10^{-6} \text{ m})^2 = 1.308 \times 10^{-6} \text{ m}^2$$

Next, convert the length to SI units.

$$l = 100 \text{ ft} \left(\frac{12 \text{ in}}{\text{ft}} \right) \left(\frac{2.54 \text{ cm}}{\text{in}} \right) \left(\frac{1 \text{ m}}{100 \text{ cm}} \right) = 30.48 \text{ m}$$

From Table 3-1, the resistivity of copper is 17.2×10^{-9} Ω·m. Now use Equation 3-1 to calculate the resistance of this wire.

$$R = \rho \frac{l}{A} = (17.2 \times 10^{-9} \Omega \cdot \text{m}) \times \left(\frac{30.48 \text{ m}}{1.308 \times 10^{-6} \text{ m}^2} \right) = 0.4 \ \Omega$$

Practice

Find the resistance of a 1000 foot piece of aluminum wire with a diameter of 10 mils.

Answer: 170 Ω

Insulators

Insulators have a very high resistivity value. Insulators cannot easily create free electrons and have a very high resistance to charge flow. Insulating material is used to cover wire to electrically isolate it from the chassis and other wires. Insulating material is used to encapsulate components, insulating them from neighboring components and wires. Insulating material is also used as an essential ingredient in the construction of components, for example, capacitors and transformers.

Example 3-2

A Teflon shim is used to insulate between two metal plates, forming a capacitor. The Teflon shim is 2 mm thick. The area is 1 cm by 1 cm. What is its resistance?

Solution

From Table 3-1, the resistivity of Teflon is $3000 \times 10^9 \, \Omega \cdot m$. Use Equation 3-1 to calculate the resistance of this insulating material.

$$R = (3000 \times 10^9 \, \Omega \cdot m) \times \left(\frac{0.002 \text{ m}}{0.01 \text{ m} \times 0.01 \text{ m}} \right) = 60 \times 10^{12} \, \Omega$$

This very small, very thin piece of Teflon has 60 Teraohms (TΩ) or 60,000,000,000,000 ohms of resistivity. If this material were a copper conductor it would have a resistance of only 143 nanoohms (nΩ) or 0.000000143 Ω. Wow! What a difference resistivity makes.

Practice

A layer of Mica insulation is used as an insulator between two conducting plates to form a capacitor. It is 6 mm thick and has an equivalent area of 1 square meter. Find its resistance.

Answer: 3 GΩ

Semiconductors

From Table 3-1, it is apparent that carbon, germanium, and silicon are neither conductors nor insulators. These materials fall between the conductor and the insulator. They create and allow carriers to flow but not as easily as a conductor. Carbon is used as a basic material in a resistive type of component classified as the carbon-composite resistor. We shall look at this resistor type later in this chapter.

As noted previously, germanium and silicon are very special materials used in the construction of devices such as transistors, integrated circuits (ICs), and diodes. The name **semiconductor** is used to denote a material of germanium or silicon even though there are other materials such as carbon that generally fit the semiconductor category (not a conductor and not an insulator). Material composed of pure germanium or pure silicon is called an **intrinsic semiconductor**.

Example 3-3

A cylindrical germanium component has a length of 5 mm and a radius of 2 mm. Find its resistance.

Solution

The cross-sectional area is

$$A = \pi r^2 = \pi (2 \times 10^{-3} \text{ m})^2 = 12.6 \times 10^{-6} \text{ m}^2$$

From Table 3-1, the resistivity of germanium is 0.47 Ω·m. Now use Equation 3-1 to calculate the resistance of this germanium material.

$$R = (0.47 \ \Omega \cdot \text{m}) \times \left(\frac{5 \times 10^{-3} \text{ m}}{12.6 \times 10^{-6} \text{ m}^2} \right) = 187 \ \Omega$$

Practice

A cylindrical silicon component has a length of 4 mm and a radius of 2 mm. Find its resistance.

Answer: 204 kΩ

3.2 Conductance

Conductance characterizes how easily charge flows in a material. The easier charge flows in a material, the more conductive a material is. Conductance is the *opposite* of resistance. The more conductive a material is, the less resistive the material.

The parameter symbol for electrical conductance is the letter **G**. The unit of electrical conductance is the **siemens**. The unit symbol for siemens is the uppercase letter **S**. Recall that the lowercase s is reserved for the unit of seconds. For example, a wire with a conductance value of 2 siemens is written as $G = 2$ S.

Conductance is the reciprocal of resistance as shown in Equation 3-2.

$$G = \frac{1}{R} \qquad \qquad \textbf{(3-2)}$$

where

G = conductance in siemens (S)
R = resistance in ohms (Ω)

G = conductance
Siemens = unit
S = unit symbol

An archaic unit of conductance is the mho (ohm spelled backwards). Its unit symbol is an upside-down omega. You may encounter this terminology elsewhere but this text consistently uses the standard unit of siemens for conductance.

Example 3-4

> The resistance of 1000 feet of wire is measured to be 3.31 Ω. Find the wire's conductance.

Solution

> The conductance is the reciprocal of resistance. Thus,

$$G = \frac{1}{R} = \frac{1}{3.31\ \Omega} = 0.302\ \text{S}$$

> This can also be written in millisiemens as 302 mS.

Practice

> A carbon resistance material has a conductance of 10 μS. Find its resistance.

Answer: 100 kΩ

3.3 Characteristic Curve

It is sometimes convenient or necessary to use a graph to characterize the resistance of a component or a circuit by relating its current and voltage. This is especially true for nonlinear resistances, where the resistance value changes with different values of current through the component. It is *standard practice* to plot the current I *through* the component as a function of voltage V *across* the component. Thus the current I is plotted on the *vertical axis* as a function of the voltage V on the *horizontal axis* as shown in Figure 3-2(c). The axes are labeled with their associated units in parenthesis, for this example I (mA) and V (V).

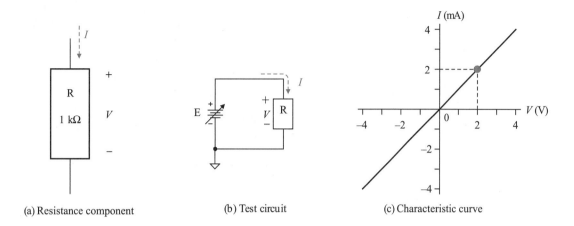

(a) Resistance component   (b) Test circuit   (c) Characteristic curve

Figure 3-2 (a,b,c) Characteristic curve of a fixed 1 kΩ resistance

Figure 3-2 represents (a) the component under test, (b) the test circuit used to find corresponding operating points of voltage and current for the component, and (c) the resulting **characteristic curve** of the measured operating data points of the component. The experimental operating data points are given in Table 3-2. Each data point is plotted on the characteristic curve and then the data points are connected with a continuous line. In the test circuit of Figure 3-2(b), the voltage source E is applied across the resistance R, causing current to flow in the resistance. The *voltage across* the resistance creates a corresponding *current through* the resistance.

The **reference voltage polarity** and the **reference current direction** for resistance R are established in Figure 3-2(b). If the measured voltage polarity and the measured current direction are the *same as their reference*, then the voltage and current values are considered to be *positive* (+). For example, with E set to 2 V, the voltage drop across R is measured to be 2 V and the resulting current through component R is measured to be 2 mA. This generates the positive data operating point of (2 V, 2 mA) as shown in Figure 3-2(c).

If the measured values are *opposite* their references, the values of current and voltage are considered to be *negative* (–). Reversing the polarity of the voltage source E in the test circuit of Figure 3-2(b) produces a negative supply voltage. This in turn reverses the voltage polarity and the current direction in the component R. For example, if supply E is set to –1 V, this creates a –1 V drop across resistance R with a corresponding meas-

Table 3-2
Measurements of the circuit of Figure 3-2

V (V)	I (mA)
4	4
2	2
0	0
–2	–2
–4	–4

ured current of –1 mA in resistance R. Figure 3-2(d) shows this example test circuit with the corresponding operating point indicated on the characteristic curve of Figure 3-2(f). The measured voltage polarities and current direction are shown in **red** while the references are shown in **black**. Again, *negative indicates that the polarity of the voltage or the direction of the current is opposite the given reference.*

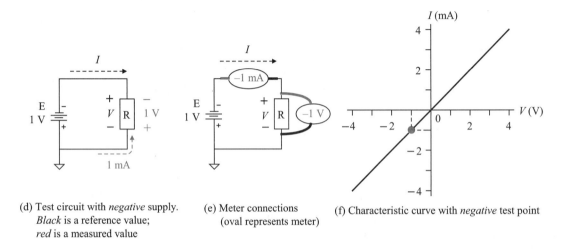

(d) Test circuit with *negative* supply.
    Black is a reference value;
    red is a measured value

(e) Meter connections
    (oval represents meter)

(f) Characteristic curve with *negative* test point

Figures 3-2(d,e,f) Test circuit with a *negative* data point

To obtain an accurate voltage polarity reading of a measured voltage, you must place the *red lead* of the voltmeter on the corresponding positive (+) schematic reference and the *black lead* on the corresponding negative (–) schematic reference as shown in Figure 3-2(e). If the voltage reading is *positive* then the actual drop of the component is the *same polarity* as the reference. In Figure 3-2(e) the voltage is negative, indicating that the measured polarity is opposite the reference.

To obtain an accurate reading of the current direction, you must place the *red lead* on the corresponding tail of the schematic reference arrow and the *black lead* on the corresponding head of the schematic reference arrow as shown in Figure 3-2(e). Recall that you must break the circuit and insert the ammeter so that the measured current flows through the meter. If the current reading is *positive* then the actual circuit current flow is the *same direction* as the reference arrow. In Figure 3-2(e) the current measurement is *negative*, indicating that the actual current direction is *opposite* the reference.

Table 3-2 lists the measured data points derived from the test circuit of Figure 3-2(b), both positive and negative. To obtain each data point, the power supply E is set to the desired voltage value to create the prescribed voltage drop across the 1 kΩ fixed resistance. And then the corresponding current is measured. All the data points from Table 3-2 are then plotted to create the *characteristic curve* of Figure 3-2(c) for this component. The characteristic curve represents the possible operating points of this component over the range tested.

A MultiSIM simulation of the (−1 V, −1 mA) data point affirms this data point for a 1 kΩ resistance as shown in Figure 3-2(g). Note that the ammeter direction and voltmeter polarity are consistent with the original circuit configuration where + is equivalent to a red test lead and − is equivalent to a black test lead.

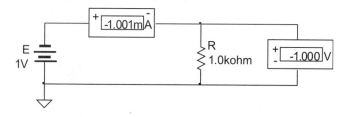

multiSIM

Figure 3-2(g) MultiSIM simulation of (−1 V, −1 mA) operating point

Simulators have the capability of performing a **dc sweep** to generate a characteristic curve. A simulator analysis is set up to sweep from one value to another value. Figure 3-2(h) shows a dc sweep of the circuit of Figure 3-2(b). A Cadence PSpice dc voltage sweep of supply E from −4 V to +4 V produces the circuit current I_R versus V_R characteristic curve of Figure 3-2(h). This simulated characteristic curve verifies the characteristic curve displayed in Figure 3-2(c). In Cadence PSpice a current probe is inserted into the circuit to measure the corresponding current at each supplied voltage. The current probe is a feature of Cadence PSpice; it is similar to a current probe that can be placed around a wire, and the current is measured by the strength of its electromagnetic field. Recall that for typical current measurements, you must break the circuit and insert the ammeter to replace the break (two points of contact). In this case, only one point of contact is shown in the simulation, equivalent to a magnetic current probe.

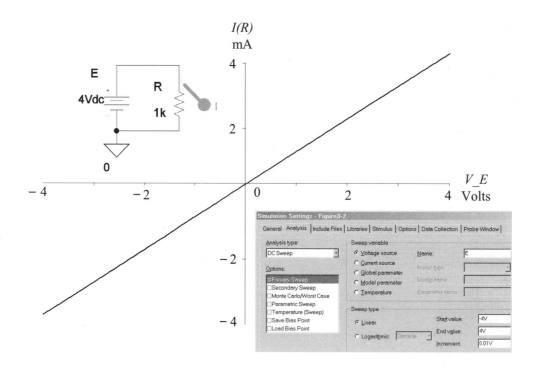

Figure 3-2(h) Cadence PSpice dc sweep of current I_R versus voltage E or V_R

Static Resistance–Ohm's Law

Look at the data of Table 3-2. Is there a pattern? Can you determine a relationship between R, I, and V? Recall that R is a fixed 1 kΩ resistance, while current varies proportionally with the voltage. Note that this characteristic curve is a straight line, suggesting a linear relationship exists between the current and voltage for this fixed 1 kΩ resistance. If the voltage is doubled, the current is doubled. That is, the resistor current is proportional to the resistor voltage drop.

A careful examination of the data shows that the following linear relationship exists.

$$R = \frac{V}{I} \qquad\qquad (3\text{-}3)$$

where

R = resistance of the component in ohms (Ω)
V = voltage across the component measured in volts (V)
I = current through the component measured in amps (A)

R is called **static resistance**. At a specific operating point, the voltage is divided by the corresponding current to obtain the component's resistance at that operating point. Note that in the characteristic curve of Figure 3-2(c), every operating point has a resistance of 1 kΩ. If you divide any operating point voltage by its corresponding current, the result is always 1 kΩ. For example, at the (2 V, 2 mA) operating point,

$$R = \frac{2\text{ V}}{2\text{ mA}} = 1\text{ k}\Omega$$

Equation 3-3 is a very important relationship in electronics that we shall study later in this book. It is known as **Ohm's law** for resistance.

Example 3-5

A voltage of 3 V is applied to the resistance of Figure 3-2. Find its corresponding current. Find its static resistance.

Solution

From the characteristic curve of Figure 3-2(c), the corresponding current I is 3 mA. The static resistance is

$$R = \frac{3\text{ V}}{3\text{ mA}} = 1\text{ k}\Omega$$

Practice

A current of 1.5 mA flows in the opposite direction of the I reference of Figure 3-2. Find its corresponding voltage. Find its static resistance.

Answer: −1.5 V, 1 kΩ.

Figure 3-3(c) is the characteristic curve for a nonlinear resistance. According to the standard, the current I is plotted on the *vertical axis* as a function of the voltage V on the *horizontal axis*.

Example 3-6

A voltage of 1.2 V is applied to the nonlinear resistance of Figure 3-3. Find its corresponding current and static resistance.

Solution

From the characteristic curve of Figure 3-3(c), the current I is estimated to be 3 mA. Its static resistance at this operating point is

$$R = \frac{1.2 \text{ V}}{3 \text{ mA}} = 400 \ \Omega$$

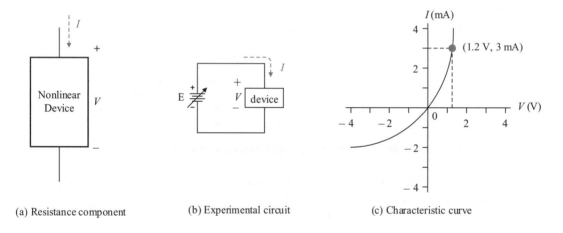

(a) Resistance component   (b) Experimental circuit   (c) Characteristic curve

Figure 3-3 (a, b, c) Nonlinear device test circuit and resulting characteristic curve

Practice

A current of –2 mA flows through the device of Figure 3-3. Find its corresponding voltage drop and static resistance.

Answer: approximately –4 V, 2 kΩ.

Exploration

Use the Cadence PSpice simulator to perform a dc sweep of a D1N4002 diode from –1 V to 1 V to produce its characteristic curve as shown in Figure 3-3(d).

Note that this is a simulated circuit only and is not to be built in the lab.

cādence®

how big can you dream?™

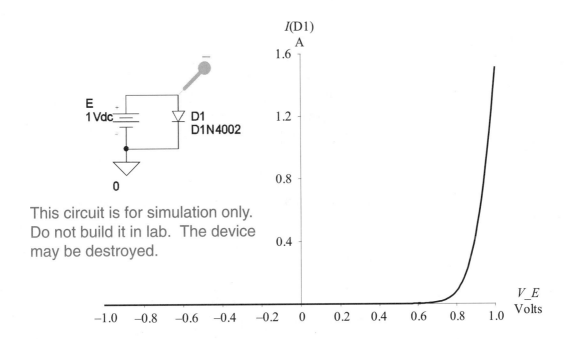

Figure 3-3(d) Simulated nonlinear curve of the D1N4002 diode

In Example 3-5, the resistance is a fixed value and thus results in a *linear* characteristic curve. In Example 3-6, the resistance of the device varies with the current and thus has a *nonlinear* curve. Notice that the nonlinear device's static resistance varies and depends upon which operating point you select. The device offers 400 Ω of resistance at the (1.2 V, 3 mA) operating point but offers 2 kΩ of resistance at the (−2 mA, −4 V) operating point. The characteristic curve is especially useful in describing and understanding the behavior of nonlinear devices.

Static Conductance

Another way to interpret an operating point on the characteristic curve is to find its **static conductance**. Substituting Equation 3-3 into Equation 3-2, we can derive a general expression for conductance in terms of its current and voltage.

$$G = \frac{1}{R} = \frac{1}{V/I} = \frac{I}{V}$$

For example, in Figure 3-3(c), the *static conductance* of the operating point (1.2 V, 3 mA) is

$$G = \frac{3 \text{ mA}}{1.2 \text{ V}} = 2.5 \text{ mS}$$

This is consistent with the value of resistance found in Example 3-6.

$$R = \frac{1}{G} = \frac{1}{2.5 \text{ mS}} = 400 \text{ } \Omega$$

3.4 Short and Open Circuits

Short Circuit

A **short** is equivalent to a very low resistance conductor. An **ideal short** is equivalent to an *ideal conductor* with 0 Ω of resistance. Its *conductance* is therefore ideally *infinite*.

Figure 3-4 demonstrates a short circuit where a wire has accidentally been placed across the lamp. In Figure 3-4(a) the current normally flows through the lamp and lights the lamp. In Figure 3-4(b) the current bypasses the lamp to flow through the very low resistive short causing what is called a **short circuit**. The lamp does not light since it has no current. Since there is very little resistance to current, a short circuit can cause *excessive current through* a circuit and damage circuit components or the power supply. In the circuit of Figure 3-4(b), the lamp is in no danger since it draws no current but the source is now drawing excessive current and is in danger unless it is protected against excessive currents. The voltage source drops close to 0 V if it is attempting to drive a pure short circuit as it is in Figure 3-4(b). Most power supplies are protected against excessive currents, but batteries are not. Batteries can overheat and possibly explode. A good rule is to never short-circuit a battery intentionally.

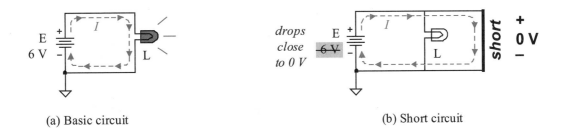

(a) Basic circuit   (b) Short circuit

Figure 3-4 Short circuit example

A characteristic of an ideal short is that it drops 0 volts. The current $V_{short} = 0\,V$ through a short can be very high and is determined by the external circuit, but a short drops 0 V.

Warning: An ammeter is a very low resistance device and is considered equivalent to a *short*. Accidentally placing an ammeter across the lamp in Figure 3-4(b) is equivalent to placing a short across the lamp. The ammeter would short out the lamp and cause excessive current, which could damage the ammeter or the voltage supply if they are not protected. *Again, note that the supply drops close to 0 V if it attempts to drive a short.*

Open Circuit

An **open** is equivalent to a break in a circuit. For example, if a wire in a circuit is cut and separated, an open is created between the two conductors. An **ideal open** is equivalent to an *ideal insulator* with infinite ohms ($\infty\ \Omega$) of resistance. Its *conductance* is therefore ideally 0 S.

Figure 3-5 demonstrates an open circuit where a wire is accidentally disconnected from the lamp creating an open. In Figure 3-5(a) the current flows normally through the lamp and lights the lamp. In Figure 3-5(b) there is no continuous path for current to flow. The lamp does not light since it does not have current. Note that the 6 V supply is dropped across the open and not across the lamp. An open circuit reduces the total circuit current and does *not* cause supply or circuit component damage.

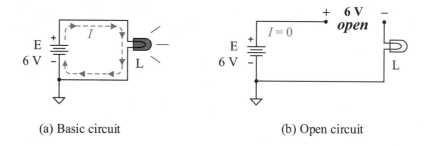

(a) Basic circuit   (b) Open circuit

Figure 3-5 Open circuit example

$I_{open} = 0\ A$

A characteristic of an open is that it always has 0 amps of current. The voltage dropped across an open is determined by the external circuit, but an open always draws 0 A. *Current cannot flow through an open.*

3.5 Switches

A simple **switch** has two possible states: **open** and **closed**. In the *open* position the switch is literally an *open*, and current cannot flow through it. In the *closed* position, the switch is a conductor, ideally a short. Figure 3-6 shows the schematic symbol of a simple switch. It is shown in the *open position*.

Figure 3-7 is the schematic of a lamp circuit controlled by a simple switch.

Figure 3-6
Simple switch schematic symbol

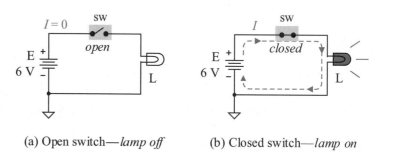

(a) Open switch—*lamp off*   (b) Closed switch—*lamp on*

Figure 3-7 A simple switch in (a) *open position* and (b) *closed position*

Figure 3-7(a) demonstrates the lamp circuit with the switch in the *open position*. Since the switch is open, no current can flow and the lamp does not light. Figure 3-7(b) demonstrates the switch in the closed position. A continuous path now exists for current to flow from the positive terminal of the supply, through the closed switch, through the lamp, and back to the negative terminal of the supply.

Multiple-pole, Multiple-throw Switches

The switch in Figure 3-8(a) is termed a **single-pole, single-throw** (SPST) switch. When closed, it connects the left switch terminal to the right switch terminal. The switch in Figure 3-8(b) is **single-pole, double-throw** (SPDT) switch. It connects the left switch terminal to one of two right switch terminals.

SPST   SPDT

(a)   (b)

Figure 3-8 (a) Single-pole, single-throw; (b) single-pole, double-throw

The two SPDT switches in Figure 3-9 are used to control a single lamp. A stairway lamp with switches at the bottom and top of the stairs is a good example for the use of such a circuit.

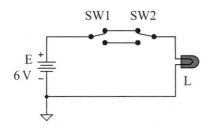

Figure 3-9 Two SPDT switches controlling a single lamp

Example 3-7

Use the circuit of Figure 3-9 to determine if the lamp is lit or dark for all possible switch positions.

Solution

SW1 up, SW2 up:
A continuous path exists. The lamp is lit.
SW 1 down, SW 2 up:
A broken path exists. The lamp is dark.
SW 1 down, SW 2 down:
A continuous path exists. The lamp is lit.
SW 1 up, SW 2 down:
A broken path exists. The lamp is dark.

Notice that throwing either switch changes the state of the lamp. If the lamp is *on*, it is switched *off*. If the lamp is *off*, it is switched *on*. This is very convenient if one switch is at the top of the stairs and the other switch is at the bottom of the stairs.

Practice

Use the circuit of Figure 3-10 and determine which switch positions light which lamp.

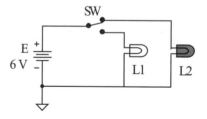

Figure 3-10 Dual lamp circuit controlled by a SPDT switch.

Answer: SW down, L1 lights; SW up, L2 lights.

Other examples of multiple-pole, multiple-throw switches are shown in Figure 3-11.

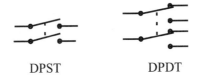

DPST   DPDT

Figure 3-11 (a) Double-pole, single-throw; (b) double-pole, double-throw

The double-pole, single-throw (DPST) switch of Figure 3-11(a) creates two separate SPST switch connections with a single throw of a switch. The dotted line indicates that the switches are **ganged** together and that both switches are opened or closed simultaneously.

The double-pole, double-throw (DPDT) switch of Figure 3-11(b) creates two separate SPDT switch connections with a single throw of a switch. Again, the dotted line indicates that the switches are ganged together and that both switches are moved simultaneously.

Figure 3-12
Rotary switch schematic symbol

Switch Types

Many types of switches exist. Common examples include toggle switches, rocker switches, push-button switches, rotary switches, stacked rotary switches, and DIP (dual in-line package) switches.

The toggle switch uses a toggle arm; for example, the home wall switch is an example of a toggle switch. A rocker switch uses a rocker arm. A rotary switch uses a knob on a shaft to rotate through several switch connections. A schematic symbol of a rotary switch is shown in Figure 3-12. The left terminal can be connected to one of four possible right terminals.

DIP (dual in-line package) switch packages include several internal switches and are used extensively in printed circuit boards. Computer boards contain DIP switches to make or break connections depending upon the computer's configuration. Figure 3-13 shows the schematic of a DIP switch with 8 internal SPST switches. Each switch is manually and individually set, typically with a slide switch or a rocker arm switch.

Switches can be metal-to-metal contacts, or they can be liquid metal connecting, for example mercury switches. Mercury is an electricity conducting liquid metal. When the pool of liquid mercury makes contact with two contacts of the switch, the switch is closed. Mercury switches are quiet switches that do not arc, whereas metal-to-metal contacts wear out, are noisy (click into place), and can arc. Arcing causes oxidation of the

Figure 3-13
DIP switch schematic symbol

metal or burn marks. Excessive arcing causes pitting and damage to the metal contacts and eventually destroys the continuity of the switch.

The switching action itself can be controlled by different mechanisms. In the home you manually move the light switch toggle to switch the lamp on and off. Your heating system uses a thermostat to detect changes in air temperature and to automatically open and close a switch to turn off or turn on your furnace. Switches can be designed to be controlled by such environmental conditions as temperature, pressure, flow, liquid level, time, inertia, sound, and light. These environmental sensors are designed to translate a physical phenomenon into a switching action. The closing or opening of a switch then controls an electronic circuit action. For example, a room goes dark. A light sensor receives insufficient light. That signals an electronic switch to close, which then provides voltage to a lamp. The lamp is lit and the room now has light.

3.6 Diodes

The **ideal diode** is a *two-terminal, semiconductor* device that allows current to flow in one direction. The diode is an electronic switch: *off* (ideally an *open*) or *on* (ideally a *short*). The diode, its schematic symbol, and its basic construction are shown in Figure 3-14.

Since the diode is polarity sensitive, the terminals are distinguished as the **anode** and the **cathode**. To distinguish the anode (positive material) from the cathode (negative material) on the component, a band is used to represent the cathode terminal as shown in Figure 3-14(a).

The schematic symbol of the diode is shown in Figure 3-14(b). The arrow represents the positive *p*-type material of the semiconductor while the line represents the negative *n*-type material.

Figure 3-14(c) illustrates the basic construction of the diode. A more detailed description of semiconductor devices can be found in a later section of this textbook. The fundamental material of the semiconductor diode is either *germanium* or *silicon*. The pure or intrinsic semiconductor (valence of +4) is doped with acceptor atoms (valence of +3) to create the *p*-type material with hole carriers (a positive carrier). The other end of the intrinsic semiconductor is doped with donor atoms (valence of +5) to create the *n*-type material with electron carriers (a negative carrier). During manufacturing, the *p* and *n* materials form a *pn* junction as the hole carriers migrate into the *n*-type material and the electron carriers migrate into the *p*-type material. The *pn* junction is an ion region devoid of carriers. A potential voltage barrier is formed by the *pn* junction: 0.2 V for germanium and 0.7 V for silicon.

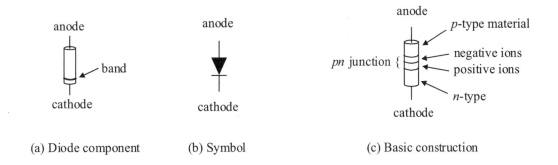

anode

band

cathode

(a) Diode component

anode

cathode

(b) Symbol

anode

pn junction {

cathode

p-type material

negative ions
positive ions

n-type

(c) Basic construction

Figure 3-14 Switching or rectifier diode

If an external voltage is applied such that the voltage at the terminal of the *p*-type material is *higher* relative to the voltage at the *n*-type terminal, the diode is considered **forward biased**. If the diode is forward biased with sufficient voltage to overcome the *pn* junction barrier potential, then it conducts and the diode can be modeled ideally as a *short*.

If insufficient external voltage is applied across the diode, that is, less than 0.2 V for germanium or less than 0.7 V for silicon, the diode does not conduct and the diode can be modeled ideally as an *open*.

If an external voltage is applied such that the voltage at the *p*-type terminal is *lower* than that of the *n*-type terminal, the diode is considered **reverse biased** and can be modeled ideally as an *open*.

Switch or Rectifier Diode

A **switch** or **rectifier diode** is designed to act as an electronic switch. If it is forward biased, the diode turns on, conducts, and ideally acts like a *short*. Figure 3-15 demonstrates a circuit with a *forward biased* diode.

The anode is at a higher voltage than the cathode. The power supply creates a voltage polarity to forward bias the diode as shown in Figure 3-15(a). A *forward biased diode* conducts current and can be ideally modeled as a *short* (dropping 0 V) as shown in Figure 3-15(b). The conducting diode completes the circuit and the lamp is lit.

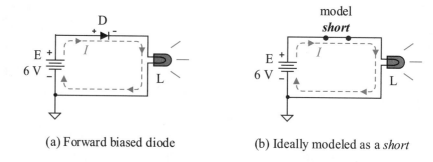

(a) Forward biased diode   (b) Ideally modeled as a *short*

Figure 3-15 A circuit with a forward biased diode

If the diode is *reverse biased*, that is, the voltage potential at the cathode is higher than that of the anode, then the diode acts ideally like an *open* and does not conduct current. Figure 3-16 demonstrates the effects of a reverse biased diode.

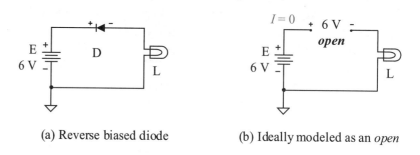

(a) Reverse biased diode   (b) Ideally modeled as an *open*

Figure 3-16 A circuit with a reverse biased diode

The power supply creates a voltage polarity on the diode to reverse bias the diode as shown in Figure 3-16(a). A higher voltage is impressed upon the cathode relative to the anode. The *reverse biased diode* acts ideally like an *open* ($I = 0$ A) causing a break in the circuit as shown in Figure 3-16(b). The circuit does not have a continuous path; thus, the lamp does not light. The reverse biased diode drops all 6 V.

From the above description, it would be convenient for circuit analysis to ideally model a forward biased switch diode as a short and a reversed biased switch diode as an open as pictured in Figure 3-17.

The characteristic curve of an ideal switch diode is shown in Figure 3-18. The vertical operating line represents the forward biased *on* diode conducting current and acting like a short (dropping 0 V). The horizontal operating line represents the reverse biased *off* diode, which acts like an open (drawing 0 A).

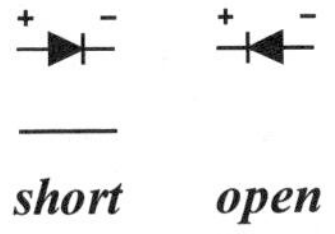

Figure 3-17
Switching diode ideal
models

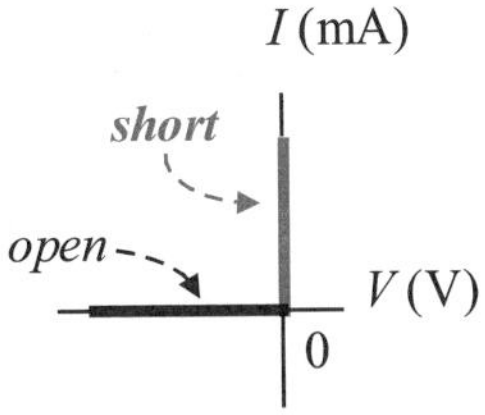

Figure 3-18
Characteristic curve
of an ideal diode

Example 3-8
Redraw the schematics of Figure 3-19 with their equivalent diode models. Assume these are ideal diodes. State whether the lamp is lit or dark.

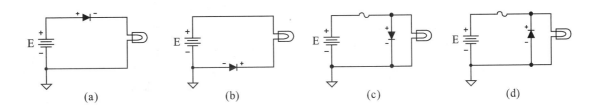

Figure 3-19 Example circuits

Solution

Each of the diodes is replaced with its appropriate ideal model. Forward biased diodes are ideally replaced with a *short*. Reverse biased diodes are ideally replaced with an *open*. See Figure 3-20 for the modeled circuits. Then analyze each circuit based upon its modeled circuit.

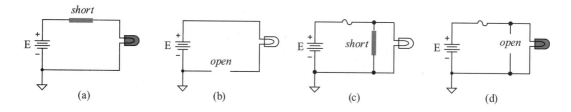

Figure 3-20 Example solutions

a. Diode is forward biased on. Diode modeled as a short. There is a continuous path. The lamp is lit.
b. Diode is reverse biased off. Diode is modeled as an open. The circuit path is broken. The lamp is dark.
c. Diode is forward biased on. Diode modeled as a short. It shorts out the lamp. Excessive current opens the fuse. The lamp is dark.
d. Diode is reverse biased off. Diode is modeled as an open. There is a continuous path through the lamp. The lamp is lit. Note: this is a protection diode; it protects the lamp from a reverse E voltage polarity.

Practice

If all the diodes of Figure 3-19 were reversed, would the lamp be lit or dark?

Answer: (a) dark, (b) lit, (c) lit, (d) dark

Switch Diode—Practical Considerations

A diode requires that the **potential barrier voltage** of its *pn* junction must be overcome before it can be forward biased *on*. This barrier voltage and the actual forward bias voltage drop of a diode depend upon the material and construction of the diode.

The barrier potential and the actual voltage drop in a live circuit of a *germanium diode is about 0.2 V* and a *silicon diode is about 0.7 V*. Figure 3-21 compares the characteristic curves for the (a) ideal diode, (b) germanium diode, and (c) silicon diode. The germanium diode must have at least 0.2 V applied to it by the circuit to turn it on and conduct, otherwise it draws 0 A (acting like an open). The silicon must have at least 0.7 V applied to it for the diode to turn on and conduct.

Once the diode turns on, it continues to drop the barrier voltage across the diode. Thus, once a silicon diode turns on, it maintains a drop of about 0.7 V. The voltage of the *on* diode remains fairly constant, with relatively large swings in diode current. For example, an *on* silicon diode current may double from 10 mA to 20 mA but still maintain a voltage drop of about 0.7 V. If the voltage across the diode drops much below 0.7 V, the diode turns off and becomes an ideal open.

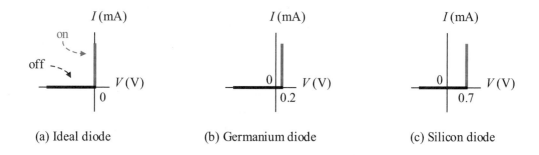

(a) Ideal diode   (b) Germanium diode   (c) Silicon diode

Figure 3-21 Characteristic curves of a switching or rectifier diode

If the actual voltage drop of the diode is significant compared to the voltage being applied to the diode circuit, the *on* diode is modeled as a voltage source as shown in Figure 3-22. Note that the voltage polarity of the model is the *same polarity as the forward biased diode*. The value of the voltage source model depends on the material of the diode: germanium is 0.2 V and silicon is 0.7 V.

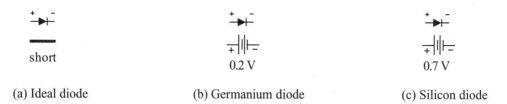

| (a) Ideal diode | (b) Germanium diode | (c) Silicon diode |

Figure 3-22 Models of forward biased *on* diode

Suppose a silicon diode is used with a supply of 7 V as shown in Figure 3-23(a). Then the *on* diode drop of 0.7 V causes a 10% error in the calculated circuit values if the ideal model is used.

$$\% \ error = \frac{0.7 \ V}{7 \ V} \times 100\% = 10\%$$

If this error is not significant for your application, then the forward biased *on* silicon diode can be modeled as an ideal diode (a short) as shown in Figure 3-23(b). If a 10% error is too much calculated error, the forward biased *on* diode must be modeled practically as a 0.7 V supply as shown in Figure 3-23(c). As you will discover later in this book, all 7 V of the supply are delivered to the lamp based upon the ideal model of Figure 3-23(a), while a more realistic 6.3 V is delivered to the lamp based upon the practical model of Figure 3-23(c).

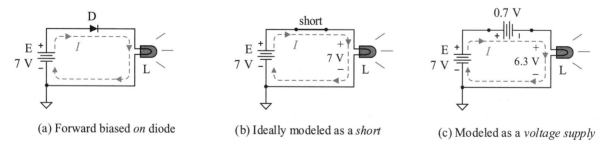

| (a) Forward biased *on* diode | (b) Ideally modeled as a *short* | (c) Modeled as a *voltage supply* |

Figure 3-23 Forward biased *on* silicon diode with its
ideal short model and its *voltage supply model*

Light Emitting Diode (LED)

The **light emitting diode** (LED) is a diode that emits light when it is *forward biased* and *conducting*. The LED must not only be forward biased but requires a minimum forward biased voltage drop to draw enough current to light. The voltage requirement varies based upon the material of the diode. The typical voltage dropped across an *on* LED ranges from 1.5 V to 5 V. A typical voltage used throughout this book is 2 V. The light brightness is proportional to the current, that is, the higher the current through the LED, the brighter the LED.

Different materials and construction create different colors of emitted light. LEDs are available with emitted colors throughout the color spectrum with a variety of light intensities. Typical colors include red, green, yellow/amber, and blue. A sample LED package and its schematic symbol are shown in Figure 3-24. LEDs come in a variety of physical packages and Figure 3-24(a) is a packaging example. The two arrows pointing away from the diode symbol in Figure 3-24(b) indicate that light waves are being emitted, thus making this an LED. Figure 3-24(c) shows the characteristic curve of an LED that drops 2 V when forward biased *on*. If a 2 V LED is *turned on* it acts like (and can be modeled as) a 2 V supply as shown in Figure 3-24(d). If the voltage across the LED drops much below 2 V, the LED turns off and acts like an open (does not emit light).

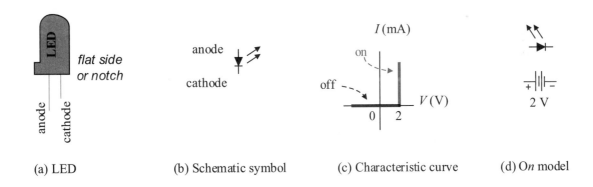

(a) LED   (b) Schematic symbol   (c) Characteristic curve   (d) *On* model

Figure 3-24 LED—light-emitting diode

LEDs have a variety of uses. They are used as *on* (green), *caution* (yellow/amber), and *off* (red) indicators. They are also used as inexpen-

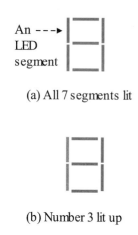

(a) All 7 segments lit

(b) Number 3 lit up

Figure 3-25
7-segment LED display

sive fuses. If a circuit is to be protected at a maximum current, an LED indicator with that maximum forward current rating can be used as both a visible display and an inexpensive semiconductor fuse. A familiar application is the use of LEDs in a 7-segment display that creates numbers 0 through 9 by lighting the appropriate segments as shown in Figure 3-25. For example, if all seven (7) LED segments are forward biased *on* as shown in Figure 3-25(a), the number eight (8) is displayed. If the two left LED segments are turned off as shown in Figure 3-25(b), then the number three (3) is displayed. Since *liquid crystal displays* (LCD) have been introduced for lower power applications such as calculators, the LED 7-segment display is more typically found in electrically powered circuits because of its higher energy consumption.

Diode—Maximum Forward Current

As just noted, an *on* diode has a maximum forward current that it can tolerate before it overheats and possibly self-destructs. This diode specification is known as the **maximum forward current**. For example, a silicon rectifier diode may be rated at 2 A. If the diode experiences this maximum current continuously, the diode overheats and self-destructs (typically opens or shorts).

Diode—Maximum Reverse Voltage

A reverse biased *off* diode has a maximum reverse voltage that it can tolerate before it breaks down and begins conducting current through the reverse direction. This diode specification is known as the **maximum reverse voltage** specification. For example, a silicon rectifier diode may be rated at 100 V. If the reverse bias *off* diode experiences this maximum reverse voltage, the diode does not necessarily self-destruct but does start conducting current in the reverse direction. This is an unwanted state for a switching diode, rectifier diode, or LED.

3.7 Fixed Resistors—Color Code

R

Figure 3-26
Resistor schematic symbol

Fixed resistors are two-terminal components designed to offer a single or fixed value of *resistance*. As noted previously, resistance is measured in units of ohms (Ω). The schematic symbol for a resistor with a fixed resistance is shown in Figure 3-26. The letter **R** is used to represent a resistor. For example, R = 10 Ω states that a resistor R has a resistance of 10 ohms.

Resistors are made of various materials. For example, wire-wound resistors and carbon composite resistors are extensively used in electrical and electronic systems. Other types of resistors include carbon film, metal oxide, and integrated circuit packages.

Wire-wound Resistors

Wire, although a conductor, offers some resistance to current. An excellent example of a wire resistance that is commonly used is that in an electric toaster. A toaster uses specially designed metal wire that conducts current, becomes very hot, and emits heat. The emitted heat then toasts the bread. Wire-wound resistors typically have low resistance values. Their resistance values can be accurately manufactured.

A wire-wound resistor is simply a resistor in which the wire of the proper gauge and length is wound up and encased in an insulating body as shown by Figure 3-27. Recall that a higher AWG number has a smaller cross-section area and thus a higher resistance per given length. Also, the longer the wire is, the greater its resistance.

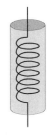

Figure 3-27
Wire-wound resistor

Carbon Composite Resistors

Carbon composite resistors are used extensively in electronics. The basic construction of a carbon composite resistor is shown in Figure 3-28(a). The external view of the resistor is shown in Figure 3-28(b).

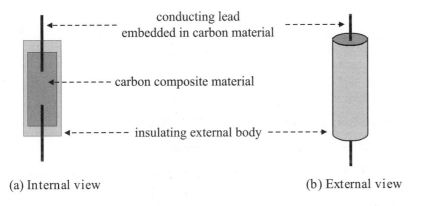

(a) Internal view   (b) External view

Figure 3-28 Carbon composite resistor basic construction

The larger the resistor, the more heat it can dissipate. Figure 3-29 shows the standard sizes of the carbon composite resistor. The largest resistor is rated at 2 watts (W), dissipating up to 2 joules per second of heat energy. The smallest resistor is 1/8 watt, dissipating up to 1/8 joule per second of heat energy. *Warning, when operating near a resistor's rated value, the resistor does become hot to the touch.*

These are enclosed ratings. That is, if the resistor is in a closed environment (like an unvented stereo cabinet), then a 2 W rated resistor must not be operated above this rating. In an open-air environment, such as an experimenter board test circuit, you can drive these resistors up to twice their rating. That is, you could operate a 2 W resistor at 4 watts, dissipating 4 joules per second of heat energy safely.

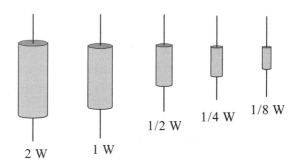

Table 3-3
Bands 1, 2, and 3 color code

Figure 3-29 Carbon composite resistor—approximate sizes

Resistor Color Code

The carbon composite resistors use a set of color coded bands to indicate the nominal (rated) value of the resistor as shown in Figure 3-30.

Color	Digit
black	0
brown	1
red	2
orange	3
yellow	4
green	5
blue	6
violet	7
gray	8
white	9

Band 1	1st digit	1 through 9
Band 2	2nd digit	0 through 9
Band 3	Multiplier	0.01, 0.1, 1, 10 … 100000
Band 4	Tolerance (5%, 10%, or 20%)	

Bands 1, 2, and 3 are used to create the **nominal** (rated) resistance value and use the colors of Table 3-3. *Bands 1 and 2* form a 2-digit number. For example, a combination of orange (3) and white (9) forms the 2-digit number of 39. *Band 3* is the multiplier value. The multiplier value is a power of 10 multiplied times the above 2-digit number to form the nominal resistance value. For example, suppose the 3rd band is red (value of 2). The value 2 is the exponent of the power of 10^n. Thus the multiplier is 10^2 or 100. Using the results from bands 1, 2, and 3, the nominal resistance of our example *orange, white, and red* is 3900 Ω, preferably stated as 3.9 kΩ.

Table 3-4
Band 4 color code

Color	Digit
gold	5%
silver	10%
none	20%

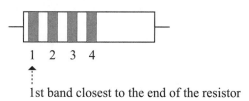

1 2 3 4

1st band closest to the end of the resistor

Figure 3-30 Resistor color bands

$$R_{nominal} = 39 \times 100 \ \Omega = 3900 \ \Omega = 3.9 \text{ k}\Omega$$

Another way of thinking of the *red multiplier* is to *add two (2) zeros* to the 2-digit number. For example, 39 with two zeros added is 3900.

Band 4 is the tolerance rating of the resistor and uses the colors of Table 3-4. It establishes the range of resistance values as guaranteed by the manufacturer. For example, a rating of 3900 $\Omega \pm 10\%$ specifies that the resistance can range from 10% *below* the nominal value to 10% *above* the nominal rating of 3900 Ω. For this example, the resistance range is 3900 $\Omega \pm 390 \ \Omega$, i.e., a range from 3510 Ω to 4290 Ω. This is the range of possible values expected for this resistor.

Additional band(s) are sometimes used to provide information on failure rate and reliability.

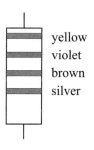

yellow
violet
brown
silver

Figure 3-31
Resistor color code problem

Example 3-9

The color code of a carbon composite resistor is yellow, violet, brown, and silver as shown in Figure 3-31. Find its (a) nominal value and (b) range of possible values.

Solution

a. Find the nominal value of the resistor.

The *2-digit number* is 47.

Band 1 (closest band to the end of the resistor)

Band 1 ➜ yellow = 4

Band 2 ➜ violet = 7

The *multiplier* is 10.

Band 3 ➜ brown = 1 ⟹ $10^1 = 10$

(or add one zero to the 2-digit number)

The *nominal resistance* is 470 Ω.

$$R_{nominal} = 47 \times 10 \ \Omega = 470 \ \Omega$$

b. Find the range of possible resistance values

Band 4 ➜ silver ⟹ ± 10%

So, the rated tolerance is 10%.

10% ⟹ 0.1 decimal equivalent

The *minimum rated value* is 423 Ω

$$R_{min} = 470 \ \Omega - (0.1) \times (470 \ \Omega) = 423 \ \Omega$$

The *maximum rated value* is 517 Ω

$$R_{max} = 470 \ \Omega + (0.1) \times (470 \ \Omega) = 517 \ \Omega$$

The *range of possible resistance values* is 470 Ω ± 47 Ω.

R_{range} is 423 Ω through 517 Ω.

Practice:

The color code of a carbon composite resistor is brown, black, orange, and gold. Find its (a) nominal value and (b) range of possible values.

Answer: (a) 10,000 Ω or 10 kΩ, (b) 9.5 kΩ through 10.5 kΩ

Standard Resistor Table

For economical reasons, a standard set of nominal values with standard tolerances have been established for commercially available resistors. Appendix C lists the standard color code resistors that are commercially available. Note that resistors are listed *vertically* based upon their *2-digit value* (color bands 1 and 2) and *horizontally* by their *multiplier* (color band 3). In Appendix C, the bolded rows are considered the values more typically used.

       All the resistors in Appendix C are available with ±5% tolerance. Resistors with the 2-digit combination of 10, 12, 15, 18, 22, 27, 33, 39, 47, 56, 68, and 82 (shown in bold print in Appendix C) are also available with ±10% tolerance. Resistors with the 2-digit combination of 10, 15, 22, 33, 47, 68 are available with ±20% tolerance. By utilizing these standard nominal values and ranges the entire spectrum of resistance is essentially covered. For example, 10 Ω plus 20% is 12 Ω, while 15 Ω less 20% is 12 Ω. The full resistance range is covered from lowest to highest in each of the tolerance ranges: 5%, 10%, and 20%.

       Note that the lower the tolerance value is, the closer the resistance is guaranteed to be to its nominal value. But lower tolerance resistors are more expensive. If your company is purchasing high volumes of components, a fraction of a cent can make a big difference in overall costs. Cost must be compared to the reliability of the component values.

       In today's competitive market, the tolerances of resistors tend to be much better than rated. It is not unusual to have ±5% tolerance resistors typically measure within 1%.

Example 3-10

A resistor in a voltage power supply is used to limit current. The minimum value it can be is 350 Ω and the maximum it can be is 550 Ω. Select the most appropriate standard resistor value and tolerance value so that it satisfies the power supply specification:

$$350 \ \Omega \le R_{\text{supply}} \le 550 \ \Omega$$

Solution

A good starting point is the value that is midway between the supply's maximum and minimum resistor specification, i.e., the average value.

$$R_{\text{average}} = \frac{350 \ \Omega + 550 \ \Omega}{2} = 450 \ \Omega$$

The closest ±20% tolerance resistor is 470 Ω. Let's check out the 470 Ω rated range to see if it meets the minimum to maximum requirement of the power supply of 350 Ω to 550 Ω.

$$376 \, \Omega \leq R_{470 \, \Omega \pm 20\%} \leq 564 \, \Omega$$

No, the resistor's rated range exceeds the power supply's 550 Ω maximum specification.

Let's look at the 470 Ω ± 10% rated range.

$$423 \, \Omega \leq R_{470 \, \Omega \pm 10\%} \leq 517 \, \Omega$$

Yes, its rated range is well within the power supply's resistor specification.

Another consideration is the **manufacturer's reliability rating** for the resistor. That is the percent of time that the manufacturer guarantees that a resistor is within the rated range. If this is not sufficient, then you may wish to choose a 470 Ω ± 5% resistor or a resistor with a higher reliability rating.

Practice:

A resistor is to be used to limit current through a light emitting diode (LED). The resistor must be at least 1 kΩ but less than 2 kΩ. Select the most appropriate standard resistor value and tolerance value so that it satisfies the specification.

Answer: 1.5 kΩ ± 20%

3.8 Ohmmeter Measurement

The **ohmmeter** is a test instrument used to measure resistance. It can be used to measure the resistance of a resistor, a conductor, or a semiconductor. It can also be used to check the continuity of connecting wire or traces as well as to test for opens and shorts.

Measuring the Resistance of a Resistor

Figure 3-32 demonstrates how to connect the ohmmeter to measure resistance. Always measure a resistor's resistance before using it in a circuit to ensure that is actually the value you intend to use. If possible, measure the resistor by itself without any other components attached to it. Polarity is not an issue with the resistor. You can reverse the leads and the measurement will be the same for a resistor. However, some devices, like diodes, are polarity sensitive.

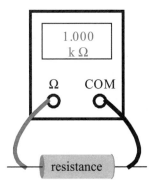

Figure 3-32 Ohmmeter measurement of a resistor

Basic Ohmmeter Concept

The ohmmeter does not directly measure resistance. Internally, the ohmmeter generates a voltage or a current, measures the current through the test leads and the voltage across the test leads, and then calculates the resistance. Equation 3-4 expresses the resistance value of an ohmmeter reading in terms of the voltage output across the ohmmeter and the current through the ohmmeter test leads.

$$R_{measured} = \frac{V_{ohmmeter}}{I_{ohmmeter}} \tag{3-4}$$

Recall Equation 3-3. This is the same formula that we discovered when looking at the characteristic curve of a resistance. This is a measure of the static resistance of the device at a particular operating point. Figure 3-33 demonstrates this concept using an ohmmeter to measure a 1 kΩ resistor. The resistance being measured completes the ohmmeter circuit. A current is generated through the resistance and a voltage is dropped across the resistance. The current generated is dependent upon the ohmmeter range selected. When the ohmmeter is connected to the 1 kΩ resistor: (1) a continuous circuit path is created, (2) a current of 1 mA flows from the ohmmeter red lead, through the resistor, and into the ohmmeter black lead, and (3) 1 V is dropped across the resistor. Applying Equation 3-4, the measured resistance of the resistor is calculated by the ohmmeter to be 1 kΩ.

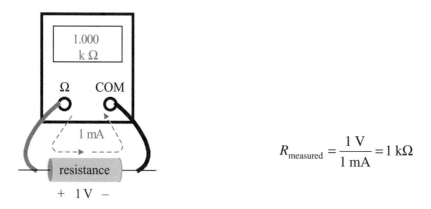

$$R_{measured} = \frac{1\ V}{1\ mA} = 1\ k\Omega$$

Figure 3-33 Ohmmeter applies a current, drops a voltage, and calculates R

This is a fixed resistance, so the resistance measures 1 kΩ on each valid meter range. Recall that in a *nonlinear* device the static resistance depends upon the operating point. If you change the ohmmeter range, you may change the voltage being applied and thus the resulting static resistance of a nonlinear device. A fixed resistance should measure the same resistance on all valid ranges. *The lowest measurable range is the most accurate range to use.*

Note, this ohmmeter example applied a fixed current on a given range and generated a voltage drop based upon the resistance being measured. Some ohmmeters apply a fixed voltage and develop a current through the resistance under test. Both types of ohmmeters use the voltage and current measurements to calculate the attached resistance being measured based on Equation 3-4.

Continuity Test the Ohmmeter

Before using the ohmmeter, ensure that the ohmmeter itself is working properly. In the lowest ohmmeter range, touch the two lead contacts together as shown in Figure 3-34(a). What should the resistance be? The reading must be zero ohms (0 Ω) or very near 0 Ω.

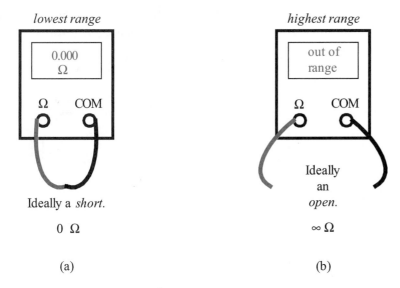

Figure 3-34 Ohmmeter (a) continuity test on lowest range, (b) open test on highest range

If the ohmmeter is not zeroed, the ohmmeter reads the wire resistance of the test leads plus the contact resistance of the two test leads touching. You must have good metal-to-metal contact of the test leads or else you increase the contact resistance. A *zero reading on the lowest scale on the ohmmeter* is interpreted as a *short*.

If you separate the leads as shown in Figure 3-34(b), what should you measure? If the leads are separated, a broken circuit is created and $I_{measure}$ is 0 A (indicating an open). There is only air between the leads and the resistance of air is too high to measure with an ohmmeter. The ohmmeter should register *out of range even on its highest range*. Out of range indicates that the resistance being measured is greater than the scale selected. For example, a 2.2 kΩ resistor measured on a 1 kΩ scale measures out of range, but is not an open.

Continuity Test a Conductor

Figure 3-35(a) demonstrates measuring the resistance of a conductor (wire or trace). Ideally, a conductor measures with a value of 0 Ω; practically, it is typically a very low resistance value. You must use the *lowest range* to make this test. Why? If you are on the highest ohmmeter range, a low resistance reading may not register a meaningful value and may falsely appear as a short. For example, a 1 kΩ resistor measured on the 1 MΩ scale looks to be approximately 0 Ω. Again, *use the lowest measurable scale possible for the most accuracy.*

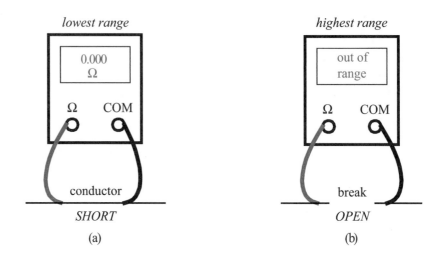

Figure 3-35 Ohmmeter measurements of (a) a conductor,
(b) a conductor break

Figure 3-35(b) demonstrates measuring the resistance of a break in a conductor. The ohmmeter measures *out of range*. On the highest ohmmeter range, this is interpreted to be an *open*. Why must you use the highest ohmmeter range to make this measurement? Suppose you set an ohmmeter to its lowest resistance range, for example 100 Ω. Then you measure a 1 kΩ resistor and it measures out of range. Does that indicate an open? No. It indicates that the reading is greater than 100 Ω.

The range setting is critical when checking for shorts, opens, or resistance values. The general rule to follow is to use the lowest range possible to produce an accurate reading.

Continuity Test a Switching or Rectifier Diode

The diode is polarity sensitive and requires two test conditions to determine if it is good: (1) a forward bias ohmmeter test and (2) a reverse bias ohmmeter test as shown in Figure 3-36.

The *forward bias check* is shown in Figure 3-36(a). Note that the red ohmmeter lead (+) is connected to the diode's *anode* and the black ohmmeter lead (−) is connected to the diode's *cathode*. Sufficient voltage must be applied by the ohmmeter to overcome the barrier potential voltage of the diode, namely, 0.2 V for germanium and 0.7 V for silicon. The actual ohmmeter voltage being applied for the test reading depends upon the ohmmeter range selected. Most ohmmeters have a diode checker or put the diode schematic symbol on the ohmmeter ranges designed to test silicon and germanium diodes. If an appropriate ohmmeter range is selected and the ohmmeter leads properly applied with the correct polarity, a nominal resistance reading validates the forward bias direction. This reading could be anywhere from a few ohms to several thousand ohms. In reality, a forward biased *on* diode is a nonlinear device, and thus has a nonlinear resistance curve. Therefore each valid ohmmeter range gives a different resistance if the ohmmeter is applying a different test voltage. You are not looking for a particular value; you are looking for a reasonable value of resistance for a forward biased diode. If you use an appropriate diode range and the diode measures as a short on the lowest resistance range or as an open on the highest resistance range, it fails the ohmmeter test.

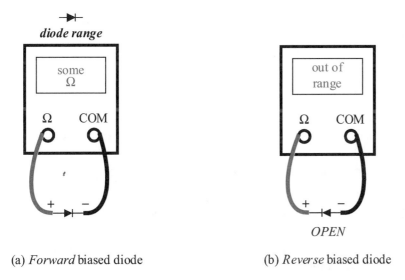

(a) *Forward* biased diode   (b) *Reverse* biased diode

Figure 3-36 Ohmmeter measurement of a diode

The *reverse bias check* is shown in Figure 3-36(b). Note that the red ohmmeter lead (+) is connected to the diode's *cathode* and the black ohmmeter lead (−) is connected to the diode's *anode*. Since the diode is reverse biased, it should read like an open on the highest resistance range.

If the forward bias check is good and the reverse bias check is good, the diode test is good. *Warning, some small signal diodes have very low maximum current ratings; checking a low current diode with an ohmmeter could destroy the diode.*

In-circuit Ohmmeter Measurement

Figure 3-37 demonstrates the technique of measuring resistance in a circuit. Figure 3-37(a) demonstrates measuring a resistance of a single resistor. Follow the current path of the ohmmeter to determine the resistance included in the measurement. In this case, *the current flows out of the ohmmeter Ω lead, through R1, and back into the ohmmeter COM (black) lead.* Thus, only R1 is included in the measurement.

Figure 3-37(b) demonstrates measuring the resistance of a resistor combination. Again, follow the ohmmeter current to see the resistance included in the ohmmeter reading. *The ohmmeter current flows out of the ohmmeter Ω (red) lead, through R1, through R2, through R3, and back into the ohmmeter COM (black) lead.* Thus, the R1-R2-R3 resistor combination is being measured. The ohmmeter only measures the resistance for which a current path is established. If the ohmmeter current does not flow through a resistance, that resistance is not included in the measurement.

You must perform the following three steps to properly measure resistance in a circuit with an ohmmeter.

1. Turn off the power supply to make an ohmmeter measurement. This is a dead circuit measurement, with no power applied. The ohmmeter provides its own internal voltage or current supply. Warning, analog ohmmeters can be damaged if connected to a live circuit.

2. Disconnect one end of the resistance you are measuring to prevent a false reading through other circuit paths. You must break an appropriate point in the circuit. For example, you could lift out one end of a resistor (a resistor lead) or a connecting wire.

3. Connect the ohmmeter test leads across the resistance to be measured. Resistors are not polarity sensitive, so the polarity of the ohmmeter test leads is not important.

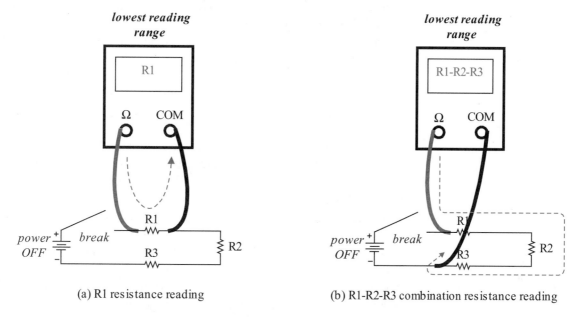

(a) R1 resistance reading   (b) R1-R2-R3 combination resistance reading

Figure 3-37 Ohmmeter resistance measurement of (a) a single resistor, (b) a resistor combination

Summary

Material exhibits electrical *resistance* based upon its *resistivity*, length, and cross-sectional area. The symbol for resistance is the letter R. Its unit is the *ohm*, and the unit symbol for the ohm is the Greek letter Ω. Resistivity is very low for conductors and very high for insulators. The resistivity of semiconductors falls between these two extremes. The higher the resistivity of a material is, the greater its resistance to current flow.

Conductance characterizes how easily charge flows in a material. It is the opposite of resistance. The conductance of a material is the *reciprocal* of its resistance. Conductors have very low resistance and very high conductance. The symbol for conductance is the letter G. Its unit is the *siemens* and the unit symbol is the capital letter S.

A *characteristic curve* of a component is a plot of its current I versus voltage V operating points. The curve represents all the possible operating points of the component. A fixed resistance has a linear characteristic curve. The characteristic curve is most useful for nonlinear devices.

A *short* is a very low resistance path, ideally 0 Ω. An unintended short in a circuit typically causes increased circuit currents. This in turn can be potentially hazardous for other circuit components. *The key char-*

acteristic of a short is that it drops very low voltage, ideally 0 V. A short drops 0 V but can draw current.

An *open* is equivalent to a break in the circuit. An open reduces the circuit current and thus does not typically cause a hazard. *The key characteristic of an open is that it draws 0 A. An open draws 0 A, but can drop voltage.*

Switches are used to turn circuits *on* and *off.* An open switch (ideally ∞ ohms) does not allow current to flow. A closed switch (ideally 0 ohms) completes a circuit to allow current to flow.

A switching or rectifier *diode* is a two-terminal semiconductor component that is voltage polarity sensitive. Its *p*-type material is labeled the *anode* (positive). Its *n*-type material is labeled the *cathode* (negative). When the diode is forward biased with sufficient voltage to overcome its *pn*-junction barrier potential (0.2 V for germanium and 0.7 V for silicon), the diode turns on and conducts current. If insufficient forward bias voltage is applied or if the diode is reverse biased, the diode is off and does not conduct current. An ideal *on* diode can be modeled as a *short*. An ideal *off* diode can be modeled as an *open*. If the externally applied voltage is relevant to the diode's barrier potential voltage, then the *pn* junction drop must be included in the circuit analysis calculations.

The *light emitting diode* (LED) emits light when forward biased with sufficient current. When turned on, the *pn* junction drops 1.5 V to 5 V.

Resistance can be fixed or variable. *Fixed resistors* are two-terminal components that provide a fixed resistance value. Carbon composite resistors are used extensively in electronics and are labeled with a standard color code that indicates the resistor's nominal resistance value, percent tolerance, and reliability.

The *ohmmeter* is the test instrument used to typically measure resistance values. The ohmmeter is a dead circuit measurement (no external voltage applied). The ohmmeter actually applies a voltage (or a current) to the component being tested, and then measures its resulting current (or voltage). Then the ohmmeter calculates the resistance being measured by dividing its voltage by its current measurement. This is a very important relationship for resistance, called Ohm's law, that we shall study further later in this textbook.

$$R = \frac{V}{I}$$

The ohmmeter measurement of resistance is not polarity sensitive. Measurement of diodes is polarity sensitive. Special ranges are provided for diodes to produce enough test voltage to forward bias them when testing them.

The next chapter expands upon this chapter, looking at resistance applications. Major topics include (1) a more detailed look at conductors and the American Wire Gauge standard, (2) circuit protection through the use of fuses and breakers, (3) rheostat and potentiometer variable resistors, (4) the effects of temperature on conductors, semiconductors, and insulators, and (5) sensors that utilize the external environment to change component resistance.

Problems

Resistance Fundamentals

3-1   Find the resistance of 1000 meters of copper wire with a diameter of 2.05 mm.

3-2   Glass is used as an insulator. It is 3 mm thick and the area is 4 cm by 2 cm. Find its resistance.

Conductance

3-3   Find the conductance of a material with 3.9 kΩ of resistance.

3-4   Find the resistance of a wire that has a conductance of 250 mS.

Characteristic Curve

3-5   For the circuit and characteristic curve of Figure 3-38.

    a. Find the voltage across R if the current through R is 20 mA.

    b. Find the static resistance of R if $I = 20$ mA.

    c. Find the static conductance of R if $I = 20$ mA.

    d. Find the current through R if 0 V is applied to R.

    e. Find the current through R if –3 V is applied to R.

    f. Find the static resistance of R if $V = 3$ V.

    g Is this a linear or nonlinear characteristic curve?

    h. With the polarities shown in the test circuit, which values of V and I for resistor R would be generated: the positive or negative values?

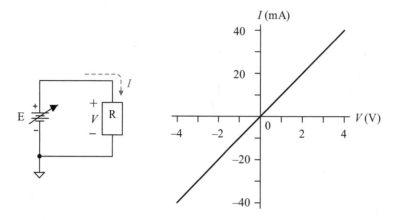

Figure 3-38

3-6 For the circuit and characteristic curve of Figure 3-39.

   a. Find the voltage across R if the current through R is 0 mA.

   b. Find the static resistance of R if $I = 40$ mA.

   c. Find the static conductance of R if $I = 40$ mA.

   d. Find the current through R if –4 V is applied to R.

   e. Find the static resistance of R if $V = 4$ V.

   f. Is this a linear or nonlinear characteristic curve?

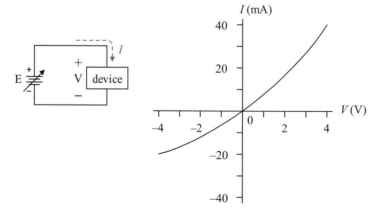

Figure 3-39

Short and Open Circuits

3-7   What is the key characteristic of a short circuit: (a) its current is 0 A, or (b) its voltage drop is 0 V?

3-8   What is the key characteristic of an open: (a) its current is 0 A, or (b) its voltage drop is 0 V?

Switches

3-9   From the circuit of Figure 3-40, what must the switch positions be for lamp L1 to light?

3-10   From the circuit of Figure 3-40 what must the switch positions be for lamp L2 to light?

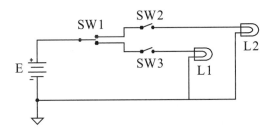

Figure 3-40

Diodes

3-11   Using the circuit of Figure 3-41(a) and assuming an ideal diode:

    a. Is the diode forward or reverse biased?

    b. Is the diode on or off?

    c. Is the lamp lit?

3-12   Using the circuit of Figure 3-41(b) and assuming an ideal diode:

    a. Is the diode forward or reverse biased?

    b. Is the diode on or off?

    c. Is the lamp lit?

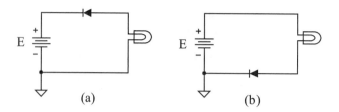

Figure 3-41

Fixed Resistors–Color Code

3-13 Find the nominal resistance, maximum nominal resistance, and minimum nominal resistance for each of the following color coded resistors:

  a. red, violet, yellow, gold

  b. brown, gray, red, silver

  c. brown, black, black

3-14 Find the nominal resistance, the maximum nominal resistance, and the minimum nominal resistance for each of the following color coded resistors:

  a. brown, black, green, gold

  b. orange, white, orange, silver

  c. green, blue, brown

Ohmmeter Measurements

3-15 Is the ohmmeter to be used in a live circuit?

3-16 Does the ohmmeter actually measure resistance directly? Explain.

3-17 If an ohmmeter range applies 1.5 V and measures 15 mA, what resistance is measured?

3-18 If an ohmmeter range applies 1.5 V to a silicon diode, will the diode turn on for a forward resistance measurement? Explain.

4

Resistance Applications

Introduction

In this chapter you are going to learn about applications of resistance. You will study conductors (wires) and understand the American Wire Gauge system of categorizing wire. You will examine methods of protecting circuits from excessive currents by using fuses and circuit breakers. You will see how rheostats and potentiometers (variable resistors) work and can be modeled. You will study components specifically designed to produce specific resistive effects based upon their external environment (for example, light sensors). You will study the effects of temperature on conductors, semiconductors, insulators, and resistance.

Objectives

Upon completion of this chapter you will be able to:

- Interpret and apply American Wire Gauge (AWG) table values.

- Calculate the resistance of wire and circuit board traces.

- Relate current capacity of wire to its AWG number.

- Apply circuit current protection to a circuit using a fuse and the circuit breaker.

- Identify, model, and apply variable resistors: rheostats and potentiometers.

- State the temperature effects on electrical material: conductors, semiconductors, and insulators.

- Calculate the resistance change of a resistor due to a temperature change.

- State the characteristics of specialized resistance components, especially transducers, which are sensors that affect component resistance.

4.1 Wire Table

Wire is produced in standard sizes. Primarily this is done for two reasons. One is to produce a set of standard sized wires so that other products that use wire can be engineered and manufactured based upon known, standard sizes. Also, it is more economical to produce a standard set of wires than provide an unlimited number of sizes of wire.

The American Wire Gauge (AWG) is the standard for wire sizes in the United States. Appendix D lists the more popular AWG wire sizes with a few associated key characteristics. Wire sizes that are typically used for high current electrical and electrical control circuits range from AWG 0000 through AWG 18, respectively. For example, home wiring typically uses 10, 12, and 14 gauge wire. AWG 16 and 18 are used for control circuits. AWG 20 and above are used for low current connecting wire. For example, 24 gauge solid wire is typically used to connect components on an experimental board. Note that Appendix D is not a complete AWG wire table but does list the more popular wire sizes.

Cross-section Area

The lowest numbered AWG size is designated 0000, which is also labeled 4/0 to represent four zeros. It is the *lowest size number* but physically represents the *largest wire*. AWG 0000 has the *largest cross-section area* of 107.2 mm^2 and is about one-half inch diameter. Compare that with AWG 40 wire that has a cross-sectional area of 0.00487 mm^2 (a wire approximately the size of a strand of hair). Figure 4-1 compares a cross-section of AWG 0000 solid wire and AWG 18 solid wire.

Figure 4-1
Cross-section view of wire

Example 4-1

From Appendix D, find the cross-section areas of AWG 12 and AWG 22 and compare them.

Solution

AWG 12 wire has a cross-section area of 3.31 mm^2.

AWG 22 wire has a cross-section area of 0.324 mm^2.

$$\frac{\text{AWG 12 area}}{\text{AWG 22 area}} = \frac{3.31 \text{ mm}^2}{0.324 \text{ mm}^2} = 10.2$$

AWG 12 has approximately 10 times the cross-section area as AWG 22.

Exploration

From Appendix D, compare the cross-section areas of 0, 10, 20, 30, and 40 gauge wire. Is there a pattern?

Answer: Increasing the AWG size by 10 *decreases* the cross-section area by a factor of about 10.

Wire Resistance

Appendix D lists the approximate resistance rating of each AWG wire size. The resistance of a wire is rated in ohms per 1000 feet (Ω/1000 ft) of wire. For example, AWG 0000 has a resistance of 0.0490 ohms per 1000 feet, while 40 gauge wire has a rating of 1,080 ohms per 1000 feet. The 4/0 gauge wire is much more conductive (less resistive) than the significantly smaller 40 gauge wire.

Example 4-2

A 60 foot 14 gauge wire is run from the electrical control box in the basement of a house to a second floor bedroom. Find the resistance of the wire.

Solution

From Appendix D, AWG 14 wire is rated at 2.53 Ω/1000 ft. The total resistance is found by multiplying the wire's length by its resistance rating.

$$R_{total} = 60 \text{ ft} \times \left(2.53 \ \frac{\Omega}{1000 \text{ ft}} \right) = 0.152 \ \Omega$$

It is important that wire that draws high current have very low resistance to prevent excessive heating and voltage loss.

Practice

The resistance of a spool of 40 gauge wire is measured to be 540 Ω. How long is it?

Answer: 500 ft

The next example compares the resistance ratings of wire in the AWG table to explore if there is a pattern.

Example 4-3

From Appendix D, find the resistance rating of AWG 12 and AWG 22 wire and compare them.

Solution

AWG 12 wire has a resistance of 1.588 Ω per 1000 feet.

AWG 22 wire has a resistance of 16.14 Ω per 1000 feet.

$$\frac{\text{AWG 22 resistance rating}}{\text{AWG 12 resistance rating}} = \frac{16.14\,\Omega/1000\,\text{ft}}{1.588\,\Omega/1000\,\text{ft}} = 10.2$$

AWG 22, the smaller diameter wire, has approximately 10 times the resistance of AWG 12 wire.

Exploration

From Appendix D, compare the resistance ratings of 0, 10, 20, 30, and 40 gauge wire. Is there a pattern?

Answer: Increasing the AWG size by 10 *increases* the resistance by a factor of about 10.

Wire—Safe Current Capacity

Appendix D lists the *safe current rating* of selected AWG wire sizes in amperes. This rating is based upon the heat produced by the current through the specified wire. For example, AWG 12 wire can safely handle 20 amperes of continuous current. If more than 20 amperes of current flows continuously in a 12 gauge wire, the wire becomes very hot and overheats its surrounding insulation and creates a possible fire hazard. You need to understand that the 12 gauge wire itself can handle much more than 20 amps of current, but currents higher than its rated value can cause excessive heat and result in serious heat damage or a fire. Current protection devices called fuses and circuit breakers are inserted into a wired circuit and designed to melt or trip, respectively, based upon the current flowing through them. This prevents the wire from overheating. These protection devices are discussed later in this chapter.

Table 4-1

Wire resistance
per unit length

AWG	μΩ/inch
0	8
2	13
4	21
6	33
8	52
10	83
12	132
14	210
16	335
18	532
20	846
22	1,345
24	2,139
26	3,401
28	5,408
30	8,600
32	13,675

NFPA No. SPP-6C,
National Electrical
Code ®, copyright ©
1996, National Fire
Protection Association,
Quincy, MA 02269

Example 4-4

From Appendix D, which gauge of copper wire can safely carry the most current?

Solution

AWG 0000 or AWG 4/0 is rated at 230 A.

Practice

A house circuit requires a wire to carry up to 18 A of current. Which wire should be used to safely carry this current?

Answer: AWG 12 should be used. It can handle up to 20 A safely.

Connector Wire

When dealing with inches of conductor length instead of hundreds of feet of wire, Table 4-1 is a more convenient form for resistance calculations. This table expresses the conductor resistance rating in micro-ohms per inch for corresponding AWG wire sizes 0 through 32.

Example 4-5

A 4 inch length of 22 gauge wire is used as a jumper (connecting wire) on an experimenter board (also referred to as a breadboard). Find its resistance. Find its conductance.

Solution

Based upon Table 4-1, a 22 gauge wire has a resistance rating of 1,345 μΩ/inch or 1.345 mΩ/inch. The 4 inch wire's total resistance is

$$R = 1345 \frac{\mu\Omega}{\text{in}} \times 4 \text{ in} = 5.38 \text{ mΩ}$$

Its conductance is

$$G = \frac{1}{R} = \frac{1}{5.38 \text{ mΩ}} = 185 \text{ S}$$

Practice

A 1 foot length of 24 gauge wire is used as a jumper wire on an experimenter board. Find its resistance. Find its conductance.

Answer: 25.67 mΩ, 38.96 S

Circuit Board Traces

Electronic circuits are mounted on **printed circuit boards** (PCB). A printed circuit board may contain anywhere from a few components to hundreds of components. Rather than using bulky wires to connect these components, the PCB utilizes metallic conductive traces to connect component leads. Copper is the metal typically used for PCB interconnectivity. It may be from only a few microns (micrometers) to 500 microns thick.

Example 4-6

The copper circuit trace of Figure 4-2 is 3 cm long, 2 mm wide, and 100 microns thick. What is the resistance of this conductor? What is its conductance?

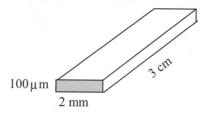

Figure 4-2 Copper trace

Solution

Use the basic definition of resistance of Equation 3-1 and the resistivity of copper from Table 3-1.

$$R = \rho \frac{l}{A} = (17.2 \times 10^{-9} \ \Omega \cdot m) \times \left(\frac{3 \ cm}{100 \ \mu m \times 2 \ mm} \right) = 0.00258 \ \Omega$$

This is an excellent conductor with a very low resistance. The conductance of this trace is thus very high.

$$G = \frac{1}{R} = \frac{1}{0.00258\ \Omega} = 388\ S$$

Practice

A 6 cm copper trace that is 2 mm wide is to have a resistance of 0.00129 Ω. Find its thickness in microns.

Answer: 400 microns

Stranded Wire

Solid wire has the disadvantage of being brittle and is subject to failure if bent or subject to vibration. Stranded wire is used in situations where flexibility and vibration are an issue. Rather than being one solid wire, stranded wire is several strands of smaller wire twisted together to provide the same effective wire gauge as if it were solid. Sample uses of stranded wire include test leads (flexibility needed) and automobile electrical wire (subject to vibration).

4.2 Fuses and Circuit Breakers

As previously noted, excessive current through a device or wire causes excessive heat, which can destroy components and creates a potential fire hazard. **Fuses** and **circuit breakers** are specially designed conductors that detect excessive heat and open the circuit that is in jeopardy. Essentially, these devices act like conductors when operating and then act like opens when the current rating is reached.

Fuses and circuit breakers are *rated based upon the current* at which the fuse melts or the circuit breaker trips. Although rated in terms of maximum current, it is the temperature of the device that triggers action. The temperature of a device is proportional to the current through it; a higher current creates a higher device temperature.

It is critical that the rating of the fuse or circuit breaker match the need of the circuit. For example, if you replace a fuse in a circuit rated for 15 A with a fuse rated at 30 A, circuit protection is lost and the circuit or device will overheat if operated above the 15 A rating. Never circumvent the protection of a circuit; the consequences can be hazardous to people and property.

Fuse

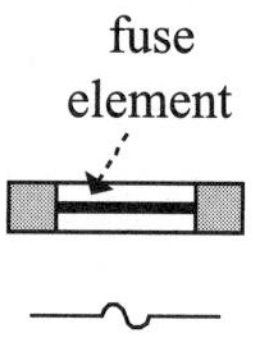

Figure 4-3
Fuse and its schematic symbol

Fuses are metal conductors designed to *melt at specific temperatures*. Figure 4-3 shows a basic fuse and its schematic symbol. The fuse element is the actual fuse. The remainder of the component is there to support the fuse element. The end-caps of the fuse body are conductors to connect the fuse element to the rest of the circuit.

Again, the temperature of the fuse element is directly related to the current through it. For example, a fuse rated at 1 A is hot enough to melt the fuse with 1 A of current flowing through it.

A **blown fuse** is slang for a fuse that has opened. Once a fuse has blown, it is destroyed and must be discarded and replaced with a new fuse. Fuses are typically used to protect such equipment as electronic circuits, stereo systems, test instruments, and electronic toys.

A fuse may be a **fast blow** fuse, which opens quickly when the rated current is reached. Or a fuse may be a **slow blow** fuse, which must be at the rated current for several seconds before it opens. The slow blow fuse is used where current spikes are allowed, and it requires a sustained excessive current to actually blow the fuse open. Typically the fast blow fuse is a flat piece of metal. The slow blow fuse is typically a corrugated, ribbon shape like that shown in Figure 4-4.

Figure 4-4
Slow blow fuse

Figure 4-5 shows a lamp circuit with fused protection. Suppose the lamp is rated for a maximum current of 1.2 A. Usually components are not operated at their maximum capacities; this tends to age components faster. In this case a 1 A fuse is selected to protect the lamp. If the current reaches 1 A, the fuse melts and the lamp is saved. If you accidentally replace the fuse with a 2 A fuse, the lamp is now vulnerable to any currents exceeding 1.2 A.

The 1 A fuse also protects the voltage supply from excessive current. If the lamp were shorted out, the supply would attempt to produce excessive current, which could damage the supply.

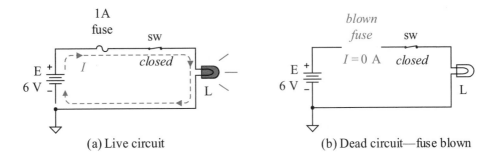

Figure 4-5 Circuit protected by fuse

Example 4-7

The lamp of Figure 4-5 is rated to draw 0.5 A with 6 V applied. Use the schematic of Figure 4-5 to determine the lamp current, lamp voltage, fuse voltage, and switch voltage for the following situations: (a) SW open, (b) SW closed, (c) SW closed and lamp shorts, and (d) SW closed and lamp burns open.

Solution

Figure 4-6 shows the schematic for each of these situations.

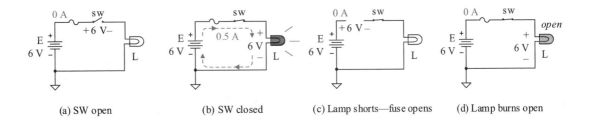

Figure 4-6 Example solutions

a.   SW open
   - no continuous path
   - $I_{lamp} = 0$ A
   - $V_{lamp} = 0$ V

- $V_{fuse} = 0$ V
- $V_{SW} = 6$ V (*open*)

b. SW closed
- continuous path
- $I_{lamp} = 0.5$ A (*lit*)
- $V_{lamp} = 6$ V
- $V_{fuse} = 0$ V
- $V_{SW} = 0$ V

c. SW closed and lamp shorts
- excessive current through the short opens the fuse
- I_{lamp} excessive and then 0 A after fuse opens
- $V_{lamp} = 0$ V
- $V_{fuse} = 0$ V initially then fuse opens and drops 6 V
- $V_{SW} = 0$ V

d. SW closed and lamp burns open
- goes from continuous path to open
- I_{lamp} goes from 0.5 A to 0 A
- $V_{lamp} = 6$ V when lit and then 6 V across the open
- $V_{fuse} = 0$ V
- $V_{SW} = 0$ V

Circuit Breaker

Figure 4-7
Circuit breaker schematic
symbol

Circuit breakers are designed to trip when the current rating is exceeded but, unlike the fuse, the circuit breaker may be reset once the fault is corrected and the circuit breaker reused. Old electrical systems used fuses for protection. For example, old homes with old electrical systems would use a 15 A fuse to protect a 14 gauge wire running out of the electrical panel to the house electrical outlets. A modern electrical system uses a 15 A circuit breaker to protect a 14 gauge wire circuit. If the circuit breaker trips, just correct the fault and reset the circuit breaker. The schematic symbol for a circuit breaker is shown in Figure 4-7 in the *open* position.

4.3 Potentiometers and Rheostats

A **variable resistor** provides a range of resistance values or set of resistance values. Variable resistors are typically controlled by turning a knob (or a screw), which in turn is connected to a shaft that rotates a *wiper arm*. The shaft in turn moves a resistor contact. As the shaft is rotated the variable resistor changes its value of resistance between its terminals.

Variable resistors are common in our technological society. A dimmer light switch, once switched on, is a variable resistor that changes resistance and dims or brightens the lights. Variable resistors are *linear* or *nonlinear* depending upon their application. The volume control on your radio and your TV are variable resistors. We hear logarithmically, therefore, volume controls on your TV and your stereo are nonlinear variable resistances so that volume changes sound linear to the ear. Two prevalent variable resistors are the rheostat and the potentiometer.

Rheostat

A **rheostat** is a two-terminal device that goes from 0 Ω to some maximum resistance. Figure 4-8(a) shows the essential elements of a rheostat. Figure 4-8(b) shows the two schematic symbols used for rheostats.

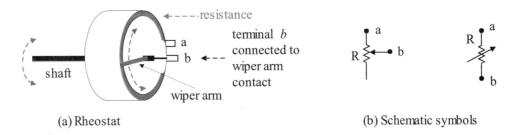

(a) Rheostat (b) Schematic symbols

Figure 4-8 Rheostat

The typical rheostat features a:
- Shaft with a knob or screw head
- Wiper arm connected to and rotating with the shaft (note the shaft could be a direct shaft as shown or another form such as a worm screw)

- Wiper arm contact that smoothly glides over the resistance surface (resistive material could be metal, carbon, or other type of resistive material)
- Wiper contact connected to an external terminal (labeled b)
- Terminal connected at one end of the variable resistance material (labeled a)
- Resistance range from $0 \text{ W} \leq R_{ab} \leq R_{max}$
- Fixed resistance of R_{max} (e.g., a 10 kΩ rheostat)

Example 4-8

A 10 kΩ rheostat is placed in three positions: (a) shaft not turned, (b) shaft turned ¼ position clockwise, and (c) shaft turned fully clockwise. Find the corresponding rheostat resistance values.

Solution

Figure 4-9 demonstrates the solution to this example.

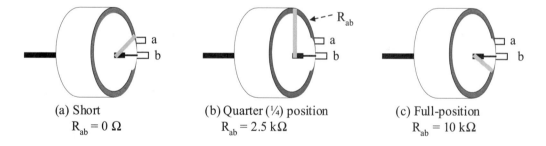

| (a) Short | (b) Quarter (¼) position | (c) Full-position |
| $R_{ab} = 0 \ \Omega$ | $R_{ab} = 2.5 \text{ k}\Omega$ | $R_{ab} = 10 \text{ k}\Omega$ |

Figure 4-9 10 kΩ rheostat in three different positions

a. Figure 4-9(a) shows the wiper arm not rotated and resting at the a contact position. Follow the path from terminal a to terminal b. What do you encounter? Terminal a connects to the wiper arm and then directly to the b connector. This is strictly a conductive path without any resistive material encountered. The ideal resistance of this conductive path is 0 Ω (or a short). Thus, $R_{ab} = 0 \ \Omega$.

b. Figure 4-9(b) shows the wiper arm in the quarter position. Follow the path from terminal a, through a fourth of the resistance, to the wiper arm and then to terminal b. A fourth of the resistive material is encountered. Thus,

$$R_{ab} = \frac{1}{4} \times 10 \text{ k}\Omega = 2.5 \text{ k}\Omega$$

c. Figure 4-9(c) shows the wiper arm fully turned to its maximum position. Follow the path from terminal a, through all of the resistance, to the wiper arm and then to terminal b. All of the resistive material is encountered. Thus,

$$R_{ab} = R_{max} = 10 \text{ k}\Omega$$

Practice

A 10 kΩ rheostat is placed in the following positions: (a) not turned, (b) mid ($\frac{1}{2}$) position, and (c) $\frac{3}{4}$ position. Find the corresponding rheostat resistances.

Answer: 0 Ω, 5 kΩ, 7.5 kΩ

Potentiometer

A **potentiometer** is a three-terminal device that has three distinct resistance values (one fixed and two variable). Figure 4-10(a) shows the essential elements of a potentiometer. Figure 4-10(b) shows the potentiometer schematic symbol. The three resistance values are R_{ac} (fixed resistance value between the outside terminals), R_{ab} (variable), and R_{bc} (variable). Note that this is the same basic construction as a rheostat except that there are two end terminals connected to the resistive material as well as the wiper arm terminal. The resistance between the two outside terminals R_{ac} is always a fixed resistance, and the potentiometer is referred to by its fixed resistance value. If there is 10 kΩ of fixed resistive material between the two outside terminals, then it is labeled a 10 kΩ potentiometer (or commonly referred to as a 10 kΩ *pot*).

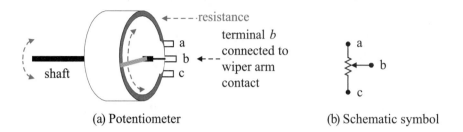

(a) Potentiometer   (b) Schematic symbol

Figure 4-10 Potentiometer

Note that $R_{ab} + R_{bc} = R_{ac}$. For example, if this is a 10 kΩ potentiometer and R_{ab} is 2.5 kΩ, then R_{bc} must be 7.5 kΩ.

Example 4-9

A 100 kΩ potentiometer shaft is rotated ¼ turn clockwise. Find its three resistance values. Draw its schematic symbol and an equivalent two-resistor model.

Solution

Figure 4-11 demonstrates the solutions to this example.

a. Figure 4-11(a) shows that the fixed resistance between terminal a and terminal c is always a fixed 100 kΩ, regardless of the placement of the wiper arm contact. Therefore, the fixed resistance of this potentiometer is always

$$R_{ac} = 100 \text{ k}\Omega$$

b. Figure 4-11(b) shows the wiper arm in the quarter turn position. The resistance between fixed terminal a and wiper terminal b is

$$R_{ab} = \frac{1}{4} \times 100 \text{ k}\Omega = 25 \text{ k}\Omega$$

c. Figure 4-11(c) highlights the resistance between wiper arm terminal b and fixed terminal c, which is

$$R_{bc} = 100 \text{ k}\Omega - 25 \text{ k}\Omega = 75 \text{ k}\Omega$$

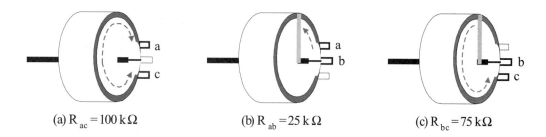

(a) $R_{ac} = 100\,k\Omega$ (b) $R_{ab} = 25\,k\Omega$ (c) $R_{bc} = 75\,k\Omega$

Figure 4-11 (a, b, c) 100 kΩ potentiometer shaft rotated ¼ turn clockwise

d. Figures 4-41(d) is the schematic symbol of the 100 kΩ potentiometer.

e. Figure 4-11(e) shows the potentiometer with equivalent resistance values between its terminals:

  $R_{ab} = 25\ k\Omega$, $R_{bc} = 75\ k\Omega$, and $R_{ac} = 100\ k\Omega$.

f. Figure 4-11(f) shows an equivalent fixed resistor model that is especially helpful when analyzing a circuit.

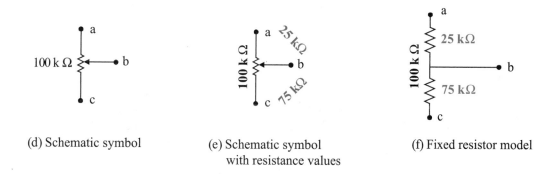

(d) Schematic symbol   (e) Schematic symbol with resistance values   (f) Fixed resistor model

Figure 4-11 (d, e, f) 100 kΩ potentiometer symbol and models

Practice

The shaft of the 100 kΩ potentiometer of Figure 4-11 is rotated a fifth of a turn clockwise. Find its three available resistances R_{ac}, R_{ab}, and R_{bc}.

Answer: 100 kΩ, 20 kΩ, 80 kΩ

Potentiometer Connected as a Rheostat

If a potentiometer is used as a rheostat where only two of its three terminals are connected to the circuit, the unused outside terminal is shorted to the wiper arm terminal as shown in Figure 4-12. This practice prevents the unused portion of the potentiometer from hanging open and acting as an antenna that can transmit and receive unwanted signals.

(a) Potentiometer   (b) Rheostat

Figure 4-12 A potentiometer connected as a rheostat

4.4 Temperature Effects

Temperature can have a significant effect on the resistance of conductors, semiconductors, resistors, and insulators. This is a major consideration in the design of an electronic system if it must operate over a wide temperature range, for example, a car radio operating from very cold to very warm temperatures.

Temperature Coefficient

Figure 4-13 demonstrates the two types of major effects of temperature on a resistance. The slope of the line is defined to be the change along the vertical axis resulting from a change along the horizontal axis. In this case, the slope is

$$m = \frac{\Delta R}{\Delta T}$$

where

    m = slope
    ΔR = change in resistance (Ω)
    ΔT = change in temperature (°C)

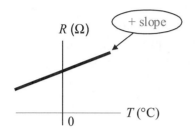

(a) *Positive* temperature coefficient

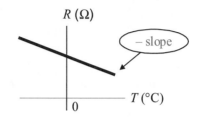

(b) *Negative* temperature coefficient

Figure 4-13 Resistance as a function of temperature

The **temperature coefficient** characterizes the resistance sensitivity of a material to its temperature at a specific resistance and is defined as

$$\alpha = \frac{m}{R}$$

where

    α = temperature coefficient (°C^{-1})
    R = a selected resistance value (Ω)
    m = the slope at the selected resistance value of R (Ω/°C)

The symbol for the temperature coefficient is the Greek letter α. A resistance has a *positive* temperature coefficient if it has a *positive slope* as shown in Figure 4-13(a), and a *negative* temperature coefficient if it has a *negative slope* as shown in Figure 4-13(b). A larger magnitude of the temperature coefficient (α) creates a greater temperature effect.

In general, conductors *increase* resistance with temperature increases while semiconductors *decrease* resistance with temperature increases.

- Conductor—*positive* α
- Semiconductor—*negative* α
- Insulator—*negative* α

Conductors

Conductors, because of their atomic structure, have an enormous supply of free electrons. Heating a conductor does not generate many more free electrons. Thus the number of current carriers does *not* increase with an increase in temperature. The free electrons do absorb more energy and become more excited. This in turn causes these carriers to have more random collisions, essentially increasing the electronic friction experienced by the carriers. The result is that an increase in temperature causes an increase in resistance. A general graph of the resistance of a conductor as a function of temperature is shown in Figure 4-14.

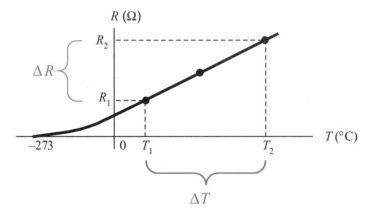

Figure 4-14 Conductor temperature effect

The curve is linear over most of its useful range. An interesting phenomenon occurs as the temperature approaches **absolute zero** (−273°C), the resistance of a conductor approaches an ideal value of 0 Ω nonlinearly. A conductor that approaches an ideal value of 0 Ω is called a **superconductor**.

Example 4-10

An incandescent light bulb uses a tungsten filament wire. The resistance of the wire is measured to be 10 Ω at room temperature (20°C). The temperature coefficient of tungsten is $0.0045°C^{-1}$. After the bulb is turned on, the temperature is measured to be 2620°C. Find the resistance of the wire when the bulb is turned on.

Solution

Use Figure 4-14. First, find the slope of the curve (assuming a linear relationship of temperature and resistance). From the equation for the temperature coefficient

$$\alpha = \frac{m}{R}$$

solve for the slope.

$$m = \alpha \ R = (0.0045 \, ^\circ C^{-1}) \times (10 \, \Omega) = 0.045 \, \frac{\Omega}{^\circ C}$$

Next, use the definition of the slope for this example.

$$m = \frac{R_2 - R_1}{T_2 - T_1}$$

Next, rearrange the above equation to solve for R_2, the upper resistance value. R_2 is the only unknown quantity.

$$R_2 = R_1 + m(T_2 - T_1)$$

Next, substitute the known values into the equation and solve for R_2.

$$R_2 = 10 \ \Omega + \left(0.045 \, \frac{\Omega}{^\circ C} \right) (2620 \, ^\circ C - 20 \, ^\circ C) = 127 \ \Omega$$

Note that this is an estimated value. The resistance versus temperature curve is only approximately linear.

Practice

A light bulb with a tungsten filament has a cold resistance of 10 Ω at room temperature (20°C) and 100 Ω resistance when lit. Find its temperature when it is lit.

Answer 2020°C

Exploration

Why do incandescent bulbs tend to burn out when you first turn them on? Hint: look at the preceding example's results: cold resistance and hot resistance.

Answer:

In the preceding example, the cold resistance of the bulb is only 10 Ω as compared to the hot resistance of the bulb of 127 Ω. The hot resistance is nearly 13 times as great as the cold resistance. Initially, when first turned on, the lamp has a very low resistance allowing a surge of current (called a **surge current**) before the lamp goes to a higher resistance that produces a much lower current. The *surge current* stresses the tungsten filament, thus causing the lamp to typically burn out when initially turned on rather than after it has warmed up.

Semiconductors

An increase in temperature significantly increases the number of intrinsic carriers in a semiconductor, namely silicon and germanium. Semiconductors create additional carriers, which are available to flow and create additional current. This effect is much more significant than the increased electronic friction created by the thermally excited carriers. Thus an increase in temperature causes an increase in current, which indicates that resistance is reduced.

Thus, for a semiconductor, an increase in temperature results in a decrease in resistance. Therefore, semiconductors have a negative temperature coefficient.

Resistors

In a carbon resistor, an increase above or a decrease below room temperature causes an increase in resistance. Figure 4-15 demonstrates this behavior. Resistance at the extremes can vary as much as 10% of a resistor's nominal value at room temperature. A system that must operate over a wide range of temperatures can be significantly impacted by the change in resistor values. Typically resistors operate at or above room temperature, where they have a positive temperature coefficient.

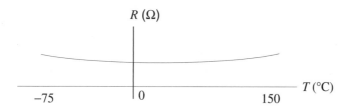

Figure 4-15 Carbon resistor temperature curve

Resistors—PPM/°C

A common rating scheme for resistance temperature sensitivity is PPM/°C, that is, **parts per million per degree Celsius**.

The parts per million is a relative scale. For example, 1000 PPM represents 1000 parts out of a million of the nominal resistance value, or 0.1% of its nominal value. The following equation demonstrates this.

$$PPM = \frac{10^3}{10^6} = 10^{-3} = 0.001 = 0.1\ \%$$

Example 4-11

A resistor's nominal value is 1 kΩ at room temperature; it is rated at 2000 PPM/°C. The resistor must operate from 20°C (room temperature) to 100°C (boiling water). Find (a) the per cent relative change in resistance, (b) its actual change in resistance, and (c) its resistance value at 100°C.

Solution

a. The change in resistance temperature is

$$\Delta T = 100°C - 20°C = 80°C$$

The relative change in resistance is

$$\frac{\Delta R}{R} = 80°C \times \left(\frac{2000}{10^6}\ °C^{-1} \right) = 0.16 = 16\%$$

b. The corresponding change in resistance is

$$\Delta R = 0.16 \times R = 0.16 \times \left(1\ kΩ \right) = 160\ Ω$$

c. The resulting resistance at 100°C

$$R_{100°C} = 1 \text{ k}\Omega + 160 \text{ }\Omega = 1.16 \text{ k}\Omega$$

Practice

A 2.2 kΩ resistor at room temperature is rated at 200 PPM/°C. The resistor must operate from 20°C (room temperature) to 80°C. Find (a) its relative change in resistance, (b) its actual change in resistance, and (c) its resistance value at 80°C.

Answer: (a) 0.012 or 1.2%, (b) 26.4 Ω, (c) 2.226 kΩ

Insulators

An increase in the temperature modestly increases the number of carriers in an insulator. This effect is more significant than the increased electronic friction created by the thermally excited carriers. Thus an increase in the temperature causes an increase in current, which signifies that resistance is reduced.

Thus, for an insulator, an increase in temperature results in a decrease in resistance. Insulators have a negative temperature coefficient.

4.5 Sensors Utilizing Resistance

A **transducer** is a device that transforms one form of energy into another form of energy. For example, a loudspeaker converts electrical energy from an electronic amplifier into mechanical energy, moving the loudspeaker's cone, which compresses the air, creating sound waves. The human ear is a transducer than converts sound waves into human electrical signals interpreted by the brain as sound.

Sensors are special transducers that are used in electronics to *sense* an external environment (for example, light, temperature, position, pressure, torque, acceleration, and strain) and translate the external physical quantity into an electrical entity, namely, resistance, voltage, or current. The resistance, voltage, or current then serves as an input to an electronic circuit, representing the measured external quantity. This electronic signal can then be used electronically, for example, to sense the position of a liquid in a tank, which is then interpreted by an electronics circuit and displayed by a digital display. Note that converting from electrical to light energy for the digital display requires another transducer; for example, an LED could serve this purpose.

This section examines a few sensors that convert an external environmental measurable quantity into a resistance value.

Thermistor

Some devices, called **temperature sensors**, actually take advantage of a device's temperature sensitivity to detect and measure a component's temperature. This type of device is also classified as a **resistance temperature detector** (RTD). For example, a temperature sensor or RTD could detect the temperature in a furnace that is then electronically interpreted and processed to display the furnace temperature. This electronic input, representing the furnace temperature, could also be used to control the furnace temperature.

The **thermistor** is a two-terminal semiconductor sensor. Its schematic symbol is shown in Figure 4-16(a) along with a representative temperature characteristic curve in Figure 4-16(b). As the temperature goes up, its resistance goes down; thus, the thermistor has a negative temperature coefficient. The symbol label is $-t^o$ to indicate a negative temperature coefficient. The logarithmic scales of Figure 4-16(b) allow characteristics to be shown over a wide temperature range.

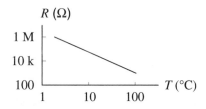

(a) Schematic symbol   (b) Characteristic temperature curve

Figure 4-16 Thermistor

Photoconductive Cell

A **photoconductive cell** is a two-terminal device that senses light and converts light intensity, measured in lumens per unit area, into a corresponding resistance value.

Its schematic symbol is shown in Figure 4-17(a) along with a representative temperature characteristic curve in Figure 4-17(b). As the light intensity increases, its resistance goes down. The arrows pointing into the component's schematic symbol indicate that it is receiving light energy and creating a corresponding resistance. Recall the LED component schematic has the arrows going out of the LED, thus indicating that it is emitting light.

(a) Schematic symbol

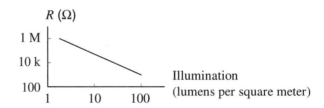

(b) Characteristic resistance curve

Figure 4-17 Photoconductive cell

Photodiode

Figure 4-18
Photodiode schematic symbol

The **photodiode** is also a two-terminal device. Its schematic symbol is shown in Figure 4-18. The diode is used in the reverse biased mode. In this mode the diode has a very high resistance. As the light intensity increases, the diode's resistance goes down. So its characteristic light curve is similar to the photoconductive cell of Figure 4-17(b).

Strain Gauge

The **resistive strain** gauge is another two-terminal device. A simple representation is shown in Figure 4-19(a). Its resistance is proportional to the stress or strain applied to it. Flexing or elongating the resistive strain gauge changes its resistance.

Another type of strain gauge is a four-terminal device that utilizes four resistors in what is called a **bridge configuration**. See Figure 4-19(b). A dc voltage is applied across one pair of terminals and the other pair of terminals, called the **output** terminals, are used to drive an electronic component that detects a voltage difference and amplifies it.

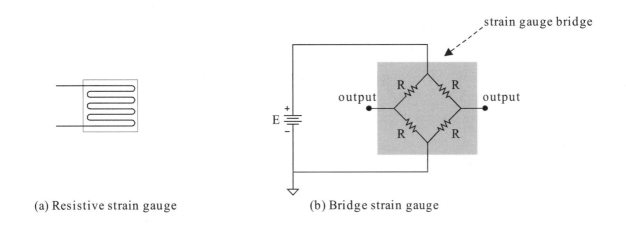

(a) Resistive strain gauge   (b) Bridge strain gauge

Figure 4-19 Strain gauge

Varistor

The **varistor** is still another two-terminal device. Its schematic symbol and characteristic curve are shown in Figure 4-20. The varistor, like the diode, is sensitive to the voltage across it. However, the varistor is designed to conduct at a specified forward and reverse biased voltage. Ideally, it acts like an open if the voltage across the varistor is less than the specification voltage. If the varistor voltage is exceeded with either polarity, it turns on and holds its rated voltage across itself and any connected component. The schematic symbol of Figure 4-20(a) reflects the diode nature of the device. The two opposing diodes reflect that the device is designed to turn *on* in both directions as shown by its schematic symbol and by its characteristic curve of Figure 4-20(b). This figure shows this varistor turning *on* when its voltage drop exceeds 100 V in either direction. If its voltage drop is less than 100 V, the varistor is *off* and acts like an open.

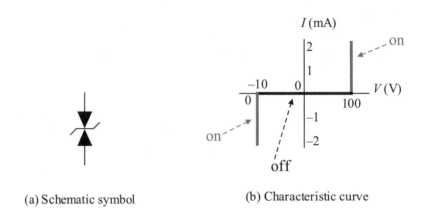

(a) Schematic symbol   (b) Characteristic curve

Figure 4-20 Varistor with a sample characteristic curve

Typically a varistor is placed across another electrical component to protect it from excessive voltage drops. If the voltage drop across the varistor is less than the specified voltage, the varistor acts like an *open* and plays no part in the circuit. If the voltage drop across the varistor exceeds its specification, then the varistor turns *on* and maintains a fixed voltage drop. In the example of Figure 4-20(b), if the varistor turns on, the voltage across it and its companion component is limited to 100 V. Thus the voltage drop across the component placed across the varistor could not exceed 100 V.

Summary

Wire is a conductor that connects electronic components and circuits. Its ideal resistance is 0 Ω, but wire does exhibit some resistance. Wire is manufactured to standard sizes based upon the American Wire Gauge (AWG) standard. Lower AWG numbers have larger diameters and have less resistance. Wires are also rated by the National Fire Protection Association (NFPA) as to the current that they can safely carry. Larger diameter wires can handle larger currents.

Fuses and *circuit breakers* are used to protect circuits from excessive currents. A fuse uses a metallic element, which melts (self-destructs) if the rated current is reached. A circuit breaker trips when the rated current is reached, but can be reset and reused.

Resistance can be fixed or variable. *Rheostats* (two terminal devices) and *potentiometers* (three terminal devices) are variable resistance components. A wiper arm that makes smooth contact with a resistance surface provides the resistance variation.

The resistance of material is temperature sensitive. If the temperature is increased,

- *Conductor* resistance increases (*positive* temperature coefficient).

- *Semiconductor* resistance decreases (*negative* temperature coefficient).

- *Insulator* resistance decreases (*negative* temperature coefficient).

Carbon composite resistors increase resistance as temperature increases or decreases from room temperature. A specification for resistance change due to temperature change is PPM/°C (parts per million per degree centigrade). This specification can be used to calculate resistance change with a change in temperature.

Transducers are devices that transform energy from one form to another, for example, electrical to mechanical. *Sensors* are devices used in electronics to sense an external measurable quantity and translate the measured value into a resistance, voltage, or current.

The next three chapters examine the three fundamental laws of circuit analysis called Kirchhoff's current law, Kirchhoff's voltage law, and Ohm's law, respectively. Kirchhoff's current law relates the currents flowing into and out of a circuit node.

Problems

Wire Table

4-1   Find the cross-section areas of 14 gauge and 24 gauge wire and compare them.

4-2   Find the resistance of 300 feet of 14 gauge and 24 gauge wire and compare them.

4-3   An electrical line must supply a current of 25 A. What gauge wire should be used?

Fuses and Circuit Breakers

4-4   What is the major advantage of a circuit breaker as compared to a fuse?

4-5   What is meant by the term slow blow fuse?

4-6   What is most likely to cause a fuse to blow or a circuit breaker to trip: an open or a short?

Potentiometers and Rheostats

4-7   A 50 kΩ linear rheostat is turned 30% of its range. Find its resistance.

4-8   A 50 kΩ potentiometer is turned 40% of its range.

    a.   Find its three available resistances.

    b.   Sketch a fixed resistor model of this potentiometer.

4-9   If only two terminals of a potentiometer are to be used (one end and the wiper terminals), what should you do with the unused, end terminal? Why?

Temperature Effects

4-10   Is the temperature coefficient positive or negative for the following materials?

    a.   Conductors

    b.   Semiconductors (germanium and silicon)

    c.   Insulators

4-11   The tungsten filament wire of an incandescent light bulb is measured to be 20 Ω at room temperature (20°C). When lit, the filament heats up to 2820°C. Find the resistance of the wire when the light is turned on.

5

KCL—Kirchhoff's Current Law

Introduction

Kirchhoff's current law (KCL) describes how currents must behave at a circuit node. *It is absolutely critical that you understand, master, and be able to apply this fundamental law of circuit analysis. Do not proceed in this book until you have mastered this chapter's materials.*

Kirchhoff's current law is based upon the principle that what flows in must flow out. The current that flows into a node must flow out of that node. Even if there are several currents flowing into the node, the sum of those currents must be the same as the sum of the currents flowing out of that node.

Objectives

Upon completion of this chapter you will be able to:

- State and apply Kirchhoff's current law (KCL).

- Define and discern the difference between a simple node, a voltage node, and a super node.

- Apply Kirchhoff's current law to simple nodes, voltage nodes and super nodes.

- Draw a current view diagram of current flowing through a circuit.

- Apply Kirchhoff's current law to several circuits including electronic circuits.

- Draw the schematic symbol of, identify the terminals of, and show the major currents of a bipolar junction transistor.

- Draw the schematic symbol of, identify the terminals of, and show the major currents of an operational amplifier.

5.1 KCL–Kirchhoff's Current Law

Kirchhoff's current law simply states that the sum of the currents into a node must equal the sum of currents coming out of that node. Mathematically

$$\sum I_{\text{into node}} = \sum I_{\text{out of node}} \tag{5-1}$$

Figure 5-1(a) graphically demonstrates this principle with a simple node, a single current into the node and a single current out of the node. Kirchhoff's current law states that the current into the node (I_{in}) must equal the current coming out of the node (I_{out}).

$$I_{\text{in}} = I_{\text{out}}$$

Figure 5-1(b) presents a water pipe analogy. The water flowing into the pipe must equal the water flowing out of that pipe. Kirchhoff's current law makes common sense. *What flows in must flow out.*

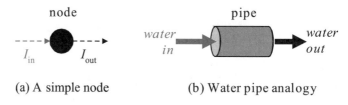

(a) A simple node   (b) Water pipe analogy

Figure 5-1 A simple node and a water pipe analogy

Figure 5-2 demonstrates this law using sample current values for the simple node and *gallons per second* (gal/s) for the water pipe analogy. Common sense tells you that if 2 gal/s is flowing into the pipe, there must be 2 gal/s flowing out of the pipe. Likewise if there is a 2 A current flowing into a node, there must be 2 A current flowing out of that node. There is no other place for the current to go, and electric charge does not accumulate within the node.

Recall that an amp (A) is equivalent to a coulomb per second (C/s). In our analogy, we can equate a gallon of water flowing in a pipe to a coulomb of charge flowing in a wire. That is, coulombs flowing per second are equivalent to gallons flowing per second. The coulomb consists of lots of flowing electrons and the gallon consists of lots of flowing water drops.

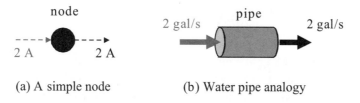

(a) A simple node   (b) Water pipe analogy

Figure 5-2 A simple node example

Figure 5-3 extends this thought process to multiple currents flowing into and out of a simple node.

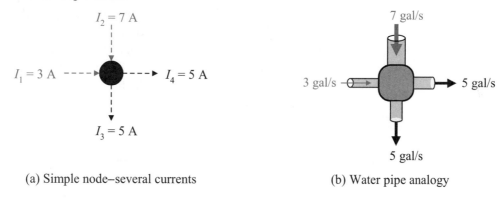

(a) Simple node–several currents   (b) Water pipe analogy

Figure 5-3 Simple node with several currents and a water pipe analogy

Figure 5-3(a) shows two currents flowing into the node and two currents flowing out of the node. Based upon Kirchhoff's current law the sum of the currents into the node must equal the sum of currents out of the node. Check it out.

$$\sum I_{\text{into node}} = \sum I_{\text{out of node}} \qquad \text{(KCL)}$$

$$I_1 + I_2 = I_3 + I_4$$

$$3\,\text{A} + 7\,\text{A} = 5\,\text{A} + 5\,\text{A}$$

$$10 \text{ A} = 10 \text{ A} \hspace{4cm} \text{(verified)}$$

The water pipe analogy is shown in Figure 5-3(b). The total flow into the pipe is the sum of 3 gal/s plus 7 gal/s, for a total of 10 gal/s. The total flow out of the pipe is the sum of 5 gal/s plus 5 gal/s, for a total of 10 gal/s. A total of 10 gal/s flows in and a total of 10 gal/s flows out. Again, this is common sense. *What flows in must flow out.*

$I_{\text{into R}} = I_{\text{out of R}}$

We can extend this concept to a two-terminal component such as a resistor, a voltage source, or a diode. The current flowing into a two-terminal device must be equal to the current flowing out of it as shown in Figure 5-4(a). The current flowing into resistor R must equal the current flowing out of it. The current flowing into supply E must equal the current flowing out of it. As shown in Figure 5-4(b), we refer to current through R as I_R, which *flows into, through, and out of* R. And we refer to the current through E as the supply current I_{supply}, which *flows into E, through E, and out of E.*

$I_{\text{into supply}} = I_{\text{out of supply}}$

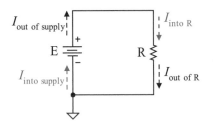 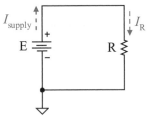

(a) Current through a two-terminal device   (b) Simplified view

Figure 5-4 KCL for a two-terminal component

5.2 KCL and the Simple Node

Let's look at several circuits and apply Kirchhoff's current law. Many of these circuits are used extensively in electronics, and we shall study them in detail later. For now we want to focus on the currents in the circuit and applying Kirchhoff's current law (KCL). You need to understand each example completely.

The first example is called a *series circuit*, where components are tied end-to-end to complete the circuit.

Example 5-1

Using the circuit of Figure 5-5(a), find the currents I_2, I_3, and I_4. Generalize about the currents in this circuit.

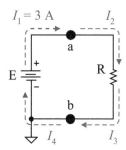

Figure 5-5(a) Example circuit–a series circuit

Solution

First, the current I_1 flowing into node a must equal the current I_2 flowing out of node a. I_2 is also the same current flowing into resistor R.

$$I_2 = I_1 = 3 \text{ A} \qquad \text{(KCL)}$$

Next, the current I_2 flowing into resistor R must equal the current I_3 flowing out of R. This is also the current flowing into node b.

$$I_3 = I_2 = 3 \text{ A} \qquad \text{(KCL)}$$

Next, the current I_3 flowing into node b must equal the current I_4 flowing out of node b. This is also the current I_4 flowing into supply E.

$$I_4 = I_3 = 3 \text{ A} \qquad \text{(KCL)}$$

Last, the current I_4 flowing into supply E must equal the current I_1 flowing out of supply E. We have come full circle and are back where we started. All the currents have been found.

$$I_1 = I_4 = 3 \text{ A} \qquad \text{(KCL)}$$

The generalization that can be made for the *series circuit* is that the *current is the same everywhere in the circuit*, in this case, 3A.

All components and wires have the same current flowing through them in a series circuit. Figure 5-5(b) shows a current view of the circuit. When working with circuits you should be able to visualize how current is flowing in the circuit.

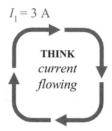

$I_1 = 3$ A

THINK
current
flowing

Figure 5-5(b) Series current of Example 5-1

Practice

In the circuit of Figure 5-6, the supply current I_{supply} is 20 mA. Find the resistor current, the diode current, and the wire current. Also, redraw the schematic of Figure 5-6 in bubble notation.

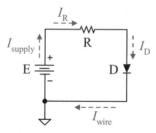

Figure 5-6 Practice circuit

Answer: All the currents are 20 mA because of KCL.
The bubble notation circuit is shown in Figure 5-7.

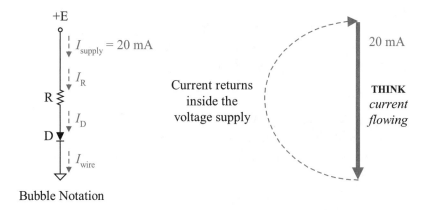

Bubble Notation

Figure 5-7 Practice circuit–bubble notation

The next example is called a *parallel circuit* because of its appearance. More than one component is connected between the same pair of nodes. Kirchhoff's current law applies to all circuits, no matter what configuration they have.

Example 5-2

Using the schematic of Figure 5-8 and KCL, state the relationship of the currents at node a, and the relationship of the currents at node b. Note, at node a, that the supply current I_{supply} splits into two *branch currents*, I_{R1} and I_{R2}. Then, these two branch currents rejoin at node b to reform the supply current I_{supply}.

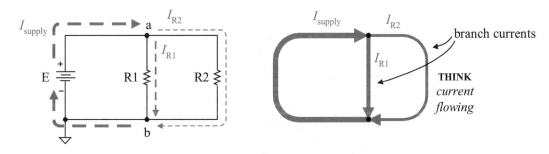

Figure 5-8 Example problem–parallel circuit with two branches

Solution

At node a, the current I_{supply} flowing into node a must equal the sum of the two currents I_{R1} and I_{R2} flowing out of node a. The supply current I_{supply} splits into two *branch currents*.

$$I_{supply} = I_{R1} + I_{R2}$$

At node b, the sum of the two branch currents I_{R1} and I_{R2} flowing into node b must equal the supply current I_{supply} flowing out of node b. The two branch currents rejoin to reform the supply current I_{supply}.

$$I_{R1} + I_{R2} = I_{supply}$$

These two nodes yield the same KCL result.

Practice

Using the circuit of Figure 5-9(a), find the branch current I_{R1}. Sketch the current view of the circuit (THINK *current flowing*).

Answer: 5A. Figure 5-9(b) shows the current view of the circuit.

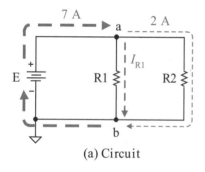

(a) Circuit

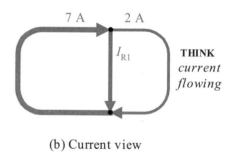

(b) Current view

Figure 5-9 Practice circuit and current view

5.3 KCL and the Voltage Node

The next example of Figure 5-10 adds a third branch to the parallel circuit. We continue to use Kirchhoff's current law to analyze the currents in the circuit; but first, let's expand our thinking of what a node really is. In the preceding section, we examined a simple node (a single point) with two or more wires connected to it. Figure 5-10(a) is a three-branch parallel circuit. Figure 5-10(b) highlights the *voltage nodes* of this circuit. The top node is all at the *same voltage potential* of +E. The bottom node is all at the *same voltage potential* of common, which is 0 V. Therefore these are called *voltage nodes*. Notice that both the top node and the bottom node each contain simple nodes.

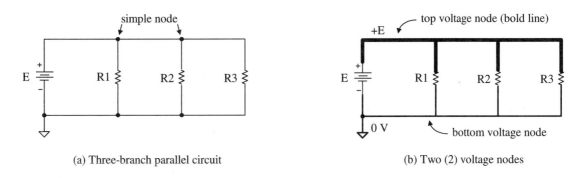

(a) Three-branch parallel circuit   (b) Two (2) voltage nodes

Figure 5-10 Node voltage view of a three-branch parallel circuit

You can apply KCL to a voltage node or a simple node. So you can visualize this circuit as seen in Figure 5-11, with two major voltage nodes. Both nodes must satisfy Kirchhoff's current law.

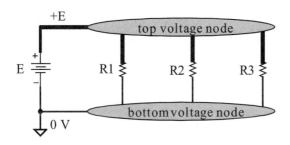

Figure 5-11 Simplified view of voltage nodes

Figure 5-12 shows the circuit with *current references*. Before solving a circuit you must always establish proper references. Let's see if we can anticipate the actual current directions and establish the appropriate reference currents. In the circuit of Figure 5-11, the supply current exits the supply's positive (+) terminal and flows *into the top voltage node*. The supply current then splits at the top voltage node into three branches. The branch currents flow from the top voltage node *down* to the bottom voltage node. Then the three branch currents rejoin into the supply current that flows *out of* the bottom node and into the supply's negative (−) terminal.

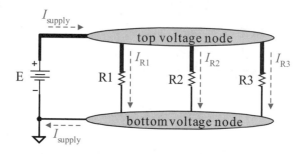

Figure 5-12 Reference currents for three-branch parallel circuit

Example 5-3

Using the schematic of Figure 5-12 and KCL, find the relationship of the currents at the top node. Then find the expression for the currents at the bottom node. Compare the results. Draw the current view diagram.

Solution

At the *top voltage node*, the supply current I_{supply} flowing into the voltage node must equal the sum of the three (3) branch currents flowing out of the top voltage node.

$$I_{supply} = I_{R1} + I_{R2} + I_{R3} \qquad \text{(KCL)}$$

At the *bottom voltage node*, the sum of the three (3) branch currents flowing into the voltage node must equal the supply current I_{supply} flowing out of the bottom voltage node.

$$I_{R1} + I_{R2} + I_{R3} = I_{supply} \qquad \text{(KCL)}$$

The resulting equations are the same whether you use the top or the bottom voltage node. The current view diagram of this circuit is shown in Figure 5-13.

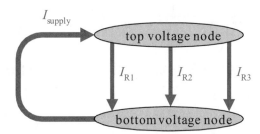

Figure 5-13 Example circuit–current view diagram

Practice

Use the circuit of Figure 5-12. If the supply current is 9 A, and the first two branch currents are 5 A and 3 A, respectively, find the third branch's current.

Answer: 1 A.

In Practice 5-3 you should have used the voltage node approach to find the unknown current with one KCL equation. You could have solved this problem by applying the simple node KCL approach, but that would have required more equations. Before solving any circuit, first examine the circuit to look for the best strategy, that is, an approach with the fewest number of steps. The more straightforward the approach, with the fewest equations, offers fewer opportunities to make calculation errors.

Next, we examine the same three-branch circuit as Practice 5-3, looking at the simple node approach.

Example 5-4

Figure 5-14 shows the supply and branch currents of the three-branch parallel circuit above. Use node a to find I_X and then node b to find I_Y.

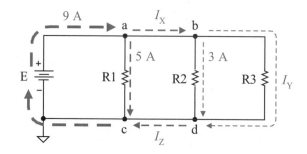

Figure 5-14 Example circuit–simple node view

Solution

At node *a*.

$$9\,\text{A} = 5\,\text{A} + I_X \qquad \text{(KCL)}$$

Solve for I_X.

$$I_X = 4\,\text{A}$$

At node *b*.

$$I_X = 3\,\text{A} + I_Y \qquad \text{(KCL)}$$

$$4\,\text{A} = 3\,\text{A} + I_Y \qquad \text{(Substitution)}$$

Solve for I_Y.

$$I_Y = 1\,\text{A}$$

Practice

Use the circuit with the values of Figure 5-14 and node *c* to find the current I_Z. Verify this answer using node *d*.

Answer: 4 A, 4 A.

The Bridge Circuit

Next, we shall explore the application of KCL to a *bridge circuit*. The circuit of Figure 5-15 shows a resistor bridge network of R1, R2, R3, and R4, with R_{load} (considered the *load resistance*). The individual branch currents are named to reflect their branch component; for example, I_{R1} corresponds to the current through R1.

All the reference currents are drawn in their correct direction, that is, the actual direction of the current. The one exception is I_{load}. We do not really know which way current flows in R_{load}. It depends upon the relative resistance values of the bridge. The three possibilities of load current are:

• Current flows from node b to node c.

• Current flows from node c to node b.

• No current flows ($I_{load} = 0$ A). This is called a **balanced bridge**.

For our example, we *assume* the reference current of I_{load} as shown in Figure 5-15(b). If we have chosen the reference direction correctly, the numerical answer for I_{load} has a positive (+) value. If the actual direction is opposite the reference, the numerical answer has a negative (−) value.

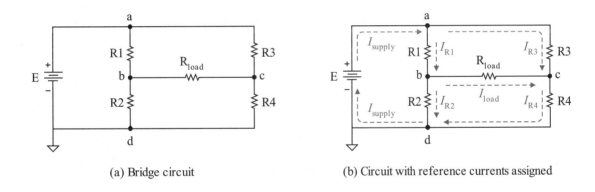

(a) Bridge circuit   (b) Circuit with reference currents assigned

Figure 5-15 Bridge circuit and *assigned reference currents*

Example 5-5

Using the circuit of Figure 5-15, use KCL and find the current relationships of simple node a and then those of simple node b.

Solution

First, apply KCL to node a.

$$I_{supply} = I_{R1} + I_{R3} \qquad \text{(KCL)}.$$

Next, apply KCL to node b.

$$I_{R1} = I_{R2} + I_{load} \qquad \text{(KCL)}$$

Practice

Using the circuit of Figure 5-15, use KCL and find the current relationships of node c and of node d.

Answer: Node c, $I_{R3} + I_{load} = I_{R4}$;   Node d, $I_{R2} + I_{R4} = I_{supply}$.

The next example is a *bridge circuit* with some currents given and we must find the remaining currents. Recall from the preceding example that all the reference current directions are in the correct direction except possibly I_{load}. The actual direction of the load current I_{load} depends upon the values of the bridge resistors (R1, R2, R3, and R4). Thus, the value of I_{load} could be positive (+) or negative (–) depending upon the reference direction chosen and the actual current direction.

Example 5-6

Using the circuit of Figure 5-16, use KCL and find the unknown currents. Redraw the schematic showing all the current values and their actual direction.

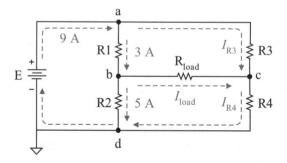

Figure 5-16 Example circuit–bridge circuit

Solution

Node a has two known currents and only *one unknown current* I_{R3}. We can solve for the unknown current I_{R3}.

$$9 \text{ A} = 3 \text{ A} + I_{R3} \qquad \text{(KCL)}$$

Solve for I_{R3}.

$$I_{R3} = 6 \text{ A}$$

Write it on the schematic after you find it so you can visualize it and use it later. Node *c* has two unknown currents so we cannot use this node yet. Let's move on to node *b*. There is only one unknown at node *b*, namely I_{load}, so let's solve that for that unknown current.

$$3 \text{ A} = 5 \text{ A} + I_{\text{load}} \qquad \text{(KCL)}$$

Solve for current I_{load}.

$$I_{\text{load}} = -2 \text{ A} \qquad \text{(flowing } out\ of \text{ node } b)$$

Note the negative (–) answer. Thus, the current really flows the *opposite direction* of the reference (*into* node b, not out of it).

$$I_{\text{load}} = +2 \text{ A} \qquad \text{(flowing } into \text{ node } b)$$

The only current remaining to find is I_{R4}. We could use either node *c* or node *d* to solve for I_{R4}. Let's use node *d*.

$$5 \text{ A} + I_{R4} = 9 \text{ A} \qquad \text{(KCL)}$$

Solve for current I_{R4}.

$$I_{R4} = 4 \text{ A}$$

Figure 5-17 shows the circuit with *actual current values and directions.* Note that I_{load} is actually flowing opposite its original reference.

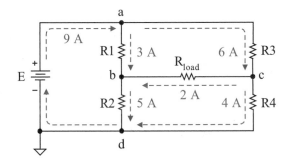

Figure 5-17 Example circuit with actual current values and directions

Practice

Use the circuit of Figure 5-17 and verify KCL at each node. This is an excellent insurance check to see if you have solved the problem correctly.

Answer: Verification of KCL at each node: *a*, *b*, *c*, and *d*.

5.4 KCL and the Super Node

The **super node** is a node that includes components or circuits within it. Previously, we have looked at the *simple node* (a single point) and the *voltage node* (continuous conductors at the same voltage potential). Now we are expanding the node concept to include components or circuits within it. The best way to demonstrate this is with an example. The previous circuit is used to demonstrate this concept of a *super node*.

Example 5-7

Figure 5-18 is the previous circuit with a super node defined. The shaded gray area is a super node because it contains components. Verify that Kirchhoff's current law works for this super node. Note that currents from the previous solution are provided.

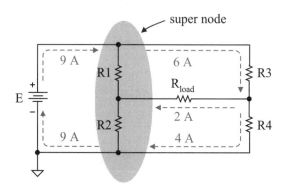

Figure 5-18 Example circuit–bridge circuit with a super node defined

Solution

This super node has three currents *flowing into it*, namely, 9 A (top left), 2 A (middle right), and 4 A (bottom right). The super node has 2 currents flowing out of it, namely, 6 A (top right) and 9 A (bottom left). Apply KCL and see if it works for this super node.

$$\sum I_{\text{into node}} = \sum I_{\text{out of node}} \qquad \text{(KCL)}$$

$$9\ \text{A} + 2\ \text{A} + 4\ \text{A} = 6\ \text{A} + 9\ \text{A} \qquad \text{(super node)}$$

$$15\ \text{A} = 15\ \text{A} \qquad \text{(KCL verified)}$$

Practice

Use the circuit of Figure 5-19. Find the current I_{R4} using the super node shown in gray. Then find I_{load} and I_{supply}.

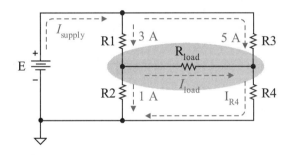

Figure 5-19 Practice circuit with a super node

Answer: 7 A, 2 A, 8 A.

Warning: if you are working with bubble notation and the super node includes a power supply or signal lead, you must first convert the node bubbles to the standard voltage supply notation to include the common returns.

5.5 KCL and a BJT Circuit

The next example introduces the **bipolar junction transistor** (BJT) in a current controlling circuit, but first let's look at the BJT component. The BJT was invented in 1947 at Bell Telephone Laboratories and launched the semiconductor revolution of electronics. The BJT is still a component used in today's technology because it is a nearly linear current amplifier and can handle relatively high currents.

The schematic symbol of an *npn* BJT and its three terminal currents are shown in Figure 5-20. The symbol Q is used as a part label, for example Q1 would represent transistor one. The BJT *arrow* is always associated with the *emitter terminal*. The *middle* terminal is always the *base*. The remaining terminal is always the *collector*. The BJT device is described in greater detail later in this book.

Figure 5-20 shows the schematic symbol of the BJT and demonstrates the direction of currents when the transistor is turned *on* and conducting. The three terminals are labeled emitter (E), base (B), and collector (C). The BJT device must satisfy KCL, that is, that the sum of the currents into the BJT must equal the current I_E out of the BJT.

$$I_E = I_B + I_C \tag{5-2}$$

A characteristic of the BJT, when the dc voltages are properly applied, is that it is a current amplifier. The smaller base current I_B is multiplied by a constant β (Greek letter beta) to create a much larger collector current I_C. **Beta** is called the *current gain* of the BJT. Beta typically has a value from 20 to several hundred.

$$I_C = \beta \, I_B \tag{5-3}$$

A small change in base current creates a big change in collector current.

Another symbol for β is h_{FE}. So Equation 5-3 can also be written as

$$I_C = h_{FE} \, I_B$$

You will find both symbols used in data sheets, but typically h_{FE} is specified. This text typically uses β for ease of mathematical manipulation.

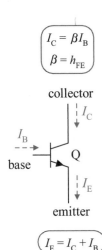

$$I_C = \beta I_B$$
$$\beta = h_{FE}$$

collector

I_C

I_B

base

Q

I_E

emitter

$$I_E = I_C + I_B$$

Figure 5-20
npn BJT–
bipolar junction
transistor

Example 5-8

Figure 5-21 is an LED light control circuit. Increasing the E_{in} supply increases the base current I_B. This increases the collector current I_C, and thus the LED current, causing the LED to glow brighter. Figure 5-21(a) shows the bubble notation for the power supplies. This is very convenient when dealing with a crowded or a complex schematic, while Figure 5-21(b) shows the circuit with its voltage supply schematic symbols. This is sometimes a more convenient form for circuit analysis. You must become familiar with both forms and should be able to solve this example using either figure. Also, you should be able to go back and forth between bubble notation and supply schematic representation. Using the relationships of Equations 5-2 and 5-3, find I_C and I_E. What is the advantage of this circuit over directly controlling the LED current without the transistor?

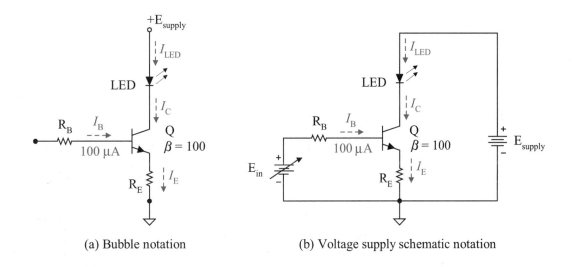

(a) Bubble notation   (b) Voltage supply schematic notation

Figure 5-21 Example circuit–LED light control

Solution

First, use the transistor β factor to find the I_C current.

$$I_C = \beta\, I_B = 100 \times 100\ \mu A = 10\ mA$$

The LED current and BJT collector current are the same by KCL.

$$I_{LED} = I_C = 10\ mA$$

Next, apply KCL on the BJT device and solve for the emitter current I_E.

$$I_E = I_B + I_C = 100\ \mu A + 10\ mA = 10.1\ mA$$

The advantage is that we are using a *very small base current* (100 μA) *to control a much larger collector current* (10 mA).

Practice

Using the circuit of Figure 5-21, the voltage supply E_{in} is adjusted to produce an LED current of 20 mA. Find the BJT's collector, base and emitter currents.

Answer: 20 mA, 200 µA, 20.2 mA. Doubling I_B doubles I_C.

In the last example circuit, if E_{in} is continuously varied, the light intensity varies proportionally. If supply E_{in} were actually a varying signal (e.g., from a microphone circuit), it would vary the light intensity of the lamp. The LED light would indicate the sound intensity and we could use fiber optic cable to detect the light changes and send a corresponding signal via the fiber cable to another circuit.

5.6 KCL and an Op Amp Circuit

The next example introduces a circuit using the *operational amplifier* (abbreviated, *op amp*). The op amp is an integrated circuit (IC) device. The integrated circuit was the next major breakthrough in electronic devices. Thousands of semiconductor components can fit into an area the size of a small pencil dot. The op amp is an integrated circuit that has widespread use throughout electronics.

The symbol of the op amp and its associated signal currents are shown in Figure 5-22. The basic symbol is a triangle shape but it represents a device that has hundreds of components. Two pins are used to connect supply voltages and/or a supply voltage and common. The two signal input pins are labeled the **inverting** (inv) pin and the **noninverting** (ni) pin. A current into the inverting (inv) terminal is *inverted*; that is, a positive (+) valued current into the inverting terminal produces a negative (−) valued current out. A current into the noninverting (ni) terminal is not inverted.

A key current characteristic is that the *input resistance* of an op amp is so high that the two input currents to the op amp (namely, I_{inv} and I_{ni}) are *ideally estimated* to be 0 A.

$$I_{inv} = 0 \text{ A}$$

$$I_{ni} = 0 \text{ A}$$

This is a very useful simple fact that we shall apply in performing circuit analysis of op amp circuits. Most op amp circuits are very easy to

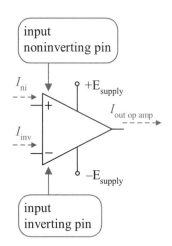

Figure 5-22

Op amp symbol and associated signal currents

analyze because of simplifying assumptions that can be made. In this book, we shall use the ideal models of the op amp for circuit analysis. In practice, there are some very, very small currents through the input terminals, but these can typically be ignored and the ideal currents of 0 amps can be assumed.

Example 5-9

Figure 5-23(a) is a very popular op amp circuit, called an *inverting summer amplifier* because it sums input signal currents. It also inverts the output current since the signal is fed to the inverting pin. That is, a positive current in produces a negative current out. The inputs to this example summer include signal currents from (1) a vocalist microphone, (2) a guitar microphone, and (3) a drummer's microphone. The microphone has the label of M, therefore, the microphones are labeled M1, M2, and M3, respectively.

At first this looks like a complex circuit, but it is really simple to apply Kirchhoff's current law to find the needed currents.

Apply KCL to find I_{in} (the net current into the amplifier). Then find I_{Rf} (the feedback resistor current). Finally find $I_{out\ op\ amp}$ (the op amp output current).

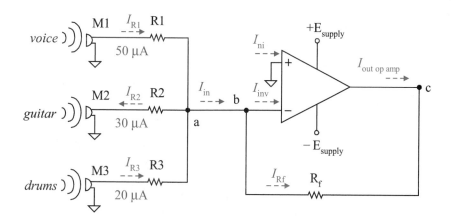

Figure 5-23(a) Example circuit–inverting summer amplifier circuit

Solution

First, apply KCL to the simple node a and find the current I_{in}. Currents I_{R1} and I_{R3} are into node a while currents I_{R2} and the net input current I_{in} are out of node a.

$$I_{R1} + I_{R3} = I_{R2} + I_{in} \qquad \text{(KCL)}$$

$$I_{in} = I_{R1} + I_{R3} - I_{R2} \qquad \text{(solve for unknown)}$$

$$I_{in} = 50\ \mu A + 20\ \mu A - 30\ \mu A = 40\ \mu A$$

Next apply KCL to the simple node b and find the current I_{Rf}. Current I_{in} is into node b while reference current I_{Rf} and current I_{inv} are out of node b.

$$I_{in} = I_{Rf} + I_{inv} \qquad \text{(KCL)}$$

$$I_{Rf} = I_{in} - I_{inv}$$

$$I_{Rf} = 40\ \mu A - 0\ \mu A = 40\ \mu A$$

Apply KCL to the simple node c and find the output op amp current $I_{out\ op\ amp}$. In this case there are two (2) reference currents flowing into node c and no reference current flowing out. Thus, the current flowing out of the node is equivalent to zero (0) amps.

$$I_{Rf} + I_{out\ op\ amp} = 0\ \mu A \qquad \text{(KCL)}$$

$$I_{out\ op\ amp} = -I_{Rf} = -40\ \mu A$$

Note that the op amp current $I_{out\ op\ amp}$ has a negative (−)value with the assumed reference direction. That implies that the current is really 40 μA *flowing into the op amp*, not out of it.

The net input current I_{in} is positive (+) while the output op amp current $I_{out\ op\ amp}$ is negative (−). This is what we expected. A positive (+) current into the inverting pin produces a negative (−) op amp output current. That is, the output is inverted as expected.

Practice

The guitar signal is boosted to 95 μA in the circuit of Figure 5-23(a). Find the feedback resistor current and the op amp output current.

Answer: −25 μA (opposite reference direction), +25 μA

Exploration

Use MultiSIM to simulate the circuit of Figure 5-23(a). Set up your MultiSIM simulated circuit as shown in Figure 5-23(b) to create the desired input currents. This type of circuit is completely analyzed in a later chapter in this textbook. Then double the signal voltages E1, E2, and E3 and observe the current measurements. Does KCL still work?

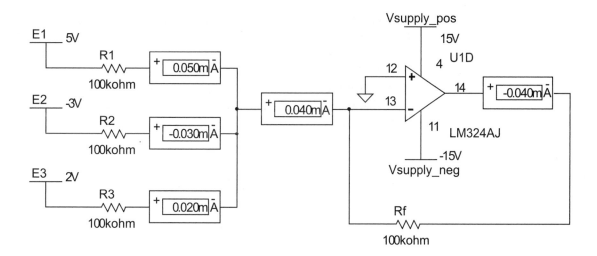

Figure 5-23(b) MultiSIM simulation of example circuit

5.7 KCL Alternate Form

We have been working with the KCL form of

$$\sum I_{\text{into node}} = \sum I_{\text{out of node}}$$

where the sum of currents into a node must equal the sum of currents out of that node.

If we algebraically subtract the right side of the equation from both sides of the equation, we end up with an **alternate form** of Kirchhoff's current law.

$$\sum I_{\text{into node}} - \sum I_{\text{out of node}} = 0 \text{ A} \tag{5-4}$$

This states that that the sum of the currents into a node minus the sum of currents out of that node must equal 0 amps. You saw this situation in the last example problem where reference currents I_{Rf} and I_{out} were both into the node and there were no currents out of the node.

Another way to view this approach is to assign the currents *into* the node as (+) positive and the currents *out of* the node as (−) negative; then, add them up and equate them to 0 amps.

Example 5-10

Using the node of Figure 5-24, find I_3 using the approach that the sum of currents into the node equals the sum of currents out of that node (Equation 5-1). Then solve for I_3 using the approach that the sum of the currents into a node less the sum of the currents out of that node must sum to 0 (Equation 5-4).

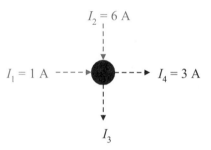

Figure 5-24 Example node

Solution

Using Equation 5-1, find I_3.

$$\sum I_{\text{into node}} = \sum I_{\text{out of node}}$$

$$1 \text{ A} + 6 \text{ A} = I_3 + 3 \text{ A} \qquad\qquad \text{(substitute)}$$

$$I_3 = 4 \text{ A} \qquad\qquad \text{(simplify)}$$

Using Equation 5-4, find I_3.

$$\sum I_{\text{into node}} - \sum I_{\text{out of node}} = 0$$

$$1\text{A} + 6 \text{ A} - I_3 - 3 \text{ A} = 0 \qquad \text{(substitute)}$$

$$I_3 = 4 \text{ A} \qquad\qquad \text{(simplify)}$$

The results are the same.

Practice

First use Equation 5-1 and then use Equation 5-4 to find the unknown current I_X of Figure 5-25. Both forms must yield the same answer.

Answer: –4 A (*note: opposite reference direction*)

Therefore current is actually *into the node* with a value of 4 A

Both forms are equally valid but, depending on the application, one form may be preferable. Both forms work for simple nodes, voltage nodes, and super nodes.

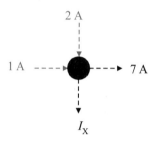

Figure 5-25
Practice node

Summary

This chapter introduced the first of three fundamental laws of circuit analysis, Kirchhoff's current law.

Kirchhoff's current law (KCL) simply states that the sum of the currents going into a node must equal the sum of currents coming out of that node. Mathematically,

$$\sum I_{\text{into node}} = \sum I_{\text{out of node}}$$

Physically it seems obvious that what flows in must flow out. KCL formalizes this for currents into and out of a *node*. The node can be a *simple node* consisting of a single point on a conductor. Or the node can be a *voltage node* that has the same voltage on continuous conductors with one or more simple nodes. Or it can be a *super node* that contains a circuit or components.

Devices must also satisfy KCL. The sum of the currents flowing into a device must equal the sum of currents flowing out of that device. We applied this to the BJT (bipolar junction transistor), a semiconductor component. The BJT is a current amplifier; its collector current is much larger than its input base current by a factor of beta (β).

The operational amplifier (op amp) is an *integrated circuit* (IC) consisting of hundreds of semiconductor components. The op amp was used to demonstrate KCL in an inverting summer amplifier, an electronic circuit with many applications.

An alternate mathematical form of KCL is

$$\sum I_{\text{into node}} - \sum I_{\text{out of node}} = 0 \text{ A}$$

This simply states that the sum of the currents into the node less the currents flowing out of the node must sum to 0 amps. Sometimes this is a more appropriate form to solve circuits. Another way to view this approach is to assign the currents into the node as (+) positive and the currents out of the node as (−) negative and add them up.

Kirchhoff's current law, Kirchhoff's voltage law, and Ohm's law are the fundamental circuit analysis laws for circuit analysis. They are the law and can never be violated. In the next chapter we shall study and apply Kirchhoff's voltage law to find node voltage drops across components and circuits. Then we shall incorporate KVL with node voltage analysis to find node voltages as well as voltage drops in a circuit.

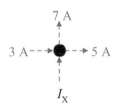

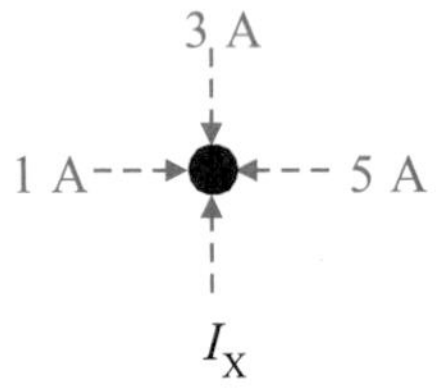

Figure 5-26

Problems

KCL—Kirchhoff's Current Law

5-1 For the node of Figure 5-26, find current I_X.

    a. Use Equation 5-1 approach (what flows *in* must flow *out*)

    b. Use Equation 5-4 approach (currents *in* less currents *out* equals 0 A)

    c. Use Equation 5-4 approach (currents *in* assigned +, currents *out* assigned −, and sum of all currents at node must be 0 A)

5-2 Repeat Problem 5-1 for the node of Figure 5-27.

Figure 5-27

KCL and the Simple, Voltage, and Super Node

5-3 Use the circuit of Figure 5-28.

    a. Find I_1, I_2, I_3, I_4, I_5, and I_6. Draw current values on your schematic as you find them.

    b. Draw a super node on your schematic that proves that $I_2 = I_5$.

    c. Draw a super node on your schematic that proves that $I_1 = I_6$.

    d. Draw the current view of this circuit with the larger currents being thicker lines.

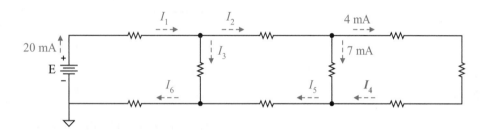

Figure 5-28

5-4 Use the circuit of Figure 5-29.

    a. Find I_1, I_2, I_3, I_4, and I_{supply}.

    b. Draw a super node on your schematic that proves that $I_1 = I_2$.

    c. Draw a super node on your schematic that proves that $I_3 = I_4$.

    d. Draw the current view of this circuit with the larger currents being thicker lines.

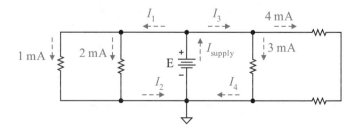

Figure 5-29

5-5   Use the circuit of Figure 5-30.

    a. Use simple nodes to find I_X and I_{lamp}.

    b. Use the top voltage node to find I_{lamp} without using I_X.

    c. Use the bottom voltage node to find I_{lamp}.

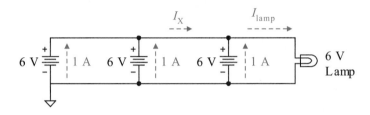

Figure 5-30

5-6   Use the circuit of Figure 5-31, a schematic of an auto's headlight (HL) and taillight (TL) lamp circuit. When turned on, a headlight draws 8 A and a taillight draws 4 A.

    a. When the switch is closed, what is the supply current I_{supply} to the 4-lamp circuit?

    b. If fuses come in 5 A increments, what is the lowest rated fuse that can be used in this circuit?

    c. Sketch the current flow diagram of this schematic with the switch closed.

    d. If the switch or the fuse is open, how much current flows through the right taillight?

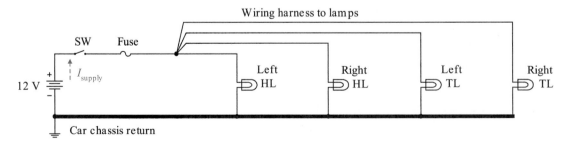

Figure 5-31

KCL and a BJT Circuit

5-7   Use the circuit of Figure 5-32.

    a. Find the base current.

    b. Find the emitter current.

    c. Find the LED current.

    d. If E_{in} is dialed down to 0 V, then $I_B = 0$ A. What do you think
the collector current and then the emitter current would be?
Does the LED light?

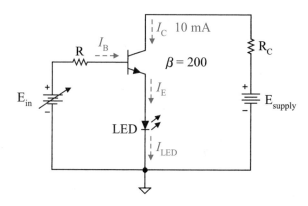

Figure 5-32

KCL and an Op Amp Circuit

5-8   Use the circuit of Figure 5-33. $I_{out\ op\ amp}$ is measured to be
−30 μA.

    a. Find the guitar current I_{R2}.

    b. Is the actual I_{R2} current into or out of node a?

    c. If the voice microphone goes bad, resulting in I_{R1} being
0 μA from its microphone, what does the output current
$I_{out\ op\ amp}$ value become?

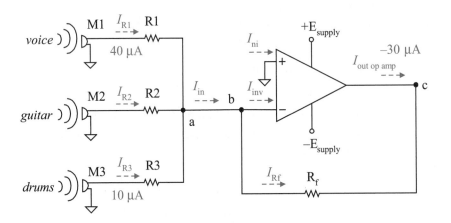

Figure 5-33

5-9 Use the circuit of Figure 5-34. Recall that the input currents to the op amp are so small that they are ideally 0 mA.

    a. Use KCL to find I_{Rf}.

    b. Use KCL to find $I_{out\ op\ amp}$.

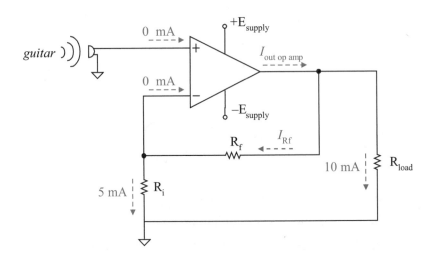

Figure 5-34

6

KVL—Kirchhoff's Voltage Law

Introduction

The next circuit law you will study and master is Kirchhoff's voltage law (KVL). Kirchhoff's voltage law describes how voltages must behave in a closed circuit loop. *It is absolutely critical that you understand, master, and be able to apply this fundamental law of circuit analysis. Do not proceed in this book until you have mastered this chapter's materials.*

Kirchhoff's voltage law essentially states that as you traverse any closed loop of a circuit (adding voltage rises and subtracting voltage falls), the net result is a 0 V difference.

Let's take a journey equivalent to a KVL journey. Imagine that you are hiking. You go up and down hills and end up right back where you started. If you add the uphill distances and subtract the downhill distances, the net result is zero difference. Just like your walk, in a closed loop walk around the circuit you end up back at the same starting point. The net result is that voltage difference between your starting point and ending point (which are the same) must be 0 V.

Objectives

Upon completion of this chapter you will be able to:

- State and apply Kirchhoff's voltage law.

- Calculate component/circuit drops between nodes and voltages at a node relative to common.

- Apply KVL mathematically and visually in a voltage walkabout procedure to several circuits.

- Draw the schematic symbol of, identify the terminals of, and show the major currents and voltages of an enhancement mode MOSFET.

- Apply KVL to electronic circuits and devices including a BJT amplifier, a MOSFET amplifier, and an op amp amplifier circuit.

6.1 KVL—Kirchhoff's Voltage Law

Kirchhoff's voltage law states that the sum of voltage rises minus the sum of the voltage falls around a closed circuit loop must equal zero (0) volts.

$$\sum V_{\text{rises}} - \sum V_{\text{falls}} = 0 \text{ V} \qquad\qquad \textbf{(6-1)}$$

Figure 6-1 graphically demonstrates this principle with a single loop circuit. Let's start at the tail of the arrow and walk around the circuit clockwise. There are four devices (numbered 1 through 4) in the circuit loop. All four device voltages must be included in the KVL expression. In Figure 6-1, we encounter a *rise* of V_1, a *fall* of V_2, a *fall* of V_3, and a *fall* of V_4. The KVL expression for the loop in Figure 6-1 is

$$+V_1 - V_2 - V_3 - V_4 = 0 \text{ V} \qquad\qquad \text{(KVL)}$$

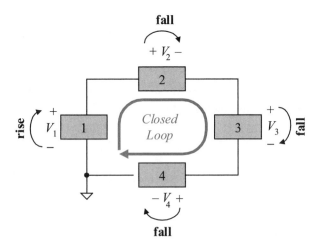

Figure 6-1 KVL with a single closed loop of four devices

The bold red line represents a closed loop path that is traversed clockwise around the circuit as we *add voltage rises* and *subtract voltage falls*. The closed loop can be clockwise or counterclockwise; both work. However, more advanced techniques that you will learn later are greatly simplified if the clockwise standard is followed. *The loop must be closed; you must include every voltage rise or fall encountered in the loop.*

Logically, Kirchhoff's voltage law makes sense. If you start at a simple node on a circuit, walk around the circuit and return to the same simple node you started at, what is the net voltage between your starting point and your ending point? It must be zero (0) volts. The net difference of a simple voltage node is 0 V. For example, if you place your voltmeter probes on the same simple node (same point) on a wire, you must read zero (0) volts difference.

Figure 6-2 demonstrates Kirchhoff's voltage law for a single closed loop circuit with voltage values.

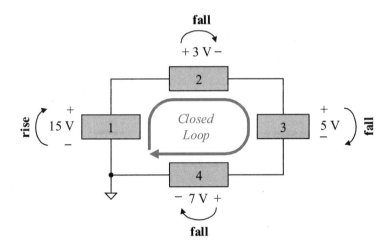

Figure 6-2 KVL applied to a single loop circuit with voltage values

Write the KVL expression for Figure 6-2.

$$+15 \text{ V} - 3 \text{ V} - 5 \text{ V} - 7 \text{ V} = 0 \text{ V}$$

Simplify the expression.

$$15 \text{ V} - 15 \text{ V} = 0 \text{ V}$$

KVL is satisfied. KVL must always be satisfied.

$$0 \text{ V} = 0 \text{ V}$$

6.2 KVL Analogy

Figure 6-3 shows an analogy of an elevator and stairwell system. If you get on the elevator on the ground floor, take it to the 7th floor, get off the elevator, walk down the stairwell to the ground floor, and walk back to the elevator, you have completed a **closed loop**. Also, you ended up where you started. Thus the rise of the elevator minus the fall of the stairwell gives you a net result of zero (0) floors. You ended up right back where you began. KVL for a circuit is basically the same principle. Since the loop is closed, you always end up where you started, and therefore the net difference must always be zero (0) volts.

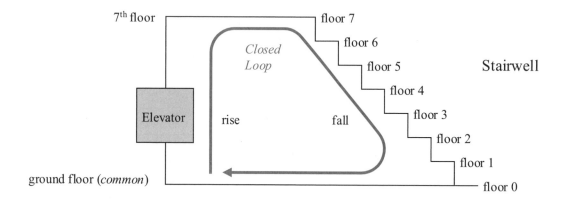

Figure 6-3 Elevator-stairwell analogy of KVL

If you start on the 3rd floor on the stairwell and travel the closed loop path, you will still end up on the 3rd floor where you started. That is a net change of 0 floors. KVL also works from any starting point.

Figure 6-4 expands the elevator analogy to include floors below ground level just like voltages can be below common level (negative valued voltages).

- Ground is equivalent to common

- 3rd floor below ground is like −3 V below common

- 3rd floor above ground is like 3 V above common

- An elevator rise of 6 floors is like a supply rise of 6 V

- A floor fall of 1 floor is like a 1 V fall (drop)

- The floor number is like the node voltage value, for example, the 2nd floor is equivalent to the 2 V node.

- Traveling a closed loop (up the elevator, down the stairs, and back to the elevator) produces a difference of 0 floors just like walking about the circuit from the –3 V node up the supplies and down the resistor drops back to the –3 V node produces a net voltage drop of 0 V.

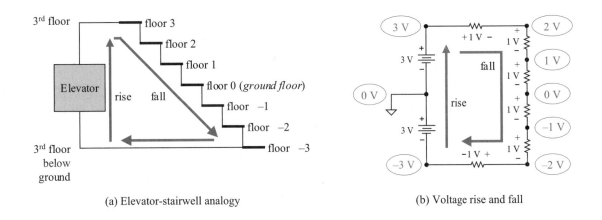

(a) Elevator-stairwell analogy   (b) Voltage rise and fall

Figure 6-4 Elevator-stairwell analogy to voltage rise-fall in a circuit

The temperature scale is another useful analogy. Above 0 degrees is equivalent to node voltage values above common (for example, 6 degrees above 0 is like +6 V). Below 0 degrees is equivalent to voltage node values below common (for example, 6 degrees below 0 is like –6 V). Use whatever visual aid helps you understand the relationship of voltage rises/falls as compared to node voltage values.

Example 6-1
     Use the circuit of Figure 6-5 and find V_4.

Solution
     Use Equation (6-1) and find V_4.

$$\sum V_{\text{rises}} - \sum V_{\text{falls}} = 0 \text{ V} \qquad\qquad \text{(KVL)}$$

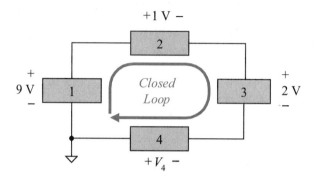

Figure 6-5 Example circuit

Figure 6-5 shows you how to visualize the rises and falls as you proceed around the loop. Start at the tail of the closed loop arrow and walk to the head of the arrow. Add voltage rises and subtract voltage falls as you proceed. *Include the voltage rise or fall of every component in the loop as you proceed around the loop. Do not leave out any component.*

- start at the tail of the loop

- *include all components in the loop*

- rise 9 V ➔ +9 V

- fall 1 V ➔ −8 V

- fall 2 V ➔ −6 V

- rise V_4 ➔ +V_4 (*based upon its reference polarity*)

- end at the arrow head of the closed loop

- double-check that *all components in the loop* have been included

Now let's visualize this process and overlay it on the schematic as shown in Figure 6-6.

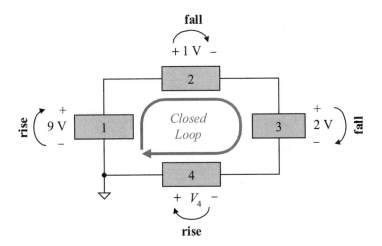

Figure 6-6 Visualize closed loop with voltage rises and falls

Use Figure 6-6, walk around the closed loop, and write the KVL equation. Starting at common and walking about clockwise,

$$+9\,\text{V} - 1\,\text{V} - 2\,\text{V} + V_4 = 0\,\text{V} \qquad \text{(KVL expression)}$$

Simplify the expression, solving for V_4.

$$V_4 = -6\,\text{V}$$

V_4 is calculated to be a negative value with the given polarity reference. Thus, V_4 is 6 V with a polarity *opposite* to that of the given reference.

Practice

Use the circuit of Figure 6-7 and find V_5. Note, the actual polarity of V_5 is not known but a reference polarity has been assumed. The voltmeter is connected to measure based upon the reference (assumed) polarity.

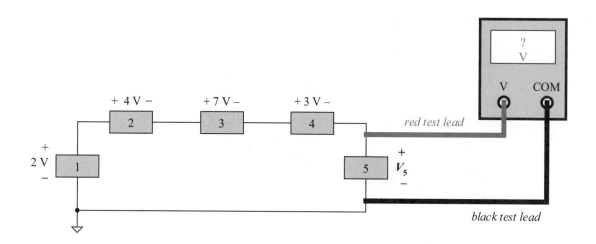

Figure 6-7 Practice problem

Answer: −12 V (voltmeter display −12 V, opposite the reference polarity)

6.3 KVL Multiple Loop Circuits

Kirchhoff's voltage law is valid for all circuit configurations. It is valid for a multiple loop circuit as well as a single loop circuit. The next example demonstrates this principle.

Instead of generic devices, we shall now start including components in the circuit and deciding (or assuming if unknown) a voltage polarity. If a polarity is not known, assume a reference polarity and be faithful to that reference.

Example 6-2

Use the circuit of Figure 6-8 and find V_1, V_2, and V_3. How many possible closed loops are there in this circuit (hint, more than 3)? Find a new loop to solve for V_3 directly.

Note, this circuit is said to have *three windows* due to its similar appearance to three windows located next to each other.

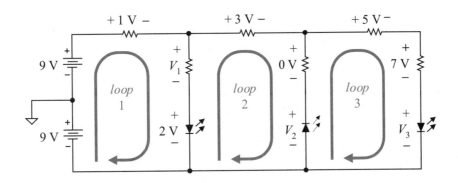

Example 6-8 Example problem–KVL and the multi-loop circuit

Solution

First, we need to select a loop that has only one unknown in it. *Loop* 1 has one unknown voltage while *loop* 2 and *loop* 3 both have two unknowns. So, we must start with loop 1 first and solve for V_1.

Write the *loop* 1 KVL expression paying attention to add a voltage *rise* (−/+) or subtract a voltage *fall* (+/−) as you proceed around the loop clockwise.

$$+9 \text{ V} +9 \text{ V} -1 \text{ V} -V_1 -2 \text{ V} = 0 \text{ V} \qquad (loop\ 1)$$

$$V_1 = 15 \text{ V} \qquad \text{(simplified)}$$

Now, we can write the *loop* 2 KVL expression since we now know V_1 to be 15 V. Note, the 2nd diode is off (it is reversed biased) and therefore open. KVL works with opens and shorts in the circuit. KVL always works; it is the law.

$$+2 \text{ V} +15 \text{ V} -3 \text{ V} -0 \text{ V} -V_2 = 0 \text{ V} \qquad (loop\ 2)$$

$$V_2 = 14 \text{ V} \qquad \text{(simplified)}$$

Now, write the *loop* 3 KVL expression since we now know V_2 to be 14 V.

$$+14 \text{ V} +0 \text{ V} -5 \text{ V} -7 \text{ V} -V_3 = 0 \text{ V} \qquad (loop\ 3)$$

$$V_3 = 2 \text{ V} \qquad \text{(simplified)}$$

How many closed loops are there? There are a total of 6 closed loops. Figure 6-9 shows three additional closed loops that could have been used to solve this problem.

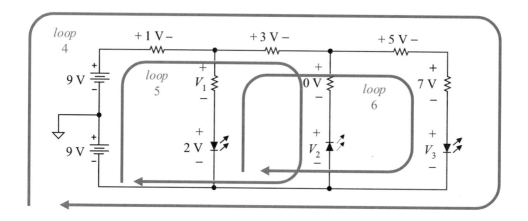

Figure 6-9 Example circuit–three additional closed loops

Let's use the outside loop (*loop 4*) KVL expression to find V_3 directly.

$$+9\ V + 9\ V - 1\ V - 3\ V - 5\ V - 7\ V - V_3 = 0\ V$$

$$V_3 = 2\ V \hspace{3cm} \text{(simplified)}$$

Always look for the best possible KVL loop to find the single unknown voltage in the loop you are solving for.

Practice

Use the circuit of Figure 6-10 and find V_1, V_2, and E using the basic KVL approach. Note, the actual polarities are not known; they have been assumed as a reference. Verify your answers by doing a voltage walkabout from the (−) voltage reference to the (+) voltage reference of V_1, V_2, and E.

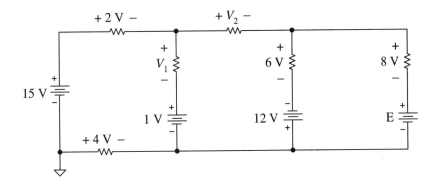

Figure 6-10 Practice circuit

Answer: 16 V, 23 V, −14 V (so, actually the E symbol is upside down)

6.4 Node Voltage Notation Review

KVL is useful in developing some convenient rules to find node voltages and voltage drops between nodes. First, let's use Figure 6-11 to review the basic node voltage terminology. These nodes are labeled *a*, *b*, and *common*. Common is a reference which is always 0 V. V_a represents the voltage of node *a* with respect to common. V_b represents the voltage at node *b* with respect to common. V_{ab} represents the voltage at node *a* with respect to node *b*.

If only the 1st node subscript appears such as V_a, then the 2nd node subscript is assumed to be common. Another way to think of a node is that it is a *voltage test point* relative to common. To measure V_a in a circuit, the voltmeter red lead is connected to node *a* and the voltmeter black lead is connected to the *common* node.

To measure V_{ab} in a circuit, the voltmeter red lead is connected to node *a* (1st subscript), and the voltmeter black lead is connected to the node *b* (the 2nd subscript).

V_{ab}, the voltage at node *a* relative to node *b*, is found by subtracting the node voltage at *b* from the node voltage at node *a*.

$$V_{ab} = V_a - V_b$$

Figure 6-11
Voltage nodes

In this case, the voltmeter *red lead* is connected to node *a* and its *black lead* is connected to node *b* (the reference node for this measurement). The voltmeter black lead is always connected to the reference node (the 2nd subscript), which if missing is assumed to be *common*.

Example 6-3

Using the node voltages of Figure 6-12, find the voltage at node *a*, the voltage at node *b*, and the voltage at node *a* relative to node *b*.

Solution

The analog common is at 0 V always; it is a reference voltage. Voltages at node *a* and *b* are given in Figure 6-12 relative to common.

$$V_a = 7 \text{ V} \qquad\qquad\qquad \text{(given)}$$

$$V_b = -3 \text{ V} \qquad\qquad\qquad \text{(given)}$$

Use node *a* and *b* voltages to find V_{ab}.

$$V_{ab} = V_a - V_b = 7 \text{ V} - (-3 \text{ V}) = 10 \text{ V} \quad \text{(definition)}$$

Figure 6-12
Node voltages

Practice

Use the node voltages of Figure 6-12 and find V_{ba} (the voltage at node *b* with respect to node *a*).

Answer: −10 V

You must be in command of voltage terminology and easily find voltages at nodes, across components, and across circuits.

The next example uses a known node voltage and circuit drop, to calculate the unknown node voltage.

Example 6-4

Use Figure 6-13(a) and find the node voltage V_a.

Solution

Inspection of Figure 6-13(a) shows that

$$V_b = -3 \text{ V} \qquad\qquad\qquad \text{(definition)}$$

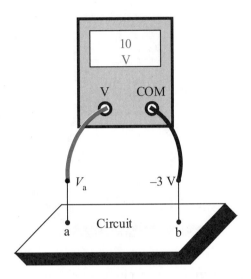

Figure 6-13(a) Example circuit–node voltages, find V_a

Next determine V_{ab} from the voltmeter reading. The *red lead* is on node *a* and the *black lead* is on node *b*, therefore the voltmeter is measuring V_{ab}.

$$V_{ab} = 10 \text{ V}$$ (voltmeter reading)

Use the definition for V_{ab}. Algebraically rewrite this expression to solve for V_a.

$$V_{ab} = V_a - V_b$$ (definition)

$$V_a = V_b + V_{ab}$$ (solve for V_a)

$$V_a = (-3 \text{ V}) + (10 \text{ V})$$ (substitute)

$$V_a = 7 \text{ V}$$ (simplify)

The following is a practical visual approach to quickly solve the same problem. Figure 6-13(b) shows the same measurement, but the +/– polarity drop of the voltmeter has been added for clarity.

- Start at node *b* with the known node voltage of –3 V.
- Rise 10 V in going from node *b* to node *a*.
- The result is 7 V at node *a*.

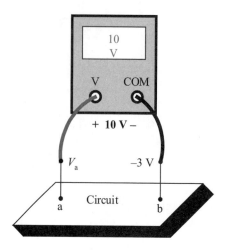

Figure 6-13(b) Example circuit–node voltages, find V_a

Practice

Use Figure 6-14 and find V_b. Use both the algebraic approach and the practical thought process.

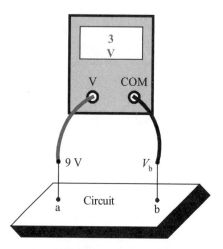

Figure 6-14 Practice circuit–node voltages, find V_b

Answer: 6 V

6.5 Voltage Walkabout

The next example is a complete circuit that incorporates a modified form of the KVL process to find the voltage drop between two nodes. In this book it is called a **voltage walkabout**. For example, to find V_{ab}, start at node b (the 2nd subscript), walk to node a (the 1st subscript), add voltage rises and subtract voltage falls.

Example 6-5

Use Figure 6-15 and KVL to find the voltage V_{bc}. First the standard KVL approach is used to find the voltage. Then a simplified KVL visual *voltage walkabout* process is demonstrated to find the same voltage more quickly and more easily.

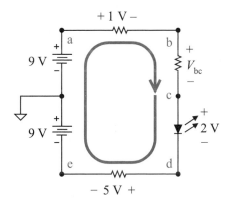

Figure 6-15 Example circuit–finding V_{bc} using KVL

Solution

Write the KVL closed loop expression.

$$-2\ \text{V} - 5\ \text{V} + 9\ \text{V} + 9\ \text{V} - 1\ \text{V} - V_{bc} = 0\ \text{V} \qquad \text{(KVL)}$$

Reorganize the equation with V_{bc} on the left. This is the *voltage walkabout* form.

$$V_{bc} = -2\ \text{V} - 5\ \text{V} + 9\ \text{V} + 9\ \text{V} - 1\ \text{V}$$

Walk from node c to node b and include all the rises and falls along the way.
Simplify and solve for V_{bc}.

$$V_{bc} = 10 \text{ V}$$

The following voltage walkabout process is very straightforward and requires only adding voltage rises and subtracting voltage falls. It is based upon KVL. Use the circuit in Figure 6-16 to visualize this process.

- Draw an unclosed loop from the *reference node c* to the *target node b* over a path that has all *known* component voltage drops. Node c is the *reference node* because it is the 2nd subscript of V_{bc}. Refer to Figure 6-16.

- Now walk the path from node c to node b, adding voltage rises and subtracting voltage falls.

- *Voltage walkabout:* $-2\text{ V} - 5\text{ V} + 9\text{ V} + 9\text{ V} - 1\text{ V}$

- The result is 10 V. This is the same V_{bc} result we derived from the formal KVL process. *The above circuit voltage walkabout process saves a few algebraic steps and always yields the voltage difference between two nodes. Again, you must know all the voltage rises and falls in the voltage walkabout path.*

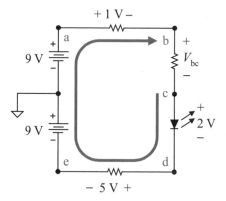

Figure 6-16 Circuit voltage walkabout (practical solution to find V_{bc})

Practice

Use Figure 6-16 and the *circuit voltage walkabout* technique to find V_a, V_e, V_{ae}, and V_{bd}. Hint, what is the implied 2nd subscript of V_a?

Answer: 9 V, –9 V, 18 V, 12 V

6.6 Node Voltages and Voltage Drops

The next example combines the basic techniques used to find node voltages, component voltage differences, and node voltage differences.

Example 6-6

Use the circuit of Figure 6-17 and find all of its node voltages. Draw the circuit on scratch paper and as you find a node voltage write it in its node voltage oval. Then find V_{bc} using the basic definition of a voltage difference between two nodes. Finally, find V_{bc} using the circuit *voltage walkabout* technique.

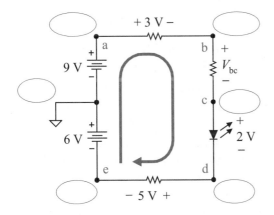

Figure 6-17 Example circuit–finding V_{bc} using KVL

Solution

The only known voltage node is common; it is 0 V. Write 0 V in the common oval.

Find V_a. Start at common (0 V) and rise 9 V to get to node a. The result is 9 V. Write 9 V in the node a oval.

$$V_a = 0 \text{ V} + 9 \text{ V} = 9 \text{ V}$$

Find V_b. Start at node a (9 V) and fall 3 V to get to node b. The result is 6 V. Write 6 V in the node b oval.

$$V_b = 9 \text{ V} - 3 \text{ V} = 6 \text{ V}$$

Find V_c. We cannot go any further around the clockwise loop since we do not know the voltage drop V_{bc}. So, let's return to common and walk around the circuit counterclockwise.

Find V_e. Start at common (0 V) and fall 6 V to get to node e. The result is –6 V. Write –6 V in the node e oval.

$$V_e = 0 \text{ V} - 6\text{V} = -6 \text{ V}$$

Find V_d. Start at node e (–6 V) and rise 5 V to get to node d. The result is –1 V. Write –1 V in the node d oval.

$$V_d = -6 \text{ V} + 5 \text{ V} = -1 \text{ V}$$

Find V_c. Start at node d (–1 V) and rise 2 V to get to node c. The result is 1 V. Write 1 V in the node c oval.

$$V_c = -1 \text{ V} + 2 \text{ V} = 1 \text{ V}$$

All the node voltages have been calculated and written in their respective node voltage ovals. See Figure 6-18.

If you missed any values, go back and redo this example. These are fundamental techniques that you must understand. You can always calculate the voltage difference between any circuit nodes if you have their node voltages. For example, the total supply voltage being applied to this circuit is between node a and node e. The total voltage supply as seen by the circuit is

$$E_{total} = V_a - V_e = 9 \text{ V} - (-6 \text{ V}) = 15 \text{ V}$$

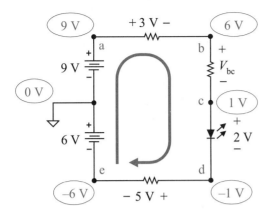

Figure 6-18 Example circuit–node voltage ovals filled in

Let's find the voltage drop V_{bc} across a component by applying the basic definition of a voltage difference of two nodes.

$$V_{bc} = V_b - V_c = 6 \text{ V} - 1 \text{ V} = 5 \text{ V}$$

Let's verify that answer by doing a *voltage walkabout* starting at node *c* (the *reference* node) and walking clockwise to node *b* (the *target* node).

$$V_{bc} = -2 \text{ V} - 5 \text{ V} + 6 \text{ V} + 9 \text{ V} - 3 \text{ V} = 5 \text{ V}$$

Practice

Use the circuit of Figure 6-19 and find all of its node voltages.

As you find a node voltage write it in its node voltage oval. Then find V_1 and V_2 using the node voltage values. Then find V_1 and V_2 using the circuit *voltage walkabout* technique to verify them.

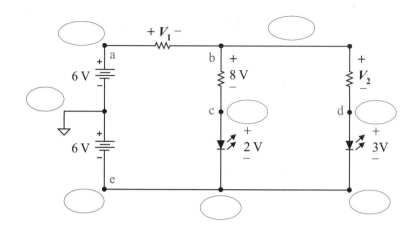

Figure 6-19 Practice circuit–fill in the node voltages

Answer: Nodes a through e (6 V, 4 V, –4 V, –3 V, –6 V); V_1 (V_{ab}, 2 V), V_2 (V_{bd}, 7 V)

6.7 KVL and a BJT Circuit

Example 6-7 explores the BJT circuit that we studied earlier, the one that was used to drive an LED. Previously we looked at the currents of this circuit. This example focuses on the voltage drops and the node voltages of that circuit.

Example 6-7

Use the circuit of Figure 6-20(a) and find all of its node voltages. The input is switched between a minimum *digital high value* of 2.4 V and *digital low value* of 0 V (common).

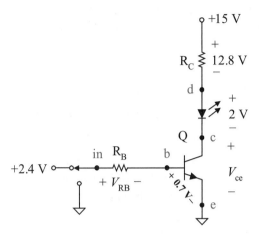

Figure 6-20(a) Example circuit–*npn* BJT circuit and its voltages

As you find a node voltage write it next to its node. Then find V_{RB}. Then find V_{ce} of the transistor. Recall that if a silicon transistor is turned on it typically drops about 0.7 V from base to emitter ($V_{be} = 0.7$ V); therefore, this voltage drop is shown on the schematic. Note that all the nodes are shown in red on the schematic.

Solution

Find V_{in}. The *in* node is connected directly to a +2.4 V supply. Thus,

$$V_{in} = 2.4 \text{ V}$$

Find V_e (the emitter node voltage). This node is tied directly to common. Thus,

$$V_e = 0 \text{ V}$$

Find V_b (the base node voltage). We know that the *e* node is tied to common (0 V). Starting at node *e* (0 V) and walking to node *b* is a 0.7 V rise. Thus,

$$V_b = 0 \text{ V} + 0.7 \text{ V} = 0.7 \text{ V}$$

Find V_d. We know that the voltage supply is set to 15 V and that the R_C resistor drops 12.8 V. Starting at the 15 V power supply bubble, fall 12.8 V to node d. Thus,

$$V_d = 15\ V - 12.8\ V = 2.2\ V$$

Find V_c (the collector node voltage). Starting at node d with a node voltage of 2.2 V, drop 2 V across the LED to node c. Thus,

$$V_c = 2.2\ V - 2\ V = 0.2\ V$$

Find V_{RB}, first using the difference in node voltages. We know V_{in} (+ node of R_B) and V_b (− node of R_B), so just apply the basic definition of a voltage difference. Thus,

$$V_{RB} = V^+ - V^- = V_{in} - V_b = 2.4\ V - 0.7\ V = 1.7\ V$$

Second, check V_{RB} using the *voltage walkabout* technique as shown in Figure 6-20(b). Note we have converted the base-emitter loop to its voltage supply schematic notation to facilitate the walkabout. Start at the reference sign (V_{RB} −) and walk clockwise to the target sign (V_{RB} +). This voltage walkabout includes a fall of 0.7 V and a rise of 2.4 V.

$$V_{RB} = -0.7\ V + 2.4\ V = 1.7\ V$$

Third, check V_{RB} using a closed loop KVL expression as shown in Figure 6-20(c). Solution of the following KVL expression confirms that V_{RB} is 1.7 V.

$$-0.7\ V + 2.4\ V - V_{RB} = 0\ V$$

Find V_{ce} (the BJT collector to emitter voltage drop) using its *node voltages*. We know V_c and V_e, so just apply the basic definition of a voltage difference between two nodes. Thus,

$$V_{ce} = V_c - V_e = 0.2\ V - 0\ V = 0.2\ V$$

Confirm V_{ce} using a *voltage walkabout* of the collector-emitter voltage loop as shown in Figure 6-20(d). For this approach we must replace the +15 V bubble with its voltage supply schematic. This voltage walkabout starts at node e (the 2nd subscript) and proceeds counterclockwise to node c (the 1st subscript). This walkabout includes a rise of 15 V for the supply, a fall of 12.8 V across the collector resistor, and a fall of 2 V for the LED for a net voltage of 0.2 V, confirming our earlier calculation.

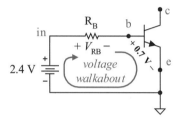

Figure 6-20(b)
Voltage walkabout

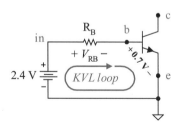

Figure 6-20(c)
KVL closed loop

$$V_{ce} = 15 \text{ V} - 12.8 \text{ V} - 2 \text{ V} = 0.2 \text{ V} \quad \text{(walkabout)}$$

Confirm V_{ce} using the *KVL closed loop* shown in Figure 6-20(e).

$$V_{ce} + 2 \text{ V} + 12.8 \text{ V} - 15 \text{ V} = 0 \text{ V} \qquad \text{(KVL)}$$

This expression simplifies to V_{ce} of 0.2 V.

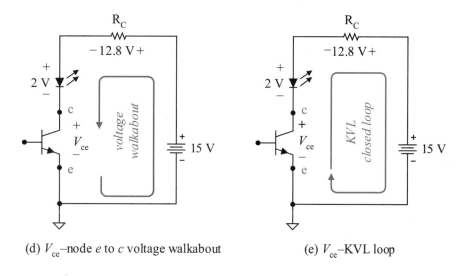

(d) V_{ce}–node e to c voltage walkabout   (e) V_{ce}–KVL loop

Figure 6-20(d,e)   Finding V_{ce} with a voltage walkabout or a KVL loop

Which method should you use: (1) node voltage difference, (2) voltage walkabout, or (3) KVL closed loop? This depends upon the problem and what you already know. In this example, since we already knew V_c and V_e, the quickest, most straightforward approach is the node voltage difference method. Also, both the voltage walkabout and KVL closed loop method forced us to redraw the circuit to include the voltage supply schematic symbols. You must look at a problem and decide which is the most straightforward approach.

Practice

Use the circuit of Figure 6-21. Voltage V_{RB} is measured and found to be 5 V; V_{RC}, 8 V; and V_{ce}, 10 V. Find all of its node voltages (nodes *e*, *b*, *in*, and *c*) and write them on the schematic. Find the supply voltages E_{in} and E_{supply} using three techniques: (1) node voltage differences, (2) voltage walkabout technique, and (3) KVL closed loop and equation.

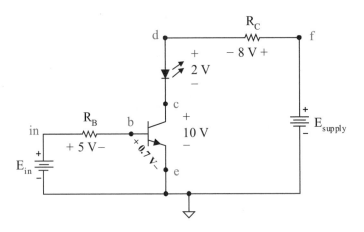

Figure 6-21 Practice circuit–*npn* BJT circuit and its voltages

---

Answer: Node voltages (*e*, 0 V; *b*, 0.7 V; *in*, 5.7 V; *c*, 10 V; *d*, 12 V; *f*, 20 V), E_{in} (5.7 V), E_{supply} (20 V)

6.8 KVL and a MOSFET Circuit

An **enhancement mode MOSFET** is a semiconductor device with three terminals: (D) drain, (S) source, and (G) gate. MOSFET stands for *metal-oxide semiconductor field effect transistor*. The schematic symbol, key variables, and associated meaningful voltages and currents are shown in Figure 6-22. Since it is a semiconductor, it is also given the schematic label of Q like the BJT.

An *enhancement* mode MOSFET is a specially constructed MOSFET that does not begin conducting until its gate-to-source is sufficiently forward biased to turn the MOSFET on. Figure 6-22(c) illustrates the currents of this device once it is turned on.

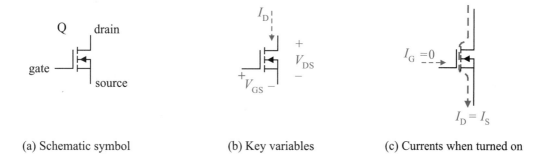

(a) Schematic symbol   (b) Key variables   (c) Currents when turned on

Figure 6-22 Schematic symbol and currents of enhancement mode *n*-channel MOSFET

This is also called an **insulated gate** (IG) device because its gate is insulated from the main body or **channel** from drain to source. Therefore the gate current I_G is always 0 A whether the device is on or off. Looking into the gate is like looking into an insulator (an open). The drain current and source current are always the same value (Kirchhoff's current law for a device requires this). When the MOSFET is turned on, the current flows from the drain, through the MOSFET channel, to the source. There are three major observations for the enhancement mode MOSFET currents.

$$I_G = 0\,\text{A} \qquad \text{(when *on* or *off*)}$$

$$I_D = 0\,\text{A} \qquad \text{(when *off*)}$$

$$I_D = I_S \qquad \text{(when *on* or *off*)}$$

The input characteristic curve of Figure 6-23 plots the *output* drain current (I_D) with respect to the *input* gate-to-source voltage (V_{GS}), demonstrating the behavior of an enhancement mode MOSFET and the relationship between the output current and the input voltage. This is a *voltage-controlled, current output devic*e.

This is a nonlinear device; thus, a characteristic curve is very helpful to better understand the device's behavior. A subsequent textbook in this series describes the MOSFET in greater detail.

This MOSFET's drain current is 0 A until the gate-to-source voltage reaches 3.7 volts. That is, the MOSFET is *off* until the **threshhold voltage** of 3.7 V is applied. Once turned *on*, very small changes in V_{GS} create very large changes in I_D. So before the MOSFET is turned on it acts like an open ($I_D = 0$ A). Once it turns on, it acts like a very low resistance device from drain to source.

V_{GS} volts	I_D amps
0.0	0.000
1.0	0.000
2.0	0.000
3.0	0.000
3.2	0.000
3.4	0.000
3.6	0.000
3.7	0.050
3.8	0.244
4.0	0.945
4.2	1.968
4.4	3.198
4.6	4.592
4.8	6.115
5.0	7.745

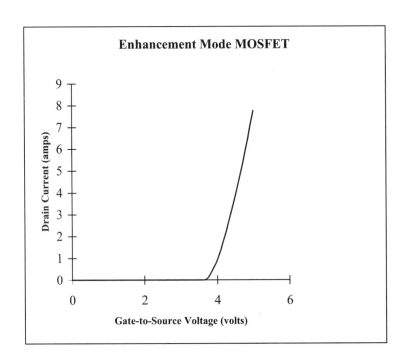

Figure 6-23 Enhancement mode *n*-channel MOSFET I_D versus V_{GS} characteristic curve

That's all you need to know about the MOSFET device at this time. Later we shall explore the MOSFET further. It is a significant component in today's electronics technology. Now let's look at a MOSFET circuit as an example of KVL and KCL.

Example 6-8

Use the circuit of Figure 6-24(a) and the MOSFET input characteristics curve of Figure 6-23. Find the value of E_{in} that produces a load current of 2 A and a load voltage of 6 V. (b) What is the value of V_{DS} at this operating point? (c) What is the static resistance of the MOSFET at this operating point?

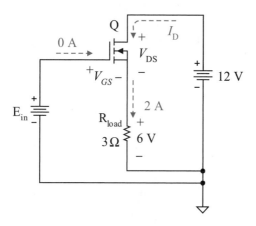

Figure 6-24(a) Example circuit–MOSFET current

Solution

a. The load current is given as 2 A. Because of KCL, the load current is the same current as source current, which is the same current as the drain current. Thus the drain current must be 2 A.

$$I_D = I_S = I_{load} = 2 \text{ A} \qquad \text{(KCL)}$$

From the input characteristic curve and associated data table, the drain current I_D closest to 2 A is 1.968 A. It has a corresponding operating point voltage V_{GS} of 4.21 V.

Write a KVL closed loop expression for the input loop shown in Figure 6-24(b). We know V_{GS} is 4.21 V and V_{load} is 6 V (given). The only unknown in this loop is E_{in}.

$$E_{in} - 4.21 \text{ V} - 6 \text{ V} = 0 \text{ V} \qquad \text{(KVL)}$$

Solve for E_{in}.

$$E_{in} = 10.21 \text{ V}$$

Double check E_{in} by doing a *voltage walkabout* from node $E_{in}(-)$ to $E_{in}(+)$.

$$E_{in} = +6 \text{ V} + 4.21 \text{ V} = 10.21 \text{ V} \qquad \text{(checks)}$$

So, an input supply voltage of 10.21 V develops a gate-to-source voltage of 4.21 V, which develops a load (drain) current of 1.968 A, which develops a load voltage drop of 6 V.

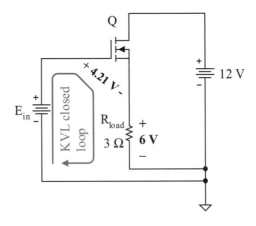

Figure 6-24(b) Solve input loop to solve for E_{in}

b. Now let's solve for the drain-to-source voltage drop V_{DS}. Let's write a KVL closed loop expression for the output loop shown in Figure 6-24(c).

$$+6 \text{ V} + V_{DS} - 12 \text{ V} = 0 \text{ V} \qquad \text{(KVL)}$$

Solve for V_{DS}.

$$V_{DS} = 6 \text{ V}$$

Double check V_{DS} by doing a voltage walkabout from node s to node d.

$$V_{DS} = -6 \text{ V} + 12 \text{ V} = 6 \text{V} \qquad \text{(walkabout check)}$$

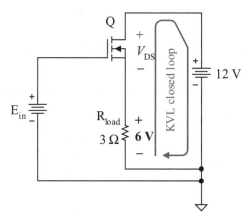

Figure 6-24(c) Solve KVL closed output loop for V_{DS}

c. The *static resistance*, as discovered earlier, is found by dividing a component's voltage by its current. In this case, we are working with the MOSFET's output, which is the drain current, and its drain-to-source voltage drop. Resistance of the MOSFET for this circuit is

$$R_{DS} = \frac{V_{DS}}{I_D} = \frac{6 \text{ V}}{2 \text{ A}} = 3 \text{ }\Omega$$

As noted earlier, when the MOSFET is turned on, it is a very low resistance device. When turned off, it is virtually an open. Since this is a nonlinear curve, the value of the static resistance varies and depends upon the operating point. The more drain current the MOSFET draws, the lower the MOSFET resistance.

Practice

Use the circuit of Figure 6-24(a) and the MOSFET input characteristics curve of Figure 6-23. (a) Find the value of E_{in} that produces a load current of 3.2 A (instead of 2 A) and a load voltage of 9.6 V (instead of 6 V). (b) What is the value of V_{DS} at this operating point? (c) What is the static resistance of the MOSFET at this operating point?

Answer: (a) 14 V, (b) 2.4 V, (c) 0.75 Ω or 750 mΩ

6.9 KVL and an Op Amp Circuit

The last example is that of a popular op amp circuit called a noninverting voltage amplifier.

Example 6-9
  Use the op amp noninverting voltage amplifier circuit of Figure 6-25. For this amplifier, the op amp input terminals drop a very, very small voltage (microvolts); therefore, this voltage drop is assumed to be ideally 0 V. The guitar microphone produces a 1 V input signal at the input *in* node. First find the voltage at node *a*. Then find the voltage at the output *out* node.

Solution
  Find V_a, the voltage at node *a*. The input signal to the op amp is 1 V at the *in* node. Thus V_{in} is 1 V. The op amp input terminals drop ideally 0 V. Since we know that the *in* input node voltage is 1 V and the op amp drops 0 V, then node *a* voltage is

$$V_a = 1\text{ V} - 0\text{ V} = 1\text{ V}$$

Find V_{out} (the output voltage). The voltage at node *a* is at 1 V. There is a 3 V rise to the *out* node. Thus,

$$V_{out} = 1\text{ V} + 3\text{ V} = 4\text{ V}$$

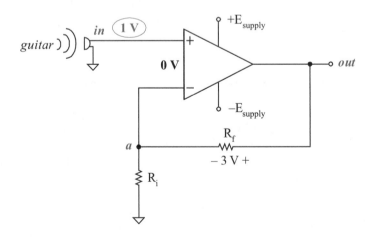

Figure 6-25 Example circuit–op amp noninverting voltage amplifier

The input is 1 V and the output is 4 V. The output voltage is greater than the input voltage. That is why this is called a voltage amplifier. If the input signal voltage is increased, the output signal voltage is increased proportionally. The following practice demonstrates this principle.

Practice

The input signal of the circuit in Figure 6-25 is increased to 2 V. This causes an increase of the drop across R_f to 6 V. Find the new V_a and the new V_{out}.

Answer: V_a (**2** V), V_{out} (8 V). Doubling the input, doubled the output.

6.10 KVL Alternate Form

We have been working with the KVL form that says the sum of voltage rises less the sum of the voltage falls in a closed circuit loop must equal zero.

$$\sum V_{rises} - \sum V_{falls} = 0 \text{ V}$$

If we algebraically add voltage falls to both sides of the equation, then an alternate form of Kirchhoff's voltage law is

$$\sum V_{rises} = \sum V_{falls} \qquad \textbf{(6-2)}$$

This states that the sum of the voltage rises must equal the sum of voltage falls in a closed circuit loop.

Example 6-10

The circuit of Figure 6-26 is the same as that of Figure 6-25. In the previous example we used the node voltage approach to solve the circuit. Let's repeat the same example but use the KVL approach of Equation 6-1 and then the alternate KVL Equation 6-2 to find V_{Ri} and V_{out}.

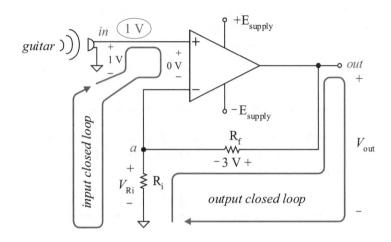

Figure 6-26 Example circuit using KVL alternate form

Solution

Set up V_{Ri} using the *input loop* and KVL Equation 6-1.

$$1 \text{ V} - 0 \text{ V} - V_{Ri} = 0 \text{ V}$$

Solve for V_{Ri}.

$$V_{Ri} = 1 \text{ V}$$

Set up V_{out} using the *output loop*, KVL Equation 6-1, and the V_{Ri} value you just found.

$$1 \text{ V} + 3 \text{ V} - V_{out} = 0 \text{ V}$$

Solve for V_{out}.

$$V_{out} = 4 \text{ V}$$

Now let's re-solve this circuit using the alternate form of Equation 6-2. Set up V_{Ri} using the *input closed loop* and KVL Equation 6-2.

$$1 \text{ V} = +0 \text{ V} + V_{Ri}$$

Solve for V_{Ri}.

$$V_{Ri} = 1 \text{ V}$$

Set up V_{out} using the *output closed loop*, KVL Equation 6-2, and the V_{Ri} value you just found.

$$V_{Ri} + 3 \text{ V} = V_{out}$$

$$1 \text{ V} + 3 \text{ V} = V_{out}$$

Solve for V_{out}.

$$V_{out} = 4 \text{ V}$$

Practice

Verify V_{Ri} by doing a walkabout from V_{Ri} (−) to V_{Ri} (+). Then verify V_{out} by doing a walkabout from common to the *out* node.

Answer: 1 V, 4 V

The technique or techniques you choose to find voltages in a circuit depends upon the circuit to be analyzed and which voltages you already know. You may also have a preference, but preference must not be your only criteria. For the op amp circuit of Example 6-9 and Example 6-10, we used the node voltage approach, the KVL forms of Equations 6-1 and 6-2, and the voltage walkabout techniques. All these techniques worked equally well, but which one was the quickest solution with the fewest mathematical steps and produced the clearest understanding? For this circuit, the best approach is that of Example 6-9 (node voltage approach). For a different circuit, a different approach might be more efficient. Keep your options open; do not just use one technique.

Summary

This chapter introduced the second of the three most fundamental laws of circuit analysis, Kirchhoff's voltage law.

Kirchhoff's voltage law (KVL) states that the sum of voltage rises minus the sum of the voltage falls around a closed loop in a circuit must equal zero (0) volts.

$$\sum V_{rises} - \sum V_{falls} = 0 \text{ V}$$

A closed loop path of the circuit must be traversed, *adding voltage rises* and *subtracting voltage falls*. The closed loop can be clockwise or counterclockwise; both work. However, more advanced techniques that you will learn later are greatly simplified if the clockwise standard is followed. The loop must be closed; you must include every voltage rise or fall encountered in the loop. Every component in the closed loop must be included.

Again this law is intuitively reasonable. If you are out for a walk and you end up back at your starting point, what is the net distance from your starting point to your finish point? It is the same point so the net difference is 0 feet. Likewise in a circuit, if you start at a simple node, add the voltage rises and subtract the voltage falls of every component (circuit) encountered in a closed loop, and end up back at the starting node, the difference in voltage from the starting node to the ending node must be zero (0) V.

We developed a *voltage walkabout* technique, based upon KVL but quicker than KVL, to find the voltage drop between two nodes if there was a path with all known voltage rises and falls. This is an unclosed loop to find an unknown voltage drop such as V_{ab}. Always walk from the reference node (*b*, the 2nd node subscript) to the target node (*a*, the 1st node subscript). If the second node subscript is missing, then it is assumed to be common. When working with the (+/−) polarity notation, the negative (−) sign is considered the reference node and the positive (+) sign is considered the target node. We applied KVL to single loop and multiple loop circuits. We also applied KVL to BJT and op amp electronic circuits.

An alternate form of the KVL equation equates the sum of the voltages rises to the sum of the voltage falls.

$$\sum V_{\text{rises}} = \sum V_{\text{falls}}$$

We have studied Kirchhoff's current law (KCL) and Kirchhoff's voltage law (KVL). These are two fundamental circuit analysis laws for circuit analysis. They are the law and can never be violated.

In the next chapter we shall study *Ohm's law* for resistance. KVL and KCL are circuit laws, whereas Ohm's law is a component law and a circuit law. Ohm's law for resistance relates the voltage, current, and resistance of a resistive component or a resistive circuit. We have already been introduced to this component law in Chapter 3. The current and voltage were related to a resistance through the characteristic curve and the measurement of a static (dc) resistance using an ohmmeter. Chapter 7 will formally introduce Ohm's law for resistance, the power rule, the power rule's alternate forms as it applies to resistance, and energy consumed by supplies and resistances.

Problems

Kirchhoff's Voltage Law

6-1   Use the circuit of Figure 6-27.

    a. Use KVL Equation 6-1 and find V_{R2}.

    b. Use the circuit walkabout technique to find and verify V_{R2}.

    c. Find node voltages V_a, V_b, and V_c.

    d. Verify V_{R2} by finding V_{bc} using the node voltages.

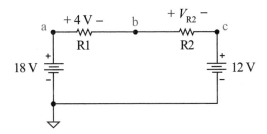

Figure 6-27

6-2   Use the circuit of Figure 6-28.

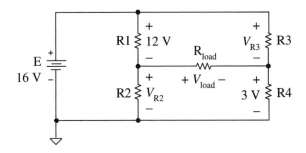

Figure 6-28

   a. Using the left window, find V_{R2}.

   b. Using the bottom-right window, find V_{load}.

   c. Using the top-right window, find V_{R3}.

   d. Starting with the original schematic with only the original known voltages, verify your answer for V_{load} by doing a voltage walkabout from the $-V_{load}$ terminal to the $+V_{load}$ terminal. What components did you include in your walkabout?

6-3   Use the circuit of Figure 6-29.

   a. Find V_a (hint: voltage walkabout starting at common)

   b. Find V_b

   c. Find V_{open} by solving for V_{ab}

   d. Find V_{open} using a voltage walkabout from node b to node a

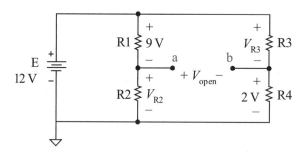

Figure 6-29

KVL and a BJT Circuit

6-4   Use the circuit of Figure 6-30. The LED drops 2 V. The BJT base-to-emitter voltage drop V_{be} is 0.7 V.

   a. Write V_{LED} and V_{be} with proper polarity on your schematic.

   b. Find V_e

   c. Find V_b

   d. Find V_{in}

   e. Find V_{RB} from its node voltages

f. Verify V_{RB} with a voltage walkabout: from node b, to node e, to *common*, to node *in*.

g. Find V_c

h. Find V_{ce}

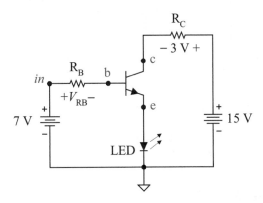

Figure 6-30

KVL and a MOSFET Circuit

6-5 Use the circuit of Figure 6-31 and the MOSFET input characteristics curve of Figure 6-23. Find the value of E_{in} that produces a load current of 3.2 A (which produces a load voltage of 9.6 V). Also, what is the value of V_{DS} at this operating point? What is the MOSFET resistance R_{DS}?

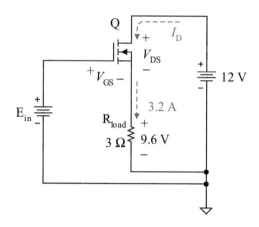

Figure 6-31

KVL and an Op Amp Circuit

6-6   Use the circuit of Figure 6-32. Recall that the voltage drop across the op amp input terminals of this inverting voltage amplifier is approximately 0 V. Node voltages are shown in the ovals.

    a.   What is the node voltage V_a? Hint: do a voltage walkabout starting at *common* and walking to node *a*.

    b.   Find the input resistor voltage drop V_{Ri}.

    c.   Find the output node voltage V_{out}.

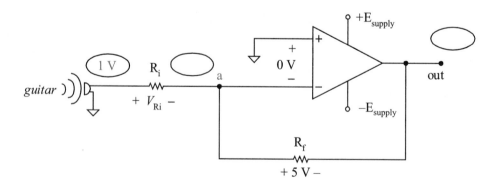

Figure 6-32

7

Ohm's Law, Power, and Energy

Introduction

This chapter examines the third major law, **Ohm's law of resistance**. Ohm's law of resistance relates resistance to its voltage and current values in a dc circuit. This law applies to both component resistances and circuit resistances (a combination of one or more resistances).

The relationship between *current direction* and *voltage polarity* is fixed. Since resistance dissipates energy, voltage potential is lost as current flows through resistance. Thus the voltage must drop +/– as current flows through a resistance.

The power rule relates voltage, current, and power of a device or a circuit. The power rule and Ohm's law of resistance can be combined to relate power, voltage, current, and resistance for resistive components. Finally, power, energy, and the cost of energy usage are examined.

Objectives

Upon completion of this chapter you will be able to:

- State and apply Ohm's law for resistance in a dc circuit.

- Relate voltage polarity with current direction through a resistance.

- State and apply the power rule for power supplies and resistances based upon the device's current and voltage.

- Relate power, current, voltage, and resistance of a resistive component or circuit using Ohm's law with the power rule.

- Calculate the maximum safe voltage or current for a resistor based upon its power rating and resistance.

- Calculate the power of and energy consumed by power supplies and resistive components and circuits.

- Calculate the cost of using electrical energy.

7.1 Ohm's Law—An Experiment

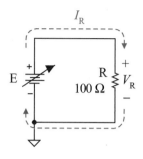

Figure 7-1(a) Ohm's law test circuit

Ohm's law for resistance is both a fundamental component law and a circuit law. The resistance could be of a component (for example, a resistor) or of a circuit of components that produce a net resistance.

Let's explore the relationship of Ohm's law for a resistance with an experiment. Figure 7-1(a) is a dc test circuit used to apply a test voltage from supply E to resistor R with a fixed resistance of 100 Ω. The test voltage E is set to a specific voltage and applied to the completed circuit. The current flows up from the positive (+) supply terminal, through the conducting wire, down through the resistor dropping +/– voltage, through the conducting wire, and back to the supply's negative (–) terminal. Ideally the conductors do not drop voltage and all of the voltage supply E is dropped across the resistor R. We can verify this with Kirchhoff's voltage law (KVL). The alternate KVL form is the most straightforward for this circuit (the sum of the voltage rises must equal the sum of voltage falls in a closed loop).

$$V_R = E \qquad\qquad \text{(KVL)}$$

Figure 7-1(b) is the test circuit schematic with the ammeter and the voltmeter properly connected.

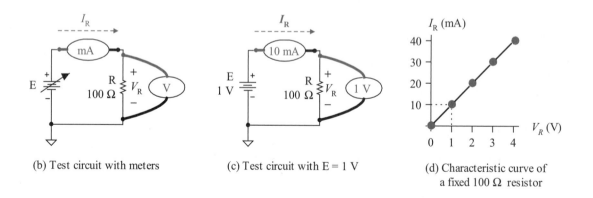

(b) Test circuit with meters

(c) Test circuit with E = 1 V

(d) Characteristic curve of a fixed 100 Ω resistor

Figure 7-1(b,c,d) Test circuit for Ohm's law experiment

The circuit is broken and the ammeter is inserted into the circuit so that the circuit current *flows into the red lead, through the ammeter, and* out the black lead to produce a positive current reading. Visualize the current reference arrow *flowing into* the red and *out of* the black lead to make proper meter connections to the circuit.

The ammeter acts ideally like a short, and current flows through the ammeter as if it were a conductor. The ammeter measures the current flowing *through* the circuit and the resistor R.

Question: is the current through the circuit, the conductors, the ammeter, and the resistance the same current? *Yes*. Recall that this is a series circuit that we studied earlier as an example. We proved by Kirchhoff's current law (KCL) that the current is everywhere the same throughout a series circuit. So the current through the ammeter is the same current that flows throughout the circuit and through the resistor.

$$I_R = I_{supply} = I_{meter} = I_{wire} \qquad \text{(KCL)}$$

The voltmeter is connected *across* the resistor R to measure its corresponding voltage drop. The voltmeter ideally acts like an open and draws no current. Note that the voltmeter polarity is connected to measure relative to its schematic +/− reference. That is, its *red lead* is connected to the corresponding positive (+) reference and the *black lead* is connected to its corresponding negative (−) reference. This is the anticipated polarity since we expect the resistor to lose voltage as the current flows through the resistance.

Now, let's set a specific supply value of E, then measure the voltage drop V_R and its corresponding current I_R through resistance R. First set supply E to 0 V. The voltmeter measures 0 V and the current meter measures 0 mA (recorded in Table 7-1). If E is set to 1 V, as shown in Figure 7-1(c), then 1 V is dropped across resistance R and 10 mA of current flows through resistance R. Table 7-1 summarizes the measurements for supply voltages of E = 0 V, 1 V, 2 V, 3 V, and 4 V with their corresponding resistor voltage measurements and current measurements. A sketch of these positive test data points is shown in Figure 7-1(d).

Next, the voltage supply E is reversed for this experiment to determine the effect of reverse polarity on the resistance's current. The voltage and current measurements now have negative values. The experiment is repeated with the reversed supply, and the measured values are also recorded in Table 7-1.

Combining all the data points, the complete *characteristic curve* of the 100 Ω resistor is generated as shown in Figure 7-1(e). This characteristic curve should look familiar. It is a linear curve that indicates that the component has a fixed resistance value, in this case, 100 Ω. The resistance does *not* vary with the current through it. The results indicate that the

Table 7-1

Measurements for the resistance of the circuit in Figure 7-1

V_R (V)	I_R (mA)
4	40
3	30
2	20
1	10
0	0
−1	−10
−2	−20
−3	−30
−4	−40

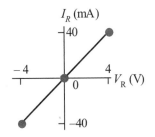

Figure 7-1(e)
100 Ω characteristic curve

resistor current is a linear function of its applied voltage. If you double the voltage, you double the current.

7.2 Ohm's Law for Resistance

Ohm's law of resistance states that current is directly proportional to its applied voltage and inversely proportional to its resistance.

Ohm's law

$$I_R = \frac{V_R}{R} \tag{7-1}$$

where

I_R = current through the resistance in amperes (A)
R = resistance in ohms (Ω)
V_R = voltage across the resistance in volts (V)

Let's verify a few of the test data points to see if this formula really works. Verify the test points shown in Figure 7-1(e),

$$I_R = \frac{4\ V}{100\ \Omega} = 40\ mA$$

$$I_R = \frac{0\ V}{100\ \Omega} = 0\ mA$$

$$I_R = \frac{-4\ V}{100\ \Omega} = -40\ mA$$

Yes, Ohm's law for resistance really works and it always works. It is a law, not an option.

Voltage polarity is fixed with current direction

How about the resistance's *voltage polarity* relative to its *current direction*? Again, since resistance dissipates energy as current flows through it, voltage potential is lost across the resistance as current flows through the resistance. Therefore, as noted in Chapter 2, the direction of current flow is *into the resistance's relatively positive (+) potential terminal* and *out of its relatively negative (−) potential terminal*. Figure 7-2 demonstrates this relationship of current direction and voltage polarity. Use this relationship when establishing reference currents and their respective voltage drops. *The current direction and voltage polarities must be consistent.*

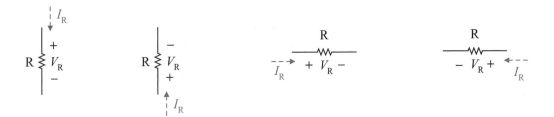

Figure 7-2 Current *enters* the relatively higher (+) voltage potential terminal

Before solving for or measuring voltages or currents in a circuit, always establish the reference voltage polarities and any corresponding reference current directions. Then you must be faithful to your established references as you solve for circuit values.

Establish reference voltage polarities and current directions

Example 7-1

Use the schematic of Figure 7-3 to do the following. (a) Establish that the circuit is continuous. (b) Establish the current direction and the voltage polarity across the 1 kΩ resistor. (c) Find the voltage drop across the 1 kΩ resistor. (d) Find the current through the 1 kΩ resistor. (e) Draw the test circuit with instrumentation to measure these values, showing the current and voltmeter connections. (f) Sketch the resistor's characteristic curve with the resistor's operating point. (g) Redraw the schematic of Figure 7-3 using bubble notation for the power supply.

Solution

a. First, establish that the circuit is *continuous* and can generate a current. Inspection of the schematic shows a continuous path out of the positive (+) supply terminal, down through the resistor, and back to the negative (−) supply terminal. The circuit has a continuous path and can therefore generate a current.

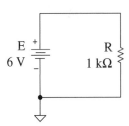

Figure 7-3
Example circuit

b. Next, establish the current direction and corresponding resistor voltage polarity. From the previous step you should note that the current flows out of the supply's positive (+) terminal, clockwise in the circuit, and drops (+/−) as it flows down through the resistor as shown in Figure 7-4(a). In this simple case, you can establish the references to be consistent with the expected results. In more complex cases, the actual current directions and voltage polarities cannot always be determined immediately and an arbitrary choice of reference current direction and corresponding voltage polarity must be assigned.

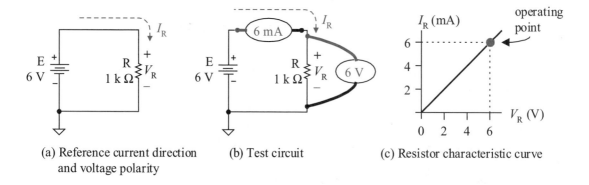

(a) Reference current direction and voltage polarity   (b) Test circuit   (c) Resistor characteristic curve

Figure 7-4(a,b,c) Example solution

c. Next, inspect the schematic and note that all of the 6 V from the supply is dropped across the 1 kΩ resistor. From the KVL alternate form

$$V_R = E = 6 \text{ V} \qquad\qquad \text{(KVL)}$$

d. Now, apply Ohm's law for resistance to calculate the resistor current and thus the circuit current.

$$I_R = \frac{V_R}{R} = \frac{6 \text{ V}}{1 \text{ k}\Omega} = 6 \text{ mA} \qquad\qquad \text{(Ohm's law)}$$

e. Figure 7-4(b) demonstrates the circuit configuration to properly measure the resistor current and voltage drop. Note that the meters are connected properly based upon the references established in Figure 7-4(a).

f. Figure 7-4(c) is a sketch of the resistor's characteristic curve. Since it is a fixed resistance, the curve is linear (or a straight line). The line must go through two operating points, namely, (0 V, 0 mA) and (6 V, 6 mA). Draw these two data points and then draw the line connecting them to create the characteristic curve.

g. Figure 7-4(d) shows the circuit schematic using the power supply bubble notation. *The bubble voltage is the voltage at that supply's node relative to common.* Note that the power supply negative (–) terminal is connected to analog common, so this is the 0 V reference. Going to the positive (+) terminal of the supply is a rise of 6 V in potential. Thus, this is a +6 V power supply, which is represented by a +6 V *bubble*.

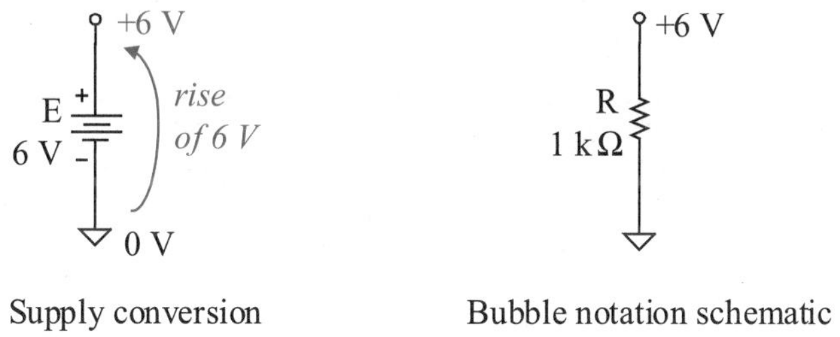

Supply conversion   Bubble notation schematic

Figure 7-4(d) Example solution–node schematic

Practice

Repeat Example 7-1 with the polarity of the 6 V supply reversed as shown in Figure 7-5(a). Maintain the same reference current direction and reference voltage polarity as in Example 7-1.

Answer: The actual current is counterclockwise and flows up through the resistor, creating a +/– polarity drop from bottom to top (opposite the references). The drop across the resistor is –6 V and the current through the resistor is –6 mA. Figure 7-5 shows (a) the reference circuit, (b) the test circuit with current meter and voltmeter connected showing measured values, (c) the complete characteristic curve for the 1 kΩ resistor with corresponding operating point, and (d) the bubble schematic notation for the negative voltage supply.

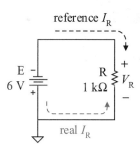

(a) Reference circuit

Figure 7-5(a)
Practice circuit

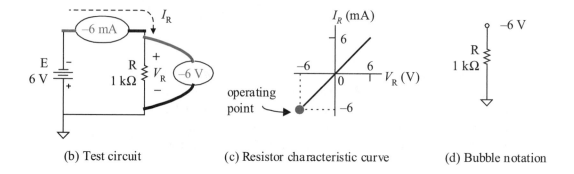

(b) Test circuit   (c) Resistor characteristic curve   (d) Bubble notation

Figure 7-5 (b, c, d) Practice solution

___

Exploration

Simulate Figure 7-1(a) using MultiSIM as shown in Figure 7-5(e) or Cadence PSpice as shown in Figure 7-5(f) and verify Table 7-1(a) values.

multi**SIM**

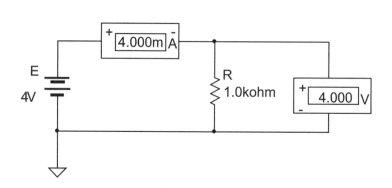

Figure 7-5(e) MultiSIM simulation at E = 4 V

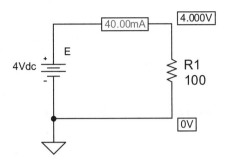

how big can you dream?™

Figure 7-5(f) Cadence PSpice simulation at E = 4 V

Notice that the MultiSIM simulation shows the actual drop across the resistance R while the Cadence PSpice simulation shows the node voltages at the top and bottom of resistance R. The actual voltage drop across the resistance R is the difference between its node voltages.

Ohm's Law–Basic Forms

Ohm's law is an algebraic equation. If you know two of the three variables (I_R, V_R, and R), then you can find the third variable. Ohm's law can be algebraically written in the following three forms.

Solve for I (the basic form).

$$I_R = \frac{V_R}{R}$$

Solve for V_R.

$$V_R = R I_R \qquad (7\text{-}2)$$

Solve for R.

$$R = \frac{V_R}{I_R} \qquad (7\text{-}3)$$

Ohm's law is used so frequently in circuit analysis that you must know and be able to quickly apply all three forms.

An interesting observation from Equation 7-2 is that a resistance must have a current to produce a voltage drop. If the current in a resistance is 0 A, then its voltage drop must be 0 V. This is a very handy troubleshooting rule. The only exception is that an open resistance (for example, a resistor that has burned open) can drop voltage depending upon the circuit.

Example 7-2

A 220 Ω resistor draws a current of 4 mA. Find its voltage drop.

Solution

Use the Ohm's law form to solve for the resistor's voltage drop. Equation 7-2 is appropriate.

$$V_R = R \quad I_R = 220 \ \Omega \times 4 \ \text{mA} = 880 \ \text{mV}$$

Practice

A resistor drops 5.4 V while drawing a current of 20 mA. Find the resistance of the resistor.

Answer: 270 Ω

Exploration

Use MultiSIM or Cadence PSpice to simulate this condition. Simulate a circuit similar to Figures 7-5(e) and 7-5(f) but change the voltage supply to 5.4 V. Then increase the 100 Ω resistance value until the current is 20 mA. Is the final resistance value 270 Ω?

Answer: Yes, it is 270 Ω

Let's apply Ohm's law of resistance in a circuit.

Example 7-3

For the circuit of Figure 7-6, find V_{R1}, I_{R2} and R4 using the three different forms of Ohm's law.

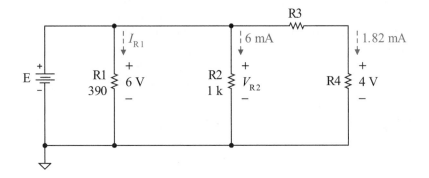

Figure 7-6 Example circuit

Solution

$$I_{R1} = \frac{V_{R1}}{R1} = \frac{6 \text{ V}}{390 \text{ }\Omega} = 15.4 \text{ mA}$$

$$V_{R2} = R2 \times I_{R2} = 1 \text{ k}\Omega \times 6 \text{ mA} = 6 \text{ V}$$

$$R4 = \frac{V_{R4}}{I_{R4}} = \frac{4 \text{ V}}{1.82 \text{ mA}} = 2.20 \text{ k}\Omega$$

Practice

For the circuit of Figure 7-7, find the total resistance of the circuit as seen by the supply. Apply Ohm's law.

Answer: 900 Ω

When working with a resistive component, a resistive circuit, or a resistive subcircuit, you can apply Ohm's law of resistance. If you know two of the three variables (current, voltage, or resistance), you can find the third algebraically.

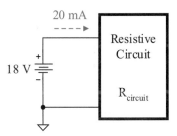

Figure 7-7
Practice circuit

7.3 Nonlinear Resistance

Some components are nonlinear (for example, diodes, LEDs, and MOS-FETs). The component's resistance depends upon its current. For a specific operating point, the device has a static (or dc) resistance, but if the current is changed then the device's resistance changes.

A characteristic curve is a very useful aid in understanding the operating behavior of a nonlinear component. For example, Figure 7-8(a) shows the characteristic curve of a silicon diode. When forward biased on, the silicon diode drops about 0.7 V even with significant changes in current. Between 0.7 V and –100 V, the diode is off and acts ideally like an open. With a reverse voltage of 100 V, the diode breaks down, holding 100 V reverse voltage across it and conducting current in the reverse direction.

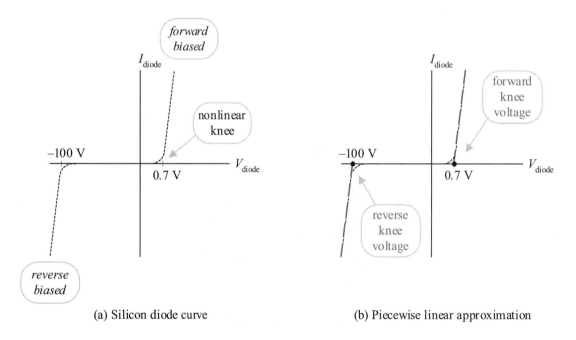

(a) Silicon diode curve   (b) Piecewise linear approximation

Figure 7-8 Silicon diode–forward biased characteristic curve

If you extend the linear portion of the forward bias curve as a straight line, it intersects the voltage axis at 0.7 V as shown in Figure 7-8(b). The voltage at which a forward bias curve intersects the voltage axis is called the **forward bias knee voltage**. Thus, when the forward biased silicon diode is turned on, it drops about 0.7 V. As the current increases in the forward biased diode, its voltage drop increases slightly due to internal diode resistance. A large swing in current provides a relatively small swing in diode voltage drop. Therefore, we can approximate the forward biased on diode to have a drop of 0.7 V.

If you extend the linear portion of the reverse bias curve as a straight line, it intersects the voltage axis at −100 V. This is the voltage at which the diode breaks down and begins conducting current in the reverse direction. It is called the **reverse knee voltage** or the **reverse breakdown voltage**. The rated reverse breakdown voltage of a diode varies depending upon the diode model. Reverse voltage ratings such as 100 V, 200 V, and 300 V for switching diodes are typical.

If the diode is being used as a switch (an *on-off* device), we want it to be *forward biased on* and *reverse biased off*. We do not want it to reach breakdown and start conducting in the reverse direction. So when picking a switching diode, be sure that its reverse breakdown rating is sufficiently high enough that breakdown cannot be achieved in its circuit.

If the actual characteristic curve of Figure 7-8(a) is replaced with straight-line approximations, then it is called a **piecewise linear** approximation curve as seen in Figure 7-8(b). If we are interested in one region, for example the forward biased on region, we could model the diode based upon the straight-line approximation of that region.

Depending upon the application, we can use the actual curve of Figure 7-8(a) or we can model the forward biased on diode as a:

- Short (ideal or first approximation)

- 0.7 V voltage supply (second approximation)

- 0.7 V voltage supply with an internal resistance (3rd approximation), a piecewise linear curve

---

Example 7-4

The circuit of Figure 7-9 uses an LED with the given characteristic curve. The LED is to operate at 20 mA. Find the value of R that limits the LED current to 20 mA. What is the static resistance of the LED?

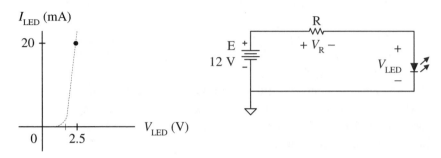

(a) LED characteristic curve   (b) Current limiting resistance

Figure 7-9(a,b) Example problem circuit–LED indicator circuit

Solution

First let's examine what we know:

a.   The LED current is 20 mA.

$$I_{LED} = 20 \text{ mA}$$

b.   The LED voltage drop is 2.5 V from its characteristic curve.

$$V_{LED} = 2.5 \text{ V}$$

Redraw the schematic and place all known values on it, as shown in Figure 7-9(c).

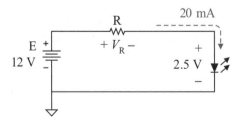

Figure 7-9(c) Circuit prepared for solution

Put on your thinking cap. This requires a three-step process for solution. We need to know I_R and V_R to solve for R.

First, what is I_R? We need to use KCL. This is a series circuit. What do we know about a series circuit? The current is the same throughout a series circuit.

$$I_R = I_{LED} = 20 \text{ mA} \qquad\qquad \text{(KCL)}$$

Next, what is V_R? We need to use KVL. Let's do a KVL voltage walkabout from the V_R (−) terminal to the V_R (+) terminal.

$$V_R = -V_{LED} + E = -2.5 \text{ V} + 12 \text{ V} = 9.5 \text{ V} \qquad \text{(KVL)}$$

Now we know R's current and we know R's voltage drop. We can use Ohm's law of resistance to find R.

$$R = \frac{V_R}{I_R} = \frac{9.5 \text{ V}}{20 \text{ mA}} = 475 \text{ } \Omega \qquad\qquad \text{(Ohm's law)}$$

Referring to the resistor table of Appendix C, the closest standard value is 470 Ω. Is this the best choice? Careful. The problem implies that we need to limit the LED current to a maximum of 20 mA. If we use a resistance of less than 475 Ω, the current goes up and we are exceeding the maximum current requirement of 20 mA. It is best to select the next higher value of resistance from the standard value table and produce a current slightly less than 20 mA. In this case, the next table value is 560 Ω.

The static resistance of the LED at its operating point of 2.5 V and 20 mA is

$$R_{LED} = \frac{2.5 \text{ V}}{20 \text{ mA}} = 125 \text{ } \Omega$$

Practice

Your company only has 560 Ω resistors to use for the circuit of Figure 7-9(b) but can change the voltage source E to any value. Find the value of E needed to produce the 20 mA operating point shown on the characteristic curve of Figure 7-8(a). Hint: you need to use KCL, then Ohm's law of resistance, and then KVL to find E.

Answer: 13.7 V

7.4 Power Rule

Current flowing through a device either generates energy (a supply) or dissipates energy (a resistor, lamp, etc.). The power supplied to a circuit must be consumed; therefore, it must be equal to the power dissipated by the circuit.

Recall that power is the rate at which energy is generated or consumed. The **power rule** relates a device's power to its voltage drop and its current.

$$P = IV \qquad\qquad (7\text{-}4)$$

where

P = power in watts (W)
I = dc current in amperes (A)
V = dc voltage in volts (V)

Proof of the Power Rule

As you learned in Chapter 2, the fundamental definition of current is defined in terms of charge flowing per unit time.

$$I = \frac{Q}{t} \qquad\qquad (2\text{-}2)$$

where

I = current in amperes (A)
Q = charge in coulombs (C)
t = time in seconds (s)

Also in Chapter 2, the fundamental definition of voltage is defined in terms of energy (or work) per unit charge.

$$V = \frac{W}{Q} \qquad\qquad (2\text{-}5)$$

where

V = voltage in volts (V)
W = energy or work in joules (J)
Q = charge in coulombs (C)

Charge Q is inherent in the definition of current and voltage. Without charge, there is no current or voltage. Substituting I and V from Equations 2-2 and 2-5 into Equation 7-4 (and canceling charge Q) verifies that current times voltage is power.

$$P = IV = \left(\frac{Q}{t}\right)\left(\frac{W}{Q}\right) = \frac{W}{t}$$

This expression simplifies to the fundamental definition of power.

$$P = \frac{W}{t}$$

Thus, dc current times dc voltage produces power. The *power rule* is verified.

Example 7-5

  A 10 V power supply delivers 200 mA of current to a circuit. How much power is it generating?

Solution

  Use the power rule of Equation 7-4 to find the power generated.

$$P = I\,V = 200\ \text{mA} \times 10\ \text{V} = 2\ \text{W}$$

Practice

  An RV operates with a 12 V dc electrical system. Its taillight is rated at 100 W. How much current does the taillight draw?

Answer: 8.33 A

Power Dissipation–Practical Application

Device specifications often rate the capacity of a device in terms of the maximum power it can dissipate. For example, a quarter watt resistor is manufactured to dissipate up to 250 mW of power safely in a closed environment. If that rating is exceeded, then the resistor becomes very hot and can be damaged or destroyed.

Example 7-6

  If an LED drops 2 V and draws 20 mA of current, find the power dissipated by this LED. If the LED is rated at 100 mW, is it safe?

Solution

Use the power rule of Equation 7-4 to find the power dissipated by the LED.

$$P = I\,V = 20 \text{ mA} \times 2 \text{ V} = 40 \text{ mW}$$

The power dissipated is less than the rated value of 100 mW. The device should not overheat. It is safe.

Practice

A half watt resistor drops 5.4 V while drawing a current of 200 mA. Find the power dissipated by the resistor. Is it safe?

Answer: 1.08 W, no it is not safe (dissipates more than 0.5 W)

7.5 Power Rule–Resistance Forms

The *power rule* can be combined with *Ohm's law for resistance* to produce very useful forms relating power, voltage, current and resistance. From Equation 7-4, the power is found by multiplying a device's or circuit's current times its voltage.

$$P = I\,V$$

Power Rule for Resistance with *V* and *R*

If the power rule current I is substituted into its Ohm's law form of (V/R) from Equation 7-1, then the *power rule for resistance* becomes

$$P = (I)V = \left(\frac{V}{R}\right)V = \frac{V^2}{R}$$

or simply

$$P = \frac{V^2}{R} \tag{7-5}$$

Using algebra, you can rewrite Equation 7-5 and solve for R.

$$R = \frac{V^2}{P}$$

You can also rewrite Equation 7-5 and solve for V^2.

$$V^2 = PR$$

Then, taking the square root of both sides of this equation, V can be found.

$$V = \sqrt{PR} \qquad (7\text{-}6)$$

Example 7-7

A 27 Ω resistor dissipates 12 W of power. Find its voltage drop and its current.

Solution

Use Equation 7-6 to find the voltage dropped across the resistor.

$$V = \sqrt{PR} = \sqrt{12\ \text{W} \times 27\ \Omega} = 18\ \text{V}$$

You could have substituted into Equation 7-5 and algebraically solved for the voltage instead of using Equation 7-6. There is no need to really memorize Equation 7-6.

Now that you have the voltage drop of the resistor, you can use Ohm's law to find the current through the resistor.

$$I = \frac{V}{R} = \frac{18\,\text{V}}{27\,\Omega} = 667\ \text{mA}$$

Practice

A 1 kΩ resistor dissipates 100 mW of power. Find its voltage drop and its current.

Answer: 10 V, 10 mA

Power Rule for Resistance with *I* and *R*

If voltage V is substituted into its Ohm's law form of (IR) from Equation 7-2 into the power rule formula, then the power rule for resistance becomes

$$P = I(V) = I(IR) = I^2 R$$

or simply

$$P = I^2 R \tag{7-7}$$

Using algebra, you can rewrite Equation 7-7 and solve for R.

$$R = \frac{P}{I^2}$$

You can also rewrite Equation 7-7 and solve for I^2.

$$I^2 = \frac{P}{R}$$

Then, taking the square root of both sides of this equation, I can be found.

$$I = \sqrt{\frac{P}{R}} \tag{7-8}$$

Example 7-8

A 27 Ω resistor dissipates 12 W of power. Find its current and its voltage drop.

Solution

Use Equation 7-8 to find the current through the resistor.

$$I = \sqrt{\frac{P}{R}} = \sqrt{\frac{12 \text{ W}}{27 \text{ } \Omega}} = 667 \text{ mA}$$

Again you could have substituted values into Equation 7-7 and algebraically found the current without using Equation 7-8.

Now use Ohm's law to find the voltage dropped across the resistor.

$$V = IR = 667 \text{ mA} \times 27 \text{ } \Omega = 18 \text{ V}$$

Note that these results agree with Example 7-7, the same problem but the order of the solution is different.

Practice

A 1 kΩ resistor dissipates 100 mW of power. Find its current and voltage drop.

Answer: 10 mA, 10 V

Summarizing, the power rule has three useful forms for a resistance.

$$P = IV$$

$$P = \frac{V^2}{R}$$

$$P = I^2 R$$

For a resistance, if you know any two variables of power (P), current (I), voltage (V), and resistance (R), you can find the other two variables. You may use these three basic forms to solve resistance related power values. You do not need to memorize the alternate forms such as Equations 7-6 and 7-8. Just use the three basic forms of Equations 7-4, 7-5, and 7-7 and solve them algebraically. Find the appropriate equation with only the one unknown in it, plug in the values, and solve for the unknown algebraically.

Resistor Ratings—Using the Power Rule

The Ohm's law forms of the power rule are very useful when determining maximum voltages and currents that can be safely applied to a resistor. Carbon composite resistors typically have power ratings of ¼ W, ½ W, 1 W and 2 W.

Given the power rating of a resistor and its resistance value, Equation 7-6, Equation 7-8, and Ohm's law can be used to find the corresponding maximum values of voltage drop and current allowed.

Example 7-9

Find the maximum safe current and the maximum safe voltage allowed for a quarter watt, 100 Ω resistor.

Solution

Use Equation 7-8 to find the maximum safe current for this resistor.

$$I = \sqrt{\frac{P}{R}} = \sqrt{\frac{0.25 \text{ W}}{100 \text{ }\Omega}} = 50 \text{ mA}$$

Then use Ohm's law to find the voltage dropped across the resistor.

$$V = IR = 50 \text{ mA} \times 100 \ \Omega = 5 \text{ V}$$

Or use Equation 7-6 to find the maximum safe voltage drop.

$$V = \sqrt{PR} = \sqrt{0.25 \text{ W} \times 100 \ \Omega} = 5 \text{ V}$$

Note that by solving for the voltage with two different methods and obtaining the same answer, you have confirmed your result.

Practice

Find the maximum safe current and the maximum safe voltage allowed for a quarter watt, 1 kΩ resistor.

Answer: 15.81 mA, 15.81 V

7.6 BJT Circuit Application

The next example examines a BJT circuit, applying the power rule to voltage sources, an LED, a resistor, and a BJT.

Example 7-10

For the BJT circuit of Figure 7-10, find the following:

a. Power generated by the 5 V supply

b. Power dissipated by R_B

c. Power dissipated by the BJT base-emitter junction

d. Power generated by the 9 V supply

e. Power dissipated by R_e

f. Power dissipated by the LED

g. Power dissipated by the transistor collector-emitter output

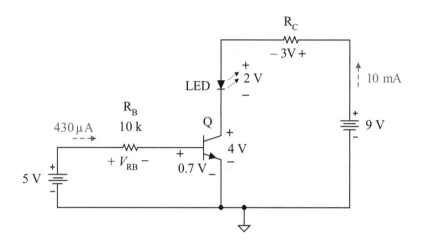

Figure 7-10 BJT example circuit

Solution

a. Power generated by the 5 V supply

$$P_{5\,\mathrm{V\,supply}} = I_{5\,\mathrm{V\,supply}}\ E_{5\,\mathrm{V\,supply}} = 430\ \mu\mathrm{A} \times 5\ \mathrm{V} = 2.15\ \mathrm{mW}$$

b. Power dissipated by R_B

$$P_{RB} = (I_{RB})^2\,R_B = (430\ \mu\mathrm{A})^2 \times 10\ \mathrm{k\Omega} = 1.85\ \mathrm{mW}$$

c. Power dissipated by the BJT base-emitter junction

$$P_{\mathrm{BJT\,be}} = I_{\mathrm{BJT\,be}}\ V_{\mathrm{BJT\,be}} = 430\ \mu\mathrm{A} \times 0.7\ \mathrm{V} = 301\ \mu\mathrm{W}$$

d. Power generated by the 9 V supply

$$P_{9\,\mathrm{V\,supply}} = I_{9\,\mathrm{V\,supply}}\ E_{9\,\mathrm{supply}} = 10\ \mathrm{mA} \times 9\ \mathrm{V} = 90\ \mathrm{mW}$$

e. Power dissipated by R_C

$$P_{RC} = I_{RC}\ V_{RC} = 10\ \mathrm{mA} \times 3\ \mathrm{V} = 30\ \mathrm{mW}$$

f. Power dissipated by the LED

$$P_{\mathrm{LED}} = I_{\mathrm{LED}}\ V_{\mathrm{LED}} = 10\ \mathrm{mA} \times 2\ \mathrm{V} = 20\ \mathrm{mW}$$

    g. Power dissipated by the transistor collector-emitter output

$$P_{\text{BJT ce}} = I_c \, V_{ce} = 10 \text{ mA} \times 4 \text{ V} = 40 \text{ mW}$$

Practice

For Example 7-10:

    a. Verify that the input circuit power supplied is equal to the input circuit power dissipated.

    b. Verify that the output circuit power supplied is equal to the output circuit power dissipated.

    c. Compare the supplied input circuit power to the supplied output circuit power.

    d. Compare the BJT input power (base-to-emitter) to the BJT output power (collector-to-emitter).

Answer: (a) 2.15 mW supplied and dissipated, (b) 90 mW supplied and dissipated, (c) collector supply power generated is about 42 times that of the base supply power, (d) BJT CE power is 133 times that of the BJT BE power (a major multiplication of power by the BJT).

7.7 MOSFET Circuit Application

The next example is the enhanced mode MOSFET circuit we saw in a previous example. Last time you were given the voltage across the load resistor. Now you can apply Ohm's law to find the load voltage drop.

Example 7-11

For the circuit of Figure 7-11(a) and its associated MOSFET characteristic curve of Figure 7-11(b), determine the E_{in} value needed to set the load voltage to 3 V. For this load voltage, calculate the load current, the load power, the supply power, and the MOSFET output power.

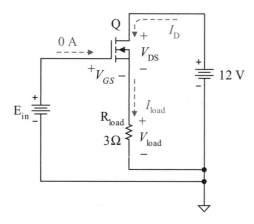

Figure 7-11(a) MOSFET voltage controlled current output

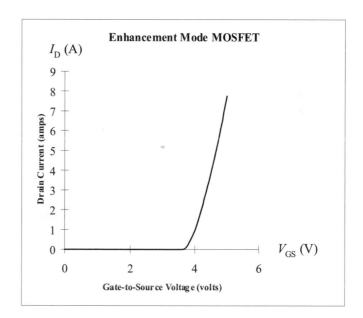

V_{GS} volts	I_D amps
0.0	0.000
1.0	0.000
2.0	0.000
3.0	0.000
3.2	0.000
3.4	0.000
3.6	0.000
3.7	0.050
3.8	0.244
4.0	0.945
4.2	1.968
4.4	3.198
4.6	4.592
4.8	6.115
5.0	7.745

Figure 7-11(b) MOSFET characteristic curve

Solution

By Ohm's law, the load current is

$$I_{\text{load}} = \frac{V_{\text{load}}}{R_{\text{load}}} = \frac{3 \text{ V}}{3 \text{ }\Omega} = 1 \text{ A}$$

By KCL, the drain current equals the load and supply currents.

$$I_{\text{D}} = I_{\text{load}} = I_{\text{supply}} = 1 \text{ A}$$

From the MOSFET's characteristic curve, at an I_{D} of 1 A

$$V_{\text{GS}} = 4.0 \text{ V}$$

By a voltage walkabout or the alternate KVL expression of the input loop, the supply voltage E_{in} is

$$E_{\text{in}} = V_{\text{load}} + V_{\text{GS}} = 3 \text{ V} + 4.0 \text{ V} = 7 \text{ V}$$

By the power rule, the load power and drainpower are

$$P_{\text{load}} = I_{\text{load}} \ V_{\text{load}} = 1 \text{ A} \times 3 \text{ V} = 3 \text{ W}$$

$$P_{\text{supply}} = I_{\text{supply}} \ E_{\text{supply}} = 1 \text{ A} \times 12 \text{ V} = 12 \text{ W}$$

The MOSFET output power must be the difference between the supplied power and the load power. That is, the power supplied must be dissipated in the two components in the output loop.

$$P_{\text{MOSFET}} = P_{\text{supply}} - P_{\text{load}} = 12 \text{ W} - 3 \text{ W} = 9 \text{ W}$$

Another way to verify this power is to find V_{DS} and apply the power rule to the MOSFET component.

$$V_{\text{DS}} = E_{\text{supply}} - V_{\text{load}} = 12 \text{ V} - 3 \text{ V} = 9 \text{ V} \qquad \text{(KVL)}$$

$$P_{\text{MOSFET}} = P_{\text{DS}} = I_{\text{D}} \ V_{\text{DS}} = 1 \text{ A} \times 9 \text{ V} = 9 \text{ W}$$

Practice

For the circuit of Figure 7-11(a) and its associated MOSFET characteristic curve of Figure 7-11(b), E_{in} is set to 10.2 V. Find the load voltage, the load current, the load power, the supply power and the MOSFET output power.

Answer: 6 V, 2 A, 12 W, 24 W, 12 W

7.8 Op Amp Circuit Application

The next example is an op amp amplifier circuit that we have seen before. Last time you were given the voltage across the feedback resistor. Now you can apply Ohm's law to find this voltage drop.

Example 7-12

For the circuit of Figure 7-12, find all the node voltages, the resistor voltage drops, resistor currents, and resistor power dissipations.

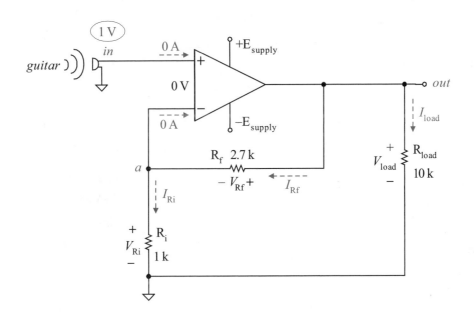

Figure 7-12 Op amp noninverting voltage amplifier

Solution

Critical. As we find values, write them on your schematic. Write node voltages by their nodes, voltage drop values with polarities by their resistors, and current values by their reference arrows. This helps you visualize the circuit values and helps you proceed in solving the rest of the circuit values.

The input voltage at node *in* is 1 V. So the op amp's noninverting (ni, +) pin is also at 1 V.

$$V_{in} = V_{ni} = 1\,V$$

The voltage drop across the op amp terminals is 0 V. So the inverting (inv, −) op amp input terminal must also be at 1 V. This terminal is tied to node *a*. So,

$$V_a = V_{inv} = V_{ni} - 0\,V = 1\,V$$

Since node *a* is at 1 V and the bottom of R_i is tied to common, the voltage drop across R_i must be 1 V.

$$V_{Ri} = V_a - 0\,V = 1\,V$$

By Ohm's law, the current through R_i is

$$I_{Ri} = \frac{V_{Ri}}{R_i} = \frac{1\,V}{1\,k\Omega} = 1\,mA$$

By Kirchhoff's current law at node *a*, the current through R_f is

$$I_{Rf} = I_{Ri} + I_{op\,amp\,inv} = 1\,mA + 0\,mA = 1\,mA$$

By Ohm's law for resistance, the voltage drop across V_{Rf} is

$$V_{Rf} = I_{Rf}\,R_f = 1\,mA \times 2.7\,k\Omega = 2.7\,V$$

Do a voltage walkabout (KVL) from common, a voltage rise across R_i, and a voltage rise across R_f to the *out* node. The output voltage V_{out} is

$$V_{out} = V_{common} + V_{Ri} + V_{Rf} = 0\,V + 1\,V + 2.7\,V = 3.7\,V$$

The voltage drop across R_{load} is the difference between the *out* node and the *common* node.

$$V_{load} = V_{out} - 0\,V = 3.7\,V - 0\,V = 3.7\,V$$

By Ohm's law, the load current is

$$I_{\text{load}} = \frac{V_{\text{load}}}{R_{\text{load}}} = \frac{3.7 \text{ V}}{10 \text{ k}\Omega} = 370 \text{ }\mu\text{A}$$

The resistor power dissipations can now be found by using the power rule.

$$P_{\text{Ri}} = I_{\text{Ri}} \text{ } V_{\text{Ri}} = 1 \text{ mA} \times 1 \text{ V} = 1 \text{ mW}$$

$$P_{\text{Rf}} = I_{\text{Rf}} \text{ } V_{\text{Rf}} = 1 \text{ mA} \times 2.7 \text{ V} = 2.7 \text{ mW}$$

$$P_{\text{load}} = I_{\text{load}} \text{ } V_{\text{load}} = 370 \text{ }\mu\text{A} \times 3.7 \text{ V} = 1.369 \text{ mW}$$

Practice

Repeat Example 7-12 with the input voltage doubled to 2 V.

Answer: All voltages and currents are doubled. All the powers are 4 times greater. (Why?)

7.9 Electrical Energy

Energy is *generated* (created) and energy is *absorbed* (dissipated or converted). The energy generated must always be equal to the energy absorbed.

DC Voltage Supplies

DC voltage supplies are designed to provide electrical energy that is consumed by the circuit components. For example, a flashlight battery generates electrical energy and the flashlight lamp absorbs energy. That absorbed energy is converted into light energy and dissipated as heat energy. The energy generated must equal the energy absorbed.

A voltage supply is designed to generate or deliver energy to a circuit. But some voltage supplies are also designed to absorb energy, for example, a rechargeable battery. A battery being recharged is absorbing energy and chemically recreating charge. Figure 7-13 distinguishes the key difference between these two applications. Figure 7-13(a) shows the source generating current *out of* its positive (+) terminal as is expected for a supply generating current when placed in a circuit. Figure 7-13(b) shows the current flowing into its positive terminal, indicating that this supply is absorbing energy. Again, an energy generating supply has a current *out of* its positive terminal and an energy absorbing supply has a current *into* its positive terminal.

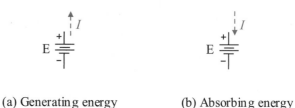

(a) Generating energy   (b) Absorbing energy

Figure 7-13 Voltage sources generate and sometimes can absorb energy

Figure 7-14 shows a simplified schematic of a battery charging circuit. The source on the left is generating energy and delivering it to the charging battery on the right.

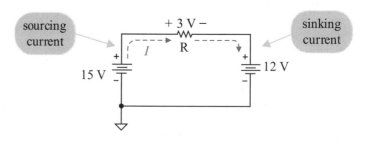

Figure 7-14 Battery charging circuit

The resistor in the schematic represents the natural resistance of the batteries. The generating voltage is greater than the charging battery voltage to create current flowing in the clockwise direction, forcing current to flow into the positive (+) terminal of the charging battery. The current flowing into the charging battering chemically recharges the battery so that it can in turn produce a flowing charge when placed into a circuit. Since the supply on the left is delivering charge to the circuit, it is said to be **sourcing** (creating) current. The source on the right is **sinking** (receiving) current. *Warning*, not all supplies are designed to sink current. Be very cautious about forcing a supply to sink current inappropriately.

Supply Energy

Recall the basic expression for power is defined in terms of energy per unit time.

$$P = \frac{W}{t}$$

This expression can be rewritten in terms of energy as

$$W = Pt$$

where
$\quad W$ = energy in joules (J)
$\quad P$ = power in watts (W)
$\quad t$ = time in seconds (s)

These basic units of measure are really too small to be practical. Therefore, other units of practical measure have been adopted when working with energy. Energy is

$$W = Pt \qquad\qquad\qquad (7\text{-}9)$$

where
$\quad W$ = energy in kilowatt hours (kWh)
$\quad P$ = power in kilowatts (kW)
$\quad t$ = time in hours (h)

Let's convert 1 kWh to its basic unit equivalent of joules (k is 1000, W is J/s, and an hour is 3600 seconds). Notice that 1 kWh is equivalent to 3.6 million joules of energy. Electrical energy companies charge per kWh as the most practical unit of energy.

$$1 \text{ kWh} = 1 \times (1000) \times \left(\frac{J}{s}\right) \times (3600 \text{ s}) = 3.6 \text{ MJ}$$

The power of the source can be found using the power rule and the known values of its voltage and current.

$$P = I\,E$$

Example 7-13

A 12 V supply generates 3 A of current for 2000 hours. Find the power of the source and the energy generated.

Solution

First find the power being generated by the supply.

$$P = I\ E = 12\ \text{V} \times 3\ \text{A} = 36\ \text{W}$$

Next, convert the power to kilowatts

$$P = 36\ \text{W} \times \left(\frac{\text{kW}}{1000\ \text{W}} \right) = 0.036\ \text{kW}$$

Now apply Equation 7-9 to find the energy produced for 2000 hours of operation.

$$W = Pt = 0.036\ \text{kW} \times 2000\ \text{h} = 72\ \text{kWh}$$

Practice

A 12 V supply generates 5 A of current for 1 year. Find the power of the source and the energy generated.

Answer: 60 W, 525.6 kWh

Resistive Energy Dissipation

Resistive components dissipate energy in the form of heat. Energy may also be converted to other forms such as light energy. The energy dissipated in a resistance can also be calculated from the basic expression for energy in terms of its power.

$$W = Pt$$

Example 7-14

A 100 Ω resistance draws 100 mA of current for 3 years. Find the power it dissipates and the energy consumed.

Solution

First find the power being dissipated by the resistance.

$$P = I^2R = (100 \text{ mA})^2 \times 100 \ \Omega = 1 \text{ W}$$

Next, convert the power to kilowatts

$$P = 1 \text{ W} \times \left(\frac{\text{kW}}{1000 \text{ W}}\right) = 0.001 \text{ kW}$$

Next convert the 3 years time interval to hours.

$$t = 3 \text{ years} \times \left(\frac{365 \text{ days}}{\text{year}}\right) \times \left(\frac{24 \text{ h}}{\text{day}}\right) = 26{,}280 \text{ h}$$

Now apply Equation 7-9 to find the energy produced for 26,280 hours of operation.

$$W = Pt = 0.001 \text{ kW} \times 26{,}280 \text{ h} = 26.28 \text{ kWh}$$

Practice

A 200 W light bulb draws 2 A of current for 2 years. Find the energy dissipated by the light bulb.

Answer: 3504 kWh

Cost of Energy

Electrical energy companies produce electricity consumed by industry and homes. Energy is measured commercially in kilowatt-hours (kWh). The price of electrical energy is charged per kilowatt-hour.

The familiar electrical meter measures the energy consumed in the home in kilowatt-hours. In the United States, this electrical energy is relatively inexpensive. Reasonably priced and consistently available electrical energy is the foundation of a vibrant technical society and the trademark of the world's industrialized countries.

Example 7-15

A 200 W bulb burns for 250 days. Find the energy consumed and the cost of that energy if the electric energy company charges 10 cents per kilowatt-hour.

Solution

First, convert the power to kilowatts.

$$P = 200 \text{ W} \times \left(\frac{\text{kW}}{1000 \text{ W}} \right) = 0.2 \text{ kW}$$

Next, convert the 250 days to hours.

$$t = 250 \text{ days} \times \left(\frac{24 \text{ h}}{\text{day}} \right) = 6000 \text{ h}$$

Now, find the energy produced for 6000 hours of operation.

$$W = Pt = 0.2 \text{ kW} \times 6000 \text{ h} = 1200 \text{ kWh}$$

Now, calculate the total electrical cost of operating the light.

$$\text{total cost} = 1200 \text{ kWh} \times \left(10 \frac{\text{cents}}{\text{kWh}} \right) = 12000 \text{ cents}$$

It cost $120 to operate a 200 W light bulb for 250 days.

Practice

An electronic circuit requires 12 V and operates at 2 A. It operates continuously for 500 days. Find the energy consumed and the cost of that energy if the electric energy company charges 10 cents per kilowatt-hour.

Answer: 288 kWh, 2880 cents or $28.80

Summary

Ohm's law for resistance is one of the most fundamental laws of electronics. Ohm's law relates the voltage and current to its resistance.

The three forms of Ohm's law are

$$I = \frac{V}{R}$$

$$V = IR$$

$$R = \frac{V}{I}$$

You must commit these to memory so that you can immediately apply Ohm's law whenever needed.

The direction of current is always into the relatively positive (+) voltage potential terminal of the resistance and out the negative (–). When establishing reference directions for a resistive component, always be consistent in establishing your current directions and corresponding voltage polarities. If a calculated or measured value is *negative*, the current direction or the voltage polarity is the *opposite* of the reference.

The power rule states that the power of a component is equal to the product of its voltage and current.

$$P = IV$$

This applies to dc calculations for any component, for example, a dc power supply, an LED, or a resistor.

For a resistive component, the power rule can be combined with Ohm's law to produce power expressions related to its resistance.

$$P = \frac{V^2}{R}$$

$$P = I^2 R$$

These formulas *cannot* apply to a voltage supply since its internal resistance is ideally 0 Ω. You must use its voltage and current values to calculate a voltage supply's power.

$$P = I E$$

Power is the energy generated or dissipated per unit time.

$$P = \frac{W}{t}$$

This expression can be rewritten in terms of its energy.

$$W = P t$$

The useful electrical unit of energy is the kilowatt-hour (kWh). This is the unit used by electrical energy companies to charge customers.

A voltage supply normally generates power. If this is the case, its current is *out of* its positive (+) terminal and said to be *sourcing current*. If a voltage supply is absorbing power, for example a charging battery, then current flows *into* its positive (+) terminal and it is said to be *sinking current*.

Problems

Ohm's Law for Resistance

7-1   Find the current through a 2.2 kΩ resistor that drops 12 V.

7-2   Find the voltage drop across a 1.8 kΩ resistor that has a current of 3 mA.

7-3   Find the resistance of a resistor that drops 18 V and has a current of 10 mA.

7-4   For the circuit of Figure 7-15.

    a. Is there circuit continuity?

    b. Redraw the schematic with the current meter and voltmeter properly connected to reflect the references. Indicate the red and black leads of the test meters.

    c. Find the resistor voltage drop and current. Be faithful to your references.

    d. Redraw the schematic into its bubble notation form.

    e. Draw the characteristic curve of the 220 Ω resistor. Include both positive and negative voltages on a scale from −15 V to +15 V. Locate the operating point of the circuit on the characteristic curve.

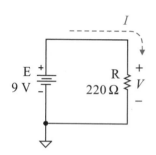

Figure 7-15

7-5   For the circuit of Figure 7-16.

    a. Redraw the schematic with the current meter and voltmeter properly connected. Indicate the red and black leads of the test meters.

    b. Find the resistor voltage drop and current. Be faithful to your references.

    c. Redraw the schematic into its bubble notation form.

    d. Draw the characteristic curve of the 1.2 kΩ resistor. Include both positive and negative voltages on a scale from −10 V to + 10 V. Locate the operating point of the circuit on the characteristic curve.

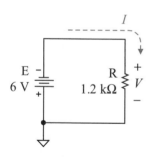

Figure 7-16

Power Rule

7-6 A 12 V bulb draws 3 A of current. Find its power dissipation

7-7 A 2.3 V light emitting diode dissipates 46 mW of power. How much current does it have?

7-8 A battery generates 10 W of power and draws 500 mA of current. What is the battery voltage?

Power Rule–Resistance Forms

7-9 A quarter watt 100 Ω resistor draws 200 mA of current. Find its power dissipation. Is it safe?

7-10 A quarter watt 3.3 kΩ resistor drops 10 V. Find its power dissipation. Is it safe?

7-11 A 1 kΩ resistor dissipates 10 W of power. Find its current and voltage drop.

7-12 A 39 Ω resistor dissipates 2 W of power. Find its current and voltage.

7-13 Find the maximum current that a quarter watt 180 Ω resistor can safely draw.

7-14 Find the maximum voltage that a quarter watt 180 Ω resistor can safely drop.

BJT Circuit Application

7-15 For the circuit of Figure 7-17, find the following values.

    a. Voltage drop across R_B.

    b. Current through R_B.

    c. Power dissipated by the BJT base-emitter junction.

    d. BJT collector-emitter voltage drop.

    e. Power dissipated by the BJT, collector to emitter.

    f. Current gain beta β of the BJT.

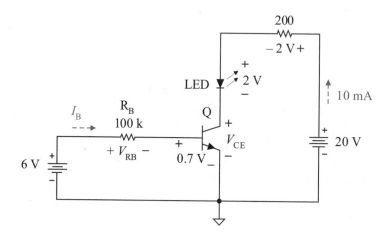

Figure 7-17

MOSFET Circuit Application

7-16   Use the MOSFET characteristic curve of Figure 7-11(b). In the circuit of Figure 7-18, the lamp drops 2 V and V_{DS} is measured to be 10 V. Find the following:

    a. Voltage across the load.
    b. Current through the load.
    c. MOSFET gate-to-source voltage.
    d. Input supply E_{in}.
    e. Power dissipated by the load.
    f. Power dissipated by the MOSFET.
    g. Power dissipated by the lamp.
    h. Power generated by the drain supply.
    i. Prove that the power generated by the drain voltage supply is dissipated by the lamp, the MOSFET, and the load resistor.

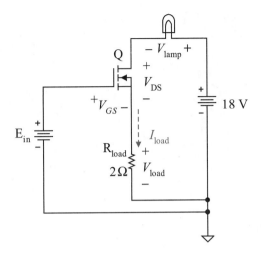

Figure 7-18

Op Amp Circuit Application

7-17 For the circuit of Figure 7-19, find the following values:

    a. Voltage at node *a*.

    b. Voltage drop across V_{Ri}.

    c. Current through R_i.

    d. Current through R_f

    e. Voltage drop across R_f.

    f. Voltage at node *out*.

    g. Load voltage.

    h. Load current.

    i. Op amp output current. Is it into or out of the op amp?

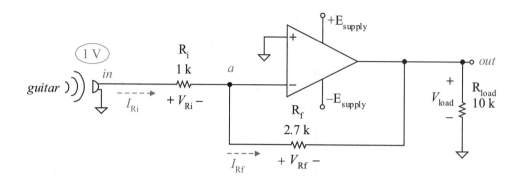

Figure 7-19

Electrical Energy

7-18   A 100 W bulb operates for 1 year at a cost of 9 cents per kilowatt-hour. Find the energy consumed and the cost to operate it.

7-19   A 12 V supply draws 200 mA and operates for 125 days. Find the power generated by the source and the cost to operate it if electrical energy costs 10 cents per kilowatt-hour.

7-20   Does a voltage supply generating electrical energy require that current exit the positive or negative terminal? Is it sourcing or sinking current?

7-21   Does a recharging battery require that electrical current enter the positive or negative terminal? Is it sourcing or sinking current?

8

Series Circuits

Introduction

The two most basic and simple forms of circuit configurations are the series circuit and the parallel circuit. We have seen both as examples in applying Kirchhoff's current law and Kirchhoff's voltage law.

   This chapter formally introduces the series circuit through the three laws of electronics. The concept of a series circuit is expanded beyond the simple circuit concept and applied to electronic circuits that exhibit series circuit behavior, even though at first glance the circuit does not appear to be a series circuit.

Objectives

Upon completion of this chapter you will be able to:

- Identify and apply circuit laws to the series circuit.

- Find total resistance of a series resistive circuit.

- Analyze the loading effect of current meters in low resistance circuits.

- Apply the voltage divider rule (VDR) and discern when it is applicable.

- Apply the VDR to BJT and MOSFET dc biasing electronic circuits.

- Model a real voltage supply and calculate its parameters and calculate its percent load voltage regulation.

- Determine a Thévenin model of a circuit based upon "What's in the Box" modeling and apply it for different circuit loads.

- Analyze and apply multiple voltage sources in series.

- Apply series circuits analysis techniques to electronic circuits.

- Apply ideal and second and third approximation modeling to switching and zener diodes in series circuit configurations.

8.1 Series Circuits and the Laws

A **series circuit** is a circuit configuration in which all the components are *two-terminal components connected end-to-end* such that each component terminal is connected to one and only one other component terminal. Figure 8-1 demonstrates a series circuit configuration. Each voltage node has only two component terminals connected to it.

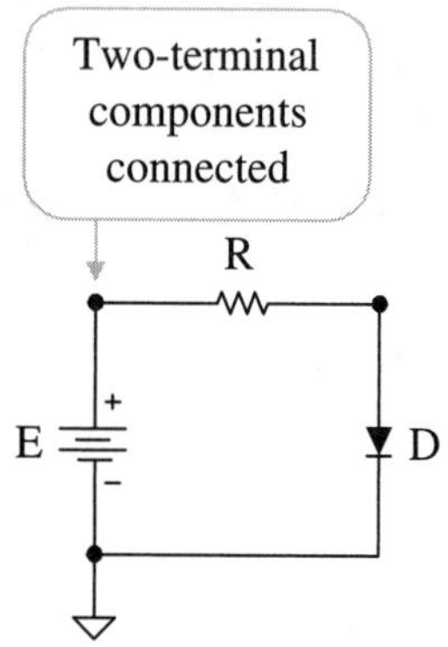

Two-terminal components connected

R

E

D

Figure 8-1
Series circuit configuration

KCL–Kirchhoff's Current Law

By Kirchhoff's current law, if there are only two currents associated with a simple node, the current flowing into that node must be equal to the current flowing out of that node. *What flows in must flow out.* Also, each component has only two terminals, so the current that flows into a component must equal the current that flows out of that component. This principle is true for the entire series circuit. Each simple node and each component has only two currents associated with it: one current flowing in and one current flowing out. *Thus the current flowing through a series circuit must be the same throughout the circuit. Each component (supply, resistor, LED, diode, wire, etc.) has the same current flowing through it.*

$$I = I_{\text{supply}} = I_{\text{R}} = I_{\text{D}} = I_{\text{wire}} \qquad \text{(KCL)}$$

Figure 8-2 offers a visual image of current flowing through a series circuit.

Series circuit current is the same throughout the circuit.

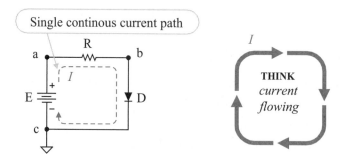

Figure 8-2 A series circuit and its current diagram

The next example utilizes the three basic laws of circuit analysis to solve for voltage and current. You must understand this example completely before moving forward in this textbook. These are fundamental techniques that will also be applied to more complex circuits. Not so amazingly, complex circuits become reasonably simple to analyze if your fundamental circuit analysis techniques are solid. *Sketching schematics with known values is critical to visualization and understanding of circuit operation and circuit analysis.* Always draw circuit schematics and place voltages, currents, and resistances on the schematic as you proceed. This significantly enhances your understanding of circuit behavior and aids in solving for other circuit values.

Example 8-1

For the circuit of Figure 8-3(a), draw the proper direction of current with associated component voltage polarities. Next find the voltage drop across R. Then find the current in the LED. The LED specification is that it drops about 2 V.

Solution

Draw the schematic of the circuit with proper current direction, component voltage polarities, and known values. The schematic of Figure 8-3(b) has been properly prepared for further circuit analysis. *This is a critical first step in circuit analysis.*

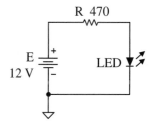

Figure 8-3(a)
Example circuit

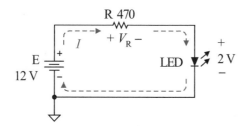

Figure 8-3(b) Preparation of schematic for circuit analysis

These steps are required *before* circuit analysis is begun:

• Draw the schematic with known values.

- Write reference current direction I on the schematic with proper direction based upon supply E.

- Establish the resistor voltage polarity based upon the current I direction written on the schematic.

- Establish the LED voltage polarity based upon the current I direction written on the schematic.

- The LED voltage polarity causes the LED to be forward-biased; that is, a relative (+) voltage on the anode compared to a (−) on the cathode.

- The supply E is greater than the 2 V needed to turn on the LED; therefore, draw the 2 V drop across the LED with its proper polarity.

Now think, what can we easily find next based upon the three laws and available known values? (1) KCL does not help us. There is one current and it is unknown. (2) Ohm's law for resistance cannot be used yet since only the resistor's resistance is known and its voltage and current are unknown. (3) KVL does help us. In the closed circuit loop, there is only one unknown voltage. Only the resistor voltage drop is not known, so we can solve for it using KVL.

Use KVL to find V_R. Look at the circuit of Figure 8-3(b). You can use the *basic KVL equation* with a closed circuit loop, or you can use the straightforward *voltage walkabout* technique from $V_R(-)$ to $V_R(+)$. Both are valid, but the KVL voltage walkabout technique provides the answer in one step. Start at the reference $V_R(-)$ terminal and walk to the $V_R(+)$ target terminal along the known voltage path. There is a fall of 2 V and a rise of 12 V in our walk. So,

$$V_R = -2 \text{ V} + 12 \text{ V} = 10 \text{ V} \qquad \text{(KVL walkabout)}$$

Critical: let's add this information to our schematic to help us visualize the next step. Figure 8-3(c) shows the updated schematic with the 10 V resistor drop written on the schematic. *Now think,* what can we do next to solve the circuit? We know all of the voltage drops. What component can we use to find the circuit current? We cannot use supply E; there is no direct relation with the circuit current. We cannot use the LED since we do not know the relationship between its voltage and current.

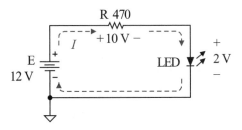

Figure 8-3(c) Example circuit with V_R of 10 V added

What about the resistor R? Yes, we know its resistance and its voltage drop, so we can use *Ohm's law for resistance* and calculate the resistor current I_R.

$$I_R = \frac{V_R}{R} = \frac{10 \text{ V}}{470 \text{ }\Omega} = 21.28 \text{ mA} \qquad \text{(Ohm's law)}$$

Critical: put the current value for the resistor on the schematic as shown in Figure 8-3(d). You found it, so write it on the schematic. *Always, always put values on your schematic as you find them.*

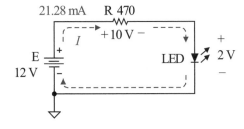

Figure 8-3(d) Example schematic with current added

Since this is a series circuit, by KCL the current is the same through every component and wire in the circuit. So, the LED current can be found from the resistor current.

$$I_{LED} = I_R = 21.28 \text{ mA} \tag{KCL}$$

Go back and review the entire process that we followed in solving this problem. Every step is important, especially establishing voltage polarities and current directions as well as circuit values.

Practice

For the circuit of Figure 8-3(a), the current through the diode is to be limited to 40 mA. Find the resistance value of the resistor R that limits the LED current to 40 mA.

The resistor R is actually called a **current limiting resistor** because it limits and determines the current through the LED.

Answer: 250 Ω (270 Ω is the closest *higher* nominal standard value)

Exploration

Simulate the example problem using MultiSIM to confirm the example solution. (See Figure 8-4.)

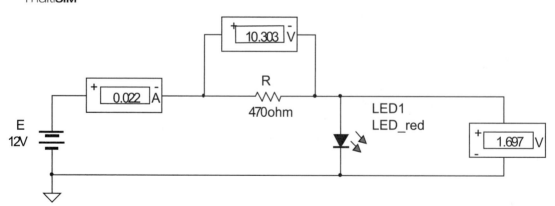

Figure 8-4 MultiSIM simulation of example circuit

Note that the LED simulated voltage drop is 1.7 V instead of 2.0 V, thus the resistive voltage drop is slightly increased and the resistor (circuit) current is slightly increased.

8.2 Resistive Series Circuit Analysis

In this section, we shall examine a series circuit with components that are resistive in nature, for example, carbon composite resistors. We shall use the laws of circuit analysis, namely, KCL, KVL, and Ohm's law for resistance, to solve for currents, voltages, and total circuit resistance.

Single Series Resistance

Figure 8-5(a) is a simple series circuit with a voltage source E and a single resistance R. We know from Section 8.1 (from KCL) that the current is the same throughout the series circuit. That is,

$$I = I_{supply} = I_R = I_{wire} \qquad \text{(KCL)}$$

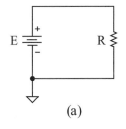 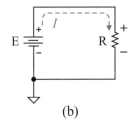 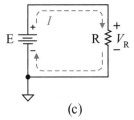

(a) (b) (c)

Figure 8-5 Simple series circuit with one resistor

The current flows out from the E(+) supply terminal down through resistor R, dropping (+/−) voltage, and returns to the E(−) supply terminal as shown in Figure 8-5(b). E volts are supplied, therefore, E volts must be dropped across the circuit components as shown in Figure 8-5(c). Thus, all E volts are dropped across R.

Or, apply Kirchhoff's voltage law around the closed loop (clockwise loop: rise E and fall V_R must sum to 0 V).

$$E - V_R = 0 \text{ V} \qquad \text{(KCL)}$$

Either way you look at it,

$$V_R = E$$

Next, we can use Ohm's law to find the current in the resistor.

$$I_R = \frac{V_R}{R} = \frac{E}{R} \qquad \text{(Ohm's law \& substitution)}$$

Now we have the circuit current and all of the component currents as well.

$$I = I_{\text{supply}} = I_{\text{wire}} = I_R \qquad\qquad \text{(KCL)}$$

Example 8-2

For the circuit of Figure 8-6(a), draw the circuit with the current direction I and the voltage polarity of V_R. Then find the voltage drop V_R, the current I_R, and the current I_{supply}.

Solution

Figure 8-6(b) shows the schematic with the circuit current and *corresponding* resistor voltage polarity drawn on the schematic. Remember that as current flows through an energy-dissipating component, such as a resistor, voltage is lost across the device, thus the $(+/-)$ polarity drop.

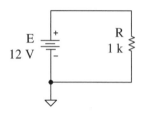

Figure 8-6(a)
Example circuit

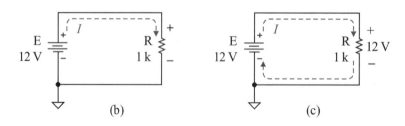

(b)   (c)

Figure 8-6(b,c) Example circuit with defined current and voltage references

Using KVL or a KVL walkabout, find the voltage drop V_R across the resistor. Let's do a KVL walkabout. Start at $V_R(-)$ reference and walk counterclockwise to $V_R(+)$. All we encounter is a rise of 12 V from the supply E. So,

$$V_R = E = 12 \text{ V} \qquad \text{(KVL walkabout)}$$

Draw this on the schematic as shown in Figure 8-6(c). Now we see that 12 V is dropped across the 1 kΩ resistor. Since we know the voltage drop across a known resistance, we can use Ohm's law to find the current flowing through the resistor.

$$I_R = \frac{V_R}{R} = \frac{12 \text{ V}}{1 \text{ k}\Omega} = 12 \text{ mA} \qquad \text{(Ohm's law)}$$

This is also the circuit current and the supply current since this is a series circuit and the current is the same throughout a series circuit.

$$I = I_{\text{supply}} = I_{\text{wire}} = I_R = 12 \text{ mA} \qquad \text{(KCL)}$$

Now draw this result on the schematic. Figure 8-6(d) shows the example circuit with all component voltages and currents drawn.

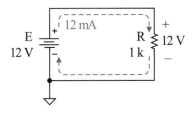

Figure 8-6(d) Example circuit voltages and currents

Practice

For the circuit of Figure 8-7, find the voltage drop V_R and the circuit current I. Be faithful to the references given.

Answer: −9 V, −90 mA

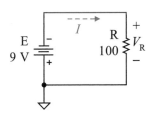

Figure 8-7
Practice circuit

Multiple Series Resistances

Next, an example is used to see the effects of a circuit with several resistors (resistances) in series. The basic laws are used to find the circuit current, the total circuit resistance as viewed by the supply, the individual resistor voltage drops, and the node voltages. Several steps are required to solve this problem, but each step is a simple, straightforward use of basic circuit laws.

Example 8-3

For the circuit of Figure 8-8(a), draw the circuit with the current direction I and the voltage polarity of each resistor. Then find the circuit current, the total circuit resistance as seen by the supply current, each resistor voltage drop, and each node voltage.

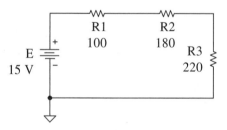

Figure 8-8(a) Example circuit

Solution

Figure 8-8(b) shows the schematic with the circuit current direction and *corresponding* resistor voltage polarities drawn on the schematic. Remember that as current flows through an energy-dissipating component such as a resistor, voltage is lost across the device, thus the (+/−) polarity drops.

First solve for circuit current I. Several steps are required using KVL, KCL, and Ohm's law.

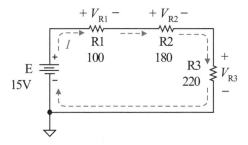

Figure 8-8(b) Example circuit with current and voltage drop references

First, using Ohm's law for resistance, we can express the voltage drops across each of the resistors.

$$V_{R1} = (100\ \Omega)\,I_{R1} \qquad\qquad \text{(Ohm's law)}$$

$$V_{R2} = (180\ \Omega)\,I_{R2}$$

$$V_{R3} = (220\ \Omega)\,I_{R3}$$

Since this is a series circuit (a single closed loop), the currents are everywhere the same.

$$I = I_{R1} = I_{R2} = I_{R3} \qquad\qquad \text{(KCL)}$$

Substituting I for the resistor currents in the Ohm's law expressions, the resistor voltage drops become

$$V_{R1} = (100\ \Omega)\,I \qquad\qquad \text{(substitution)}$$

$$V_{R2} = (180\ \Omega)\,I$$

$$V_{R3} = (220\ \Omega)\,I$$

Next write the KVL loop expression for this single closed loop circuit. Start at common and walk clockwise around the circuit.

$$+15\ \text{V} - V_{R1} - V_{R2} - V_{R3} = 0\ \text{V} \qquad \text{(KVL loop)}$$

Simplify this expression with unknowns on the left side and the known voltage on the right side of the expression.

$$V_{R1} + V_{R2} + V_{R3} = 15\ \text{V}$$

Substituting the Ohm's law expressions for each resistor into the KVL loop expression, the KVL loop expression becomes

$$(100\ \Omega)I + (180\ \Omega)I + (220\ \Omega)I = 15\ \text{V}$$

The current I on the left side of this expression can be factored out of all three terms. The resulting equation is

$$(100\ \Omega + 180\ \Omega + 220\ \Omega)I = 15\ \text{V}$$

Rewrite this expression with the coefficient simplified.

$$(500\ \Omega)I = 15\ \text{V}$$

Now we can solve for the circuit current I.

$$I = \frac{15\ \text{V}}{500\ \Omega} = 30\ \text{mA}$$

Find the total circuit resistance R_T as seen by the supply E.

Look at the last expression where we solved for the current I. Writing this expression in literal terms, you can think of it as an Ohm's law expression for the circuit where E is the voltage across the total circuit resistance R_T.

$$I = \frac{\text{E}}{R_T}$$

So, the total circuit resistance as seen by the supply is the sum of the individual series resistances of R1, R2, and R3.

$$R_T = 500\ \Omega$$

Write the known current on the schematic as shown in Figure 8-8(c) and then solve for each of the resistor voltage drops.

Substitute the known current value in each resistor's Ohm's law expression to find its voltage drop.

$$V_{R1} = 100\ \Omega \times 30\ \text{mA} = 3\ \text{V} \qquad \text{(substitution)}$$

$$V_{R2} = 180\ \Omega \times 30\ \text{mA} = 5.4\ \text{V}$$

$$V_{R3} = 220\ \Omega \times 30\ \text{mA} = 6.6\ \text{V}$$

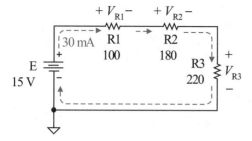

Figure 8-8(c) Example circuit with calculated current

Figure 8-8(d) shows a MultiSIM simulation with the complete solution of the example series circuit with the circuit current value and all of its resistor voltage drops.

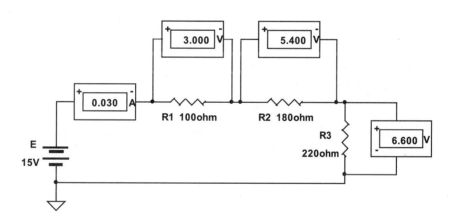

Figure 8-8(d) Example circuit with current and resistor voltages

Notice that KVL is satisfied in this single voltage loop circuit. Writing the KVL clockwise closed loop expression

$$+15 \text{ V} - 3 \text{ V} - 5.4 \text{ V} - 6.6 \text{ V} = 0 \text{ V} \qquad \text{(KVL)}$$

$$0 \text{ V} = 0 \text{ V} \qquad \text{(KVL satisfied)}$$

Finally, *find all the node voltages*. First label the nodes so that there are practical references as shown in Figure 8-8(e). The current and resistor designators have been removed from this figure for the sake of clarity and simplification.

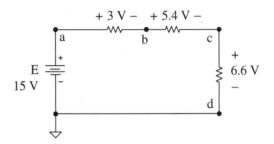

Figure 8-8(e) Sample circuit nodes defined

Start at the only known node voltage. *Common* is a reference node and is assigned the node voltage of 0 V. Write 0 V on your schematic at the common node.

$$V_d = 0 \text{ V} \qquad\qquad \text{(common)}$$

You may walk clockwise or counterclockwise around the voltage loop to find the rest of the node voltages. Let's walk around clockwise to find the node voltages. Write the node voltages on your schematic as you proceed; this will help you visualize the node voltages and assist in finding the next node voltage.

Find V_a. Start at common (0 V) and walk to node *a*. There is a 15 V rise across supply E.

$$V_a = V_{common} + E = 0 \text{ V} + 15 \text{ V} = 15 \text{ V}$$

Find V_b. Starting at node *a* (15 V) and walking to node *b*, there is a 3 V fall across resistor R1.

$$V_b = V_a - V_{R1} = 15 \text{ V} - 3 \text{ V} = 12 \text{ V}$$

Find V_c. Starting at node *b* (12 V) and walking to node *c*, there is a 5.4 V fall across resistor R2.

$$V_c = V_b - V_{R2} = 12 \text{ V} - 5.4 \text{ V} = 6.6 \text{ V}$$

Find V_d. Starting at node c (6.6 V) and walking to node d, there is a 6.6 V fall across resistor R3.

$$V_d = V_c - V_{R3} = 6.6 \text{ V} - 6.6 \text{ V} = 0 \text{ V}$$

Note that we have walked completely around the loop and the node voltage at common (V_d) is verified to be 0 V in this last calculation.

The resulting node voltages are shown in Figure 8-8(f) in the shaded ovals.

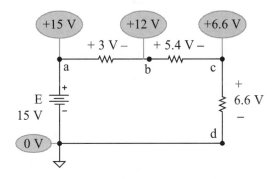

Figure 8-8(f) Node voltages of example circuit

Personally verify the node voltages by starting at common and walking counterclockwise around the circuit. You must get the same node voltage values whether you walk clockwise or counterclockwise. And you must always end up with 0 V at common.

Practice

For the circuit of Figure 8-9, find the circuit current, the total circuit resistance as viewed from the supply, the resistor voltage drops, and the node voltages.

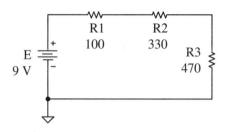

Figure 8-9 Practice circuit

Answer 10 mA; 900 Ω; 1 V, 3.3 V, 4.7 V; 0 V, 9 V, 8 V, 4.7 V, 0 V

8.3 Resistance of a Series Circuit

In the last section we used the basic laws to solve a relatively simple series resistive circuit with three resistors. Several steps were required to solve this relatively simple series circuit. Sometimes a circuit exhibits a characteristic, based upon the circuit laws, which can be used to streamline the circuit solution process.

In a series circuit, the *total circuit resistance* R_T as viewed from the supply can be used to streamline the calculations process. The circuit of Figure 8-10(a) is a generic series circuit with three resistors in series, and Figure 8-10(b) is its modeled total circuit resistance.

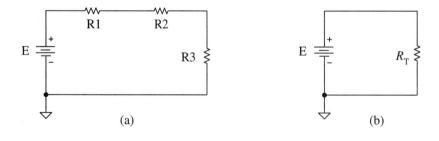

Figure 8-10 Series circuit and equivalent circuit model to find R_T

We want to find an expression for the total circuit resistance R_T in terms of its individual resistors R1, R2, and R3. Use Figure 8-11 and the basic circuit laws to solve this problem just as we did in the previous example. Figure 8-11(b) has been modeled as a single resistance R_T with a voltage drop of E since it is the only component in this series circuit.

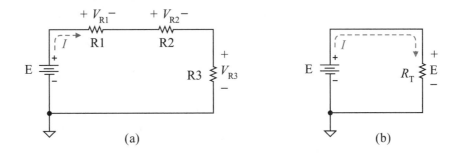

Figure 8-11 Circuits used to find total circuit resistance R_T

First, let's apply Ohm's law to the resistors of Figure 8-11(a).

$$V_{R1} = (R1)I_{R1} \qquad \text{(Ohm's law)}$$

$$V_{R2} = (R2)I_{R2}$$

$$V_{R3} = (R3)I_{R3}$$

Since this is a series circuit (a single closed loop), the current must be the same everywhere throughout the circuit.

$$I = I_{R1} = I_{R2} = I_{R3} \qquad \text{(KCL)}$$

Substituting I for the resistor currents in the Ohm's law expressions, the resistor voltage drops become

$$V_{R1} = (R1)I \qquad \text{(substitution)}$$

$$V_{R2} = (R2)I$$

$$V_{R3} = (R3)I$$

Next write the KVL loop expression for this single closed loop circuit. Start at common and walk clockwise around the circuit.

$$E - V_{R1} - V_{R2} - V_{R3} = 0 \text{ V} \qquad \text{(KVL)}$$

Reorder this KVL expression with resistor voltage drops on the left side and the supply voltage E on the right side of the equation.

$$V_{R1} + V_{R2} + V_{R3} = E \qquad \text{(KVL)}$$

Substituting the Ohm's law expressions for each resistor into the KVL loop expression, the KVL loop expression becomes

$$(R1)I + (R2)I + (R3)I = E$$

The current I on the left side of this expression can be factored out of all three terms. The resulting equation is

$$(R1 + R2 + R3)I = E$$

Rewrite this expression with only the resistance on the left for the circuit of Figure 8-11(a).

$$R1 + R2 + R3 = \frac{E}{I} \qquad \text{(Ohm's law)}$$

Now write the similar Ohm's law expression for the circuit of Figure 8-11(b).

$$R_T = \frac{E}{I} \qquad \text{(Ohm's law)}$$

Since both resistances are equal to the same E/I expression on the right side of the equation, they are equal to each other.

Series circuit resistances add

$$R_T = R1 + R2 + R3 \qquad \textbf{(8-1)}$$

So, a major characteristic or feature of a series circuit is that *series resistances add to create the total circuit resistance*. Now that we know that, we can apply this characteristic to any series circuit.

How does that help us? Look back at the several steps using basic laws that were used to prove this statement. We can use this result to streamline solving series circuits. Let's solve for the circuit current of Example 8-3 again but this time use the relationship of Equation 8-1. The previous example required many, many steps to find the current I. This method uses basically only two steps to find the circuit current.

Example 8-4

For the circuit of Figure 8-12(a), find the current I using the total circuit resistance R_T approach. Draw the equivalent R_T model of the circuit.

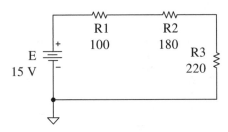

Figure 8-12(a) Example series circuit

Solution

Since this is a series circuit, the total circuit resistance R_T is the sum of its individual resistances.

$$R_T = R1 + R2 + R3$$

or

$$R_T = 100\ \Omega + 180\ \Omega + 220\ \Omega = 500\ \Omega$$

Figure 8-12(b) is the R_T circuit model of this series circuit. Since there is only one resistance in this series circuit, the voltage drop across R_T must be the supply voltage E.

$$E = 15\ V$$

Now we can simply use Ohm's law to find the circuit current I.

$$I = \frac{E}{R_T} = \frac{15\ V}{500\ \Omega} = 30\ mA$$

This is the same result, and it was achieved much more quickly, with fewer steps than the previous example. Always select the

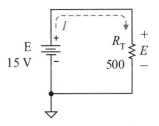

Figure 8-12(b)
Series circuit model
with R_T

most streamlined process to solve a circuit. Fewer steps introduce fewer opportunities to make mistakes.

Practice

For the circuit of Figure 8-13, find total circuit resistance, then the circuit current.

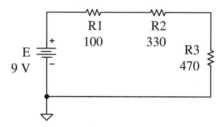

Figure 8-13 Practice circuit

Answer: 900 Ω, 10 mA

Effect of Changing Series Circuit Resistance

What is the overall effect if resistance is changed in a series circuit, assuming the supply voltage remains fixed? Ohm's law provides the answer to this question.

$$I = \frac{E}{R_T}$$ (Ohm's law)

Based upon Ohm's law, if the resistance R_T of a series circuit is increased, the circuit's current must decrease. So, as R_T goes up, circuit current I must come down.

Adding additional resistors in series causes the circuit current to decrease. This reduced current, in turn, causes individual resistor voltage drops to be reduced. *As resistors are added to a series circuit, the circuit current is reduced and the voltage drop across each resistor is reduced.*

Example 8-5

The resistance of a series circuit is doubled. What happens to its circuit current?

Solution

Assume that the original circuit's resistance is R_T. Then the new circuit's resistance is twice as much, or $2R_T$. The new circuit current is

$$I_{new} = \frac{E}{2R_T} = \frac{1}{2}\left(\frac{E}{R_T}\right) = \frac{1}{2}I$$

So, if the series resistance doubles, the current is halved.

Practice

The resistance of a series circuit is halved. What happens to its circuit current?

Answer: Circuit current doubles

8.4 Current Meter Loading

If a current meter is placed in a circuit and it actually changes the current reading, then there is current meter loading.

Current meters are very, very low resistance instruments. Ideally the current meter can be modeled as an ideal conductor with $0\ \Omega$ of resistance (equivalent to a *short*). Since current flows inside the wire, the circuit must be broken at a component terminal or wire terminal and the current meter inserted in series with the circuit as shown in Figure 8-14. Writing the expression for the total circuit resistance R_T of this series circuit, including the meter resistance R_{meter}, the total circuit resistance is

$$R_T = R_{meter} + R1 + R2 \tag{8-2}$$

Assuming the current meter has a very low resistance, under what conditions could the meter itself start affecting the circuit resistance and cause measuring errors? If the value of R1 plus R2 is much greater than R_{meter}, then the total circuit resistance is not greatly affected by the meter resistance.

$$R1 + R2 \gg R_{meter}$$

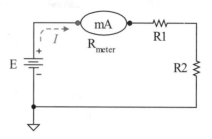

Figure 8-14 Series circuit with milliameter

Thus, assuming no loading effect, Equation 8-2 can be approximated as

$$R_T \approx R1 + R2$$

However, if the value of R1 plus R2 is very low and comparable to R_{meter}, then R_{meter} must be included. In this case, there is a real possibility that the current meter is loading (changing) the circuit current.

Example 8-6

The resistance of the current meter in Figure 8-14 is 1 Ω. The supply voltage E is 12 V and resistors R1 and R2 are 1 kΩ each. Find the circuit current including the meter resistance and then excluding the meter resistance. Calculate the percent error and compare. Is the current meter loading the circuit?

Solution

Including the meter resistance, the total circuit resistance is

$$R_T = 1\ \Omega + 1\ k\Omega + 1\ k\Omega = 2.001\ k\Omega$$

By Ohm's law, the circuit current is

$$I = \frac{E}{R_T} = \frac{12\ V}{2.001\ k\Omega} = 5.997\ mA$$

Excluding the meter resistance by assuming an ideal current meter of 0 Ω, the total circuit resistance is

$$R_T \approx 1\ k\Omega + 1\ k\Omega = 2\ k\Omega$$

So, by Ohm's law the circuit current is ideally

$$I = \frac{E}{R_T} = \frac{12 \text{ V}}{2 \text{ k}\Omega} = 6 \text{ mA}$$

The percent error using an ideal current meter for this circuit is very, very small and reasonable, much less than 1%.

$$Percent\ error = \frac{I_{approximate} - I_{ideal}}{I_{ideal}} \times 100\ \% = \frac{5.997 \text{ mA} - 6 \text{ mA}}{6 \text{ mA}} \times 100\ \% = -0.05\ \%$$

The tolerance of the resistors causes more error in calculating the expected circuit values. There is no significant current meter loading effect for this circuit since

$$R1 + R2 \gg R_{meter}$$

Practice

The resistance of the current meter in Figure 8-14 is 1 Ω. The supply voltage E is 12 V and resistors R1 and R2 are both 10 Ω each. Find the circuit current including the meter resistance and then excluding the meter resistance. Calculate the percent error and compare. Is the current meter loading the circuit?

Answer: 571.4 mA, 600 mA, − 4.77% (meter loading is a problem)

For low resistance circuits, beware of possible current meter loading effects.

8.5 VDR–Voltage Divider Rule

In a series circuit, the current is the same through every component and connecting wire. What about the voltage drops across the resistors? They are different for different resistances based upon Ohm's law for resistance.

$$V_R = R I \qquad \text{(Ohm's law)}$$

In a series resistive circuit, where the current *I* is the same, the supply voltage *divides its voltage among the individual resistances*.

- Equal resistances drop the same voltage.

- Larger resistances drop more voltage.

- Smaller resistances drop less voltage.

How does the voltage divide up among series resistors? A resistor's voltage drop depends upon the applied supply voltage, the total circuit resistance, and the resistor's resistance. Let's use the circuit of Figure 8-15 to quantify this *voltage dividing effect* and find the fraction of the supplied voltage E dropped across R1.

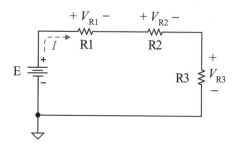

Figure 8-15 Series circuit resistors divide up the applied voltage

First, write the Ohm's law expression for R1.

$$V_{R1} = (R1)\, I_{R1} \qquad \text{(Ohm's law for } V_{R1})$$

Since this is a series circuit, the current is the same and can be written in terms of the circuit current I.

$$V_{R1} = (R1)\, I \tag{8-3}$$

We also know from a circuit perspective that Ohm's law for the total circuit resistance R_T can be rewritten to solve for the circuit current I

$$I = \frac{E}{R_T} \tag{8-4}$$

where $R_T = R1 + R2 + R3$.

Substituting the circuit current I of Equation 8-4 into Equation 8-3 for the resistor, the resistor voltage expression for V_{R1} becomes

$$V_{R1} = (R1)\frac{E}{R_T} \qquad \text{(substitution)}$$

Rearrange the equation to express the resistance as a fraction of the supply voltage E.

$$V_{R1} = \left(\frac{R1}{R_T}\right)E \qquad\qquad (8\text{-}5)$$

VDR is valid only for series circuits where the current is the same

Equation 8-5 is called the **voltage divider rule** (VDR) for *series circuits*. It expresses the voltage across a resistance as a resistive fraction of the supplied voltage. The advantage of this rule is that the resistor voltage drop can be found in one step without finding the resistor current first.

This rule can be applied to a set of resistors in a series circuit as well. For example, the voltage drop across the series combination of R1 and R2 in Figure 8-15 can be found as

$$V_{R12} = \left(\frac{R1+R2}{R_T}\right)E \qquad\qquad (8\text{-}6)$$

where $R_T = R1 + R2 + R3$.

The major advantage of the voltage divider rule is that the voltage drops across series circuit resistor combinations can be found by using only the known resistances and the applied voltage. The circuit current does not need to be explicitly calculated. The current calculation is actually buried inside the voltage divider rule with the term E/R_T but it is not explicitly and separately calculated.

The current direction is still used to determine voltage polarities of the resistor drops. The VDR only determines magnitude. It does not determine polarity.

Example 8-7

Using the voltage divider rule, find the voltage drop across each of the resistors in the circuit of Figure 8-16.

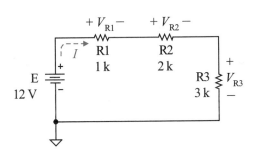

Figure 8-16 Example circuit

Also, use VDR to find the voltage drop across the series combination of R2 and R3. Note, the current I is used to establish resistor voltage polarities, but the current is not explicitly calculated when applying VDR.

Solution

Find the total circuit resistance R_T.

$$R_T = 1\,k\Omega + 2\,k\Omega + 3\,k\Omega = 6\,k\Omega$$

Use Equation 8-5 to find the voltage drop across each resistor.

$$V_{R1} = \frac{R1}{R_T} \times E = \left(\frac{1\ k\Omega}{6\ k\Omega}\right) 12\ V = 2\ V$$

$$V_{R2} = \frac{R2}{R_T} \times E = \left(\frac{2\ k\Omega}{6\ k\Omega}\right) 12\ V = 4\ V$$

$$V_{R3} = \frac{R3}{R_T} \times E = \left(\frac{3\ k\Omega}{6\ k\Omega}\right) 12\ V = 6\ V$$

Note: 1/6 of E dropped across R1; 1/3, across R2; 1/2, across R3. *All the fractions must add up to 1 to account for all of the drops.* Use Equation 8-6 to find the voltage drop across R2 and R3.

$$V_{R23} = \frac{R2 + R3}{R_T} \times E = \left(\frac{2\ k\Omega + 3\ k\Omega}{6\ k\Omega}\right) 12\ V = 10\ V$$

Practice

For the circuit of Figure 8-16, use the VDR to find the drop across the series circuit combination of R1 and R2.

Answer: 6 V

Refer back to the last example and the voltage drops across the resistances. A simple pattern emerges. If the resistance is doubled, the voltage drop is doubled. If the resistance is tripled, the voltage drop is tripled. *The voltage drops across series circuit resistances is linearly proportional to the individual resistances.*

- 1 kΩ dropped 2 V

- 2 kΩ dropped 4 V

- 3 kΩ dropped 6 V

Example 8-8

In a series circuit a 1 kΩ resistor drops 2 V. Find the voltage dropped by a series 3.3 kΩ resistor.

Solution

The 3.3 kΩ resistor is 3.3 times as great as the 1 kΩ resistance. So the voltage drop across the 3.3 kΩ resistor is

$$V_{3.3k} = \left(\frac{3.3 \ k\Omega}{1 \ k\Omega} \right) 2 \ V = 6.6 \ V$$

Practice

In a series circuit a 2.2 kΩ resistor drops 4 V. Find the voltage dropped by a series 3.3 kΩ resistor.

Answer: 6 V

The next example applies the voltage divider rule to the bubble notation form. Again, the bubble notation is sometimes a convenient way to represent dc supplies in a more complex schematic.

Example 8-9

For the circuit of Figure 8-17, find the total voltage E_{total} being applied to the circuit. Use VDR to find the voltage dropped across R1. Then use VDR to find the voltage dropped across the series combination of R1 and R2.

Solution

The net voltage applied to the circuit is the voltage difference between node a and node d (+12 V and –12 V supplies).

$$E_{net} = V_a - V_d = 12 \ V - (-12 \ V) = 24 \ V$$

So, a total or net of 24 V is applied to the three resistors in series. This voltage is divided up among the three resistors based upon the voltage divider rule. The total circuit resistance R_T is

$$R_T = 1 k\Omega + 2 \ k\Omega + 3 \ k\Omega = 6 \ k\Omega$$

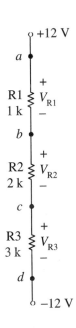

Figure 8-17
Example circuit

The voltage across R1 is

$$V_{R1} = \left(\frac{1\ k\Omega}{6\ k\Omega}\right) 24\ V = 4\ V$$

The voltage across the series combination of R1 and R2 is

$$V_{R12} = \left(\frac{1\ k\Omega + 2\ k\Omega}{6\ k\Omega}\right) 24\ V = 12\ V$$

Practice

For the series circuit of Figure 8-17 find V_{R2}, then use VDR to find the voltage across the combination of R2 and R3 in series.

Answer: 8 V, 20 V

BJT Voltage Divider Bias

The next example is a very popular BJT circuit called the voltage divider bias circuit. This circuit is shown in Figure 8-18. The β (current gain) of a BJT can vary widely (for example, from 50 to 250) for the same BJT model. To offset this variation and nullify its effect, the voltage divider bias circuit is typically designed so that the base current I_B is very small compared to its feeder current I_{R1} coming from the supply as shown in Figure 8-18(a). This design, combined with a BJT $\beta > 50$, creates a β **independent circuit**.

$$I_{R1} \gg I_B \qquad \text{(design criteria)}$$

Kirchhoff's current law relates the three currents at node b.

$$I_{R1} = I_{R2} + I_B \qquad \text{(KCL)}$$

Therefore, we can make the following approximation.

$$I_{R1} \approx I_{R2}$$

Since these currents are approximately the same, this is *essentially a series circuit* with the supply voltage splitting between R1 and R2. To start analyzing this circuit the first step is to find V_B, the dc voltage at the base of the BJT relative to common. Using the voltage divider rule,

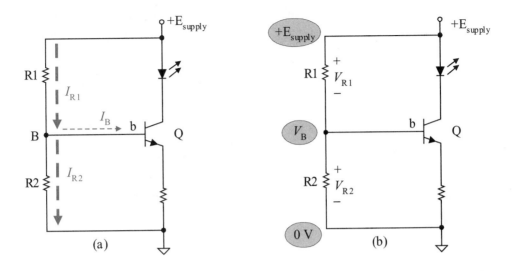

Figure 8-18 Voltage divider bias BJT circuit

$$V_B = V_{R2} \approx \left(\frac{R2}{R1+R2} \right) E_{supply}$$

Once we have V_B, the rest of the circuit values can be found using fundamental circuit analysis techniques.

Example 8-10

The circuit of Figure 8-19(a) has been designed to be β independent with $I_{R1} \gg I_B$ and $\beta > 50$. Find the base voltage V_B using the voltage divider rule. Then find the emitter voltage V_E, the emitter current I_E, and finally the LED and collector currents. Assume that this is a silicon BJT.

Solution

Since $I_{R1} \gg I_B$, then $I_{R1} \approx I_{R2}$. Thus, R1 and R2 form an *essentially series circuit* and we can apply the voltage divider rule.

$$V_B = V_{R2} \approx \left(\frac{R2}{R1+R2} \right) E_{supply} = \left(\frac{3.3 \text{ k}\Omega}{3.3 \text{ k}\Omega + 10 \text{ k}\Omega} \right) 20 \text{ V} = 5 \text{ V}$$

Now that we know V_B, we can find the rest of the circuit values using basic circuit analysis. Refer to Figure 8-19(b).

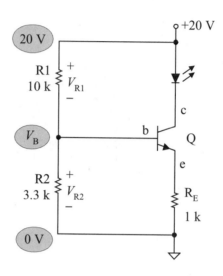

Figure 8-19(a) Example circuit

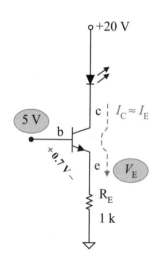

Figure 8-19(b)
Example circuit–
find V_E, V_{RE}, I_E, I_C, I_{LED}

Next we can find V_E. Since this is a silicon BJT, V_{BE} must be about 0.7 V. Start with V_B of 5 V and drop 0.7 V in going from the base to the emitter,

$$V_E = 5 \text{ V} - 0.7 \text{ V} = 4.3 \text{ V}$$

The top of R_E is at 4.3 V (emitter) and the bottom of R_E is at 0 V (analog common), so

$$V_{RE} = 4.3 \text{ V} - 0 \text{ V} = 4.3 \text{ V}$$

By Ohm's law the current through R_E is

$$I_{RE} = \frac{4.3 \text{ V}}{1 \text{ k}\Omega} = 4.3 \text{ mA}$$

By KCL (series current)

$$I_E = I_{RE} = 4.3 \text{ mA}$$

Since $\beta > 50$, then $I_B \ll I_C$. So,

$$I_C \approx I_E = 4.3 \text{ mA}$$

By KCL (series current),

$$I_{LED} = I_C \approx 4.3 \text{ mA}$$

Notice that we met our goal of designing a β *independent circuit*. We calculated circuit values without calculating the base current or knowing the BJT current gain β. The circuit design was successful, but what about our assumptions? Let's check them out. Let's assume the worst case with a BJT β of only 50. Then

$$I_B < \frac{I_C}{\beta} \approx \frac{4.3 \text{ mA}}{50} = 86 \text{ μA}$$

The voltage divider network current is

$$I_{R1} \approx I_{R2} = \frac{5 \text{ V}}{3.3 \text{ k}\Omega} = 1.51 \text{ mA}$$

Yes, our original assumption is satisfied.

$$I_{R1} \gg I_B$$

$$1.51 \text{ mA} \gg 86 \text{ μA}$$

Figure 8-19(c) is a simulation of the example problem. Table 8-1 shows that an error of only about 2% exists between the calculated and simulated values. The only major difference is the base current I_B because we assumed a BJT β of 50 while MultiSIM assumed 200. However, the circuit is proven to be β independent and thus the output collector current is solidly set at about 4.3 mA in spite of the differences in the assumed β.

Practice

The supply voltage of the circuit in Figure 8-19(a) is changed from 20 V to 28 V. Assuming the circuit is β independent, find V_B, V_E, V_{RE}, I_{RE}, I_E, I_C, and I_{LED}.

Answer: 7 V, 6.3 V, 6.3 V, 6.3 mA, 6.3 mA, 6.3 mA, 6.3 mA

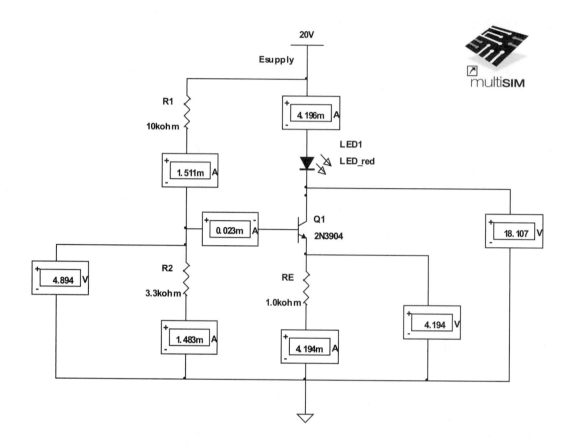

Figure 8-19(c) MultiSIM simulation of example BJT circuit

Table 8-1 Compare calculated and simulated calculations

Voltage or Current	Calculated Value	Simulated Value
V_B (V)	5	4.894
V_E (V)	4.3	4.194
$I_E \approx I_C$ (mA)	4.3	4.19
I_B (mA)	0.086	0.023
I_{R1} (mA)	1.51	1.511
I_{R2} (mA)	1.51	1.483

MOSFET Voltage Divider Bias

The next example uses a voltage divider network to bias the gate-to-source voltage of a MOSFET.

Example 8-11

For the circuit of Figure 8-20, find the relationship of R1 and R2 needed to produce a load current of 1 A. From the MOSFET's characteristic curve, I_D is 1 A when V_{GS} is 4 V. Suggest a set of values for R1 and R2 to produce the 1 A load current.

Solution

First, the drain current I_D and the load current must be the same since they are in series (by KCL). So,

$$I_{load} = I_D = 1 \text{ A}$$

Based upon the MOSFET's characteristic curve, a drain current of 1 A is produced from a gate-to-source voltage of

$$V_{GS} = 4 \text{ V}$$

Using a KVL walkabout from $V_{R2}(-)$ to $V_{R2}(+)$ shows that

$$V_{R2} = V_{GS} = 4 \text{ V}$$

Recall that a key characteristic of an E-MOSFET is that it is an insulated gate device and that its gate current I_G is for all practical purposes 0 A.

$$I_G = 0 \text{ A}$$

So, by KCL,

$$I_{R1} = I_{R2}$$

Thus, R1 and R2 form an *essentially series circuit*. Therefore we can apply the voltage divider rule to it. Applying the VDR to R1, R2, and the supply voltage produces the following expression.

$$V_{R2} = \left(\frac{R2}{R1 + R2} \right) E_{supply}$$

Substitute the known voltages.

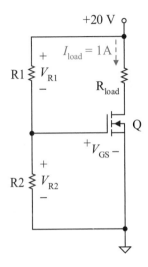

Figure 8-20
Example circuit

$$4\,V = \left(\frac{R2}{R1+R2}\right)20\,V$$

This simplifies to

$$R1 + R2 = 5(R2)$$

Which simplifies to

$$R1 = 4(R2)$$

Thus, we must select two resistances with R1 being 4 times the resistance of R2. Also, to prevent loading effects, the resistances should be reasonably high. One example from the Appendix C resistor color code table is

$$R1 = 39\ k\Omega \quad \text{and} \quad R2 = 10\ k\Omega$$

Practice

Repeat the example with the supply voltage changed to 24 V.

Answer: R1 is 5 times the value of R2; example, 51 kΩ and 10 kΩ

8.6 Real Voltage Supply Model

We have been working with only ideal voltage supplies to this point. An ideal voltage supply supplies a fixed dc voltage under various load conditions (that is, with various load currents). An ideal voltage supply is assumed to have an internal resistance of 0 Ω. But, in the practical world, all supplies have some internal resistance even though it may be very low. The lower the internal resistance, the better the supply and the closer it is to the ideal supply model.

Ideal Voltage Supply Model

Figure 8-21 demonstrates the effects of an ideal voltage supply. The ideal 12 V supply *delivers all* 12 V to the load resistance whether the supply is

    a. *Unloaded* at 0 A (*no load,* the load is an open)

    b. *Lightly loaded* at 1 mA (with a 12 kΩ load resistance)

    c. *Heavily loaded* at 1 A (with a 12 Ω load resistance)

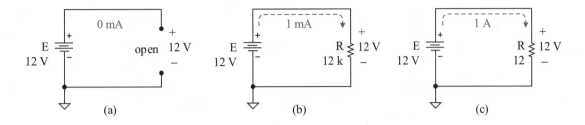

Figure 8-21 Ideal voltage supply delivers fixed voltage

In all three cases, Kirchhoff's voltage law confirms that all 12 V are delivered to the load.

$$V_R = E = 12 \text{ V} \qquad\qquad \text{(KVL)}$$

The circuit current is then calculated based upon Ohm's law for resistance.

$$I = \frac{12 \text{ V}}{\infty \ \Omega} = 0 \text{ A} \qquad \text{(no load or unloaded)}$$

$$I = \frac{12 \text{ V}}{12 \text{ k}\Omega} = 1 \text{ mA} \qquad \text{(lightly loaded)}$$

$$I = \frac{12 \text{ V}}{12 \ \Omega} = 1 \text{ A} \qquad \text{(heavily loaded)}$$

Never Short a Voltage Supply

Never place a short or extremely low resistance across a voltage supply. Why not? Figure 8-22 demonstrates the problem of trying to drive a short or a very low resistance with a voltage supply.

Initially, all 12 V of the supply appears to drop across the conducting wire and short (ideally the circuit resistance is 0 Ω). Applying Ohm's law for resistance, the circuit current appears to be infinite.

$$I = \frac{12 \text{ V}}{0 \ \Omega} = \infty \text{ A} \qquad \text{(shorted power supply)}$$

In reality the conductor has a very low resistance which would generate amps of current, not infinite current. For example, suppose the connect-

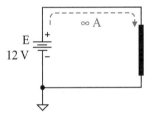

Figure 8-22
Never drive a short with a voltage supply

A voltage supply drops to 0 V when driving a *short circuit*

ing wire and short combined for a resistance of 0.1 Ω; then initially the current would attempt to go to 120 A. That current is still excessive though not infinite. *This excessive current can be dangerous.* For example, shorting out a car battery could cause excessive current, battery overheating, and a possible dangerous explosion of acid. Electronic voltage supplies can be designed to limit current to a maximum value to protect against excessive circuit currents. Such supplies are called **current limited supplies**.

Another feature of driving a short circuit is that the voltage supply output drops its output voltage down close to 0 V. The power supply cannot really support such high currents and typically its *voltage drops close to 0 V*.

Real Voltage Supply Model

The ideal voltage supply has 0 Ω of internal resistance. Again, more realistically, every voltage supply has some internal resistance.

Figure 8-23 is a model of a real voltage supply that incorporates its two main features, its *ideal voltage supply* (E_{supply}) and its *internal resistance* (R_{supply}).

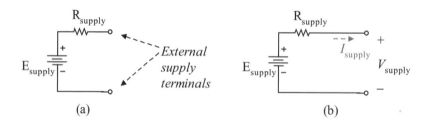

(a)   (b)

Figure 8-23 Model of a real voltage supply

Figure 8-23(a) emphasizes that the only access to the supply is its external terminals. We cannot go inside a voltage supply and directly measure its E_{supply} and R_{supply}. For example, you cannot go inside a D cell battery and measure these values. Our only access to the D cell is its positive and negative terminals. Figure 8-23(b) provides the terminology we associate with a real voltage supply. V_{supply} represents the voltage at the supply's terminals and I_{supply} represents the current delivered by the supply to an attached load.

Unloaded or No Load Voltage

Figure 8-24 represents an *unloaded* power supply, that is, no load is attached to the power supply. Without a load, the supply sees an open circuit as its load. In this case, the voltage supplied V_{supply} is the same as its ideal supply E_{supply}. Why?

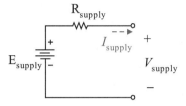

Figure 8-24   Unloaded power supply

$$I_{supply} = 0 \text{ A} \qquad \text{(open circuit)}$$

$$V_{Rsupply} = 0 \text{ A} \times R_{supply} = 0 \text{ V} \qquad \text{(Ohm's law)}$$

$$V_{supply} = E_{supply} \qquad \text{(KVL walkabout)}$$

To identify this special *unloaded* condition, we label the supply voltage as the **voltage no load** V_{NL} condition. This is also labeled the *open circuit* voltage V_{OC} since the load is an open.

$$V_{NL} = V_{OC} = E_{supply}$$

V_{NL} represents the voltage no load condition for a power supply

The next example is a practical example of measuring the no load voltage.

Example 8-12

You take the D cell battery out of your flashlight and measure it with a DMM voltmeter. It measures 1.55 V. What is the battery's unloaded or open circuit voltage? What is the value of the ideal voltage supply model E_{supply}?

Solution

Since you are measuring the battery without a load (assuming an ideal DMM voltmeter), your measurement is the unloaded or the open circuit voltage. It is the same value as the supply's ideal voltage supply E_{supply}.

$$V_{NL} = V_{OC} = E_{supply} = 1.55 \text{ V}$$

Practice

(a) In lab, you want to set up your electronic dc supply so that its ideal voltage E_{supply} is 12 V. Should you adjust the supply's voltage with the load attached or unattached? (b) Suppose you want the actual voltage delivered to the load to be 12 V. Should you adjust the supply's voltage with the load attached or unattached?

Answer: (a) unattached, (b) attached

Loaded Power Supply

Figure 8-25 represents a power supply with a resistive load. The power supply is modeled as an ideal source E_{supply} in series with the supply's internal resistance R_{supply}. R_{load} is attached to the supply, and its associated current and voltage are shown. Note that this is a simple series circuit with two resistances in series. All we need to do to solve these types of circuits is apply the basic laws of circuit analysis and the characteristics of a series circuit.

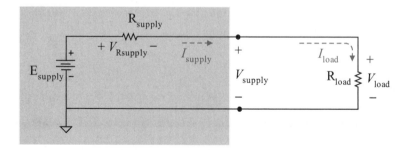

Figure 8-25 Loaded power supply

Since this is a series circuit (or by KCL) the current through every component is the same.

$$I_{supply} = I_{load}$$

The supply voltage actually delivered to the load is now V_{supply}. It is no longer the ideal supply voltage E_{supply}. Since the supply has internal resistance, the voltage supplied to the load must be less than the ideal voltage supply E_{supply}. Do a clockwise KVL voltage walkabout from the $V_{supply}(-)$ terminal to the $V_{supply}(+)$ terminal.

$$V_{supply} = +E_{supply} - V_{Rsupply}$$

Applying Ohm's law to R_{supply}, the voltage delivered by a real supply is

$$V_{supply} = E_{supply} - I_{supply} R_{supply}$$

Another way to view this circuit is that the ideal supply voltage must voltage divide (VDR) between the load resistance and the supply resistance. Thus only a fraction of E_{supply} is delivered to the load.

$$V_{supply} = \left(\frac{R_{load}}{R_{load} + R_{supply}} \right) E_{supply}$$

Ideally $R_{load} \gg R_{supply}$, then $V_{supply} \approx E_{supply}$

Example 8-13

A power supply is specified to have an unloaded voltage of 10 V and an internal resistance of 0.1 Ω. Draw the voltage supply model of this power supply with an attached load resistance of 100 Ω. Then find the load current and load voltage. How much voltage is dropped internally in the voltage supply?

Solution

The modeled voltage supply with the attached 100 Ω load is shown in Figure 8-26.

Find the total circuit resistance of this series circuit.

$$R_T = R_{supply} + R_{load} \qquad \text{(series circuit)}$$

$$R_T = 0.1\,\Omega + 100\,\Omega = 100.1\ \Omega$$

Find the load current using Ohm's law with the ideal supply voltage and total circuit resistance.

$$I_{load} = I_{supply} = \frac{10\text{ V}}{100.1\,\Omega} = 99.90\text{ mA}$$

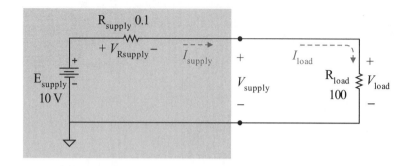

Figure 8-26 Example circuit

Find the load current using Ohm's law for the load resistance.

$$V_{load} = R_{load}\, I_{load}$$

$$V_{load} = 100\ \Omega \times 99.90\ \text{mA} = 9.99\ \text{V}$$

The voltage lost inside the voltage supply across R_{supply} is the difference between the ideal supply voltage and the delivered load voltage.

$$V_{Rsupply} = 10\ \text{V} - 9.99\ \text{V} = 0.01\ \text{V}$$

This internally lost supply voltage can also be found using Ohm's law.

$$V_{Rsupply} = 0.1\ \Omega \times 99.90\ \text{mA} = 0.00999\ \text{V} = 0.01\ \text{V}$$

As the supply current increases, the supply's internal R_{supply} voltage drop increases, decreasing the supply's output voltage.

Practice

Repeat the previous example but with an internal supply resistance of 10 Ω. Find total circuit resistance, load current, load voltage, and the supply's internal drop.

Answer: 110 Ω, 90.91 mA, 9.091 V, 0.909 V

Maximum Load Current–Full Load

A voltage supply is rated to have a maximum supply current that it can provide. The maximum current is limited naturally (e.g., a battery), may be protected by a fuse or circuit breaker, or may have an electronic circuit that limits the supply current to a maximum while reducing the voltage supplied automatically (current limiting).

The term **voltage full load** V_{FL} is used to indicate the voltage of a supply under full load (maximum supply current) condition.

V_{FL} represents the voltage full load condition (supply at its maximum current rating)

Example 8-14

A power supply is rated for a maximum of 18 V at 1 A. Find the minimum resistance that can be connected to this power supply with an 18 V output. Also, find the power dissipated by that resistor.

Solution

Using Ohm's law for resistance, find this minimum resistance that produces the maximum load current.

$$R_{minimum} = \frac{V_{supply}}{I_{maximum}} \qquad \text{(Ohm's law)}$$

$$R_{minimum} = \frac{18 \text{ V}}{1 \text{ A}} = 18 \text{ } \Omega$$

We can find either the power delivered by the supply or the power dissipated by the resistor; they are the same. The power delivered by the supply is

$$P_{delivered} = V_{supply} \, I_{supply} \qquad \text{(power rule)}$$

$$P_{delivered} = 18 \text{ V} \times 1 \text{ A} = 18 \text{ W}$$

The power dissipated by the load resistance is

$$P_{load} = V_{load} \, I_{load}$$

$$P_{dissipated} = 18 \text{ V} \times 1 \text{ A} = 18 \text{ W}$$

or

$$P_{dissipated} = \frac{(18 \text{ V})^2}{18 \text{ } \Omega} = 18 \text{ W}$$

or

$$P_{dissipated} = (1 \text{ A})^2 \ 18 \ \Omega = 18 \text{ W}$$

Practice

The above 18 V power supply is current limited to 1 A. If a resistance lower than 18 Ω is attached, the power supply drops its output voltage to maintain a current of 1 A in the load.

(a) A 10 Ω resistor is attached to the above power supply. What does the supply voltage become to maintain the 1 A maximum current? (b) A 1 Ω resistor is attached to the above power supply. What does the supply voltage become to maintain the 1 A maximum current? (c) A short is attached to the above power supply. What does the supply voltage become to maintain the 1 A maximum current?

Answer: (a) 10 V (b) 1 V (c) 0 V

Note that in the last practice problem the output supply voltage dropped to 0 V if a *short* was attached to the power supply. If a voltage supply drives a short circuit, its output voltage drops to 0 V since it cannot produce infinite current.

Percent Load Voltage Regulation

How good is a power supply? Ideally the supply would maintain the same output voltage under varying load conditions. That is, the no load output voltage V_{NL} would equal its full load output voltage V_{FL}. In our previous example, the output supply ideally maintains an 18 V output whether no load is attached (0 A) or the supply is fully loaded (1 A).

$$V_{FL} = V_{NL} \qquad \text{(ideal voltage supply)}$$

A practical measure of a voltage supply's performance with varying load conditions is to compare the *no load voltage* V_{NL} and *full load voltage* V_{FL}. The **percent load voltage regulation** of a power supply is defined to be

$$\% \ VR = \frac{V_{NL} - V_{FL}}{V_{FL}} \times 100 \ \% \qquad \textbf{(8-7)}$$

Thus, the percent load voltage regulation for an *ideal voltage supply* is 0%. The larger the *% VR*, the greater the variation in the voltage delivered to load under varying load conditions.

Example 8-15

A power supply delivers 18.00 V unloaded and 17.98 V fully loaded at 1 A. Find its percent load voltage regulation. Also, find the minimum load resistance, the supply's ideal supply voltage E_{supply}, and its internal resistance R_{supply}.

Solution

Use Equation 8-7 to find the percent load voltage regulation.

$$\% \ VR = \frac{18 \ V - 17.98 \ V}{17.98 \ V} \times 100 \ \% = 0.11 \ \%$$

The relative change in supply voltage over the extremes of its operating load conditions (from no load to full load) is very low. Whether this is good enough depends upon the application. If this is a power supply for an inexpensive radio, this is more than adequate. If this is a power supply for a precision circuit that is navigating a space shuttlecraft around the moon, then it may not be adequate. You must evaluate the impact of the variation in the dc voltage supply on the circuits affected.

Use Ohm's law to find the minimum load resistance. We are given the full load current and full load voltage drop.

$$R_{load} = \frac{V_{load}}{I_{load}} = \frac{17.98 \ V}{1 \ A} = 17.98 \ \Omega$$

To find the internal resistance of the power supply, we need to draw the modeled circuit of the power supply under (a) no load conditions to find E_{supply} and then (b) full load conditions to perform circuit analysis and find R_{supply}.

The schematic of Figure 8-27(a) depicts this power supply under no load conditions. The *unloaded output voltage* is given as 18 V. The supply current must be 0 A since the load is an open circuit. The internal supply resistance R_{supply} must drop 0 V since its current is 0 A (Ohm's law). Perform a counterclockwise KVL voltage walkabout from $E_{supply}(-)$ to $E_{supply}(+)$ to verify that E_{supply} must be 18 V. That is,

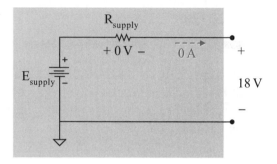

Figure 8-27(a) Example circuit–unloaded power supply model

$$E_{supply} = V_{NL} = 18 \text{ V} \qquad \text{(unloaded)}$$

This is always true. The no load voltage is the ideal supply voltage since the internal resistance does not drop any voltage when the supply is unloaded.

Now that we have the ideal supply voltage E_{supply} we can draw the power supply circuit model under full load conditions to find the supply's internal resistance. Figure 8-27(b) represents the voltage supply under full load conditions: I_{load} is 1 A and V_{load} is 17.98 V.

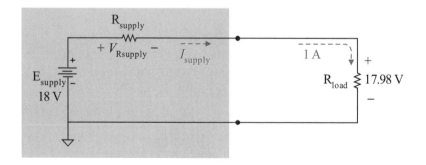

Figure 8-27(b) Example circuit–fully loaded supply

First, use a KVL voltage walkabout to find the voltage drop across the supply's internal resistance $V_{Rsupply}$.

$$V_{\text{Rsupply}} = -17.98 \text{ V} + 18 \text{ V} = 0.02 \text{ V} \qquad \text{(walkabout)}$$

Second, note that the supply current is equal to the load current since this is a series circuit.

$$I_{\text{supply}} = I_{\text{load}} = 1 \text{ A} \qquad \text{(series circuit)}$$

Now we can apply Ohm's law for resistance to find the internal supply resistance.

$$R_{\text{supply}} = \frac{V_{\text{Rsupply}}}{I_{\text{supply}}} = \frac{0.02 \text{ V}}{1 \text{ A}} = 0.02 \text{ }\Omega$$

A supply resistance in the lower milliohm range is very typical of a good voltage supply.

Practice

An aging D cell battery has an unloaded voltage measurement of 1.6 V and a full load voltage measurement of 1.4 V at 1 A of current. Find its percent load voltage regulation. Also, find the load resistance R_{load} needed to produce the 1 A current. Find the supply's ideal voltage E_{supply} and its internal resistance R_{supply}. Is a no load test sufficient to determine if a battery is good?

Answer: 14.3% (not good), 1.4 Ω, 1.6 V, 0.2 Ω, no (you must test a battery under load conditions)

8.7 Thévenin's Theorem

Thévenin's theorem could be appropriately nicknamed "What's in the box?" For our purposes at this juncture, Thévenin's theorem essentially states that *any linear bilateral network may be modeled as an ideal voltage supply in series with a resistance*. A linear network consists of linear components such as resistors and dc voltage supplies. A bilateral network operates in the same manner regardless of the supplied voltage polarity or current direction. Again, resistors and supplies satisfy this criterion, while a diode is polarity sensitive and is not bilateral. Figure 8-28 demonstrates this theorem. Figure 8-28(a) represents a complicated circuit and Figure 8-28(b) represents its simple Thévenin model equivalent.

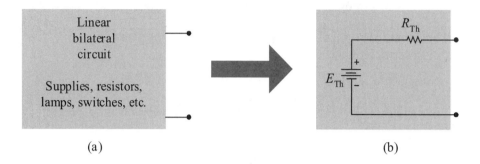

Figure 8-28 Thévenin model equivalent of a circuit

Does the Thévenin model look familiar? Absolutely. The Thévenin model is the same model we have been using to represent the real voltage power supply, that is, an ideal voltage supply in series with an internal resistance. A real voltage supply is a circuit of many components, but we have simplified it down to its Thévenin model equivalent to perform circuit analysis. The model represents the actual circuit in terms of all possible load conditions. Once the model is established, it is easier to work with under various load conditions than an original complicated circuit.

What's in the Box?

This is a very powerful theorem. Circuits that are very complicated can be reduced to a simple two-terminal circuit with an ideal voltage supply in series with a resistance. For now, we are going to take a "What's in the box?" approach. Later in this textbook, when appropriate, circuits are converted to their Thévenin model to facilitate analysis.

Measuring E_{Th}

So there is an electronic gizmo with a pair of output terminals where the load is attached (like the power supply). How do we find its Thévenin voltage and Thévenin resistance so we can model it? First, how would you go about finding the Thévenin voltage E_{Th} of Figure 8-28(a)? You would use a voltmeter if you want to measure voltage. You cannot go inside the circuit and measure this voltage; you can only work at the terminals of the original circuit of Figure 8-28(a). Since the model is an equivalent circuit, how would you measure the E_{Th} for the circuit of Figure 8-28(b)? As we did for the real power supply measurement, just measure the open circuit voltage at the terminals. Since there is no current with the unloaded circuit, the voltmeter measures the value of E_{Th} as shown in Figure 8-29. This assumes that a reasonably good voltmeter is used that does not load the circuit. Good voltmeters have very high internal resistances (1 MΩ or higher). If R_{Th} is a high resistance on the order of the voltmeter's internal resistance, then the voltmeter could load the circuit and cause a measurement error.

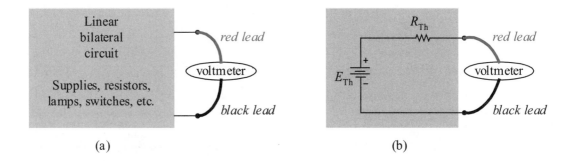

Figure 8-29 Measuring E_{Th}

The real measurement of E_{Th} requires that you connect the voltmeter to the actual circuit as shown in Figure 8-29(a). Figure 8-29(b) demonstrates with the model that you are in fact measuring the corresponding equivalent E_{Th} value of the real circuit. Figure 8-29(b) clarifies that the open circuit measurement produces no current through R_{Th} and thus no voltage drop across R_{Th}. Thus by KVL, the voltmeter is truly measuring E_{Th}.

Thus to measure the E_{Th} of a circuit, identify the pair of terminals that you wish to model without the load attached (open circuit) and measure the voltage across the terminals. This is a no load or open circuit voltage measurement (V_{NL} or V_{OC}).

$$E_{Th} = V_{NL} = V_{OC}$$

Measuring/Calculating R_{Th}

Measuring R_{Th} requires that we attach a load and draw a load current through R_{Th}. We need to measure two of the following: (1) the load voltage, (2) the load current, or (3) the load resistance. If you know two of these measurements, you can calculate the third using Ohm's law for resistance. Figure 8-30 represents this loaded measurement.

Again the real measurement is made on the real circuit as shown in Figure 8-30(a). The model in Figure 8-30(b) is used to understand and relate to the value of R_{Th} based upon the measurements of Figure 8-30(a). Typically you would measure the load voltage V_{load} and the load resistance R_{load}, and then calculate R_{Th} based upon E_{Th}, R_{load}, and V_{load} using basic circuit analysis techniques. This is the same technique we used to analyze real voltage supplies.

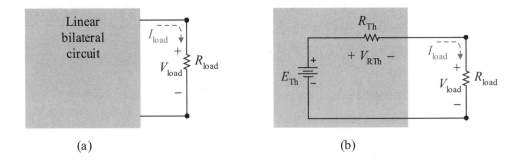

(a)   (b)

Figure 8-30 Finding R_{Th} with loaded circuit and model

Example 8-16

A bridge circuit is analyzed in the lab. The load is removed and its open circuit voltage is measured to be 10 V. The load resistance of 4 kΩ is reattached and the load voltage is measured to be 8 V. Find and draw the Thévenin model for this circuit. If a 1 kΩ resistor were attached as a load instead of the 4 kΩ resistor, what would the load voltage and current be?

Solution

Based upon the no load measurement,

$$E_{Th} = 10 \text{ V}$$

Figure 8-31(a) represents the Thévenin model for this circuit with known values: open circuit voltage, loaded circuit voltage, and load resistance. Use circuit analysis and find the value of R_{Th} based upon the measured values and the Thévenin model.

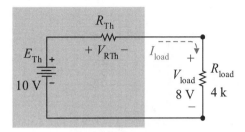

Figure 8-31(a) Model loaded with 4 kΩ load resistor

Find the voltage drop across R_{Th} by doing a KVL voltage walka-bout clockwise from V_{RTh} (−) to V_{RTh} (+).

$$V_{RTh} = -8\text{ V} + 10\text{ V} = 2\text{ V}$$

Next, calculate the load current I_{load}, which is also the current through R_{Th} since this is a series circuit.

$$I_{RTh} = I_{load} = \frac{8\text{ V}}{4\text{ k}\Omega} = 2\text{ mA}$$

You know the drop across R_{Th} and the current through R_{Th}, so you can find R_{Th} using Ohm's law for resistance.

$$R_{Th} = \frac{2\text{ V}}{2\text{ mA}} = 1\text{ k}\Omega$$

Figure 8-31(b) represents the Thévenin model of the bridge cir-cuit with a 1 kΩ load attached for analysis. Find the total circuit resistance of the series circuit.

$$R_{total} = 1\text{ k}\Omega + 1\text{ k}\Omega = 2\text{ k}\Omega$$

Find the series circuit current, which is also the load current.

$$I_{load} = \frac{10\text{ V}}{2\text{ k}\Omega} = 5\text{ mA}$$

Find the load voltage using Ohm's law (or VDR).

$$V_{load} = 5\text{ mA} \times 1\text{ k}\Omega = 5\text{ V}$$

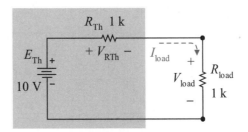

Figure 8-31(b) Model loaded with a 1 kΩ load resistor

So, the original bridge circuit with a 1 kΩ load resistor instead of a 4 kΩ resistor would have a load current of 5 mA and a load voltage drop of 5 V. Notice that we never analyzed the original circuit. We characterized the original bridge circuit as its Thévenin model and then used the model to find the original circuit's output with a different load resistance.

Practice

An unknown circuit in a box is analyzed in the lab. The load is removed and its open circuit voltage is measured to be 12 V. The load resistance of 3 kΩ is reattached and the load voltage is measured to be 9 V. Find the Thévenin model for this circuit. If a 2 kΩ resistor were attached as a load for the original circuit, what would the load voltage and current be?

Answer: model (12 V in *series* with 1 kΩ), 8 V, 4 mA

8.8 Voltage Supplies in Series

Many applications require that voltage supplies or voltage cells be placed in series. This arrangement raises the applied voltage to the load.

For example, flashlights typically require two batteries in series to create enough voltage to drive a 3-volt lamp.

Example 8-17

A flashlight requires 2 D cell batteries in series as represented by the circuit of Figure 8-32. Each battery operates at 1.5 V. Find the voltage applied to the lamp of the flashlight when the switch is closed.

Solution

Assume the switch is closed to complete the circuit. Find the voltage applied to the lamp by doing a KVL voltage walkabout from the lamp (−) to the lamp (+) reference. The closed switch drops 0 V.

$$V_{lamp} = 1.5 \text{ V} + 1.5 \text{ V} + 0 \text{ V} = 3 \text{ V}$$

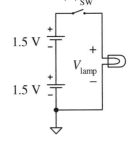

Figure 8-32
Example circuit

Practice

You accidentally put one of the flashlight batteries in backward in the above flashlight. Find the voltage delivered to the lamp if the switch is closed.

Answer: 0 V (the cells oppose each other)

The batteries in a flashlight, assuming they are put in correctly, are said to be series aiding. That is, supplies are **series aiding** if they produce current in the same direction as shown in Figure 8-33(a); their voltages add (E1 + E2). Supplies are **series opposing** if they produce current in opposite directions as shown in Figure 8-33(b); their voltages subtract from each other (E1 − E2).

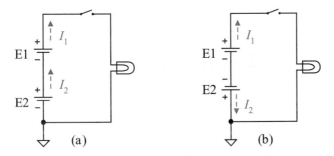

Figure 8-33 (a) *Series aiding* supplies   (b) *Series opposing* supplies

Battery Charger

The next example takes advantage of two opposing supplies. The example circuit is a battery charging circuit (or generator). One supply is the generator and the other supply is the battery being charged. As the battery is charged its voltage increases until it is fully charged. The generator must have a greater dc supply voltage than the battery being charged so that the resulting current is into the (+) supply terminal of the charging battery. Thus, the generator is sourcing current and the charging battery is sinking current.

Example 8-18

The circuit of Figure 8-34 is a schematic of a battery charging circuit, modeling each device with its voltage supply or Thévenin model equivalent. The supply on the left is the generator. The battery on the right is the battery being charged. Note that these are opposing supplies but that the generator's 14 V is greater than the battery's 12 V, thus the resulting current is out of the generator (+) terminal and into the battery (+) terminal.

Find the charging current I_{gen} and the generator voltage V_{gen}. How much charge capability is regenerated in the battery over a period of 5 minutes?

Solution

This is a series circuit, so the total series circuit resistance is

$$R_T = 0.06 \ \Omega + 0.94 \ \Omega = 1 \ \Omega$$

The net voltage being applied to this circuit resistance is the difference of the two supplies. Since we are assuming a clockwise current in the circuit, walk clockwise around the circuit adding supply voltage rises and subtracting supply voltage falls. *Include only the supply voltages; do not include the resistor voltage drops.*

$$E_{net} = 14 \ \text{V} - 12 \ \text{V} = 2 \ \text{V}$$

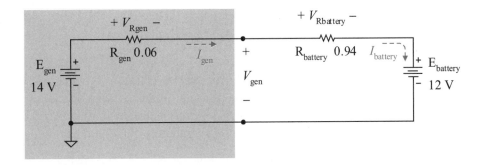

Figure 8-34 Example circuit–battery charger modeled circuit
first approximation model

The KVL supply loop result is positive (+) indicating that clockwise is the correct choice of current direction. The net supply voltage forces current clockwise. If the result had been negative, the current would have been flowing in the other direction. But we chose correctly and the current is flowing clockwise.

Now we have the net voltage being applied to the two circuit resistances, so we can use Ohm's law for resistance to calculate the current flowing through the resistances.

$$I_{RT} = \frac{E_{net}}{R_T} = \frac{2\text{ V}}{1\ \Omega} = 2\text{ A}$$

Since this is a series circuit, the current is the same throughout the circuit.

$$I_{gen} = I_{battery} = 2\text{ A}$$

We can find the generator voltage V_{gen} by doing a walkabout from the $V_{gen}(-)$ reference to the $V_{gen}(+)$ reference. We can walk counterclockwise or clockwise. The result is the same.

The clockwise walkabout for V_{gen} is

$$V_{gen} = +E_{gen} - I_{gen}\,R_{gen}$$

$$V_{gen} = +14\text{ V} - (2\text{ A} \times 0.06\ \Omega) = 13.88\text{ V}$$

The counterclockwise walkabout for V_{gen} is

$$V_{gen} = +E_{battery} + I_{battery}\,R_{battery}$$

$$V_{\text{gen}} = +12 \text{ V} + (2 \text{ A} \times 0.94 \text{ }\Omega) = 13.88 \text{ V}$$

Again, both walkabouts must produce the same result. The generator output is 13.88 V. Note that this voltage lies between 14 V and 12 V. As the battery charges, its voltage rises and its resistance decreases. Thus, this is a dynamic circuit with the battery voltage, the battery resistance, the circuit current, and the generator output voltage gradually changing over time.

To find the amount of charge rejuvenated in the battery by the charging circuit over a 5 minute period, we need to go back to the fundamental definition of current.

$$I = \frac{Q}{t}$$

Rewriting the current definition to solve for charge and expressing 5 minutes in basic units as 300 seconds, the charge is

$$Q = It = 2 \text{ A} \times 300 \text{ s} = 600 \text{ C}$$

Practice

After charging awhile, the above battery has charged to 12.5 V and its internal resistance has decreased to 0.44 Ω. Find the charging current I_{gen} and the generator voltage V_{gen}. How much charge capability is regenerated in the battery over a period of 5 minutes?

Answer: 3 A, 13.82 V, 900 C

Diode Models

The diode was previously introduced as basically an on-off device. When off, the diode behaves like an open and can be modeled as an open. When forward-biased on, the diode acts *ideally* like a conductor or *more realistically* as a constant voltage drop (0.2 V for germanium and 0.7 V for silicon). Figure 8-35 demonstrates the modeling of a silicon diode as ideal (first approximation model), second approximation model, and third approximation model. The second approximation model is more realistic than the ideal model. And the third approximation model is more realistic than the second. Which model do you choose to use for circuit analysis?

That depends upon the circuit and which diode characteristics significantly impact the analysis.

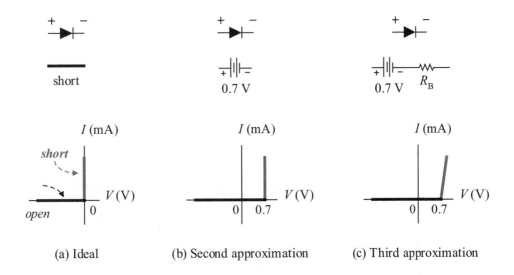

Figure 8-35 Silicon diode models and characteristic curves

The **ideal model** is called the **first approximation model** of the component. This is the minimum model of a component that still retains its fundamental characteristics. Figure 8-35(a) demonstrates the ideal or first approximation model of the silicon diode. For the purposes of circuit analysis, we can replace the diode with its equivalent model in the schematic. The ideal model works well for higher voltage circuits (for example, in a 100 V circuit the ideal model produces less than 1% error).

A **second approximation model** includes an additional attribute(s) that makes the model more realistic. Figure 8-35(b) shows the second approximation model of a silicon diode that includes its 0.7 V *pn* junction barrier potential voltage. Once the silicon diode is turned on, the diode drops and maintains a voltage drop of about 0.7 V. The second approximation model exhibits this major characteristic. If the voltages in the circuit are small enough (comparable to the 0.7 V) then a diode model including the 0.7 V supply must be used. For example, a 10 V circuit would produce an error of about 7% if the ideal model were used instead of the second approximation model.

Likewise a **third approximation model** includes an additional attribute(s) that yields a more accurate model. A more detailed model produces more accurate results but increases the difficulty level in analyzing its circuit. Figure 8-35(c) shows the third approximation model of a silicon diode that includes its 0.7 V *pn* junction barrier potential voltage and R_B (its natural resistance to dc current). As the characteristic curve indicates, as current is increased in a forward-biased on diode, its voltage drop increases linearly. This is due to the natural resistance of the diode's *p* and *n* type materials and is called the **bulk resistance** R_B. This natural resistance is very low, a few ohms or less. Usually this characteristic can be ignored but in *low resistance circuits* or *high current circuits*, the third approximation model must be used.

Example 8-19

The circuit of Figure 8-36(a) is a simple series circuit with a resistance and a forward-biased on silicon diode. The diode's resistance is about 0.2 Ω. Use (a) the ideal model, then (b) the second approximation model, and then (c) the third approximation model to find the circuit current, the diode voltage drop, and the resistor voltage drop. (d) Make an observation about the most appropriate diode model for this example based upon your results.

Solution

a.   The schematic of Figure 8-36(b) uses the ideal (first approximation) model of the diode (an *ideal conductor* or a *short*).

From KVL, the net voltage applied to the load resistor is

$$V_R = E_{supply} = 6 \text{ V}$$

From Ohm's law, the circuit current is

$$I = I_R = I_D = \frac{6 \text{ V}}{10 \text{ }\Omega} = 0.6 \text{ A} = 600 \text{ mA}$$

The diode voltage drop (across a short) is

$$V_D = 0 \text{ V}$$

b.   The schematic of Figure 8-36(c) uses the second approximation model of the diode (*an ideal* 0.7 V *voltage supply*).

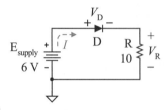

Figure 8-36(a)
Example circuit

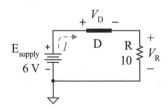

Figure 8-36(b)
Example circuit–diode
first approximation model

From KVL or a KVL walkabout, the net voltage applied to the load resistor is

$$V_R = E_{supply} - V_D = 6 \text{ V} - 0.7 \text{ V} = 5.3 \text{ V}$$

From Ohm's law, the circuit current is

$$I = I_R = I_D = \frac{5.3 \text{ V}}{10 \text{ }\Omega} = 0.53 \text{ A} = 530 \text{ mA}$$

The diode voltage drop is its model value.

$$V_D = 0.7 \text{ V}$$

c. The schematic of Figure 8-36(d) uses the third approximation model of the diode (*an ideal 0.7 V voltage supply in series with its internal resistance*).

The total circuit resistance is

$$R_T = 10 \text{ }\Omega + 0.2 \text{ }\Omega = 10.2 \text{ }\Omega$$

From a clockwise KVL walkabout of the supplies only, the net voltage supplied to the total circuit resistance is

$$E_{net} = E_{supply} - V_D = 6 \text{ V} - 0.7 \text{ V} = 5.3 \text{ V}$$

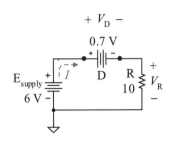

Figure 8-36(c)
Example circuit–diode *second approximation model*

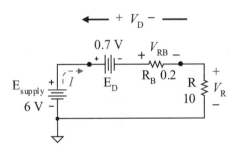

Figure 8-36(d) Example circuit using the third approximation model

From Ohm's law, the circuit current in a series circuit is the net supply voltage divided by the total circuit resistance.

$$I = I_R = I_D = \frac{E_{net}}{R_T} = \frac{5.3\ V}{10.2\ \Omega} = 0.52\ A$$

The diode voltage drop across its terminals is a series addition of its rated 0.7 V and the diode's internal resistance voltage drop based upon Ohm's law. Do a KVL walkabout from $V_D(-)$ across the diode to $V_D(+)$ to include both voltages and to find the total drop across the diode.

$$V_D = 0.7\ V + (0.52\ A \times 0.2\ \Omega) = 0.804\ V$$

By Ohm's law, the resistor voltage drop is

$$V_R = 0.52\ A \times 10\ \Omega = 5.2\ V$$

d.   Let's look at the results of the three diode models for this circuit and compare the results in Table 8-2.

Table 8-2 Example diode circuit results

Diode model	I (A)	V_R (V)	V_D (V)
Ideal	0.60	6.0	0.0
2nd approximation	0.53	5.3	0.7
3rd approximation	0.52	5.2	0.8

Going from the ideal model to the second approximation model, the circuit current goes from 0.60 A to 0.53 A. This is about a 13% error.

Going from the second approximation model to the third approximation model, the circuit current goes from 0.53 A to 0.52 A. This is about a 2% error.

If you are looking for a very rough estimate, the ideal model provides this with about a 15% error. If you needed more accuracy, then the second approximation model yielded reasonable results with about a 2% error. In retrospect, we could have forecasted the model selection impact.

How about the second approximation model? Compare the circuit supply with the model diode supply. Is 0.7 V meaningful as compared to the supply voltage of 6 V seen by the diode? Ignoring the diode voltage drop causes a calculation error greater than 10%. Is this significant in this particular instance?

How about the third approximation model? Does the diode's internal resistance to dc current need to be included? Is the diode's resistance of 0.2 Ω meaningful as compared with the circuit resistance it sees of 10 Ω? It is about a 50 to 1 ratio. It looks like ignoring the diode's internal resistance produces a calculation error of about 2%. Is this calculation error significant?

In the final analysis, the allowable calculation error depends upon how accurate your calculation of expected circuit values needs to be. That is a judgment call.

Practice

Using the diode circuit of this example, the load resistance is changed to 1 kΩ. Find the load current using (a) the ideal diode model, (b) the second diode approximation model, and (c) the third diode approximation model?

Answer: 6 mA, 5.3 mA, 5.299 mA

Zener Diode Voltage Regulator

The **zener diode** is a diode specially designed to work in the reverse-biased region. Normally a reverse-biased diode is off and acts like an open. However, *if a diode is reverse-biased with sufficient voltage*, the diode turns on and conducts current in the reverse direction while maintaining a reasonably constant voltage drop. Figure 8-37 shows the characteristic curve of a zener diode designed to start conducting in the *reverse-bias on region* at V_Z. Its forward bias curve is very similar to any forward-biased silicon diode curve with a forward knee voltage of 0.7 V.

Two types of reverse voltage breakdown can occur in a diode, *avalanche breakdown* or *zener breakdown* (greater detail is provided later in this textbook). Even though two types of breakdown can occur, all diodes of this type are called zener diodes. They are also known as *voltage regu-*

lator diodes since they produce a fixed value of voltage that can be used to produce a regulated fixed dc voltage supply or reference under varying load conditions. Such supplies have a very low percent voltage variation, approaching 0%.

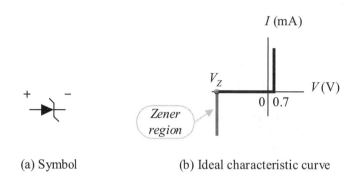

(a) Symbol   (b) Ideal characteristic curve

Figure 8-37 Zener diode symbol and ideal characteristic curve

Breakdown does not harm the diode, and we can take advantage of this phenomenon to provide a fixed dc voltage drop of V_Z. The value of V_Z depends upon the diode's material and construction. Excessive current in either the forward or reverse direction can harm the diode, so diodes are rated for maximum forward current and maximum reverse current. If these currents are exceeded, the diode is attempting to dissipate too much power and self-destructs.

As shown in Figure 8-38, the zener diode operating in its zener region has two models that are used in circuit analysis. The ideal model is an ideal voltage supply with the value of V_Z, the reverse knee voltage. The second approximation model includes the zener diode's internal resistance R_z to dc current as well as its ideal zener voltage supply V_Z.

The next example examines the zener diode in a simple series circuit using both the ideal and second approximation models.

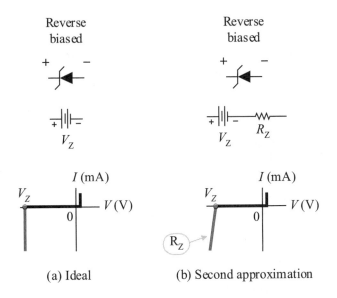

Figure 8-38 Zener diode models

(a) Ideal

(b) Second approximation

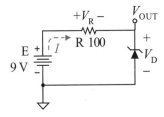

Figure 8-39(a)
Example circuit
V_{OUT} across zener diode

Example 8-20

The circuit of Figure 8-39(a) is a simple series circuit with a resistance (to limit current in the zener diode) and a reverse-biased zener (voltage regulator) diode. The zener is rated at 6 V, its resistance is about 2 Ω, and its maximum reverse current rating is 100 mA. The resistor is rated at a quarter watt. Use (a) the ideal zener diode model then (b) the second approximation model to find the zener current and the output voltage V_{OUT} across the diode. Check that the zener diode and the resistor are safe. (c) Make an observation about the most appropriate diode model for this example based upon your results.

Solution

a. Figure 8-39(b) is the schematic of the ideal (first approxima-tion) model of the zener diode (an ideal voltage supply). In this case the output voltage is just the voltage of the zener diode. Do a walkabout from common and rise 6 V to the V_{OUT} node.

$$V_{\text{OUT}} = V_D = V_Z = 6 \text{ V}$$

From a clockwise KVL walkabout from $V_R(-)$ to $V_R(+)$, the voltage drop across R is

$$V_R = -6 \text{ V} + 9 \text{ V} = 3 \text{ V}$$

Using Ohm's law for resistance, the circuit current

$$I = I_R = I_Z = \frac{3 \text{ V}}{100 \text{ }\Omega} = 30 \text{ mA}$$

Is the zener diode safe? Yes it is operating at less than 100 mA. Is the resistor safe? First we need to calculate its power dissipation.

$$P_R = I_R V_R = 30 \text{ mA} \times 3 \text{ V} = 90 \text{ mW}$$

Yes, it is dissipating less than a quarter watt.

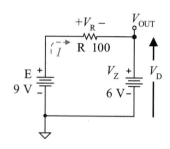

Figure 8-39(b)
Example circuit
zener diode–
ideal model

b.   Figure 8-39(c) is the schematic of the second approximation model of the zener diode (an ideal 6 V supply in series with an internal zener resistance of 2 Ω).

The total series circuit resistance is

$$R_T = 100 \text{ }\Omega + 2 \text{ }\Omega = 102 \text{ }\Omega$$

From a clockwise KVL walkabout of the *supplies only*, the net voltage applied to the total circuit resistance is

$$E_{\text{net}} = E - V_Z = 9 \text{ V} - 6 \text{ V} = 3 \text{ V}$$

From Ohm's law, the circuit current in a series circuit is the net supply voltage divided by the total circuit resistance.

$$I = I_R = I_Z = \frac{E_{\text{net}}}{R_T} = \frac{3 \text{ V}}{102 \text{ }\Omega} = 29.4 \text{ mA}$$

The zener diode voltage drop across its terminals is a series addition of its rated 6 V plus its internal resistance voltage drop based upon Ohm's law. Do a KVL walkabout from $V_D(-)$ up the diode to $V_D(+)$ to include both voltages and to find the total drop across the zener diode. This is also the output voltage V_{OUT}.

$$V_{\text{OUT}} = V_D = 6 \text{ V} + (29.4 \text{ mA} \times 2 \text{ }\Omega) = 6.059 \text{ V}$$

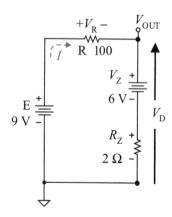

Figure 8-39(c)
Example circuit
zener diode–
*second approximation
model*

The current decreased slightly in this model since a very small additional series resistance was added to the circuit. So, both the resistor and the zener diode are safe.

c.   Let's look at the results of the two zener diode models for this circuit and compare the results.

Using the ideal model

$$V_{OUT} = V_D = V_Z = 6 \text{ V}$$

Using the second approximation model with the internal zener resistance added about 59 mV to the drop across the zener and thus to the output voltage.

$$V_{OUT} = V_D = 6 \text{ V} + 0.059 \text{ V} = 6.059 \text{ V}$$

The internal zener resistance voltage drop becomes an issue only if the zener diode is drawing significant current creating a larger relevant drop internally, creating a noticeably higher zener diode voltage drop. Comparing the V_{OUT} values for each model yields a 1% error between the calculations. *The ideal model would have been sufficient.*

Practice

The zener diode of the example is replaced with a 5 V zener that has an internal resistance of 5 Ω, with a maximum reverse current rating of 200 mA. Repeat the example with this zener diode.

Answer: (a) 40 mA, safe, 5 V, 160 mW, safe
(b) 38.1 mA, safe, 5.19 V, 145 mW, safe
(c) 3% error (ideal model sufficient)

Summary

A *series circuit* is one in which the *same current* passes through all the components. The components are two-terminal components that are connected end-to-end such that only two components are joined at each connection.

If there are two or more resistances in series, the total circuit resistance of the series circuit is *the sum of the individual resistances.*

$$R_T = R1 + R2 + R3 + ... + Rn$$

Ohm's law applies to total circuit or equivalent resistances as well as to individual resistances.

$$I_{\text{circuit}} = \frac{V_{\text{circuit}}}{R_{\text{circuit}}}$$ (Ohm's law)

Adding a resistance in series increases the total circuit resistance, decreases the circuit current, and thus decreases the drop across each of the series resistances.

To measure current, the circuit must be broken and a current meter inserted into the circuit as if it were a series component. Ideally the current meter acts like an ideal conductor (a short) and therefore does not drop any voltage. However, in reality, the current meter has some resistance although very low. Thus in low resistive circuits, the current meter may impact the measurement or produce a current meter loading effect by increasing the circuit resistance and thus producing a lower than expected current measurement.

In a *series resistive circuit*, the applied voltage divides across the series resistances proportionately based upon their respective resistance values as compared to the total series resistance. This is known as the *voltage divider rule* (VDR).

$$V_{R1} = \left(\frac{R1}{R_T}\right)E$$ (VDR)

The voltage V_{R1} receives ($R1/R_T$) of the applied voltage E. The advantage of using the voltage divider rule is that it eliminates the need to directly calculate the circuit current. The series circuit current calculation is buried within the rule as (E/R_T) but is not separately or explicitly calculated.

The voltage divider rule also applies to the drop across multiple resistors in series. For example, the voltage drop across R1 and R2 in a series circuit is

$$V_{R12} = \left(\frac{R1 + R2}{R_T}\right)E$$ (VDR)

Series and *essentially series* circuits are used to set up proper dc voltages (bias) components such as BJTs and MOSFETs. These are called *voltage divider bias circuits*. An *essentially series circuit* is one that appears at first glance *not* to be a series circuit but the currents that branch off the main current branch are relatively so small as to be negligible. Thus there appears to be one main current even though there are very small currents that branch off it.

Ideal voltage sources produce a fixed dc output under various load conditions. The ideal voltage source has an internal resistance of 0 Ω. A realistic voltage has an internal resistance although typically very, very low. The supply's output voltage varies somewhat with varying load conditions, although typically not very much, based upon the relationship of the internal supply resistance and the attached load resistance.

$$V_{\text{supply}} = \left(\frac{R_{\text{load}}}{R_{\text{load}} + R_{\text{supply}}} \right) E_{\text{supply}}$$

The actual voltage delivered from the supply is a supplied voltage divided between the load resistance and supply's internal resistance. Typically for a good voltage supply, $R_{\text{load}} \gg R_{\text{supply}}$.

A measure of the quality of a power supply is its percent load voltage regulation

$$\% \ VR = \frac{V_{\text{NL}} - V_{\text{FL}}}{V_{\text{FL}}} \times 100 \ \%$$

where V_{NL} is its output voltage under no load conditions (0 A load current) and V_{FL} is its output voltage under full load conditions (maximum allowable load current).

Thévenin's theorem could be appropriately nicknamed "What's in the box?" Thévenin's theorem essentially states that *any linear bilateral network may be modeled as an ideal voltage supply in series with a resistance.* An ideal voltage supply in series with a resistance is the model we used to model a real power supply. Thus, we have already used a Thévenin model to represent a real dc voltage supply. The Thévenin voltage is just the *open circuit voltage* (output voltage without the circuit's load attached). The Thévenin voltage and resistance can be determined by measurement and calculation. For now, we are approaching this as a pragmatic approach (based upon measurements) of characterizing a circuit by a Thévenin model.

Voltage supplies in series occur naturally, for example, flashlight batteries and battery chargers. Series supplies attempting to produce current in the same direction are *series aiding* (e.g., flashlight batteries). Series supplies attempting to produce currents in opposite directions are *series opposing* (e.g., battery charger). Devices such as a forward-biased on diode or LED naturally act like a voltage supply. They provide a fixed dc output voltage. Ideally these are fixed dc supplies, but like the real voltage supply, these devices do have very low internal resistances. Thus, their output dc voltages can vary slightly with changing load currents.

Device modeling provides a way of modeling a device in terms of simpler circuit components such as ideal voltage supplies and resistance that can be more easily analyzed in circuit analysis. The simplest model that retains the main characteristic(s) of a device is known as the *ideal* or *first approximation model*. The next simplest model that adds another feature is the *second approximation model*; then comes the *third approximation model*, and so forth. Adding additional characteristics produces a more accurate model but usually increases the complexity of analyzing the circuit.

The ideal model of a forward-biased on diode is a conductor (a short). The second approximation model includes its *pn* junction barrier voltage as an ideal voltage supply (0.2 V for germanium and 0.7 V for silicon). The third approximation model adds the diode's internal resistance (similar to a real voltage supply). This is basically the Thévenin model. Be careful. The diode does not really have a voltage supply in it. But once the diode turns on, it drops a relatively fixed voltage like a voltage supply does.

The zener diode utilizes the reverse breakdown voltage to provide a fixed, regulated dc across it. When a load is placed across the reverse-biased on zener diode, the zener provides a relatively fixed supply voltage with varying load conditions. The ideal model of a zener diode is an ideal voltage supply. The second approximation model of a zener diode is a real voltage supply (an ideal voltage supply with a series resistance), the Thévenin model.

Problems

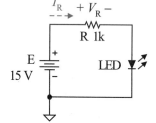

Figure 8-40

Series Circuits and the Laws

8-1   For the circuit of Figure 8-40:

   a. Draw the proper direction of current with associated component voltage polarities.

   b. Then find the voltage drop across R.

   c. Then find the current in the LED.

   d. Draw the corresponding bubble notation schematic for this circuit. The LED specification is that it drops about 2 V.

8-2 Repeat Problem 8-1 for the circuit of Figure 8-41. Be faithful to the resistor reference current direction and the reference voltage polarity.

Resistive Series Circuit Analysis

8-3 For the circuit of Figure 8-42:

  a. Write the KVL expression in terms of the circuit current I.

  b. Solve the KVL expression for the current I.

  c. Find the voltage drop across each resistor.

  d. Find the node voltages V_a, V_b, V_c, and V_d.

  e. Find voltage drops V_{ac} and V_{ca}.

  f. Using the supply voltage and the circuit current, find the total circuit resistance.

  g. Draw the bubble notation schematic for this circuit.

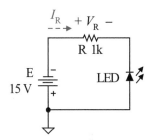

Figure 8-41

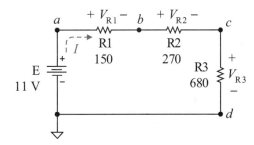

Figure 8-42

Resistance of a Series Circuit

8-4 For the circuit of Figure 8-43:

  a. Find the total circuit resistance from the individual resistances.

  b. Find node voltages V_a and V_d.

  c. Find the supplied voltage V_{ad}.

  d. From the total circuit resistance and the supplied voltage, find the circuit current I.

e. Find each resistor voltage drop.

f. Find node voltages V_b and V_c.

g. Find voltage drops V_{ac}, V_{bd}, and V_{db}.

h. Draw the bubble notation schematic for this circuit.

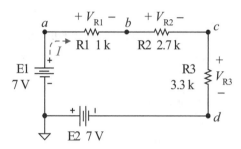

Figure 8-43

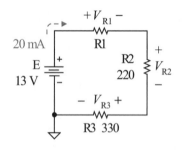

Figure 8-44

8-5   For the circuit of Figure 8-42:

a. Find the total circuit resistance from the individual resistances.

b. From the total circuit resistance and the supplied voltage, find the circuit current.

8-6   For the circuit of Figure 8-43:

a. Find the total circuit resistance from the individual resistances.

b. From the total circuit resistance and the supplied voltage, find the circuit current.

8-7   For the circuit of Figure 8-44:

a. Find the total circuit resistance from the supply voltage and current.

b. Find the resistance value of R1 using the total circuit resistance.

c. Find the power generated by the voltage supply.

    d. Find the power dissipated by R3. If this is a quarter watt rated resistor, is it safe?

8-8 For the circuit of Figure 8-44:

    a. Find the voltage drops across R2 and R3.

    b. Find the voltage drop across R1.

    c. Using the voltage drop across R1 and the current through R1, find its resistance. Is this the same value of R1 as you found in Problem 8-7?

Current Meter Loading

8-9 For the circuit of Figure 8-45:

    a. Find the circuit current if the ammeter is ideal.

    b. Find the circuit current if the ammeter has an internal resistance of 1 Ω.

    c. What is the percent error of this measurement? Is it significant?

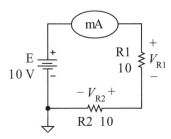

Figure 8-45

8-10 The ammeter is accidentally placed across R2. Redraw the schematic and model the ammeter as an ideal short. Find the milliameter reading, the current through R1, and the current through R2. Which devices are in possible jeopardy of destruction?

VDR–Voltage Divider Rule

8-11 For the circuit of Figure 8-42, use the voltage divider rule to find:

    a. The voltage drop across each resistor.

    b. The voltage drop of V_{ac}.

    c. The voltage drop of V_{bd}.

    d. The voltage drop of V_{db}.

8-12 For the circuit of Figure 8-43, use the voltage divider rule to find:

    a. The voltage drop across each resistor.

    b. The voltage drop of V_{ac}.

    c. The voltage drop of V_{bd}.

    d. The voltage drop of V_{db}.

8-13 For the β independent circuit of Figure 8-46, assuming a silicon BJT with a β of 250:

    a. Use the voltage divider rule to find base voltage V_B.

b. Find V_E.

c. Find I_E.

d. Find I_C.

e. Find V_C.

f. Find V_{CE}.

g. Assuming β is 250, find the base current.

h. Test your assumption. Is this circuit β independent? That is, is $I_{R1} \gg I_B$?

8-14 Assume the β independent circuit of Figure 8-46.

a. If the BJT goes bad and its β becomes 10, find the base current. Note: you can assume the current from Problem 8-13 as a starting point.

b. Test your assumption. Is this circuit still β independent? That is, is $I_{R1} \gg I_B$?

8-15 Assuming a β independent circuit, redesign the circuit of Figure 8-46 by changing the resistance of R1 to produce a base voltage of 5.9 V.

8-16 For the circuit of Figure 8-47, use the MOSFET characteristics of Figure 7-11(b) on page 267. Assume the load drops 6 V. Find:

a. V_{GS}

b. I_D

c. V_{DS}

d. The power dissipated by the MOSFET.

8-17 Redesign the circuit of Figure 8-47 to produce a 2 A load current. Change the value of R2 to produce this new operating point.

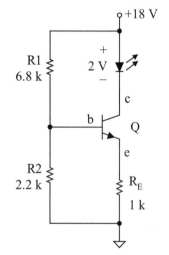

Figure 8-46

Real Voltage Supply Model

8-18 The open circuit voltage of a voltage supply measures 18 V. The fully loaded supply voltage is 17.9 V at 1 A.

a. What is the supply's ideal internal voltage supply value?

b. Find the load resistance attached to produce the stated load values.

c. Find the drop across the supply's internal resistance.

   d. Find the supply's internal resistance.

   e. What is the maximum power that this supply can deliver?

   f. Find the supply's percent load voltage regulation.

   g. Using the model of the voltage supply, find the load current if a short were attached as the load. Is this a concern?

8-19   The open circuit voltage of a D cell battery measures 1.6 V. The fully loaded supply voltage is 1.45 V at 1 A.

   a. What is the battery's ideal internal voltage supply value?

   b. Find the load resistance attached to produce the stated load values.

   c. Find the drop across the battery's internal resistance.

   d. Find the battery's internal resistance.

   e. Find the battery's percent load voltage regulation.

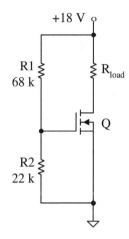

Figure 8-47

Thévenin's Theorem

8-20   A circuit's open circuit voltage measures 12 V. Under normal load conditions, the load voltage is 8 V and load current is 1 mA.

   a. What is its Thévenin voltage?

   b. Find its Thévenin resistance.

   c. Draw the Thévenin model.

   d. Using the Thévenin model, attach a load resistance of 2 kΩ and find the load voltage and load current.

8-21   A circuit's open circuit voltage measures 20 V. Under normal load conditions, the load voltage is 5 V and load current is 5 mA.

   a. What is its Thévenin voltage?

   b. Find its Thévenin resistance.

   c. Draw the Thévenin model.

   d. Using the Thévenin model, attach a load resistance of 2 kΩ and find the load voltage and load current.

Voltage Supplies in Series

8-22   A flashlight uses four 1.5 D cell batteries in series.

   a. Draw the flashlight schematic including the supply, the switch, and the lamp.

b. Find the required voltage rating of the lamp.

c. If the bulb is rated at 15 W, find the current through the lamp when it is turned on.

d. What is the resistance of the bulb when it is turned on?

8-23 A flashlight uses four 1.5 D cell batteries in series. One is accidentally reversed.

a. Draw the flashlight schematic including the supply (with the reversed battery), the switch, and the lamp.

b. Find the net supply voltage driving the lamp.

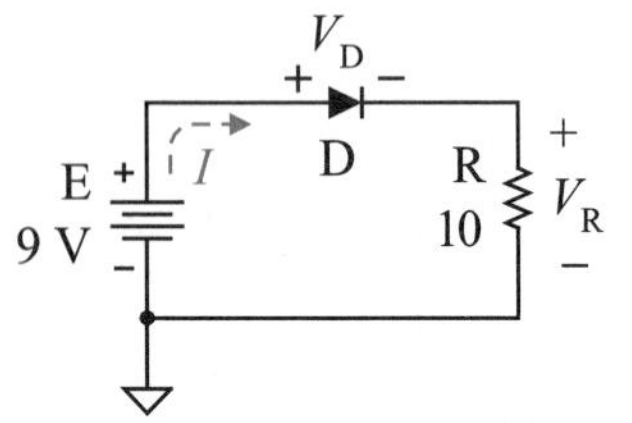

8-24 For the circuit of Figure 8-48, assume a silicon diode with an internal resistance of 0.4 Ω. Use the proper model to find the resistor voltage drop, the circuit current, and the diode voltage drop.

a. Ideal model: draw the modeled circuit and use it.

b. Second approximation model: draw the modeled circuit and use it.

c. Third approximation model: draw the modeled circuit and use it.

8-25 For the circuit of Figure 8-49, assume a 5 V zener diode with an internal resistance of 8 Ω. Use the proper model to find the resistor voltage drop, the zener current, and the zener diode voltage drop (the output voltage).

a. Ideal zener diode model: draw the modeled circuit and use it.

b. Second approximation model: draw the modeled circuit and use it.

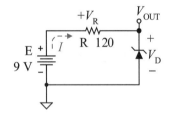

Figure 8-48

Figure 8-49

9

Essentially Series

Introduction

This chapter introduces the operational amplifier (op amp) as an ideal component. The ideal op amp is examined in several popular circuits that are *essentially series* in nature.

First you will study the op amp as a comparator. The op amp itself is a natural comparator, comparing two inputs and producing either a high or low output voltage. Next you will examine positive feedback comparators (created by feeding part of the output signal back to the input with the same polarity).

Then you will take a more detailed look at the two popular op amp amplifier circuits you have seen before: the inverting and noninverting voltage amplifiers. You will prove that these amplifiers have negative feedback (that is, part or all of the output signal is fed back to the input signal with the opposite polarity) to produce a voltage amplifier.

In all these cases, the primary circuit analysis is simplified to an essentially series circuit analysis using the ideal characteristics of the op amp.

Objectives

Upon completion of this chapter you will be able to:

- Specify the ideal op amp characteristics including its input limits, its voltage supply limitations, and its output limits.

- Calculate output voltages with given input voltages and supply voltages.

- Analyze an op amp as an inverting or noninverting comparator.

- Analyze an op amp with positive (series circuit) feedback as an inverting or noninverting comparator with hysteresis to prevent comparator jitters.

- Analyze an op amp with negative (series circuit) feedback with both an inverting and noninverting amplifier circuit.

9.1 The Ideal Operational Amplifier

The **operational amplifier (op amp)** is a nearly ideal voltage amplifier device. Its output voltage $V_{\text{out op amp}}$ is much greater than its input voltage V_{error}, its input resistance is very high (ideally ∞ Ω), and its output resistance is very low (ideally 0 Ω). Figure 9-1 shows the op amp symbol along with its associated key voltages.

The triangle symbol actually represents a circuit that may have hundreds of components internally. It is an **integrated circuit (IC)**, and these hundreds of components (resistors, diodes, BJTs, etc.) are constructed into a semiconductor chip of microscopic size. The IC, in turn, is housed inside a package that brings its connecting terminals to pins that can be physically connected to a circuit board. The terminals shown in Figure 9-1 are the essential pin connections that are available externally. Other pins may be available (for example, to balance the chip), but for now focus your attention on the primary terminals of the chip.

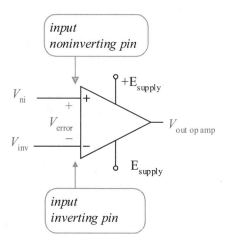

Figure 9-1 Op amp symbol and associated voltages

First look at the $\pm E_{\text{supply}}$ dc voltages. These are the supply voltages that power the chip and bias the internal components (such as diodes, BJTs, and MOSFETs). It is *critical* that these supply voltages be connected properly to the chip. If a supply is connected to the wrong pin, there is a good chance that you will destroy the chip. The $-E_{\text{supply}}$ terminal can be connected directly to a negative supply or, for some op amps, it can be connected to common to provide a configuration which is called a **single-supply** op amp.

Op Amp Input Voltage

The op amp has two voltage signal input pins, labeled V_{inv} (–, inverting input) and V_{ni} (+, noninverting input). Signals can be applied to either or both input terminals. V_{error} is the *net input* to the op amp and is defined to be

$$V_{error} = V_{ni} - V_{inv}$$

The "error" terminology comes from the control and instrumentation fields where an error control signal is fed back to mix with the original signal to provide correction signaling. The op amp was first associated with error correction and the name V_{error} has stuck ever since. Just think of V_{error} as being the net *input voltage* across the input terminals of the op amp.

$$V_{error} = V_{input\ op\ amp}$$

As noted earlier, the input currents to the op amp are considered, for all practical purposes, to be 0 A since the input resistance of the op amp is very high. For the ideal op amp, consider the input currents (I_{inv} and I_{ni}) to be 0 A.

Example 9-1

For each of the op amps shown in Figure 9-2, find the net op amp input voltage V_{error}.

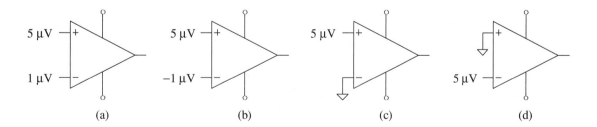

(a) (b) (c) (d)

Figure 9-2 Example op amp inputs

Figures 9-2(a) and 9-2(b) have dual inputs (input voltages on both the input terminals). Figures 9-2(c) and 9-2(d) have a single ended input and require the other input terminal to be connected to common (0 V).

Solution

a. See Figure 9-2(a).

$$V_{error} = V_{ni} - V_{inv}$$

$$V_{error} = (5 \ \mu V) - (1 \ \mu V) = 4 \ \mu V$$

b. See Figure 9-2(b).

$$V_{error} = (5 \ \mu V) - (-1 \ \mu V) = 6 \ \mu V$$

c. See Figure 9-2(c).

$$V_{error} = (5 \ \mu V) - (0 \ V) = 5 \ \mu V$$

d. See Figure 9-2(d).

$$V_{error} = (0 \ V) - (5 \ \mu V) = -5 \ \mu V$$

Practice

For each of the op amps shown in Figure 9-3, find the net op amp input voltage V_{error}.

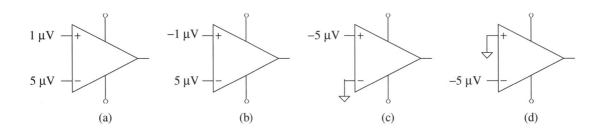

(a)   (b)   (c)   (d)

Figure 9-3 Practice problem

Answer: (a) $-4 \ \mu V$ (b) $-6 \ \mu V$ (c) $-5 \ \mu V$ (d) $+5 \ \mu V$

Op Amp Input Voltage Limitations

The input voltages to the op amp must not exceed the supply voltages of $\pm E_{supply}$. That is, V_{inv} and V_{ni} must lie between the supply voltage values.

$$-E_{supply} < V_{inv} \ \& \ V_{ni} < +E_{supply}$$

If either input voltage is outside the supply range, some of the components (namely, diodes and BJTs) would be biased improperly and possibly destroyed. Practically, the input can usually exceed these limits by about 0.3 V and still not cause op amp damage.

Example 9-2

Which of the following circuits in Figure 9-4 are unsafe based upon the op amp signal input voltages relative to the supplies?

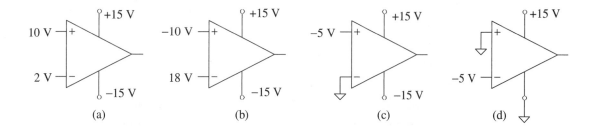

Figure 9-4 Examples of safe and unsafe inputs

Solution

a. See Figure 9-4(a). Both inputs (10 V and 2 V) are within the supply voltage range (–15 V to +15 V). The op amp is safe for the given inputs relative to the supply voltages.

b. See Figure 9-4(b). The +18 V input lies outside the supply voltage range (–15 V to +15 V), therefore, this op amp is not safe.

c. See Figure 9-4(c). Both inputs (–5 V and 0 V) are within the supply voltage range (–15 V to +15 V). The op amp is safe for the given inputs relative to the supply voltages.

d. See Figure 9-4(d). The –5 V input is outside the supply range (0 V to +15 V), therefore, this op amp is not safe.

Practice

For each of the op amps shown in Figure 9-5, determine if the input voltages are within the appropriate range for the op amp with its respective supplies.

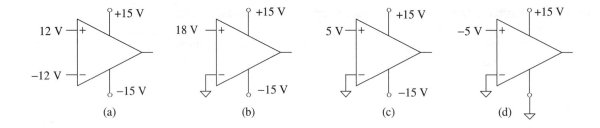

Figure 9-5 Practice input voltages

Answer: (a) safe, (b) not safe, (c) safe, (d) not safe

Op Amp Output Voltage–Rail Voltage

The op amp multiplies the input voltage V_{error} (the difference between the input voltage pins) by a very large constant to produce its output voltage $V_{out\ op\ amp}$.

$$V_{out\ op\ amp} = A V_{error}$$

where

$V_{out\ op\ amp}$ = output voltage of the op amp
A = voltage gain of the op amp
V_{error} = input voltage to the op amp ($V_{ni} - V_{inv}$)

The output voltage of the op amp is limited to the range of its supply voltages of $\pm E_{supply}$ less some headroom voltage needed to operate the op amp. That is, $V_{out\ op\ amp}$ cannot exceed the supply voltage values.

$$-E_{supply} < V_{out\ op\ amp} < +E_{supply}$$

These are also referred to as the op amp's **rails** or **rail voltages**. In practice, the actual output range is less than $\pm E_{supply}$. The circuits internal to the op amp may require some *headroom* to operate. For example, supplies of ± 15 V may produce an output voltage range of

$$-14\ V \leq V_{out\ op\ amp} \leq +13\ V$$

In this case, the op amp requires 2 V of *headroom* for the $+E_{supply}$ and 1 V of *headroom* for the $-E_{supply}$ to operate. Op amps require anywhere from millivolts to a couple of volts of headroom, depending upon the op amp model. The +13 V rail is referred to as the *upper rail*. And –14 V is referred to as the *lower rail*.

$$V_{lower\ rail} \leq V_{out\ op\ amp} \leq V_{upper\ rail}$$

Once the output hits one of the rails, the op amp is said to be **saturated**. *Any additional increase in the input voltage will not increase the output voltage; the output will stay at the rail voltage.*

Example 9-3

Find the output voltages for the op amps of Figure 9-6. The dc voltage gain of the op amp is 100,000 ($A=10^5$). Check that the input voltages are within the appropriate range before solving. Assume that the real rail voltages are +13 V and –14 V.

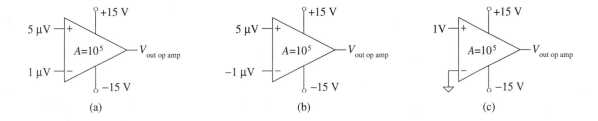

(a) (b) (c)

Figure 9-6 Example inputs, gains, and outputs

Solution

In all three cases the input voltages are appropriate and fall between ±15 V.

a. See Figure 9-6(a). The net input voltage to the op amp is

$$V_{error} = V_{ni} - V_{inv} = (5\ \mu V) - (1\ \mu V) = 4\ \mu V$$

The op amp output voltage is

$$V_{out\ op\ amp} = AV_{error} = (10^5)(4\ \mu V) = 400\ mV$$

This is well within the rail voltage range.

b. See Figure 9-6(b). The net input voltage to the op amp is

$$V_{error} = (5\ \mu V) - (-1\ \mu V) = 6\ \mu V$$

The op amp output voltage is

$$V_{out\ op\ amp} = (10^5)(6\ \mu V) = 600\ mV$$

This is well within the rail voltage range.

c. See Figure 9-6(c). The net input voltage to the op amp is

$$V_{error} = (1\ V) - (0\ V) = 1\ V$$

The op amp output voltage is

$$V_{out\ op\ amp} = (10^5)(1\ V) = 100,000\ V$$

No. This is not possible. The maximum output voltage is limited to the op amp's upper rail, that is, +13 V. Therefore, the actual output voltage is its upper rail voltage.

$$V_{out\ op\ amp} = 13\ V$$

Practice

Find the output voltages for the op amps of Figure 9-7. The gain of the op amp is 100,000 ($A = 10^5$). Check that the input voltages are within the appropriate range before solving. Assume that the real rail voltages are +13 V and −14 V.

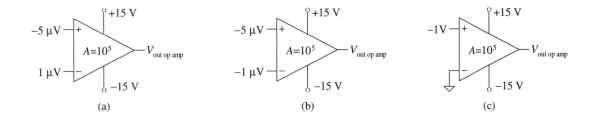

(a)   (b)   (c)

Figure 9-7 Practice problem

Answer: (a) –600 mV (b) –400 mV (c) –14 V (lower rail, saturated)

Supply ±E$_{supply}$ Limitations

Each op amp has a maximum limitation of the difference in supply voltages that it can handle. The difference in supplies is the net supply voltage being applied to the chip.

$$E_{supply\ difference} = (+E_{supply}) - (-E_{supply})$$

If this voltage difference exceeds the manufacturer's specifications, then the op amp is in jeopardy. For example, the LM324 has a maximum supply difference specification of 32 V. If this specification is exceeded, the op amp can be damaged.

Example 9-4

An LM324 op amp has a positive supply +E$_{supply}$ of +20 V and a negative –E$_{supply}$ of –20 V. Is the op amp safe?

$$E_{supply\ difference} = (20\ V) - (-20\ V) = 40\ V$$

Solution

No. The specification of 32 V is exceeded.

Practice

An LM324 op amp has a positive supply +E$_{supply}$ of +30 V and the negative –E$_{supply}$ is connected to common. Is the op amp safe?

Answer: Yes (difference less than the maximum specification of 32 V)

Op Amp Maximum Output Current

The **short circuit** current is the current through a short attached to the op amp. The op amp output is designed to be **current limited**. The short circuit current of the LM741 is about 20 mA and the LM324 is about 40 mA. *However, the maximum LM741 and LM324 output op amp current under normal operating conditions is typically only a few milliamps.* Once that current is exceeded the op amp output degrades significantly.

9.2 Op Amp Comparator

Due to the very high voltage gain of the op amp (for example, $A = 10^5$ for the LM324 op amp), the op amp slams to either the *upper rail* (based upon $+E_{supply}$) or the *lower rail* (based upon $-E_{supply}$). Very little input voltage is needed to slam to one rail or the other.

For the LM324 op amp, the upper rail voltage is approximately its positive supply voltage less about 1.5 V (worst case).

$$V_{upper\ raill} \approx +E_{supply} - 1.5\ V \qquad \text{(maximum output)}$$

For the LM324 op amp, the lower rail voltage is approximately its negative supply voltage plus about 0.8 V (worst case).

$$V_{lower\ rail} \approx -E_{supply} + 0.8\ V \qquad \text{(minimum output)}$$

The LM324 is designed to work with digital as well as analog electronics. When used with digital electronic circuits, +5 V is applied to the $+E_{supply}$ pin and the $-E_{supply}$ pin is connected to common. The rail voltages fall within the required *digital high* and *digital low*. The LM741 cannot work with digital circuits since it requires about 2 V of headroom.

Example 9-5

An LM324 op amp with a dc voltage gain of $A = 10^5$ has an upper rail voltage of +13 V. What is the minimum op amp input voltage required to drive this op amp to its upper rail?

Solution

$$V_{out\ op\ amp} = A V_{error}$$

Solving for V_{error}

$$V_{error} = \frac{V_{out\ op\ amp}}{A} = \frac{13\ V}{10^5} = 130\ \mu V \approx 0\ V$$

Only +130 μV drives this op amp to its upper rail. For all practical purposes, this input voltage is so small that you can assume that anything above 0 V will drive the op amp to its upper rail.

Practice

An LM324 op amp with a dc voltage gain of $A = 10^5$ has a lower rail voltage of −14 V. What is the minimum op amp input voltage required to drive this op amp to its lower rail?

Answer: −140 μV (approximately 0 V, so ideally assume that anything less than 0 V drives the op amp to its lower rail).

Transfer Curve

A transfer curve relates an output variable to an input variable and gives a visual view of the interaction of the output as a function of the input. In the case of the op amp, its output voltage $V_{\text{out op amp}}$ is related to and controlled by its input voltage V_{error}. The next example demonstrates a test circuit that can be used to generate a transfer curve for an op amp.

Example 9-6

The circuit of Figure 9-8 is a test circuit that changes the input voltage to an LM741 op amp with a dc voltage gain of $A = 10^5$.

Assume the op amp needs about 2 V of headroom for its low and high outputs so that its rails are ±10 V. (a) First analyze the potentiometer input circuit to determine the range of the input voltage to the op amp. (b) Based upon the input range, sketch a transfer curve of the output voltage $V_{\text{out op amp}}$ as a function of its input voltage V_{error}.

Solution

a. The potentiometer circuit is essentially a series circuit. Why? Since the op amp input current is ideally 0 A, the resistor circuit forms an *essentially series* circuit. The current flows from the +12 V supply down through R1, R$_{\text{pot}}$, R2, and into the −12 V supply. The net voltage drop across this resistive circuit is

$$E_{\text{net}} = E_{\text{supply difference}} = +12\,\text{V} - (-12\,\text{V}) = 24\,\text{V}$$

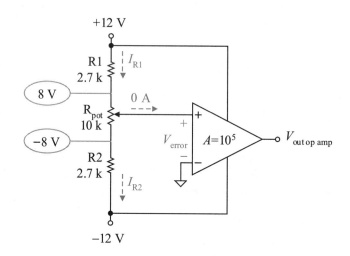

Figure 9-8 Test circuit to generate the op amp's transfer curve

The essentially series circuit resistance of R1, R_{pot}, and R2 has a total resistance of

$$R_{series} = 2.7 \text{ k}\Omega + 10 \text{ k}\Omega + 2.7 \text{ k}\Omega = 15.4 \text{ k}\Omega$$

The essentially series circuit current is

$$I_{series} = \frac{24 \text{ V}}{15.4 \text{ k}\Omega} = 1.56 \text{ mA}$$

The node voltage at the top of the 10 kΩ pot is its +12 V node voltage (bubble voltage) less the fall across the top 2.7 kΩ resistor.

$$V_{top \text{ of pot}} = +12 \text{ V} - (1.56 \text{ mA}) (2.7 \text{ k}\Omega) = 7.8 \text{ V}$$

The node voltage at the bottom of the 10 kΩ pot is its −12 V node voltage (bubble voltage) plus the rise across the bottom 2.7 kΩ resistor.

$$V_{bottom \text{ of pot}} = -12 \text{ V} + (1.56 \text{ mA}) (2.7 \text{ k}\Omega) = -7.8 \text{ V}$$

The op amp input voltage range of V_{error} is approximately

$$-8 \text{ V} \leq V_{error} \leq +8 \text{ V}$$

b. Assuming a headroom of 2 V for the op amp, the op amp output slams the −10 V lower rail for $V_{error} < 0$ V. And the op amp output slams the +10 V upper rail for $V_{error} > 0$ V. *The input voltage at which the comparator trips from one state to the other is approximately 0 V.*

$$V_{error} = \frac{V_{out\,op\,amp}}{A} = \frac{\pm 10 \text{ V}}{10^5} = \pm 100 \text{ μV} \approx 0 \text{ V}$$

As you sweep the potentiometer wiper arm *from the bottom to the top* of the potentiometer (−8 V to +8 V), you generate the transfer curve of Figure 9-9(a).

For $V_{error} < 0$ V, $V_{out\,op\,amp} = -10$ V

For $V_{error} = 0$ V, $V_{out\,op\,amp}$ trips to +10 V.

For $V_{error} > 0$ V, $V_{out\,op\,amp} = +10$ V

You obtain the same results if you sweep the potentiometer wiper arm from *the top to the bottom* (from +8 V to −8 V) as shown in Figure 9-9(b).

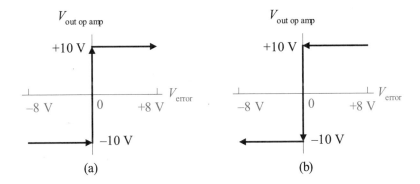

(a) (b)

Figure 9-9 Noninverting comparator transfer curve

This is called a **noninverting comparator** since a *positive* input drives the output to the *upper rail voltage* and a negative input drives the output to the lower rail voltage. This noninverting result is created since the *input signal is applied to the noninverting terminal of the op amp.*

Practice

Repeat this example with the signal applied to the inverting input pin and the noninverting input pin connected to common as shown in Figure 9-10(a).

Answer: Figure 9-10(b) is the transfer curve generated as the potentiometer sweeps from –8 V to +8 V or +8 V to –8 V. A positive input signal drives the op amp to its lower rail. Thus this circuit is termed an **inverting comparator**. This becomes intuitively obvious if you notice that *the input signal is driving the inverting terminal of the op amp*.

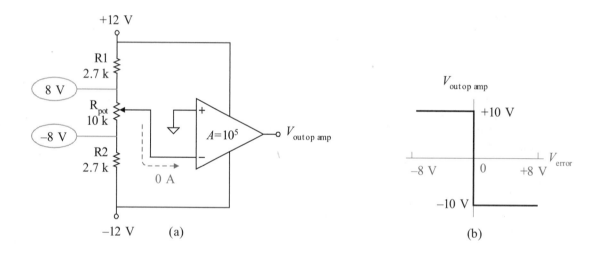

Figure 9-10 Inverting comparator test circuit and transfer curve

The comparator used in the above example is called a **zero-crossover detector** or comparator since the op amp switches output rails when the op amp input crosses through 0 V.

For example, the noninverting comparator of the last example uses the inverting input pin as the fixed *reference voltage* (in this example it is a 0 V reference). As the input voltage on the noninverting pin goes from –8 V to +8 V, the op amp slams to the opposite rail *when the input voltage approximately equals the reference voltage of* 0 V.

Suppose you need a crossover voltage other than 0 V, then a reference voltage needs to be set up at one of the input op amp pins. First determine which op amp input terminal is to receive the input signal V_{in}. Is this to be a noninverting or an inverting comparator? The other op amp input pin becomes the reference voltage V_{ref} pin. For example, if you need a noninverting comparator with a crossover voltage of +5 V, then the signal V_{in} is fed to the noninverting pin and a fixed 5 V dc voltage is applied to the inverting input terminal. Figure 9-11(a) demonstrates this reference voltage circuit and Figure 9-11(b) shows its corresponding transfer curve.

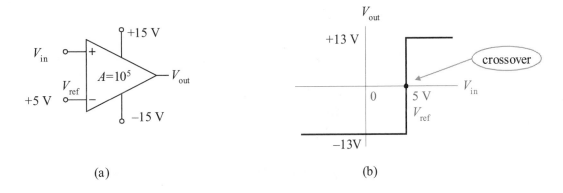

(a)   (b)

Figure 9-11 Noninverting comparator with a +5 V crossover (reference voltage)

The next example and practice problem provide reference voltages with two different approaches. The example problem uses a voltage divider resistor circuit to establish a reference voltage, while the practice problem uses a zener diode circuit. Both are essentially series circuits since the op amp's input current is ideally 0 A.

Example 9-7

The circuit of Figure 9-12(a) is a noninverting comparator that uses a voltage divider circuit to establish a reference voltage at the inverting input pin. (a) Find the value of R1 needed to establish a reference voltage of 6.2 V. (b) Draw its transfer curve. Assume that the op amp requires ±2 V of headroom.

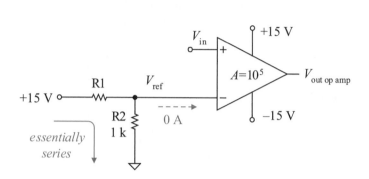

Figure 9-12(a)   Example of establishing a reference voltage

Solution

a. R1 and R2 form a simple voltage divider essentially series circuit. The simplest approach is to use the voltage divider rule. The reference voltage V_{ref} is the same as the drop across the R2 resistor.

$$V_{ref} = \left(\frac{R2}{R1+R2}\right)(+E_{supply})$$

Substitute the known values into this equation.

$$6.2\ V = \left(\frac{1\ k\Omega}{R1+1\ k\Omega}\right)(+15\ V)$$

Set up to solve for R1. Note that the voltage units cancel and the answer is in ohms.

$$6.2\ V\ (R1+1\ k\Omega) = (1\ k\Omega)(+15\ V)$$

Solving for R1

$$R1 = \frac{15\ k\Omega - 6.2\ k\Omega}{6.2\ k\Omega} = 1.42\ k\Omega$$

Unfortunately this is a nonstandard value. The closest possible standard value is 1.5 kΩ, and that produces a V_{ref} of 6.4 V, not the required 6.2 V. This situation is exacerbated when you take into account the tolerance of the resistors being used. For example, using ±5% tolerance resistors, the error range could vary from 0 to 10%. Another possible solution is to replace R1 with a custom precision resistor, but that is a very expensive solution. Another solution is to replace R1 with a potentiometer and accurately adjust it to 1.42 kΩ. But, this is a very labor intensive, expensive, unrealistic solution.

A better solution is to use a zener reference diode to provide the desired reference voltage. The practice problem uses a reverse biased zener diode to accurately establish a 6.2 V reference.

b. Figure 9-12(b) shows the transfer curve for this noninverting comparator with a 6.2 V reference voltage.

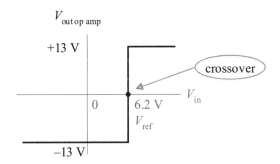

Figure 9-12(b) Transfer curve of example problem

Practice

The circuit of Figure 9-13 uses a resistor and a 6.2 V zener (essentially in series) to establish a 6.2 V reference voltage at the inverting input pin.

The zener current is very important. This reverse biased zener drops 6.2 V at 20 mA. Find the resistance R needed to produce a 6.2 V zener drop with a current close to 100 mA. A slight variation in zener current does not significantly affect the zener voltage drop. However, zener voltages can also vary slightly even with the same zener diode model.

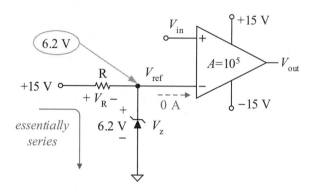

Figure 9-13 Establishing a reference voltage using a zener diode

---

Answer: 440 Ω (possible standard values are 430 Ω, or 470 Ω)

The next example is a practical circuit that uses a temperature transducer that senses temperature and converts it to a signal voltage. This signal voltage is compared to a reference voltage (essentially representing a reference temperature). The comparator output then controls two LEDs that indicate the temperature status (above or below the temperature reference).

Example 9-8

The circuit of Figure 9-14(a) is a temperature sensor and status display circuit. The LM34 is a temperature sensor. It is rated at 10 mV per °F with 0 V being equivalent to 0°F. So 10 mV represents 1°F, 20 mV represents 2°F, etc. The potentiometer controls the reference voltage to the inverting op amp input pin. The LM324 op amp is specified to need 1.5 V of upper rail headroom and 0.8 V of lower rail headroom. The LEDs are specified to be 2 V each.

$R_{pot\ bottom}$ is set to 1.28 kΩ. (a) Find the reference voltage established at the op amp's inverting input pin. (b) Find the corresponding temperature for that reference voltage. What significance does this temperature have? (c) If the sensor temperature is 20°F, find V_{error}. Find the op output voltage. Which LED is lit? (d) If the sensor temperature is 35°F, find V_{error}. Find the op amp output voltage. Which LED is lit? (e) Find the current flowing through each LED when it is lit.

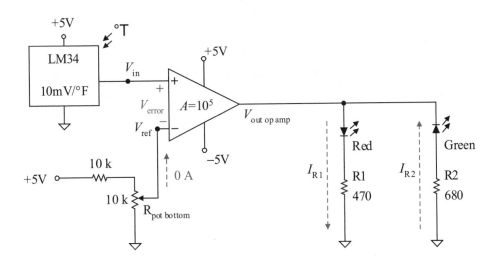

Figure 9-14(a)   Temperature sensor and status display

Solution

a. The potentiometer wiper arm connects to the input of the op amp which has ideally 0 A of current flowing into it. Thus, the potentiometer is part of an *essentially series* circuit with the 10 kΩ fixed resistance in series with the 10 kΩ potentiometer. Apply the voltage divider rule to this series circuit to find the voltage drop across the bottom of the potentiometer. This is the same voltage as the reference voltage V_{ref}.

$$V_{ref} = V_{pot\ bottom} = \left(\frac{1.28\ k\Omega}{20\ k\Omega}\right)(+5\ V) = 320\ mV$$

b. Using the specification of the LM34 temperature sensor of 10 mV per °F, the corresponding temperature of the reference is

$$T_{ref} = \frac{320\ mV}{10\ mV/\ °F} = 32°F$$

The reference voltage is set to correspond to the temperature at which water freezes.

c. The sensor senses a temperature of 20°F. The voltage input V_{in} to the op amp from the sensor is

$$V_{in} = (10 \text{ m V/}°F) \ (20°F) = 200 \text{ mV}$$

The op amp input voltage V_{error} is the difference between its op amp input terminals.

$$V_{error} = V_{ni} - V_{inv} = V_{in} - V_{ref}$$

$$V_{error} = 200 \text{ mV} - 320 \text{ mV} = -120 \text{ mV}$$

The LM324 has a typical voltage gain of 100,000 (that is, 10^5). If possible, the op amp output voltage would climb to

$$V_{out \ op \ amp} = AV_{error} = (10^5)(-120 \text{ mV}) = -12,000 \text{ V}$$

No, this is not possible. The op amp slams to its lower rail. Allowing for 0.8 V of headroom for the lower rail, its lower rail voltage is

$$V_{out \ op \ amp} = -5 \text{ V} + 0.8 \text{ V} = -4.2 \text{ V}$$

This forward biases the green LED and reverse biases the red LED. Thus, the green LED is lit indicating that the temperature is below freezing.

Figure 9-14(b) shows the circuit with the above voltages.

d. The sensor senses a temperature of 35°F. The voltage input V_{in} to the op amp from the sensor is

$$V_{in} = (10 \text{ mV} / °F)(35°F) = 350 \text{ mV}$$

The op amp input voltage V_{error} is the difference between its op amp input terminals.

$$V_{error} = 350 \text{ mV} - 320 \text{ mV} = 30 \text{ mV}$$

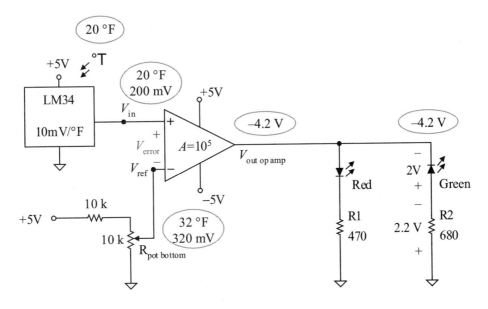

Figure 9-14(b) Temperature sensor with green LED lit

The LM324 has a typical voltage gain of 10^5, so the op amp output voltage is

$$V_{\text{out op amp}} = A\,V_{\text{error}} = (10^5)\,(30 \text{ mV}) = 3000 \text{ V}$$

No. The op amp slams to its upper rail. Allowing for 1.5 V of headroom for the upper rail, its upper rail voltage is

$$V_{\text{out op amp}} = 5 \text{ V} - 1.5 \text{ V} = 3.5 \text{ V}$$

So the op amp slams to its upper rail of +3.5 V. This forward biases the red LED and reverse biases the green LED. Thus, the red LED is lit, indicating that the temperature is above freezing. Figure 9-14(c) shows the circuit with these corresponding voltages.

e. It is important that the op amp output current be only a few milliamps to prevent the op amp from going into current limiting, which causes its output voltage to decrease. The red LED is in series with R1, which is 470 Ω. The red LED is lit when the op amp is slamming its upper rail of +3.5 V. Since the LED drops 2 V, the R1 resistor must drop the remaining 1.5 V. Using Ohm's law for resistance.

$$I_{R1} = I_{red\ LED} = \frac{1.5\ V}{470\ \Omega} = 3.2\ mA$$

The green LED is in series with R2, which is 680 Ω. The green LED is lit when the op amp is slamming its lower rail of –4.2 V. Since the LED drops 2 V, the R2 resistor must drop the remaining 2.2 V. Using Ohm's law for resistance.

$$I_{R1} = I_{green\ LED} = \frac{2.2\ V}{680\ \Omega} = 3.2\ mA$$

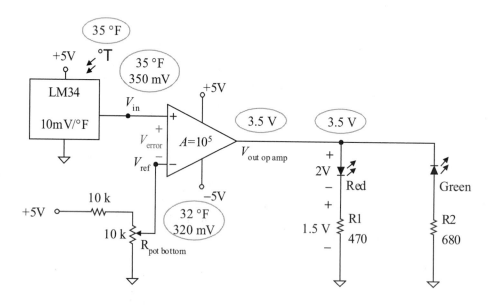

Figure 9-14(c)   Temperature sensor with red LED lit

Practice

The supply voltages for the circuit of Figure 9-14(a) are changed to ±9 V. (a) Determine the value of $R_{pot\ bottom}$ needed to produce a 70°F reference voltage. (b) Find the new op amp output rail voltages. (c) Find the standard values of R1 and R2 needed to limit LED currents to about 15 mA.

Answer: (a) 1.56 kΩ (b) 7.5 V, –8.2 V (c) 367 Ω (standard 390 Ω), 413 Ω (standard 430 Ω)

The problem with the above comparator is that very little noise voltage (in the microvolt range) could cause the op amp output to **jitter** between its upper and lower rails. If the input voltage is very close to its reference voltage, a little noise could cause this *jitter* effect. For some circuits, this would be an intolerable situation. For example, if this were a comparator sending a signal to the on-off control circuitry of a heavy-duty compressor for a freezer, jittering on and off could be potentially damaging to the compressor and at minimum an annoyance.

The next section of this chapter provides a solution to the jitter problem. Positive feedback (feeding part of the output signal back to add to the input signal) eliminates *jitter*.

9.3 Positive Feedback Comparators

Positive feedback can be used to eliminate the jitter around the reference voltage. *With positive feedback, a part of the output signal is fed back to add with the original signal to increase the magnitude of the signal to the op amp input terminals.* V_{error} grows, causing V_{out} to grow, causing V_{error} to grow, etc. Very quickly the op amp is driven to a rail voltage. The output becomes fixed at the rail voltage. The feedback signal (which is a fraction of the output voltage) is then also fixed. So the entire circuit becomes stable with fixed voltages and currents *with the op amp output voltage fixed at the upper or lower rail depending upon the value of* V_{error}.

Positive feedback
$V_{\text{out op amp}} = \pm V_{rail}$

As noted earlier, an input signal can be fed to either the noninverting op amp input pin (creating a noninverting comparator) or to the inverting op amp input pin (creating an inverting comparator). First you will examine a noninverting positive feedback comparator example and then an inverting positive feedback comparator circuit. In both cases you will use circuit laws and basic circuit analysis techniques to evaluate these circuits. *The analysis is centered on the essentially series circuit nature of the feedback circuit since the op amp input currents are ideally* 0 A.

Noninverting Positive Feedback Comparator

The circuit of Figure 9-15 is a **noninverting positive feedback compara-tor**. The input signal is connected to the noninverting op amp input pin, so the output signal is not inverted (**noninverting**).

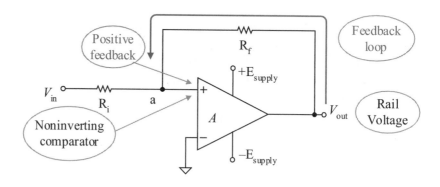

Figure 9-15 Noninverting positive feedback comparator

For a simple op amp circuit, the type of feedback (positive or nega-tive) is very easy to determine. With simple feedback, the feedback signal is going from V_{out} to the *noninverting op amp input pin* (+); therefore, it is **positive feedback**.

To *formally* prove positive feedback for this circuit, assume V_{in} is a positive (+) input test signal. The op amp output is still a (+) test signal. That (+) test signal is fed back through R_f to the noninverting terminal as a (+) feedback signal. The input test signal and the feedback signal meet at node *a*; and both signals are positive (+). *The feedback signal is the same sign as the original test signal; therefore this is positive feedback.*

This is called a **positive feedback circuit** since the feedback signal *adds to* (increases the magnitude of) the V_{error} signal. That larger V_{error} voltage is amplified by the op amp to create a larger output signal, which creates a larger feedback signal. This positive feedback continues with the signal growing larger and larger *until the output slams the op amp's rail* (either the lower or upper rail) depending on the polarity of V_{error}. The only *two possible outputs* of this circuit are $\pm V_{rail}$. Since this is a nonin-verting comparator, a positive V_{error} drives the op amp to the upper rail.

$$V_{out} = V_{out\ op\ amp} = V_{upper\ rail} \qquad\qquad (V_{error} > 0)$$

$$V_{out} = V_{out\ op\ amp} = V_{lower\ rail} \qquad\qquad (V_{error} < 0)$$

It is critical that you understand the above thought process to prove positive feedback. Later circuits with negative feedback, called amplifiers, will be examined. The results are completely different.

Example 9-9

The circuit of Figure 9-16 has an op amp with rail voltages of ±10 V. The input has been fixed at 3 V, which drives the output to the +10 V rail.

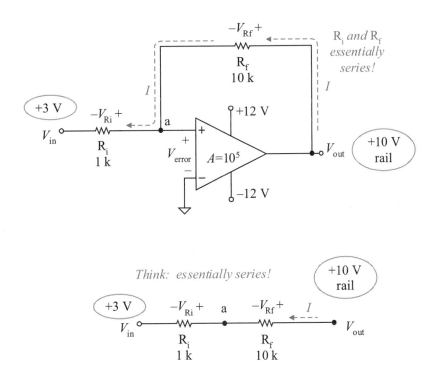

Figure 9-16 Slamming upper rail with positive input

(a) Verify that this is a noninverting circuit. (b) Verify that this is a positive feedback circuit. (c) Find the feedback current I. (d) Find the voltage drop across R_i. (e) Find the voltage at node a. (f) Find the op amp input voltage V_{error}. (g) Does the value of

V_{error} support the supposition that the op amp is slamming the upper rail?

Solution

a. The input signal V_{in} is being fed to the noninverting op amp input pin, so this is a *noninverting circuit.*

b. The feedback signal is being fed back to the op amp's (+) noninverting input pin so this is *positive feedback.*

c. Since the input current to the op amp is assumed to be 0 A, R_i and R_f form an *essentially series circuit.* Note that the current I is flowing from the higher potential V_{out} (+10 V) to the lower potential V_{in} (+3 V). First find the voltage drop V_{RT} across the two resistors that are essentially in series.

$$V_{RT} = V_{out} - V_{in} = (10 \text{ V}) - (3 \text{ V}) = 7 \text{ V}$$

Now find the feedback current I using Ohm's law.

$$I = \frac{7 \text{ V}}{11 \text{ k}\Omega} = 0.636 \text{ mA}$$

d. Next find the voltage drop across R_i using Ohm's law.

$$V_{Ri} = (0.636 \text{ mA})\,(1 \text{ k}\Omega) = 0.636 \text{ V}$$

e. Now find the node voltage V_a. Start at the known node voltage at V_{in} of 3 V. Walk to node a, raising the node voltage by V_{Ri} of 0.636 V.

$$V_a = 3 \text{ V} + 0.636 \text{ V} = 3.636 \text{ V}$$

f. Next find V_{error} which is always the difference between the op amp's input pins.

$$V_{error} = V_{ni} - V_{inv} = 3.636 \text{ V} - 0 \text{ V} = 3.636 \text{ V}$$

g. Since $V_{error} > 0$ V, the op amp does slam to its upper rail of +10 V. The output is confirmed to be +10 V.

Practice

The circuit of Figure 9-17 has an op amp with rail voltages of ± 10 V. The input has been fixed at –3 V, which drives the output to the –10 V rail.

(a) Verify that this is a noninverting circuit. (b) Verify that this is a positive feedback circuit. (c) Find the feedback current I. (d) Find the voltage drop across R_i. (e) Find the voltage at node a. (f) Find the op amp input voltage V_{error}. (g) Does the value of V_{error} support the supposition that the op amp is slamming the lower rail?

Note that the current direction and voltage polarities are the reverse of the example problem. Since V_{in} (–3 V) is at a higher potential voltage than V_{out} (–10 V), the essentially series current flows from V_{in} to V_{out}.

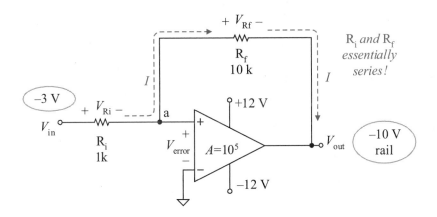

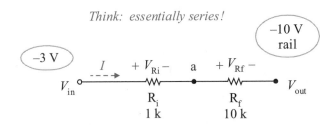

Figure 9-17 Slamming lower rail with negative input

Answer: (a) input signal fed to the noninverting op input pin (b) feedback signal fed back to the noninverting op amp input pin (c) 0.636 mA (d) 0.636 V (e) –3.636 V (f) –3.636 V (g) yes, $V_{error} < 0$ V

In the previous example and practice, the op amp changes from +3 V to –3 V causing a change in the output from +10 V to –10 V. At what V_{in} voltage does the output trip from its upper rail voltage (+10 V) to its lower rail voltage (–10 V)?

Example 9-10

Find the input voltage V_{in} required to trip the circuit of Figure 9-18(a) from its upper rail (+10 V) to its lower rail (–10 V).

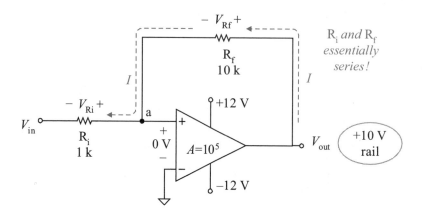

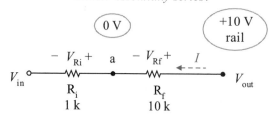

Figure 9-18(a) Find V_{in} that trips the circuit to the –10 V rail

Solution

As noted earlier in this chapter, the op amp slams to its other rail as V_{error} crosses through 0 V. Again, back to basics. Redraw the schematic of Figure 9-16, assign V_{error} equal to 0 V and solve for V_{in}.

Since V_{error} is 0 V, the node voltage V_a must also be at 0 V.

$$V_a = V_{\text{common}} + V_{\text{error}} = 0 \text{ V} + 0 \text{ V} = 0 \text{ V}$$

Op amp always trips at
$V_{\text{error}} \cong 0 \text{ V}$

Now find V_{Rf} using the difference between its node voltages.

$$V_{\text{Rf}} = V_{\text{out}} - V_a = 10 \text{ V} - 0 \text{ V} = 10 \text{ V}$$

Apply Ohm's law for resistance R_f. This is also the current that flows through R_i since the op amp input current is 0 A.

$$I = I_{\text{Ri}} = I_{\text{Rf}} = \frac{10 \text{ V}}{10 \text{ k}\Omega} = 1 \text{ mA}$$

Apply Ohm's law for resistance R_i.

$$V_{\text{Ri}} = 1 \text{ mA} \times 1 \text{ k}\Omega = 1 \text{ V}$$

The last step is to find the node voltage V_{in} (which is the input voltage that trips the op amp). The right node of R_i is at 0 V. So starting at the right node of 0 V and falling 1 V across R_i, the input node voltage V_{in} must be –1 V.

$$V_{\text{in}} = V_a - V_{\text{Ri}} = 0 \text{ V} - (1 \text{ V}) = -1 \text{ V}$$

Summarizing, the input voltage to this comparator V_{in} must go from a positive voltage, through 0 V, to –1 V before the op amp trips from its upper rail to its lower rail (–10 V). Figure 9-18(b) shows the transfer curve for this comparator as the input V_{in} goes from positive input voltages to negative input voltages.

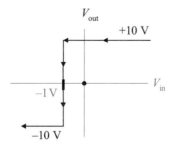

Figure 9-18(b) Example circuit transfer curve (–1 V trip point)

Practice

Find the input voltage V_{in} required to trip the circuit of Figure 9-19(a) from its lower rail of (–10 V) to its upper rail (+10 V). This is similar to the example except in this case the input voltage is going from a negative value to a positive value.

Answer: +1 V (the input voltage V_{in} starts negative, passes through 0 V, and trips the op amp at +1 V to its +10 V rail). See Figure 9-19(b) for its transfer curve.

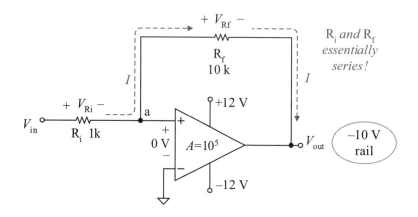

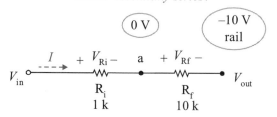

Figure 9-19(a) Find V_{in} that trips the circuit to +10 V rail

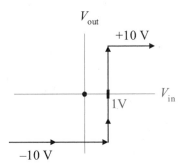

Figure 9-19(b) Example circuit transfer curve (+1 V trip point)

The preceding example and practice problem produced two trip points. When the op amp output is at its upper rail (+10 V) the trip point is −1 V. When the op amp output is at its lower rail (−10 V) the trip point is +1 V. Since there are two trip points, distinguish them by calling them the lower trip point (LTP) and the upper trip point (UTP).

$$V_{\text{LTP}} = -1 \text{ V} \qquad \text{(lower trip point)}$$

$$V_{\text{UTP}} = +1 \text{ V} \qquad \text{(upper trip point)}$$

Figure 9-20 combines the two transfer curves overlaid on one graph to help us visualize how this circuit operates and demonstrates how jitter around the reference is eliminated. If the op amp is slammed against the upper rail (+10 V), the input voltage must go past 0 V down to −1 V to trip. Once the op amp does trip, a new trip point of +1 V is created. Now the input voltage must go past 0 V to the upper trip point before it trips again.

This effect where two different traces are created based upon the starting voltage is called the **hysteresis effect**. The trip point is determined by the current state of the op amp output voltage. An important feature of this positive feedback comparator is that it does not jitter around 0 V. The input voltage must go significantly past 0 V before it trips to the other state. In this circuit, a 1 V threshold is created before the comparator trips. If this is too insensitive, you must change the feedback voltage dividing resistor R_i and R_f to produce a lower feedback voltage. For example, if the feedback resistor R_f were changed to a 100 kΩ resistor, the trip point voltages would be changed to ±0.1 V and the comparator would be much more sensitive. This circuit has a special name; it is called a **noninverting Schmitt trigger**.

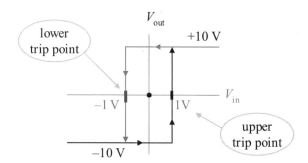

Figure 9-20 Example circuit complete transfer curve

Inverting Positive Feedback Comparator

The circuit of Figure 9-21 is an inverting positive feedback circuit. The input signal is being fed directly to the inverting input, so it is an **inverting comparator**.

The feedback signal in this simple op amp circuit is being fed back to the noninverting (+) op amp input pin, thus it is **positive feedback**.

More formally, assume V_{in} is a positive (+) input test signal. The op amp output signal becomes a (−) test signal. That (−) test signal is fed back through R_f to the op amp noninverting terminal as a (−) signal. That feedback signal appears as a (−) at the op amp output pin. The feedback signal meets the test signal at the op amp output. *Both signals have the same sign; they are both negative (−). Therefore, this is positive feedback.*

Since it is positive feedback, the op amp is driven to either the upper or lower rail voltage. The only *two possible outputs* of this circuit are $\pm V_{rail}$. Since this is an inverting comparator, a negative V_{error} drives the op amp to the upper rail.

$$V_{out} = V_{out\,op\,amp} = V_{upper\,rail} \qquad (V_{error} < 0)$$

and

$$V_{out} = V_{out\,op\,amp} = V_{lower\,rail} \qquad (V_{error} > 0)$$

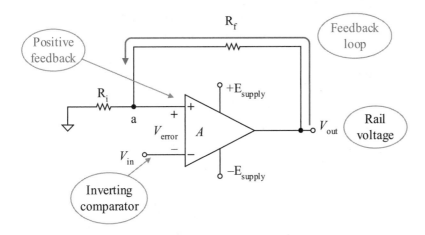

Figure 9-21 Inverting positive feedback comparator

Example 9-11

The circuit of Figure 9-22 has an op amp with rail voltages of ±10 V. The input has been fixed at 3 V, which drives the output to the −10 V rail.

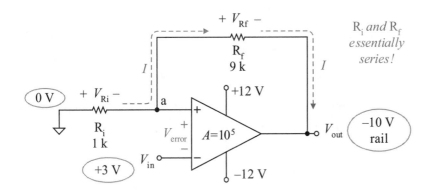

Figure 9-22 Slamming lower rail with positive input

(a) Verify that this is an inverting circuit. (b) Verify that this is a positive feedback circuit. (c) Find the feedback current I. (d) Find the voltage drop across R_i. (e) Find the voltage at node a. (f) Find the op amp input voltage V_{error}. (g) Does the value of V_{error} support the supposition that the op amp is slamming the upper rail?

Solution

a. The input signal V_{in} is driving the op amp's inverting input pin so this is an *inverting circuit*.

b. The feedback signal is being fed back to the op amp's (+) noninverting input pin so this is *positive feedback*.

c. Since the input current to the op amp is assumed to be 0 A, R_i and R_f form an *essentially series circuit*. Note that the current I is flowing from the higher potential *common* (0 V) to the lower potential V_{out} (–10 V) First find the voltage drop V_{RT} across these two resistors.

$$V_{RT} = V_{common} - V_{out} = (0\ V) - (-10\ V) = 10\ V$$

Now find the feedback current I using Ohm's law.

$$I = \frac{10\ V}{10\ k\Omega} = 1\ mA$$

d. Next find the voltage drop across R_i using Ohm's law.

$$V_{Ri} = 1\ mA \times 1\ k\Omega = 1\ V$$

e. Next, find V_a. Start at common and walk across R_i, falling in voltage by 1 V.

$$V_a = V_{common} - V_{Ri} = 0\ V - 1\ V = -1\ V$$

f. Now find V_{error} to determine the state of the op amp's output. V_{error} is the difference between V_{ni} and V_{inv}, where

$$V_{ni} = V_a = -1\ V$$

and

$$V_{inv} = V_{in} = 3\ V$$

So by definition

$$V_{error} = V_{ni} - V_{inv} = -1\ V - 3\ V = -4\ V$$

g. Since V_{error} is less than 0 V, the op amp does slam its lower rail of -10 V. The output is confirmed to be -10 V.

Practice

The circuit of Figure 9-23 has an op amp with rail voltages of ± 10 V. The input has been fixed at -3 V and output at $+10$ V.

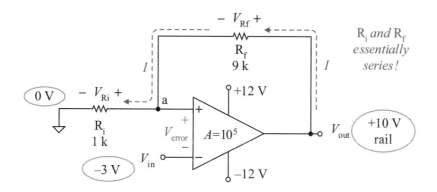

Figure 9-23 Slamming upper rail with negative input

(a) Verify that this is an inverting circuit. (b) Verify that this is a positive feedback circuit. (c) Find the feedback current I. (d) Find the voltage drop across R_i. (e) Find the voltage at node a. (f) Find the op amp input voltage V_{error}. (g) Does the value of V_{error} support the supposition that the op amp is slamming the upper rail?

Answer: (a) The input signal is fed to the inverting op amp input pin (b) The output voltage is being fed back to the noninverting (+) op amp input pin (c) 1 mA (d) 1 V (e) 1 V (f) 4 V (g) yes, $V_{error} > 0$ V

Based upon the previous example and practice problem, the op amp must change rails as its input goes from a $+3$ V to a -3 V. At what V_{in} voltage does the output trip from its lower rail voltage (-10 V) to its upper rail voltage ($+10$ V)? The next example finds this *trip voltage.*

Example 9-12

Find the input voltage V_{in} required to trip the circuit of Figure 9-24(a) from its lower rail (−10V) to its upper rail (+10 V).

Solution

Again, back to basics. Assign V_{error} equal to 0 V, and solve for V_{in}.

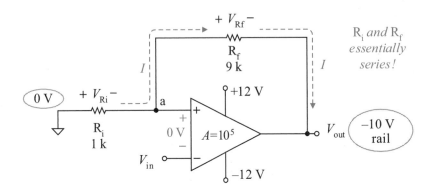

Figure 9-24(a) Find the input voltage V_{in} that trips output to +10 V

V_a serves as the *fixed reference voltage*. The op amp trips to its upper rail when V_{in} is approximately equal to the reference voltage established at V_a (the voltage applied to the op amp noninverting input pin).

Recall that the voltage drop V_{RT} across these two resistors is

$$V_{RT} = V_{common} - V_{out} = (0 \text{ V}) - (-10 \text{ V}) = 10 \text{ V}$$

Next use the voltage divider rule of this *essentially series* circuit of R_f and R_i to find the voltage drop across R_i. This saves the step of calculating the feedback current I.

$$V_{Ri} = \left(\frac{1 \text{ k}\Omega}{1 \text{ k}\Omega + 9 \text{ k}\Omega} \right) 10 \text{ V} = 1 \text{ V}$$

Next find V_a. Start at common and walk across R_i, falling by 1 V. So,

$$V_a = V_{common} - V_{Ri} = 0 \text{ V} - 1 \text{ V} = -1 \text{ V}$$

Summarizing, the input voltage to this comparator V_{in} must go from a positive input voltage V_{in}, through 0 V, to –1 V before the op amp trips from its lower rail (–10 V) to its upper rail (+10 V). Figure 9-24(b) shows the transfer curve for this comparator as the input V_{in} goes from positive voltages to negative voltages.

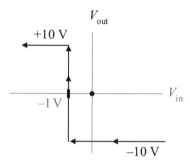

Figure 9-24(b) Example circuit transfer curve (–1 V trip point)

Practice

Find the input voltage V_{in} required to trip the circuit of Figure 9-25(a) from its upper rail of (+10 V) to its lower rail, (–10 V).

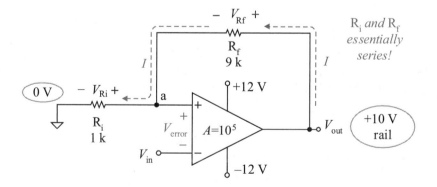

Figure 9-25(a) Find the input voltage V_{in} that trips output to –10 V

This is similar to the example except in this case the input voltage V_{in} is going from a negative value to a positive value.

Answer: +1 V (the input voltage V_{in} starts negative, passes through 0 V, and trips the op amp at +1 V). See Figure 9-25(b) for its transfer curve.

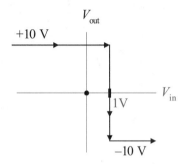

Figure 9-25(b) Example circuit transfer curve (+1 V trip point)

The preceding example and practice problem produced two trip points. When the op amp output is at its upper rail (+10 V) the trip point is +1 V. When the op amp output is at its lower rail (−10 V) the trip point is −1 V. Thus the two trip points for circuit are

$$V_{LTP} = -1 \text{ V} \hspace{3cm} \text{(lower trip point)}$$

$$V_{UTP} = +1 \text{ V} \hspace{3cm} \text{(upper trip point)}$$

Figure 9-26 combines the two transfer curves overlaid on one graph to help us visualize how this circuit operates and demonstrates how jitter around the reference is eliminated.

If the op amp is slammed against the upper rail (+10 V), the input voltage must go past 0 V to the upper trip point of +1 V. Once the op amp does trip to −10 V, a new trip point of −1 V is created. Now the input voltage must go past 0 V to the lower trip point of −1 V before it trips again.

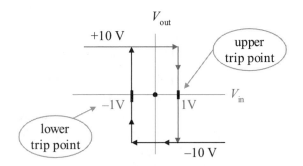

Figure 9-26 Example circuit complete transfer curve

Again the hysteresis effect of tracing two different paths desensitizes the comparator from jitter. This circuit is called an *inverting, positive feedback comparator circuit* and has a special name. It is called an **inverting Schmitt trigger**.

9.4 Negative Feedback Amplifiers

Negative feedback is used to produce a controlled circuit output as a function of its input. The op amp has a very large, uncontrolled voltage gain that varies significantly from op amp to op amp. With negative feedback, a part of the output signal is fed back to subtract from the magnitude of V_{error}, the signal to the op amp input terminals. V_{error} decreases, causing V_{out} to decrease, causing V_{error} to decrease, and so forth. Very quickly the op amp's input voltage V_{error} is driven to a very low value (in the microvolt range).

$$V_{\text{error}} \approx 0 \text{ V} \qquad \text{(negative feedback)}$$

Ideally assume it is 0 V.

$$V_{\text{error}} = 0 \text{ V} \qquad \text{(ideally)}$$

The output voltage depends upon the circuit components, but the analysis is based upon an essentially series feedback circuit. Again, the input signal can be fed to either the noninverting op amp input pin (creating a noninverting amplifier) or to the inverting op amp input pin (creating an inverting amplifier).

Negative feedback
$V_{\text{error}} \approx 0 \text{ V}$

There is one major requirement of the op amp. The op amp must not be in saturation. That is, the output voltage must be inside the rail voltages.

$$V_{\text{lower rail}} < V_{\text{out}} < V_{\text{upper rail}}$$ (restriction)

You have seen the negative feedback circuit in earlier chapters while studying KCL, KVL, and Ohm's law for resistance. You assumed the op amp's V_{error} was 0 V and proceeded to solve the circuits using basic laws.

First you will review the noninverting negative feedback voltage amplifier and then the inverting negative feedback voltage amplifier circuit. In both cases you will use circuit laws and basic circuit analysis techniques to evaluate these circuits. Much of the analysis will be centered on the essentially series circuit nature of the feedback circuits.

Noninverting Negative Feedback Amplifier

The circuit of Figure 9-27 is a *noninverting negative feedback amplifier*. The input signal is connected to the noninverting op amp input pin, so the output signal is not inverted (noninverting).

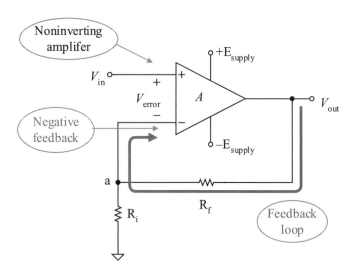

Figure 9-27 Noninverting negative feedback amplifier

The feedback path of this simple op amp circuit is connected to the op amp's inverting input pin at node a. Therefore, this is a **negative feedback** circuit. This causes V_{error} to be equal to 0 V, ideally.

To more formally prove negative feedback, assume V_{in} is a positive (+) input test signal. The op amp output is still a (+) test signal. That (+) test signal is fed back through R_f to the op amp inverting terminal as a (–) feedback signal. That feedback signal appears as a (–) at the op amp output pin. At the op amp output pin, the test signal appears as a (+) and the feedback signal appears as a (–). *Where the feedback signal meets the original signal, they are of opposite signs. Therefore, this is negative feedback. The feedback signal subtracts from the original signal, creating a smaller net signal.*

Voltage Follower–Buffer Amplifier

Since the input currents to the op amp are ideally 0 A, the resistors R_f and R_i form an essentially series circuit and simple circuit analysis can be applied.

Example 9-13

The circuit of Figure 9-28 is called a **voltage follower** circuit. The output voltage follows the input voltage. Ideally,

$$V_{out} = V_{in}$$

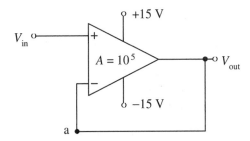

Figure 9-28 Example circuit—voltage follower

(a) Verify that this is a noninverting circuit. (b) Verify that this is a negative feedback circuit so the assumption that V_{error} is approximately 0 V can be made. (c) Find V_a, the voltage at feedback node a. (d) Find the op amp output voltage V_{out}. (e) Does the circuit name of *voltage follower* make sense? (f) Find the voltage gain A_v of this circuit (that is, the output voltage divided by the input signal voltage). (g) Find the amplifier input resistance and its potential loading impact on the signal source V_{in}.

Solution

a. The input signal is fed to the noninverting pin, so this is a noninverting circuit.

b. The output is fed back to the op amp's inverting pin, so this is negative feedback. Since this is negative feedback, the ideal assumption is made that

$$V_{error} = 0 \text{ V} \qquad\qquad \text{(ideally)}$$

c. Find the voltage at node a by starting with node voltage V_{in} and subtracting V_{error}.

$$V_a = V_{in} - V_{error} = V_{in} - 0 \text{ V} = V_{in}$$

d. Since node a and node out are the same voltage node, they have the same voltage value.

$$V_{out} = V_a$$

Substituting,

$$V_{out} = V_{in}$$

e. The circuit name *voltage follower* is logical, since the output voltage equals (or follows) the input voltage. *The output voltage follows the input voltage in both magnitude and sign.* Note the op amp must not be saturated for this circuit to work properly. That is, V_{out} must be between the rail voltages.

f. Since $V_{out} = V_{in}$, the voltage gain of the circuit A_v is

$$A_v = \frac{V_{out}}{V_{in}} = 1$$

If the op amp is not saturated, the voltage gain of this circuit is always unity (1). So what good is this amplifier? It does not really amplify the signal voltage.

This amplifier can be used as a *buffer amplifier* between two circuits. If a voltage supplied signal is delivered to a load circuit that produces heavy loading on the original circuit, the buffer amplifier can be inserted between these two circuits to eliminate the loading effect. The buffer amplifier prevents an excessive loss of the signal, so a gain of unity is a very desirable outcome.

g. Ideally the input op amp current is 0 A, therefore, *the amplifier looks like an open to the signal source* V_{in}. Practically, the input resistance to this amplifier is in the gigaohm (GΩ) range. Thus, this amplifier does not load down the signal supply V_{in}.

$$R_{amplifier\ input} = \frac{V_{in}}{I_{in}} = \frac{100\ mV}{0\ A} \approx \infty\ \Omega$$

Practice

Use the circuit of Figure 9-28. Assume that the rail voltages are ± 13 V. Find the output voltage V_{out}, if (a) V_{in} is 2 V, (b) V_{in} is 15 V, and (c) V_{in} is 20 V.

Answer: (a) 2 V (b) $V_{in} > +V_{upper\ rail}$ therefore $V_{in} = V_{upper\ rail} = +13V$ (c) $V_{in} > +E_{supply}$. The input voltage is greater than the supply voltage. This could damage the op amp. If the op amp survives, then V_{out} is limited to the $V_{upper\ rail}$ of $+13$ V.

Noninverting Voltage Amplifier–Lightly Loaded

The next example utilizes a negative feedback resistor circuit intended to produce a voltage gain greater than unity. That is, the signal voltage at the output is greater than the signal voltage at the input. Again, the circuit voltage gain of a circuit A_v is defined to be

$$A_v = \frac{V_{out}}{V_{in}}$$

This circuit in turn drives a load. *If the op amp output current is within specifications and the op amp output voltage is between the rail voltages, the circuit is truly a voltage amplifier, linearly multiplying the input signal voltage by its constant voltage gain A_v. That is,*

$$V_{out} = A_v \ V_{in}$$

Example 9-14

Use the circuit of Figure 9-29(a).

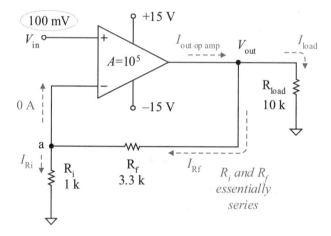

Figure 9-29(a) Example circuit–noninverting voltage amplifier

(a) Verify that this is a noninverting circuit. (b) Verify that this is a negative feedback circuit. (c) Use basic circuit analysis and find the following: amplifier input resistance, node voltage V_a, voltage drop V_{Ri}, feedback current I_f, voltage drop V_{Rf}, output voltage V_{out}, op amp output voltage $V_{out \ op \ amp}$, load voltage V_{load}, load current I_{load}, op amp output current $I_{out \ op \ amp}$. (d) Verify that the output voltage is within the rail voltages and that the op amp output current is less than its rated maximum (assume a rating of 10 mA). (e) Find the circuit voltage gain A_v.

Solution

a. The input signal is fed to the *noninverting pin*, so this is a noninverting circuit.

b. A noninverted op amp output is fed back to the op amp's *inverting pin* through the R_f and R_i feedback circuit, so this is *negative feedback*. Since this is negative feedback, the ideal assumption is made that

$$V_{error} = 0 \text{ V} \qquad \text{(ideally)}$$

c. Find all circuit voltages and currents with the following assumptions.

$$I_{in \text{ op amp}} = 0 \text{ A} \qquad \text{(nature of op amp)}$$

$$V_{error} = 0 \text{ V} \qquad \text{(negative feedback)}$$

The following is a solution to all the remaining voltages and currents using basic circuit analysis techniques. The above two assumptions are included on the schematic as shown in Figure 9-29(b) to help aid in circuit analysis. As you proceed, write values, voltage polarities, and current directions on the schematic to better visualize circuit values and aid in solving the rest of the circuit.

First, find the input resistance of the amplifier. Ideally the input op amp current is 0 A, therefore, *the amplifier looks like an open to the signal source* V_{in}. Practically, the input resistance to this amplifier is in the gigaohm (GΩ) range.

$$R_{amplifier \text{ input}} = \frac{V_{in}}{I_{in}} = \frac{100 \text{ mV}}{0 \text{ A}} = \infty \ \Omega$$

Node voltage V_a is the input signal voltage less the V_{error} drop.

$$V_a = V_{in} - V_{error}$$

$$V_a = 100 \text{ mV} - 0 \text{ V} = 100 \text{ mV}$$

The voltage drop across R_i is the difference between its two nodes.

$$V_{Ri} = V_a - V_{common}$$

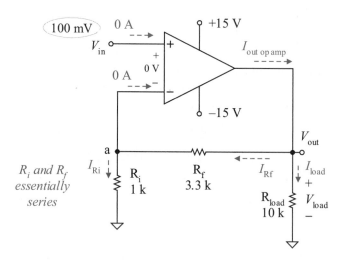

Figure 9-29(b) Example circuit redrawn for analysis

$$V_{Ri} = 100 \text{ mV} - 0 \text{ V} = 100 \text{ mV}$$

Apply Ohm's law to R_i and find current I_{Ri}.

$$I_{Ri} = \frac{V_{Ri}}{R_i}$$

$$I_{Ri} = \frac{100 \text{ mV}}{1 \text{ k}\Omega} = 100 \text{ μA}$$

Use KCL at node a and find the feedback current I_{Rf}.

$$I_{Rf} = I_{Ri} + 0 = 100 \text{ μA}$$

Find V_{Rf} by using Ohm's law.

$$V_{Rf} = I_{Rf} R_f$$

$$V_{Rf} = 100 \text{ μA} \times 3.3 \text{ k}\Omega = 330 \text{ mV}$$

Find V_{out} doing a walkabout from node a to the *out* node. This is also the output voltage of the op amp.

$$V_{out} = V_{out \, op \, amp} = V_{a'} + V_{Rf}$$

$$V_{\text{out}} = 100 \text{ mV} + 330 \text{ mV} = 430 \text{ mV}$$

Find V_{error} (the op amp input voltage) by taking the output op amp voltage $V_{\text{out op amp}}$ and dividing by the op gain A.

$$V_{\text{error}} = \frac{V_{\text{out op amp}}}{A}$$

$$V_{\text{error}} = \frac{430 \text{ mV}}{10^5} = 4.3 \text{ μV}$$

Note that V_{error} is very small compared to V_{in} of 100 mV. So approximating V_{error} as 0 V is a reasonable assumption.

The load voltage drop across R_{load} is the difference between its two nodes.

$$V_{\text{load}} = V_{\text{out}} - V_{\text{common}}$$

$$V_{\text{load}} = 430 \text{ mV} - 0 \text{ V} = 430 \text{ mV}$$

Apply Ohm's law to R_{load} and find the load current I_{load}.

$$I_{\text{load}} = \frac{V_{\text{load}}}{R_{\text{load}}}$$

$$I_{\text{load}} = \frac{430 \text{ mV}}{10 \text{ k}\Omega} = 43 \text{ μA}$$

Apply KCL to find the op amp output current $I_{\text{out op amp}}$.

$$I_{\text{out op amp}} = I_{\text{Rf}} + I_{\text{load}}$$

$$I_{\text{out op amp}} = 100 \text{ μA} + 43 \text{ μA} = 143 \text{ μA}$$

d. Inspect the results to ensure that the op amp is within its operating range. The op amp output voltage is 430 mV, well within the op amp's rail voltages of approximately ±13 V. The op amp output current of 143 μA is well below the maximum specified op amp current rating of 10 mA. The op amp is operating within its limits. This circuit is performing as a linear voltage amplifier.

e. Since the op amp output voltage is less than the op amp rail voltage, the circuit voltage gain is

$$A_v = \frac{V_{out}}{V_{in}}$$

$$A_v = \frac{430 \text{ mV}}{100 \text{ mV}} = 4.3$$

Note this is significantly less than the op amp gain of 10^5. The circuit (or system) voltage gain is actually being determined by the voltage divider circuit R_i and R_f.

Practice

Using the circuit of Figure 9-30, find the input signal voltage V_{in} required to produce an output voltage of 500 mV. Find the op amp's input voltage V_{error}. Find the op amp output current and confirm that it is a reasonable value. Find the circuit voltage gain.

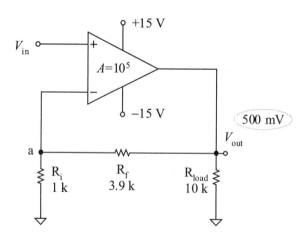

Figure 9-30 Problem circuit

Answer: 102 mV; 5 µV; 152 µA (within specifications), 4.9

Notice that the last example and practice had circuit voltage gains A_v of 4.3 and 4.9, respectively. This is significantly less than the op amp's *open loop* voltage gain A of 10^5. Let's analyze the example circuit using literal symbols to determine just what is creating the *circuit voltage gain*, which is also called the **system voltage gain** or the **closed loop voltage gain**. Use the circuit of Figure 9-30. If the output voltage is less than the op amp rail voltage, then the op amp voltage across its input pins (i.e., V_{error}) must be very small or approximately 0 V.

$$V_{error} = \frac{V_{out}}{A} \approx 0 \text{ V} \qquad \text{high value of } A$$

Thus, by Kirchhoff's voltage law the voltage at node a is approximately equal to the input signal voltage V_{in}, and the voltage drop across R_i is equal to V_{in}.

$$V_a = V_{in} - V_{error} \approx V_{in} = V_{Ri}$$

Use Ohm's law to find the current through R_i.

$$I_{Ri} = \frac{V_{Ri}}{R_i}$$

Since the input currents to the op amp are very small, approximately 0 A, Kirchhoff's current law dictates that the essentially series current I is

$$I = I_{Ri} = I_{Rf}$$

The output voltage is the sum of voltage drops across R_i and R_f by Kirchhoff's voltage law.

$$V_{out} = V_{Ri} + V_{Rf}$$

Apply Ohm's law for each resistor voltage drop using the essentially series circuit current I.

$$V_{out} = IR_i + IR_f$$

The voltage drop across R_i is the same as the input signal V_{in}; and by Ohm's law,

$$V_{in} = V_{Ri} = IR_i$$

The circuit voltage gain is the circuit output voltage divided by the circuit input voltage.

$$A_v = \frac{V_{out}}{V_{in}}$$

Substituting

$$A_v = \frac{IR_i + IR_f}{IR_i}$$

Simplify by canceling out the term I. The circuit voltage gain is

$$A_v = \frac{R_i + R_f}{R_i}$$

Note that the circuit voltage gain is the reciprocal of the voltage divider circuit of R_i and R_f. This is called the **feedback factor** B, which must be less than or equal to 1 (that is, $0 \le B \le 1$).

$$B = \frac{R_i}{R_i + R_f} = \frac{1}{A_v}$$

Dividing out the R_i term, another algebraic form for circuit voltage gain A_v of the noninverting voltage amplifier is

$$A_v = 1 + \frac{R_f}{R_i}$$

Let's verify the voltage gain and calculate the feedback factor of the previous example using R_f and R_i.

$$A_v = \frac{1 \ k\Omega + 3.3 \ k\Omega}{1 \ k\Omega} = 4.3$$

Its feedback factor is

$$B = \frac{1 \ k\Omega}{1 \ k\Omega + 3.3 \ k\Omega} = 0.233 \qquad \text{(or 23.3\%)}$$

If the op amp output is not hitting its rails, the feedback factor and the voltage gain are fixed by the R_i and R_f resistance values. You can design a circuit with a specific voltage gain by proper selection of R_i and R_f.

Noninverting Amplifier–Excessive Loading

The next example uses the same amplifier as the previous example but the load resistance is changed to be a very low resistive value creating a high load current, which excessively loads the output of the op amp. If the output op amp current is too high, the op amp output voltage is degraded and the linear amplifier is seriously compromised.

Example 9-15

Use the circuit of Figure 9-31(a). The amplifier output solution V_{out} for this circuit is the same as the last example except for finding the load current and the op amp output current. Using V_{out} of the last example, find the load current and output op amp load current of this circuit. What is implication of the result?

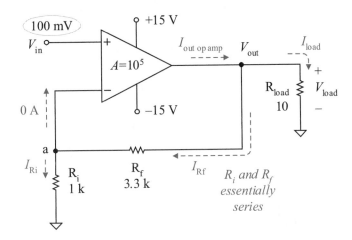

Figure 9-31(a) Example circuit–noninverting voltage amplifier

Solution

The output voltage from the previous example with the amplifier lightly loaded is 430 mV. However, the load current for the 10 Ω load is substantial.

$$I_{load} = \frac{430 \text{ mV}}{10 \text{ }\Omega} = 43 \text{ mA}$$

Apply KCL to find the op amp output current $I_{out\ op\ amp}$.

$$I_{out\ op\ amp} = 100 \text{ }\mu\text{A} + 43 \text{ mA} = 43.1 \text{ mA}$$

Inspection of the results shows that the op amp's output current could be a problem. The op amp selected must support a relatively high output current. The LM741 and LM324 op amp output current specifications are on the order of 10 to 40 mA. Neither meets the demands of the circuit. *This circuit does not work with these op amps. Either select an op amp that can handle this level of output current or create a composite amplifier using a BJT or MOSFET current booster in the feedback loop as shown in the next set of examples.*

Practice

Simulate the example circuit using MultiSIM. Simulate with an LM741 op amp and observe the op amp and circuit voltages and currents.

Answer: See Figure 9-31(b). Note that the op amp can only support 25 mA in this simulator, reflecting the op amp short circuit current rating of 20 to 25 mA. Since the load current is significantly reduced, the load voltage drops correspondingly by Ohm's law to 0.25 V.

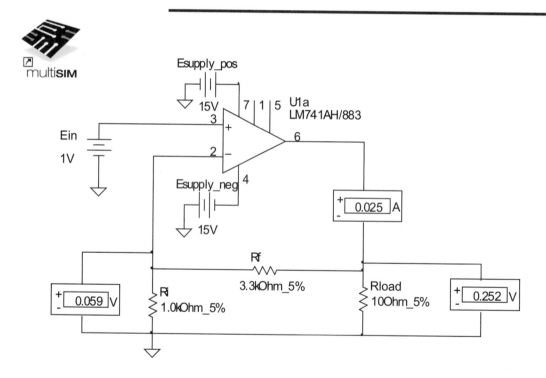

Figure 9-31(b) Practice circuit using LM741 op amp

Composite Amplifiers

Composite amplifiers are circuits composed of and using two or more different amplifiers to optimize their individual strengths and minimize their weaknesses. For example, a linear negative feedback amplifier with a nonlinear BJT or a very nonlinear MOSFET in its feedback circuit can create a linear amplifier that provides high load currents.

Composite Amplifier–BJT Current Booster

The circuit of Figure 9-31(a) has an inherent flaw. The op amp is attempting to drive a load of 43.1 mA. Op amps like the LM741 and the LM324 typically cannot typically produce such high op amp output currents. The circuit of Figure 9-32 remedies this problem by using a BJT to boost the current from the op amp output to the load. The signal out of the op amp is fed into the base of the BJT. The BJT boosts its base current by a factor of beta (β) to the collector and ($\beta+1$) to the emitter. This is a *composite amplifier* since it is composed of and uses two different amplifiers to optimize their strengths and minimize their weaknesses. The linear, low current op amp amplifier circuit contains a high current, somewhat nonlinear amplifier (the BJT) within its feedback loop to produce a high current, linear voltage amplifier.

Example 9-16

The **current booster** circuit of Figure 9-32 uses an op amp and an *npn* BJT as a current booster. Verify that this is a negative feedback circuit. Using the values of Example 9-14, find the values of I_E, I_B (same as $I_{out\ op\ amp}$), and $V_{out\ op\ amp}$. Is the op amp operating at less than 10 mA? Is the op amp output voltage within the rail voltages?

Solution

First verify that this is a negative feedback system. The input signal proceeds from op amp noninverting (+) input through the op amp, out of the op amp output, across the base-emitter junction of the BJT. The signal at the *emitter follows the signal at the base* (there is *not* a signal inversion). The feedback signal proceeds across R_f to the op amp inverting (–) input pin. The result is *negative feedback*; the op amp is performing as an amplifier.

See the previous two examples to find the following circuit values.

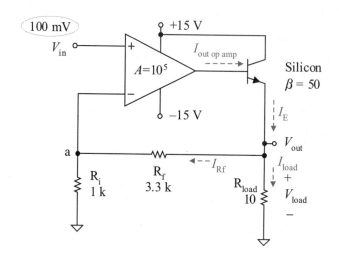

Figure 9-32 Example circuit–op amp amplifier with BJT current booster

$$V_{Ri} = V_a = 100 \text{ mV}$$

$$I_{Ri} = I_{Rf} = 100 \text{ μA}$$

$$V_{Rf} = 330 \text{ mV}$$

$$V_{load} = V_{out} = 430 \text{ mV}$$

$$I_{load} = 43 \text{ mA}$$

Apply KCL to find the BJT emitter current I_E.

$$I_E = I_{Rf} + I_{load}$$

$$I_E = 100 \text{ μA} + 43 \text{ mA} = 43.1 \text{ mA}$$

Since the BJT $\beta \geq 50$, this is a *good BJT beta* and it can be approximated that $I_C \approx I_E$. Thus,

$$I_B = \frac{I_E}{\beta + 1} \approx \frac{I_E}{\beta}$$

$$I_B \approx \frac{43.1 \text{ mA}}{50} = 862 \text{ μA}$$

Since I_B and $I_{\text{out op amp}}$ are the same current,

$$I_{\text{out op amp}} = I_B = 862 \ \mu A$$

$I_{\text{out op amp}}$ is now well within the op amp's current capability. This circuit works, whereas the circuit of Figure 9-31(a) cannot work with the overloading of its op amp output capability.

What about the op amp output voltage? Is that within the rail voltages? The BJT base voltage is approximately 0.7 V higher than its emitter.

$$V_E = V_{\text{out}} = 430 \ \text{mV}$$

$$V_B = V_E + 0.7 \ \text{V}$$

$$V_B = 430 \ \text{mV} + 0.7 \ \text{V} = 1.13 \ \text{V}$$

The op amp output voltage is at the same potential as the base of the BJT. This op amp output is well within the rail voltages of approximately ± 13 V.

$$V_{\text{out op amp}} = V_B = 1.13 \ \text{V}$$

Practice

The input signal voltage of the example problem is changed to 500 mV. Find the circuit output voltage, voltage gain, the op amp output current, and the op amp output voltage. Can the op amp handle this output current and voltage if it has a maximum usable output current of 3 mA? Is the op amp output voltage within the rail voltages?

Answer: 2.15 V, 4.3, 4.3 mA (No, output op amp current is too high), 2.85 V (output op amp voltage is within the rail voltages)

Composite Amplifier–MOSFET Current Booster

The MOSFET can be used to boost current in place of the BJT for higher current applications in the ampere range.

Example 9-17

Use MultiSIM and simulate the **MOSFET current booster** circuit of Figure 9-33. Calculate and verify the load voltage and

MOSFET current. Find the amplifier system voltage gain. Based upon the simulation, are the op amp output voltage and current acceptable?

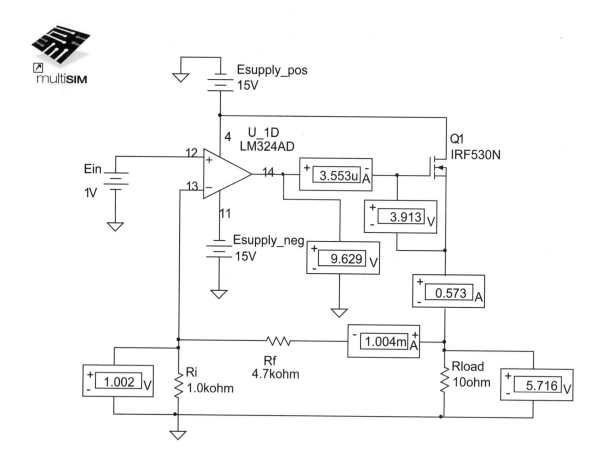

Figure 9-33 Example circuit–op amp amplifier with MOSFET current boost

Solution

First verify that this is a negative feedback system. The input signal proceeds from op amp input pin #12 through the op amp, out of op amp pin #14, across the gate-to-source of the MOSFET. The signal at the source follows the signal at the gate; thus there is *not* a signal inversion. The feedback signal proceeds across R_f

to op amp pin #13 (the inverting op amp pin). The result is negative feedback.

Assuming that the op amp output is not at a rail voltage, the circuit is behaving as a linear voltage amplifier. Therefore, the error voltage across the signal op amp input terminals is approximately 0 V, and the two input signal op amp pins are at approximately the same voltage.

$$V_{\text{pin }13} \approx V_{\text{pin }12} = 1 \text{ V}$$

The simulation does include a realistic 2 mV **offset voltage** difference between the input signal pins (pins #12 and #13) caused by the internal differences of the LM324 op amp. In this simulation, pin #13 is simulated to be 2 mV above pin #12. Thus the voltage at pin #13 is more practically represented to be 1.002 V instead of 1.000 V. This 2 mV offset voltage must be taken into account for small signal operation and does appear in amplifier simulations.

$$V_{\text{Ri}} = V_{\text{pin }13} - V_{\text{common}}$$

$$V_{\text{Ri}} = 1 \text{ V} - 0 \text{ V} = 1 \text{ V}$$

$$I_{\text{Ri}} = \frac{V_{\text{Ri}}}{R_{\text{i}}}$$

$$I_{\text{Ri}} = \frac{1 \text{ V}}{1 \text{ k}\Omega} = 1 \text{ mA}$$

$$I_{\text{Rf}} = I_{\text{Ri}} + 0 = 1 \text{ mA}$$

$$V_{\text{Rf}} = I_{\text{Rf}} R_{\text{f}}$$

$$V_{\text{Rf}} = 1 \text{ mA} \times 4.7 \text{ k}\Omega = 4.7 \text{ V}$$

V_{out} is the voltage at the top of the load relative to common.

$$V_{\text{out}} = V_{\text{pin }13} + V_{\text{Rf}}$$

$$V_{\text{out}} = 1 \text{ V} + 4.7 \text{ V} = 5.7 \text{ V}$$

$$V_{\text{load}} = V_{\text{out}} - V_{\text{common}}$$

$$V_{\text{load}} = 5.7 \text{ V} - 0 \text{ V} = 5.7 \text{ V}$$

Note that the simulated output voltage is 5.716 V due to the 2 mV input offset voltage.

$$I_{load} = \frac{V_{load}}{R_{load}}$$

$$I_{load} = \frac{5.7 \text{ V}}{10 \text{ }\Omega} = 570 \text{ mA}$$

Apply KCL at the output node to find the MOSFET source current I_S.

$$I_S = I_{Rf} + I_{load}$$

$$I_S = 1 \text{ mA} + 570 \text{ mA} = 571 \text{ mA}$$

The overall circuit voltage gain of the amplifier is

$$A_v = \frac{V_{out}}{V_{in}} = \frac{5.7 \text{ V}}{1 \text{ V}} = 5.7$$

Rewriting the voltage gain expression to solve for V_{out}, this is a linear amplifier with

$$V_{out} = 5.7 \, V_{in}$$

as long as the op amp output voltage is not at a rail voltage.

Based upon the simulation, the op amp output current is 3.55 µA which is well within specifications. The op amp output voltage is 9.63 V, which is well within the op amp rail voltages.

Practice

Simulate the example circuit with an input signal of 4 V. What problem is encountered?

Answer: The output op amp voltage is simulated to be 15.5 V (the op amp output signal exceeds the upper rail voltage; the output is the upper rail voltage)

Composite Amplifier–Active Feedback

Composite amplifiers may also include an amplifier as part of the negative feedback loop. Look at Figure 9-34. The current through the load is sampled by R_{sense}, amplified by the difference amplifier U2, and fed back to U1. That op amp drives the load through the small signal transistor Q1 and power transistor Q2.

Example 9-18

Calculate all currents and voltages in the circuit shown in Figure 9-34 when $E_{in} = 4$ V. (Example courtesy of Professor J. Michael Jacob of Purdue University.)

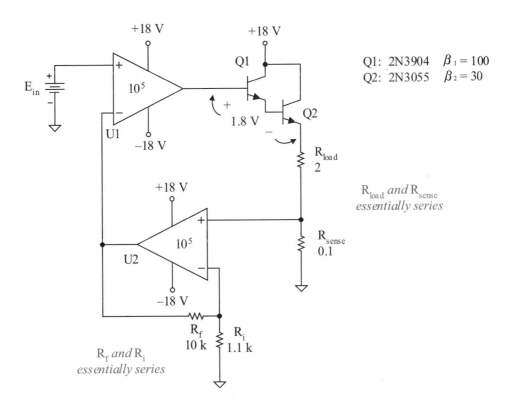

Figure 9-34 Active feedback within a composite amplifier

Solution

Note that this is negative feedback. A signal enters U1 (+) pin, exits U1, enters Q1 base, exits Q1 emitter (no inversion), enters Q2 base, exits Q2 emitter (no inversion), enters U2 (+) pin, exits U2, and enters U1 (−) pin.

1. Noninverting input voltage, U1

$$V_{\text{ni U1}} = E_{\text{in}} = 4 \text{ V}$$

2. Inverting voltage, U1

$$V_{\text{inv U1}} = V_{\text{ni U1}} = 4 \text{ V}$$

$$V_{\text{out U2}} = V_{\text{inv U1}} = 4 \text{ V}$$

3. Input and feedback resistors current, U1

$$I_{\text{Rf}} = I_{\text{Ri}} = \frac{4 \text{ V}}{10 \text{ k}\Omega + 1.1 \text{ k}\Omega} = 360.4 \text{ }\mu\text{A}$$

At this point you have to modify the analysis technique a little. But you still only need Ohm's law and Kirchhoff's laws.

$$V_{\text{inv U2}} = I_{\text{Ri}} R_{\text{i}} = 360.4 \text{ }\mu\text{A} \times 1.1 \text{ k}\Omega$$

$$V_{\text{inv U2}} = 396.4 \text{ mV}$$

Since U2 has negative feedback, its two input voltages are ideally equal.

$$V_{\text{ni U2}} = V_{\text{inv U2}} = 396.4 \text{ mV}$$

This is the voltage across R_{sense}. Therefore, by Ohm's law the current through the sensing resistor is

$$I_{\text{Rsense}} = \frac{396.4 \text{ mV}}{0.1 \Omega} = 3.964 \text{ A}$$

This is also the current through the load.

$$I_{\text{load}} = I_{\text{Rsense}} = 3.964 \text{ A}$$

An input voltage of 4 V produces a load current of 4 A. The composite amplifier uses its negative feedback to sense the current through the load, convert that to voltage and apply the voltage as negative feedback to U1.

Transistors Q1 and Q2 are power drivers, providing the 4 A required load current, which in turn requires a much smaller current from the op amp. The base current of Q2 is

$$I_{B2} = \frac{I_{E2}}{\beta_2 + 1} = \frac{3.964 \text{ A}}{31} = 127.8 \text{ mA}$$

The emitter current of Q2 is

$$I_{E1} = I_{B2}$$

The base current of Q1 is

$$I_{B1} = \frac{I_{E1}}{\beta_1 + 1} = \frac{127.8 \text{ mA}}{101} = 1.278 \text{ mA}$$

The output op amp current is within specifications.

$$I_{out\ op\ amp} = I_{B1} = 1.278 \text{ mA}$$

Find the voltage drop across the load using Ohm's law.

$$V_{load} = 3.964 \text{ A} \times 2 \text{ } \Omega = 7.928 \text{ V}$$

The voltage at the emitter of Q2 is a voltage rise across R_{sense} and R_{load}.

$$V_{Q2\ emitter} = 396.4 \text{ mV} + 7.928 \text{ V} = 8.324 \text{ V}$$

A power transistor may drop 1 V or more between its base and emitter leads. So for the **Darlington pair** of Q1 and Q2 a drop of 1.8 V is reasonable. The op amp output voltage of U1 is well within its rail voltages. The op amp output voltage of U2 is the same as the input signal voltage. The op amps are operating within specifications.

$$V_{out\ U1} = 8.32 \text{ V} + 1.8 \text{ V} = 10.1 \text{ V}$$

$$V_{out\ U2} = 4 \text{ V}$$

multiSIM

Practice

Perform a MultiSIM simulation to verify this example problem.

Inverting Negative Feedback Amplifier

The circuit of Figure 9-35 is an **inverting negative feedback amplifier**. Compare this circuit with that of Figure 9-27, the noninverting amplifier just examined. The only difference is that the input voltage V_{in} and common have been swapped going into the op amp input pins. The input signal V_{in} is now driving the inverting terminal through resistor R_i. This causes an inverted output signal, so this is an inverting circuit.

The feedback circuit is connected to the op amp's inverting input pin at node a, and so it is a negative feedback circuit. Again, negative feedback forces the op amp's input voltage V_{error} to be ideally equal to 0 V. Negative feedback creates an amplifier. So this is an *inverting voltage amplifier*.

To prove negative feedback more formally, assume V_{in} is a positive (+) input test signal. The op amp inverts this signal and produces a (−) signal at the op amp output pin. That (−) signal is fed back through R_f to node a where the feedback signal meets the original signal. At node a, the signal appears as a (+) and the feedback signal appears as a (−). *Where the feedback signal meets the original signal, they are of opposite signs. Therefore, this is negative feedback.*

Since the input currents to the op amp are ideally 0 A, the resistors R_f and R_i again form an essentially series circuit and simple circuit analysis can be applied.

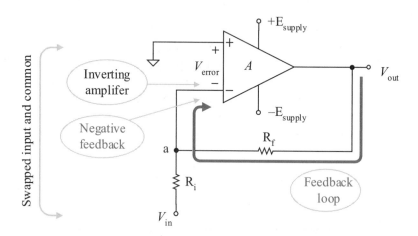

Figure 9-35 Inverting negative feedback amplifier

Example 9-19

The circuit of Figure 9-36(a) uses an LM324 op amp. (a) Verify that this is an inverting circuit. (b) Verify that this is a negative feedback circuit. (c) Use basic circuit analysis to find: V_a, V_{Ri}, I_{Ri}, I_{Rf}, V_{Rf}, V_{out}, V_{load}, I_{load}, and $I_{out\ op\ amp}$. (d) Find the input resistance of this amplifier system. (e) Are there any problem areas (too much input voltage, too much op amp output current, or a saturated op amp hitting the rails)? (f) Find the circuit voltage gain.

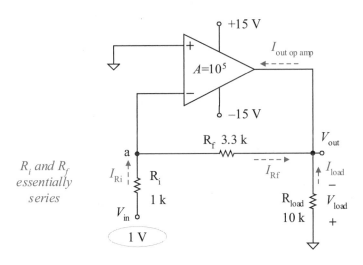

Figure 9-36(a) Example circuit with an inverting voltage amplifier

Solution

a. The input signal is fed to the *inverting pin* through R_i, so this is an inverting circuit.

b. The op amp output is fed back to the op amp's *inverting pin* through the R_f and R_i feedback circuit. Thus this is *negative feedback*. Since this is negative feedback, the ideal assumption is made that

$$V_{error} = 0\ V \qquad\qquad (ideally)$$

c. Find all circuit voltages and currents with the following assumptions.

$$I_{\text{in op amp}} = 0 \text{ A} \qquad\qquad \text{(nature of op amp)}$$

$$V_{\text{error}} = 0 \text{ V} \qquad\qquad \text{(negative feedback)}$$

Figure 9-36(b) shows the schematic with these assumptions. Again, place values with voltage polarities and current directions on your schematic as you solve the problem.

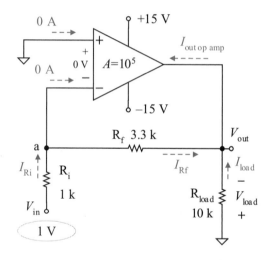

Figure 9-36(b) Inverting voltage amplifier with assumptions

$$V_{\text{a}} = V_{\text{common}} - V_{\text{error}} = 0 \text{ V} - 0 \text{ V} = 0 \text{ V}$$

$$V_{\text{Ri}} = V_{\text{in}} - V_{\text{a}} = 1 \text{ V} - 0 \text{ V} = 1 \text{ V}$$

$$I_{\text{Ri}} = \frac{V_{\text{Ri}}}{R_{\text{i}}} = \frac{1 \text{ V}}{1 \text{ k}\Omega} = 1 \text{ mA}$$

$$I_{\text{Rf}} = I_{\text{Ri}} + 0 \text{ mA} = 1 \text{ mA}$$

$$V_{\text{Rf}} = I_{\text{Rf}} \, R_{\text{f}} = 1 \text{ mA} \times 3.3 \text{ k}\Omega = 3.3 \text{ V}$$

$$V_{\text{out}} = V_{\text{out op amp}} = V_{\text{a}} - V_{\text{Rf}} = 0 \text{ V} - 3.3 \text{ V} = -3.3 \text{ V}$$

$$V_{\text{load}} = V_{\text{common}} - V_{\text{out}} = 0 \text{ V} - (-3.3 \text{ V}) = 3.3 \text{ V}$$

$$I_{load} = \frac{V_{load}}{R_{load}} = \frac{3.3 \text{ V}}{10 \text{ k}\Omega} = 0.33 \text{ mA}$$

$$I_{out \text{ op amp}} = I_{Rf} + I_{load} = 1 \text{ mA} + 0.33 \text{ mA} = 1.33 \text{ mA}$$

d. The amplifier input resistance is found with Ohm's law using the input voltage and input current.

$$R_{input} = \frac{V_{in}}{I_{in}} = \frac{1 \text{ V}}{1 \text{ mA}} = 1 \text{ k}\Omega$$

This makes sense. The 1 kΩ input resistance is tied to common, thus the resistance seen from input to common should be 1 kΩ as verified by the Ohm's law calculation.

e. Everything is within specifications: input op amp voltages within the rails, supply difference within specification, output op amp voltage within the rail voltages, and the op amp output current less than a few mA. Therefore, this is a valid output.

f. Since the op amp output voltage is within the op amp rail voltages, the circuit voltage gain is

$$A_v = \frac{V_{out}}{V_{in}}$$

$$A_v = \frac{-3.30 \text{ V}}{1 \text{ V}} = -3.3$$

Note that the negative sign indicates that the output signal is inverted relative to its input. This input signal is positive; therefore the amplifier output signal must be negative.

Practice

The feedback resistor R$_f$ in the circuit of Figure 9-36(a) is changed to 10 kΩ. Find the circuit output voltage and the op amp output current. Are any specifications violated? Find the circuit voltage gain.

Answer: −10 V, 2 mA, no specifications violated (output valid), −10

Use the circuit of Figure 9-36(a) to analyze the example circuit using literal symbols to determine just what is creating its *circuit voltage gain*. If the output voltage is less than the op amp rail voltage, the op amp input voltage (or V_{error}) must be very small or approximately 0 V.

$$V_{error} = \frac{V_{out}}{A} \approx 0 \text{ V}$$

By Kirchhoff's voltage law the voltage at node a is approximately at common potential (0 V); so the voltage drop across R_i is equal to V_{in}.

$$V_a \approx 0 \text{ V}$$

$$V_{R1} \approx V_{in}$$

Use Ohm's law to find the current through R_i.

$$I_{Ri} = \frac{V_{Ri}}{R_i}$$

The input currents to the op amp are approximately 0 A, therefore Kirchhoff's current law dictates that the essentially series current I is

$$I = I_{Ri} = I_{Rf}$$

Since node a is a *virtual common* (≈ 0 V), the output voltage is the voltage fall across R_f.

$$V_{out} \approx 0 \text{ V} - V_{Rf} = -V_{Rf}$$

Apply Ohm's law using the essentially series current I.

$$V_{out} = -IR_f$$

The voltage drop across R_i is approximately the same as the input signal V_{in}; and by Ohm's law,

$$V_{in} \approx V_{Ri} = IR_i$$

The circuit voltage gain is the circuit output voltage divided by the circuit input voltage.

$$A_v = \frac{V_{out}}{V_{in}}$$

Substituting

$$A_v = -\frac{IR_f}{IR_i}$$

Simplify by canceling out the term I. The circuit voltage gain is

$$A_v = -\frac{R_f}{R_i}$$

The negative sign indicates that the output is inverted relative to the input signal. Let's confirm the previous example's voltage gain using R_f and R_i.

$$A_v = -\frac{3.3 \text{ k}\Omega}{1 \text{ k}\Omega} = -3.3$$

If the op amp output is not hitting its rails, the voltage gain is fixed by the R_i and R_f resistance values. You can design this circuit for a specific voltage gain by proper selection of R_i and R_f.

The circuit input resistance is defined by Ohm's law.

$$R_{in} = \frac{V_{in}}{I_{in}}$$

Substituting the Ohm's law form of V_{in} and the essentially series feedback current I, this equation reduces to an intuitive result. This input resistance is a load consideration for the source circuit.

$$R_{in} \approx \frac{IR_i}{I} = R_i$$

Zero and Span Amplifier

The next example uses the LM34 temperature sensor. You used this temperature sensor in an earlier example of a comparator. In the next example, the circuit converts the Fahrenheit temperature reading to its corresponding Celsius temperature.

Example 9-20

Verify that the circuit of Figure 9-37(a) converts Fahrenheit temperature (10 mV per °F) to centigrade (10 mV per °C). (a) Assume the sensor temperature is 32°F (freezing) and find V_{out}. Does this match with 0°C? (b) Assume the sensor temperature is 212°F (boiling) and find V_{out}. Does this match with 100°C? *Critical: place values on the schematic as you find them: node voltages, resistor voltage drops, and resistor currents.*

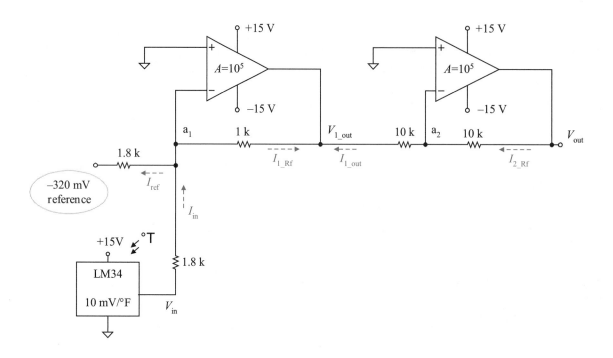

Figure 9-37(a) Example zero and span circuit

This is called a zero and span circuit since it establishes a new reference 0 V (as 0°C instead of 0°F) and changes the span (a 180°F span to a 100°C span).

Solution

This circuit contains two stages. Both are inverting negative feedback voltage amplifiers. Therefore, V_{error} is ideally 0 V for both op amps and the node voltages at node a_1 and node a_2 are both 0 V. You also know that the input currents of the op amp are ideally 0 A. Write these values on your schematic and write voltages and currents as you find them. You must do this if you are to fully understand and properly solve these circuits. These circuits are easily solved if you follow the technique of writing all values on the schematic as you find them. Conversely, if you do not write values on the schematic as you find them, circuits become very difficult to solve.

a. Redraw the schematic assuming the sensor temperature is 32°F. Write values on the schematic as you find them. The output of the LM34 is

$$V_{in} = (10 \text{ mV/°F})(32°F) = 320 \text{ mV}$$

Apply Ohm's law to the *input resistance* to find I_{in}.

$$I_{in} = \frac{320 \text{ mV}}{1.8 \text{ k}\Omega} = 177.8 \text{ μA}$$

Apply Ohm's law to the *reference resistance* to find I_{ref}.

$$I_{ref} = \frac{320 \text{ mV}}{1.8 \text{ k}\Omega} = 177.8 \text{ μA}$$

The 1st stage is the inverting summer amplifier used in an earlier example to verify KCL. Applying KCL to node a_1,

$$I_{in} = I_{ref} + I_{1\_Rf}$$

or

$$I_{1\_Rf} = I_{in} - I_{ref}$$

$$I_{1\_Rf} = 177.8 \text{ μA} - 177.8 \text{ μA} = 0 \text{ A}$$

Therefore the drop across the 1st stage feedback resistor is 0 V (Ohm's law) and the output of the first op amp is 0 V.

Thus, the input to the 2nd stage is 0 V, which develops 0 A input current through the two 10 kΩ resistors of the 2nd stage. This produces 0 V out of the 2nd stage.

Check to be sure that this is the right temperature for freezing water. The output voltage conversion factor should now be 10 mV per °C.

$$T_{out} = \frac{0 \text{ V}}{10 \text{ mV} / °C} = 0°C$$

Yes. Freezing on the Fahrenheit sensor scale translated to freezing on the Celsius scale.

b. Redraw the schematic assuming the sensor temperature is 212°F. See Figure 9-37(b). The output of the LM34 is

$$V_{in} = \left(10 \frac{mV}{°F}\right)(212°F) = 2.12 \text{ V}$$

Apply Ohm's law to the *input resistance* to find I_{in}.

$$I_{in} = \frac{2.12 \text{ V}}{1.8 \text{ k}\Omega} = 1.1778 \text{ mA}$$

I_{ref} remains unchanged as found earlier by Ohm's law.

$$I_{ref} = \frac{320 \text{ mV}}{1.8 \text{ k}\Omega} = 177.8 \text{ μA}$$

Applying KCL to node a_1,

$$I_{1\_Rf} = I_{in} - I_{ref}$$

$$I_{1\_Rf} = 1.1778 \text{ mA} - 177.8 \text{ μA} = 1 \text{ mA}$$

Find the voltage drop across the 1st stage feedback resistor.

$$V_{1\_Rf} = 1 \text{ mA} \times 1 \text{ k}\Omega = 1 \text{ V}$$

The output voltage of the 1st stage is

$$V_{1\_out} = 0 \text{ V} - 1 \text{ V} = -1 \text{ V}$$

The output current of the 1st stage, which is the input current to the 2nd stage, is

$$I_{1\_out} = \frac{1 \text{ V}}{10 \text{ k}\Omega} = 100 \text{ μA}$$

The current in the feedback of the 2nd stage is by KCL

$$I_{2\_Rf} = I_{1\_out} = 100 \text{ μA}$$

The voltage across the feedback resistor of the 2nd stage is

$$V_{2\_Rf} = 100 \text{ μA} \times 10 \text{ k}\Omega = 1 \text{ V}$$

The output voltage of the 2nd stage is the 1 V raised across its feedback resistor.

$$V_{out} = 0\ V + 1\ V = 1\ V$$

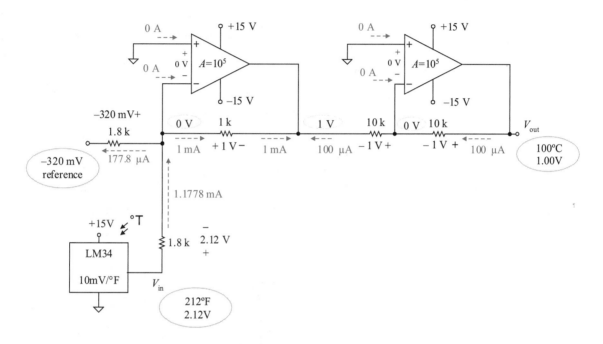

Figure 9-37(b) Example zero and span circuit–conversion from 212°F to 100°C

The 2nd stage merely inverts the signal voltage coming from the 1st stage. The –1 V into the 2nd stage is inverted to 1 V. Check that this is the correct temperature for boiling water. The output voltage conversion factor should now be 10 mV per °C.

$$T_{out} = \frac{1V}{10\ mV/°C} = 100°C$$

Yes. This zero and span circuit composed of two inverting voltage amplifiers does convert from the Fahrenheit scale to the Celsius scale. Figure 37(b) shows all the node voltages, voltage drops, and currents for the temperature sensor at

212°F. *Your final schematic drawing should have the same values displayed if you properly solved and documented your schematic.*

Practice

For the zero and span circuit of Figure 9-37(a), find the output voltage V_{out} at a room temperature of 72°F. Redraw the original schematic and document all node voltages, voltage drops, and currents on your schematic.

Answer: 222 mV

Difference Amplifier

A **difference amplifier** as shown in Figure 9-38 is designed to amplify

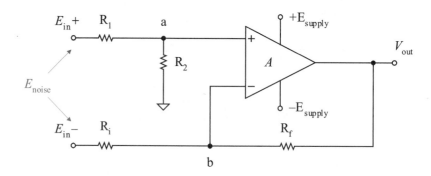

Figure 9-38 Difference amplifier

the *difference* between the input signals on each op amp input terminal. Identical external noise mixed in with the signal produces a net ideal noise output of 0 V while the *signal difference* is amplified. This produces the desired output signal with a significant reduction of noise in the output voltage. Note the two *essentially series* circuits (R1 with R2, R_i with R_f).

Example 9-21

For the difference amplifier circuit of Figure 9-39(a) find the signal output voltage, the amplifier voltage gain, and the noise output voltage. Assume an ideal op amp.

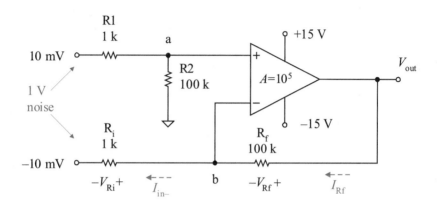

Figure 9-39(a) Example difference amplifier

Solution

First analyze the signal voltages and currents. The noninverting input signal voltage divides at node a. Since the current into the op amp is essentially 0, the two resistors at the noninverting input circuit of R1 and R2 form an *essentially series* circuit.

$$V_a = \frac{100 \text{ k}\Omega}{101 \text{ k}\Omega} \times 10 \text{ mV} = 9.901 \text{ mV}$$

Since this is a negative feedback amplifier (and assuming the op amp output is not at a rail voltage), the signal difference between the op amp input terminals is ideally 0 V.

$$V_{error} = 0 \text{ V}$$

Therefore,

$$V_b = V_a = 9.901 \text{ mV}$$

The voltage drop across V_{Ri} is the difference of its two node voltages.

$$V_{\text{Ri}} = 9.901 \text{ mV} - (-10 \text{ mV}) = 19.901 \text{ mV}$$

By Ohm's law, the inverting input signal current is

$$I_{\text{Ri}} = \frac{19.901 \text{ mV}}{1 \text{ k}\Omega} = 19.901 \text{ }\mu\text{A}$$

By Kirchhoff's current law at node b and assuming the op amp input current is 0, the feedback resistor current is

$$I_{\text{Rf}} = I_{\text{Ri}} = 19.90 \text{ }\mu\text{A}$$

By Ohm's law, the voltage drop across the feedback resistor is

$$V_{\text{Rf}} = 19.90 \text{ }\mu\text{A} \times 100 \text{ k}\Omega = 1.990 \text{ V}$$

The output signal voltage is equal to the voltage at node b plus the rise of voltage across feedback resistor R_{f}.

$$V_{\text{out}} = V_{\text{b}} + V_{\text{Rf}}$$

$$V_{\text{out}} = 9.901 \text{ mV} + 1.990 \text{ V} = 2.00 \text{ V}$$

The voltage gain of this amplifier is the output voltage 2.00 V divided by the net input voltage to the amplifier system of 20 mV (that is, the difference input signal 10 mV less −10 mV).

$$A_{\text{v}} = \frac{2.00 \text{ V}}{20 \text{ mV}} = 100$$

Perform the same analysis as above for the *noise signal*.

$$V_{\text{a}} = \frac{100 \text{ k}\Omega}{101 \text{ k}\Omega} \times 1 \text{ V} = 0.9901 \text{ V}$$

$$V_{\text{b}} = V_{\text{a}} = 0.9901 \text{ V}$$

$$V_{\text{Ri}} = 1 \text{ V} - 0.9901 \text{ V} = 9.9 \text{ mV}$$

$$I_{\text{Ri}} = \frac{9.9 \text{ mV}}{1 \text{ k}\Omega} = 9.9 \text{ }\mu\text{A}$$

$$I_{Rf} = I_{Ri} = 9.9\ \mu A$$

$$V_{Rf} = 9.9\ \mu A \times 100\ k\Omega = 0.99\ V$$

$$V_{out\ noise} = V_b - V_{Rf}$$

$$V_{out\ noise} = 0.99\ V - 0.99\ V = 0\ V$$

Note the voltage polarity drop of R_f. The noise common to both input signals cancels itself out in the amplifier, reducing the common noise to 0 V output. *The result is that the output signal voltage is 2 V and the output noise voltage is ideally 0 V.*

Practice

Repeat the example problem if R2 and R_f are changed to 50 kΩ.

Answer 1.0 V, 50, 0 V

Exploration

Suppose the example problem resistors are ±5 % tolerance resistors and the actual values are those shown in the circuit of Figure 9-39(b). Calculate the output signal voltage and output noise voltage. Then calculate the voltage signal-to-noise ratio, that is, the output signal voltage divided by the output noise voltage.

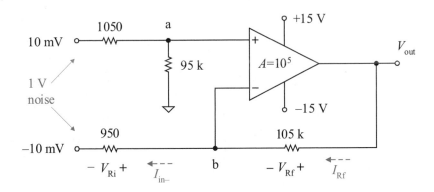

Figure 9-39(b) Difference amplifier with ± 5% tolerance resistors (worst case)

Answer: signal: 2.21 V; noise: 218 mV; voltage signal-to-noise ratio: 10 to 1. The resistors were selected to create the worst case scenario. The signal voltage now has an error of 10%. The noise voltage went from an ideal value of 0 V to a value of 218 mV. Slight variations in resistance values of the difference amplifier can create significant errors in output signal and noise voltages. Ideally the signal-to-noise ratio is infinity (∞) since the ideal value of noise for the output signal is 0 V.

Instrumentation Amplifier

The difference amplifier in the previous example has two significant shortcomings. First, the input resistance seen by E_{in-} depends on the voltage at E_{in+}. So it is not fixed, and does not match the resistance at the other input. Second, and much worse, is that to adjust the voltage gain you have to adjust both R_f resistors while keeping them exactly equal. Even a 1% difference can seriously degrade the amplifier's performance.

The instrumentation amplifier using two op amps shown in Figure 9-40 solves both of these problems. Each input is connected directly into an op amp's noninverting input. These amplifier input resistances are typically in the GΩ range. The single resistor R_g sets the gain. So there are no matching problems. (This section and example provided courtesy of Professor J. Michael Jacob of Purdue University.)

Example 9-22

Analyze the circuit in Figure 9-40 for all currents and voltages.

Solution

As you find current and voltage values, place them on the schematic with proper current direction and voltage polarity references. Be sure you understand each value and its reference direction before proceeding to the next calculation.

1. Noninverting input voltages:

$$V_{ni\ U1} = 2.45 \text{ V} \qquad\qquad V_{ni\ U2} = 2.55 \text{ V}$$

2. Inverting voltages:

$$V_{inv\ U1} = 2.45 \text{ V} \qquad\qquad V_{inv\ U2} = 2.55 \text{ V}$$

3. Input resistor currents, U1:

$$I_{Ri1} = \frac{2.45 \text{ V}}{9 \text{ k}\Omega} = 272.2 \text{ }\mu\text{A} \leftarrow$$

$$I_{Rg} = \frac{2.55 \text{ V} - 2.45 \text{ V}}{200 \text{ }\Omega} = 500 \text{ }\mu\text{A} \leftarrow$$

4. Feedback resistor current, U1:

$$I_{Rf1} = 500 \text{ }\mu\text{A} - 272 \text{ }\mu\text{A} = 228 \text{ }\mu\text{A} \rightarrow$$

5. Feedback resistor voltage, U1:

$$V_{Rf1} = 228 \text{ }\mu\text{A} \times 1 \text{ k}\Omega = 228 \text{ mV} \quad + \text{ to } -$$

6. Output voltage, U1:

$$V_{\text{out U1}} = 2.45 \text{ V} - 0.228 \text{ V} = 2.222 \text{ V}$$

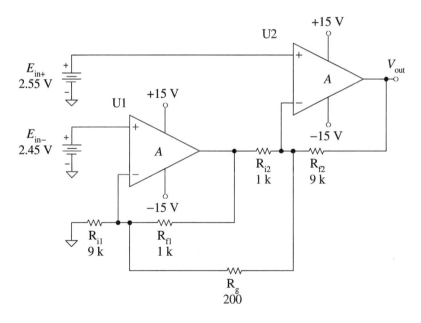

Figure 9-40 Two op amp instrumentation amplifier

Now, go back through the steps 3 through 6 for U2.

1. Input resistor current, U2:

$$I_{Ri2} = \frac{2.55 \text{ V} - 2.222 \text{ V}}{1 \text{ k}\Omega} = 328 \text{ μA} \leftarrow$$

2. Feedback resistor current, U2:

$$I_{Rf2} = 328 \text{ μA} + 500 \text{μA} = 828 \text{ μA} \leftarrow$$

3. Feedback resistor voltage, U2:

$$V_{Rf2} = 828 \text{ μA} \times 9 \text{ k}\Omega = 7.45 \text{ V} \quad - \text{ to } +$$

4. Output voltage, U2:

$$V_{out} = 7.45 \text{ V} + 2.55 \text{ V} = 10.00 \text{ V}$$

Practice

   a. Repeat the manual analysis with $R_g = 400 \ \Omega$.

   b. Does R_g directly or inversely affect the output?

   c. Verify your calculations with a simulation.

Answer: $V_{out} = 5.50$ V

Current-to-Voltage Converter

To convert a current supply or current signal to a voltage, a **current-to-voltage converter** can be used. A current-to-voltage converter is shown in Figure 9-41(a).

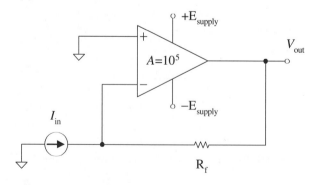

Figure 9-41(a) Current-to-voltage amplifier

The current supply or current input signal is naturally a high resistance source and needs to drive a very low resistive input amplifier circuit to prevent loss at the input. The output of the amplifier is a voltage source, which is naturally a low resistance source.

Example 9-23

For the circuit of Figure 9-41(a), find the voltage V_{out} in terms of the input current I_{in}.

Solution

First, establish known and reference currents and voltages as shown in Figure 9-41(b). Then use basic laws and circuit analysis to establish the relationship of V_{out} to I_{in}.

Since this is a negative feedback amplifier, the voltage drop across the op amp input signal terminals is ideally 0 V (assuming the output is not at its rail voltages). Therefore,

$$V_a = 0 \text{ V}$$

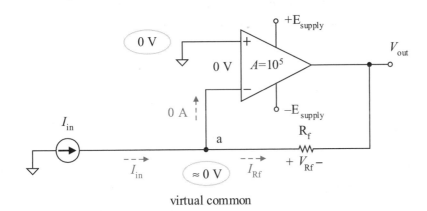

Figure 9-41(b) Setting appropriate references

Since the op amp device has very high input resistance, the current into the op amp signal terminals is ideally 0 A. Apply Kirchhoff's current law at node a.

$$I_{Rf} = I_{in}$$

Apply Ohm's law to the feedback resistor R_f.

$$V_{Rf} = R_f \, I_{Rf}$$

$$V_{Rf} = R_f \, I_{in}$$

The voltage at V_{out} is found by starting at node a (approximately 0 V) and fall V_{Rf} voltage to the output node.

$$V_{out} = 0 \text{ V} - V_{Rf} = -V_{Rf}$$

Substitute the Ohm's law form for V_{Rf}.

$$V_{out} = -R_f \, I_{in}$$

Note that the output voltage is inverted as expected since the input signal is driving the inverting input terminal of the op amp.

Practice

A 4–20 mA current loop, used in an industrial control loop, must be converted to 1–5 V control signal to control an industrial circuit. Find the value of the feedback resistor R_f of Figure 9-42 needed to convert the amplifier output to a 1–5 V output.

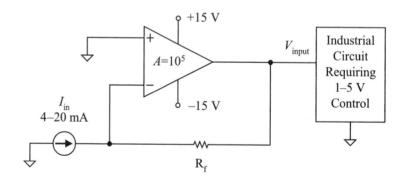

Figure 9-42 Current-to-voltage amplifier practice problem

Answer: 250 Ω

Voltage-to-Current Converter

A voltage-to-current converter is shown in Figure 9-43(a). The voltage input signal E_{in} is converted to an output current signal I_{out}.

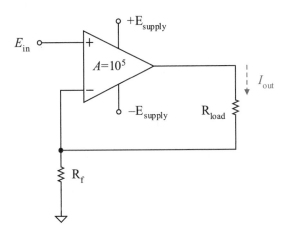

Figure 9-43(a) Voltage-to-current converter

To convert a voltage supply or voltage signal to a current, a **voltage-to-current converter** can be used. The voltage supply or voltage input signal is naturally a low resistance source and needs to drive a high resistive input amplifier circuit to prevent loss on the input. The output of the amplifier is a current source, which is naturally a high resistance source that naturally needs to drive a low resistance load.

Example 9-24

For the circuit of Figure 9-43(a), find the current I_{out} in terms of the input voltage E_{in}.

Solution

First, establish known and reference currents and voltages as shown in Figure 9-43(b). Then use basic laws and circuit analysis to establish the relationship of I_{out} to E_{in}.

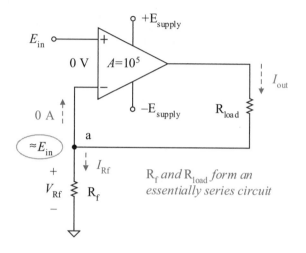

Figure 9-43(b) Setting appropriate references

Since this is a negative feedback amplifier, the voltage drop across the op amp input signal terminals is approximately 0 V (assuming the output is not at its rail voltages). Therefore, the input op amp signal terminals are about the same voltage relative to common. Ideally,

$$V_a = E_{in}$$

The voltage drop across feedback resistor R_f is therefore approximately the input voltage E_{in}.

$$V_{Rf} = E_{in}$$

Apply Ohm's law to the feedback resistor R_f and substitute E_{in}.

$$I_{Rf} = \frac{V_{Rf}}{R_f} = \frac{E_{in}}{R_f}$$

Since the op amp device has very high input resistance, the current into the op amp signal terminals is ideally 0 A. Apply Kirchhoff's current law at node a.

$$I_{Rf} = I_{out}$$

R_{load} and R_f *form an essentially series circuit.* Substitute I_{out} for I_{Rf} and the output current in terms of the input voltage is

$$I_{out} = \frac{E_{in}}{R_f}$$

This configuration looks very similar to the noninverting voltage amplifier. The major difference is the placement of the load. In this case, the op amp current is feeding the load directly, that is, the load is in series with the op amp output current.

Practice

A current meter movement is to be used to measure voltage. Design a voltage-to-current converter amplifier so that a voltage from 0 V to 1 V drives a current meter movement from 0 μA to 100 μA.

Answer: See Figure 9-44. Note: resistor R_f is selected to be 10 kΩ to meet specifications.

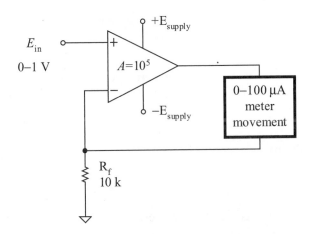

Figure 9-44 Current-to-voltage amplifier practice problem

Active Diode–Ideal Diode

The *pn* junction barrier voltage of a germanium diode is 0.2 V and the silicon diode is 0.7 V. This barrier voltage can be reduced to approximately 0 V by placing the diode in the feedback path of an op amp amplifier.

Example 9-25

For the circuits of Figure 9-45(a), find the output voltage V_{out} for the passive and active rectifier circuits and compare your results.

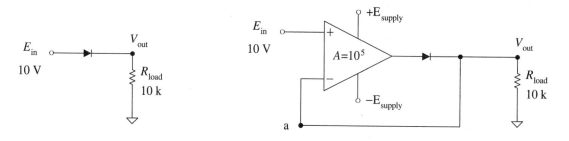

Passive rectifier   Active rectifier

Figure 9-45(a) Example problem of passive and active rectifiers

Solution

For the passive rectifier circuit of Figure 9-45(a), the output voltage is

$$V_{out} = 10 \text{ V} - 0.7 \text{ V} = 9.3 \text{ V}$$

The active rectifier circuit is essentially a voltage follower with the output voltage following the input voltage.

$$V_a = E_{in} - V_{error}$$

Ideally,

$$V_a = 10 \text{ V} - 0 \text{ V} = 10 \text{ V}$$

The output voltage is ideally at the same potential as node *a*.

$$V_{out} = V_a = 10 \text{ V}$$

The op amp output voltage is 0.7 V higher due to the 0.7 V drop across the diode. This voltage is within the op amp rail voltages.

$$V_{out\ op\ amp} = 10 \text{ V} + 0.7 \text{ V} = 10.7 \text{ V}$$

Check out the value of V_{error} now required to turn on the silicon diode. The gain of the amplifier system is 10^5 until the diode turns on. Once the diode is turned on, the system becomes a negative feedback amplifier. A passive silicon diode requires 0.7 V to turn it on. This active diode only requires 7 μV to turn on.

$$V_{error} = \frac{V_D}{A} = \frac{0.7 \text{ V}}{10^5} = 7 \text{ μV}$$

Practice

Simulate the example circuits and verify their resulting outputs.

Answer: See Figure 9-45(b). The simulated answers are approximately the same as the calculated answers.

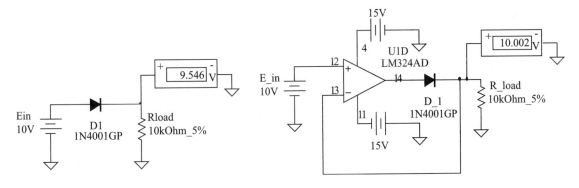

Passive diode

Active diode

Figure 9-45(b) Simulation of passive and active diode circuits

Summary

The ideal operational amplifier (op amp) was introduced and used in several types of circuits. The analysis of these circuits used circuit laws and fundamental circuit analysis techniques. Several circuits associated with the op amp turn out to be *essentially series* circuits since the op amp's input currents are ideally 0 A.

The net input voltage to the op amp is called V_{error}. V_{error} is the difference between the op amp's noninverting input pin (V_{ni}) and its inverting op amp pin (V_{inv}).

$$V_{error} = V_{ni} - V_{inv}$$

The op amp input voltages must not exceed its supply voltages else the op amp could be damaged.

$$-E_{supply} < V_{inv} \ \& \ V_{ni} < +E_{supply}$$

The op amp input voltage V_{error} is multiplied by a very large constant A to produce its output voltage $V_{out\ op\ amp}$.

$$V_{out\ op\ amp} = A V_{error}$$

If the input signal is fed to the op amp's noninverting input pin, its output is not inverted. If the input signal is fed to the op amp's inverting input pin, its output is inverted. The op amp output is limited by its rail voltages (the supply voltages less any headroom voltage needed to internally operate the op amp circuits).

$$V_{lower\ rail} \leq V_{out\ op\ amp} \leq V_{upper\ rail}$$

Each op amp has a maximum limitation of the difference in supply voltages that it can handle. The difference in supplies is the net supply voltage being applied to the chip.

$$E_{supply\ difference} = (+E_{supply}) - (-E_{supply})$$

The op amp can be used as a comparator. For all comparators,

$$V_{out} = V_{upper\ rail} \qquad\qquad \text{for } V_{error} > 0V$$

$$V_{out} = V_{lower\ rail} \qquad\qquad \text{for } V_{error} < 0V$$

The op amp is said to trip from one rail to the other as it crosses through an input voltage of 0 V. A reference voltage can be established other than 0 V by impressing it on one of the op amp input pins.

To eliminate jitter, positive feedback can be added to the op amp circuit. Positive feedback feeds part of the output signal back to the op amp's noninverting input pin. This feedback signal increases the magnitude of V_{error}, which increases V_{out}, which increases V_{error}, etc. until the op amp drives to one of its rail voltages. This action creates an upper trip point voltage and a lower trip point voltage that eliminates comparator jitter.

Negative feedback is used to produce an amplifier with a stable gain. Negative feedback feeds part or all of the output signal back to the op amp's inverting input pin. This feedback signal decreases the magnitude of V_{error}, which decreases V_{out}, which decreases V_{error}, etc. until the op amp's V_{error} approaches 0 V.

The positive and negative feedback circuits use an essentially series circuit of two resistors that voltage divide the output voltage. These two resistors are tied to one of the op amp input pins. Since the op amp input current is 0 A, the two resistors are essentially in series.

Problems

The Ideal Operational Amplifier

9-1   For each of the op amps shown in Figure 9-46, find the net amp input voltage V_{error}.

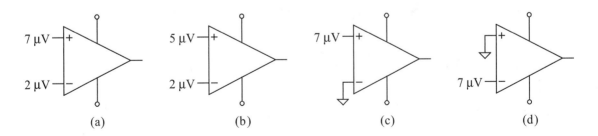

Figure 9-46

9-2   Repeat Problem 9-1 for the circuits of Figure 9-47.

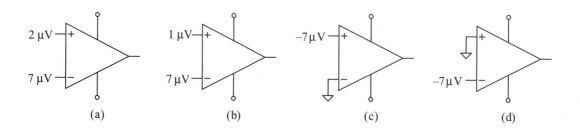

(a)   (b)   (c)   (d)

Figure 9-47

9-3   Which of the circuits in Figure 9-48 violate the criterion that the signal input voltages must not exceed the op amp's supply voltages and are therefore unsafe?

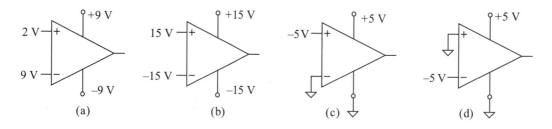

(a)   (b)   (c)   (d)

Figure 9-48

9-4   Repeat Problem 9-3 for the circuits of Figure 9-49.

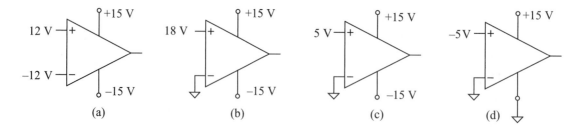

(a)   (b)   (c)   (d)

Figure 9-49

9-5   Find the output voltages for the op amps of Figure 9-50. The op
amp voltage gain is 100,000 ($A=10^5$). Check that the input voltages are within the appropriate range before solving. Assume the
rail voltages are +13 V and −13 V.

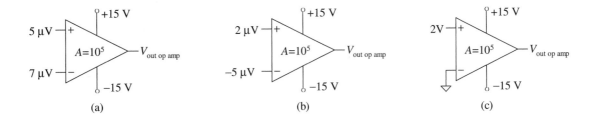

(a)   (b)   (c)

Figure 9-50

9-6   Find the output voltages for the op amps of Figure 9-51. The op
amp voltage gain is 100,000 ($A=10^5$). Check that the input voltages are within the appropriate range before solving. Assume the
rail voltages are +13 V and −13 V.

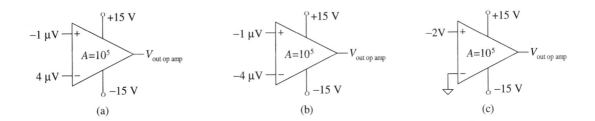

(a)   (b)   (c)

Figure 9-51

Op Amp Comparator

9-7   The circuit of Figure 9-52 is a test circuit that changes the input
voltage to an op amp with a voltage gain of $A=10^5$. Assume the
headroom required by the op amp is 2 V.

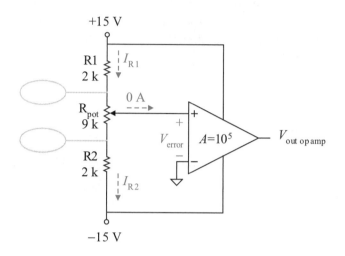

Figure 9-52

a. First analyze the potentiometer input circuit to determine the range of the input voltage to the op amp.

b. Based upon the input range, sketch a transfer curve of the output voltage $V_{\text{out op amp}}$ as a function of its input voltage V_{error}.

9-8   Repeat Problem 9-7 with the potentiometer wiper arm connected to the inverting op amp input pin and common connected to the noninverting pin.

9-9   For the circuit of Figure 9-53, find the resistance R1 needed to establish a reference voltage of 5 V. Is this a noninverting or inverting comparator? Draw the transfer curve for this comparator.

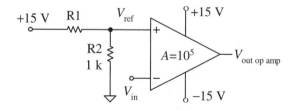

Figure 9-53

9-10   For the circuit of Figure 9-54, find the resistance R needed to establish a reference zener current of 10 mA. Is this a noninverting or an inverting comparator? Draw the transfer curve for this comparator.

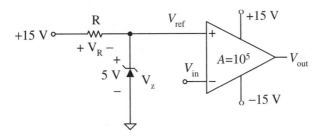

Figure 9-54

9-11   For the circuit of Figure 9-55:

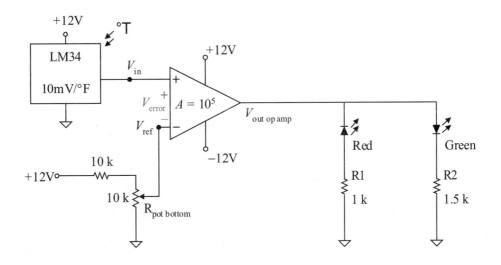

Figure 9-55

a. Find the reference voltage needed to produce a reference temperature of 100°F.
b. Find the value of $R_{pot\ bottom}$ needed to establish this reference voltage.
c. Assuming an op amp rail headroom of ±2V, find the currents flowing through the LEDs when they are turned on.
d. For what temperature range is the green LED turned on?

Positive Feedback Comparator

9-12   For the circuit of Figure 9-56, the output voltage is currently at its upper rail. Assume the op amp rail headroom needed is 2 V for the upper and lower rails.

a. If V_{in} is 3 V, find the current flowing through R_i, the voltage drop across R_i, and the op amp's input voltage V_{error}.
b. Find the input voltage V_{in} required to trip this circuit to its lower rail.

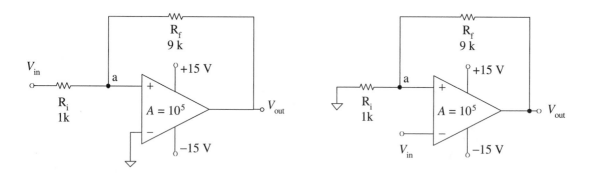

Figure 9-56   **Figure 9-57**

9-13   For the circuit of Figure 9-56, the output voltage is currently at its lower rail. Assume the op amp rail headroom needed is 2 V for the upper and lower rails.

a. If V_{in} is –3 V, find the current flowing through R_i, the voltage drop across R_i, and the op amp's input voltage V_{error}.
b. Find the input voltage V_{in} required to trip this circuit to its upper rail.

9-14 For the circuit of Figure 9-57, the output voltage is currently at its upper rail. Assume the op amp rail headroom needed is 2 V for the upper and lower rails.
   a. If V_{in} is −3 V, find the current flowing through R_i, the voltage drop across R_i, and the op amp's input voltage V_{error}.
   b. Find the input voltage V_{in} required to trip this circuit to its lower rail.

9-15 For the circuit of Figure 9-57, the output voltage is currently at its lower rail. Assume the op amp rail headroom needed is 2 V for the upper and lower rails.
   a. If V_{in} is +3 V, find the current flowing through R_i, the voltage drop across R_i, and the op amp's input voltage V_{error}.
   b. Find the input voltage V_{in} required to trip this circuit to its upper rail.

Negative Feedback Amplifiers

9-16 Identify the circuit type of Figure 9-58. Find V_a, V_{out}, and V_{error}.

9-17 Identify the circuit type of Figure 9-59. Assume that the rail voltages are ±13 V and the maximum op amp output current is 5 mA. Find the ideal amplifier input resistance. Find V_a, I_{Ri}, I_{Rf}, V_{Rf}, V_{out}, V_{load}, I_{load}, and $I_{out\ op\ amp}$. Are any op amp limitations reached?

9-18 The load resistance R_{load} of the circuit in Figure 9-59 is changed to 10 Ω. Find V_a, I_{Ri}, I_{Rf}, V_{Rf}, V_{out}, V_{load}, I_{load}, and $I_{out\ op\ amp}$. Are any op amp limitations reached?

9-19 Assume that the load resistance of the circuit of Figure 9-59 is changed to 10 Ω. Use a BJT with a β of 30 to current boost the circuit load. Find the op amp output current. Is it reasonable?

9-20 Identify the circuit type of Figure 9-60. Assume that the rail voltages are ±13 V and the maximum op amp output current is 10 mA. Find V_a, I_{Ri}, I_{Rf}, V_{Rf}, V_{out}, V_{load}, I_{load}, and $I_{out\ op\ amp}$. Find the amplifier input resistance. Are any op amp limitations reached?

9-21 The input voltage V_{in} of the circuit in Figure 9-60 is changed to be 2 V. Find V_a, I_{Ri}, I_{Rf}, V_{Rf}, V_{out}, V_{load}, I_{load}, and $I_{out\ op\ amp}$. Find the amplifier input resistance. Are any op amp limitations reached?

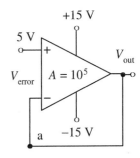

Figure 9-58

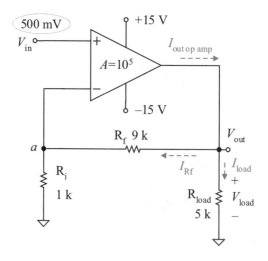

Figure 9-59

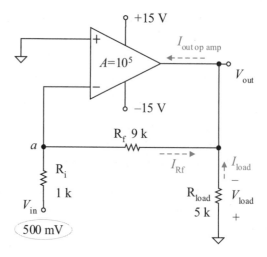

Figure 9-60

9-22 The circuit of Figure 9-61 utilizes conductive foam as a variable conductance or resistance. Its resistance varies from 60 kΩ with no pressure applied to the conductive foam. When maximum pressure is uniformly applied to the conductive foam, its resistance is 30 kΩ. Find the amplifier's output voltage range.

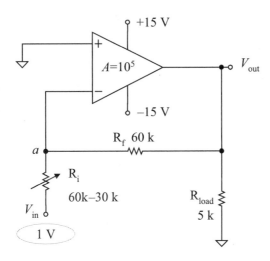

Figure 9-61

9-23 For the circuit of Figure 9-62, find the 1st stage output voltage V_{1_out}, the output voltage V_{out}, and the corresponding output temperature reading (in degrees Celsius) if the input temperature is 82°F.

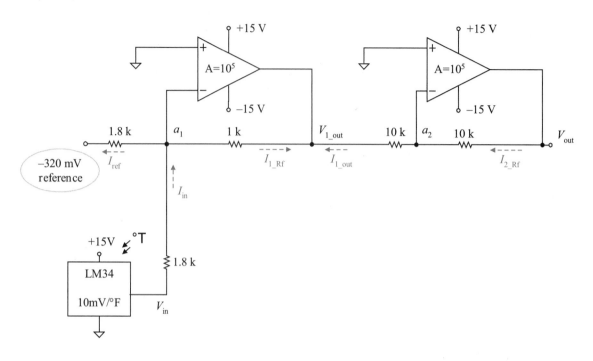

Figure 9-62

9-24 For the circuit of Figure 9-63, find the signal voltage output and the noise signal output assuming an ideal op amp. What is the signal-to-noise ratio?

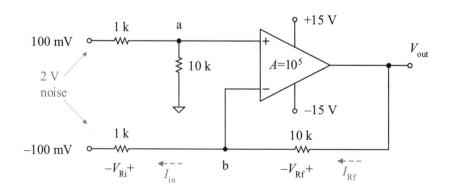

Figure 9-63

10

Parallel Circuits

Introduction

This chapter formally introduces the parallel circuit, a fundamental circuit configuration in electrical and electronics circuits. The characteristics and features of the parallel circuit are discovered using the basic laws and circuit analysis techniques. The current supply is introduced and applied to circuit analysis techniques.

Objectives

Upon completion of this chapter you will be able to:

- Identify parallel circuits and apply circuit laws to parallel circuits.

- Find total resistance and conductance of a parallel resistive circuit.

- Apply the current divider rule (CDR) and discern when applicable.

- Apply an ideal current supply to component modeling and circuit analysis.

- Model a real current supply and calculate its parameters.

- Determine a Norton model of a circuit based upon "What's in the Box" modeling and apply it for different circuit loads.

- Analyze circuits with parallel voltage and current supplies.

- Determine and apply the Norton model to circuit analysis.

- Use supply conversion technique to analyze circuits.

10.1 Parallel Circuits and the Laws

A **parallel circuit** is a circuit configuration that has two distinct voltage nodes with components (or circuits) connected to each node. A terminal from each component (or circuit) is connected to one of two voltage nodes and another terminal from each component (or circuit) is connected to the other voltage node. Recall that all terminals that are connected to the same voltage nodes are at the same potential. A voltage node essentially consists of all of the metal conductors (wire, component leads, etc.) that connect two or more components together.

Figure 10-1 shows a parallel circuit configuration. Each component (or circuit) connected between these two voltage nodes is called a **branch**. A branch may consist of a single two-terminal component or it may be a circuit with two different connections to each of the two nodes.

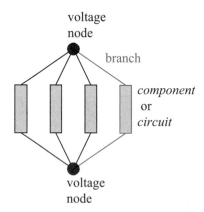

Figure 10-1 Parallel circuit configuration

Figure 10-2(a) demonstrates a simple voltage node configuration with one connecting point for each node. Figure 10-2(b) shows a voltage node configuration with several connecting wires but demonstrates the node voltage concept. The gray oval represents that this is virtually a single node at the same voltage potential.

Figure 10-3 shows the corresponding current paths in these two configurations from the node voltage viewpoint. You must be able to readily identify parallel circuits based upon their configuration and visualize how current flows into and out of their two nodes.

Visualize the current flowing in the parallel circuit of Figure 10-2(a) as demonstrated in Figure 10-3(a). The current exits the (+) supply terminal. The current then flows up and to the right into the top simple node. The current enters the top node and splits into three branch currents. The branch currents then recombine at the bottom node and flow into the (−) supply terminal. The top node is the *start* of the current split; the bottom node is the *end* of the current split. You have uniquely identified the *start* of the current split and the *end* of the current split for the three branches, thus verifying that these three branches are in parallel.

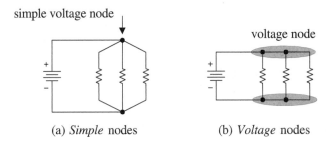

(a) *Simple* nodes   (b) *Voltage* nodes

Figure 10-2 Basic parallel circuit configurations

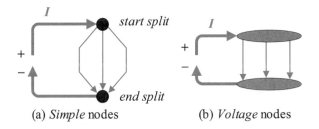

(a) *Simple* nodes   (b) *Voltage* nodes

Figure 10-3 Visualize current flowing into and out of nodes

Example 10-1

Figure 10-4 is the schematic of a three-branch load circuit. The three branch currents are I_1=10 mA, I_2=20 mA, and I_3=30 mA. Find the supply current I_{supply}. Then find the current I_x.

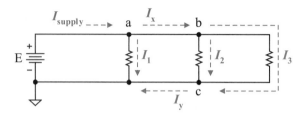

Figure 10-4 Example three-branch circuit

Solution

Using the *top voltage node* as demonstrated in Figures 10-2(b) and 10-3(b), apply KCL.

$$I_{supply} = I_1 + I_2 + I_3$$

$$I_{supply} = 10 \text{ mA} + 20 \text{ mA} + 30 \text{ mA} = 60 \text{ mA}$$

To find the current I_x, apply KCL to simple node b.

$$I_x = I_2 + I_3$$

$$I_x = 20 \text{ mA} + 30 \text{ mA} = 50 \text{ mA}$$

Or apply KCL to simple node a to find the current I_x.

$$I_{supply} = I_1 + I_x$$

$$I_x = I_{supply} - I_1$$

$$I_x = 60 \text{ mA} - 10 \text{ mA} = 50 \text{ mA}$$

Practice

For the circuit of Figure 10-4, the supply current is 100 mA, the second branch current I_2 is 15 mA, and the current I_y is 25 mA. Find the currents I_1, I_3, and I_x.

Answer: 75 mA, 10 mA, 25 mA

A key feature of the parallel circuit is that the voltage across each parallel branch must be the same. Apply KVL to any closed loop in the parallel circuit of Figure 10-5 and you can prove that each branch voltage in a parallel circuit must be the same. This is expressed by Equation 10-1.

$$V_1 = V_2 = V_3 = E \qquad \textbf{(10-1)}$$

Parallel branches must drop the *same voltage.*

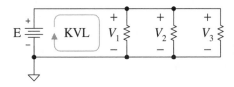

Figure 10-5 Parallel branches have the same voltage drop

Example 10-2

Figure 10-6 is the schematic of a recreational vehicle lamp circuit (excluding switches and fuses). Find the lamp voltages.

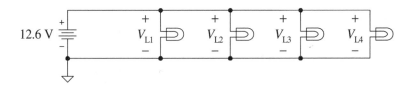

Figure 10-6 Recreational vehicle headlight and taillight schematic

Solution

Apply KVL to each closed loop that includes the 12.6 V supply for each light. For example, lamp L1 drops 12.6 V since

$$12.6 \text{ V} - V_{L1} = 0 \text{ V}$$

Or perform a walkabout from $V_{L1}(-)$ across the voltage supply with a 12.6 V rise to $V_{L1}(+)$ to find this voltage directly.

$$V_{L1} = 12.6 \text{ V}$$

Using KVL closed loop analysis or the KVL walkabout technique,

$$V_{L1} = V_{L2} = V_{L3} = V_{L4} = 12.6 \text{ V}$$

Practice

Figure 10-7 is a schematic of the headlamp circuit and taillight circuit for a car (without fuses or switches). The headlamps L1 and L2 each draw 15 A and the taillights L3 and L4 each draw 5 A. Find the voltage dropped by each lamp and the total current supplied to the four lamps by the voltage supply. Find the total circuit resistance.

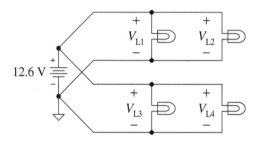

Figure 10-7 Practice problem

Answer: 12.6 V across each lamp, 40 A, 0.315 Ω or 315 mΩ

10.2 Resistance and Conductance

What is the equivalent resistance of a parallel combination of resistances? For example, what is the equivalent circuit or total resistance R_T of three resistors in parallel as shown in Figure 10-8? In other words, if an ohmmeter were connected to the three resistors in parallel (between nodes a and b), what R_T would the ohmmeter read?

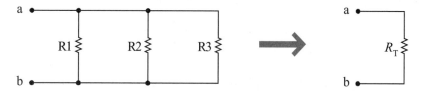

Figure 10-8 Total or equivalent parallel resistance

Apply the basic laws to a parallel resistive circuit to determine the effect of parallel resistors. First let's examine this with an example circuit.

Example 10-3

For the circuit of Figure 10-9(a), find (a) the resistor voltage drops, (b) the branch currents and the supply current, and (c) the total circuit resistance.

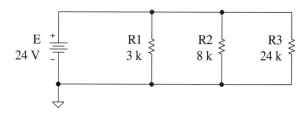

Figure 10-9(a) Example circuit

Solution

 a. Use Kirchhoff's voltage law or apply the knowledge that parallel branches have equal voltage drops because of KVL.

$$V_{R1} = V_{R2} = V_{R3} = 24 \text{ V}$$

 b. Use Ohm's law to find each branch current.

$$I_{R1} = \frac{24 \text{ V}}{3 \text{ k}\Omega} = 8 \text{ mA}$$

$$I_{R2} = \frac{24 \text{ V}}{8 \text{ k}\Omega} = 3 \text{ mA}$$

$$I_{R3} = \frac{24 \text{ V}}{24 \text{ k}\Omega} = 1 \text{ mA}$$

Use Kirchhoff's current law to find the supply current.

$$I_{\text{supply}} = 8 \text{ mA} + 3 \text{ mA} + 1 \text{ mA} = 12 \text{ mA}$$

Figure 10-9(b) shows the resulting circuit values.

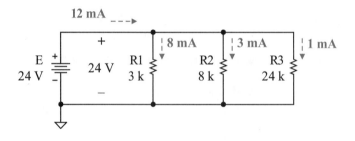

Figure 10-9(b) Example circuit with voltages and currents

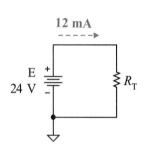

Figure 10-9(c)
Example cicuit–total
resistance view

c. Visualize the supply voltage and the supply current driving the equivalent total resistance of the circuit as shown in Figure 10-9(c). This is the circuit as viewed by the supply looking into an equivalent total resistance. Now, apply Ohm's law to this circuit equivalence and find the equivalent total resistance R_T of this parallel circuit.

$$R_T = \frac{24 \text{ V}}{12 \text{ mA}} = 2 \text{ k}\Omega$$

This is an interesting result. *The total resistance of the parallel circuit is less than any of the branch resistances.*

Practice

Figure 10-10 is a simulation of a parallel circuit. Use basic laws and circuit analysis techniques to verify (a) the resistor voltage drops, (b) the branch currents and the supply current, and (c) the total circuit resistance. Note that the supply current meter is an ammeter with 1 mA resolution while the branch current meters are milliammeters with 0.001 mA resolution.

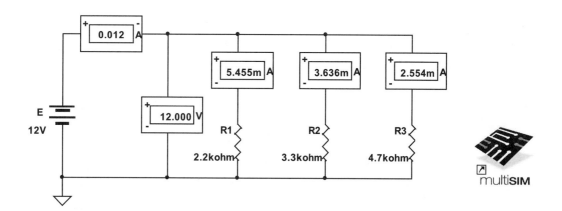

Figure 10-10 Practice problem–MultiSIM simulation

Answer: (a) 12 V each; (b) 5.455 mA, 3.636 mA, 2.554 mA, 11.65 mA; (c) 1.03 kΩ. The branch currents are less than 0.01 A, so MultiSIM creates milliammeters which produce 4 significant figure accuracy. Note the round off error of the supply ammeter of 0.012 A.

Now, using the circuit of Figure 10-11, apply the same approach as above using the laws and basic circuit analysis techniques; but, use literals instead of circuit values and find the equivalent circuit total resistance of the parallel combination.

First, note that the voltage drop across each branch and thus each resistor is the same as the supplied voltage E based upon KVL.

$$V_{R1} = V_{R2} = V_{R3} = E$$

Use Ohm's law to express each branch current and the supply current.

$$I_{R1} = \frac{E}{R1}$$

$$I_{R2} = \frac{E}{R2}$$

$$I_{R3} = \frac{E}{R3}$$

$$I_{supply} = \frac{E}{R_T}$$

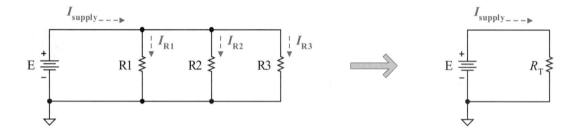

Figure 10-11 Equivalent circuit parallel resistance

Use Kirchhoff's current law to express the supply current.

$$I_{supply} = I_{R1} + I_{R2} + I_{R3}$$

Substitute the Ohm's law form for each of these currents.

$$\frac{E}{R_T} = \frac{E}{R1} + \frac{E}{R2} + \frac{E}{R3}$$

Now, divide out the common term E from the numerator of each term.

$$\frac{1}{R_T} = \frac{1}{R1} + \frac{1}{R2} + \frac{1}{R3}$$

This relationship can be extended to any number n of resistors in parallel as expressed in Equation 10-2. This is a quick solution technique to find the total resistance of a parallel resistance combination without doing extensive circuit analysis.

$$\frac{1}{R_T} = \frac{1}{R1} + \frac{1}{R2} + \frac{1}{R3} + ... + \frac{1}{Rn} \qquad \text{(10-2)}$$

Example 10-4

For the circuit of Figure 10-12 (the same as Figure 10-9 of Example 10-3), find the total circuit resistance R_T and then find the supply current.

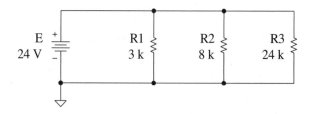

Figure 10-12 Example circuit

Solution

Find the total parallel resistance R_T by substituting the branch resistance values into Equation 10-2 and solving for R_T.

$$\frac{1}{R_T} = \frac{1}{3\ k\Omega} + \frac{1}{8\ k\Omega} + \frac{1}{24\ k\Omega}$$

This can be solved as follows. First, invert both sides of the equality, then reduce the right side of the expression.

$$R_T = \frac{1}{\dfrac{1}{3\ k\Omega} + \dfrac{1}{8\ k\Omega} + \dfrac{1}{24\ k\Omega}}$$

or

$$R_T = \left(\frac{1}{3\ k\Omega} + \frac{1}{8\ k\Omega} + \frac{1}{24\ k\Omega} \right)^{-1}$$

or

$$R_T = \left((3\ k\Omega)^{-1} + (8\ k\Omega)^{-1} + (24\ k\Omega)^{-1} \right)^{-1}$$

This last form is especially useful for a calculator solution. The calculator keystrokes are:

3 EE 3 x⁻¹ + 8 EE 3 x⁻¹ + 2 4 EE 3 x⁻¹ Enter x⁻¹ Enter

where

EE	*Power of ten* key
x⁻¹	*Reciprocal* key
Enter	*Enter* or = key

The resulting equivalent total parallel resistance is

$$R_T = 2 \text{ k}\Omega$$

For this example, a shortcut is to assume that all of the resistance units are in kΩ and thus the answer can be assumed to be in units of kΩ.

3 x⁻¹ + 8 x⁻¹ + 2 4 x⁻¹ Enter x⁻¹ Enter

The resulting numerical answer of 2 is interpreted as 2 kΩ.

Now find the supply current by applying Ohm's law.

$$I_{supply} = \frac{E}{R_T} = \frac{24 \text{ V}}{2 \text{ k}\Omega} = 12 \text{ mA}$$

Practice

For the circuit of Figure 10-13, find the total circuit resistance R_T and then find the supply current.

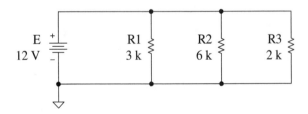

Figure 10-13 Practice problem

Answer: 1 kΩ; 12 mA

Two-Branch Resistance Rule

If only two branches are in parallel, the general expression for total parallel resistance can be simplified. Start with the general expression for two parallel branches with resistances R1 and R2.

$$\frac{1}{R_T} = \frac{1}{R1} + \frac{1}{R2}$$

Form the common denominator of (R1 × R2) and simplify.

$$\frac{1}{R_T} = \frac{R2}{R1 \times R2} + \frac{R1}{R1 \times R2} = \frac{R1 + R2}{R1 \times R2}$$

Take the reciprocal of both sides of the equality. The total resistance of two branches in parallel is expressed as Equation 10-3.

$$R_T = \frac{R1 \times R2}{R1 + R2} \qquad \text{(10-3)}$$

<u>product</u>
sum

The total resistance of two parallel branches is the product of their branch resistances divided by the sum of their two branch resistances. This only works for two parallel branches.

Example 10-5

Find the total resistance of the circuit in Figure 10-14.

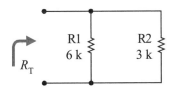

Figure 10-14 Example problem

Solution

The general rule for finding the total parallel resistance is

$$\frac{1}{R_T} = \frac{1}{6\ k\Omega} + \frac{1}{3\ k\Omega}$$

Solving for R_T

$$R_T = 2\ k\Omega$$

Using the two-branch rule of Equation 10-3, total resistance is

$$R_T = \frac{R1 \times R2}{R1 + R2} = \frac{6\ k\Omega \times 3\ k\Omega}{6\ k\Omega + 3\ k\Omega} = 2\ k\Omega$$

Typically the general rule is more useful when dealing with real circuits with less friendly resistance values and with the use of the calculator. The practice problem compares and demonstrates this observation.

Practice

Find the total resistance of two parallel branches with 470 Ω and 1.2 kΩ of resistance, respectively. First calculate the total resistance using the general rule of Equation 10-2. Then calculate the total resistance using the two branch rule of Equation 10-3.

Answer: 338 Ω

Equal Resistances in Parallel

The equivalent total resistance of equal resistances in parallel can be simplified. Assume that three resistors of equal resistance value R are in parallel. Then the general expression of Equation 10-2 for total parallel resistance becomes

$$\frac{1}{R_T} = \frac{1}{R} + \frac{1}{R} + \frac{1}{R} = \frac{3}{R}$$

Taking the reciprocal of both sides of the equation, the total resistance is

$$R_T = \frac{R}{3}$$

That is, the total resistance of three equal parallel resistances is the branch resistance value divided by three, the number of branches. Extending this to the case of n equal resistances in parallel, the total resistance is expressed by Equation 10-4.

$$R_T = \frac{R}{n}$$

(10-4)

Example 10-6

    Find the total resistance of five 10 kΩ resistors in parallel.

Solution

    The general rule of Equation 10-2 for finding total parallel resistance is

$$\frac{1}{R_T} = \frac{1}{10 \text{ k}\Omega} + \frac{1}{10 \text{ k}\Omega} + \frac{1}{10 \text{ k}\Omega} + \frac{1}{10 \text{ k}\Omega} + \frac{1}{10 \text{ k}\Omega} = \frac{5}{10 \text{ k}\Omega}$$

Solving for R_T

$$R_T = 2 \text{ k}\Omega$$

Using the equal resistance branch rule of Equation 10-4, the total resistance is

$$R_T = \frac{R}{n} = \frac{10 \text{ k}\Omega}{5} = 2 \text{ k}\Omega$$

Practice

    Ten 330 Ω resistors are in parallel. Find the equivalent resistance of this parallel circuit.

Answer: 33 Ω

Conductance

Sometimes conductance is more useful than resistance in analyzing parallel circuits. Recall that conductance is the reciprocal of resistance. Thus, the resistance form of Equation 10-2 can be expressed in the conductance form of Equation 10-5.

$$\frac{1}{R_T} = \frac{1}{R1} + \frac{1}{R2} + \frac{1}{R3} + ... + \frac{1}{Rn}$$

$G = \frac{1}{R}$

$$G_T = G1 + G2 + G3 + ... + Gn \qquad \text{(10-5)}$$

Obviously the conductance equivalent Equation 10-5 is a much easier calculation than its parallel resistance equivalent. More complex analysis may be simplified by using the conductance form. The next example demonstrates this principle in solving for an unknown branch resistance by first finding its conductance.

Example 10-7

In a three-branch parallel circuit, the first branch is 470 Ω and the second branch is 330 Ω. Find the resistance of the third branch required to produce a total resistance of 103 Ω.

Solution

The total resistance form of this parallel circuit is relatively messy to solve, even using the calculator.

$$\frac{1}{103\ \Omega} = \frac{1}{470\ \Omega} + \frac{1}{330\ \Omega} + \frac{1}{R3}$$

However, its equivalent conductance form of Equation 10-5 is relatively easy to solve. First, simplify each resistance to its equivalent conductance form.

$$9.709\ \text{mS} = 2.128\ \text{mS} + 3.030\ \text{mS} + G3$$

Recall that the conductive unit is the siemens (uppercase S). Next, solve for the unknown conductance $G3$.

$$G3 = 9.709\ \text{mS} - 2.128\ \text{mS} - 3.030\ \text{mS} = 4.551\ \text{mS}$$

Resistance $R3$ is then the reciprocal of conductance $G3$.

$$R3 = \frac{1}{G3} = \frac{1}{4.551\ \text{mS}} = 220\ \Omega$$

Practice

Five 1 kΩ resistors are placed in parallel. Find the conductance of each branch. Find the conductance of the parallel circuit. Find the circuit's resistance from its conductance.

Answer: 1 mS; 5 mS; 200 Ω

10.3 CDR–Current Divider Rule

Current entering the node of a parallel circuit splits up or divides into smaller currents. Of course, the sum of those currents must sum up to be the current entering the node (Kirchhoff's current law). In the circuit of Figure 10-15, the current flowing into the node is 6 A. The current *divides* into three branch currents, 1 A, 2 A, and 3 A, respectively. One-sixth of the current (1A/6A) flows into the left branch. Two-sixths (2A/6A) flows into the middle branch. Three-sixths (3A/6A) flows into the right branch. Note that one-sixth plus two-sixths plus three-sixths must sum up to be the whole current. The whole is equal to the sum of its parts.

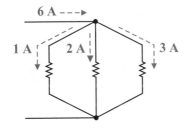

Figure 10-15 Current divides into parallel branches

This concept of current dividing leads to the current divider rule (CDR) for parallel resistive branches. Let's use the parallel circuit of Figure 10-16 to develop the current divider rule (CDR).

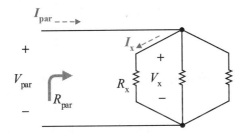

Figure 10-16 CDR–current divider rule for resistive branches

The parallel circuit voltage is V_{par}, the total current into the parallel circuit is I_{par}, and the total resistance of the parallel circuit is R_{par}. The voltage drop across each branch is equal to the voltage across the parallel circuit. Thus, for branch x,

$$V_{par} = V_x$$

Substitute the Ohm's law form for V_{par} and for V_x.

$$R_{par} \cdot I_{par} = R_x \cdot I_x$$

Solve for the branch current I_x. This is the resistance form of the **current divider rule** (*resistive* CDR) to find the current in parallel branch x.

$$I_x = \frac{R_{par}}{R_x} I_{par} \tag{10-6}$$

where

I_x	current flowing into branch x
I_{par}	current flowing into the parallel circuit
R_x	x branch resistance
R_{par}	parallel circuit resistance

Convert the resistances to their respective conductances. The conductance form of the current divider rule (*conductance* CDR) for branch x is

$$I_x = \frac{G_x}{G_{par}} I_{par} \tag{10-7}$$

where

I_x	current flowing into branch x
I_{par}	current flowing into the parallel circuit
G_x	x branch conductance
G_{par}	parallel circuit conductance

Equation 10-7 seems reasonable. *The more conductive a branch is, the larger its fraction of the total current.*

Example 10-8

Use the resistive current divider rule to find each of the branch currents in Figure 10-17. Then verify Kirchhoff's current law for the top voltage node.

Solution

First find the total resistance of the parallel circuit, R_{par}.

$$R_{par} = \left((3\ k\Omega)^{-1} + (8\ k\Omega)^{-1} + (24\ k\Omega)^{-1} \right)^{-1} = 2\ k\Omega$$

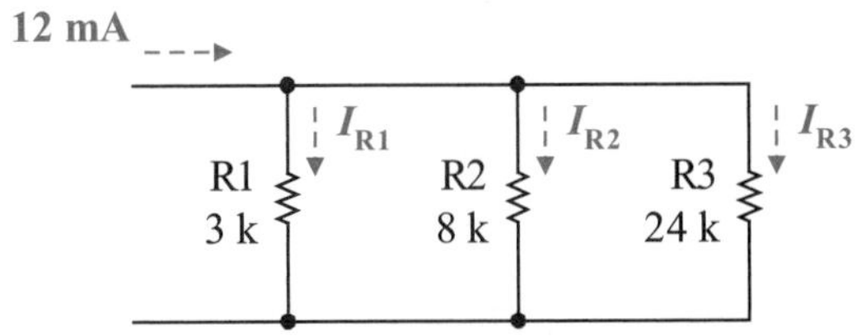

Figure 10-17 Example circuit

Now it is convenient to apply Equation 10-6 for each branch. Note that the resistor ratio must be less than one so that the branch current is less than the total current supplied to the parallel circuit.

$$I_{R1} = \left(\frac{R_{par}}{R1} \right) I_{par} = \left(\frac{2\ k\Omega}{3\ k\Omega} \right) 12\ mA = 8\ mA$$

$$I_{R2} = \left(\frac{R_{par}}{R2} \right) I_{par} = \left(\frac{2\ k\Omega}{8\ k\Omega} \right) 12\ mA = 3\ mA$$

$$I_{R3} = \left(\frac{R_{par}}{R3} \right) I_{par} = \left(\frac{2\ k\Omega}{24\ k\Omega} \right) 12\ mA = 1\ mA$$

If these branch currents are correct, Kirchhoff's current law must be satisfied.

$$I_{par} = 8\ mA + 3\ mA + 1\ mA = 12\ mA \qquad (KCL\ \checkmark)$$

Practice For the circuit in Figure 10-17,
(a) Find the conductance of each branch and the conductance of the parallel circuit. (b) Then find the currents for each branch using the conductance form of the CDR.

Answer: (a) 333 μS, 125 μS, 41.7 μS; 500 μS (b) 8 mA, 3 mA, 1 mA

Two-Branch Current Divider Rule

If only two branches are in parallel, the general expression for the resistive CDR is simplified. Start with the general expression for two parallel branches with resistances R1 and R2. From Equation 10-6

$$I_{R1} = \left(\frac{R_{par}}{R1} \right) I_{par}$$

From Equation 10-3, the parallel resistance of two resistors R_{par} is

$$R_{par} = \frac{R1 \times R2}{R1 + R2}$$

Substituting R_{par} into the equation for I_{R1} and simplifying,

$$I_{R1} = \frac{\dfrac{R1 \times R2}{R1 + R2}}{R1} I_{par}$$

$$I_{R1} = \left(\frac{R2}{R1 + R2} \right) I_{par} \qquad \textbf{(10-8)}$$

Two-branch CDR

In a two-branch parallel circuit, the current in a given branch is equal to the resistance of the *other branch* divided by the sum of the two branch resistances times the supplied current to the parallel circuit.

Example 10-9

Use the two-branch CDR of Equation 10-8 to find the 6 kΩ current in the circuit of Figure 10-18.

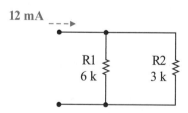

Figure 10-18 Example problem

Solution

$$I_{6k\Omega} = \left(\frac{3 \text{ k}\Omega}{6 \text{ k}\Omega + 3 \text{ k}\Omega} \right) 12 \text{ mA} = 4 \text{ mA}$$

Practice

Use the two-branch CDR of Equation 10-8 to find the 3 kΩ current in the circuit of Figure 10-18.

Answer: 8 mA

10.4 Current Supplies

A **dc current supply** produces a fixed current in a closed circuit that flows in one direction. Figure 10-19 shows the schematic symbol of an ideal current supply. The arrow indicates the direction of current flow if the current supply is connected to a circuit with a continuous path for current flow. The ideal current supply produces the same fixed current under various load conditions.

I

Figure 10-19
Current supply
symbol

Example 10-10

For the circuit of Figure 10-20, find the load current and load voltage for the following loads: short, 1 Ω, 1 kΩ, and an open.

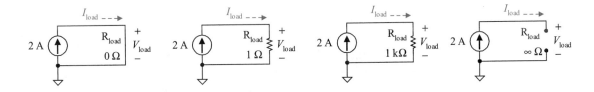

Figure 10-20 Example of varying load resistances

Solution

R$_{load}$ = short: $I_{load} = 2 \text{ A}$

$V_{load} = 2 \text{ A} \times 0 \ \Omega = 0 \text{ V}$

$R_{load} = 1\ \Omega$: $I_{load} = 2\ A$

$V_{load} = 2\ A \times 1\ \Omega = 2\ V$

$R_{load} = 1\ k\Omega$: $I_{load} = 2\ A$

$V_{load} = 2\ A \times 1\ k\Omega = 2\ kV$

$R_{load} = $ open: $I_{load} = 0\ A$

$V_{load} = 0\ A \times \infty\ \Omega \rightarrow$ undefined

In all cases, except the open circuit, the load current remains fixed at 2 A even though the load voltage is changing. However, care must be taken that the current supply is driving a reasonable load resistance. The following practice demonstrates the potential problem. Also, that last case violates Kirchhoff's current law that in a series circuit the current must be the same throughout.

Practice

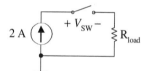

Figure 10-21
Practice problem

Use the circuit of Figure 10-21. (a) The switch is closed. Find the voltage drop V_{SW} across the switch. (b) The switch is opened suddenly. When the switch is just barely opened, an equivalent resistance of 1 MΩ exists between the switch terminals due to the resistance of the air. Assuming the current of 2 A still *momentarily exists* and the switch now acts like a 1 MΩ resistance, find the voltage drop across the switch terminals. What do you suppose happens under this condition?

Answer: (a) 0 V (b) 2 MV, the air breaks down and the terminals momentarily arc (equivalent to an electronic lightning strike eventually leading to the burning and pitting of the switch contacts)

The basic laws and circuit analysis techniques apply to current supplied circuits as well as voltage supplied circuits. The next example demonstrates driving a parallel circuit with a dc current supply.

Example 10-11

For the circuit of Figure 10-22(a), find the supply current I_{supply}, each branch current (I_{R1}, I_{R2}, and I_{R3}), and the voltage drop across the circuit (the supply voltage, V_{supply}).

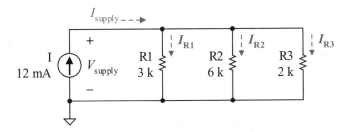

Figure 10-22(a) Example problem

Solution

KCL dictates that the supply current I_{supply} must be the same as the ideal supply current I.

$$I_{supply} = I = 12 \text{ mA}$$

To further analyze the circuit, either the equivalent total circuit resistance R_T or conductance G_T must be found. Let's find the total circuit resistance of the parallel circuit.

$$\frac{1}{R_T} = \frac{1}{3 \text{ k}\Omega} + \frac{1}{6 \text{ k}\Omega} + \frac{1}{2 \text{ k}\Omega}$$

Thus, $R_T = 1 \text{ k}\Omega$

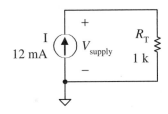

Figure 10-22(b)
Example problem–
reduced circuit

Apply Ohm's law to the reduced equivalent circuit shown in Figure 10-22(b). The supply current drives a total circuit resistance of 1 kΩ.

$$V_{supply} = V_{RT} = 12 \text{ mA} \times 1 \text{ k}\Omega = 12 \text{ V}$$

Since the circuit of Figure 10-22 is a parallel circuit, each branch drops the same voltage, 12 V. Each branch current can now be calculated by using Ohm's law.

$$I_{R1} = \frac{12 \text{ V}}{3 \text{ k}\Omega} = 4 \text{ mA}$$

$$I_{R2} = \frac{12 \text{ V}}{6 \text{ k}\Omega} = 2 \text{ mA}$$

$$I_{R3} = \frac{12\ V}{2\ k\Omega} = 6\ mA$$

A quick check of the top node shows that KCL is satisfied.

$$I_{supply} = I_{R1} + I_{R2} + I_{R3} = 4\ mA + 2\ mA + 6\ mA = 12\ mA$$

Practice

Solve Example 10-11 using the current divider rule approach. Using the total resistance found in the above solution, use the resistive CDR to find each branch current.

Answer: 4 mA, 2 mA, 6 mA

Real Current Supply

An ideal dc current supply delivers a fixed current under varying load conditions. However, a real current supply delivers less load current as the load resistance is increased. To account for this characteristic, the **real current supply** is modeled as *an ideal current supply with an internal parallel resistance* as shown in Figure 10-23.

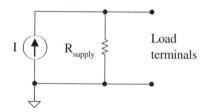

Figure 10-23 Real current supply model

The following example demonstrates the effect of real loading for a current supply.

Example 10-12

For the circuit of Figure 10-24, find the load current I_{load} and the corresponding load voltage V_{load} for the following load resistances: a short (0 Ω), 1 kΩ, 10 kΩ, and an open (∞ Ω).

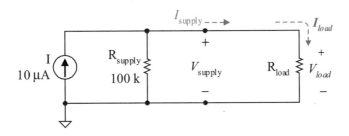

Figure 10-24 Example using a more realistic current supply

Solution

Use the two-branch CDR to find each current and then Ohm's law to find the corresponding load voltage drop.

$$I_{load} = \left(\frac{R_{supply}}{R_{load} + R_{supply}} \right) I$$

R_{load} = short:

$$I_{load} = \left(\frac{100 \text{ k}\Omega}{100 \text{ k}\Omega + 0 \text{ k}\Omega} \right) 10 \text{ } \mu A = 10 \text{ } \mu A$$

$$V_{load} = 10 \text{ } \mu A \times 0 \text{ } \Omega = 0 \text{ V}$$

R_{load} = 1 kΩ:

$$I_{load} = \left(\frac{100 \text{ k}\Omega}{100 \text{ k}\Omega + 1 \text{ k}\Omega} \right) 10 \text{ } \mu A = 9.90 \text{ } \mu A$$

$$V_{load} = 9.90 \text{ } \mu A \times 1 \text{ k}\Omega = 9.9 \text{ mV}$$

R_{load} = 10 kΩ:

$$I_{load} = \left(\frac{100 \text{ k}\Omega}{100 \text{ k}\Omega + 10 \text{ k}\Omega} \right) 10 \text{ } \mu A = 9.09 \text{ } \mu A$$

$$V_{load} = 9.09 \text{ } \mu A \times 10 \text{ k}\Omega = 90.9 \text{ mV}$$

$$R_{\text{load}} = \text{open:} \quad I_{\text{load}} = \left(\frac{100 \text{ k}\Omega}{100 \text{ k}\Omega + \infty \text{ k}\Omega} \right) 10 \text{ μA} = 0 \text{ μA}$$

$$V_{\text{load}} = 10 \text{ μA} \times 100 \text{ k}\Omega = 1 \text{ V} \qquad (R_{\text{supply}} \text{ drop})$$

If the load is a short, all of the ideal current supply is delivered to the load. *A current supply can, and most efficiently, drives a short circuit.*

As the load resistance is increased, the current dividing circuit delivers less current to the load. Even with major increases in the load resistance, the load current decreases slightly while the load voltage changes dramatically. This is the characteristic of a current supply.

If there is no load, that is an open load, then the voltage dropped across the supply's load terminals is essentially the drop across the supply's internal resistance.

Practice

Using the circuit of Figure 10-24 and the two branch CDR, find the load current and load voltage for load resistances of 0 Ω, 1 kΩ, 2 kΩ, 4 kΩ, and 8 kΩ. Make an observation about the changing voltage and current. Does your observation confirm that this supply is indeed acting like a current supply?

Answer: I_L, V_L: 10 μA, 0 V; 9.90 μA, 9.90 mV; 9.80 μA, 19.6 mV; 9.615 μA, 38.5 mV; 9.26 μA, 74.1 mV. The load resistance and voltage are doubling while the load current is decreasing very slowly. Yes, it acts like a real current supply.

Batteries are a physical source of voltage. These voltage supplies are readily available at local stores. But, have you ever heard anyone ask the local merchant for a current supply? Unlike the voltage supply, the current supply is not a stand-alone supply that you buy like a battery. However, many devices exhibit the characteristic of acting like a current supply.

BJT Constant Current Supply

The BJT transistor dc model includes a base-emitter barrier voltage supply model (0.7 V for silicon) and a collector current supply model as shown in Figure 10-25.

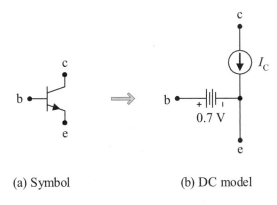

(a) Symbol　　　　　(b) DC model

Figure 10-25 A silicon *npn* BJT transistor

Example 10-13

Redraw the schematic of Figure 10-26(a), converting its silicon BJT symbol to its dc model equivalent. Then find the emitter voltage V_E, the voltage drop across R_E, the emitter current I_E and the collector current I_C. Assume the BJT current gain β is very large.

Solution

Figure 10-26(b) is the schematic redrawn with the BJT dc model. Begin the analysis at the base (*b*) of the BJT. The base is at common potential so the base node voltage is at 0 V.

$$V_B = 0 \text{ V}$$

There is a 0.7 V fall from the base (*b*) to the emitter (*e*).

$$V_E = 0 \text{ V} - 0.7 \text{ V} = -0.7 \text{ V}$$

The drop across R_E is

$$V_{RE} = -0.7 \text{ V} - (-10 \text{ V}) = 9.3 \text{ V}$$

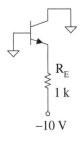

Figure 10-26(a)
Example problem

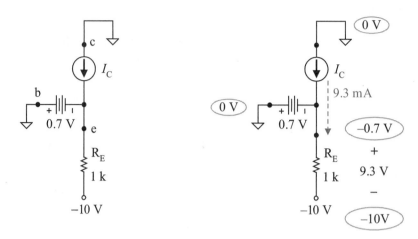

(b) With BJT dc model   (c) Modeled schematic with circuit values

Figure 10-26(b,c) Modeled dc BJT

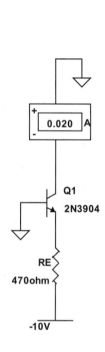

Figure 10-26(d)
Practice problem–
simulation

multi**SIM**

The current flowing out of the emitter (e) and down through R_E is found using Ohm's law.

$$I_E = I_{RE} = \frac{9.3\ V}{1\ k\Omega} = 9.3\ mA$$

The BJT current gain is very high, so the collector current is approximately equal to the emitter current (KCL).

$$I_C \approx I_E = 9.3\ mA$$

Figure 10-26(c) shows the modeled circuits with the calculated circuit values.

Practice

Assume that emitter resistor R_E in the circuit of Figure 10-26 is changed to 470 Ω. Find the collector current supplied. Support your calculation with a computer simulation as shown in Figure 10-26(d).

Answer: 19.8 mA

Zener Regulated BJT Constant Current Supply

The next example incorporates a zener diode in the BJT biasing circuit to provide fixed, more accurate collector current to the load. This circuit is BJT β independent if the BJT current gain β is a reasonably high value, namely, 50 or higher. Assuming a reasonable β, the base current is assumed to be ideally 0 and the collector and emitter currents approximately the same value.

Example 10-14

For the zener regulated bias circuit of Figure 10-27(a), apply basic laws and circuit analysis techniques and find the load current. Then find the zener current. The zener minimum current of operation is 1 mA, its maximum rated current is 60 mA, and its power rating is 500 mW. Is the zener diode in its operating range and safe? The BJT is a silicon transistor with a current gain β of 100.

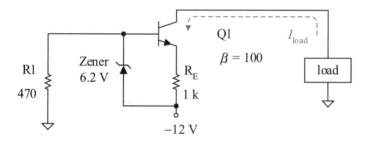

Figure 10-27(a) Zener biased BJT current supply

Solution

Redraw the schematic and include references as shown in Figure 10-27(b). The node voltages have been placed at the BJT base and emitter to help visualize the circuit solution. Starting with the negative supply of –12 V, there is a 6.2 V rise to the base of the transistor.

$$V_B = -12 \text{ V} + 6.2 \text{ V} = -5.8 \text{ V}$$

Starting at the base, the base-emitter junction drops 0.7 V.

$$V_E = -5.8 \text{ V} - 0.7 \text{ V} = -6.5 \text{ V}$$

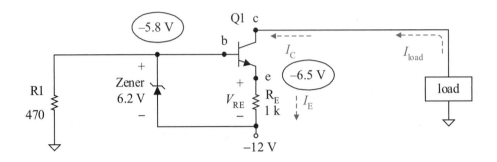

Figure 10-27(b) Example circuit set up for analysis

The drop across R_E is the difference between its node voltages, that is, its higher potential node voltage of −6.5 V and its lower potential node voltage of −12 V.

$$V_{RE} = -6.5 \text{ V} - (-12 \text{ V}) = 5.5 \text{ V}$$

Use Ohm's law to find the current flowing out of the emitter (e) and down through R_E.

$$I_E = I_{RE} = \frac{5.5 \text{ V}}{1 \text{ k}\Omega} = 5.5 \text{ mA}$$

The BJT current gain is reasonably high, so the collector current is approximately equal to the emitter current (KCL). And the load current is the same collector current.

$$I_{load} = I_C \approx I_E = 5.5 \text{ mA}$$

Figure 10-27(c) shows the circuit with references established to find the base current, the R1 voltage and current, and the zener current. The base current is the collector current divided by the BJT current gain β.

$$I_B = \frac{55 \text{ mA}}{100} = 5.5 \text{ μA}$$

The drop across R1 is the difference between its node voltages.

$$V_{R1} = 0 \text{ V} - (-5.8 \text{ V}) = 5.8 \text{ V}$$

Use Ohm's law to find the current in R1.

$$I_{R1} = \frac{5.8 \text{ V}}{470 \text{ } \Omega} = 12.3 \text{ mA}$$

Use KCL to find the zener current.

$$I_Z = I_{R1} - I_B$$

The base current, as expected, is very small compared to the remaining circuit currents due to the large BJT β.

$$I_Z = 12.3 \text{ mA} - 55 \text{ } \mu\text{A} \approx 12.3 \text{ mA}$$

The zener current is operating above its minimum requirement of 1 mA and below its maximum allowable current of 60 mA. The power dissipation of the zener diode is its current times its voltage drop.

$$P_Z = 12.3 \text{ mA} \times 6.2 \text{ V} = 76.3 \text{ mW}$$

The zener diode is operating safely.

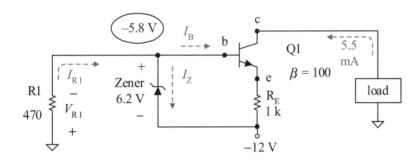

Figure 10-27(c) Example circuit–solving for the zener current

---

Practice

The zener diode of Figure 10-27 is changed to a 5 V zener. Find the load current.

Answer: 4.3 mA

Exploration

Using a 5 V, 500 mW zener diode, find the lowest value of R1 that limits the zener to its maximum current rating.

Answer 70 Ω

Photovoltaic Current-to-Voltage Converter

The next example utilizes a photovoltaic diode that converts light to current as shown in Figure 10-28. This is a negative feedback amplifier circuit; therefore, the differential input signal V_{error} is approximately 0 V. The diode symbol includes an incoming wave symbol to indicate incoming light to the diode. The diode operates in the reverse bias mode and drops very low voltage (a few tenths of a volt); therefore, one end of the diode can be connected to common while the other end is connected to virtual common. This circuit arrangement provides the unique advantage that the input is at virtual common. The input current from the photovoltaic diode is then fed through the feedback resistor R_f, creating an output voltage. As the input light varies, the diode current varies causing the current through the feedback resistor to vary, which causes the output voltage to vary.

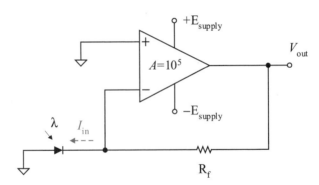

Figure 10-28 Photovoltaic diode current-to-voltage converter

Example 10-15

For the circuit of Figure 10-29, the input current from the photovoltaic cell is 10 μA. Find the output voltage.

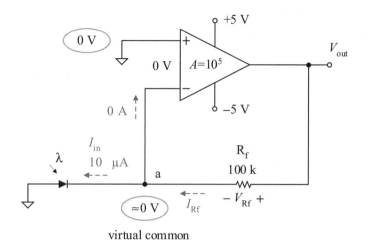

Figure 10-29 Example light-to-voltage converter circuit

Solution

Note that node *a* is at *virtual common* (very close to 0 V). Also, the current into the op amp is ideally 0; therefore, the diode and R_f form an essentially series circuit with the same current, namely I_{in}. Use Ohm's law to find the voltage drop across R_f.

$$V_{Rf} = 10 \ \mu A \times 100 \ k\Omega = 1 \ V$$

There is a 1 V rise from virtual common (0 V) to the output node. Therefore, the output voltage is

$$V_{out} = 0 \ V + 1 \ V = 1 \ V$$

Practice

The light intensity in the example problem increases, causing the photovoltaic current to double. Find the output voltage for this current.

Answer: 2 V

IC Current Supply

A more direct approach to produce a current supply is to use an IC (*integrated circuit*) that is designed specifically for that purpose. The LM334

IC chip is used in the circuit of Figure 10-30 as a current supply. It is a three-terminal device (pins V^+, V^-, and R). The supplied current I_{SET} is determined by the external resistance R_{SET}. Based upon the manufacturer's application notes, at room temperature (25°C) the supplied current is

$$I_{SET} = \frac{67.7 \text{ mV}}{R_{SET}}$$

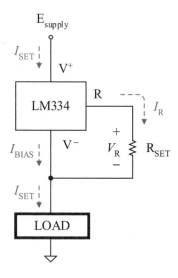

Figure 10-30 Example IC current supply

Example 10-16

    The LM334 IC of the circuit in Figure 10-30 is operating at room temperature. If R_{SET} is 10 Ω, find the current supplied to the load.

Solution

    Substituting into the manufacturer's application notes formula,

$$I_{SET} = \frac{67.7 \text{ mV}}{10 \text{ }\Omega} = 6.77 \text{ mA}$$

Practice

Find the resistance RSET needed to produce a current of 1 mA.

Answer: 67.7 Ω

Voltage-to-Current Industrial Transmitter

Another major application of a current supply circuit is in the area of industrial controls. Control signals are either voltage supplied signals of 1–5 V or current supplied signals of 4–20 mA. Control circuits are often located far from the controlling signal supplies. This requires long wire runs with terminal connections along the wired circuit. If a voltage supply is used to signal the control circuits, voltage can be lost across long wire runs and across terminal connections. Recall that all wire exhibits some resistance and the longer the wire, the greater the resistance and voltage loss. A current supplied signal does not have this problem. If a 20 mA current is generated as a signal, 20 mA is flowing throughout this current loop. Voltages are still dropped but the controlling 20 mA signal is not degraded. Figure 10-31 demonstrates these different approaches.

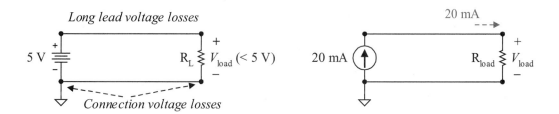

Figure 10-31 Voltage and current control loops

Example 10-17

The circuit of Figure 10-32 converts a 1 V control voltage signal to a current supplied signal for a control loop. Find the control loop current produced. Also, find the output current of the op amp. Note, the BJT is provided as a current booster. The op amp is limited to a very low current output that is amplified by the BJT to support the 4–20 mA current loop.

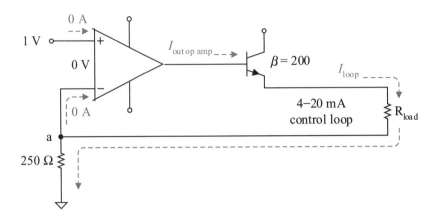

Figure 10-32 Voltage-to-current converter

Solution

This is a negative feedback amplifier with a BJT placed in the feedback loop to boost the op amp output current to support the 4–20 mA current control loop. Assuming the op amp is not slamming its rails, the op amp input voltage is ideally 0 V as shown on the schematic of Figure 10-32. The input signal is 1 V to the non-inverting pin of the op amp. Therefore V_a, the voltage at node a, must be 1 V. This drops 1 V across the 250 Ω resistor. Use Ohm's law and calculate the current flowing *down* through the 250 Ω resistor.

$$I_{250\Omega} = \frac{1\ V}{250\ \Omega} = 4\ mA$$

Since the current flowing into the op amp is ideally 0 A, the control loop current is the same as the current flowing through the 250 Ω resistor. And this is the same as the emitter current of the BJT.

$$I_{load} = I_E = 4\ mA$$

Calculate the base current of the BJT by dividing by its current gain β. The BJT base current is the same as the op amp output current.

$$I_{\text{out op amp}} = I_B = \frac{4 \text{ mA}}{200} = 20 \text{ μA}$$

A general purpose op amp, such as the LM741 or the LM324, can support several mA of current, so 20 μA is well within typical op amp output current specifications.

Practice

Suppose the control signal input is changed to 5 V. Find the control loop current and the op amp output current.

Answer: 20 mA, 100 μA

10.5 Parallel Current Supplies

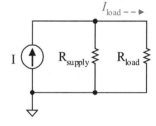

Figure 10-33
Real current supply

Before we look at parallel current supplies, you must first understand the resistive nature of the current supply as it appears to the rest of the circuit.

Ideal Internal Resistance of a Current Supply

Figure 10-33 represents the model of a real current supply with an *ideal current supply* I and a *parallel internal supply resistance* R_{supply}. The real current supply drives a load represented by R_{load} with a load current of I_{load}. The load current can be expressed using the two-branch current

$$I_{\text{load}} = \left(\frac{R_{\text{supply}}}{R_{\text{load}} + R_{\text{supply}}} \right) I$$

divider rule.

The greater the internal resistance of the real current supply, the more ideal the real current supply becomes. Note, as the internal resistance of R_{supply} increases, the resistive fraction approaches unity (1) and the load current I_{load} approaches the ideal supply current *I*. Thus, a real current supply approaches an ideal current supply if $R_{\text{supply}} \gg R_{\text{load}}$. The real current supply becomes ideal if the supply's internal resistance approaches an *open*, that is, $R_{\text{supply}} \rightarrow \infty \ \Omega$. An ideal current supply is considered to have infinite ohms of internal resistance. Expressed another way, the *ideal current supply* looks like an *open* to the rest of the circuit.

An ideal current supply has infinite internal resistance– looks like an *open*.

An alternate way to understand this is to zero the ideal current supply and deduce its resistance. Let's assume that we have an ideal current supply with 0 A of current. If any component has 0 A of current, what does this imply? It implies that the component is open since it is not drawing current. Therefore, we can deduce that the natural resistance of an ideal current supply is an *open* as perceived by the rest of the circuit.

This analogy also holds for voltage supplies. If we zero a voltage supply, then it is producing 0 V. A component that drops 0 V in a live circuit is considered to have 0 Ω of resistance, equivalent to a *short*.

An ideal voltage supply has zero internal resistance–looks like a *short*.

$$+ \quad E = 0$$

SHORT

Parallel Current Supplies

Current supplies with different current values can be placed in parallel as demonstrated by Figure 10-34. Since each ideal current supply appears as an open to the other current supplies, the current supplies do not have a path for current to flow. That is, I2 and I3 appear as opens to I1, therefore, I1 has no completed path for current to flow. Once a load is attached to the load terminals then each current supply produces a current through the load.

Figure 10-34 Parallel current supplies and their equivalent model

To simplify circuit analysis, ideal current supplies in parallel can be combined to produce a single equivalent ideal current supply as shown in Figure 10-34. Let's view this circuit from the top node. I1, I2, and I3 all flow into the top node if a load is connected. As a standard, currents flowing into a node are considered positive (+) and currents flowing out of a node are considered negative (−). Applying Kirchhoff's current law (KCL) for these current supplies at the top node, the net current flowing into the top node (if there were a load attached) is

$$I_{net} = I1 + I2 + I3 \qquad \text{(into top node)}$$

Suppose the rightmost current supply's direction were reversed as shown in Figure 10-35. What would the effect be? Viewing this from the top node's perspective again, I1 and I2 are flowing into the top node and I3 is flowing out of the top node. Thus, the net current supplied as viewed from the top node (if a load were connected) is

$$I_{net} = I1 + I2 - I3 \qquad \text{(into top node)}$$

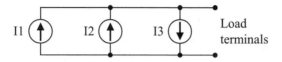

Figure 10-35 Parallel current supplies

Example 10-18

Reduce the three current supplies of Figure 10-36(a) to one equivalent current supply. Find and draw the single supply model.

Figure 10-36(a) Example problem

Solution

All three ideal current supplies flow *into* the top node, so the net current *into the top node* is the sum of these current supplies. See Figure 10-36(b).

$$I_{net} = 1 \text{ A} + 2 \text{ A} + 3 \text{ A} = 6 \text{ A}$$

Figure 10-36(b) Example equivalent current supply model

Figure 10-36(c)
Practice equivalent
current supply model

The next example examines real current supplies in parallel. Essentially
the ideal current supplies are combined in parallel and their internal
resistances are combined in parallel to produce a single real supply
model.

Example 10-19

Reduce the three real current supplies of Figure 10-37(a) to one
equivalent real current supply. Then draw this single current supply model with the load attached, and find the load current, the
load voltage, and the load power.

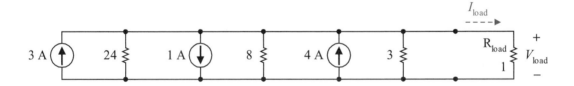

Figure 10-37(a) Example of parallel currents and resistances

Solution

The net current *into the top node* from the three ideal current supplies is

$$I_{net} = 3 \text{ A} - 1 \text{ A} + 4 \text{ A} = 6 \text{ A}$$

The three supply resistors are in parallel and can be reduced to a
single equivalent resistance, R_{supply}.

$$\frac{1}{R_{supply}} = \frac{1}{24 \ \Omega} + \frac{1}{8 \ \Omega} + \frac{1}{3 \ \Omega}$$

$$R_{supply} = 2 \ \Omega$$

The resulting single supply modeled circuit is shown in Figure 10-37(b).

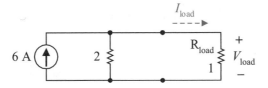

Figure 10-37(b) Single real supply equivalent model

Use the two-branch current divider rule to find the load current.

$$I_{load} = \left(\frac{2\ \Omega}{2\ \Omega + 1\ \Omega} \right) 6\ A = 4\ A$$

Use Ohm's law to find the load voltage.

$$V_{load} = 4\ A \times 1\ \Omega = 4\ V$$

Use the power rule to find the load power.

$$P_{load} = 4\ A \times 4\ V = 16\ W$$

Practice

The middle current supply of Figure 10-37(a) is changed from 1 A to 10 A. Find the equivalent single current supply model. Then use that model to find the load current, load voltage, and load power. Be faithful to the references of Figure 10-37(a).

Answer: Model: 3 A *out of the top node*, 2 Ω; load: −2 A, −2 V, 4 W

10.6 Norton's Theorem

Norton's theorem is the current supply version of Thévenin's theorem. Thévenin's theorem produced a real voltage supply model of a circuit. Norton's theorem produces a **Norton model** (*a real current supply model*)

of a circuit. Figure 10-38 demonstrates the overall concept of an equivalent Norton model of a circuit.

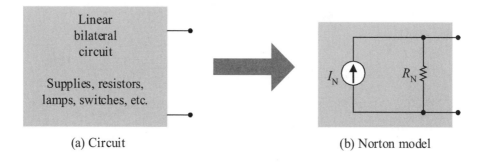

(a) Circuit   (b) Norton model

Figure 10-38 Norton model equivalent of a circuit

Norton's theorem essentially states that *any linear bilateral network may be modeled as an ideal current supply in parallel with a resistance.* The ideal current supply is called the **Norton current supply** I_N and its parallel resistance is called its **Norton resistance** R_N. The Norton model lends itself to modeling components or circuits that produce a current output. For example, the BJT collector is modeled as current supply. Ideally it is an ideal current supply. More accurately, the BJT collector can be modeled as a real current supply with a parallel, high valued resistance.

What's in the Box–Measuring I_N

Look at Figure 10-39. The model must emulate the circuit under all load conditions including a short circuit load.

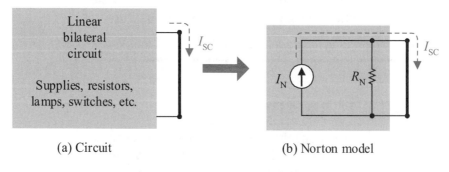

(a) Circuit   (b) Norton model

Figure 10-39 Determining (or measuring) I_N directly

What can be connected to the load terminals to produce and determine (measure) the ideal Norton current supply I_N? A conductor (a short) connected across the load terminals produces the **short circuit current**, labeled I_{SC}. Figure 10-39 demonstrates this technique.

Look at Figure 10-39(b). Note that the Norton current totally *bypasses* R_N and flows through the attached short. Thus, the Norton current is the same as the short circuit current.

$$I_N = I_{SC}$$

Note, it requires a Norton current *flowing upward* to produce a short circuit current *flowing downward* through the short.

To directly measure the Norton current I_N of a circuit, identify the pair of terminals that you wish to model without the load attached, connect a short (an ammeter) as the load, and measure the current through the short. This is the short circuit current measurement (I_{SC}). This concept is demonstrated in Figure 10-40. Recall that the internal resistance of an ammeter is very close to 0 Ω (a short).

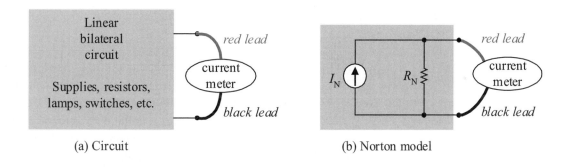

(a) Circuit   (b) Norton model

Figure 10-40 Measuring short circuit current (*warning, short circuits cause higher currents*)

Warning, do not attach a short as a load unless you are absolutely sure that the circuit can handle a shorted output. Shorts produce higher currents that the circuit may or may not be able to handle. Also, some circuits are characteristically a voltage supply by nature. If their outputs are shorted, it drives the voltage toward 0 V and the short circuit current is not a valid measurement.

What's in the Box–Measuring/Calculating R_N

To determine R_N requires that we attach a load and draw a load current. We need to measure two of the following: (1) the load voltage, (2) the load current, or (3) the load resistance. If you know two of these measurements, you can calculate the third using Ohm's law for resistance. Figure 10-41 represents this loaded measurement.

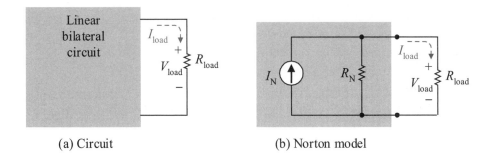

(a) Circuit   (b) Norton model

Figure 10-41 Finding R_N with loaded circuit and model

Again the real measurement is made on the real circuit as shown in Figure 10-41(a). The model in Figure 10-41(b) is used to understand and relate to the value of R_N based upon the measurements of Figure 10-37(a). Typically you would measure the load voltage V_{load}, load current I_{load}, and/or the load resistance R_{load}, and then calculate R_N based upon I_N, V_{load}, I_{load}, and R_{load} using basic circuit analysis techniques.

Example 10-20

A bridge circuit is analyzed in the lab. The load is removed and replaced with a short (an ammeter). The ammeter measures 10 mA. A load resistance of 4 kΩ is attached and the load current is measured to be 2 mA. Find and draw the Norton model for this circuit. If a 1 kΩ resistor were attached as a load instead of the 4 kΩ resistor, what would the load current and voltage be?

Solution

Based upon the short circuit current measurement,

$$I_N = I_{SC} = 10 \text{ mA}$$

Figure 10-42(a) represents the Norton model for this circuit with known values: short circuit current, loaded circuit current and load resistance. Use circuit analysis and find the value of R_N based upon the measured values and the Norton model.

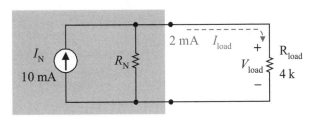

Norton model

Figure 10-42(a) Model loaded with 4 kΩ load resistor

Find the load voltage using Ohm's law.

$$V_{load} = 2 \text{ mA} \times 4 \text{ k}\Omega = 8 \text{ V}$$

R_N and R_{load} are in parallel; therefore, they have the same voltage drop.

$$V_{RN} = V_{load} = 8 \text{ V}$$

The current flowing through R_N is found using KCL.

$$I_{RN} = 10 \text{ mA} - 2 \text{ mA} = 8 \text{ mA}$$

You know the drop across R_N and the current through R_N, so you can find R_N using Ohm's law for resistance.

$$R_N = \frac{8 \text{ V}}{8 \text{ mA}} = 1 \text{ k}\Omega \qquad \text{(Ohm's law)}$$

Figure 10-42(b) represents the Norton model of the bridge circuit with a 1 kΩ load attached for analysis. We can now find the load current and the load voltage from the Norton model. These are the same load values that would be found from analyzing the original circuit with a 1 kΩ load resistance.

Find the load current using the two-branch CDR.

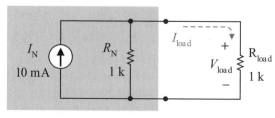

Norton model

Figure 10-42(b) Model loaded with a 1 kΩ load resistor

$$I_{load} = \left(\frac{1 \text{ k}\Omega}{1 \text{ k}\Omega + 1 \text{ k}\Omega} \right) 10 \text{ mA} = 5 \text{ mA}$$

Find the load voltage using Ohm's law.

$$V_{load} = 5 \text{ mA} \times 1 \text{ k}\Omega = 5 \text{ V}$$

So, the original bridge circuit with a 1 kΩ load resistor instead of a 4 kΩ resistor would have a load current of 5 mA and a load voltage drop of 5 V. Notice that we never analyzed the original circuit. We characterized the original bridge circuit as its Norton model and then used the model to find the original circuit's output with a different load resistance.

Practice

An unknown circuit in a box is analyzed in the lab. The load is removed and its short circuit current is measured to be 12 mA. The load resistance of 3 kΩ is attached and the load voltage is measured to be 9 V. Find the Norton model for this circuit. If a 2 kΩ resistor is attached as a load for the original circuit, what would the load voltage and current be?

Answer: model (12 mA with parallel 1 kΩ); 8 V, 4 mA

10.7 Supply Conversion Analysis

A real current supply can be converted to a real voltage supply and vice versa. A real current supply is represented by the Norton model of Figure 10-43(a). A real voltage supply is represented by the Thévenin model of Figure 10-43(b). These models must produce the same load voltage and load current with various loads.

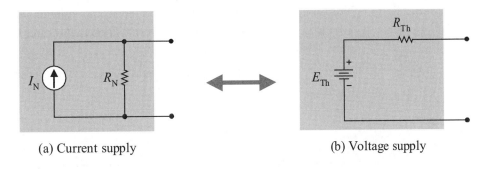

(a) Current supply   (b) Voltage supply

Figure 10-43 Supply conversion from Norton model to Thévenin model

First let's examine these supplies without a load and equate their open circuit voltages V_{OC} as shown in Figure 10-44(a,b). The open circuit of both models must be equal if these two models are equivalent. Ohm's law is applied to the current supply of Figure 10-44(a), while the open circuit voltage of Figure 10-44(b) must be equal to E_{Th}.

$$V_{OC} = E_{Th}$$
$$R_N \times I_N = E_{Th}$$

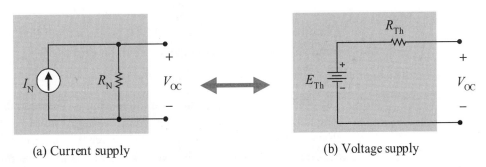

(a) Current supply   (b) Voltage supply

Figure 10-44(a,b) Supplies must have the same open circuit voltage to be equivalent models

Rewrite this equation to express R_N.

$$R_N = \frac{E_{Th}}{I_N} \tag{10-9}$$

Next let's examine the short circuit current of these supplies. These supplies must produce the same short circuit current if they are equivalent models. Figures 10-40(c,d) show the supplies with shorted outputs. The short circuit current of the current supply of Figure 10-44(c) is the ideal Norton current I_N. Ohm's law is applied to the voltage supply of Figure 10-44(d) to find its short circuit current. These short circuit load currents must be equal to each other.

$$I_N = \frac{E_{Th}}{R_{Th}}$$

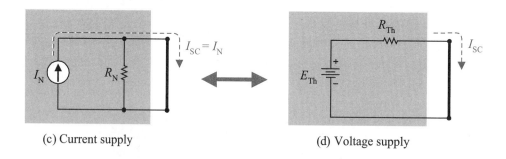

(c) Current supply   (d) Voltage supply

Figure 10-44(c,d) Supplies must have the same short circuit current to be equivalent models

Rewrite this equation to express R_{Th}.

$$R_{Th} = \frac{E_{Th}}{I_N} \tag{10-10}$$

Examining Equations 10-9 and 10-10, it is apparent that R_{Th} must be the same value as R_N if these models are to be equivalent.

$$R_{Th} = R_N \tag{10-11}$$

The conversion between supply models (that is, between the Norton current supply model and the Thévenin voltage supply model) is very straightforward. The supply resistances have the same value as expressed by Equation 10-11, and the Norton current and Thévenin voltage are related by Ohm's law as expressed by Equations 10-9 and 10-10.

Example 10-21

Convert a current supply with an ideal current supply of 10 mA and a parallel supply resistance of 2 kΩ to its equivalent voltage supply.

Solution

The internal supply resistances must have the same value.

$$R_{Th} = R_N = 2 \text{ k}\Omega$$

Use Ohm's law to find the ideal voltage supply value.

$$E_{Th} = R_N \times I_N = 2 \text{ k}\Omega \times 10 \text{ mA} = 20 \text{ V}$$

Note *polarity* of
drop across R_N

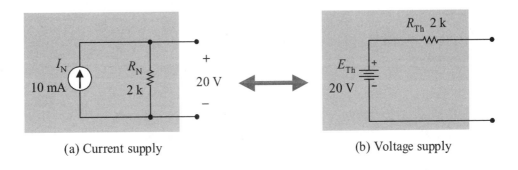

(a) Current supply   (b) Voltage supply

Figure 10-45 Example problem solution

Figure 10-45 shows the equivalent supply models. Note, that the E_{Th} polarity is established by Figure 10-45(a) with the drop across R_N. The real current supply model always has a parallel supply resistance, and the real voltage supply model always has a series circuit resistance.

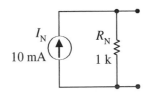

Figure 10-46
Practice solution

Practice

A voltage supply model has an ideal voltage supply of 10 V and a series supply resistance of 1 kΩ. Find and draw its current supply equivalent model.

Answer: 10 mA with a parallel resistance of 1 kΩ. See Figure 10-46.

Converting supply forms can be a useful tool in solving some types of circuits. The next example demonstrates this technique.

Example 10-22

Find the two resistor currents and the two resistor voltage drops for the circuit of Figure 10-47(a). Redraw the circuit with circuit values.

Solution

This circuit has two supplies that cannot be directly combined. Series voltage supplies can be combined. Parallel current supplies can be combined. We could (1) convert the current supply on the left to a voltage supply and solve a series circuit, or (2) we could convert the voltage supply on the right to a current supply and solve a parallel circuit.

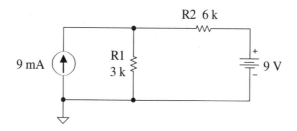

Figure 10-47(a) Example problem

Option (1) is a little more straightforward so let's convert the current supply to a voltage supply and solve a series circuit.

First, the resistances must be the same to be equivalent models.

$$R_{Th} = R_N = 3 \text{ k}\Omega$$

Note *polarity* of drop across R1

Second, Ohm's law must be satisfied.

$$E_{Th} = R_N \times I_N = 3 \text{ k}\Omega \times 9 \text{ mA} = 27 \text{ V}$$

Figure 10-47(b) shows the converted voltage supply. Figure 10-47(c) shows the original circuit with the current supply converted to its Thévenin model equivalent.

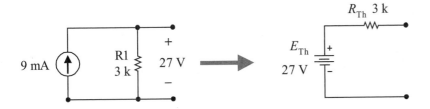

Figure 10-47(b) Convert current supply to a voltage supply

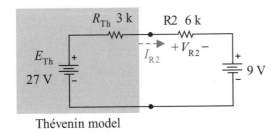

Figure 10-47(c) Circuit with current supply replaced with model

Figure 10-47(c) is now a series circuit. Use this circuit to solve for the current and voltage of resistor R2. You cannot use this circuit to solve for the current and voltage of R1. The Thévenin model is just that, a model. R1 and the ideal current supply have been modeled away. Only the voltages and currents that can be associated with the original circuit and its components can be resolved with this modeled circuit.

Going clockwise around the series loop of Figure 10-47(c), find the net voltage applied.

$$E_{net} = +27 \text{ V} - 9 \text{ V} = 18 \text{ V} \qquad \text{(CW)}$$

Sum up the series resistances.

$$R_{\text{total}} = 3 \text{ k}\Omega + 6 \text{ k}\Omega = 9 \text{ k}\Omega$$

Use Ohm's law to find the series circuit current, which is the same as the R2 current.

$$I_{R2} = \frac{18 \text{ V}}{9 \text{ k}\Omega} = 2 \text{ mA}$$

Use Ohm's law and find the R2 resistor voltage drop.

$$V_{R2} = 6 \text{ k}\Omega \times 2 \text{ mA} = 12 \text{ V}$$

Stop! You can do no more with the modeled circuit. You must now take the current and voltage of R2 back to the original (unmodeled) circuit to find the current of R1 and voltage of R1. Figure 10-47(d) shows the original circuit with the resistor R2 voltage and current values placed on the schematic. Now solve this circuit for remaining voltages and currents.

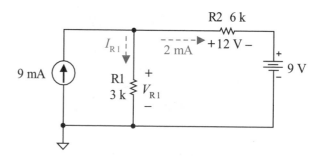

Figure 10-47(d) Back to original circuit to find I_{R1} and V_{R1}

Use KCL and find the current I_{R1}.

$$I_{R1} = 9 \text{ mA} - 2 \text{ mA} = 7 \text{ mA}$$

Use Ohm's law and find V_{R1}.

$$V_{R1} = 3 \text{ k}\Omega \times 7 \text{ mA} = 21 \text{ V}$$

The current supply voltage drop is the same as its resistor's voltage drop V_{R1} since they are in parallel.

$$V_{\text{current supply}} = V_{R1} = 21 \ \text{V}$$

Verify KVL with a clockwise loop around the right window of the circuit (21 V rise across R1, 12 V fall across R2, and 9 V supply fall).

$$+21 \ \text{V} - 12 \ \text{V} - 9 \ \text{V} = 0 \ \text{V} \qquad \qquad \text{(KVL ✓)}$$

Practice

> Repeat the example problem but convert the voltage supply to a Norton model equivalent to solve for circuit values. This produces a parallel circuit from which I_{R1} and V_{R1} can be found first.

Answer: same as the example

10.8 Parallel Voltage Supplies

Typically voltage supplies are arranged in series circuits to increase the supply voltage. For example, two 1.5 V batteries in series in a flashlight provides a 3 V supply to drive a 3 volt lamp as shown in Figure 10-48(a). So why would you want to put voltage supplies in parallel as shown in Figure 10-48(b)?

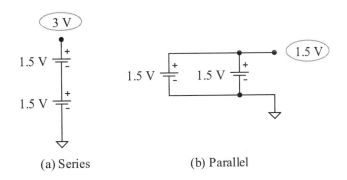

(a) Series   (b) Parallel

Figure 10-48 Series and parallel voltage supplies

Parallel voltage supplies of equal voltage produce that same voltage across the load. So the supply voltage is not increased with parallel supplies. If only one voltage supply is used, it must produce all of the current needed by the load. If, however, two equal parallel voltage supplies were used, each would produce only half of the current based upon Kirchhoff's current law. The next example demonstrates this principle.

Example 10-23

For the circuit of Figure 10-49(a), find the load current. For the circuit of Figure 10-49(b), find the load current and then find the three voltage supply currents.

Solution

The 1 kΩ load resistor drops all 3 V from the supply. So,

$$I_{load} = \frac{3 \text{ V}}{1 \text{ k}\Omega} = 3 \text{ mA}$$

In the circuit of Figure 10-49(b), the supplies are equal and in parallel with the load.

$$V_{load} = 3 \text{ V}$$

Again, find load current by Ohm's law.

$$I_{load} = \frac{3 \text{ V}}{1 \text{ k}\Omega} = 3 \text{ mA}$$

Use Kirchhoff's current law (KCL) for the top node.

$$I_{load} = I_1 + I_2 + I_3 = 3 \text{ mA}$$

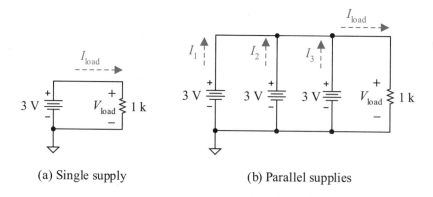

(a) Single supply   (b) Parallel supplies

Figure 10-49 Example problem

Assuming identical ideal voltage supplies, we can assume that each supply produces the same current. Divide the total load current by the number of voltage supplies to determine the current that must be supplied by each voltage supply.

$$I_1 = I_2 = I_3 = \frac{3 \text{ mA}}{3} = 1 \text{ mA}$$

The result is that each supply must produce 1 mA of current while a single supply circuit must generate all 3 mA of current. Thus the parallel voltage supply system reduces the current loading of each supply. If these were batteries, it would triple their lives.

Practice

For the circuit of Figure 10-50, find (a) the load voltage V_{load}, (b) the resistor currents of I_5 and I_6, (c) the supply current I_{supply}, (d) each supply current of I_1, I_2 and I_3, and (e) the current I_4.

The first step is to determine the total supplied current to the load. In turn, this determines the current that must be supplied by the combined three voltage supply circuit. Then that supplied current demand is shared by the three voltage supplies.

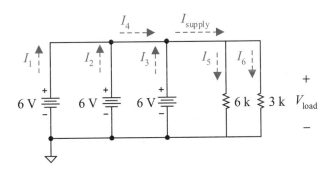

Figure 10-50 Practice circuit

Answer: (a) 6 V (b) 1 mA, 2 mA (c) 3 mA (d) 1 mA each (e) 2 mA

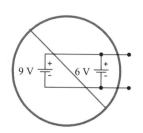

Figure 10-51
Never put unequal
voltage supplies in
parallel

Note that in all the above examples, all parallel voltage supplies are the same voltage. Never place unequal voltage supplies in parallel as shown in Figure 10-51. Not only does it violate Kirchhoff's voltage law (KVL), the lower valued voltage supply loads down a parallel higher voltage supply and the resulting voltage is somewhere between the two values. If there is an excessive difference, it can cause the lower valued supply to sink current and cause damage to either or both supplies.

Summary

Two-terminal components or circuits are connected in *parallel* when they are connected between the same two nodes. Each of these parallel components or circuits is considered a *branch*. The voltage across each parallel branch is the same based upon Kirchhoff's voltage law. The current in each parallel branch can be different.

Adding a branch in parallel increases the supplied current, which causes the total parallel circuit resistance to decrease. If resistances R1, R2, R3, ... , Rn are in parallel, then the total parallel resistance R_T is

$$\frac{1}{R_T} = \frac{1}{R1} + \frac{1}{R2} + \frac{1}{R3} + ... + \frac{1}{Rn}$$

For two branches in parallel, the total parallel circuit resistance can be simplified to the expression of the resistors' product over sum.

$$R_T = \frac{R1 \times R2}{R1 + R2}$$

The total circuit parallel resistance of n equal branch resistances in parallel is the branch resistance value divided by the number of branches.

$$R_T = \frac{R}{n}$$

The total conductance G_T of a parallel circuit with n branches is the sum of its branch conductances.

$$G_T = G1 + G2 + G3 + ... + Gn$$

Depending upon the circuit and signals being analyzed, either the resistance or conductance form may be the best to use.

Current flowing into the node of a parallel circuit divides up into branch currents and rejoins at the other node. The current divider rule

(CDR) is a formula to determine a specified branch current based upon its branch resistance, the total parallel circuit resistance, and the current flowing into the parallel circuit. The resistive form of the CDR to find the current in branch x is

$$I_x = \frac{R_{par}}{R_x} I_{par}$$

where

I_x	current flowing into branch x
I_{par}	current flowing into the parallel circuit
R_x	x branch resistance
R_{par}	parallel circuit resistance

The conductance form of the current divider rule (*conductance* CDR) for branch x is

$$I_x = \frac{G_x}{G_{par}} I_{par}$$

where

I_x	current flowing into branch x
I_{par}	current flowing into the parallel circuit
G_x	x branch conductance
G_{par}	parallel circuit conductance

In a two-branch parallel circuit, the current in a given branch is equal to the resistance of the *other branch* divided by the sum of the two branch resistances times the supplied current to the parallel circuit.

$$I_{R1} = \left(\frac{R2}{R1 + R2} \right) I_{par}$$

where

R1, R2	resistances of branch one and branch two, respectively
I_{R1}	current flowing into branch with resistance R1
I_{par}	current flowing into the two-branch parallel circuit

A *dc current supply* produces a fixed current in a closed circuit that flows in one direction under varying load conditions. The arrow in the symbol of the current supply is the direction of conventional current flow when the current supply is connected to a closed circuit.

The ideal current supply has infinite resistance. A real current supply typically has a very high resistance and is modeled as an ideal current supply with an internal parallel resistance. Ideal and real current supplies are typically used to model devices such as BJTs and MOSFETs. Parallel current supplies and resistances can be reduced to a model with a single ideal current supply and a single parallel resistance.

The Norton theorem states that a linear, bilateral circuit can be modeled as a real current supply with an ideal current supply I_N in parallel with a parallel resistance R_N. The Norton model represents the circuit for all possible loads and produces the same load current and voltage as the original circuit. This is an especially appropriate model for components such as the BJT and MOSFET, which are characterized by current supply outputs.

The Norton model can be converted to a Thévenin model and vice versa where

$$R_{\text{Th}} = R_{\text{N}} = \frac{E_{\text{Th}}}{I_{\text{N}}}$$

That is, the Thévenin and Norton resistance have the same value, and must satisfy Ohm's law with E_{Th} and I_{N}.

Voltage supplies can be placed in parallel to reduce current loading of the voltage supplies, but the voltage supplies must have the same voltage value to prevent supply loading and potential damage.

Problems

Parallel Circuits and the Laws

10-1   The lamps of Figure 10-52 each draw 6 A of current.

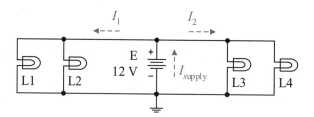

Figure 10-52

a. Find each lamp's voltage drop.

b. Find the currents I_1, I_2, and I_{supply}.

c. Find the resistance of each lamp.

d. Find the total resistance of the circuit using only the supply voltage and current.

e. Find the total conductance of the circuit using only the supply voltage and current.

10-2 For the circuit of Figure 10-52, lamps L1 and L2 are each rated at 24 W and lamps L3 and L4 are each rated at 60 W.

a. Find each lamp's voltage drop.

b. Find the currents I_{L1}, I_{L2}, I_1, I_{L3}, I_{L4}, I_2, and I_{supply}.

c. Find the resistance of each lamp.

d. Find the total resistance of the circuit using only the supply voltage and current.

e. Find the total conductance of the circuit using only the supply voltage and current.

10-3 For the circuit of Figure 10-53, draw all currents with direction and resistor voltages with polarity on your schematic as you find them.

a. Find each resistor's voltage drop.

b. Find each resistor current.

c. Find the supply current.

d. Find total circuit resistance using the supply voltage and supply current.

e. Find total circuit conductance using the supply voltage and supply current.

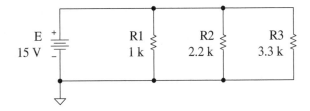

Figure 10-53

10-4   For the circuit of Figure 10-54, draw all currents with direction and resistor voltages with polarity on your schematic.

a. Find each resistor's voltage drop.

b. Find each resistor current.

c. Find I_1, I_2, I_3, and I_{supply}

d. Find total circuit resistance using the supply voltage and current.

e. Find the resistance of the R3-R4-R5 combination using the supply voltage and I_3.

f. Find the total circuit conductance using the supply voltage and current.

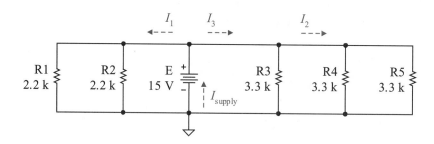

Figure 10-54

10-5   Each lamp in the circuit of Figure 10-52 draws 6 A of current.

a. Find each lamp's resistance.

b. Find the total circuit resistance using the individual lamp resistances.

c. Find the resistance seen by current I_1 using the resistances of L1 and L2.

d. Find each lamp's conductance.

e. Find the total circuit conductance using the individual lamp conductances.

f. Find the resistance seen by current I_1 using the conductances of L1 and L2.

10-6   For the circuit of Figure 10-52, lamps L1 and L2 are each rated at 24 W and lamps L3 and L4 are each rated at 60 W.

    a. Find each lamp's conductance.

    b. Find the total circuit conductance.

    c. Find the total circuit resistance.

    d. Find the supply current.

    e. Find the power delivered by the supply.

    f. Find the total power consumed by the lamps.

    g. Does the power generated equal the total power dissipated?

10-7   For the circuit of Figure 10-53:

    a. Find the total circuit resistance.

    b. Find the supply current.

    c. Find the power delivered by the supply.

    d. Find the power dissipated by each resistor.

    e. Find the total power dissipated by the resistors.

    f. Does the power generated equal the total power dissipated?

10-8   For the circuit of Figure 10-53, add a fourth parallel resistor to produce a total resistance of 257 Ω. Hint: use conductance.

CDR–Current Divider Rule

10-9   For the circuit of Figure 10-55, the current flowing into the current divider circuit $I_{\text{into split}}$ is 400 mA. Using the current divider rule, find each resistor branch current I_1, I_2, and I_3. Verify your current results using Kirchhoff's current law at the top voltage node.

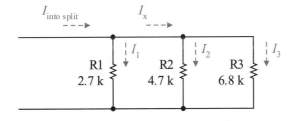

Figure 10-55

10-10 For the circuit of Figure 10-55, the current flowing into the current divider circuit $I_{\text{into split}}$ is 400 mA. Using the current divider rule, find the current I_x.

10-11 For the circuit of Figure 10-56:

    a. Find the total circuit resistance.

    b. Find the supplied current.

    c. Find the current in R1 using the current divider rule.

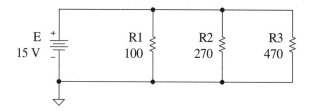

Figure 10-56

10-12 For the circuit of Figure 10-56:

    a. Find the total circuit conductance.

    b. Find the supplied current.

    c. Find the conductance of R1.

    d. Find the current in R1 using the current divider rule (conductance form).

Current Supplies

10-13 For the circuit of Figure 10-57:

    a. Find the current flowing through resistor R. Then find the voltage V_R if resistance R is:

    b. a short   **e.** 1 MΩ

    c. 1 Ω   **f.** 1 GΩ

    d. 1 kΩ   **g.** an open

    h. Are all of these voltages reasonable?

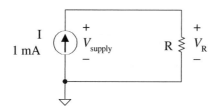

Figure 10-57

10-14   For the series circuit of Figure 10-58:

    a. Find the voltage drop across each resistor.

    b. Find the voltage drop across the current supply.

    c. Find the total power delivered by the supply.

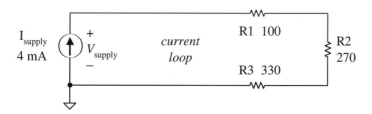

Figure 10-58

10-15   For the current supply circuit of Figure 10-59, find the supply current I_{supply} for a load resistance of:

    a. $0\ \Omega$ (a short)   **e.** $1\ k\Omega$

    b. $1\ \Omega$ **f.** $10\ k\Omega$

    c. $10\ \Omega$ **g.** $100\ k\Omega$.

    d. $100\ \Omega$

    h. At what load resistance is this supply no longer acting as a fixed dc current supply?

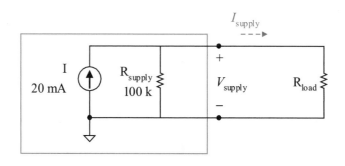

Figure 10-59

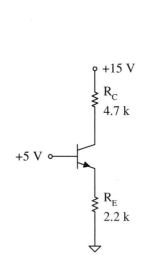

+15 V

R_C
4.7 k

+5 V

R_E
2.2 k

Figure 10-60

10-16 For the silicon BJT supply circuit of Figure 10-60, assume a very high BJT β so that $I_C \approx I_E$.

   a. Find the emitter node voltage V_E.

   b. Find the collector current I_C.

   c. Find the collector node voltage V_C.

   d. Find the collector-to-emitter voltage drop V_{CE}.

   e. Draw the BJT equivalent dc model with values.

Parallel Current Supplies

10-17 For the circuit of Figure 10-61:

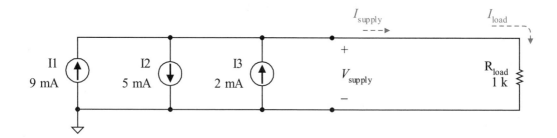

Figure 10-61

a. Redraw the circuit with a single equivalent current supply and find the supplied current I_{supply}.

b. Find the load current.

c. Find the load voltage.

d. Find the load power.

10-18 For the circuit of Figure 10-62:

a. Redraw the circuit with a single equivalent real current supply (ideal equivalent current source in parallel with its equivalent internal supply resistance).

b. Find the supplied current I_{supply} (use the current divider rule).

c. Find the load current.

d. Find the load voltage.

e. Find the load power.

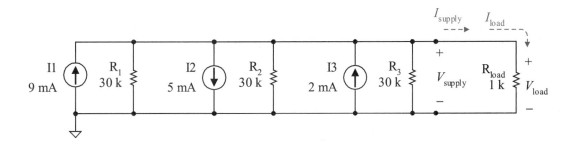

Figure 10-62

Norton's Theorem

10-19 A circuit's short circuit current is measured to be 10 mA. When loaded with a 1 kΩ resistor, its load current drops to 9.9 mA.

a. Find and draw the Norton model for this circuit.

b. A load of 4.7 kΩ is attached to the original circuit; use your Norton model to find the new load current and load voltage.

10-20   A circuit's short circuit current is measured to be 100 mA. With a load resistor attached, its load voltage is 10 V and its load current is 99 mA.

    a.   Find and draw the Norton model for this circuit.

    b.   A load of 10 Ω is attached to the original circuit; use your Norton model to find the new load current and load voltage.

Supply Conversion Analysis

10-21   Convert a Norton model of 1 mA and 10 kΩ to its equivalent Thévenin model. Draw the Thévenin model with component values.

10-22   Convert a Thévenin model of 10 V and 1 kΩ to its equivalent Norton model. Draw the Norton model with component values.

10-23   For the circuit of Figure 10-63:

    a.   Convert the current supply to its equivalent Thévenin model and redraw the circuit with the Thévenin model.

    b.   Find the current in and the voltage drop across R2.

    c.   Going back to the original circuit and using the values found for R2, find the current in and the voltage drop across R1.

    d.   What is the voltage drop across the current supply?

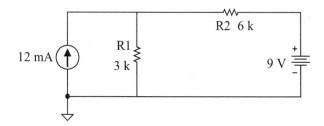

Figure 10-63

10-24   For the circuit of Figure 10-63:

    a.   Convert the voltage supply to its equivalent Norton model and redraw the circuit with the Norton model.

    b.   Find the current in and the voltage drop across R1.

c. Going back to the original circuit and using the current and voltage found for R1, find the current in and the voltage drop across R2.

d. What is the current through the voltage supply?

Parallel Voltage Supplies

10-25 For the circuit of Figure 10-64:

a. Find the resistor currents.

b. Find the supplied current I_{supply}.

c. Find the current produced from each voltage supply.

d. Find the voltage drop across each resistor.

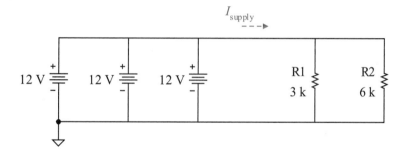

Figure 10-64

10-26 For the circuit of Figure 10-64, what happens if one of the voltage supplies drops 11 V?

Series-Parallel Circuits

Introduction

This chapter formally introduces the series-parallel circuit through the three laws of electronics. The series-parallel circuit combines series circuit and parallel circuit combinations.

Resistance reduction techniques are demonstrated to reduce a resistive series-parallel circuit to its simplest series or parallel circuit form. Voltages and currents are then calculated based upon the reduced equivalent circuit. These voltages and/or currents are then used to solve original circuit values.

Objectives

Upon completion of this chapter you will be able to:

- Identify and apply circuit laws to series-parallel circuits.

- Identify and reduce simple series and simple parallel resistance combinations to simplify the circuit for circuit analysis.

- Reduce circuits to a simple series or a simple parallel circuit to find supplied currents and voltages; and in turn use those voltages and currents to find additional circuit voltages and currents.

- Analyze ladder circuits using basic laws and circuit analysis techniques.

- Analyze an R-2R ladder circuit feeding an inverting voltage amplifier to form a simple digital-to-analog converter circuit.

- Use the Thévenin model approach to analyze the bridge circuit, a series-parallel circuit which cannot be simply reduced to a simple series circuit or a simple parallel circuit.

11.1 Series-Parallel Combinations

Series-parallel circuits are a combination or combinations of simple series and simple parallel circuits. Before examining series-parallel circuits, let's first review the fundamentals of series circuits and parallel circuits.

Series Circuits

A purely series circuit has one continuous path through its components as demonstrated in Figure 11-1. Each block represents a single two-terminal component (for example, resistor, lamp, diode, and LED).

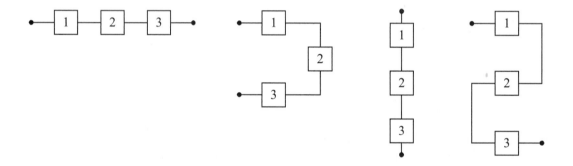

Figure 11-1 Series circuit configurations–note $\boxed{1}$ represents component 1

When connected to a supply, the current flowing into a component or wire must equal the current flowing out of that component or wire (Kirchhoff's current law). Therefore the current is the same throughout a series circuit.

$$I_1 = I_2 = I_3$$

Figure 11-2 demonstrates the current flow for each series circuit with the given polarity of applied voltage. Note, this applied voltage could be from a supply or it could be from a circuit to which this series circuit is attached. The voltage polarities of the components of Figure 11-2 are assumed to be for *voltage dropping components* such as resistors, diodes, and LEDs (*not* voltage or current supplies).

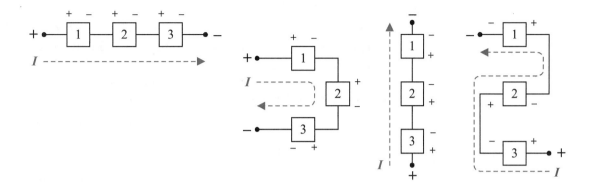

Figure 11-2 Series circuit *current* with corresponding *polarities of voltage dropping components*

Essentially Series Circuits

An *essentially series circuit* does not at first glance appear to be a series circuit. But if all branching currents are zero (0 A) and what remains is a circuit with essentially the same current, the circuit is considered to be an *essentially series circuit*. Figure 11-3 demonstrates this principle. The current branching off to component 3 is 0 A. Kirchhoff's current law dictates that the current in component 1 and component 2 must be the same.

The op amp is a classic example of this concept since the current into the op amp is virtually 0 A. Another example is the MOSFET with a gate current of 0 A. The BJT approaches this when the base current is much less than other currents in the BJT circuit (that is, $I_B \ll I_C$).

Thus in Figure 11-3, components 1 and 2 can be considered to be *essentially in series* and treated as a series circuit; this includes application of the voltage divider rule (VDR).

$$I = I_1 \approx I_2$$

If the branch current is not really 0 or not significantly less than the current I, then do not apply this concept. For example, if the current of component 1 is 100 mA and the current of component 3 is 1 mA, only a 1% error is introduced in assuming an *essentially series circuit*. However, if the current of component 3 is 10 mA, now an error on the order of 10% is introduced.

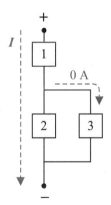

Figure 11-3
Essentially
series circuit

Parallel Circuits

A *purely parallel circuit* has two voltage nodes associated with it. Branches of components or circuits are connected between these two nodes. The ends of each branch are connected to each of these two nodes. Figure 11-4 demonstrates the parallel circuit configurations. In its simplest form, components 1, 2, and 3 are two-terminal components.

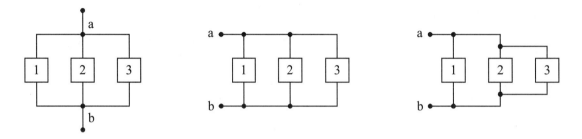

Figure 11-4 Parallel circuit configurations

Nodes *a* and *b* are the two nodes for each of the parallel circuits in Figure 11-4. Components 1, 2, and 3 form the branches of the parallel circuit. The top of the branches are connected to node *a*, and the bottom of the branches are connected to node *b*.

When connected to a supply or circuit supplying voltage or current, the voltage across each branch must be the same (Kirchhoff's voltage law). Figure 11-5 demonstrates this principle that parallel branches must have the same voltage drop.

$$V_{ab} = V_1 = V_2 = V_3$$

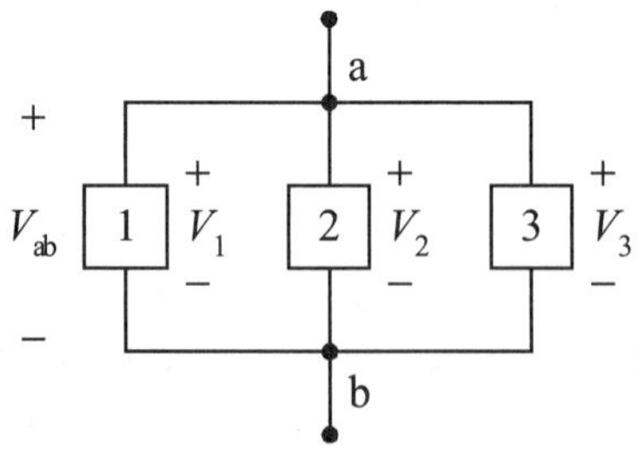

Figure 11-5 The *voltage* across parallel branches *must be the same*

Figure 11-6 demonstrates current dividing in a parallel circuit. In this example, current enters from the top of the circuit, divides at node *a*, flows separately through the individual branches, rejoins at node *b*, and then flows out of the bottom of the parallel circuit.

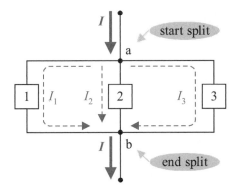

Figure 11-6 Current divides in a parallel circuit

In a parallel circuit, you must be able to uniquely identify the node at which current divides (the *start* of the current split, node *a*) and the node at which current rejoins (the *end* of the current split, node *b*). Based upon Kirchhoff's current law, the branch currents must sum up at the top (or bottom) node to be equal to the current flowing into (or out of) the parallel circuit.

$$I = I_1 + I_2 + I_3$$

Series-Parallel Circuits

Series-parallel circuits are a combination of series and parallel circuits. Figure 11-7 shows several examples of series-parallel circuit configurations.

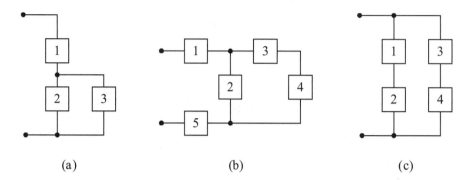

Figure 11-7 Several series-parallel circuit configurations

First look at Figure 11-8(a). Do you see any simple series or parallel combinations? Components 2 and 3 share two nodes and form a simple parallel circuit. We can combine them together and indicate that these two components are in parallel (//) by the following notation.

$$2 \mathbin{/\mkern-5mu/} 3$$

Figure 11-8(b) shows the circuit redrawn with component 2 and component 3 combined to simplify the circuit. Note: it may not be possible to physically combine 2 and 3 (for example, an LED and a resistor); however, this is a technique that demonstrates the series-parallel nature of the circuit.

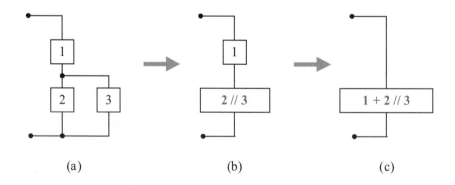

Figure 11-8 The circuit of Figure 11-7(a) reduced to a *series circuit equivalent*

Now look at Figure 11-8(b). Is anything simply in series or in parallel? Component 1 is now in series with the parallel combination of components 2 and 3 in parallel. Use the + sign to indicate that components or component combinations are in series. Figure 11-8(c) shows the complete reduction of this circuit to its simplest form. A shorthand notation for expressing this combination is

$$1 + (2 /\!/ 3)$$

where the parentheses indicate that the parallel combination is to be resolved first.

Another approach is to examine how current flows in the circuit. Assume that current is flowing into the top node and out of the bottom node. Figure 11-9 is Figure 11-7(a) redrawn showing the current flowing through the circuit. The current enters node *a*, and then flows through component 1 without branching; thus, component 1 is in *series*. The current then splits at node *b* between the two components 2 and 3 and rejoins at node *c*; therefore, these two components (branches) must be in *parallel*. The current then exits the circuit at node *d*.

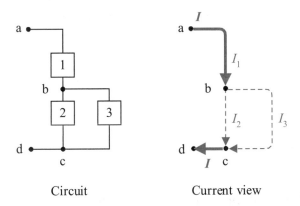

Circuit   Current view

Figure 11-9 Current view of Figure 11-7(a)

You must be able to analyze series-parallel with both techniques to determine series circuit combinations and parallel circuit combinations.

- By visual inspection, determine simple series and simple parallel combinations.

- By visualizing current flowing through the circuit, determine components that share the same current (series) and components that divide up the current (identifying the node that starts the split and the node that ends the split).

Example 11-1

Use the circuit of Figure 11-7(b). (a) Determine what is in series and parallel based upon simple combinations and circuit reduction. Write a shorthand notation expression of the reduced series-parallel combination. Draw a current view diagram of this circuit assuming the current is flowing into the top node.

Solution

Figure 11-10(a) shows the visual steps in reducing this circuit and understanding its series-parallel combinations. The components 3 and 4 are simply in series, that is (3 + 4). This is the starting point to analyze this circuit. Next, this combination is in parallel with component 2, that is, 2 // (3 + 4).

That combination is in series with components 1 and 5. The resulting combination expression is

$$1 + (2 // (3 + 4)) + 5$$

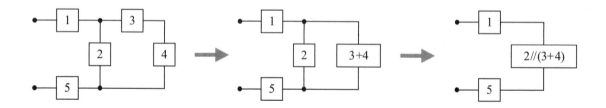

Figure 11-10(a) Example solution reducing circuit to a *series circuit equivalent*

The current view of the circuit is shown in Figure 11-10(b). Visualize the current entering node *a*, and flowing in *series* through component 1. The current then enters node *b* and *splits* into two branches. The *split starts* at node *b* and the *split ends* at node *c*. Thus branch 2 is in *parallel* with the (3 + 4) branch. After the current rejoins at node *c*, it flows in *series* through component 5.

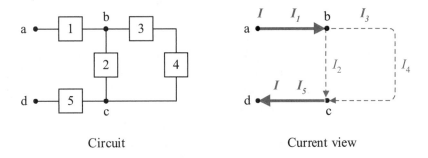

Figure 11-10(b) Example solution showing current view

Practice

Use the circuit of Figure 11-7(c). (a) Determine what is in series and parallel based upon simple combinations and circuit reduction. Write a shorthand notation expression of the reduced series-parallel combination. Draw a current view diagram of this circuit assuming the current is flowing into the top node (see Figure 11-11).

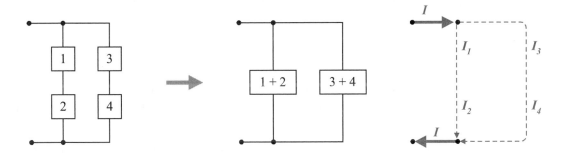

Figure 11-11 Practice problem reduced to its *parallel circuit equivalent*, and its current view

Answer: $(1 + 2) // (3 + 4)$

11.2 Resistance Reduction Technique

If the circuit components of a series-parallel circuit are resistances that form simple series, simple parallel, or simple series-parallel combinations, then the circuit can be reduced to its simplest resistance form of series or parallel. Once the circuit has been reduced to its simplest form, then basic series circuit or basic parallel circuit analysis can be used to find voltages and currents that can be applied to the original circuit. Then the remaining voltages and currents of the original circuit can be found.

As previously shown, the total resistance of the *resistive series circuit* of Figure 11-12 is found by the following equation

$$R_\mathrm{T} = R_\mathrm{ab} = R_{1-3} = R1 + R2 + R3$$

Figure 11-12
Resistive
series circuit

where the combined resistance could be written as

R_T total resistance of the circuit
R_ab resistance between nodes *a* and *b*
R_{1-3} combined resistance of R1, R2, and R3

Depending upon the specific application, one of the three symbols (R_T, R_ab, R_{1-3}) may be the most appropriate for expressing the combined resistance of a circuit.

As previously demonstrated, the total resistance of the *resistive parallel circuit* of Figure 11-13 is found by the following equation:

$$\frac{1}{R_\mathrm{T}} = \frac{1}{R_\mathrm{ab}} = \frac{1}{R_{1-3}} = \frac{1}{R1} + \frac{1}{R2} + \frac{1}{R3}$$

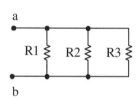

Figure 11-13
Parallel resistive
circuit

The next several examples demonstrate resistor reduction techniques. When combining resistances in series or in parallel, their equivalent resistance is found and then used in subsequent resistance reduction.

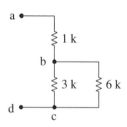

Figure 11-14(a)
Example problem

Example 11-2

Use the resistor reduction technique to reduce the circuit of Figure 11-14(a).

Solution

This circuit configuration is the same as that of Figure 11-7(a). The simplest combination is the 3 kΩ resistor in *parallel* with the 6 kΩ resistor (between nodes *b* and *c*). Reduce this parallel combination first.

$$\frac{1}{R_{bc}} = \frac{1}{3 \text{ k}\Omega} + \frac{1}{6 \text{ k}\Omega}$$

$$R_{bc} = 2 \text{ k}\Omega$$

Replace the parallel circuit with its resistive equivalent as shown in Figure 11-14(b). The circuit is now reduced to a simple series combination of the 1 kΩ in series with the R_{bc} equivalent resistance of 2 kΩ. Thus, the total resistance of this circuit is the sum of these two series resistances.

$$R_T = R_{ad} = 1 \text{ k}\Omega + 2 \text{ k}\Omega = 3 \text{ k}\Omega$$

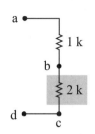

Figure 11-14(b)
Example solution

Practice
Find resistances R_{ab}, R_{bc}, and R_{ac} for the circuit of Figure 11-15.

Answer: 2 kΩ, 6 kΩ, 8 kΩ

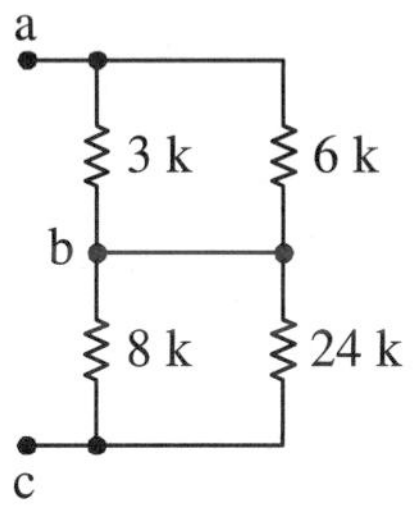

The next example has two branches in parallel. Each of these parallel branches has a series configuration that must be resolved first.

Example 11-3
Use the resistor reduction technique to reduce the circuit of Figure 11-16(a). First simplify each of the parallel branches and then find the combined parallel circuit resistance R_{bc}. Then find the total circuit resistance R_{ad}.

Figure 11-15
Practice circuit

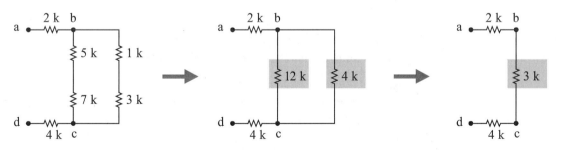

(a) Original circuit   (b) Parallel branches simplified   (c) Reduced to series circuit

Figure 11-16 Example circuit

Solution

The left parallel branch has 5 kΩ in series with 7 kΩ, forming a total branch resistance of 12 kΩ.

$$R_{\text{left branch}} = 5 \text{ k}\Omega + 7 \text{ k}\Omega = 12 \text{ k}\Omega$$

The right parallel branch has 1 kΩ in series with 3 kΩ, forming a total branch resistance of 4 kΩ.

$$R_{\text{right branch}} = 1 \text{ k}\Omega + 3 \text{ k}\Omega = 4 \text{ k}\Omega$$

Figure 11-16(b) shows each parallel branch reduced to its branch resistance. Now, the two parallel branches can be combined.

$$\frac{1}{R_{\text{bc}}} = \frac{1}{12 \text{ k}\Omega} + \frac{1}{4 \text{ k}\Omega}$$

$$R_{\text{bc}} = 3 \text{ k}\Omega$$

The reduced circuit is shown in Figure 11-16(c). It is a simple series circuit and the total circuit resistance can now be calculated.

$$R_{\text{ad}} = 2 \text{ k}\Omega + 3 \text{ k}\Omega + 4 \text{ k}\Omega = 9 \text{ k}\Omega$$

Practice

(a) In the circuit of Figure 11-16(c), a wire is accidentally placed between node b and node c. Find resistances R_{bc} and R_{ad}. Did the *short* circuit increase or decrease the total circuit resistance?

(b) In the circuit of Figure 11-16(a), the 1 kΩ resistor becomes open. Find resistances R_{bc} and R_{ad}. Did the *open* circuit increase or decrease the total circuit resistance?

Answer: (a) 0 Ω (short), 6 kΩ, decrease   (b) 12 kΩ, 18 kΩ, increase

The resistance reduction technique is next applied as an overall technique to analyze resistive circuits. Once a circuit's total circuit resistance has been determined, then the circuit laws and analysis techniques can be applied to find the circuit currents and voltages.

11.3 Basic Analysis Techniques

This section utilizes several examples to examine basic series-parallel circuit techniques. Before punching numbers into the calculator, determine what the problem really is, what values you need to find and a plan to find them. Then proceed to solve the problem.

Example 11-4

For the circuit of Figure 11-17, find the load current, load voltage and load power.

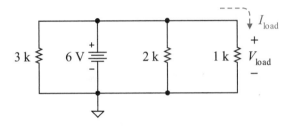

Figure 11-17 Example circuit

Solution

Look at the circuit. Is there a quick, easy solution for I_{load} or V_{load}? Yes, this is a purely parallel circuit, therefore, by KVL, all the parallel branches must have the same voltage. Once you know the load voltage, you can easily find the load current by Ohm's law and the load power by the power rule.

$$V_{\text{load}} = 6 \text{ V} \qquad\qquad \text{(parallel)}$$

$$I_{\text{load}} = \frac{6 \text{ V}}{1 \text{ k}\Omega} = 6 \text{ mA} \qquad\qquad \text{(Ohm's law)}$$

$$P_{\text{load}} = 6 \text{ V} \times 6 \text{ mA} = 36 \text{ mW} \qquad \text{(power rule)}$$

Practice

Let's solve the example problem the hard way to discourage you from jumping to problem solving without first thinking through an efficient problem solving plan. Be wise. Make a plan that is as straightforward as possible.

The hard way: (a) find total circuit resistance, (b) find the supply current, (c) use the current divider rule to find the load current, and (d) then use Ohm's law to find the load voltage V_{load}. Note, in the example, this was a one-step solution by observing that parallel circuits drop equal voltages across their branches.

Answer: (a) 545.5 Ω, 11 mA, 6 mA, 6 V (unnecessary, excessive work)

The last practice problem takes four unnecessary steps. Each step has the potential for error. The example took a few seconds to make an observation, while the practice took several steps with equations to arrive at the same result. It is always worth the time to think through the problem solution process and optimize it.

Example 11-5

For the circuit of Figure 11-18(a), find the load voltage V_{load}. *First*, examine the circuit and its nature. Visualize simple series and/or parallel combinations that can be reduced. Visualize the current flowing through the circuit and help determine what is in series, parallel, or simple series-parallel. Devise an efficient plan to find the load voltage. And, finally, execute that plan.

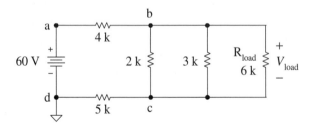

Figure 11-18(a) Example circuit

Solution

What is the nature of this circuit? Is any resistance simply in series or in parallel that could be reduced to simplify the circuit? The most obvious reduction is to combine the three parallel resistances of 2 kΩ, 3 kΩ, and 6 kΩ, as in Figure 11-18(b), into an equivalent resistance of R_{bc}.

You can also visualize the series-parallel nature of the circuit by examining the current flowing through the circuit as shown in Figure 11-18(c). The current flows up out of the supply, through the 4 kΩ resistor, splits at node b into the three branch currents, rejoins at node c, flows through the 5 kΩ resistor and back up into the supply.

Figure 11-18(d) is the resistance-reduced circuit with the parallel resistors combined into a single equivalent resistance R_{bc}. Note, that the voltage drop across all three branches and across R_{bc} must be same voltage and thus be equal to V_{load}. Therefore, if we can find the voltage drop across R_{bc}, we have found the load voltage V_{load}. What is the best approach to find the voltage drop across R_{bc}? We could find the current in Figure 11-17(d) and then multiply by R_{bc} (that's two steps). Anything better? How about a one-step voltage divider rule (VDR)?

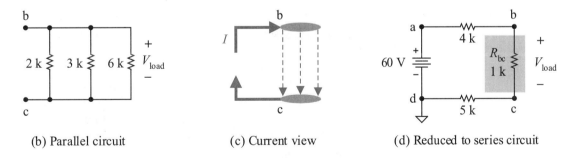

(b) Parallel circuit   (c) Current view   (d) Reduced to series circuit

Figure 11-18(b,c,d) Example circuit solution

This is the plan: find (1) the parallel combination of R_{bc}, (2) the total circuit resistance R_{ad}, and (3) the load voltage V_{load} using VDR. Now, let's execute the *minimal step plan*.

$$\frac{1}{R_{bc}} = \frac{1}{2 \ k\Omega} + \frac{1}{3 \ k\Omega} + \frac{1}{6 \ k\Omega}$$

$$R_{bc} = 1 \ k\Omega$$

$$R_{ad} = 4 \ k\Omega + 1 \ k\Omega + 5 \ k\Omega = 10 \ k\Omega$$

$$V_{load} = V_{bc} = \left(\frac{1 k\Omega}{10 k\Omega}\right) 60 \ V = 6 \ V$$

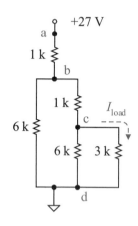

Figure 11-19
Practice circuit

Practice

For the circuit of Figure 11-19, find the load current I_{load}. *First,* examine the circuit and its nature. Visualize simple series and/or parallel combinations that can be reduced. Continue to reduce the circuit until you can calculate the total circuit resistance of R_{ad}. Visualize the current flowing down through the circuit to help determine what is in series, parallel, or simple series-parallel. Devise an efficient plan to find the load current. And, finally, execute that plan.

Answer: 4 mA

The next example incorporates LEDs. When these devices are forward biased on, they drop a relatively fixed voltage (for example, 2 V). They cannot be directly combined with resistors to simplify the circuit.

Example 11-6

Examine the circuit of Figure 11-20(a). The bias of LEDs must be checked first to see if they are *on* or *off.* Redraw the circuit schematic with the LEDs properly modeled and use that circuit to solve the rest of this problem. Find the 1 kΩ resistor voltage drop and current. Find the right LED's voltage drop. Find all of the node voltages.

Solution

The left LED is forward biased *on* (dropping 2 V) and the right LED is reverse biased *off* (acting like an open). The LEDs have been appropriately modeled in the circuit of Figure 11-20(b).

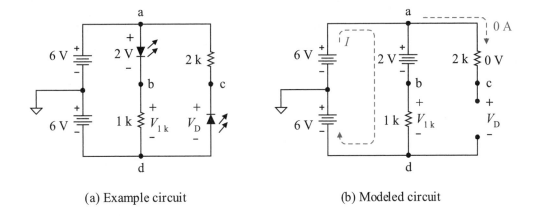

(a) Example circuit   (b) Modeled circuit

Figure 11-20 Example circuit

Since the right LED is reverse biased *off*, it is ideally an open and its branch current must be 0 A and the 2 kΩ resistor must drop 0 V. Therefore, only the left LED branch completes the circuit and draws a current. This is *an essentially series circuit*. This becomes a very straightforward circuit to analyze without further circuit reduction.

$$V_{1k} = +6 \text{ V} + 6 \text{ V} - 2 \text{ V} = 10 \text{ V} \qquad \text{(walkabout)}$$

$$I_{1k} = \frac{10 \text{ V}}{1 \text{ k}\Omega} = 10 \text{ mA} \qquad \text{(Ohm's law)}$$

$$V_D = +6 \text{ V} + 6 \text{ V} - 0 \text{ V} = 12 \text{ V} \qquad \text{(walkabout)}$$

Now find each of the node voltages starting from common and walking to node *a*, and so forth.

$$V_a = 0 \text{ V} + 6 \text{ V} = 6 \text{ V}$$

$$V_b = 6 \text{ V} - 2 \text{ V} = 4 \text{ V}$$

$$V_c = 6 \text{ V} - 0 \text{ V} = 6 \text{ V}$$

$$V_d = 0 \text{ V} - 6 \text{ V} = -6 \text{ V}$$

Practice

For the circuit of Figure 11-21, assume that the wiper arm of the potentiometer is at the top. Examine the circuit; there are two essentially series circuits. Find V_{in}, V_a, and V_{out}.

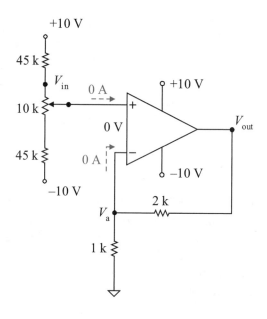

Figure 11-21 Practice circuit

Answer: 1 V, 1 V, 3 V

11.4 Ladder Circuits

Ladder circuits are called ladder circuits because they have the shape of a ladder as demonstrated in Figure 11-22. Ladders are a special form of the series-parallel circuit. To analyze the resistive ladder circuit the circuit must first be reduced to find its equivalent total circuit resistance. Then with a supplied voltage or current, the circuit currents and voltages can be calculated.

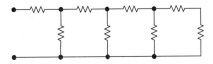

Figure 11-22 Ladder circuit

Example 11-7
Find the total resistance of the ladder circuit of Figure 11-23(a).

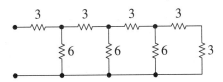

Figure 11-23(a) Example circuit

Solution
Employ the resistor reduction technique. Is there an obvious simple series or simple parallel combination? There is only one such
combination in the circuit of Figure 11-23; the right pair of 3 Ω
resistors forms a series circuit and a combined resistance of 6 Ω.
Figure 11-23(b) is the corresponding schematic of the resistor-
reduced circuit. Notice that the rightmost pair of 6 Ω resistors
now forms a parallel circuit that can be reduced to 3 Ω.

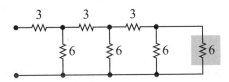

Figure 11-23(b) Example circuit: first reduction

Using Figure 11-24 shows the entire resistor reduction
process, combining simple series or simple parallel combinations

until the circuit has been totally reduced to its simplest equivalent circuit, namely, a simple series circuit. The total circuit resistance of the series circuit is 6 Ω.

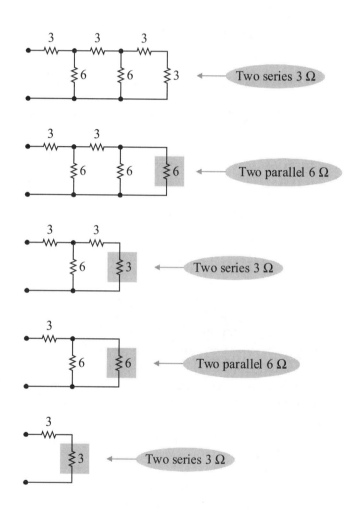

Figure 11-24 Resistor reduction of a ladder network

This is the thought process of resistor reduction of the ladder circuit but it is cumbersome, especially if you draw each reduced circuit. However, there is a shorthand notation that uses only the original schematic to simplify the circuit. Figure 11-25 uses an arrow notation to indicate equivalent *resistance looking into the rest of the circuit*. The rightmost 6 Ω arrow includes only the two right 3 Ω resistors in series. Put your hand over the circuit to the left of the arrow and block it out. What remains is the circuit that is equivalent to 6 Ω.

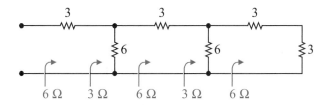

Figure 11-25 Resistor reduction: *shorthand notation*

Now slide your hand to the left of the 3 Ω arrow. The circuit that remains is the 6 Ω resistor in parallel with the two series 3 Ω resistors. The arrow represents looking into the circuit to the right with circuit to the left disconnected. Figure 11-26 demonstrates this view of the circuit.

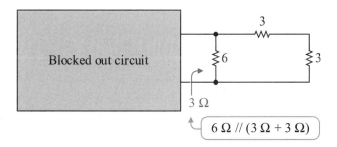

Figure 11-26 Equivalent resistance looking into the circuit

Practice

Reduce the ladder circuit of Figure 11-27. Practice both methods demonstrated in the example problem. What is the total circuit resistance?

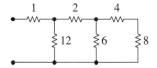

Figure 11-27
Practice circuit

Answer: 5 Ω

Current View of Ladder Circuit

Figure 11-28 demonstrates current flowing through a ladder circuit.

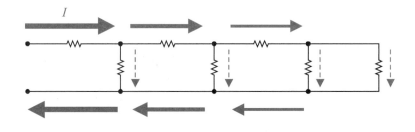

Figure 11-28 Current division in a ladder circuit

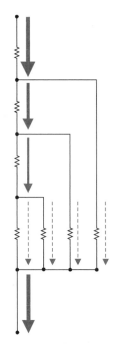

Figure 11-29
Another ladder circuit configuration

The supplied current divides as it flows into a top node. A branch current flows down through its branch (the dashed lines). The remaining current is reduced and flows right to the rest of the circuit. The currents rejoin at the bottom nodes

Another ladder circuit configuration and its associated currents are shown in Figure 11-29. This figure is analogous to a river flowing with branches breaking off and taking water away from the river. Then all four branches rejoin at the bottom of the river.

Can you visualize (in both configurations) as current enters the first node, the current splits into two parallel branches? One branch is a simple resistor branch. The other branch is the rest of the circuit (a series parallel circuit).

Visualize the node where the current split starts and the node where the two branch currents rejoin. Figure 11-30 demonstrates this concept of a parallel complex branch. The start and the end of the current split define parallel branches. You must be able to identify the defining two nodes of a parallel circuit to clearly establish parallel branches. In Figure 11-30, the incoming current I divides (splits) at node a. Branch current I_1 flows down to node b. Branch current I_2 flows to the right through a series-parallel circuit and then to node b. The branch currents I_1 and I_2 are parallel currents defined by the starting node a and the ending node b.

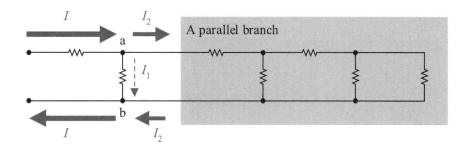

Figure 11-30 I_1 and I_2 are parallel branch currents (starting node a, ending node b)

Ladder Circuit Voltages and Currents

The following example applies a voltage supply to a ladder network and examines the currents and voltages developed in the ladder circuit.

Example 11-8

For the ladder circuit of Figure 11-31(a), find its total resistance, supply current, all resistor currents and voltages and the node voltages.

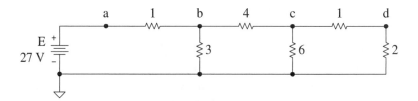

Figure 11-31(a) Example circuit

Solution

First the total circuit resistance R_T must be found by resistor reduction as shown in Figure 11-31(b). Start at the right end of the circuit and reduce the simple series and the simple parallel combinations until the entire circuit has been reduced to a single resistance.

$$R_T = 3 \ \Omega$$

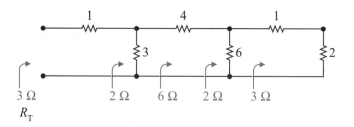

Figure 11-31(b) Resistor reduction to find total resistance R_T

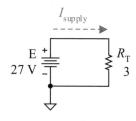

Figure 11-31(c)
Find supply current

Next find the supply current from the modeled circuit of Figure 11-31(c) using the applied voltage driving the total circuit resistance.

$$I_{supply} = \frac{27 \ V}{3 \ \Omega} = 9 \ A$$

Now apply basic circuit laws and analysis techniques to solve for all resistor currents and voltage drops. As you find these values draw them on the schematic: *voltages with polarities* and *currents with direction*. Figure 11-31(d) shows all the currents and voltages associated with this ladder circuit.

Associate each of the following calculations with the schematic of Figure 11-31(d). Start with the supply side of the circuit and work your way into the circuit using only the basic laws: Kirchhoff's current law (KCL), Kirchhoff's voltage law (KVL or KVL walkabout), and Ohm's law. Note that the currents at the bottom nodes sum up as they flow back to the supply.

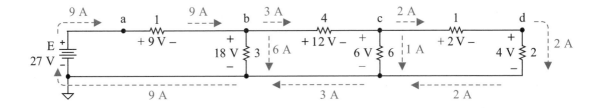

Figure 11-31(d) Example circuit currents and voltages

$$I_{1\Omega} = I_{\text{supply}} = 9 \text{ A} \qquad\qquad (\text{KCL, node } a)$$

$$V_{1\Omega} = 9 \text{ A} \times 1 \text{ } \Omega = 9 \text{ V} \qquad\qquad (\text{Ohm's law})$$

$$V_{3\Omega} = 27 \text{ V} - 9 \text{ V} = 18 \text{ V} \qquad\qquad (\text{KVL, left window})$$

$$I_{3\Omega} = \frac{18 \text{ V}}{3 \text{ } \Omega} = 6 \text{ A} \qquad\qquad (\text{Ohm's law})$$

$$I_{4\Omega} = 9 \text{ A} - 6 \text{ A} = 3 \text{ A} \qquad\qquad (\text{KCL, node } b)$$

$$V_{4\Omega} = 3 \text{ A} \times 4 \text{ } \Omega = 12 \text{ V} \qquad\qquad (\text{Ohm's law})$$

$$V_{6\Omega} = 18 \text{ V} - 12 \text{ V} = 6 \text{ V} \qquad\qquad (\text{KVL, center window})$$

$$I_{6\Omega} = \frac{6 \text{ V}}{6 \text{ } \Omega} = 1 \text{ A} \qquad\qquad (\text{Ohm's law})$$

$$I_{1\Omega} = 3 \text{ A} - 1 \text{ A} = 2 \text{ A} \qquad\qquad (\text{KCL, node } c)$$

$$V_{1\Omega} = 2 \text{ A} \times 1 \text{ } \Omega = 2 \text{ V} \qquad\qquad (\text{Ohm's law})$$

$$I_{2\Omega} = I_{1\Omega} = 2 \text{ A} \qquad\qquad (\text{KCL, node } d)$$

$$V_{2\Omega} = 2 \text{ A} \times 2 \text{ } \Omega = 4 \text{ V} \qquad\qquad (\text{Ohm's law})$$

Find the node voltages by starting at common with a known voltage of 0 V. Then proceed clockwise around the outside loop to each node, dropping the appropriate voltage as you proceed. Figure 11-31(e) shows the node voltages of this ladder circuit. The currents and resistance values have been removed to simplify the voltage visualization of the circuit.

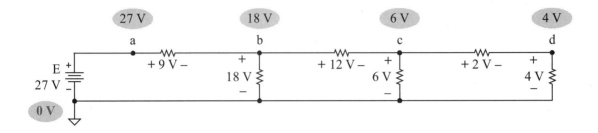

Figure 11-31(e) Example circuit node voltages

$$V_a = 27 \text{ V}$$

$$V_b = 27 \text{ V} - 9 \text{ V} = 18 \text{ V}$$

$$V_c = 18 \text{ V} - 12 \text{ V} = 6 \text{ V}$$

$$V_d = 6 \text{ V} - 2 \text{ V} = 4 \text{ V}$$

Practice

For the ladder circuit of Figure 11-32(a), find its total resistance, supply current, all resistor currents and voltages, and the node voltages. Place voltages with polarities and currents with directions on your schematic as you proceed.

Draw a separate voltage node schematic showing only the component voltages and node voltages.

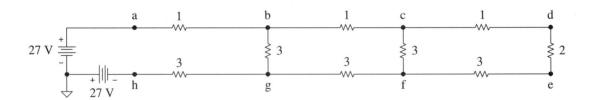

Figure 11-32(a) Practice circuit

Answer: $R_T = 6 \text{ }\Omega$. See Figures 11-32(b,c) for voltages and currents.

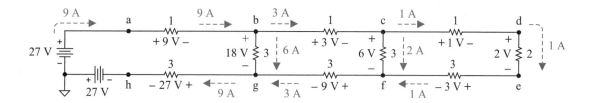

Figure 11-32(b) Practice circuit resistor currents and voltage drops

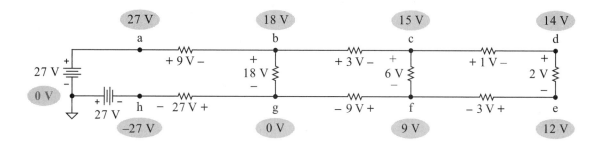

Figure 11-32(c) Practice circuit node voltages

Thinking Through the Process

You must think through a problem before attacking it with number crunching. Develop a plan before attacking a problem. Learn to think in terms of literals as well as numerical values. Use the ladder circuit of Figure 11-33 and think through the solution of the circuit parameters using literals without any numbers.

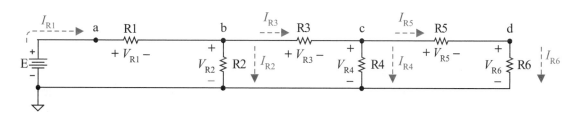

Figure 11-33 Thinking through the process

Current flow overview: The supply current flows through R1 and then splits at node *b*, placing R2 in parallel with the rest of the circuit R3 through R6 (R3–R6) circuit. The parallel R3 current I_{R3} then splits within its own series-parallel subcircuit. I_{R3} splits at node *c* into a current I_{R4} through R4 and a parallel current I_{R5} through R5 and R6 (which are in series).

Resistance analysis: Starting at the end of the ladder, perform resistor reduction of the circuit. Note: R_{56} is the combined resistances of R5 and R6. $R_{4\text{-}6}$ is the combined resistance of R4, R5, and R6.

$$R_{56} = \text{R5} + \text{R6} \qquad \text{(series)}$$

$$R_{4\text{-}6} = \text{R4} \ // \ R_{56} \qquad \text{(parallel)}$$

$$R_{3\text{-}6} = \text{R3} + R_{4\text{-}6} \qquad \text{(series)}$$

$$R_{2\text{-}6} = \text{R2} \ // \ R_{3\text{-}6} \qquad \text{(parallel)}$$

$$R_{\text{T}} = R_{1\text{-}6} = \text{R1} + R_{2\text{-}6} \qquad \text{(series)}$$

Supply current: The supply current is the supply voltage divided by the total circuit resistance.

$$I_{R1} = I_{\text{supply}} = \frac{E}{R_{\text{T}}} \qquad \text{(Ohm's law)}$$

Resistor currents and voltages: Starting with the supply voltage and supply current, work your way back into the circuit applying basic laws and circuit analysis techniques.

$$V_{R1} = I_{R1} \times \text{R1} \qquad \text{(Ohm's law)}$$

$$V_{R2} = E - V_{R1} \qquad \text{(KVL, left window)}$$

$$I_{R2} = \frac{V_{R2}}{\text{R2}} \qquad \text{(Ohm's law)}$$

$$I_{R3} = I_{R1} - I_{R2} \qquad \text{(KCL, node } b\text{)}$$

$$V_{R3} = I_{R3} \times \text{R3} \qquad \text{(Ohm's law)}$$

$$V_{R4} = V_{R2} - V_{R3} \qquad \text{(KVL, left window)}$$

$$I_{R4} = \frac{V_{R4}}{\text{R4}} \qquad \text{(Ohm's law)}$$

$$I_{R5} = I_{R3} - I_{R4} \qquad \text{(KCL, node } c\text{)}$$

$$V_{R5} = I_{R5} \times R5 \qquad \text{(Ohm's law)}$$

$$I_{R6} = I_{R5} \qquad \text{(KCL)}$$

$$V_{R6} = I_{R6} \times R6 \qquad \text{(Ohm's law)}$$

Much of your thought process in the real world will be without numerical values. Numerical voltages, currents, and resistances bring circuits to life but many problems must be solved abstractly first.

Also think in terms of the circuit action. For example, we conceptualized about the current flowing through the circuit and branching at nodes that split the current. As the current flows through a series resistor, it drops voltage to the rest of the circuit, and so forth.

R-2R Ladder Circuit

The R-2R ladder circuit is a ladder circuit that uses two resistor values. As the name implies, the ratio of the resistors is 1 to 2. For example, if R is 1.1 kΩ then 2R is 2.2 kΩ. Figure 11-34(a) is the schematic of a basic R-2R ladder circuit. Figure 11-34(b) shows the effective resistances of this ladder circuit. As a current comes into the first current splitting node, the current splits between a 2R branch down and a 2R effective branch resistance to the right. This pattern continues and produces a natural binary divider circuit. Each R-2R circuit splits the current in half.

The next example takes advantage of the R-2R ladder network to convert a three-bit binary number to an analog equivalent value. The ladder is arranged with switches. The switch has a 0 or 1 digital position.

- *Switch position 0*: The branch current is fed directly to common and does not flow through the ammeter.

- *Switch position 1*: The branch current is fed to the ammeter where it adds with other branch currents with switches in position 1.

Table 11-1 is a binary-to-decimal conversion table for the decimal numbers 0 through 7 (called the **octal number system**), showing the equivalent values between the binary system (digits 0 and 1) and the decimal system (digits 0–9). Switches 2, 1, and 0 correspond to the binary position number. Note: if the switches were digitally controlled, this circuit would form a digital-to-analog converter (DAC) circuit.

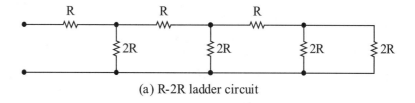

(a) R-2R ladder circuit

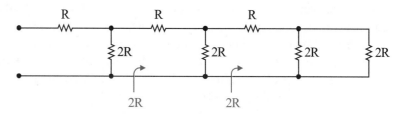

(b) R-2R ladder circuit equivalent resistances

Figure 11-34 The R-2R ladder circuit

Table 11-1 Binary-to-Decimal Conversion

SW 2	SW 1	SW 0	Binary Number	Decimal Number	Current (mA)
0	0	0	000	0	0
0	0	1	001	1	1
0	1	0	010	2	2
0	1	1	011	3	3
1	0	0	100	4	4
1	0	1	101	5	5
1	1	0	110	6	6
1	1	1	111	7	7

Example 11-9

For the circuit of Figure 11-35, find the total circuit resistance looking into the R-2R ladder circuit. Then find the supply current generated. Next find each branch current. Verify the ammeter current for each switch setting of Table 11-1. Find the circuit's node voltages.

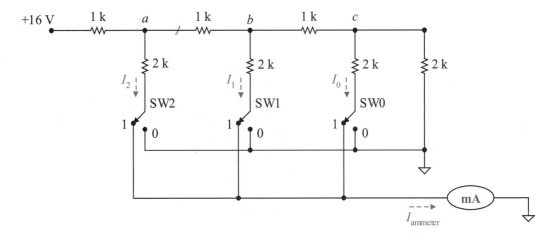

Figure 11-35 Example circuit: R-2R ladder shown in the 111 position

Solution

Examine the circuit. Notice that the 2 kΩ branches are always connected to the circuit; they are either connected directly to common or connected through the ammeter to common. The ladder is always intact independent of the switch positions.

Resistance-reduce the ladder circuit by starting with the two rightmost 2 kΩ resistors in parallel at node c (reduces to 1 kΩ). Combine that effective 1 kΩ resistance with the 1 kΩ resistance between nodes b and c (reduces to 2 kΩ). Continue this resistance reduction technique until the circuit is completely reduced to a single equivalent total resistance.

$$R_T = 2 \text{ k}\Omega$$

Find the supply current using Ohm's law.

$$I_{supply} = \frac{16 \text{ V}}{2 \text{ k}\Omega} = 8 \text{ mA} \qquad \text{(Ohm's law)}$$

The current splits equally at each node.

$$I_2 = \frac{8 \text{ mA}}{2} = 4 \text{ mA} \qquad\qquad \text{(CDR)}$$

$$I_1 = \frac{4 \text{ mA}}{2} = 2 \text{ mA}$$

$$I_0 = \frac{2 \text{ mA}}{2} = 1 \text{ mA}$$

If all the switch positions are in position 0, then none of the branch currents flow through the ammeter and the ammeter current is 0 mA. If any of the switches are in position 1, their currents sum up and flow through the ammeter (Kirchhoff's current law). The following are the calculations for all eight switch position combinations. The switch positions are given in parentheses.

$$I_{\text{ammeter}} = 0 \text{ mA} \qquad\qquad (000)$$

$$I_{\text{ammeter}} = 1 \text{ mA} \qquad\qquad (001)$$

$$I_{\text{ammeter}} = 2 \text{ mA} \qquad\qquad (010)$$

$$I_{\text{ammeter}} = 2 \text{ mA} + 1 \text{ mA} = 3 \text{ mA} \qquad\qquad (011)$$

$$I_{\text{ammeter}} = 4 \text{ mA} \qquad\qquad (100)$$

$$I_{\text{ammeter}} = 4 \text{ mA} + 1 \text{ mA} = 5 \text{ mA} \qquad\qquad (101)$$

$$I_{\text{ammeter}} = 4 \text{ mA} + 2 \text{ mA} = 6 \text{ mA} \qquad\qquad (110)$$

$$I_{\text{ammeter}} = 4 \text{ mA} + 2 \text{ mA} + 1 \text{ mA} = 7 \text{ mA} \qquad\qquad (111)$$

The node voltages also divide in a binary fashion.

$$V_a = 16 \text{ V} - (8 \text{ mA} \times 1 \text{ k}\Omega) = 8 \text{ V}$$

$$V_b = 8 \text{ V} - (4 \text{ mA} \times 1 \text{ k}\Omega) = 4 \text{ V}$$

$$V_c = 4 \text{ V} - (2 \text{ mA} \times 1 \text{ k}\Omega) = 2 \text{ V}$$

Practice

The R-2R ladder circuit of the example is connected to an inverting voltage amplifier to convert the current in the ammeter to an output voltage as shown in Figure 11-36. Find the output voltage V_{out} for all eight switch configurations (000, ... , 111).

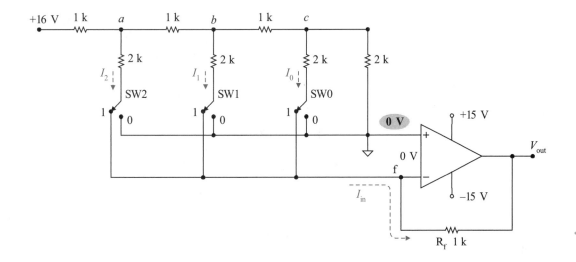

Figure 11-36 Practice problem: binary to decimal voltage converter

Answer: 0, −1 V, −2 V, −3 V, −4 V, −5 V, −6 V, −7 V. Note, to obtain a positive voltage at the output, this output voltage can be fed to an inverting voltage amplifier with a gain of unity.

Exploration

> The MX7524 is an 8-bit DAC IC. Its R-2R ladder consists of 10 kΩ and 20 kΩ resistors. If the input reference voltage to the R-2R ladder network is 10 V, find the analog output current for the digital inputs (a) 00000001, (b) 00000010, (c) 11111111.

Answer: (a) 1.953 µA = [(1/256)×500 µA],   (b) 3.906 µA = [(2/256)×500 µA],   (c) 498.1 µA = [(255/256)×500 µA]

11.5 Bridge Circuit

The basic construction of a bridge circuit is shown in Figure 11-37. A dc voltage is applied across the top and bottom terminals (nodes a and d, respectively); and the output voltage taken between nodes b and c.

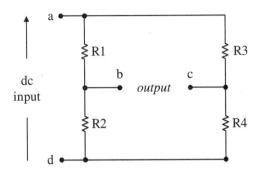

Figure 11-37 Bridge circuit

Balanced Bridge

Balanced bridge

The bridge is said to be **balanced** if R1 is to R2 as R3 is to R4.

$$\frac{R1}{R2} = \frac{R3}{R4}$$

If this is true, then the voltages at nodes b and c must be the same since each branch must form the same voltage divider fraction (VDR).

$$V_b = V_c$$

Thus, the output voltage for a balanced bridge must always be 0 V no matter what load (except a supply) is connected to the output.

Balanced bridge output
$V_{load} = 0$ V
$I_{load} = 0$ A

$$V_{out} = V_b - V_c = 0\,V$$

Since the voltage drop between node b and c is 0 V, the load voltage and the load current must be 0 even if a load is connected.

Example 11-10

For the circuit of Figure 11-38, prove that the bridge is balanced using the bridge's resistor values. Verify that node voltages V_b and V_c are the same for this bridge circuit. Find the output voltage V_{out}.

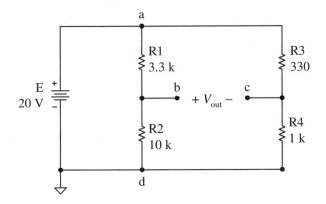

Figure 11-38 Example circuit: balanced bridge

Solution

Use the bridge resistor ratios to test to see if the bridge is balanced.

$$\frac{3.3 \text{ k}\Omega}{10 \text{ k}\Omega} = \frac{330 \text{ } \Omega}{1 \text{ k}\Omega} \qquad (?)$$

$$0.33 = 0.33 \qquad (\text{yes})$$

Next find the output node voltages and the *open circuit output voltage drop.*

$$V_b = \left(\frac{10 \text{ k}\Omega}{13.3 \text{ k}\Omega}\right) 20 \text{ V} = 15 \text{ V}$$

$$V_c = \left(\frac{1 \text{ k}\Omega}{1.3 \text{ k}\Omega}\right) 20 \text{ V} = 15 \text{ V}$$

$$V_{out} = V_{bc} = 15 \text{ V} - 15 \text{ V} = 0 \text{ V} \qquad (\text{yes})$$

The bridge has been proven to be balanced by two approaches: the bridge resistance ratios are the same and the open circuit output voltage is 0 V.

Practice

Change the 1 kΩ resistor of the bridge circuit in Figure 11-38 to a 330 Ω resistor. Then test the bridge resistor ratio to see if the bridge is balanced. Also, find the output node voltages V_b and V_c and the output voltage V_{out}. Is the bridge balanced?

Answer: $0.33 \neq 0.5$, 15 V, 10 V, 5 V, bridge is *not* balanced

Loaded Unbalanced Bridge

Circuit analysis is much more challenging if the bridge circuit is unbalanced and has a load as shown in Figure 11-39(a).

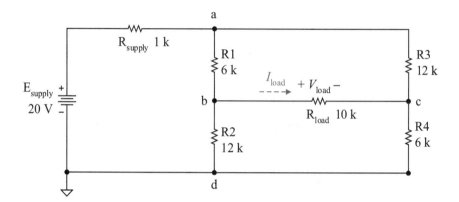

Figure 11-39(a) Unbalanced loaded bridge

This is a series-parallel circuit that cannot be reduced. This circuit cannot be resistor-reduced. The load connected between nodes *b* and *c* creates a mutual coupling between the bridge's left branch and the right branch that cannot be simplified.

What advanced technique(s) could be used to help solve this problem? First, what is the problem? The problem is R_{load}. If R_{load} could be removed, this is a relatively simple parallel series circuit. Or if the load were a short, this becomes a series-parallel circuit that can be resistor-reduced and analyzed. Let's see, *open circuit voltage* and *short circuit current*. If we can find the open circuit voltage (E_{Th}) and the short circuit current (I_N), then we can find the Thévenin resistance R_{Th}. Then we can use the Thévenin model of the bridge circuit with the load attached in series to find the load current and the load voltage.

Example 11-11

Find the load current and load voltage of the bridge circuit in Figure 11-39(a).

Solution

Form a plan to solve this problem. Several major steps are required.

Step 1: Remove the load and find the open circuit voltage (E_{Th}).
Step 2: Remove the load and replace it with a short and find the short circuit current (I_N).
Step 3: Calculate the Thévenin resistance R_{Th}.
Step 4: Draw the Thévenin model with the load attached and find the load current and voltage.

Now let's proceed with the plan. Be certain of the results of each major step before proceeding to the next major step. Verify all values of each step before proceeding to the next step.

Step 1: Remove the load, see Figure 11-39(b), and find the open circuit voltage (E_{Th}). Results: $R_T = 10$ kΩ, $I_{supply} = 2$ mA, $V_a = 18$ V, $V_b = 12$ V, $V_c = 6$ V, $E_{Th} = 6$ V.

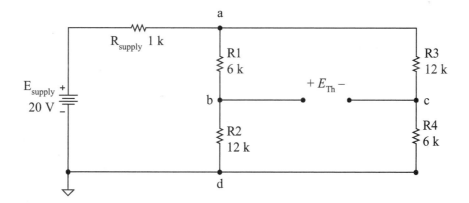

Figure 11-39(b) Find the open circuit voltage E_{Th}

Step 2: Remove the load, see Figure 11-39(c), and replace it with a short. Find the short circuit current (I_N). Results: $R_T = 9$ kΩ, $I_{supply} = 2.222$ mA, $V_a = 17.8$ V, $V_b = 8.89$ V, $I_{R1} = 1.482$ mA, $I_{R2} = 0.741$ mA, $I_N = 0.741$ mA.

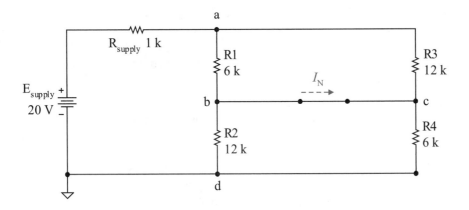

Figure 11-39(c)   Find the short circuit current I_N

Step 3: Calculate the Thévenin resistance R_{Th}.

$$R_{Th} = \frac{E_{Th}}{I_N} = \frac{6 \quad V}{0.741 \quad mA} = 8.1 \quad k\Omega$$

Step 4: Draw the Thévenin model with the load attached and find the load current and the load voltage. See Figure 11-39(d).

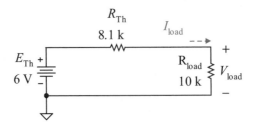

Figure 11-39(d) Thévenin model with load

$$I_{\text{load}} = \frac{6 \text{ V}}{18.1 \text{ k}\Omega} = 0.333 \text{ mA}$$

$$V_{\text{load}} = 0.333 \text{ mA} \times 10 \text{ k}\Omega = 3.33 \text{ V}$$

This was a lengthy solution in terms of the number of major steps required. Also, finding the Thévenin voltage and the Norton current were lengthy steps in and of themselves. The overall plan is critical in analyzing circuits, especially an advanced analysis like this example. *Always think through a solution before attempting the solution. Also, the manually calculated load values were verified by a MultiSIM simulation.*

Practice

The **Wheatstone bridge circuit** is used to sense static strain measurements. **Strain gauges** can be 1, 2, or 4 sensors (that is, one, two, or four of the resistances change with stress and strain). A strain causes a change in bridge resistances, which translates into a difference voltage as seen in the circuit of Figure 11-40. This circuit represents measuring the pressure from the weight of a gas tank. R3 of 350 Ω indicates an empty tank. Simulate this circuit using MultiSIM and find the values of R3 for a ¼, ½, and a full tank of gas if 20 mV represents a full tank of gas. Hint: use the demonstrated reading to linearly approximate the values needed and then fine-tune these values using the simulator.

multi**SIM**

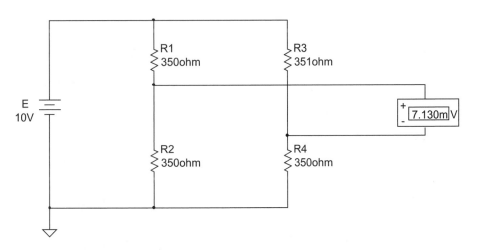

Figure 11-40 MultiSIM simulation of a strain gauge

Answer: 350.7 Ω, 351.45 Ω, 352.8 Ω (full tank)

Exploration

Simulate the example circuit, changing the strain gauge to a four sensor device. Set R1 and R4 to 349 Ω and R2 and R3 to 351 Ω.

Note, the example resistance bridge can be changed to force load current to flow in the other direction, that is, from node *c* to *b*. An easy test to determine load current direction and load voltage polarity is to examine the unloaded node voltages of nodes *b* and *c*. The higher potential node becomes the higher potential node of the loaded system as well.

11.6 Circuit Loading

A technician measures two flashlight batteries with a voltmeter. They test good, 1.5 V each. The batteries are properly put into a flashlight but the lamp is dim, not bright like it should be. What's the problem?

The problem may be that the batteries were tested without a proper load. Figure 11-41 demonstrates this principle of loading. Figure 11-41(a) demonstrates that an unloaded voltmeter test produces essentially 0 A load current. Since there is no load current, the internal resistance of the battery drops 0 V (Ohm's law).

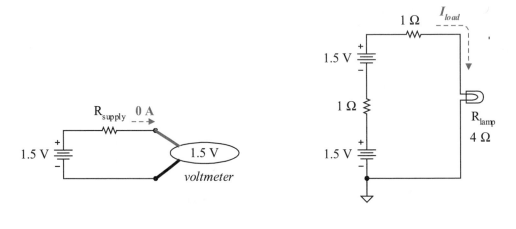

(a) Unloaded battery test   (b) Loaded circuit

Figure 11-41 Circuit and supply loading

Now the batteries are put in the flashlight, the flashlight turned on, and the series circuit is drawing current as shown in the circuit of Figure 11-41(b). The 3-V lamp is not receiving the full 3 V from the batteries. Why not? Under load conditions, the batteries' internal resistance is dropping voltage that is not available to the flashlight lamp. Each battery is dropping 0.5 V internally; this voltage is lost.

$$I_{lamp} = \frac{3\ V}{6\ \Omega} = 0.5\ A$$

$$V_{lamp} = 0.5\ A \times 4\ \Omega = 2\ V \qquad\qquad \text{(dim lamp)}$$

The open circuit voltage test is equivalent to finding the Thévenin voltage of a supply or a circuit. This is the maximum possible voltage that a circuit can produce under a no load condition. A circuit must be designed, developed, and tested under full load condition. **Full load condition** is interpreted to be *under maximum load current condition*.

Recall that voltage supplies are rated under full load conditions. Many components and circuits have full load ratings. These are not necessarily maximum current ratings that relate to the destruction of the device. Full load ratings relate to maximum normal operating conditions. If the component or circuit is overloaded, components are not necessarily destroyed. The component or circuit ceases to function normally. For example, an op amp may have a short circuit current rating of 40 mA but its voltage output may start degrading at 10 mA of load current.

Always be aware of component and circuit limitations. As you advance in your electronics education, you will learn more details and specifications about components and circuit operations. The basic laws and circuit analysis techniques are an invaluable tool in circuit analysis as long as component or circuit specifications are within range.

Voltmeter Loading

The voltmeter is a very high resistance instrument; ideally it is treated as an open. However, if the circuit that is being measured also contains high resistance values, then measurement error is possible as shown in the next example.

Example 11-12

Assuming an ideal voltmeter, find the voltmeter measurement of the circuit in Figure 11-42(a). If the voltmeter has an internal resistance of 10 MΩ, find the voltmeter measurement.

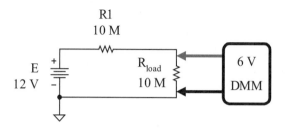

Figure 11-42(a) Ideal voltmeter reading

Solution

This is a simple series circuit. The voltage divides equally between the two circuit resistors. Assuming the voltmeter is an ideal open, find the load voltage using the voltage divider rule.

$$V_{meter} = V_{load} = \left(\frac{10 \ \text{M}\Omega}{20 \ \text{M}\Omega}\right) 12 \ \text{V} = 6 \ \text{V}$$

To include the resistance of the voltmeter, model it as a 10 MΩ resistance in parallel with the load resistance as shown in Figure 11-42(b). The load and the meter now form a parallel circuit with a combined resistance of 5 MΩ. The supply now drives a series circuit of R1 in series with a parallel load combination of the load resistance and the meter resistance. Again, applying the voltage divider rule, the load voltage is now

$$V_{meter} = V_{load} = \left(\frac{5 \ \text{M}\Omega}{15 \ \text{M}\Omega}\right) 12 \ \text{V} = 4 \ \text{V}$$

This is a false reading caused by the internal resistance of the meter as demonstrated in the circuit of Figure 11-42(c). The internal resistance of the voltmeter is on the order of the resistance of the circuit resistance under measurement. Whenever working with high resistance circuits, beware of the possibility of voltmeter loading. If this is a possibility, very high resistance specialty voltmeters can be used to prevent voltmeter loading measurement errors.

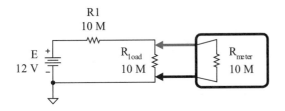

(b) Voltmeter internal resistance model

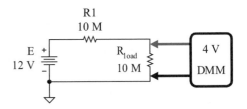

(c) Voltmeter loading error

Figure 11-42(b,c) Voltmeter loading of high resistance circuit

Practice

Replace R1 and R_{load} of the circuit of Figure 11-42(a) with 1 kΩ resistors. Ideally, what is the load voltage? If a voltmeter with an internal resistance of 10 MΩ is used to measure the load voltage, accurately find the meter reading. Is the voltmeter producing any significant loading?

Answer: 6 V, 5.997 V, no significant loading effect

Zener Voltage Regulator Loading

Previously you studied a zener diode series circuit, modeling the reverse biased zener diode to a *first approximation* as a voltage supply V_Z and to a second approximation as a voltage supply V_Z in series with its internal resistance R_Z. The next example attaches a load in parallel with the reverse biased zener diode as a **voltage regulator**. The zener diode maintains a reasonably constant voltage drop across the load even with varying supply or load conditions.

Example 11-13

For the circuit of Figure 11-43(a), find the output voltage, load current, supply current, zener current, and the power dissipated by the R_{supply} resistance. To solve this problem, model the reverse biased zener diode (a) as a 5 V supply and (b) as a 5 V supply in series with its internal zener resistance of 1 Ω. (c) Compare the results using the first and second approximation models.

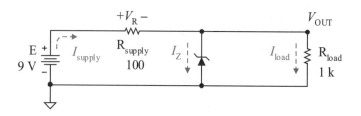

Figure 11-43(a) Example circuit of loaded zener diode circuit

Solution

a. Replace the zener diode with its first approximation model of a 5 V supply and redraw the modeled circuit as shown in Figure 11-43(b). Notice that this is a reverse biased zener diode; that is, the voltage drop is reversed and the current through the zener is reversed.

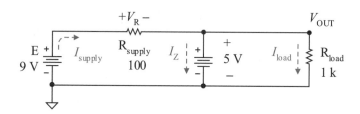

Figure 11-43(b) Zener regulator diode *first approximation model*

Examine the modeled circuit and critically observe what is hopefully apparent in starting the solution of this problem. The 5 V supply is purely in parallel with the load. The output voltage is the voltage drop across the 5 V zener supply.

$$V_{out} = V_z = 5 \text{ V}$$

Find the load current using Ohm's law.

$$I_{\text{load}} = \frac{5\ \text{V}}{1\ \text{k}\Omega} = 5\ \text{mA}$$

Find the supply resistor voltage drop using Kirchhoff's voltage law or do a walkabout from its (−) reference to its (+) reference.

$$V_{\text{Rsupply}} = -5\ \text{V} + 9\ \text{V} = 4\ \text{V}$$

Find the supply current applying Ohm's law to the supply resistor.

$$I_{\text{Rsupply}} = \frac{4\ \text{V}}{100\ \Omega} = 40\ \text{mA}$$

Find the zener current by applying Kirchhoff's current law at the output node.

$$I_Z = 40\ \text{mA} - 5\ \text{mA} = 35\ \text{mA}$$

Find the power dissipated by supply resistance R_{supply} by applying the power rule. *Warning*, in this type of circuit, R_{supply} tends to dissipate significant power and you must ensure it is sufficiently rated to handle the power dissipation.

$$P_{\text{Rsupply}} = 40\ \text{mA} \times 4\ \text{V} = 160\ \text{mW}$$

b. Replace the zener diode with its second approximation model of a 5 V supply in series with a 1 Ω zener resistance and redraw the modeled circuit as shown in Figure 11-43(c).

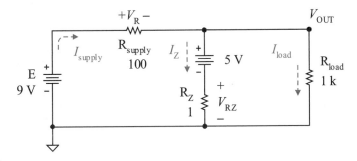

Figure 11-43(c) Zener regulator diode *second approximation model*

The circuit is more challenging. It has series supplies contained within parallel branches.

You must use an advanced circuit analysis technique to solve the circuit. You can solve this problem using the supply conversion technique. Convert the two series voltage supply branches into their current supply parallel equivalents. That is, convert the series Thévenin branches into their parallel Norton equivalents, insert the Norton models back into the circuit creating a purely parallel equivalent circuit, and solve for the load current using the current divider rule CDR.

The left branch converts to a current supply of

$$I_1 = \frac{9\text{ V}}{100\text{ }\Omega} = 90\text{ mA}$$

in parallel with a 100 Ω resistor.

The center branch converts to a current supply of

$$I_2 = \frac{5\text{ V}}{1\text{ }\Omega} = 5\text{ A}$$

in parallel with a 100 Ω resistor.

Figure 11-43(d) is the converted, modeled circuit ready to be solved for the load current and output voltage.

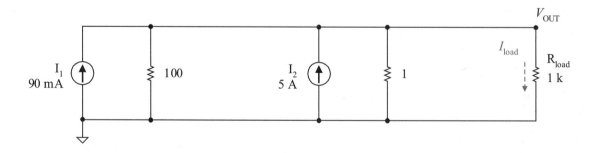

Figure 11-43(d) Zener regulator converted to current sources

The net current into the top node is the sum of the currents from the two modeled current supplies

$$I_{net} = 90 \text{ mA} + 5 \text{ A} = 5.09 \text{ A}$$

The circuit is now simplified to the modeled circuit of Figure 11-43(e) with a single net supply and three parallel resistors including the original load resistor.

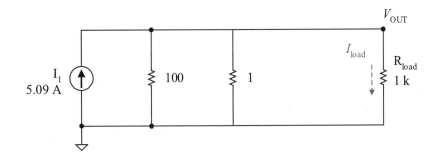

Figure 11-43(e) Zener regulator with current sources, simplified

Use the current divider rule to find the load current.

$$I_{load} = \left(\frac{R_{total}}{R_{load}} \right) I_{net}$$

R_{total} is the parallel combination of the three resistors.

$$I_{load} = \left(\frac{0.989 \ \Omega}{1 \text{ k}\Omega} \right) 5.09 \text{ A} = 5.034 \text{ mA}$$

Find the output or load resistor voltage using Ohm's law.

$$V_{out} = V_{load} = 5.034 \text{ mA} \times 1 \text{ k}\Omega = 5.034 \text{ V}$$

The supply current can now be found from the original circuit of Figure 11-42(a) as

$$I_{supply} = \frac{9 \text{ V} - 5.034 \text{ V}}{100 \ \Omega} = 39.66 \text{ mA}$$

Find the zener current using Kirchhoff's current law.

$$I_Z = 39.66 \text{ mA} - 5.034 \text{ mA} = 34.63 \text{ mA}$$

By Ohm's law

$$P_{\text{Rsupply}} = (39.66 \text{ mA})^2 \times 100 \; \Omega = 157 \text{ mW}$$

c. Ideally the output voltage was 5 V using the first approximation model. Using the second approximation model, including the zener internal resistance, the output voltage is more accurately calculated to be 5.034 V. The zener resistance is so low that it drops very little voltage and causes less than a 1% change in the output voltage. In this case, the first approximation model was sufficient to analyze the circuit.

Practice

Use the example circuit of Figure 11-44. This is the same circuit as the example problem, but the load has been changed to 10 kΩ. Find the output voltage, load current, supply current, and zener current. Solve this problem modeling the reverse biased zener diode (a) as a 5 V supply and (b) as a 5 V supply in series with its internal zener resistance of 1 Ω. (c) Compare the results using the first and second approximation models. (d) Compare the load voltage and zener current with the example problem. Even though the load resistance went up by a factor of 10, was the load voltage provided across the zener regular changed dramatically or did it properly regulate? As the load current went down, what did the zener current do?

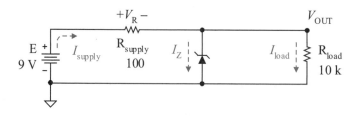

Figure 11-44 Practice circuit of loaded zener diode circuit

Answer: (a) 5 V, 0.5 mA, 40 mA, 39.5 mA (b) 5.039 V, 0.504 mA, 39.61 mA, 39.11 mA (c) the first and second approximations are very close to each other (about a 2% difference) (d) the load resistance went up by a factor of 10, the load current down by a factor of 10, the supply current remained about the same, and the zener current went up (the zener diode compensated for the change in load current, while the supply current remained relatively the same)

Summary

Series-parallel circuits consist of basic series and basic parallel circuits. Series circuits share the same current. Parallel circuits share the same branch voltage. Purely resistive circuits may be reduced to a single effective circuit resistance if the circuit consists of basic series and basic parallel resistances.

It is important to recognize simple series circuit and parallel circuit configuration. This can be observed by basic circuit configurations or by following the flow of current through a circuit.

Essentially series circuits are circuits that at first glance appear to be a series-parallel circuit. However, one or more branches have 0 A or virtually 0 A of current, leaving an essentially series circuit. Examples include circuits that use the op amp, which has a very, very small current flowing into its input terminals relative to the external voltages.

It is critical to examine a circuit and understand its behavior before starting to mathematically analyze it. This may be looking for simple resistor combinations that may be reduced, circuits that have 0 A branches to form essentially series circuits, circuits that have components that have various states based upon whether or not they are forward biased, and so forth. Before attempting to solve a circuit, form a plan or a strategy and think through the solution process.

Some circuits, like the ladder circuit, must be reduced to its total equivalent resistance before voltage and current analysis can take place. The R-2R ladder is of particular interest in converting binary to decimal (or, digital values to analog values) since it is a natural binary current divider circuit.

The bridge circuit is an example of a mutually shared resistance between two other branches. The bridge cannot be simplified by simple resistor reduction. An advanced technique of finding the Thévenin equivalent model of the bridge circuit can be used to determine the load current and voltage. This is an excellent example of thinking through the circuit analysis process before attempting a numerical solution. Also, this is an excellent example of using circuit simulators to confirm a long, complex circuit analysis solution.

When dealing with components and circuits, it is important to understand their limitations. For example, an op amp that is slamming a rail cannot be acting as a linear amplifier. An op amp that is being excessively loaded by a circuit attempting to demand too much output current may not be operating normally. Circuit analysis laws and techniques work consistently as long as component and/or circuit limits are not forcing a circuit into an abnormal state of operation.

Problems

Series-Parallel Combinations

11-1   For the circuit of Figure 11-45(a):

   a. Identify the simple series combinations.

   b. Identify the simple parallel combinations.

   c. Identify the major series combination.

11-2   For the circuit of Figure 11-45(b):

   a. Identify the simple series combinations.

   b. Identify the simple parallel combinations.

   c. Identify the major parallel combination.

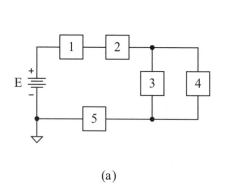

(a)

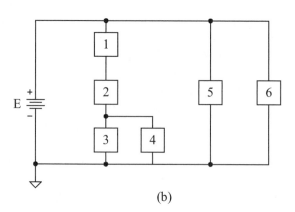

(b)

Figure 11-45

Resistance Reduction Technique

11-3 Find the combined resistance R_A of Figure 11-46(a). That is, an ohmmeter is connected to the A nodes; what does it read?

11-4 Find the combined resistance R_B of Figure 11-46(b).

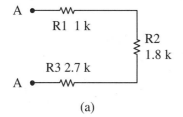

(a)

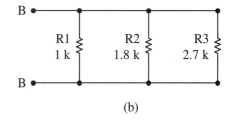

(b)

Figure 11-46

11-5 For the circuit of Figure 11-47(a):

  a. Find the combined resistance R_A.

  b. Find the combined resistance R_A if the 3 kΩ resistor is shorted out.

  c. Find the combined resistance R_A if the 4 kΩ resistor is opened.

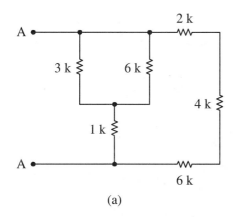

(a)

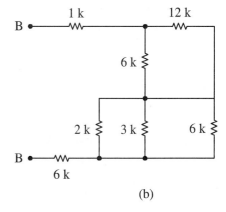

(b)

Figure 11-47

11-6   For the circuit of Figure 11-47(b):

    a.   Find the combined resistance R_B.

    b.   Find the combined resistance R_B if the 2 kΩ resistor is shorted out.

    c.   Find the combined resistance R_B if the 2 kΩ resistor is opened.

11-7   For the circuit of Figure 11-48(a):

    a.   Find the combined resistance R_A.

    b.   Find the combined resistance R_A if one of the 6 kΩ resistors is shorted out.

    c.   Find the combined resistance R_A if one of the 6 kΩ resistors opens.

11-8   For the circuit of Figure 11-48(b):

    a.   Find the combined resistance R_B.

    b.   Find the combined resistance R_B if the 24 kΩ resistor is shorted out.

    c.   Find the combined resistance R_B if the 24 kΩ resistor opens.

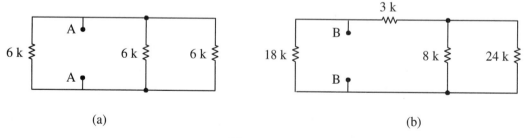

(a)   (b)

Figure 11-48

Basic Analysis Techniques

11-9   For the circuit of Figure 11-49(a):

    a.   Find the supply current.

    b.   Find the 3 kΩ resistor current and voltage drop.

    c.   Find the node voltages V_a, V_b, and V_e.

    d. Find the voltage differences V_{ae}, V_{ab}, and V_{bd}.

    e. Find the supply current if the 1 kΩ resistor is shorted out.

    f. Find the supply current if the 1 kΩ resistor is opened.

11-10 For the circuit of Figure 11-49(b):

    a. Find the supply current.

    b. Find the 3 kΩ resistor current and voltage drop.

    c. Find the node voltages V_a, V_b, and V_f.

    d. Find the voltage differences V_{af}, V_{ab}, and V_{bd}.

    e. Find the supply current if the 12 kΩ resistor is shorted out.

    f. Find the supply current if the 12 kΩ resistor is opened.

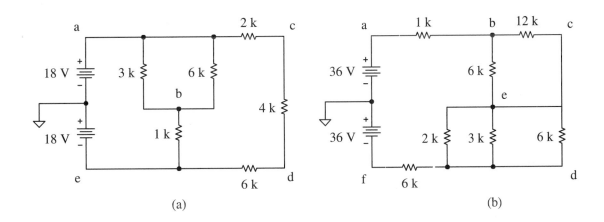

Figure 11-49

11-11 For the circuit of Figure 11-50(a):

    a. Find the supply current.

    b. Find the 4 kΩ resistor voltage and current.

    c. Find the LED currents.

    d. Find the node voltages V_a, V_b, V_c, and V_d.

    e. Find the voltage differences V_{ad}, V_{ab}, and V_{bc}.

11-12 For the circuit of Figure 11-50(b):

    a. Find the supply current.

    b. Find the 1 kΩ resistor current and voltage drop.

    c. Find the node voltages V_a, V_b, V_c, and V_d.

    d. Find the voltage differences V_{ad}, V_{ac}, and V_{ca}.

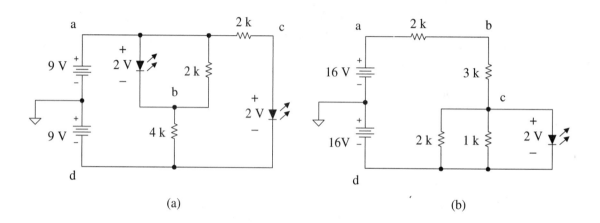

(a)   (b)

Figure 11-50

Ladder Circuits

11-13 For the circuit of Figure 11-51:

    a. Find the total circuit resistance.

    b. Find the supply current.

    c. Find the load voltage drop V_{load} and the current I_{load}.

    d. If an ammeter is accidentally placed across the 6 kΩ resistor: What does the ammeter read? Which components are in jeopardy?

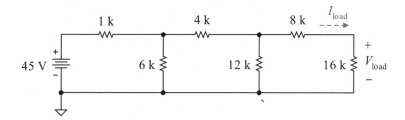

Figure 11-51

11-14 For the circuit of Figure 11-52:

   a. Find the total circuit resistance.

   b. Find the supply current.

   c. Find the load voltage drop V_{load} and load current I_{load}.

   d. An ammeter is accidentally placed across the voltage supply: What does the ammeter read? Which components are in jeopardy?

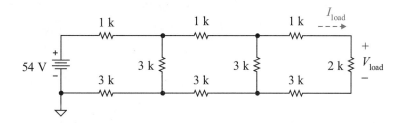

Figure 11-52

11-15 For the circuit of Figure 11-53:

   a. Find the open circuit voltage.

   b. Find the short circuit current.

   c. Find the Norton (or Thévenin) resistance.

   d. Attach a 1 kΩ resistor as the load to the circuit Thévenin model and find I_{load} and V_{load}.

e.   Attach a 1 kΩ resistor as the load to the circuit Norton model and find I_{load} and V_{load}.

f.   Attach a 1 kΩ load resistor to the actual circuit and find I_{load} and V_{load}.

g.   Did the Thévenin and Norton models produce the same load current and load voltage as the actual circuit?

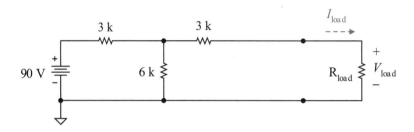

Figure 11-53

11-16   For the circuit of Figure 11-54:

a.   Find the total circuit resistance R_A.

b.   Find the branch currents I_3, I_2, I_1, I_0.

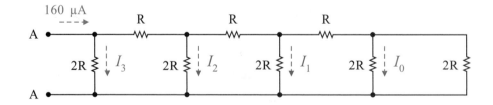

Figure 11-54

11-17   Use the digital-to-analog converter (DAC) circuit of Figure 11-55.

a.   Find the currents I_3, I_2, I_1, and I_0.

   b. Find the currents I_3, I_{32}, I_{321}, and I_{3210}.

   c. Find the output voltage V_{out}.

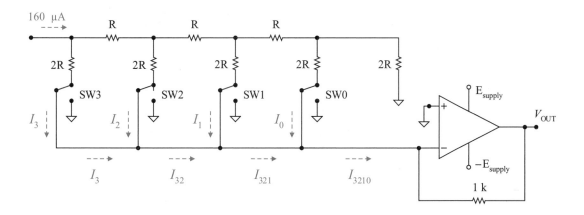

Figure 11-55

11-18 Use the DAC circuit of Figure 11-56.

   a. Find the currents I_3, I_2, I_1, and I_0.

   b. Find the currents I_3, I_{32}, I_{321}, and I_{3210}.

   c. Find the output voltage V_{out}.

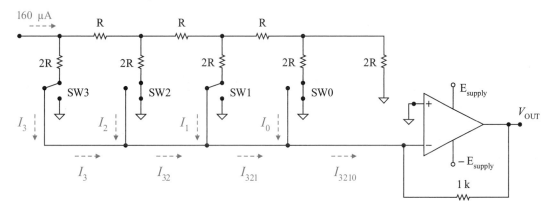

Figure 11-56

11-19 Use Figure 11-56. Currently its switches digitally read 1010. A digital sequence goes from 1010 to 1001 to 0111. What are the resulting three analog voltages?

Bridge Circuit

11-20 For the bridge circuit of Figure 11-57(a):

    a. Find the open circuit voltage.

    b. Find the short circuit current.

    c. Find the Thévenin resistance.

    d. Find the load current and the load voltage.

    e. If the load is replaced with an LED that drops 2 V, find the load current.

    f. Find the value of R2 that balances the bridge.

11-21 Repeat Problem 11-20 for Figure 11-57(b).

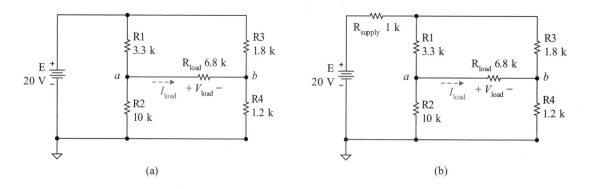

(a)   (b)

Figure 11-57

Waveforms

Introduction

Voltage or current values that vary with time produce alternating values or waveforms when plotted as a function of time. This chapter introduces the waveform (variable as a function of time), its terminology, and its application to fundamental circuit analysis and to electronic circuit analysis.

Objectives

Upon completion of this chapter you will be able to:

- Identify and apply waveform terminology.

- Identify the waveshape and key characteristics of the sine wave, square wave, rectangle wave, triangle wave, and sawtooth wave.

- Define and apply ac voltage and current supplies to circuits.

- Interpret, apply, and measure the average (dc) value of a waveform.

- Calculate the average value of a square wave, triangle wave, sine wave, and rectangle wave with and without dc offsets.

- Separate a waveform into its ac signal and dc offset; and determine the ac signal key parameters.

- Define, calculate, and apply average power of a resistive circuit.

- Interpret, apply, and measure the effective (rms) values of the fundamental waveforms.

- Calculate the rms value of a square wave, triangle wave, and sine wave with and without a dc offset.

- Superimpose (add) combinations of ac and/or dc voltages and currents in resistive and electronic circuits.

- Analyze ac signals applied to electronic circuits.

12.1 Waveform Terminology

A **waveform** graphically represents a variable as a function of time, for example, a voltage as a function of time as shown in Figure 12-1. The vertical axis represents the voltage value and the horizontal axis represents the time value. The symbol $v(t)$ represents the voltage as a function of time. The lowercase v is used to represent that this is a variable that can change with time.

A dc (direct current) voltage or current is a fixed value and does not vary with time. Figure 12-2 is a sketch of a 6 V dc voltage waveform. Its value is fixed at 6 V and can be expressed as

$$v(t) = 6 \ V$$

The voltage is always 6 V, whether it is measured at a time of 0 s or a time of 3 s. Mathematically,

$$v(0\,s) = 6 \ V$$

$$v(3\,s) = 6 \ V$$

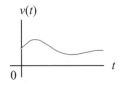

Figure 12-1
Varying voltage

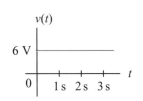

Figure 12-2
DC waveform

AC Waveform–Basic Parameters

An **ac** (**alternating current**) voltage or current varies with time. Typically it is a periodic signal with a repeating pattern that averages zero (0) over a period of time. Figure 12-3 is a sketch of an ac waveform, namely a sine wave. The generic expression for a voltage sine wave with a maximum value of 1 is given by the expression

$$v(t) = \sin(2\pi f t)$$

where

$v(t)$	voltage as a function of time with units of volts (V)
sin	sine function
π	Greek letter *pi* ≈ 3.1415927…
f	frequency (quantifies how rapidly the signal is changing)
t	time with units of seconds (s)

The scales of Figure 12-3 have been modified to include the units in the axes labels. The $v(V)$ label states that the vertical axis is a voltage v measured in units of volts (V). The $t(ms)$ label states that the horizontal axis is time t measured in units of milliseconds (ms). As time increases (goes from the origin 0 to the right), the value of the voltage varies. The voltage initially starts at 0 V, increases to a *maximum* (peak) of 1 V, back to 0 V, then to a *minimum* of –1 V, and back to 0 V. Mathematically,

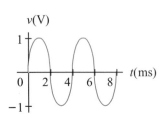

Figure 12-3
AC waveform

$$v(0\,\text{ms}) = 0\text{ V}$$
$$v(1\,\text{ms}) = 1\text{ V}$$
$$v(2\,\text{ms}) = 0\text{ V}$$
$$v(3\,\text{ms}) = -1\text{ V}$$
$$v(4\text{ ms}) = 0\text{ V}$$

The voltage waveshape then repeats this same basic waveshape over the same period of time; therefore, it is called a **periodic** waveform. The waveshape from 0 ms to 4 ms is repeated from 4 ms to 8 ms, and so forth. One repetition of the waveform is called a **cycle**. Figure 12-3 is displaying *two cycles* of the waveform. The waveform time can also be expressed in terms of the number of cycles or fraction of a cycle. For example, a *quarter of a cycle* would be equivalent to 1 ms for this waveform.

Periodic–repeating waveform

Since the waveform is repeated periodically, the voltage is repeated periodically.

$$v(0\text{ ms}) = v(4\text{ ms}) = v(8\text{ ms}) = v(12\text{ ms}) = 0\text{ V}$$
$$v(1\text{ ms}) = v(5\text{ ms}) = v(9\text{ ms}) = v(13\text{ ms}) = 1\text{ V}$$
$$v(2\text{ ms}) = v(6\text{ ms}) = v(10\text{ ms}) = v(14\text{ ms}) = 0\text{ V}$$
$$v(3\text{ ms}) = v(7\text{ ms}) = v(11\text{ ms}) = v(15\text{ ms}) = -1\text{ V}$$

The time of one cycle is called a **period** and its symbol is an *upper-case T*. The period of the waveform in Figure 12-3 is 4 ms since the waveshape repeats itself every 4 ms.

$$T = 4\text{ ms}$$

T–symbol for period

This is interpreted to be 4 ms per cycle.

We can find the number of *cycles per second* by taking the reciprocal of the *seconds per cycle*.

$$\frac{1\text{ cycle}}{4\text{ ms}} = 250\,\frac{\text{cycles}}{\text{s}}$$

The number of cycles that occur in 1 second is the reciprocal of its period, and is called the waveform's **frequency**. The symbol for frequency is a lowercase *f*.

$$f = \frac{1}{T}$$

Again, the period *T* is understood to be *time per cycle*. Thus, the frequency of the waveform in Figure 12-3 is

$$f = \frac{1}{T} = \frac{1}{4 \text{ ms}} = \frac{1}{4 \text{ ms/cycle}} = 250 \frac{\text{cycles}}{\text{second}}$$

The unit of $\left(\dfrac{\text{cycle}}{\text{second}} \right)$ is given the unit name of hertz; and the unit symbol for hertz is *Hz*. Thus, the frequency of this waveform is

$$f = \frac{1}{T} = \frac{1}{4 \text{ ms}} = 250 \frac{\text{cycles}}{\text{second}} = 250 \text{ Hz}$$

Example 12-1

Is the waveform of Figure 12-4 periodic? How many cycles are displayed? Find the waveform's period and frequency. The maximum value of the ac sine wave is called its **amplitude**. Find the amplitude of the signal and its minimum value.

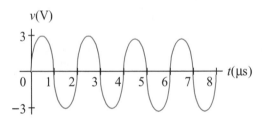

Figure 12-4 Example waveform

Solution

Yes, the waveform is *periodic* since the same basic waveshape is repeated every 2 μs. *Four cycles* are displayed since the basic waveshape form is repeated four times. By observation, the *period* T of the waveshape is 2 μs, since the basic waveshape repeats itself every 2 μs.

$$T = 2 \ \mu s$$

The *frequency* of the voltage waveshape is

$$f = \frac{1}{T} = \frac{1}{2 \ \mu s} = 500 \ \text{kHz}$$

By observation, the amplitude (maximum voltage) is 3 V_{max} and the minimum voltage is –3 V_{min}. Thus, this is a periodic waveshape with four cycles, a period of 2 μs, a frequency of 500 kHz, an amplitude of 3 V, and a minimum of –3 V.

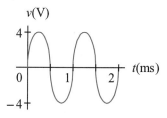

Practice

Sketch two cycles of a sine wave with a maximum of 4 V, a minimum of –4V, and a frequency of 1 kHz.

Answer: See Figure 12-5.

Figure 12-5
Practice waveform

Figure 12-6 summarizes the basic terminology of an alternating waveform. The vertical scale can be typically voltage v or current i. Upper case letters are used for V_{max} and V_{min} since these are fixed values for a given waveform and do not vary with time.

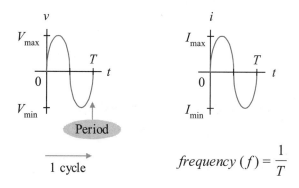

Figure 12-6 Waveform: basic terminology

Peak and Peak-to-Peak Values

Figure 12-7 displays additional waveform parameters that are commonly used, namely, **peak** and **peak-to-peak** values. This is an ac waveform that has been **offset** by a dc voltage; it is no longer centered around 0 V.

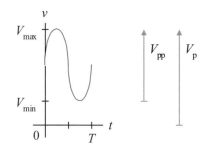

Figure 12-7 *Peak* and *peak-to-peak* values

The **peak** value is labeled with a lower case *p* subscript and is just its maximum value.

$$V_p = V_{max}$$

The **peak-to-peak** value is labeled with a lower case *pp* subscript and is the difference between its maximum value V_{max} and its minimum value V_{min}.

$$V_{pp} = V_{max} - V_{min}$$

Example 12-2

For the waveform of Figure 12-8, find its maximum value, minimum value, peak value, and peak-to-peak value.

Solution

By observation, the maximum and minimum voltages are

$$V_{max} = 4 \text{ V}_{max} = 4 \text{ V}_p$$

$$V_{min} = 1 \text{ V}_{min}$$

The *peak* voltage is

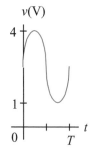

Figure 12-8
Example figure

$$V_p = V_{max} = 4 \ V_p$$

The *peak-to-peak* voltage is

$$V_{pp} = 4 \ V - 1 \ V = 3 \ V_{pp}$$

Practice

> For the waveform of Figure 12-9, find its maximum value, minimum value, peak value, and peak-to-peak value.

Answer: $5 \ V_{max}$, $-1 \ V_{min}$, $5 \ V_p$, $6 \ V_{pp}$

For clarification, when dealing with waveform values, subscripts are provided with the resulting voltage and current units.

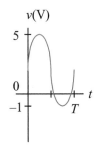

Figure 12-9
Practice waveform

Basic Waveforms

The basic waveshapes used in electronics are shown in Figure 12-10. The sine wave is a fundamental waveshape used in electrical systems and many electronic analog circuits such as audio amplifiers. The square wave and rectangle wave are used extensively in digital circuits as well as analog electronics. The triangle wave and sawtooth wave are used in wave shaping and timing circuits. A sawtooth voltage is used in televisions and oscilloscopes to control the trace of the electron beam on the surface of the screen (called a cathode ray tube, CRT). The oscilloscope is a test instrument that displays waveshapes on the CRT so they can be visibly observed. The exponential waveform is also used in timing and wave-shaping circuits.

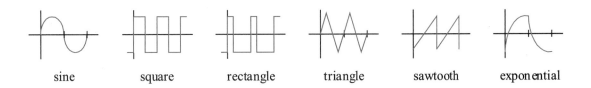

sine square rectangle triangle sawtooth exponential

Figure 12-10 Basic waveshapes

12.2 Average (DC) Value

Periodic waveforms have an average (or dc) value. In its simplest form, average is the summation of values divided by the number of values. For example, two people have a total of $30. On the average, each person has $15 even though one person may have all the money.

$$\$_{avg} = \frac{\$30}{2} = \$15$$

With respect to waveforms, the **average** or **dc** value is the *net area A_{net}* (area under the curve with respect to the horizontal 0 V axis) divided by the period T.

$$V_{dc} = \frac{A_{net}}{T}$$

If the area is *above* the 0 V axis, it naturally creates a *positive* area. If the area is *below* the 0 V axis, it naturally creates a *negative* area. In Figure 12-11(a), the positive half cycle (red area A_1) of a sine wave has the same area as its negative half cycle (black area A_2). That is,

$$A_{net} = A_1 - A_2 = 0$$

Thus, the dc (average) voltage of this waveform is

$$V_{dc} = \frac{A_{net}}{T} = \frac{A_1 - A_2}{T} = \frac{0\ V}{T} = 0\ V$$

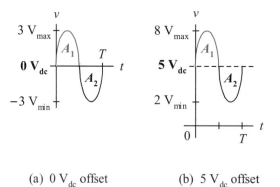

(a) 0 V_{dc} offset   (b) 5 V_{dc} offset

Figure 12-11 Sine wave (a) without dc offset and (b) with a fixed 5 V_{dc} offset

Let's add a *fixed* 5 V to this waveform; that is, add a 5 V *dc offset* to the waveform to produce an ac waveform with a nonzero dc offset. The new key parameters become

$$V_{\text{max with offset}} = 3 \text{ V} + 5 \text{ V} = 8 \text{ V}_{\text{p}}$$

$$V_{\text{dc with offset}} = 0 \text{ V} + 5 \text{ V} = 5 \text{ V}_{\text{dc}}$$

$$V_{\text{min with offset}} = -3 \text{ V} + 5 \text{ V} = 2 \text{ V}_{\text{min}}$$

Figure 12-11(b) shows the new waveform with this 5 V_{dc} *offset* included. The shape is unchanged, but each of the voltage values of the waveform has been increased by 5 V. The area A_1 above the 5 V_{dc} line is equal to the area A_2 below the 5 V_{dc}. These areas cancel each other out since they are equal. A line drawn through the dc value of a waveform has equal *areas above and below this dc line*.

Because of the symmetry of the sine wave, square wave, triangle wave, and sawtooth waves, a line drawn through the center of these waveforms (midway between the maximum and minimum values) produces an equal area above and below this centerline. Therefore, the dc value (average value) of the waveform can be simply calculated as

$$V_{\text{dc}} = \frac{1}{2}\left(V_{\text{max}} + V_{\text{min}}\right)$$

Average or
dc value

For Figure 12-11(a),

$$V_{\text{dc without offset}} = \frac{1}{2}\left(3 \text{ V} + \left(-3 \text{ V}\right)\right) = 0 \text{ V}_{\text{dc}}$$

For Figure 12-11(b),

$$V_{\text{dc with offset}} = \frac{1}{2}\left(8 \text{ V} + \left(2 \text{ V}\right)\right) = 5 \text{ V}_{\text{dc}}$$

This is the waveform's dc value as it would be measured by a *dc voltmeter*. The dc voltmeter and dc ammeter measure average waveform values.

What about the peak voltage? Does it include the dc offset? *Yes.*

$$V_{\text{max without offset}} = 3 \text{ V}_{\text{p}} \qquad \text{(Figure 12-11(a))}$$

$$V_{\text{max with offset}} = 8 \text{ V}_{\text{p}} \qquad \text{(Figure 12-11(b))}$$

What about the peak-to-peak voltage? Does it include the dc offset? *No.*

$$V_{pp \text{ without offset}} = 3 \text{ V} - (-3 \text{ V}) = 6 \text{ V}_{pp} \qquad \text{(Figure 12-11(a))}$$

$$V_{pp \text{ with offset}} = 8 \text{ V} - 2 \text{ V} = 6 \text{ V}_{pp} \qquad \text{(Figure 12-11(b))}$$

Example 12-3

For the waveform of Figure 12-12(a), find its maximum current, minimum current, dc current, peak current, and peak-to-peak current.

Solution

By observation, maximum and minimum currents are

$$I_{max} = 8 \text{ A}_p$$

$$I_{min} = -4 \text{ A}_{min}$$

By observation, maximum and peak current are the same.

$$I_p = I_{max} = 8 \text{ A}_p$$

Since this is a square wave,

$$I_{dc} = \frac{1}{2}(8 \text{ A} + (-4 \text{ A})) = 2 \text{ A}_{dc}$$

$$I_{pp} = (8 \text{ A} - (-4 \text{ A})) = 12 \text{ A}_{pp}$$

Figure 12-12(b) summarizes these results.

Figure 12-12(a)
Example waveform

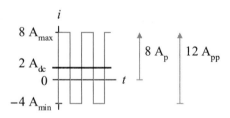

Figure 12-12(b) Example problem solution

Practice

For the waveform of Figure 12-13, find its maximum current, minimum current, dc current, peak current, and peak-to-peak current.

Answer: 1 A_{max}, −5 A_{min}, −2 A_{dc}, 1 A_p, 6 A_{pp}

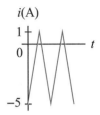

Figure 12-13
Practice waveform

Average (DC) Mathematical Definition

The official definition of average (dc) comes from calculus. The calculus definition is provided here for the sake of completeness although calculus usage is not required until a later textbook in this series. The official definition for all periodic waveshapes is

$$V_{average} = V_{dc} = \frac{1}{T} \int_0^T v(t)\, dt$$

where

$\int_0^T v(t)\, dt$	*area under the curve* (like A_{net} in previous examples)
$\int$	*summing* symbol (sum up lots of little slices of area)
T	period of the periodic waveshape
$v(t)$	function v expressed as a function of time t
dt	an infinitesimally small quantity of time

When the areas of the waveform are symmetrical (like the sine wave) or if they are simple (like rectangle waves), the areas can be calculated with relatively straightforward mathematics. However, if the waveforms' shapes are more complicated, like the exponential waveform, then more advanced mathematics (like calculus or computer numerical analysis techniques) must be employed to find the net area and thus the average value of the waveform.

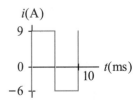

Figure 12-14
Example waveform

Example 12-4

Find the average (dc) current of a square wave in Figure 12-14. First, find the average simply using its maximum and minimum values. Then find the average (dc) current by finding the positive and negative areas under the curve and dividing by the period.

Solution

First, let's take the straightforward approach since this is square wave,

$$I_{dc} = \frac{1}{2}\left(9 \text{ A} + \left(-6 \text{ A}\right)\right) = 1.5 \text{ A}_{dc}$$

Next, let's use the basic definition of area divided by period to find the average. The net area under the curve is the difference between the positive and negative areas relative to the 0 A axis.

The positive area occurs from 0 ms to 5 ms.

$$A_{positive} = 9 \text{ A} \times \left(5 \text{ ms} - 0 \text{ ms}\right) = 45 \text{ A} \cdot \text{ms}$$

The negative area occurs from 5 ms to 10 ms.

$$A_{negative} = -6 \text{ A} \times \left(10 \text{ ms} - 5 \text{ ms}\right) = -30 \text{ A} \cdot \text{ms}$$

The net area is the difference between the positive area and the negative area.

$$A_{net} = 45 \text{ A} \cdot \text{s} - 30 \text{ A} \cdot \text{ms} = 15 \text{ A} \cdot \text{ms}$$

Find the average by dividing the net area by the period.

$$I_{dc} = I_{average} = \frac{15 \text{ A} \cdot \text{ms}}{10 \text{ ms}} = 1.5 \text{ A}_{dc}$$

Calculus tidbit: For those who have knowledge of integral calculus, the official calculus approach is provided. If you have no knowledge of calculus, then you may skip this calculus solution. The integral produces the net area under the curve A_{net}.

$\int i(t)dt$

$$I_{dc} = I_{average} = \frac{1}{T} A_{net} = \frac{1}{T} \int_{0}^{T} i(t)\,dt$$

$$I_{dc} = I_{average} = \frac{1}{10 \text{ ms}}\left[\int_{0 \text{ ms}}^{5 \text{ ms}}\left(9 \text{ A}\right)dt + \int_{5 \text{ ms}}^{10 \text{ ms}}\left(-6 \text{ A}\right)dt\right]$$

$$I_{dc} = \frac{1}{10 \text{ ms}}\left[\left(9 \text{ A}\right)t \Big|_{0 \text{ ms}}^{5 \text{ ms}} + \left(-6 \text{ A}\right)t \Big|_{5 \text{ ms}}^{10 \text{ ms}}\right]$$

$$I_{dc} = \frac{1}{10 \text{ ms}}\left[\left(9 \text{ A}\left(5 \text{ ms} - 0 \text{ ms}\right)\right) + \left(-6 \text{ A}\left(10 \text{ ms} - 5 \text{ ms}\right)\right)\right]$$

$$I_{dc} = \frac{1}{10 \text{ ms}} \left[45 \text{ A} \cdot \text{ms} + \left(-30 \text{ A} \cdot \text{ms} \right) \right]$$

$$I_{dc} = \frac{1}{10 \text{ ms}} \left[15 \text{ A} \cdot \text{ms} \right] = 1.5 \text{ A}_{dc}$$

Practice

Find the average (dc) current of the rectangle waveshape in Figure 12-15. First find the positive area, then the negative area, and finally the dc current. You must use the net area divided by the period approach since this waveshape is not symmetrical. Note that the positive area occurs from 0 ms to 2 ms, and the negative area occurs from 2 ms to 8 ms.

Answer: 18 A·ms, −36 A·ms, −2.25 A$_{dc}$

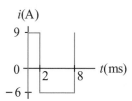

Figure 12-15
Practice waveform

Separating AC Signals with a DC Offset

An **offset signal** is an ac signal that is riding on a dc level. Frequently the dc signal is needed as a bias signal for the ac signal to ride upon, like waves on a lake. The tranquil (quiescent) lake level is equivalent to the dc value. The ripples in the lake are analogous to the ac signal. Some circuits require offset ac signals like ripples (ac) in a lake require lake water (dc).

Typically, the analysis of an *offset signal* is separated into a *dc analysis* and an *ac analysis*. Then the results of these two separate analyses are combined by a process known as *superposition* (to be studied later in this chapter). Thus, it is sometimes required to separate the offset signal into its dc signal and its ac signal as shown in Figure 12-16.

Figure 12-16(a) shows an offset ac signal that is separated into its dc component as shown in Figure 12-16(b) and its ac signal component as shown in Figure 12-16(c).

The process to separate the offset ac signal into its dc and ac components is reasonably straightforward. All that is needed is the offset signal's dc value and its peak-to-peak value. Find the dc component V_{dc} of the mixed signal and the peak-to-peak voltage V_{pp} of the offset signal as described previously in this text.

Then find the ac signal component values. *The average (dc) of the ac signal by definition must always be 0.*

$$V_{dc} = 0 \text{ V} \qquad \text{(ac signal)}$$

V_{pp} of the offset signal and the ac only signal *must be the same.* Find the V_{pp} signal of the offset signal first. Then, find the ac signal V_{max} and V_{min} by dividing V_{pp} by 2. This is the simplest approach to translate from a dc offset ac signal to its pure ac signal component.

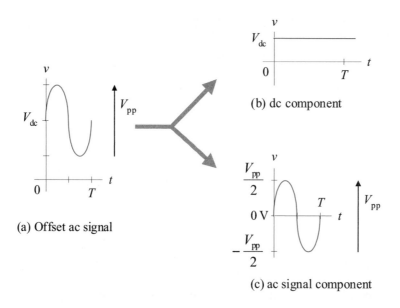

(a) Offset ac signal

(b) dc component

(c) ac signal component

Figure 12-16 A dc offset ac signal separated into its dc and ac components

$$V_{max} = V_p = \frac{V_{pp}}{2} \qquad \text{(ac signal)}$$

$$V_{min} = -\frac{V_{pp}}{2} \qquad \text{(ac signal)}$$

Recall that V_{pp} is the same for both the offset signal and the ac signal component. V_{pp} is independent of dc offset.

Another approach to find the *ac signals* V_{max} and V_{min} is to use the offset signal V_{max}, V_{min}, and V_{dc}.

$$V_{max} = V_{\text{offset signal max}} - V_{dc} \qquad \text{(ac signal)}$$

$$V_{min} = V_{\text{offset signal min}} - V_{dc} \qquad \text{(ac signal)}$$

An example of an ac component is an audio signal (for example, voice or music) riding on a fixed dc bias to operate a BJT over a linear range and prevent distortion. Visualize the ac signal in Figure 12-16(c) being a single tone in an audio amplifier. A standard technique is to use a test signal of a single frequency sine wave in an amplifier to test the amplifier's response. A single frequency (tone) is much easier to observe and analyze than a mixture of many frequencies like a voice pattern.

Another term applied to an ac only signal is the term *amplitude*. You can think of *amplitude* as analogous to *volume*. Thinking in terms of an audio signal (no dc offset), an increase in the signal peak-to-peak voltage leads to an increase in the signal maximum voltage (or peak voltage), an increase in the volume, and an increase in the amplitude.

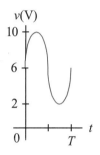

Figure 12-17(a)
Example problem–
offset signal

Example 12-5

For the waveform of Figure 12-17(a), find its offset signal dc voltage and its peak-to-peak voltage. Then find the ac signal voltage's dc voltage, maximum voltage (peak), and minimum voltage. Sketch the ac signal waveform.

Solution

By observation, the maximum and minimum voltages of the *offset signal* are

$$V_{max} = 10 \text{ V}_p \qquad \text{(offset signal)}$$

$$V_{min} = 2 \text{ V}_{min} \qquad \text{(offset signal)}$$

Since this is a sine wave, the dc voltage of the *offset signal* can be found by

$$V_{dc} = \frac{1}{2}\left(10 \text{ A}_p + (2 \text{ A}_{min})\right) = 6 \text{ V}_{dc} \qquad \text{(offset signal)}$$

and its peak-to-peak voltage by

$$V_{pp} = 10 \text{ V}_p - (2 \text{ A}_{min}) = 8 \text{ V}_{pp} \qquad \text{(offset signal)}$$

The ac signal parameters can now be calculated from the offset signal parameters of V_{dc} and V_{pp}. The *ac signal parameters* are

$$V_{dc} = 0 \text{ V}_{dc} \qquad \text{(always for an ac signal)}$$

$$V_{max} = \frac{1}{2} V_{pp} = \frac{1}{2} (8\ V_{pp}) = 4\ V_p \qquad \text{(ac signal)}$$

$$V_{min} = -\frac{1}{2} V_{pp} = -\frac{1}{2} (8\ V_{pp}) = -4\ V_{min} \qquad \text{(ac signal)}$$

The waveform solution is shown in Figure 12-17(b).

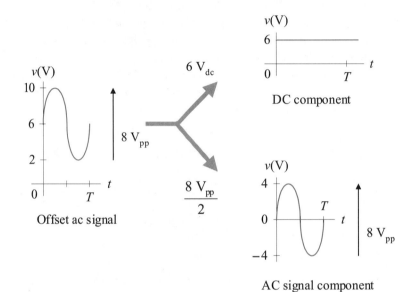

Figure 12-17(b) Example solution waveforms

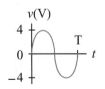

Figure 12-18
Practice solution

Practice

The dc offset of Figure 12-18 is changed to 13 V. Find the ac signal component values of V_{dc}, V_{max}, and V_{min}. Sketch the ac signal waveform. Find the total signal values of V_{dc}, V_{max}, and V_{min}. Sketch the total, offset signal waveform.

Answer: $0\ V_{dc}$, $4\ V_{max}$, and $-4\ V_{min}$; $13\ V_{dc}$, $17\ V_{max}$, and $9\ V_{min}$

12.3 AC Voltage Supplies

The symbols for ac voltage supplies are shown in Figure 12-19. The symbol is a circle with the shape of the ac signal inside the circle. If there is a plus (+) sign associated with the symbol, the plus (+) sign indicates that that terminal is considered the positive half cycle for analysis purposes. The actual voltage polarity and current direction reverse themselves every half cycle. However, we can freeze the supply in its positive half cycle and assign voltage polarities and current directions based upon the supply's positive half cycle reference (+).

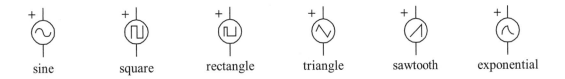

sine   square   rectangle   triangle   sawtooth   exponential

Figure 12-19 AC voltage supply symbols

Example 12-6

For the circuit of Figure 12-20(a), the potentiometer wiper arm is adjusted to 3 kΩ of resistance from wiper contact to common. First redraw the schematic with the potentiometer properly modeled into its two equivalent resistances and three terminals. Establish the supply and load resistor voltage polarities and current directions on your schematic.

Then find the total circuit resistance, the supply peak current, the load peak current, and the load peak voltage. Also, find the node voltages at nodes *a*, *b*, and *out*.

Sketch the load voltage and the load current waveforms. Properly label and scale the voltage, current, and time axes.

Solution

First draw the schematic with the potentiometer modeled as 7 kΩ and a 3 kΩ fixed resistance. Then establish the supply and load resistor voltage polarities and the current directions. Using the established positive half cycle reference (+) of the voltage supply, use the same practices as performed for dc circuits to establish voltage polarities and current directions. See Figure 12-20(b). R_x and R_y represent the top and bottom potentiometer resistances, respectively.

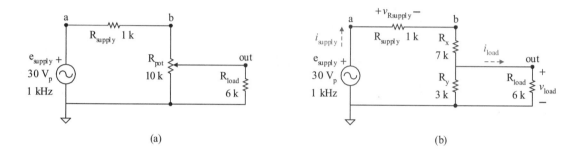

Figure 12-20(a,b) Example circuit

These voltage and current references are valid while the supply is in its positive half cycle. During the negative half cycle, all the voltage polarities and current directions will reverse. The voltage polarities and current directions reverse each half cycle, that is, they flip twice per full cycle of the supply.

Use resistor reduction techniques to reduce the circuit. First let's follow the supply current path. It flows from the supply, through the 1 kΩ resistor and down through the 7 kΩ resistance. The current has not split yet so these two resistances are in series. The current splits at the wiper contact between the 3 kΩ resistance and the 6 kΩ resistance and rejoins at the bottom node (common); placing the 3 kΩ and 6 kΩ resistances in parallel. And that parallel combination is in series with the 1 kΩ and the 7 kΩ resistances.

R_y in parallel with R_{load} is found by the general parallel resistance formula.

$$\frac{1}{R_{y,load}} = \frac{1}{3 \ k\Omega} + \frac{1}{6 \ k\Omega}$$

$$R_{y,load} = 2 \ k\Omega$$

The total circuit resistance is the sum of the 1 kΩ, the 7 kΩ, and the parallel combination 2 kΩ resistances.

$$R_{total} = 1 \text{ k}\Omega + 7 \text{ k}\Omega + 2 \text{ k}\Omega = 10 \text{ k}\Omega$$

By Ohm's law, the supply current is

$$I_{p \text{ supply}} = \frac{30 \text{ V}}{10 \text{ k}\Omega} = 3 \text{ mA}_p$$

By Ohm's law, the supply resistor drop is

$$V_{p \text{ Rsupply}} = 3 \text{ mA} \times 1 \text{ k}\Omega = 3 \text{ V}_p$$

The supply current then splits between the 3 kΩ and the 6 kΩ load resistances. Use the two-branch current divider rule to find the load current.

$$I_{p \text{ load}} = \frac{3 \text{ k}\Omega}{9 \text{ k}\Omega} \times 3 \text{ mA}_p = 1 \text{ mA}_p$$

Peak load voltage (that is the output *amplitude*) is

$$V_{p \text{ load}} = 1 \text{ mA} \times 6 \text{ k}\Omega = 6 \text{ V}_p$$

Starting at common and rising 30 V_p across the supply, the node voltage at *a* is

$$V_a = 0 \text{ V} + 30 \text{ V} = 30 \text{ V}_p$$

Falling 3 V_p across the supply resistor, the voltage at node *b* is

$$V_b = 30 \text{ V} - 3 \text{ V} = 27 \text{ V}_p$$

Starting at common (that is, 0 V_p) and rising 6 V_p across the load resistor, the voltage at node *out* is

$$V_{out} = 0 \text{ V} + 6 \text{ V} = 6 \text{ V}_p$$

Figure 12-20(c) shows the original schematic with all of the calculated voltages and currents, including the waveform voltages at nodes *a*, *b*, and *out*.

All the voltages and currents have the same frequency as the supply frequency. To sketch these waveforms, we must first calculate the period of the waveforms, which is just the reciprocal of the frequency.

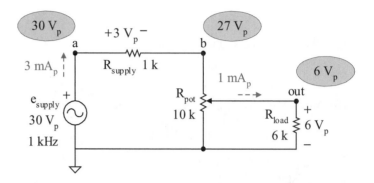

Figure 12-20(c) Example circuit voltages and currents

$$T = \frac{1}{1\ \text{kHz}} = 1\ \text{ms}$$

Figure 12-20(d) shows the sketch of the load voltage and load current waveforms.

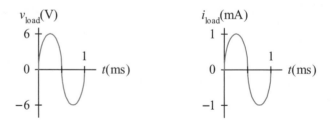

Figure 12-20(d) Load voltage and current waveforms

Practice

Repeat Example 12-6 for the circuit of Figure 12-21(a). The potentiometer wiper arm is positioned such that the resistance from node *b* to the wiper arm contact is 4 kΩ.

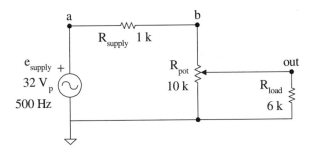

Figure 12-21(a) Practice circuit

Answer: 8 kΩ, 4 mA$_p$, 4 V$_p$, 2 mA$_p$, 12 V$_p$; node a (32 V$_p$), node b (28 V$_p$), node *out* (12 V$_p$); period (2 ms). See Figure 12-21(b) for the waveform sketches.

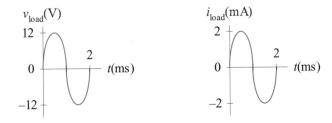

Figure 12-21(b) Practice circuit load voltage and current waveforms

12.4 Average Power–RMS Values

An ac supply is used to light a light bulb and generates a given amount of light as shown in Figure 12-22. What is the average power dissipated by the light? What is the equivalent dc supply needed to generate the same amount of light (that is, the same amount of power)?

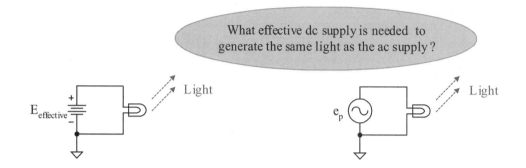

Figure 12-22 Equivalent ac and dc supplies

If the same quantity of light is being generated, then both bulbs are operating at the same average power and dissipating the same energy.

To explore these questions and find the answers, the resistive circuit of Figure 12-23 is used.

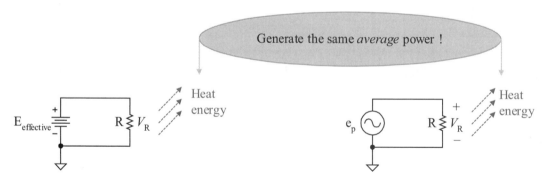

Figure 12-23 Equivalent ac and dc supplies

What is the average power dissipated by resistance R? Based on the dc power rule, the average power dissipated by the load is

$$P_{\text{average R}} = I_{\text{effective}} \, V_{\text{effective}}$$

where $I_{\text{effective}}$ and $V_{\text{effective}}$ are the resistor current and voltage, respectively.

Using Ohm's law, the power can be written as voltage squared divided by the resistance.

$$P_{\text{average R}} = \frac{V^2_{\text{effective}}}{R}$$

What is the average power being generated by the ac supply? The average ac voltage generated is 0 V, and the average ac current generated is 0 A. But certainly the average power is not 0 W, else the lamp would not be emitting light and there would be no heat dissipation in the resistance circuit. The average ac power must be the same as the average dc power generated by the dc supply to produce the same energy (heat dissipation).

The average of a periodic function is defined to be its net area divided by its period, or in calculus terms

$$P_{\text{average}} = \frac{1}{T} \int_0^T p(t)\,dt$$

where $p(t)$ is the instantaneous power and the integral generates the net area (area under the curve). The function $p(t)$ represents the value of the power at any given instant of time t. *Calculus is used to formally develop the effective voltage expression, but it is not required that the student know or use calculus to understand and apply the results.*

The average power from the ac circuit must be equal to the average power from the dc circuit, therefore,

$$P_{\text{average of dc}} = P_{\text{average of ac}}$$

Substitute the average integral expression for the ac average power on the right side of the expression.

$$P_{\text{average of dc}} = \frac{1}{T} \int_0^T p(t)\,dt$$

Convert the resistor power variables to their V^2/R forms.

$$\frac{V^2_{\text{effective}}}{R} = \frac{1}{T} \int_0^T \frac{v^2(t)}{R}\,dt$$

R is assumed to be a constant and can be eliminated from both sides of the expression.

$$V^2_{\text{effective}} = \frac{1}{T} \int_0^T v^2(t)\,dt$$

Taking the square root of both sides of the equation, the effective voltage (the dc voltage required to produce the same power as the ac signal) is

$$V_{\text{effective}} = \sqrt{\frac{1}{T} \int_0^T v^2(t)\,dt}$$

A closer look at this formula shows the following three ordered steps:
- First, *square* the function $v(t)$.

$$v^2(t)$$

- Second, take the *mean* (average) of the squared function.

$$\frac{1}{T} \int_0^T v^2(t)\,dt$$

- Third, take the square *root* of the mean of the squared function

$$\sqrt{\frac{1}{T} \int_0^T v^2(t)\,dt}$$

Because of this mathematical process (the *root* of the *mean* of the *square*), this value is named the **root-mean-square** (rms) value.

$$V_{\text{rms}} = \sqrt{\frac{1}{T} \int_0^T v^2(t)\,dt}$$

The *rms* voltage is effectively the dc voltage needed to produce the same power as the ac voltage. The term *rms* is standard vocabulary and is used from this point forward in the text instead of the term *effective*.

The rms value can also be found for the current; it has the same form.

$$I_{\text{rms}} = \sqrt{\frac{1}{T} \int_0^T i^2(t)\,dt}$$

The following examples demonstrate how to use this process for circuits with simple signals: dc, square wave, and rectangle wave.

rms is always positive

Note the rms value is always considered a *positive* (+) value. Since the function is squared, negative values become positive. Then the positive square root is taken of a positive number for a positive result.

Example 12-7

Find the rms voltage of a 10 V fixed dc voltage.

Solution

The waveform is shown in Figure 12-24(a).

$$v(t) = 10 \text{ V}_{dc}$$

Square the signal as shown in Figure 12-24(b).

$$v^2(t) = (10 \text{ V})^2 = 100 \text{ V}^2$$

Let's find its net area over 2 s. Use Figure 12-24(c).

$$A_{net} = 100 \text{ V}^2 \times 2 \text{ s} = 200 \text{ V}^2 \cdot \text{s}$$

Find the mean (average) value over that 2 s (this is a very simple case).

$$V^2_{average \ of \ the \ square} = \frac{200 \text{ V}^2 \cdot \text{s}}{2 \text{ s}} = 100 \text{ V}^2$$

Then take the square root.

$$V_{rms} = \sqrt{100 \text{ V}^2} = 10 \text{ V}_{rms}$$

The rms voltage of a positive dc is its dc voltage.

Practice

Repeat Example 12-7 but assume that the voltage is a fixed negative voltage of -10 V_{dc}. Write a general equation for the relationship of dc and rms voltage.

Answer: 10 V_{rms} (positive). $V_{rms} = |V_{dc}|$, i.e., the rms value of a dc voltage is the *absolute value of the dc voltage*.

The next simplest example is the square wave. It also has a very straightforward result.

Example 12-8

Find the rms voltage of a 10 V_p square wave with a period of 2 s.

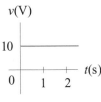

(a) voltage

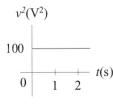

(b) Square

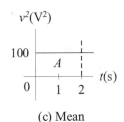

(c) Mean

Figure 12-24
Example problem

Solution

The waveform is shown in Figure 12-25(a).

$$v(t) = +10 \text{ V}_{dc} \qquad\qquad 0 \text{ s} < t < 1 \text{ s}$$

$$v(t) = -10 \text{ V}_{dc} \qquad\qquad 1 \text{ s} < t < 2 \text{ s}$$

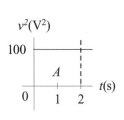

$v(V)$

(a) voltage

Square the signal as shown in Figure 12-25(b). Once both are squared, they have the same 100 V^2 value.

$$v^2(t) = (10 \text{ V})^2 = 100 \text{ V}^2 \qquad 0 \text{ s} < t < 1 \text{ s}$$

$$v^2(t) = (-10 \text{ V})^2 = 100 \text{ V}^2 \qquad 1 \text{ s} < t < 2 \text{ s}$$

Let's find its net area over 2 s. Use Figure 12-24(c).

$$A_{net} = 100 \text{ V}^2 \times 2 \text{ s} = 200 \text{ V}^2 \cdot \text{s}$$

$v^2(V^2)$

(b) Square

Find the mean (average) value over that 2 s.

$$V^2_{\text{average of the square}} = \frac{200 \text{ V}^2 \cdot \text{s}}{2 \text{ s}} = 100 \text{ V}^2$$

Then take the square root.

$$V_{rms} = \sqrt{100 \text{ V}^2} = 10 \text{ V}_{rms}$$

$v^2(V^2)$

(c) Mean

The rms voltage of a square wave is its peak voltage.

$$V_{rms} = V_p \qquad\qquad \text{(square wave)}$$

Figure 12-25
Example problem

Practice

Find the rms value of a square wave current with a maximum voltage of 7 mA$_p$.

Answer: 7 mA$_{rms}$

Example 12-9

Find the rms voltage of the square wave of Figure 12-26(a). Notice that this square wave has a -1 V$_{dc}$ offset.

Solution

The waveform is shown in Figure 12-26(a).

$$v(t) = +2 \text{ V}_{dc} \qquad\qquad 0 \text{ s} < t < 1 \text{ s}$$

$$v(t) = -4 \text{ V}_{dc} \qquad\qquad 1 \text{ s} < t < 2 \text{ s}$$

Square the signal. Results are shown in Figure 12-26(b). Notice that both areas *A*1 and *A*2 are now positive.

$$v^2(t) = (2 \text{ V})^2 = 4 \text{ V}^2 \qquad\qquad 0 \text{ s} < t < 1 \text{ s}$$

$$v^2(t) = (-4 \text{ V})^2 = 16 \text{ V}^2 \qquad\qquad 1 \text{ s} < t < 2 \text{ s}$$

Let's find its net area over 2 s. Use Figure 12-24(b).

$$A_{net} = 4 \text{ V}^2 \times 1 \text{ s} + 16 \text{ V}^2 \times 1 \text{ s} = 20 \text{ V}^2 \cdot \text{s}$$

v(V)

(a) voltage

Find the mean (average) value over that 2 s. See Figure 12-26(c).

$$V^2_{\text{average of the square}} = \frac{20 \text{ V}^2 \cdot \text{s}}{2 \text{ s}} = 10 \text{ V}^2$$

Then take the square root.

$$V_{rms} = \sqrt{10 \text{ V}^2} = 3.16 \text{ V}_{rms}$$

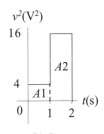

*v*²(V²)

(b) Square

Practice
Find the rms value of the rectangle wave of Figure 12-27.

Answer: 5.29 mA$_{rms}$

If an ac signal has a dc offset as in Example 12-9, then there is a quicker approach to finding the rms value. Again, we turn to calculus for the formal definition but merely use it to get to a useful result. Assume the signal *v*(*t*) is an ac signal $v_{ac}(t)$ with a dc offset of V_{dc}. Then *v*(*t*) can be expressed as

$$v(t) = v_{ac}(t) + V_{dc}$$

The general form for finding the rms value is

$$V_{rms} = \sqrt{\frac{1}{T} \int_0^T v^2(t) \, dt}$$

Substitute for *v*(*t*) and the general form becomes

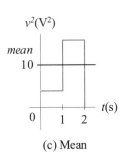

*v*²(V²)

(c) Mean

Figure 12-26
Example problem

i(mA)

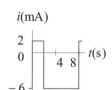

Figure 12-27
Practice example

$$V_{rms} = \sqrt{\frac{1}{T}\int_0^T \left(v_{ac}(t)+V_{dc}\right)^2 dt}$$

Square the voltage term.

$$V_{rms} = \sqrt{\frac{1}{T}\int_0^T \left(v_{ac}^2(t)+2v_{ac}V_{dc}+V_{dc}^2\right)dt}$$

Expanding the integral of each term under the radical,

$$V_{rms} = \sqrt{\frac{1}{T}\int_0^T v_{ac}^2(t)\,dt + \frac{2}{T}\int_0^T v_{ac}V_{dc}\,dt + \frac{1}{T}\int_0^T V_{dc}^2 dt}$$

This is equivalent to

$$V_{rms} = \sqrt{V_{rms\ of\ ac}^2 + 0 + V_{dc}^2}$$

The middle term is a constant times the average of a pure ac signal which is always 0 V. Simplified,

$$V_{rms} = \sqrt{V_{rms\ of\ ac}^2 + V_{dc}^2}$$

Thus, if there is an offset to a purely ac signal, we can find the rms value of the total signal (the offset ac signal) by finding the rms of the ac component signal and the dc offset value, and then applying this equation. Let's see if it really works by doing Example 12-9 but using this new approach for Example 12-10.

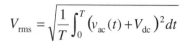

Example 12-10

Find the rms voltage of the square wave of Figure 12-28. Then compare the results with those of Example 12-9.

Solution

First split the offset signal into its ac and dc parts. Find its dc (average) voltage.

$$V_{dc} = \frac{1}{2}\left(2\text{ V}+(-4\text{ V})\right) = -1\text{ V}$$

Next find the peak-to-peak voltage.

$$V_{pp} = 2\text{ V} - \left(-4\text{ V}\right) = 6\text{ V}_{pp}$$

The ac signal maximum voltage is half of its peak-to-peak.

v(V)

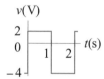

Figure 12-28
Example problem

$$V_{\text{p of ac}} = \frac{1}{2}\left(6 \text{ V}\right) = 3 \text{ V}_{\text{p}}$$

Since the ac component is a pure square wave, its rms value is equal to its peak voltage.

$$V_{\text{rms of ac}} = V_{\text{p of ac}} = 3 \text{ V}_{\text{p}}$$

Now apply the total waveform rms voltage formula

$$V_{\text{rms}} = \sqrt{V_{\text{rms of ac}}^2 + V_{\text{dc}}^2}$$

$$V_{\text{rms}} = \sqrt{\left(3 \text{ V}\right)^2 + \left(-1 \text{ V}\right)^2} = \sqrt{10 \text{ V}^2} = 3.16 \text{ V}_{\text{rms}}$$

This is the same answer obtained from Example 12-9 with the same waveform.

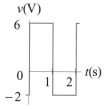

Figure 12-29
Practice problem

Practice

Find the dc, peak-to-peak, and rms value of the rectangle wave of Figure 12-29.

Answer: 2 V_{dc}, 8 V_{pp}, $4.47 \text{ V}_{\text{rms}}$

Other waveshapes such as the sine, the triangle, and the exponential waveforms require calculus to solve. Each waveform can be substituted into the general form, and its rms (effective) voltage and/or current value can be found.

$$V_{\text{rms}} = \sqrt{\frac{1}{T}\int_0^T v^2(t)\, dt}$$

$$V_{\text{rms dc}} = \left|V_{\text{dc}}\right| \qquad\qquad \text{(dc)}$$

$$V_{\text{rms square wave}} = V_{\text{p}} \qquad\qquad \text{(ac square wave)}$$

$$V_{\text{rms sine wave}} = \frac{V_{\text{p}}}{\sqrt{2}} = 0.707 \text{ V}_{\text{p}} \qquad\qquad \text{(ac sine wave)}$$

$$V_{\text{rms triangle wave}} = \frac{V_{\text{p}}}{\sqrt{3}} = 0.577 \text{ V}_{\text{p}} \qquad \text{(ac triangle or sawtooth wave)}$$

Example 12-11

The U.S. commercial electrical industry produces a 60 Hz sine wave voltage of 120 V_{rms}. Find its amplitude and peak-to-peak voltage. Sketch one cycle of this waveform.

Solution

The rms value of the sine wave is found by

$$V_{rms \ sine \ wave} = \frac{V_p}{\sqrt{2}} \qquad \text{(pure ac sine wave)}$$

Reorganize the equation to find the amplitude (ac peak voltage).

$$V_p = \sqrt{2} \ V_{rms} = \sqrt{2} \left(120 \ V_{rms} \right) = 170 \ V_p$$

The peak-to-peak voltage for a sine wave is twice its peak.

$$V_{pp} = 2 \left(170 \ V_p \right) = 340 \ V_{pp}$$

The period of U.S. electrical voltage is

$$T = \frac{1}{f} = \frac{1}{60 \ Hz} = 16.7 \ ms$$

A sketch of this voltage signal is shown in Figure 12-30.

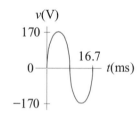

Figure 12-30
Example problem–
U.S. electrical voltage

Practice

The electrical industry of a European country produces a 50 Hz sine wave voltage of 220 V_{rms}. Find its peak, peak-to-peak voltage, and period.

Answer: 311 V_p, 622 V_{pp}, 20 ms

The next example is that of an ac ripple voltage formed in a half-wave rectifier dc power supply. Ideally a dc power supply would convert the ac voltage from the electrical utilities into a pure dc voltage output, but the electronic circuit used does not totally eliminate the unwanted ac (called ripple voltage, just like ripples in a lake).

Example 12-12

A half-wave rectifier power supply has an unwanted sawtooth ripple voltage as shown in Figure 12-31. Find its dc output voltage, peak-to-peak ripple voltage, the peak ripple voltage (ac only signal), the rms ripple voltage, the period of the ripple voltage, and the frequency of the ripple voltage.

Solution

$$V_{\text{max}} = 20 \text{ V}_{\text{p}}$$

$$V_{\text{min}} = 18 \text{ V}_{\text{min}}$$

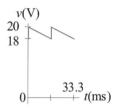

Its dc output voltage is the *average* value (note that this is midway between its minimum and maximum voltages).

$$V_{\text{average}} = \frac{1}{2}(20 \text{ V} + 18 \text{ V}) = 19 \text{ V}_{\text{dc}}$$

Its peak-to-peak voltage is the difference between its maximum and minimum voltages.

$$V_{\text{pp}} = 20 \text{ V} - (18 \text{ V}) = 2 \text{ V}_{\text{pp}}$$

Figure 12-31
Example problem–
ripple voltage

The ripple's peak voltage (ac only signal) is half of the peak-to-peak.

$$V_{\text{p ripple}} = \frac{1}{2}(2 \text{ V}) = 1 \text{ V}_{\text{p}}$$

The ripple voltage waveshape is very close to a sawtooth waveform. Thus, the ripple rms voltage is approximately

$$V_{\text{rms ripple}} = \frac{V_{\text{p}}}{\sqrt{3}} = \frac{1 \text{ V}_{\text{p}}}{\sqrt{3}} = 577 \text{ mV}_{\text{rms}}$$

Figure 12-31 is displaying 2 cycles of ripple voltage. Thus, one period of the ripple cycle is

$$T_{\text{ripple}} = \frac{33 \text{ ms}}{2} = 16.7 \text{ ms}$$

$$f_{\text{ripple}} = \frac{1}{T_{\text{ripple}}} = \frac{1}{16.7 \text{ ms}} = 60 \text{ Hz}$$

Practice

A full-wave rectifier power supply has an unwanted sawtooth ripple voltage as shown in Figure 12-32. Find its dc output voltage, peak-to-peak ripple voltage, the peak ripple voltage (ac only signal), the rms ripple voltage, the period of the ripple voltage, and the frequency of the ripple voltage.

Answer: 18 V$_{dc}$, 4 V$_{pp}$, 2 V$_p$, 1.15 V$_{rms}$, 8.35 ms, 120 Hz

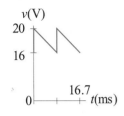

Figure 12-32
Practice problem–
full-wave ripple

DC and AC Power Calculations

To find the average power of an ac signal in a resistance or of an ac supply, use rms values. Recall that the definition of rms (effective value) came from the definition of average power in a resistive load.

To find the average power of a dc signal in a resistance or a dc supply, just use the dc voltage and current values. These are effectively the same as the rms values used to find average power.

Example 12-13

The U.S. commercial voltage of 120 V$_{rms}$ drives a 100 W light bulb. How much current is flowing through the bulb?

Solution

Use the power rule and set up to solve for the current I.

$$I_{rms\ bulb} = \frac{P_{average}}{V_{rms\ bulb}}$$

$$I_{rms\ bulb} = \frac{100\ W}{120\ V_{rms}} = 0.833\ A_{rms}$$

Practice

A 100 mV$_{pp}$ sine wave with a dc offset of 5 V drives a BJT with an input resistance of 1 kΩ looking into its base circuit. Find the dc power dissipated by the base. Find the ac rms voltage applied to the base.

Answer: 25 mW, 35.36 mV$_{rms}$

12.5 AC and DC Supplies

Many circuits mix ac and dc voltages and currents. Typically the dc and the ac are analyzed separately and then the results combined (or *superimposed*). This section utilizes examples to analyze circuits that have a mixture of ac and dc voltage and current supplies. The schematic symbols used in this textbook series are shown in Figure 12-33.

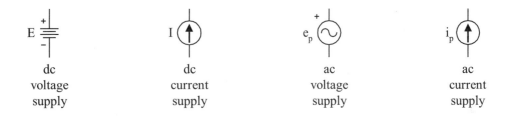

Figure 12-33 Supply schematic symbols

The ac supply values are stated in peak (p), peak-to-peak (pp), or rms. The ac waveform is either explicit with the symbol, understood, or must be stated. If the frequency is relevant, then the frequency or its period must be understood or stated. The ac voltage polarity sign (+) is a reference that is to be used throughout the analysis as a reference. That is, this supply terminal is to be assumed to be in its positive half cycle for analysis purposes.

Likewise, the ac current supply arrow indicates the reference direction of the current supply, assuming that current supply is in its positive half cycle. This current direction is to be assumed throughout the circuit analysis.

If an ac supply does not have an assigned waveform, it is assumed to be a sine wave.

Example 12-14
Identify the supply and its primary values in Figure 12-34.

Solution
    a.   dc voltage supply with a fixed voltage of 6 V_{dc}.

    b.   dc current supply with a fixed current of 7 mA_{dc}.

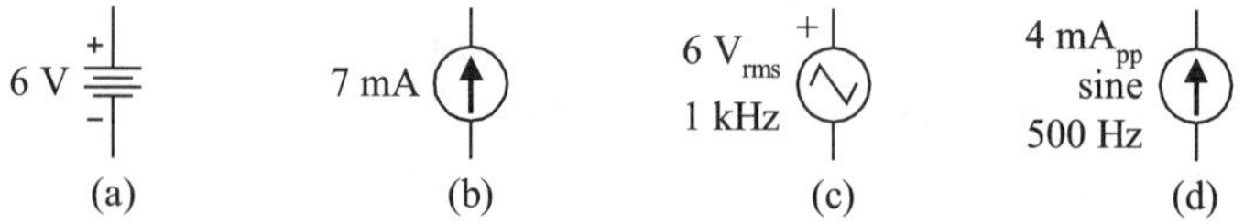

Figure 12-34 Example problem

c.   ac triangle voltage supply with a frequency of 1 kHz and a voltage of 6 V_{rms}. It has the following key values.

$$T = \frac{1}{1 \text{ kHz}} = 1 \text{ ms} \qquad \text{(period)}$$

$$V_{max} = V_p = \sqrt{3} \times 6 \ V_{rms} = 10.4 \ V_p \qquad \text{(max or peak)}$$

$$V_{pp} = 2 \times 10.4 \ V_p = 20.8 \ V_{pp} \qquad \text{(peak-to-peak)}$$

d.   ac sine current supply with a frequency of 500 Hz and a current of 4 mA_{pp}. It has the following key values.

$$T = \frac{1}{500 \text{ Hz}} = 2 \text{ ms} \qquad \text{(period)}$$

$$I_{max} = I_p = \frac{4 \text{ mA}_{pp}}{2} = 2 \text{ mA}_p \qquad \text{(max or peak)}$$

$$I_{rms} = \frac{2 \text{ mA}_p}{\sqrt{2}} = 1.414 \text{ mA}_{rms} \qquad \text{(rms)}$$

Practice

Identify the supply and its primary values in Figure 12-35.

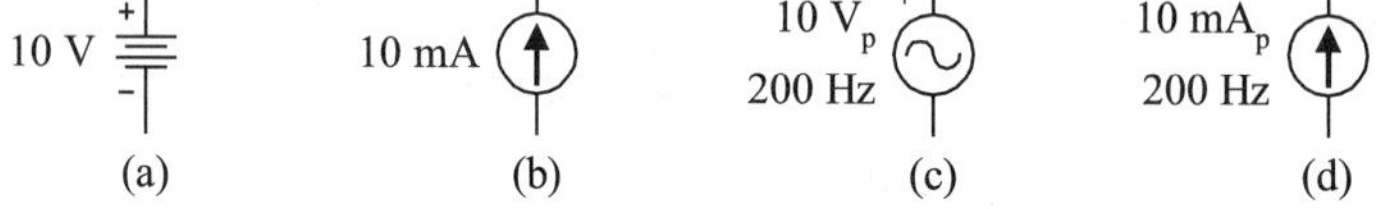

Figure 12-35 Example problem

Answer: (a) 10 V_{dc}; (b) 10 mA_{dc}; (c) sine, 10 V_p, 20 V_{pp}, 7.07 V_{rms}, 5 ms; (d) assumed sine wave, 10 mA_p, 20 mA_{pp}, 7.07 mA_{rms}, 5 ms

---

The next example combines an ac signal with a dc offset.

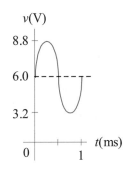

Figure 12-36(a)
Example problem

Example 12-15

For the supply of Figure 12-36(a), draw the total waveform of its output. Separate the ac signal from its dc offset. Find all key values of the signal including the total signal rms voltage. Draw an equivalent circuit with a dc supply and a separate ac supply.

Solution

The dc supplied voltage is a fixed 6 V_{dc}.

$$E_{dc} = 6 \text{ V}_{dc}$$

The ac supplied voltage is a 2 V_{rms} sinusoidal supply with a frequency of 1 kHz. The key values of the ac supply are:

$$T = \frac{1}{1 \text{ kHz}} = 1 \text{ ms}$$

$$e_{rms} = 2 \text{ V}_{rms}$$

$$V_{p \text{ ac only}} = V_{max \text{ ac only}} = \sqrt{2} \times 2 \text{ V}_{rms} = 2.828 \text{ V}_p$$

$$V_{max \text{ total signal}} = 6 \text{ V}_{dc} + 2.828 \text{ V} = 8.828 \text{ V}_p$$

$$V_{min \text{ total signal}} = 6 \text{ V}_{dc} - 2.828 \text{ V} = 3.172 \text{ V}_{min}$$

The total waveform is shown in Figure 12-36(b). The peak-to-peak voltage is always double its *ac only* signal peak value.

$$V_{pp} = 2 \times 2.828 \text{ V}_{p \text{ ac only}} = 5.656 \text{ V}_{pp}$$

Verify the peak-to-peak voltage by taking the difference of the signal's maximum and minimum voltages.

$$V_{pp} = 8.828 \text{ V}_p - 3.172 \text{ V}_{min} = 5.656 \text{ V}_{pp}$$

The total signal rms voltage is

$$V_{rms \text{ total signal}} = \sqrt{(6 \text{ V}_{dc})^2 + (2 \text{ V}_{rms})^2} = 6.32 \text{ V}_{rms}$$

Figure 12-36(b)
Example waveform

Figure 12-36(c)
Example equivalent modeled supplies

An equivalent model of the supply system with the dc and ac supplies separated is shown in Figure 12-36(c). Since the ac and dc simply add to each other, this is *equivalent to a series circuit.*

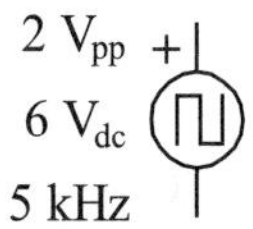

2 V_{pp} +
6 V_{dc}
5 kHz

Figure 12-37
Practice circuit

Practice

For the supply of Figure 12-37, separate the ac signal from its dc offset. Find all key values of the signal including the total signal rms voltage.

Answer: Square wave, 6 V_{dc}, 1$V_{p \, ac \, signal}$, 7 $V_{p \, total \, signal}$, 5 $V_{min \, total \, signal}$, 1 $V_{rms \, ac \, signal}$, 6.08 $V_{rms \, total \, signal}$, 200 μs period

---

12.6 Electronic Applications

Several electronic circuit examples demonstrate the use of mixed ac and dc voltages and currents.

Half-wave Rectifier Circuit

The half-wave rectifier circuit is a fundamental building block for a dc voltage power supply. An ac signal (0 V_{dc}) is fed to a rectifier circuit with a diode to allow current to flow in only one direction through the load. Since current is allowed to flow in only one direction through the load, a net dc voltage is created across the load.

Example 12-16

Use Figure 12-38(a). Assume that the ac voltage generator is at its maximum voltage (positive half cycle). Is the silicon diode *on* or *off*? What is the diode voltage drop? Find the load voltage and load current.

Assume that the ac voltage generator is at its minimum voltage (negative half cycle). Is the silicon diode *on* or *off*? Find the load current, load voltage drop, and diode voltage drop.

Sketch the load voltage and load current waveforms.

Solution

First inspect the circuit. This series circuit has a silicon diode in the circuit. The voltage polarity across the diode flips with each half cycle of the input voltage. The diode is forward biased on the positive half cycle, acting ideally like a short or to a second approximation dropping 0.7 V. Realistically, the diode does not turn on and allow current to flow until at least 0.7 V is applied to the circuit. As the voltage is increased above 0.7 V, the net voltage difference is applied to the load resistor. As the supply voltage increases, the drop across the load increases until it reaches its maximum voltage, when the supply voltage is at its maximum of $10\ V_p$.

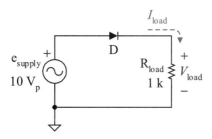

Figure 12-38(a) Example circuit of a half-wave rectifier

Visualize, as shown in Figure 12-38(b), the circuit with the ac supply fixed at its *maximum supply voltage* and solve for *maximum circuit values*. The silicon diode is forward biased and assigned a drop of 0.7 V. It is a simple series circuit. Solve for the load voltage using KVL or a KVL walkabout from $(-)\ V_{load}$ to $(+)$ V_{load}.

$$V_{load} = +10\ V - 0.7\ V = 9.3\ V_p$$

The maximum forward circuit current is found by using Ohm's law on the resistor. Since this is a series circuit, the diode, the supply, and the load experience the same maximum forward current. This is a critical number for the diode. If the circuit current exceeds the maximum forward current rating of the diode, then the diode current can be damaged.

$$I_{max\ load} = I_{max\ forward\ diode} = \frac{9.3\ V_p}{1\ k\Omega} = 9.3\ mA_p.$$

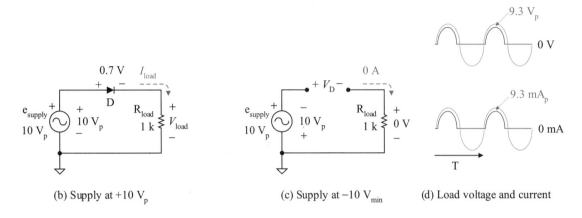

(b) Supply at +10 V$_p$ (c) Supply at −10 V$_{min}$ (d) Load voltage and current

Figure 12-38(b,c,d) Example circuit solution

The diode is reverse biased on the negative half cycle, acting ideally like an *open*. Since this is a series circuit and the diode is off (acting like an open), current cannot flow in this series circuit. Visualize the circuit as shown in Figure 12-38(c).

$$I_{load} = 0 \text{ A}$$

Therefore, the voltage drop across the load resistance is 0 V.

$$V_{load} = 0 \text{ mA} \times 1 \text{ k}\Omega = 0 \text{ V}$$

The maximum reverse voltage drop across the diode occurs when it is reverse biased with the greatest voltage applied to it. Frequently this is listed in the specifications as positive voltage even though the diode is under reverse bias conditions. If the magnitude of this voltage is exceeded in the circuit, the diode turns on and conducts in the reverse direction (undesirable result).

$$V_{\text{max reverse voltage of diode}} = 0 \text{ V} - 10 \text{ V} = -10 \text{ V}$$

The sketches of the load voltage and load current are shown in Figure 12-38(d). Note that the diode must be forward biased by at least 0.7 V before it turns *on* and conducts current. Therefore the output voltage curve is always 0.7 V less than the input signal when the diode is forward-biased *on*.

Practice

Perform a MultiSIM simulation of the example circuit. Set the sine wave signal generator to 10 V_p at 60 Hz. Use the oscilloscope test instrument to measure the input signal on channel A (labeled CH A) with a vertical sensitivity setting of 5 V per division and to measure the output across channel B (labeled CH B) with a vertical sensitivity setting of 2 V per division. Set the time scale to 5 ms per division. Set the triggering for a single sweep.

Answer: See Figure 12-38(e). Some significant observations follow:

1. The *first time marker* T1 (the left vertical line) has been placed when the input voltage is at its *maximum*. The simulation shows in the T1 box that the time T1 is 4.3 ms, the voltage of VA1 (input) is 10.0 V, and the voltage VB1 (output) is 9.4 V. The voltage difference of VA1 and VB1 is 0.6 V, which is the simulated forward drop across the diode at time T1.

2. The *second time marker* T2 (the right vertical line) has been placed when the input voltage is at its *minimum*. The simulation shows in the T1 box that the time T2 is 12.5 ms, the voltage of VA1 (input) is −10.0 V, and the voltage VB1 (output) is at −670.8 μV (approximately 0 V). The small voltage drop across the diode in the reverse direction is due to a small reverse leakage current in the diode. Assuming an ideal case, this current is considered to be 0. The voltage difference of VA2 and VB2 is about −10 V, which is the simulated reverse voltage drop across the diode at time T2. This represents the *peak inverse voltage* experienced by the diode. The diode must be rated to handle this worst-case reverse voltage, otherwise it will break down and begin conducting current in the reverse direction.

3. CH A voltage sensitivity was selected to better distinguish between the waveforms. Change the voltage sensitivity of CH A to match CH B, and observe the waveforms. They track each other very closely and you have to look closely and use the T1 and T2 markers to discern the two waveforms.

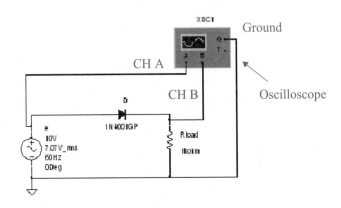

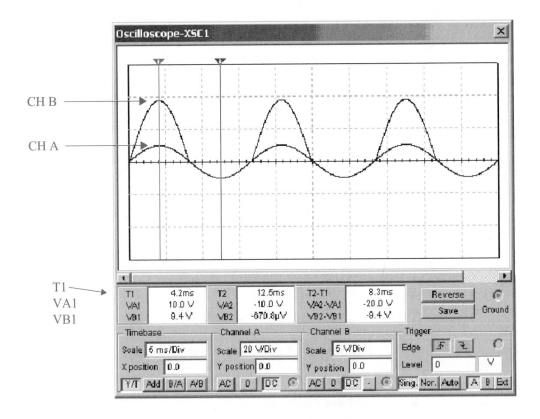

Figure 12-38(e) MultiSIM oscilloscope analysis of example half-wave rectifier

Exploration

      Reverse the diode in circuit of Figure 12-38 and repeat the practice problem.

Answer: See Figure 12-39 for expected oscilloscope outcome.

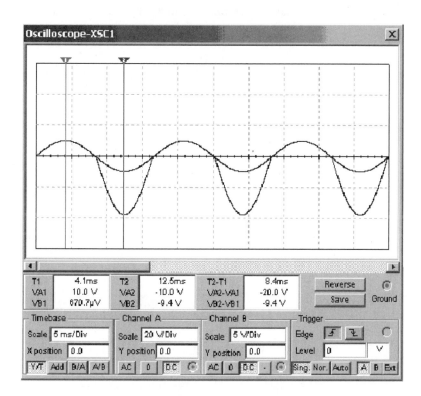

Figure 12-39 Exploration results with reversed diode

Op Amp Noninverting Voltage Amplifier Circuit

In the following example, the feedback circuit is a voltage divider circuit composed of two resistors, such that a fraction of the output voltage is fed back to the op amp inverting input terminal. This creates a voltage gain (a greater voltage) at the output; therefore, this is called a voltage amplifier.

Example 12-17

For the circuit of Figure 12-40(a), find the node *a* voltage V_a, the resistor R_i voltage drop and current, the resistor R_f current and voltage drop, and the output voltage v_{out}. Use circuit laws and basic circuit analysis techniques only to find these values. Then sketch the input and the output voltage waveforms.

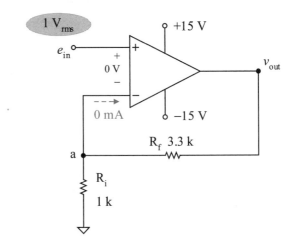

Figure 12-40(a) Example circuit: noninverting voltage amplifier

Solution

Start at e_{in}. There is a 0 V drop to node *a*.

$$V_a = 1\ V_{rms} - 0\ V = 1\ V_{rms}$$

Resistor R_i has 1 V_{rms} at its top node and 0 V (common) at its bottom node, therefore it drops 1 V_{rms}. Assuming the input is during its positive half cycle, the *current flows down* through R_i and its value is

$$I_{Ri} = \frac{1\ V_{rms}}{1\ k\Omega} = 1\ mA_{rms}$$

Since 0 mA is flowing into the op amp, all of the current flowing through resistor R_i must also be flowing through R_f (Kirchhoff's current law). The R_f current must be flowing from *right to left*.

$$I_{Rf} = I_{Ri} = 1 \text{ mA}_{rms}$$

By Ohm's law, the voltage drop across R_f is

$$V_{Rf} = 1 \text{ mA}_{rms} \times 3.3 \text{ k}\Omega = 3.3 \text{ V}_{rms}$$

To find V_{out}, start with V_a and raise the voltage by v_{Rf}.

$$V_{out} = 1 \text{ V}_{rms} + 3.3 \text{ V}_{rms} = 4.4 \text{ V}_{rms}$$

Figure 12-40(b) shows the circuit with its voltage and current values.

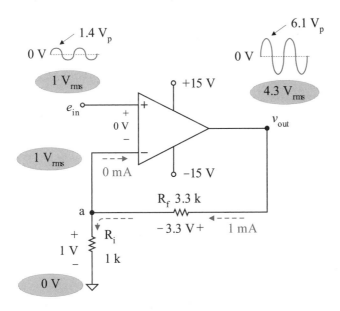

Figure 12-40(b) Example circuit with voltages and currents

Note that the voltage polarities and current directions assume that the input signal is in its positive half cycle. The rms values are always positive, but the voltage polarities and current directions are assumed to be instantaneous to analyze the circuit.

To sketch the input and output voltage waveforms, convert their rms voltages to peak voltages. These sketches are shown in Figure 12-40(b).

$$e_{p\ in} = \sqrt{2} \times 1\ V_{rms} = 1.414\ V_p$$

$$v_{p\ out} = \sqrt{2} \times 4.3\ V_{rms} = 6.081\ V_p$$

Practice

For the circuit of Figure 12-41, verify all voltages and currents *including voltage polarities and current directions*. The output voltage is *not really a negative* rms voltage; the negative sign indicates that the output is *inverted* (that is, phase shifted by 180°).

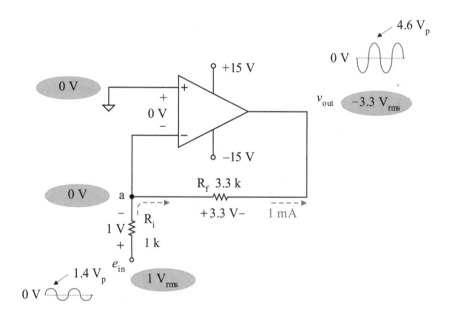

Figure 12-41 Practice circuit: inverting voltage amplifier

Op Amp Voltage Buffer Amplifier Circuit

The **voltage buffer amplifier** is a circuit designed to buffer the load resistance from the resistance of a circuit supplying the signal. If a supply resistance is much greater than the load resistance and directly connected, the load voltage is greatly reduced. The voltage buffer amplifier is placed between the supplying circuit and the load *to prevent this loss*.

Example 12-18

Find the output voltage of the circuit of Figure 12-42(a). Then insert an op amp voltage buffer amplifier and find the output voltage.

Solution

This is a series circuit with a simple voltage divider. There is a tremendous loss of signal at the output; the input signal is reduced from 2 V_{pp} to 0.2 V_{pp} at the output. The signal is **attenuated** (reduced) by a factor of 10.

$$V_{out} = \left(\frac{1\ k\Omega}{10\ k\Omega} \right) 2\ V_{pp} = 0.2\ V_{pp}$$

An op amp voltage buffer amplifier that prevents voltage loss is shown in Figure 12-42(b).

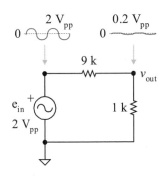

Figure 12-42(a)
Practice circuit–
Lossy circuit

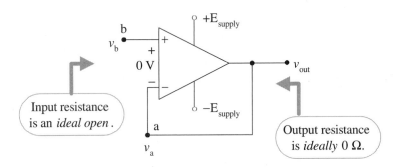

Figure 12-42(b) Op amp voltage buffer amplifier

The voltage dropped across the op amp input terminals is approximately 0 V. Starting with the node voltage at node b, 0 V is dropped to node a. Therefore the voltage at node a is the same as the voltage at node b. Since node a is connected by conductors to the output node, the output voltage must be the same as the signal at node a. Thus,

$$v_{out} = v_a = v_b$$

Since the output voltage is equal to the input voltage, this circuit is also called a **voltage follower** (that is, the output voltage *follows* the input signal voltage). Two other major characteristics of this circuit are:

- Its input resistance is *very, very high*, ideally infinite but more realistically in the gigaohm range.

- Its output resistance is very low, ideally 0 Ω. That is, the buffer amplifier output appears to have a very, very low resistance compared to the load resistance.

The circuit of Figure 12-42(c) removes the loss of the series circuit by inserting an op amp voltage buffer amplifier between the input signal source and the load.

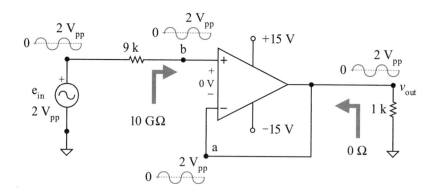

Figure 12-42(c) Example problem–op amp buffer amplifier

In this example, the input resistance to the buffer amplifier is given as 10 GΩ. The 2 V_{pp} input signal drives the 9 kΩ resistor in series with buffer amplifier input resistance of 10 GΩ. Using the voltage divider rule, the voltage at node *b* is virtually still 2 V_{pp}.

$$V_b = \left(\frac{10 \text{ G}\Omega}{9 \text{ k}\Omega + 10 \text{ G}\Omega} \right) 2 \text{ V}_{pp} = 1.999998 \text{ V}_{pp} \approx 2 \text{ V}_{pp}$$

Starting at node b with 2 V_{pp} and dropping 0 V across the op amp input terminals, the voltage at node a is

$$V_a = 2\ V_{pp} - 0\ V \approx 2\ V_{pp}$$

Now, follow this 2 V_{pp} signal along the feedback wire to the output. All the connecting wire between node a, the op amp output terminal, and the output node V_{out} are all the same voltage node and therefore at the same voltage potential. Therefore,

$$V_{out} = V_a = 2\ V_{pp}$$

All 2 V_{pp} have been delivered to the output without any loss or degradation of the originally supplied signal. The buffer amplifier worked; it isolated the supply resistance from the load resistance and maintained the original signal.

Practice

For the circuit of Figure 12-42(c), find the output signal peak voltage, rms voltage, rms current, and average output power.

Answer: 1 V_p, 0.707 V_{rms}, 0.707 mA$_{rms}$, 0.5 mW

Sine Wave Driving an Op Amp Comparator

Let's apply a sine wave to an open loop op amp comparator. Recall the ideal nature of the op amp. If the difference of the input is greater than 0, ideally the op amp slams to its upper rail voltage. If the difference of the op amp input is negative, ideally the op amp slams to its lower rail voltage.

Example 12-19

A 2 V_{pp} sine wave is applied to a noninverting comparator as shown in Figure 12-43(a). Find and draw the output waveform. Assume that the op amp needs 2 V of headroom to operate from both the positive and negative supply voltages.

Solution

With a positive supply voltage of 15 V, the upper rail voltage of the op amp is about +13 V due to its required overhead voltage of 2 V.

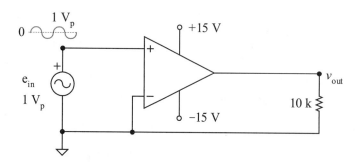

Figure 12-43(a) Example circuit–no feedback signal

$$V_{\text{upper rail}} = 15 \text{ V} - 2 \text{ V} = 13 \text{ V}$$

With a negative supply voltage of −15 V, the lower rail voltage of the op amp is about −13 V due to its required overhead voltage of 2 V.

$$V_{\text{lower rail}} = -15 \text{ V} + 2 \text{ V} = -13 \text{ V}$$

Assuming a typical op amp gain of 10^4, the positive half cycle slams the positive rail with an input voltage of only +1.30 mV and the negative half cycle slams the lower rail with an input voltage of only −1.30 mV. Note the op amp voltage gain is frequency dependent above its **critical frequency**. As the signal frequency increases above its critical frequency, the ac signal voltage gain A decreases. This is covered thoroughly in later texts in this series.

$$V_{+\text{ input}} = \frac{13 \text{ V}}{10^4} = 1.30 \text{ mV}$$

$$V_{-\text{ input}} = \frac{-13 \text{ V}}{10^4} = -1.30 \text{ mV}$$

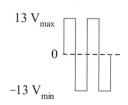

Figure 12-43(b)
Example output voltage

Very little input voltage is needed to slam either the positive or negative rail. Ideally, the comparator op amp output voltage is at either the upper rail with a positive input or the lower rail with a negative input voltage. For this circuit, a sinusoidal input produces a square wave output voltage as shown in Figure 12-43(b).

Practice

The circuit of Figure 12-43(a) is converted to an inverting comparator by switching the signal and common on its input pins. Find and draw the output voltage waveform of the inverting voltage amplifier.

Answer: See Figure 12-43(c).

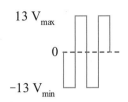

Figure 12-43(c)
Practice output voltage

12.7 Superposition

In a linear circuit with multiple supplies or signals (ac, dc, or ac-dc), the effect of each supply or signal source can be analyzed separately and then the results superimposed (added together).

For example, if a resistor has a 1 mA current due to one supply and a 2 mA current from another supply, then these currents can be superimposed (added together) for a total 3 mA. See Figure 12-44. Superposition applies to voltages and it applies to currents; that is, currents superimpose and voltages superimpose.

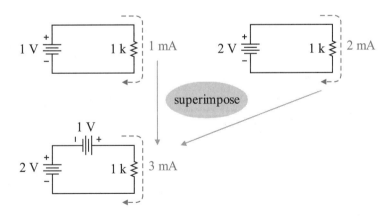

Figure 12-44 Superposition of two currents

When dealing with multiple supplies, the effect of each supply can be analyzed separately and the resulting individual currents or voltages superimposed to obtain the net result. If we are looking at the effect of one supply at a time, what do we do with the other supplies in the circuit? For example, in the circuit of Figure 12-44 when using the 2 V supply, we assumed that the 1 V supply allowed a current through it to the load.

Zeroed Voltage and Current Supply Models

When looking at the contributions of a supply, the other supplies in the circuit must be zeroed to eliminate their contributions. That is, unused current supplies are turned down to 0 A and unused voltage supplies are turned down to 0 V. Recall that a component that has 0 A is equivalent to an *open*, and a component that has 0 V is equivalent to a *short*.

The ideal internal resistance of an ac or dc voltage supply is 0 Ω, that is, ideally a short or an ideal conductor. Dialing a voltage supply to 0 V is equivalent to replacing it with an *ideal conductor* or a *short* as shown in Figure 12-45.

Figure 12-45 Zeroed voltage supply ideally modeled as a short

The ideal internal resistance of an ac or dc *current supply* is ∞ Ω, that is, ideally an *open*. Dialing a current supply to 0 A is ideally equivalent to replacing it with an *open* as shown in Figure 12-46.

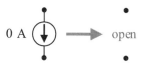

Figure 12-46 Zeroed current supply ideally modeled as an open

The next example demonstrates the superposition of two dc supplies.

Example 12-20

For the circuit of Figure 12-47(a), find the voltage supply's contribution to the load current (draw the equivalent circuit). Then find the current supply's contribution to the load current (draw the equivalent circuit). Superimpose to find the total load current.

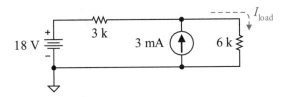

Figure 12-47(a) Example circuit with multiple supplies

Solution

Find the contribution from the 18 V voltage supply. Redraw the circuit with the *current supply zeroed* (equivalent to an *open*) as shown in Figure 12-47(b). The load current has been labeled as $I_{load\ 1}$ to identify it as the current from the voltage supply.

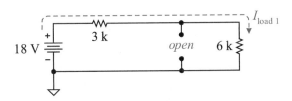

Figure 12-47(b) Example circuit to find contribution of 18 V supply

$$I_{load\ 1} = \frac{18\ V}{9\ k\Omega} = 2\ mA$$

(Ohm's law)

Now, let's look at the contribution of the 3 mA current supply to the load current. Draw the circuit with the current supply in the circuit but the *voltage supply zeroed* (replaced by a *short*) as shown in Figure 12-47(c). This is a parallel circuit, and the supply current splits between a 3 kΩ branch and a 6 kΩ branch.

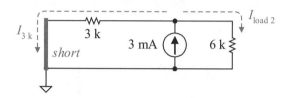

Figure 12-47(c) Example circuit to find contribution of 3 mA supply

$$I_{\text{load } 2} = \left(\frac{3\ k\Omega}{9\ k\Omega} \right) 3\ mA = 1\ mA \qquad \text{(two-branch CDR)}$$

Superimpose the contributions. Both contributions are in the same direction as the original reference current I_{load}, therefore they are positive (+) valued.

$$I_{\text{load}} = I_{\text{load } 1} + I_{\text{load } 2}$$

$$I_{\text{load}} = 2\ mA + 1\ mA = 3\ mA$$

Practice

Reverse the current supply of Figure 12-47(a) and find its contribution to the load current. Then superimpose the contributing load currents to find the new load current.

Answer: $I_{\text{load 2}}$ (−1 mA, direction opposite the reference), I_{load} (1 mA)

The next example demonstrates superimposing two ac supplies with *identical* frequencies, wave shapes, and starting times. *The resulting frequency, wave shape, and starting times are the same in a resistive circuit.* As will be discussed later in this chapter, ac signal superposition is a point-by-point summation of the signals, and the output is dramatically different if there is any difference in frequency, wave shape, or signal starting times. Thus, this example is a very special case.

Example 12-21

Assume the supplies have *identically* the same frequency, same wave shape, and same starting time. (a) For the circuit of Figure 12-48(a), find the voltage supply's contribution to the load voltage (draw the equivalent circuit). (b) Then find the current supply's contribution to the load voltage (draw the equivalent circuit). (c) Then superimpose to find the total load voltage.

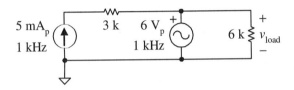

Figure 12-48(a) Example circuit with multiple ac supplies

Solution

The two contribution circuits are (a) the current supply contribution with the voltage supply zeroed and (b) the voltage supply contribution with the current supply zeroed. See Figure 12-48(b) and Figure 12-48(c), respectively.

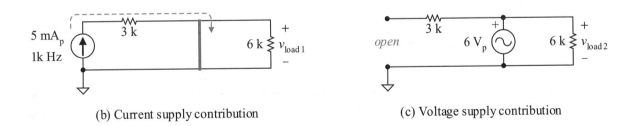

(b) Current supply contribution   (c) Voltage supply contribution

Figure 12-48(b,c) Example circuit models for superposition

    a.   From the modeled circuit of Figure 12-48(b), it is apparent that the voltage supply produces a short around the 6 kΩ resistor. None of the current supply's current flows through the load.

$$V_{\text{load 1}} = 0 \ V_p$$

b.   From the modeled circuit of Figure 12-48(c), it is apparent that the voltage supply is directly across the load.

$$V_{\text{load 2}} = 6 \ V_p$$

c.   Superimposing the contributing voltages,

$$v_{\text{load}} = v_{\text{load 1}} + v_{\text{load 2}}$$

$$V_{\text{load}} = 0 \ V_p + 6 \ V_p = 6 \ V_p$$

Practice

For the circuit of Figure 12-49, find the load voltage contribution v_{load} due to the current supply and due to the voltage supply. Then superimpose the results to find the total load voltage. Assume identical supply frequencies, starting times, and waveshapes.

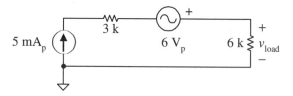

Figure 12-49 Practice circuit

Answer: Current supply (30 V_p), voltage supply (0 V), total (30 V_p)

The next circuit mixes ac and dc supplies together. But again, we can analyze the circuit with the dc and the ac supplies independently and then superimpose the results.

Example 12-22

For the circuit of Figure 12-50(a), find the dc voltage supply's contribution to the load current (draw the equivalent circuit). Then find the ac current supply's contribution to the load voltage (draw the equivalent circuit). Then superimpose to find the total load voltage. Sketch the total voltage waveform. The ac current supply is a sine wave with a frequency of 1 kHz.

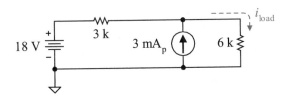

Figure 12-50(a) Example circuit with a dc supply and an ac supply

Solution

The two contribution circuits are (a) the dc voltage supply contribution with the *ac current supply zeroed* (replaced with an *open*) and (b) the ac current supply contribution with the *dc voltage supply zeroed* (replaced with a *short*). See Figure 12-50(b) and Figure 12-50(c), respectively. Note that uppercase is used for dc variables and lowercase is used for ac variables to more readily distinguish dc and ac variables.

(b)   (c)

Figure 12-50(b,c) Example circuit models for superposition

(a) The 18 V_{dc} supply produces a dc $I_{load\ 1} = 2$ mA$_{dc}$.

(b) The 3 mA$_p$ supply produces an ac $i_{load\ 2} = $ of 1 mA$_p$.

You cannot directly add the dc and ac currents. Rather you must think of this as an ac sine wave riding around a dc average current. Combining the ac and dc load currents in graphical form, the total waveform can be viewed. See Figure 12-50(d). Note the frequency of 1 kHz produces a period of 1 ms.

$$I_{\text{max total signal}} = 2 \text{ mA}_{\text{dc}} + 1 \text{ mA}_{\text{p}} = 3 \text{ mA}_{\text{p}}$$

$$I_{\text{min total signal}} = 2 \text{ mA}_{\text{dc}} - 1 \text{ mA}_{\text{min}} = 1 \text{ mA}_{\text{min}}$$

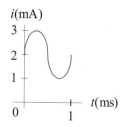

Figure 12-50(d)
Example total waveform

Practice

Find and draw the Thévenin model of the circuit left of the load in Figure 12-50(a). This model would include a dc E_{Th} in series with an ac e_{Th}.

Answer: Figure 12-51 represents the Thévenin model of Figure 12-50(a). Various loads can be attached to this model and load values derived. The dc circuit values and ac circuit values can be solved separately and then the results combined as a waveform.

Inverting Summer Amplifier–Offset Added to AC

The inverting op amp summer amplifier circuit of Figure 12-52 sums and amplifies the input signals. The input signals can be dc (labeled E_{in}) and/or ac (labeled e_{in}). That is, the input signals could be ac signals, dc signals, or ac riding on a dc offset.

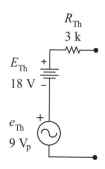

Figure 12-51
Practice solution–
Thévenin model

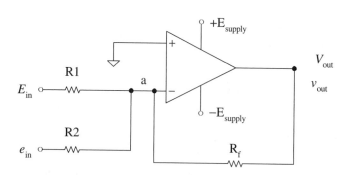

Figure 12-52 Inverting summer amplifier

The circuit shown in Figure 12-52 shows one dc input E_{in} and one ac input e_{in}. *To ease in the visualization of the variables, uppercase letters are used for dc variables and lowercase letters are used for ac variables.* A key *ideal* assumption is that the input currents to the op amp are 0 A. Another key *ideal* assumption for this negative feedback circuit is that the op amp input voltage between the inverting and noninverting terminals is 0 V. The current and voltage references are shown in the circuit of Figure 12-53 along with the ideal assumptions. The figure has been slightly modified to accommodate labeling but it is the same circuit.

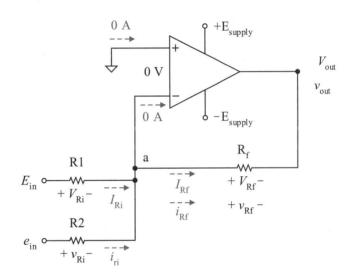

Figure 12-53 Inverting summer amplifier with references and ideal assumptions.

A few basic observations can be made. The voltage at node *a* must be 0 V since the noninverting op amp input terminal is at ground potential and the drop across the op amp is ideally 0 V. This is called a **virtual ground** or **virtual common** since it is at 0 V potential but not directly tied to common.

$$V_a = 0 \text{ V}$$

Virtual common

Since voltage at node *a* is 0 V, the dc voltage drop across R1 is E_{in} and the ac voltage drop across R2 is e_{in}. Note that uppercase is used for dc variables and lowercase is used for ac variables to more readily distinguish dc and ac variables.

$$V_{Ri} = E_{in}$$

$$v_{ri} = e_{in}$$

Apply Ohm's law to the input resistors to find the input currents.

$$I_{R1} = \frac{V_{Ri}}{R1} = \frac{E_{in}}{R1}$$

$$i_{R2} = \frac{v_{Ri}}{R2} = \frac{e_{in}}{R2}$$

Apply Kirchhoff's current law at node a. The dc current I_{R1} flowing through R1 into node a must be the same as dc current I_{Rf} flowing out of node a through R_f, since the current flowing out of node a into the op amp is assumed to be ideally 0 A. This is also true for the ac currents.

$$I_{Rf} = I_{R1}$$

$$i_{Rf} = i_{R2}$$

Apply Ohm's law to R_f for each signal to find the voltage drop across the feedback resistor R_f.

$$V_{Rf} = R_f \times I_{Rf}$$
$$v_{Rf} = R_f \times i_{Rf}$$

The respective dc and ac voltages are found by starting at the node a voltage potential of 0 V and falling by the voltage drop across R_f.

$$V_{out} = 0 \text{ V} - V_{Rf} = -V_{Rf}$$

$$v_{out} = 0 \text{ V} - v_{Rf} = -v_{Rf}$$

The ac and dc signal currents are initially superimposed at node a. However, we can also consider the ac and the dc signals independently to find the dc and the ac output voltages separately, and superimpose them.

Example 12-23

Find the dc and ac (sine wave) output voltages in the circuit of Figure 12-54(a). Sketch the output voltage ac only waveform and the total waveform. Find the average power delivered to the load.

Solution

Virtual common

First, perform dc circuit analysis.

$$V_a = 0 \text{ V}$$

$$V_{R1} = E_{in} = 1 \ V_{dc}$$

$$I_{R1} = \frac{V_{R1}}{R1} = \frac{1 \ V_{dc}}{2 \ k\Omega} = 0.5 \ mA_{dc}$$

$$I_{Rf} = I_{R1} = 0.5 \ mA_{dc}$$

$$V_{Rf} = R_f \times I_{Rf} = 10 \ k\Omega \times 0.5 \ mA_{dc} = 5 \ V_{dc}$$

$$V_{out} = 0 \ V - V_{Rf} = -V_{Rf} = -5 \ V_{dc}$$

Inverted output

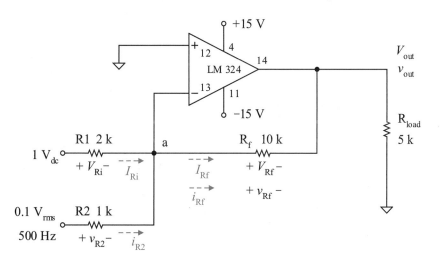

Figure 12-54(a) Example circuit

Next, perform ac circuit analysis. Note, lowercase is used for all ac variables for clarity.

$$v_a = 0 \ V$$

Virtual common

$$v_{R2} = e_{in} = 0.1 \ V_{rms}$$

$$i_{R2} = \frac{v_{R2}}{R2} = \frac{0.1 \ V_{rms}}{1 \ k\Omega} = 0.1 \ mA_{rms}$$

$$i_{Rf} = i_{R2} = 0.1 \ mA_{rms}$$

Inverted output

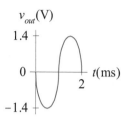

Figure 12-54(b)
Example ac waveform

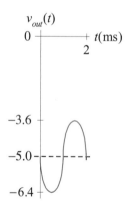

Figure 12-54(c)
Example total waveform

$$v_{Rf} = R_f \times i_{Rf} = 10 \text{ k}\Omega \times 0.1 \text{ mA}_{rms} = 1 \text{ V}_{rms}$$

$$v_{out} = 0 \text{ V} - v_{Rf} = -v_{Rf} = 1 \text{ V}_{rms} \qquad \text{(inverted)}$$

The rms value is always a positive value. The negative sign indicates that the signal at the output is inverted.

The period of the waveform is

$$T = \frac{1}{f} = \frac{1}{500 \text{ Hz}} = 2 \text{ ms}$$

To sketch the total waveform, the ac rms voltage must be converted to its peak value.

$$v_{p\,out} = 1 \text{ V}_{rms} \times \sqrt{2} = 1.414 \text{ V}_p$$

A sketch of one cycle of the ac only output voltage signal is shown in Figure 12-54(b). The total output voltage, that is the ac signal with a –5 V_{dc} offset, is shown in Figure 12-54(c).

Check that the output voltage is within the op amps rail voltages. The LM324 would produce an upper rail of about 13.5 V and a lower rail of –14.2 V. The total signal operates well within this range and is not clipped.

Find the average power delivered to the load. The power that the dc signal delivers to the load is

$$P_{dc\,load} = \frac{\left(-5 \text{ V}_{dc}\right)^2}{5 \text{ k}\Omega} = 5 \text{ mW}$$

The average power that the ac signal delivers to the load is

$$P_{average\,ac\,load} = \frac{\left(1 \text{ V}_{rms}\right)^2}{5 \text{ k}\Omega} = 0.2 \text{ mW}$$

The total average power dissipated by the load is the sum of the dc power and the ac power.

$$P_{average\,load} = 5 \text{ mW} + 0.2 \text{ mW} = 5.2 \text{ mW}$$

Another approach is to find the total average power that the signal delivers to the load by finding the total signal rms voltage and use that voltage to find the average power.

$$V_{\text{rms total signal}} = \sqrt{\left(-5\ V_{\text{dc}}\right)^2 + \left(1\ V_{\text{rms ac}}\right)^2} = 5.099\ V_{\text{rms}}$$

$$P_{\text{average load}} = \frac{\left(5.099\ V_{\text{rms}}\right)^2}{5\ k\Omega} = 5.2\ mW$$

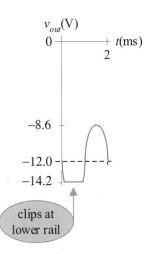

$v_{out}(V)$

clips at
lower rail

Figure 12-54(d)
Practice
total waveform

Practice

The feedback resistor of the summer circuit of Figure 12-54(a) is changed to 24 kΩ. Sketch the total output voltage waveform.

Answer: See Figure 12-54(d). The signal clips at the lower rail.

AC Signal Superposition

Superposition of individual ac signals is a bit more interesting than the simplistic ideal case given in a previous example. The circuit of Figure 12-55(a) is an inverting summing amplifier with three musical notes of middle E that have identical frequencies and identical starting times.

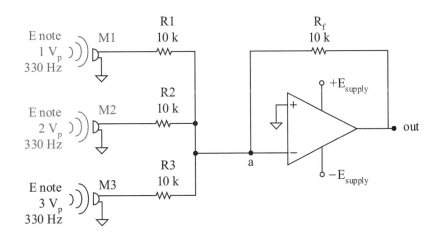

Figure 12-55(a) Inverting summing amplifier with same input frequency

Until now it has been assumed that all individual ac supplies in a circuit start at the same time and have exactly the same frequency. This is somewhat unrealistic. The same single ac supply fed to a resistive circuit does start at the same time and has the same frequency throughout the circuit, but it is unrealistic to assume that separate individual supplies would typically start at the same time and have exactly the same frequency. If the signals start at exactly the same time with identical frequencies (musical note E), then the individual outputs and the total output are as shown in Figures 12-55(b) and (c).

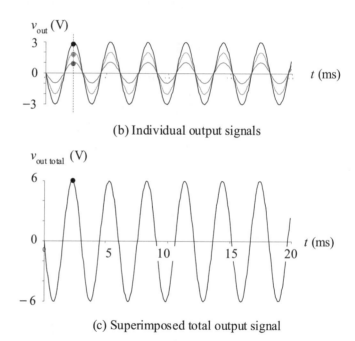

(b) Individual output signals

(c) Superimposed total output signal

Figure 12-55(b,c) Identical frequencies starting at identical times

This is a very idealistic case. This total output signal is only possible if these input signals are all coming from the same source and the source is exactly the same distance from each microphone. The total signal is the superposition of the three individual signals. Each total signal data point is the sum of the three individual data points. For example, each signal reaches its peak value at the same time. The sum of the three peak values with amplitudes of 1 V_p, 2 V_p, and 3 V_p is a sum of 6 V_p. This is a point-by-point summation. Spreadsheets and mathematics packages are designed to easily perform such signal summation tasks. These signal waveform sketches were created using a spreadsheet. The graphics production is not perfect but does give a reasonable presentation of the general wave shape especially if viewed in the spreadsheet software. Note that the output frequency is the same as the input frequencies (musical note E). All the signals are inverted since this is an inverting voltage amplifier.

Figures 12-55(d) and (e) demonstrate changing the starting times of the individual signals; that is, the time at which the signals start at 0 V. The frequencies are still identical, but signal M2 and signal M3 have been started later.

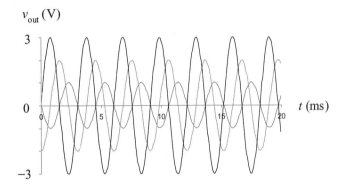

(d) Individual output signals

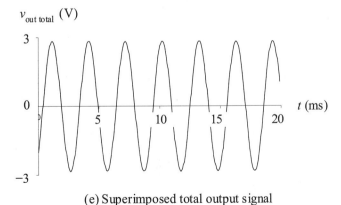

(e) Superimposed total output signal

Figure 12-55(d,e) Identical frequencies starting at *different times*

Notice the total output signal still has that same identical frequency but the amplitude is now significantly reduced and the starting time is not at 0 ms. The starting times of the three signals significantly affect the total signal amplitude; changing the starting time of one of the signals changes the total output signal amplitude and its starting time. The maximum possible output signal is the one in which all the signals start at the same time as shown in Figure 12-55(c).

Example 12-24

Find the rms value of each of the individual sine waves of Figure 12-55(b). Then find the rms value of the total signal of Figure 12-55(c). Is the sum of the individual signal rms values equal to the rms of the total signal? Next find the rms value of the total output signal of Figure 12-55(e). Does the sum of the individual signal rms values equal the rms of the total signal of Figure 12-55(e)? What conclusion must you draw?

Solution

The rms values of the individual signals are

$$V_{\text{rms M1}} = \frac{1 \text{ V}_{\text{p}}}{\sqrt{2}} = 0.707 \text{ V}_{\text{rms}}$$

$$V_{\text{rms M2}} = \frac{2 \text{ V}_{\text{p}}}{\sqrt{2}} = 1.414 \text{ V}_{\text{rms}}$$

$$V_{\text{rms M3}} = \frac{3 \text{ V}_{\text{p}}}{\sqrt{2}} = 2.121 \text{ V}_{\text{rms}}$$

The rms value of the total signal of Figure 12-55(c) is

$$V_{\text{rms total signal}} = \frac{6 \text{ V}_{\text{p}}}{\sqrt{2}} = 4.242 \text{ V}_{\text{rms}}$$

while the sum of the individual rms values is

$$V_{\text{rms ideal}} = 0.707 \text{ V}_{\text{rms}} + 1.414 \text{ V}_{\text{rms}} + 2.121 \text{ V}_{\text{rms}} = 4.242 \text{ V}_{\text{rms}}$$

These values are the same. This confirms your previous work with the ideal case of superimposing signals with identical frequencies and identical starting times.

The rms value of the total signal of Figure 12-55(e) gives a different result.

$$V_{\text{rms different}} = \frac{3\,V_p}{\sqrt{2}} = 2.121\ V_{\text{rms}}$$

Warning, in general, you cannot simply superimpose ac signals. AC signals are superimposed point-by-point. Only in the very ideal case where the frequencies are identical and the starting times are identical can you do simple superposition of ac sources.

Practice

If you have trigonometry knowledge, use a spreadsheet and generate the signal waveforms of Figures 12-55(b,c). Warning, the sine wave arguments must be in terms of radians and not degrees since mathematical operations are performed within the arguments.

The circuit of Figure 12-56(a) is the same inverting summer amplifier but with three different frequency inputs: in this case the musical chord of notes middle C-E-G of 262 Hz, 330 Hz, and 392 Hz, respectively.

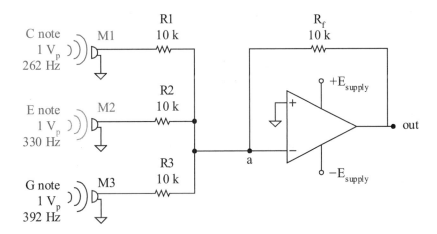

Figure 12-56(a) Inverting summing amplifier with chord input C-E-G

Figure 12-56(b,c) shows the individual output signals and their resulting superposition into the total signal. The frequencies shown have been rounded up to whole numbers. Musical notes are standard frequencies that are specific into the decimal range.

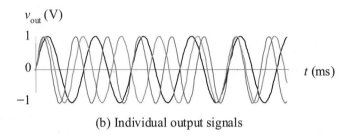

(b) Individual output signals

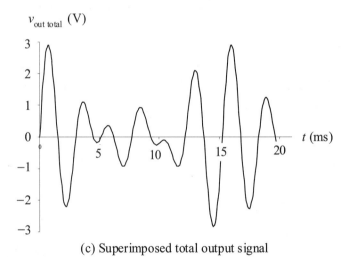

(c) Superimposed total output signal

Figure 12-56(b,c) Chord C-E-G with signals starting at the same time

Any change in signal amplitude or relative signal starting times results in a changing of the total signal wave shape.

12.8 AC and DC Measurements

Measurements of ac and dc voltages and currents can be made with the digital multimeter. In the dc position, the digital multimeter measures the dc or average value. In the ac position, the digital multimeter measures the rms value. There are three types of *rms meters*. Be sure you understand which type you are using.

Sine wave rms meter
The simplest ac meter assumes that all signals are pure ac sine waves. It filters out any dc and then detects the signal's peak voltage and multiplies the measured peak by 0.707 to obtain its rms reading. This is appropriately correct for commercial electrical voltages, which are ac sine waves.

True rms meters of the ac signal
This type of meter measures the true rms value ac voltages. Any dc signal is filtered out and only the pure ac signal remains. Then, regardless of the ac waveform, the true rms value of the pure ac signal is calculated and displayed.

True rms meters of the total signal
This type of meter measures the true rms value of the total signal, including ac signals with dc offsets. The value of this measured signal is equivalent to

$$V_{\text{rms total signal}} = \sqrt{V_{\text{rms of ac}}^2 + V_{\text{dc}}^2}$$

where

$V_{\text{rms of ac}}$ is the rms value of the ac only signal
V_{dc} is the dc value of the signal

Example 12-25

A 4 V_{pp} triangle wave is measured with the above three types of ac meters. A dc meter measures 5 V_{dc}. What would each type of meter read?

Solution

Sine wave rms meter

$$V_{\text{rms}} = \frac{\dfrac{4\ V_{\text{PP}}}{2}}{\sqrt{2}} = 1.414\ V \qquad \text{(dc filtered out)}$$

True rms ac meter

$$V_{rms} = \frac{\frac{4\ V_{PP}}{2}}{\sqrt{3}} = 1.155\ V \qquad\qquad \text{(dc filtered out)}$$

True rms meter of the total signal

$$V_{rms\ total\ signal} = \sqrt{V_{rms\ of\ ac}^2 + V_{dc}^2}$$

$$V_{rms\ total\ signal} = \sqrt{(1.155\ V)^2 + (5\ V)^2} = 5.132\ V_{rms\ total}$$

The type of meter obviously makes a difference. Know your test equipment.

Practice

A 4 V_{pp} square wave is measured with the above three types of ac meters. A dc meter measures 5 V_{dc}. What would each type of meter read?

Answer: 1.414 V_{rms}, 2 V_{rms}, 5.385 $V_{rms\ total}$

Summary

Voltage and current can be a fixed dc value or can vary with time. Typical repetitive ac waveforms include the sine wave, the triangle wave, and the square wave. Key current and voltage waveform parameters for an alternating current (ac) waveform include its maximum (or peak), minimum, and peak-to-peak values. The time for one complete cycle of the waveform is called its period. The reciprocal of the period is the frequency of the signal measured in hertz (or cycles per second).

The average value of a pure ac signal is 0 V or 0 A. If the signal has a dc offset, the average value of the total signal is its dc offset voltage or current.

The rms (root-mean-square) value of a varying signal is equivalent to the dc voltage or current required to produce the same power to the circuit. The rms value of a signal is dependent on its waveshape. The rms value for a pure sine wave is 0.707×V_p; for a triangle wave it is 0.577×V_p; and for a square wave it is 1.000×V_p. A signal can be separated into its dc and ac parts. The rms value of the total signal is found from its rms ac signal value and its dc value by

$$V_{\text{rms total signal}} = \sqrt{V^2_{\text{rms of ac}} + V^2_{\text{dc}}}$$

In resistive circuits the *average power* generated or dissipated is calculated using V_{rms} and I_{rms} values.

$$P_{\text{average}} = I_{\text{rms}} \, V_{\text{rms}}$$

Supplies can be either voltage supplies or current supplies. A multiple supply circuit can be analyzed as single supplied equivalent circuits and the resulting currents and voltages superimposed (or added up). Supplies exhibit a natural resistance: voltage supplies have very low internal resistance (ideally 0 Ω, a *short*) and current supplies have a very high internal resistance (ideally ∞ Ω, an *open*). Superposition is especially useful in analyzing a circuit that contains dc and ac signals. The dc circuit can be analyzed separately from the ac circuit, and the results superimposed to find the total (composite) signal.

AC circuit analysis techniques are the same as dc techniques in resistive circuits. All of the basic laws and circuit analysis techniques are still valid. These circuit techniques also apply to circuits where the analysis is predominantly resistive circuit analysis, such as an op amp amplifier.

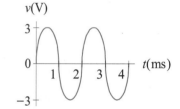

Figure 12-57

Problems

Waveform Terminology

12-1 For the waveform of Figure 12-57, find its peak, minimum, and peak-to-peak voltages. Find its period and frequency. How many cycles are displayed?

12-2 For the waveform of Figure 12-58, find its peak, minimum, and peak-to-peak currents. Find its period and frequency. How many cycles are displayed?

12-3 Find the periods of the following signal frequencies.
    a. 100 Hz
    b. 5 MHz

12-4 Find the periods of the following signal frequencies.
    a. 1 kHz
    b. 2 GHz

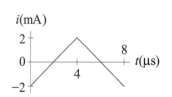

Figure 12-58

12-5 Find the frequencies of the following signal periods.
    a. 10 ms
    b. 33 ns

12-6   Find the frequencies of the following signal periods.
  a. 100 μs
  b. 500 ps

Average (DC) Value

12-7   For the signal of Figure 12-57:
  a. What is its average (dc) value?
  b. A 3 V dc offset is added to the signal; find the signal's peak, minimum, and peak-to-peak values. Sketch the new waveform.

12-8   For the signal of Figure 12-58:
  a. What is its average (dc) value?
  b. A 5 mA dc offset is added to the signal; find the signal's peak, minimum, and peak-to-peak values. Sketch the new waveform.

12-9   For the signal of Figure 12-59, find its peak, minimum, peak-to-peak, and average values.

12-10   For the signal of Figure 12-60, find its peak, minimum, peak-to-peak, and average values.

AC Voltage Supplies

12-11   For the circuit of Figure 12-61, find and sketch the output voltage for the potentiometer wiper arm at its
  a. Top
  b. Middle
  c. Bottom

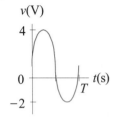

Figure 12-59

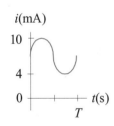

Figure 12-60

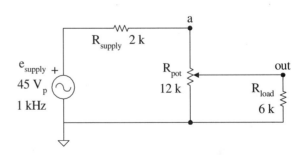

Figure 12-61

12-12 For the circuit of Figure 12-61, find and sketch the voltage at node *a* for the potentiometer wiper arm at its

   a. Top

   b. Middle

   c. Bottom

Average Power–RMS Values

12-13 For the circuit of Figure 12-61, find the average load power for the potentiometer wiper arm at its

   a. Top

   b. Middle

   c. Bottom

12-14 For the circuit of Figure 12-61, find the average power generated by the supply with the potentiometer wiper arm at its

   a. Top

   b. Middle

   c. Bottom

Electronic Applications

12-15 For the circuit of Figure 12-62:

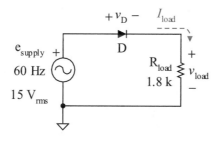

Figure 12-62

   a. Find the peak supply voltage.

   b. Find the peak load voltage.

   c. Sketch the load voltage waveform.

   d. Find the peak load current.

   e. Sketch the load current waveform.

   f. What is the maximum reverse voltage experienced by the diode?

12-16 For the circuit of Figure 12-63:

   a. Find the voltage at node *a*.

   b. Find the output voltage v_{out}.

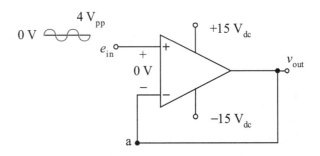

Figure 12-63

12-17 For the circuit of Figure 12-64:

   a. Is the circuit using negative or positive feedback?

   b. Is this a noninverting or inverting voltage amplifier?

   c. Find the voltage at node *a*.

   d. Find the voltage drop across R_i.

   e. Find the current through R_f.

   f. Find the voltage drop across R_f.

   g. Find the output voltage. Is the output voltage an inverted or a noninverted waveform?

   h. Find the average power dissipated by the feedback resistor R_f.

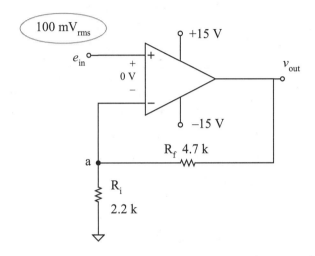

Figure 12-64

12-18 For the circuit of Figure 12-65:

   a. Is the circuit using negative or positive feedback?

   b. Is this a noninverting or inverting voltage amplifier?

   c. Find the voltage at node a.

   d. Find the voltage drop across R_i.

   e. Find the current through R_f.

   f. Find the voltage drop across R_f.

   g. Find the output voltage. Is the output voltage an inverted or a noninverted waveform?

   h. Find the average power dissipated by the feedback resistor R_f.

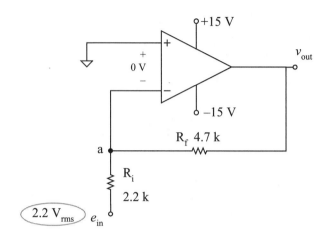

Figure 12-65

Superposition

12-19 For the dc circuit of Figure 12-66(a):

    a. Find the load voltage due to the voltage supply.

    b. Find the load voltage due to the current supply.

    c. Find the total load voltage.

    d. Find the total load current.

    e. Find the total load power.

12-20 For the dc circuit of Figure 12-66(b):

    a. Find the load current due to the voltage supply.

    b. Find the load current due to the current supply.

    c. Find the total load current.

    d. Find the total load voltage.

    e. Find the total load power.

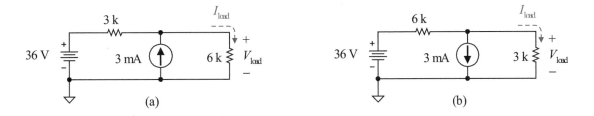

Figure 12-66

12-21 For the dc circuit of Figure 12-67:

    a. Find the load current due to the voltage supply.

    b. Find the load current due to the 6 mA current supply.

    c. Find the load current due to the 12 mA current supply.

    d. Find the total load current.

    e. Find the total load voltage.

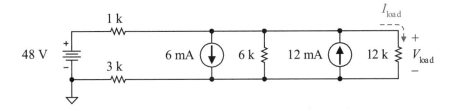

Figure 12-67

12-22 For the dc circuit of Figure 12-68(a):

    a. Find the load voltage due to the 36 V voltage supply.

    b. Find the load voltage due to the 12 V voltage supply.

    c. Find the total load voltage.

    d. Find the total load current.

12-23 For the dc circuit of Figure 12-68(b):

    a. Find the load current due to the current supply.

    b. Find the load current due to the voltage supply.

    c. Find the total load current.

    d. Find the total load voltage.

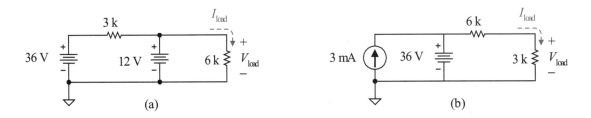

          (a)   (b)

Figure 12-68

12-24 Assume that the supplies in Figure 12-69(a) have identical frequencies, starting times, and waveshapes.

    a. Find the load voltage due to the voltage supply.

    b. Find the load voltage due to the current supply.

    c. Find the total load voltage.

    d. Find the total load current.

    e. Find the average load power.

    f. Find the peak load voltage.

    g. Find the peak load current.

12-25 Assume that the supplies in Figure 12-69(b) have identical frequencies, starting times, and waveshapes.

    a. Find the load current due to the voltage supply.

    b. Find the load current due to the current supply.

    c. Find the total load current.

    d. Find the total load voltage.

    e. Find the average load power.

    f. Find the peak load voltage.

    g. Find the peak load current.

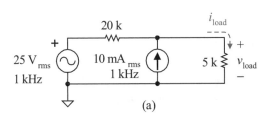

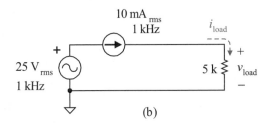

Figure 12-69

12-26 For the ac-dc circuit of Figure 12-70(a):
  a. Find the dc load voltage.
  b. Find the ac load voltage.
  c. Find the total signal rms voltage.
  d. Find the average load power.
  e. Sketch the total voltage waveform.

12-27 For the ac-dc circuit of Figure 12-70(b):
  a. Find the dc load current.
  b. Find the ac load current.
  c. Find the total signal rms current.
  d. Find the average load power.
  e. Sketch the total current waveform.

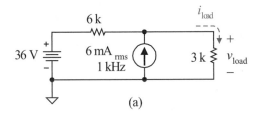

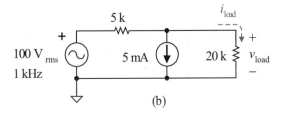

Figure 12-70

12-28 For the dc circuit of Figure 12-71, E1 is 150 mV$_{dc}$, E2 is 200 mV$_{dc}$, and E3 is 250 mV$_{dc}$.

   a. Find E1's output V_{OUT_1}.

   b. Find E2's output V_{OUT_2}.

   c. Find E3's output V_{OUT_3}.

   d. Find the net output V_{OUT} using superposition.

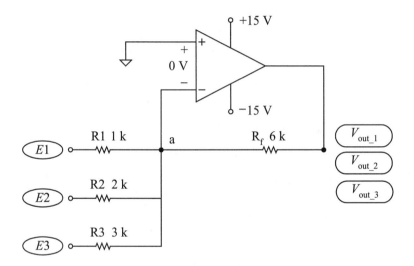

Figure 12-71

12-29 For the dc circuit of Figure 12-71, E1 is 150 mV$_{dc}$, E2 is −200 mV$_{dc}$, and E3 is 250 mV$_{dc}$.

   a. Find E1's output V_{OUT_1}.

   b. Find E2's output V_{OUT_2}.

   c. Find E3's output V_{OUT_3}.

   d. Find the net output V_{OUT} using superposition.

12-30 For the ac-dc circuit of Figure 12-72:

   a. Find the dc output voltage V_{OUT}.

  b. Find the ac output voltage v_{out}.

  c. Find the peak-to-peak voltage of the ac only output voltage.

  d. Sketch the ac only output voltage waveform.

  e. Sketch the total output voltage waveform.

  f. Find the rms value of the total output voltage waveform.

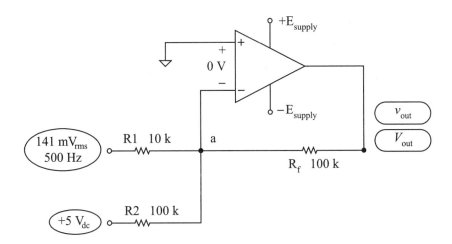

Figure 12-72

AC and DC Measurements

12-31 A 5 V_{pp} sine wave is measured with the three basic types of ac meters. A dc meter measures 5 V_{dc}. What would each type of meter read?

  a. A simple sine wave ac meter.

  b. A true rms of the ac signal.

  c. A true rms of the total signal meter.

12-32 A 5 V_{pp} triangle wave is measured with the three basic types of ac meters. A dc meter measures 10 V_{dc}. What would each type of meter read?

  a. A simple sine wave ac meter.

  b. A true rms of the ac signal.

  c. A true rms of the total signal meter.

13

Capacitance and Reactance

Introduction

Capacitance and capacitor components store energy in the form of electrical charge. Capacitance exhibits opposition to ac current. This opposition is called reactance, that is, the reaction to ac current. Capacitors in series, parallel, and series-parallel exhibit a total combined capacitance based upon their individual capacitance and the circuit configuration. The waveforms in *RC* circuits are primarily dependent upon the shape and frequency of the supplied signal.

Objectives

Upon completion of this chapter you will be able to:

- Describe the basic structure of capacitor components.

- Define and apply capacitance based upon its electrical parameters.

- Define and apply capacitance based upon its geometrical parameters.

- Describe the basic types of capacitors and their unique characteristics.

- Define and apply the capacitive reactance formula for ac signals.

- Find the total capacitance for series, parallel, and series-parallel capacitors.

13.1 Capacitors

Capacitance is created when two conductors are separated by an insulator. Figure 13-1 demonstrates several examples of natural capacitance: two conductor plates separated by an insulator, two wires separated by insulation, and an insulated wire lying on a chassis. The latter two are examples of stray (unwanted) capacitance. Capacitance is the quantitative measure of the ability to store charge.

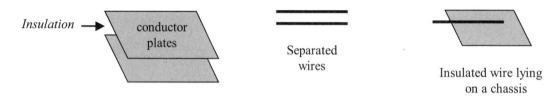

Figure 13-1 Capacitance examples

Capacitor

A **capacitor** is a component specifically designed to have two conductors separated by an insulator. Figure 13-2 demonstrates the fundamental structure of a capacitor.

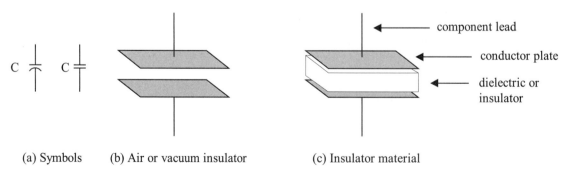

Figure 13-2 Basic capacitor structure

The capacitor symbol of Figure 13-2(a) has two parallel lines (the conductors) separated by a gap (the insulator). The insulating material separating the two plates is also called a **dielectric**. Examples of insulating materials include a vacuum, air, glass, mica, or rubber.

If a fixed voltage is applied to a capacitor, a charge builds up on each conductor plate as shown in Figure 13-3(a). An electric field is created between the two charges. An analogy that most people can relate to is a charged cloud during a thunderstorm as demonstrated in Figure 13-3(b). If someone with long hair is caught in a strong electric field just before a lighting strike, they might very well look like the caricature shown in Figure 13-3(c). If you are in an electrical storm and all of your hair starts to point upward towards a storm cloud, chances are that a lightning strike is about to occur very, very close to you. Your hair will become charged like the earth and be attracted to the cloud. The electric field becomes so intense between the earth and the cloud that the air (insulator) breaks down and becomes a conductor for charge to flow from cloud to earth or earth to cloud (that is, a lightning strike).

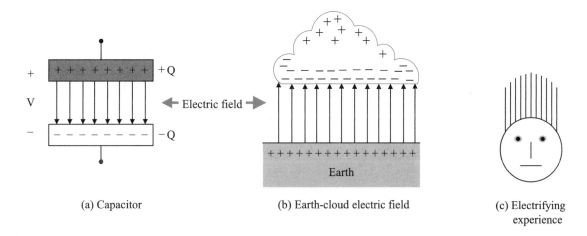

(a) Capacitor   (b) Earth-cloud electric field   (c) Electrifying experience

Figure 13-3 A charge difference creates an electric field

Capacitance Value

Referring to Figure 13-4, the capacitance value in terms of electrical units is defined to be

$$C = \frac{Q}{V} \qquad \text{(13-1)}$$

where
    C capacitance in units of farads (F)
    Q charge in units of coulombs (C)
    V voltage in units of volts (V)

Figure 13-4
Capacitance

Example 13-1

A capacitor stores 100 μC of charge on each plate with a voltage drop of 10 V_{dc}. Find the capacitance value of the capacitor.

Solution

Substitute into Equation 13-1.

$$C = \frac{Q}{V} = \frac{100 \ \mu C}{10 \ V} = 10 \ \mu F$$

Practice

A 470 μF capacitor drops 20 V_{dc}, find the charge stored on its plates.

Answer: 9.4 mC (rewrite Equation 13-1 to solve for Q)

Dielectric Constant

Inserting an insulator between the plates of a capacitor increases its capacity to store charge per unit volt dropped across the capacitor. Refer to Figure 13-5. In terms of basic geometry, capacitance can be defined as

$$C = \varepsilon \frac{A}{d} \qquad \text{(13-2)}$$

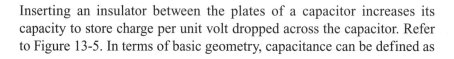

Figure 13-5
Capacitance

where

    C capacitance in units of farads (F)
    ε dielectric constant of insulator in farads per meter (F/m)
    A area of one of the plates in units of meters squared (m^2)
    d distance between the plates in units of meters (m)

Example 13-2

A capacitor has a vacuum as its insulator. Find the capacitance of a capacitor that has a plate 1 m by 1 m and a plate distance of 1 mm. The dielectric constant of a vacuum is 8.85×10^{-12} F/m.

Solution

Find the area of the plate.

$$A = (1 \text{ m})^2 = 1 \text{ m}^2$$

The distance between the plates in meters is

$$d = 1 \text{ mm} = 0.001 \text{ m}$$

Substitute the appropriate values into Equation 13-2.

$$C = \varepsilon \frac{A}{d} = \left(8.85 \times 10^{-12} \frac{F}{m} \right) \frac{1 \text{ m}^2}{0.001 \text{ m}} = 8.85 \text{ nF}$$

Frequently capacitance is expressed in microfarads, nanofarads, or picofarads.

Practice

The capacitor in the example uses mica as an insulator. Its dielectric constant is five times as great as that of a vacuum. Find its dielectric constant and its capacitance.

Answer: 44.25×10^{-12} F/m, 44.25 nF

Relative Dielectric Constant

The dielectric constant of a vacuum is the lowest dielectric constant possible. It is used as a standard (base reference), and other dielectric constants are expressed relative to that of a vacuum. Table 13-1 is a table of some common relative dielectric constants. Each dielectric constant is expressed as a multiple of the standard value of a vacuum of

$$\varepsilon_0 = 8.85 \times 10^{-12} \text{ F/m}$$

The dielectric constant of a dielectric can be expressed as

$$\varepsilon = \varepsilon_0 \, \varepsilon_r$$

Table 13-1

Relative
Dielectric
Constants

Dielectric	ε_r
Vacuum	1.0
Air	1.001
Teflon	2.0
Paper	2.5
Rubber	3.0
Mica	5.0
Bakelite	7.0
Glass	7.5

where

ε dielectric constant of insulator in farads per meter (F/m)

ε_0 dielectric constant of a vacuum

ε_r relative dielectric constant

Thus, Equation 13-2 can be written as

$$C = \varepsilon_o\, \varepsilon_r\, \frac{A}{d}$$

Example 13-3

The effective area of the plates of a capacitor is 10 m² and the distance between the plates is 1 mm. Find the dielectric constant of the paper insulator. Find the capacitance of the capacitor.

Solution

The relative dielectric constant of paper is 2.5. Therefore its dielectric constant is

$$\varepsilon = \varepsilon_0\, \varepsilon_r = 2.5\left(8.85\times10^{-12}\ \text{F/m}\right) = 22.13\times10^{-12}\ \text{F/m}$$

The capacitance is

$$C = \varepsilon\, \frac{A}{d} = \left(22.13\times10^{-12}\ \text{F/m}\right)\frac{10\ \text{m}^2}{10^{-3}\ \text{m}} = 221.3\ \text{nF}$$

Practice

The capacitor in the example uses glass as an insulator. Find its dielectric constant and its capacitance.

Answer: 66.38×10^{-12} F/m, 663.8 nF

The following observations can be made about capacitance:

- Doubling the area *doubles* the capacitance.

- Doubling the dielectric constant *doubles* the capacitance.

- Doubling the distance between the plates *halves* the capacitance.

Leakage Current

Theoretically, after a capacitor has been charged to a voltage and the capacitor is removed from the circuit, the capacitor retains the charge and the voltage drop forever. However, charge leaks between the plates through the dielectric over time due to free electrons in the dielectric producing what is termed **leakage current**. This effect is modeled and quantified by placing a resistance R_{leakage} in parallel with the plates of the capacitor. Most capacitors like mica have a very high leakage resistance, on the order of hundreds of MΩ of resistance. Such high resistance produces a very low and usually inconsequential leakage current. However there are some capacitor types like electrolytic capacitors that have a relatively lower leakage resistance and produce higher leakage currents. These capacitors do not hold charge long once disconnected from their supply source.

Measuring Capacitance

Capacitance test instruments especially designed to measure capacitance and its characteristics are needed to accurately measure capacitor parameters. An ohmmeter can be used to provide a simple test on some capacitor types to check if a discharged capacitor's resistance goes from a short (0 Ω) to an open (∞ Ω). If the capacitor reaches a steady reading of resistance, it is most likely defective. A general purpose instrument is an **RLC meter** which measures resistance, conductance, inductance (discussed later in this textbook), and capacitance.

Fixed Capacitors

A **fixed capacitor** has a fixed, single value of capacitance, for example, 100 μF. Capacitors come in a variety of shapes and sizes. Sample shapes are shown in Figure 13-6. For a given type, a physically larger capacitor has more capacitance (can store more charge per unit volt).

The capacitance value may be color coded or number coded. Capacitor nominal numeric values correspond to the resistor color code with two-digit codes of 10, 15, 22, 33, and 47. For example, a disc capacitor labeled 223 is interpreted as 22×10^3 picofarads. The resulting value is 22 nF.

Major types of capacitors include ceramic capacitors, thin film capacitors, tantalum capacitors, and electrolytic capacitors. As the names indicate, the ceramic capacitor uses a ceramic dielectric material, the thin film capacitor uses a film such as polyester, and the tantalum capacitor uses a tantalum dielectric.

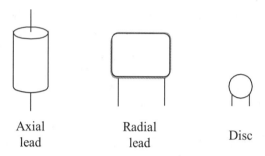

Figure 13-6 Various capacitor shapes

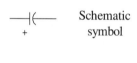

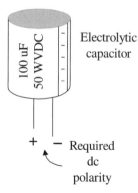

Figure 13-7
Electrolytic capacitor

Each capacitor type has its associated capacitance range and other special characteristics. For example, *approximate ranges* are: mica and ceramic capacitors from 1 pF to 100,000 pF, thin film from 1 nF to 47 μF, tantalum from 1 nF to 1000 mF, and electrolytics from 100 nF to 15,000 μF. Many capacitor types exist and new types are being invented. Check manufacturer's specifications for capacitive ranges and other essential characteristics.

Capacitors are either polarized or not polarized. Capacitors that are polarized are polarity sensitive and require the proper dc voltage polarity connection. For example, electrolytic capacitors and tantalum capacitors are polarized. If a capacitor is not polarized, for example the ceramic capacitor, then the dc polarity is of no consequence.

An **electrolytic capacitor** is a specially constructed capacitor that is polarity sensitive. See Figure 13-7. If the applied voltage polarity is accidentally reversed, the capacitor does not function properly and can explode. The electrolytic capacitor is labeled with a plus (+) sign with its schematic symbol. The capacitor component is usually large enough to be labeled with its value and its polarity. Extra caution must be taken to assure that an electrolytic capacitor is properly placed in the circuit. Also, exceptional care must be taken to not exceed the working dc voltage rating of the electrolytic capacitor. *Caution: if an electrolytic capacitor's working dc voltage is exceeded or if it is reverse biased, the electrolytic capacitor has the potential to explode.* It is designed for high values of capacitance and is used extensively in such applications as dc power supplies.

Hidden Capacitance–Stray Capacitance

Whenever two conductors are separated by an insulator, capacitance exists: wires in the same run, a wire running along a ground plane, leads of a component, and so forth. This natural capacitance, called **hidden or stray capacitance**, is typically very low (in the pF range). It only affects component and circuit behavior at very high frequencies or circuits with suddenly changing supplies or signals (namely, dc switching circuits and circuits with rectangular signals). Figure 13-8 demonstrates several of these types of hidden capacitance.

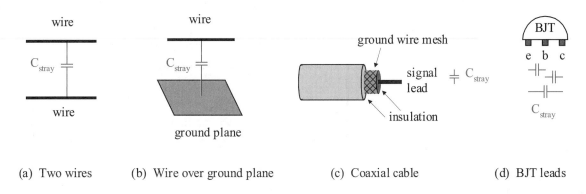

(a) Two wires   (b) Wire over ground plane   (c) Coaxial cable   (d) BJT leads

Figure 13-8 Natural hidden or stray capacitance

Figure 13-8(a) demonstrates the stray capacitance that exists between two wires running parallel to each other separated by their insulation and/or air. Figure 13-8(b) demonstrates the stray capacitance that exists between a conducting wire and a ground plane. Figure 13-8(c) demonstrates the capacitance that exists between the center signal wire and the grounded shielded cable (wire mesh) surrounding the signal wire and separated by insulation. This wire configuration is called shielded or coaxial cable. It is intended primarily for high frequency signals. The signal lead is shielded from external noise by wire mesh surrounding the signal lead. The wire mesh is typically connected to common to short external noise to common. Coaxial cable is typically connected to instrumentation with

BNC (Bayonet Naval Connector) connectors to minimize capacitive effects and the external noise distortion. Wire or cable runs are rated in terms of capacitance per unit distance. For example, coaxial cable is approximately 3 pF per inch. A 6-foot coaxial cable has a net capacitance of 18 pf. This is an accumulated parallel capacitance and, as demonstrated later in this chapter, parallel capacitance adds.

Device specifications of capacitance are rated between terminals, for example, C_{be} is a specification of the base-to-emitter capacitance as shown in Figure 13-8(d). Or the device capacitance may be specified as the *input* capacitance, *output* capacitance, and *reverse transfer* capacitance from output to input.

Figure 13-9 demonstrates the same hidden capacitance effect between traces on a printed circuit board (PCB). When laying out a printed circuit board, many layout considerations must be taken into account, including stray capacitance.

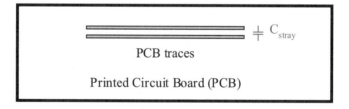

Figure 13-9 Capacitance between PCB board runs

Variable Capacitors

A **variable capacitor** has a range of values, for example, 0 to 100 μF. Figure 13-10(a) shows the schematic symbols for variable capacitors. Variable capacitors typically have a rotating shaft. As the shaft rotates, a set of capacitor plates rotates relative to a set of fixed capacitor plates. As the shaft rotates, the rotating plates rotate, and the area between the plates is increased or decreased. This effectively increases or decreases the capacitor's capacitance as shown in Figure 13-10(b).

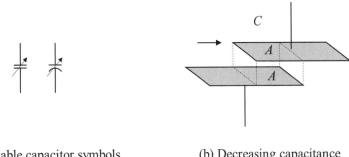

(a) Variable capacitor symbols   (b) Decreasing capacitance

Figure 13-10 Variable capacitors

Capacitance Key Characteristics

A few key characteristics that describe capacitance are that capacitance

- stores charge in the form of a voltage $V = Q/C$.

- retains voltage if disconnected from the circuit.

- returns charge (voltage) to a circuit like a voltage supply E.

- resists change in voltage.

The instantaneous relationship of the voltage and current in a capacitor is

$$i_c(t) = C \frac{d\,v_c(t)}{dt}$$

Or if working with average values,

$$i_c(t) = C \frac{\Delta v_c(t)}{\Delta t}$$

A changing capacitor voltage is required to produce an induced capacitor current. It is called an *induced capacitor current* since current does not really flow through the capacitor (equivalent to a lightning strike) but rather the changing capacitor voltage induces a current in the circuit external to the capacitor.

- An uncharged capacitor initially acts like a short to a sudden change in dc voltage.

- A charged capacitor initially acts like a dc voltage supply to a sudden change in dc voltage.

- A steady state capacitor acts like an open.

Current Response to Simple Voltage Waveforms

Since capacitor current depends on the rate of change of the capacitor voltage, the capacitor current and voltage do not have a linear resistance like a fixed resistance.

Example 13-4

Given the capacitor voltage waveforms of Figure 13-11(a,b), find the corresponding capacitor current waveforms for a 10 μF capacitor.

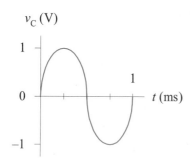

(a) Sinusoidal capacitor voltage

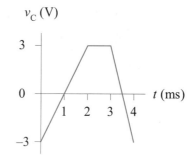

(b) Trapezoidal capacitor voltage

Figure 13-11(a,b) Example capacitor voltage waveforms

Solution

a. For the sine wave of Figure 13-11(a), substitute the voltage sine wave function of $\sin(2\pi f t)$ into the basic definition of capacitance. The capacitor current is

$$i_c(t) = C \frac{d\,v_c(t)}{dt}$$

$$i_c(t) = C \frac{d \sin(2\pi ft)}{dt}$$

If the frequency is fixed at a constant value, the derivative of the sine function is the cosine function multiplied by the constant $2\pi f$. The capacitor current becomes

$$i_c(t) = C\ (2\pi f)\ \cos(2\pi ft)$$

simplifying to

$$i_c(t) = 2\pi fC \cos(2\pi ft)$$

Thus, if a sinusoid voltage is applied to a capacitor, the current is a sinusoid. Substituting the capacitance value of 10 μF and the frequency of 1 kHz, the capacitor current expression is

$$i_c(t) = 2\pi \times 1 \text{ kHz} \times 10\mu\text{F} \times \cos(2\pi ft)$$

and reduces to

$$i_c(t) = 62.8 \cos(2\pi ft)\, \text{mA}$$

as shown in Figure 31-11(c).

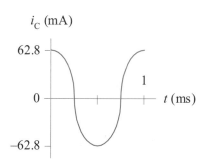

(c) Sinusoidal response

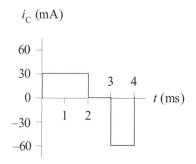

(d) Trapezoidal response

Figure 13-11(c,d) Example resulting capacitor current waveforms

b.   For the capacitor signal current of Figure 13-11(b), separate the problem into three parts for each of the distinct parts of the waveform. Results are shown in Figure 13-11(d).

For $0 \leq t < 2$ ms, the rate of change is the slope of the line (*up* 6 V and over 2 ms).

$$i_c(t) = C \frac{\Delta v_c(t)}{\Delta t} = 10\mu F \left(\frac{6 \text{ V}}{2 \text{ ms}} \right) = 30 \text{ mA}$$

For $2 \text{ ms} \leq t < 3$ ms, the rate of change of a constant capacitor voltage of 6 V is zero (0).

$$i_c(t) = C \frac{\Delta v_c(t)}{\Delta t} = 10\mu F \left(\frac{0 \text{ V}}{1 \text{ ms}} \right) = 0 \text{ mA}$$

For $3 \leq t < 4$ ms, the rate of change is the slope of the line (*down* 6 V and over 1 ms).

$$i_c(t) = C \frac{\Delta v_c(t)}{\Delta t} = 10\mu F \left(\frac{-6 \text{ V}}{1 \text{ ms}} \right) = -60 \text{ mA}$$

Practice

Given the capacitor voltage waveforms of Figure 13-12(a,b), find the corresponding capacitor current waveforms for a 1 mF capacitor.

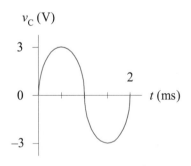

(a) Sinusoidal capacitor voltage

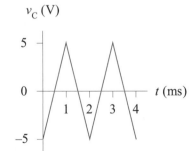

(b) Triangle wave capacitor voltage

Figure 13-12(a,b) Practice capacitor voltage waveforms

Answer: See Figure 13-12(c,d) for capacitor current response wave-forms.

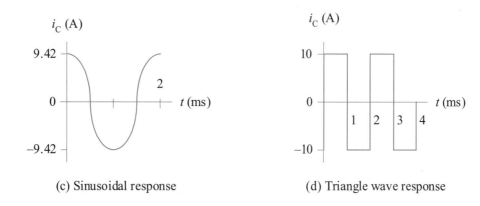

(c) Sinusoidal response (d) Triangle wave response

Figure 13-12(c,d) Practice response capacitor current waveforms

13.2 Capacitive Reactance

Capacitance exhibits an opposition to ac current. This reaction to ac current is called **reactance**. Specifically, it is called **capacitive reactance**.

This opposition to ac current is different from resistance. Resistance dissipates energy in the form of heat, light, and so forth. The capacitor stores energy in the form of charge that can be returned to the circuit. Rather than dissipating the energy, energy is absorbed and returned by the capacitor.

From Example 13-4, a capacitor with a sinusoidal voltage of

$$v_c(t) = \sin(2\pi ft)$$

creates a capacitor current of

$$i_c(t) = 2\pi f C \cos(2\pi ft)$$

Divide the capacitor voltage by its current.

$$\frac{v_c(t)}{i_c(t)} = \frac{\sin(2\pi ft)}{2\pi f C \cos(2\pi ft)}$$

Separate the coefficient.

$$\frac{v_c(t)}{i_c(t)} = \frac{1}{2\pi fC} \times \frac{\sin(2\pi ft)}{\cos(2\pi ft)}$$

The magnitude of the capacitive reactance to an ac sinusoidal signal is the formula

Capacitive reactance

$$X_C = \frac{1}{2\pi fC} \qquad (13\text{-}3)$$

where

X_C capacitive reactance with units of ohms (Ω)

f frequency of the ac sinusoidal signal with units of hertz (Hz)

C capacitance with units of farads (F)

Capacitive reactance is in units of ohms like resistance, but the reactance magnitude cannot be directly added to resistance magnitude. They are fundamentally different in that resistance dissipates energy and reactance stores energy. A subsequent textbook in this textbook series explores the combination of resistance and reactance; it is a vector sum relationship.

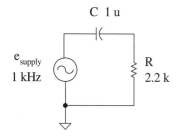

Figure 13-13
Example circuit

Example 13-5

In the circuit of Figure 13-13, find the reactance of the capacitor. Can the circuit resistance be added directly to the capacitive reactance?

Solution

Apply Equation 13-3.

$$X_C = \frac{1}{2\pi fC} = \frac{1}{2\pi(1\text{ kHz})(1\text{ }\mu\text{F})} = 159\text{ }\Omega$$

Reactance and resistance magnitudes cannot be directly added; they have a vector sum relationship.

**Reactance ohms
and resistive ohms
do *not* add directly**

Practice

The capacitance of the capacitor in the circuit of Figure 13-13 is changed to 100 µF. Find its capacitive reactance.

Answer: 1.59 Ω

Capacitive reactance is inversely proportional to capacitance and frequency. As frequency or capacitance is increased, the capacitive reactance decreases.

$f \uparrow \quad X_C \downarrow$

Example 13-6

In a series *RC* circuit, the capacitive reactance of a capacitor is 10 Ω. The capacitor is replaced by a capacitor 100 times its original value. What is the reactance of this capacitor? Would you expect the ac voltage drop across the new capacitor to increase or decrease?

$C \uparrow \quad X_C \downarrow$

Solution

The capacitance of the capacitor is *increased* by a factor of 100; therefore, its reactance is *reduced* by the same factor.

$$X_C = \frac{10 \ \Omega}{100} = 0.1 \ \Omega$$

In dc circuit analysis, a reduced resistance in a series circuit drops less voltage. Even though resistance magnitudes and reactance magnitudes do not directly add, the same concept of a reduced resistance dropping less voltage holds. Thus, it is expected that the new capacitor with less reactance will drop less voltage.

Practice

In a series *RC* circuit, the capacitive reactance of a capacitor is 10 kΩ. The capacitor is replaced by a capacitor 100 times smaller than its original value. What is the reactance of this capacitor? Would you expect the ac voltage drop across the new capacitor to increase or decrease?

Answer: 1 MΩ, drops more voltage

Quick-look Frequency Effects

The capacitor is a component that has two conducting plates separated by an insulator. Since the insulator is ideally ∞ ohms, the capacitor is natural open to dc current once the charge on the capacitor plates has stabilized (called **steady state**).

The capacitor's opposition to ac current is a function of its frequency. If the supplied signal is a sinusoid, then the capacitor current and voltage are a sinusoid.

Recall the reactance equation for a capacitance.

$$X_C = \frac{1}{2\pi fC}$$

If the frequency is relatively low, then the capacitor acts like an open and drops most of the supply voltage across it. If the signal frequency is relatively high, then the capacitor acts like a short and drops 0 V and all of the drop is across the resistor R. Figure 13-14 demonstrates these extreme frequency responses of a sinusoid in an *RC* circuit.

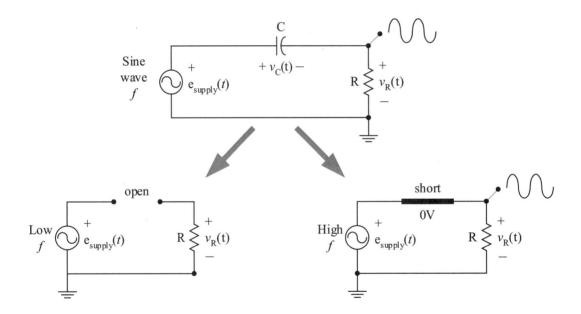

Figure 13-14 Extreme frequency responses of sinusoids in an *RC* circuit

The circuit of Figure 13-15(b) demonstrates the use of a filter capacitor in a dc power supply. This is a half-wave rectifier circuit with a filter capacitor placed across the load. The diode rectifies the ac signal to eliminate the negative half of the signal and to produce only the positive half cycle in the output, as shown by the circuit of Figure 13-15(a), from a previous example. The purpose of the filter capacitor is to charge to the peak voltage of the signal out of the rectifier diode and hold that voltage across the load at a fixed dc value. When the diode is forward biased the capacitor is charged up to the voltage e_p less the diode voltage drop. Once the capacitor is charged, it ideally maintains this fixed dc voltage across the load. Visualize the capacitor as a *dc open*. The filter capacitor also serves the purpose of shorting out any ac signals above a specific frequency. Visualize the capacitor as an *ac short* and all ac signals shorted out to common. Thus the dc is passed to the load and all ac signals shorted to common resulting in an *electronic dc power supply*. The circuit is called a half-wave, filtered dc power supply.

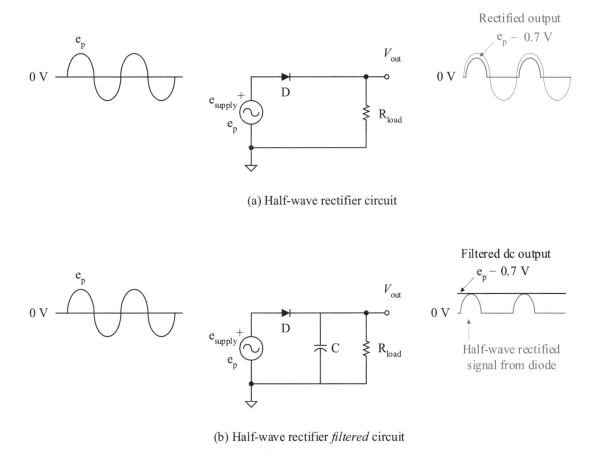

(a) Half-wave rectifier circuit

(b) Half-wave rectifier *filtered* circuit

Figure 13-15 Half-wave rectifier with filter capacitor–quick look frequency effect

Example 13-7

For the circuit of Figure 13-16(a), find the output dc voltage and the output ac voltage.

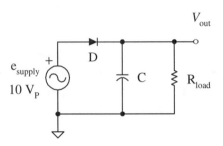

Figure 13-16(a) Example problem of half-wave, filtered rectifier

Solution

The output out of the diode is a half-wave rectified signal with a peak voltage of

$$V_{p \text{ out}} = 10 \text{ V} - 0.7 \text{ V} = 9.3 \text{ V}_p$$

The capacitor charges up to this voltage and *ideally* holds that voltage across the load. The capacitor acts as an open to dc and holds 9.3 V fixed across the load.

$$V_{dc \text{ out}} = V_{p \text{ out}} = 9.3 \text{ V}_{dc}$$

The capacitor ideally acts like an open to ac so the load voltage for ac is ideally 0 V_{rms}.

Practice

The diode in the circuit of Figure 13-16(a) is reversed. Find the dc output voltage. What is the ac output voltage?

Answer: -9.3 V_{dc}, 0 V_{rms}

Exploration

Simulate the circuit of Figure 13-16(a) using a 1N4001 diode, a 10 kΩ resistor, and a 1000 μF capacitor. Change C to 10 μF and notice the output voltage change. The output now has ac ripple voltage due to the discharging and recharging of C. Figure 13-16(b) shows the results.

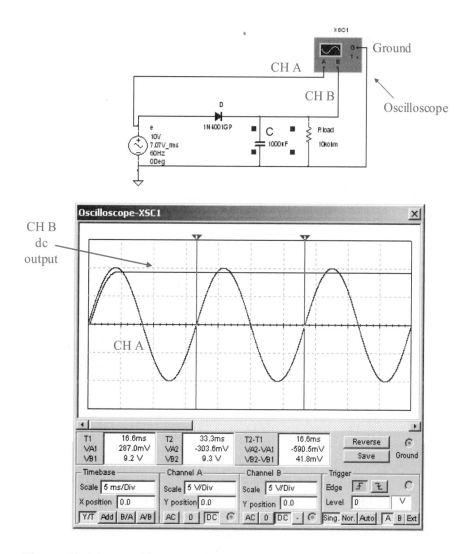

Figure 13-16(b) Half-wave rectifier with filter capacitor MultiSIM simulation

13.3 Series and Parallel Capacitance

Total Series Reactance and Capacitance

Total series capacitive reactance

Capacitive reactance in series adds like resistance in series. The total capacitive reactance of capacitors in series is the sum of their reactances. Thus, the total capacitive reactance of the three capacitors in the series circuit of Figure 13-17 is

$$X_{C\,total} = X_{C1} + X_{C2} + X_{C3} \qquad (13\text{-}4)$$

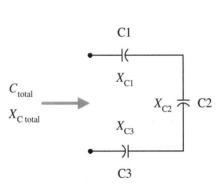

Figure 13-17 Three series capacitors

The total equivalent series capacitance can be found from substituting each individual capacitive reactance form of Equation 13-3 into the series circuit capacitive reactance equation of 13-4.

$$\frac{1}{2\pi f\, C_{total}} = \frac{1}{2\pi f\, C1} + \frac{1}{2\pi f\, C2} + \frac{1}{2\pi f\, C3}$$

The term $2\pi f$ is common in each denominator. Algebraically multiplying both sides of the equation, and thus each numerator by this term, cancels out the $2\pi f$ terms and the equation reduces to

Total series capacitance

$$\frac{1}{C_{total}} = \frac{1}{C1} + \frac{1}{C2} + \frac{1}{C3} \qquad (13\text{-}5)$$

The form of Equation 13-5 for total series circuit capacitance is due to the reciprocal relationship of X_C and C. Take the reciprocal of both sides of Equation 13-5 to solve for total capacitance.

$$C_{total} = \left(\frac{1}{C1} + \frac{1}{C2} + \frac{1}{C3} \right)^{-1}$$

Expressing the reciprocal of each capacitance in x^{-1} form, this equation becomes

$$C_{total} = \left(C1^{-1} + C2^{-1} + C3^{-1} \right)^{-1}$$

This form is very handy for calculator use. This is the same as the form for parallel resistances. *A good memory trick is to think that capacitors in series combine like resistors in parallel.*

Example 13-8

A 100 μF capacitor, a 47 μF capacitor, and a 10 μF capacitor are in series. The circuit is supplied with a 1 kHz sinusoidal supply. Find the reactance of each capacitor, the total circuit capacitive reactance, and the combined total circuit capacitance.

Solution

The individual capacitive reactances are

$$X_{C1} = \frac{1}{2\pi f \, C1} = \frac{1}{2\pi(1 \text{ kHz})(100 \text{ μF})} = 1.59 \text{ Ω}$$

$$X_{C2} = \frac{1}{2\pi f \, C2} = \frac{1}{2\pi(1 \text{ kHz})(47 \text{ μF})} = 3.38 \text{ Ω}$$

$$X_{C3} = \frac{1}{2\pi f \, C3} = \frac{1}{2\pi(1 \text{ kHz})(10 \text{ μF})} = 15.92 \text{ Ω}$$

The total circuit capacitive reactance is the sum of the individual series reactances.

$$X_{Ctotal} = 1.59 \text{ Ω} + 3.38 \text{ Ω} + 15.92 \text{ Ω} = 20.89 \text{ Ω}$$

Rewriting Equation 13-3 to solve for capacitance

$$C_{total} = \frac{1}{2\pi f \, X_{Ctotal}} = \frac{1}{2\pi(1 \text{ kHz})(20.89 \text{ Ω})} = 7.62 \text{ μF}$$

This answer may be confirmed by using the total series capacitance form of Equation 13-5.

$$C_{total} = \left(\frac{1}{100\ \mu F} + \frac{1}{47\ \mu F} + \frac{1}{10\ \mu F} \right)^{-1} = 7.62\ \mu F$$

or

$$C_{total} = \left((100\ \mu F)^{-1} + (47\ \mu F)^{-1} + (10\ \mu F)^{-1} \right)^{-1} = 7.62\ \mu F$$

Practice

Repeat the example with a sinusoidal frequency of 10 Hz. You can repeat the above solution process or use a shortcut. If the frequency is decreased by a factor of 100, what is the impact on each capacitive reactance and the total capacitance?

Answer: 159 Ω, 338 Ω, 1.592 kΩ, 2.089 kΩ, 7.62 μF (C_{total} unchanged)

Total Parallel Reactance and Capacitance

Capacitive reactance in parallel combines like resistance in parallel. Thus, the total capacitive reactance of the three capacitors in the parallel circuit of Figure 13-18 is

Total parallel capacitive reactance

$$\frac{1}{X_{C\ total}} = \frac{1}{X_{C1}} + \frac{1}{X_{C2}} + \frac{1}{X_{C3}} \qquad \text{(13-6)}$$

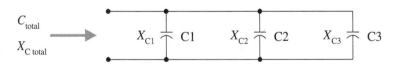

Figure 13-18 Three parallel capacitors

The total equivalent parallel capacitance can be found from substituting each individual capacitive reactance form of Equation 13-3 into the parallel circuit capacitive reactance Equation 13-6.

$$\frac{1}{\frac{1}{2\pi f\ C_{total}}} = \frac{1}{\frac{1}{2\pi f\ C1}} + \frac{1}{\frac{1}{2\pi f\ C2}} + \frac{1}{\frac{1}{2\pi f\ C3}}$$

Algebraically inverting and multiplying the denominators, this equation reduces to

$$2\pi f\ C_{total} = 2\pi f\ C1 + 2\pi f\ C2 + 2\pi f\ C3$$

Algebraically dividing each term by the common factor of $2\pi f$, this equation simplifies to

$$C_{total} = C1 + C2 + C3 \qquad\qquad (13\text{-}7)$$

Capacitance in parallel is the sum of the individual capacitance. This is the same as the form for series resistances. *A good memory trick is to think that capacitors in parallel combine like resistors in series.*

Example 13-9

A 100 µF capacitor, a 47 µF capacitor, and a 10 µF capacitor are in parallel. The circuit is supplied with a 1 kHz sinusoidal supply. Find the total circuit capacitance and reactance.

Solution

The total circuit capacitance is

$$C_{total} = 100\ \mu F + 47\ \mu F + 10\ \mu F = 157\ \mu F$$

The total circuit capacitance reactance is

$$X_{Ctotal} = \frac{1}{2\pi f\ C_{total}} = \frac{1}{2\pi(1\ kHz)(157\ \mu F)} = 1.01\ \Omega$$

Practice

The total circuit capacitance is 30 pF of three equal valued capacitors in parallel. Find the capacitance of each capacitor. Find the total capacitive reactance at a frequency of 100 Hz.

Answer: 10 pF each, 53.1 MΩ

Series-Parallel Capacitance Reduction

For series-parallel capacitor circuits, reduce simple series capacitance combinations and simple parallel capacitance combinations. These must be pure capacitance combinations. Other components such as resistors cannot be simply combined with capacitors. Resistance and reactance require vector addition.

Example 13-10

Reduce the circuit of Figure 13-19(a) to a single value.

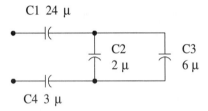

Figure 13-19(a) Circuit of example problem

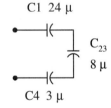

Figure 13-19(b)
Example circuit
reduced

Solution

The simplest combination is the parallel combination of C2 and C3, that is, C2//C3. Sum the parallel capacitances.

$$C_{23} = 2\ \mu F + 6\ \mu F = 8\ \mu F$$

Combining C2 and C3, the circuit is reduced to the equivalent circuit of Figure 13-19(b). Now there is a series combination of C1, C23, and C4. Combine these purely series capacitances to a single capacitance value.

$$C_{\text{total}} = \left(\frac{1}{24\ \mu F} + \frac{1}{8\ \mu F} + \frac{1}{3\ \mu F} \right)^{-1} = 2\ \mu F$$

Another approach is to combine pairs of series capacitors; namely, C1//C_{23} for a combination of C_{123} with a value of 6 μF. That 6 μF is in series with 4 μF for a series combination value of 2 μF.

Practice

Find the total capacitance for the circuit of Figure 13-20.

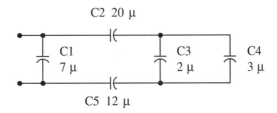

Figure 13-20 Practice circuit

Answer: 10 μF

Charge and Voltage of Multiple Capacitors

Current is defined to be the time rate of change of charge. A net charge must be flowing in the same direction to create current. In analyzing charge distribution in multiple capacitor circuits, charge can be considered to flow like current.

$i = \dfrac{dq}{dt}$

The charge in a series circuit must be the same everywhere just as current is the same throughout a series circuit. This is essentially Kirchhoff's current law of charge. Different capacitors in series have different voltage drops since $V = Q/C$. In a purely capacitive series circuit, the voltage drops sum up to be the total voltage drop (Kirchhoff's voltage law). Summarizing, *in purely series capacitive circuits*

$I = \dfrac{Q}{t}$

$$I_{\text{supplied}} = I_1 = I_2 = I_3$$

$$Q_{\text{supplied}} = Q_1 = Q_2 = Q_3$$

$$V_{\text{supplied}} = V_1 + V_2 + V_3$$

$$V_C = \dfrac{Q}{C}$$

$V = \dfrac{Q}{C}$

The voltage across parallel capacitors must be the same (Kirchhoff's voltage law). The charge on different parallel capacitances is different since $Q = CV$. The sum of charge into a parallel combination is the sum of the charge in each branch. This is essentially Kirchhoff's current law in charge form. Summarizing, *in purely parallel capacitive circuits*

$$I_{\text{supplied}} = I_1 + I_2 + I_3$$

$$Q_{\text{supplied}} = Q_1 + Q_2 + Q_3$$

$$V_{\text{supplied}} = V_1 = V_2 = V_3$$

$$V_C = \frac{Q}{C}$$

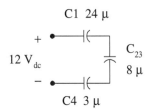

Figure 13-21
Example circuit

Example 13-11

Find the total charge supplied to the series capacitive circuit of Figure 13-21 assuming that 12 V_{dc} is applied to the circuit and that it is in steady state. Find the voltage drop across each capacitor.

Solution

First find the total series capacitance of the circuit.

$$C_{\text{total}} = \left(\frac{1}{24\ \mu F} + \frac{1}{8\ \mu F} + \frac{1}{3\ \mu F} \right)^{-1} = 2\ \mu F$$

Next apply the basic definition of capacitance to the total circuit.

$$Q_{total} = C_{total}\, V_{total} = 2\ \mu F \times 12\ V = 24\ \mu C$$

The charge is the same on each series capacitor. Now apply the basic definition of capacitance to each capacitor to its individual voltage drop.

$$V_{C1} = \frac{Q_{\text{total}}}{C1} = \frac{24\ \mu C}{24\ \mu F} = 1\ V$$

$$V_{C2} = \frac{Q_{\text{total}}}{C2} = \frac{24\ \mu C}{8\ \mu F} = 3\ V$$

$$V_{C3} = \frac{Q_{\text{total}}}{C3} = \frac{24\ \mu C}{3\ \mu F} = 8\ V$$

Verify Kirchhoff's voltage law to ensure that these are reasonable voltages.

$$V_{supplied} = 1 \text{ V} + 3 \text{ V} + 8 \text{ V} = 12 \text{ V}$$

KVL confirmed

Practice

Find the charge of each capacitor of the circuit of Figure 13-22 assuming that 12 V_{dc} is applied to the circuit and that it is in steady state. Find the total charge supplied to the parallel capacitive circuit.

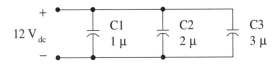

Figure 13-22 Practice circuit

Answer: 12 μC, 24 μC, 36 μC, 72 μC

13.4 Capacitive Coupler and Filter

Circuits that have a combination of ac and dc voltages use the capacitor to **filter out**, **bypass**, or **couple** these signals, separating the ac signal from the dc signal. Ideally, the capacitor acts like an open to dc (blocks dc current). In many circuits the capacitor is designed to act like a short to ac (couples or passes ac current). For now, we shall ideally assume that the capacitor is designed to act as an *ac short*. See Figure 13-23.

Capacitors couple ac current and block dc current

Figure 13-23 Ideal capacitor models for ac and dc

Figure 13-24 demonstrates that a capacitor acts as an open to dc and therefore can develop a dc voltage drop. In this circuit, the capacitor is dropping 10 V_{dc}. Capacitors are rated for maximum working dc voltage drops. If the rating is exceeded the capacitor could be damaged.

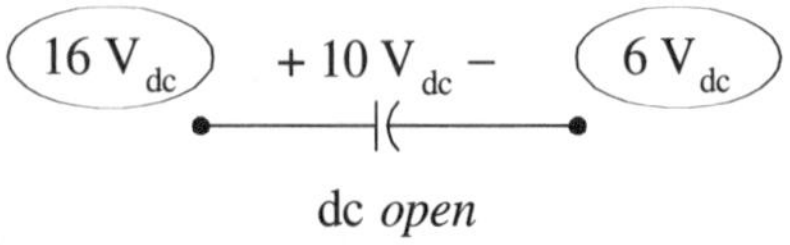

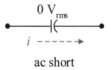

$$\boxed{16 \text{ V}_{dc}} \quad + 10 \text{ V}_{dc} - \quad \boxed{6 \text{ V}_{dc}}$$

dc *open*

Figure 13-24 The dc voltage drop across a capacitor

Figure 13-25 demonstrates that a capacitor ideally acts as an ac short at very high frequencies. The opposition to ac induced current in a capacitor is inversely proportional to its capacitance and the ac signal frequency.

$$0 \text{ V}_{rms}$$

i - - - - - →

ac short

Figure 13-25 Ideal ac coupling or bypass capacitor

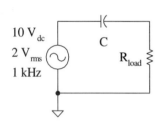

Figure 13-26(a)
Example circuit

10 V$_{dc}$
2 V$_{rms}$
1 kHz
C
R$_{load}$

Example 13-12

For the circuit of Figure 13-26(a), draw the dc model and find the dc resistor and capacitor voltage drops. Draw the ac model of the circuit treating the capacitor as an ideal ac short and find the ac resistor and capacitor voltage drops. Is this circuit coupling the ac signal to the resistor or is it bypassing (shorting) the ac signal around the resistor?

Solution

For dc analysis, model the capacitor as an open and solve the circuit. Since the capacitor is acting as an open, then dc current in the circuit is 0 A_{dc}, the resistor drops 0 V_{dc}, and the capacitor drops all 10 V_{dc}.

For ac circuit analysis, model the capacitor as an ideal short. The capacitor must then drop 0 V_{rms} and the load resistor drop all 2 V_{rms}. The load resistor drops 0 V_{dc} and 2 V_{rms}. Thus, the capacitor is *blocking dc* current to the load resistor and *passing ac* current to the load resistor; therefore this is serving as an *ac coupling capacitor*. Figure 13-26(b) shows the circuit models and resulting voltages.

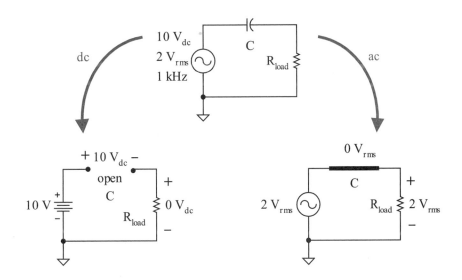

Figure 13-26(b) Example problem models and solutions

Practice

For the circuit of Figure 13-27(a), draw the dc model and find the load resistor and capacitor dc voltage drops. Draw the ac model of the circuit treating the capacitor as an ideal ac short and find the load resistor and capacitor ac voltage drops. Is this circuit coupling the ac signal to the load resistor or is it bypassing (shorting) the ac signal around the load resistor?

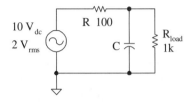

Figure 13-27(a) Practice circuit

Answer: 9.09 V_{dc}, 0 V_{rms}. See Figure 13-27(b). The ac current bypasses (shorts around) the load resistor; therefore, the capacitor is serving as an ac bypass capacitor.

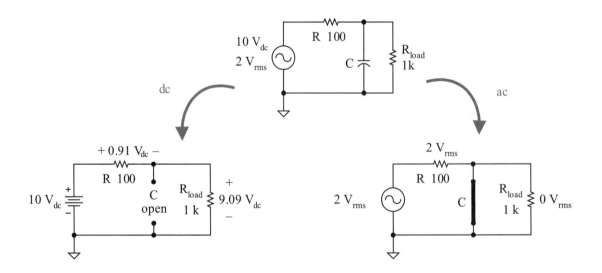

Figure 13-27(b) Practice solution

BJT Amplifier Coupling and Bypass Capacitors

The circuit of Figure 13-28(a) is a BJT amplifier stage. The input capacitor C1 and the output capacitor C2 are *ac coupling capacitors* and *dc blocking capacitors*. These capacitors allow the ac signal to be coupled from stage to stage (left to right) but isolate the dc voltage of each stage. C3 is a *bypass capacitor*, bypassing the ac signal around R_E. The dc signal sees R_E from emitter to common, whereas the ac signal sees a short from the emitter to common.

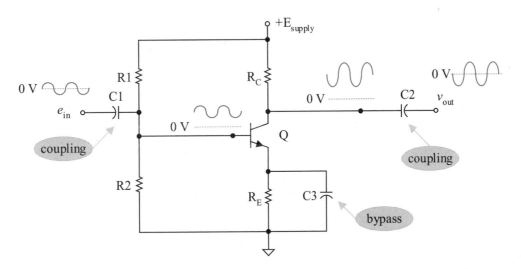

Figure 13-28(a) BJT amplifier stage using ac coupling and bypass capacitors

Figures 13-28(b,c) show the equivalent modeled circuits of dc and ac, respectively. These show the ideal modeling of the capacitor as a *dc open* and an *ac short*. In an ideal world, this is how you would like to design this amplifier. However, a quick look using the ideal dc and ac models is always an effective way to quickly look at such a circuit to gain an initial understanding of how the circuit is working. The modeled dc circuit looks completely different from the ac modeled circuit. In fact, typically the dc modeled circuit must be analyzed first to determine key dc values. Those key dc values are in turn used to determine key ac parameters before the ac analysis can continue. Look at both Figure 13-28(b) and Figure 13-28(c). Be sure you can visualize the capacitors as dc opens and ac shorts.

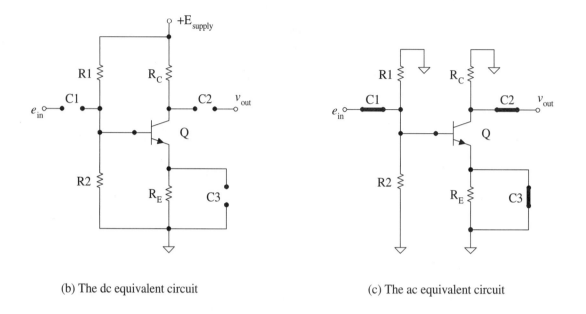

(b) The dc equivalent circuit   (c) The ac equivalent circuit

Figure 13-28(b,c) Model dc (*capacitors opened*) and ac (*capacitors shorted*)

BJT Amplifier Example

The next example is a BJT amplifier stage that has an ac signal riding on a dc bias voltage. This dc bias voltage is needed to properly bias the BJT to operate in its linear region so that the BJT output signal has the same shape as its input voltage. Figure 13-29 shows a basic BJT amplifier stage.

The coupling capacitor is a device that is frequency sensitive. For now all you need to know is that the capacitor value is selected so that it acts ideally like a short to the ac signal. By its very nature, the capacitor acts like an open to dc. Thus, this type of capacitor is called an ac coupling capacitor (couples the ac through it) and a dc blocking capacitor. The coupling capacitor allows the ac signal to pass from the left to the right through the circuit:

- The input signal e_{in} at the *in* node is coupled to the BJT base.

- The base signal is amplified to the collector of the BJT.

- The collector signal is coupled out through C_{out} to the output node.

These ac coupling capacitors at the same time act like dc blocking capacitors to isolate the biasing of BJT stages from each other.

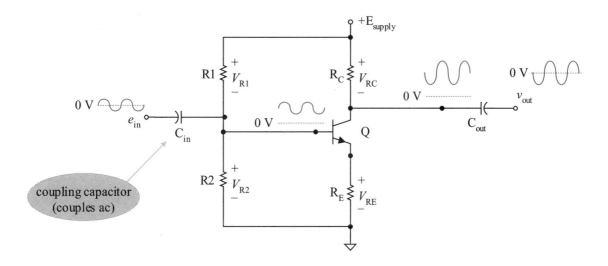

Figure 13-29 BJT amplifier stage

Example 13-13

Use the BJT amplifier circuit of Figure 13-30(a).

Find the dc bias voltage at the base V_B of the BJT. Then sketch the
total signal at the base of the BJT.

The BJT ac voltage amplifier gain is −2.2 for this amplifier
design from base to collector. The negative sign implies that the
signal is inverted from base to collector (also called a 180° phase
shift). Find and sketch the total signal at the collector.

Then find and sketch the output signal v_{out}.

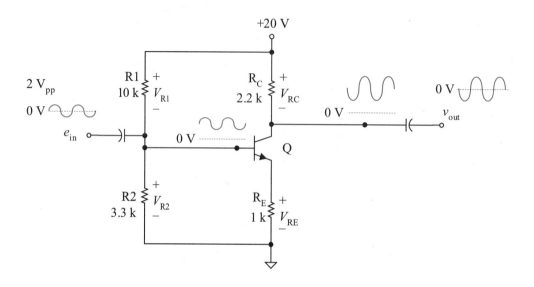

Figure 13-30(a) Example circuit of a BJT amplifier stage

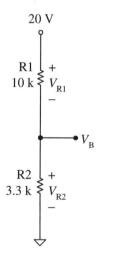

Figure 13-30(b)
Example figure–
dc bias circuit

Solution

Let's make a couple of reasonable simplifying assumptions. Assume that this dc voltage divider bias circuit has been properly designed so that the dc base current I_B is much, much less than the current flowing through R1. Also, let's assume that dc current gain β_{dc} of the BJT is greater than 50. Then this voltage divider dc bias circuit is BJT β independent and we can assume that R1 and R2 form an essentially series circuit.

$$I_{R1} \approx I_{R2}$$

Figure 13-30(b) shows this essentially series dc bias circuit. The dc base voltage V_B is the same as the voltage drop across R2. Since this is an essentially series circuit, the most efficient solution is to apply the voltage rule to the R1 and R2 series combination.

$$V_B = V_{R2} \approx \left(\frac{R2}{R1+R2} \right) E_{\text{supply}}$$

$$V_{B} \approx \left(\frac{3.3 \text{ k}\Omega}{3.3 \text{ k}\Omega + 10 \text{ k}\Omega} \right) 20 \text{ V} = 5 \text{ V}_{dc}$$

Since C_{in} couples the ac input signal to the base without any losses, the ac voltage base signal is

$$v_b = e_{in} = 2 \text{ V}_{pp}$$

$$v_b = e_{in} = \frac{1}{2} \left(2 \text{ V}_{pp} \right) = 1 \text{ V}_{p}$$

Combining the dc and ac BJT base voltages, the total base signal is shown in Figure 13-30(c). A 1 V_p ac signal is riding on a 5 V_{dc} bias.

$$v_{p \text{ base total signal}} = 5 \text{ V}_{dc} + 1 \text{ V}_{p} = 6 \text{ V}_{p}$$

To find the dc voltage of the collector V_C requires a few steps. Assuming this is a silicon transistor, the base-to-emitter dc voltage drop is 0.7 V_{dc}. Therefore the dc voltage at the BJT emitter is

$$V_E = 5 \text{ V} - 0.7 \text{ V} = 4.3 \text{ V}_{dc}$$

The voltage drop V_{RE} is the same as V_E.

$$V_{RE} = V_E = 4.3 \text{ V}_{dc}$$

The emitter resistor current I_{RE} is found by Ohm's law and is the same as the BJT emitter current I_E. Also, since the BJT β is greater than 50, we can make the assumption that the collector current is approximately the same as the emitter current (essentially series collector-emitter circuit).

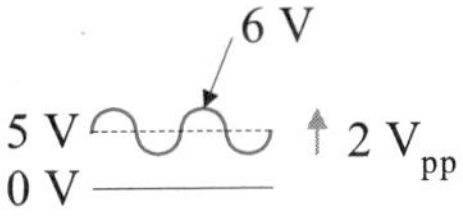

$$I_C \approx I_E = I_{RE} = \frac{4.3 \text{ V}}{1 \text{ k}\Omega} = 4.3 \text{ mA}$$

Figure 13-30(c)
Example circuit–
total signal at the
base of the BJT

The collector dc voltage can now be found as the supply voltage less the drop across the R_C resistor.

$$V_C = 20 \text{ V} - (4.3 \text{ mA} \times 2.2 \text{ k}\Omega) = 10.5 \text{ V}_{dc}$$

The ac signal collector voltage is found by taking the input base BJT ac signal voltage and multiplying by the amplifier gain given as –2.2. Again, the negative sign indicates that the ac signal is inverted in going from the base to the collector.

$$v_c = -2.2 \times 1 \text{ V}_p = -2.2 \text{ V}_p$$

$$v_c = 2.2 \times 2 \text{ V}_{pp} = 4.4 \text{ V}_{pp}$$

The total waveform at the BJT collector is shown in Figure 13-30(d), including the inverted signal effect.

$$v_{p \text{ collector total signal}} = 10.5 \text{ V} + 2.2 \text{ V} = 12.7 \text{ V}_p$$

The C_{out} capacitor filters out the dc and couples through the ac signal from the collector of the BJT.

$$V_{OUT} = 0 \text{ V}_{dc}$$

$$v_{out} = v_c = 4.4 \text{ V}_{pp}$$

The total output signal is shown in Figure 13-30(e).

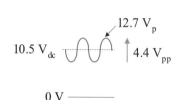

Figure 13-30(d)
Example circuit–
total signal at the collector
of the BJT

Practice

The input signal e_{in} is doubled to 4 Vpp. Find and sketch the total signals at the BJT base, the BJT collector, and the output.

Answer: The dc bias values remain the same. All of the ac signal values are doubled. e_{in} and v_b are 4 V_{pp}. v_c and v_{out} are 8.8 V_{pp}. That's the nature of a linear amplifier; double the input and the output ac signal voltage should double (e.g., same sound but louder).

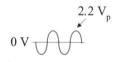

Figure 13-30(e)
Example circuit–
total signal at the
output

Single Supply Op Amp

Op amps can be operated with a split supply voltage system ($\pm E_{supply}$) or as a single supply system ($+E_{supply}$ and common). For example, a portable radio uses a 9 V battery producing a 9 volt power supply. Figure 13-31(a) demonstrates the problem of feeding an ac audio signal to a single-sided op amp supply. Assume that rail headroom required is 1 V for the upper and lower rails. A volt below the upper rail is 8 V. A volt above the lower rail is 1 V. The output voltage range is limited from 1 V to 8 V. Thus the output voltage can only swing between 1 V and 2 V instead of the desired –2 V to +2 V. The audio signal has been seriously distorted and does not sound the same as the original signal. In this case, the peak-to-peak signal has been diminished as well as distorted.

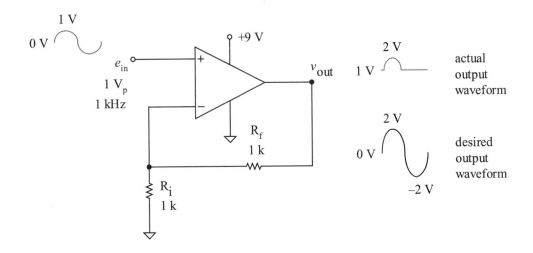

Figure 13-31(a) Distortion caused by single-sided supply of op amp

The next example addresses this problem by providing a dc offset input voltage that produces a dc offset output that prevents hitting the rails. This requires an input ac coupling capacitor, an ac output coupling capacitor, and a capacitor in the feedback circuit that creates the original ac signal gain but a dc voltage gain of unity.

Example 13-14

Redesign the circuit of Figure 13-31(a) to produce the desired ac signal output without distortion.

Solution

First, place an ac coupling capacitor C1 and a voltage divider on the input to raise the input signal offset from 0 V_{dc} to 4.5 V_{dc}, as shown in Figure 13-31(b). The voltage divider of 100 kΩ in series with 100 kΩ provides an essentially series circuit, voltage dividing the supply voltage. Recall that there is 0 A_{dc} going into the op amp and 0 A_{dc} coming from the coupling capacitor. The ac coupling cap is absolutely needed. Why? Without the coupling capacitor the dc supply would have a shorted path through the ac supply, bypassing the bottom 100 kΩ resistor. The ac input signal remains a 2 V_{pp} signal but is now riding on an offset of 4.5 V_{dc}.

High resistance values are selected for the voltage divider to prevent wasted power dissipation.

$$V_{\text{dc input op amp}} = \left(\frac{100 \text{ k}\Omega}{200 \text{ k}\Omega} \right) 9 \text{ V} = 4.5 \text{ V}$$

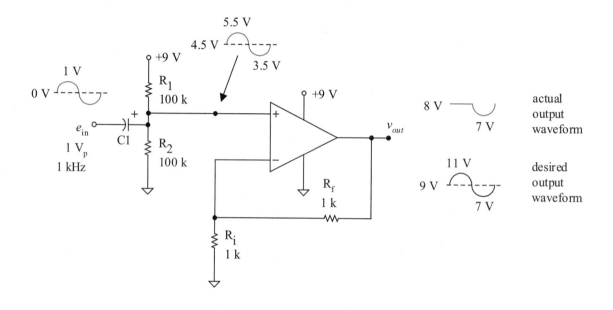

Figure 13-31(b) Input signal offset by 4.5 V_{dc}

The input signal to the op amp has been appropriately raised to 4.5 V_{dc}, however, this 4.5 V_{dc} is now amplified by the amplifier and produces a dc voltage offset of 9 V_{dc}, slamming the signal into the upper rail. The output signal attempts to vary from 7 V_{min} to 9 V_{max} but slams the rail at 8 V.

The ac voltage gain must remain two (2) but at the same time the dc voltage must be one (unity). Use the C3 capacitor in the feedback loop as shown in Figure 13-31(c). For the ac signal, visualize capacitor C3 as a short, creating the same ac circuit as the original ac circuit. For the dc voltage, visualize capacitor C3 as an open, forming a dc voltage buffer amplifier with a gain of unity. Also, an ac output coupling capacitor C2 has been added to filter

out the 4.5 V_{dc} to produce the desired 4 V_{pp} output voltage without a dc offset and without clipping distortion by the op amp rails.

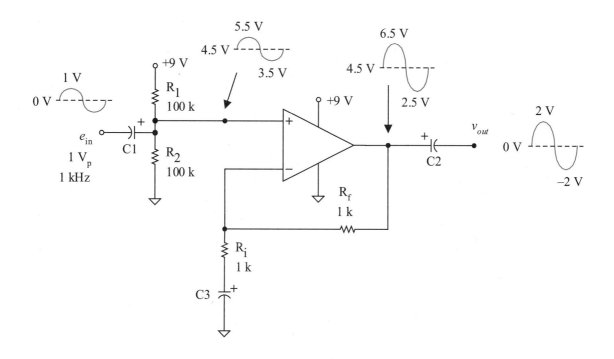

Figure 13-31(c) Single-sided op amp amplifier using dc offset and capacitors

---

Practice

Draw the dc model and the ac model of the circuit and verify the
dc and ac values of the example circuit of Figure 13-31(c).

Answer: Figures 13-31(d,e), dc and ac models, respectively.

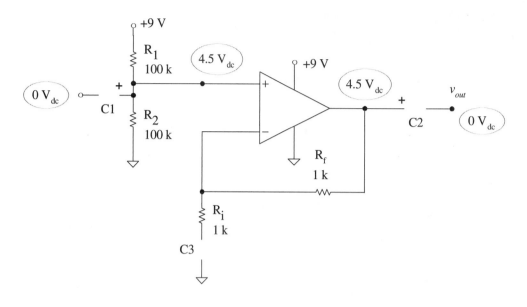

Figure 13-31(d) The dc model of circuit–*capacitors open*

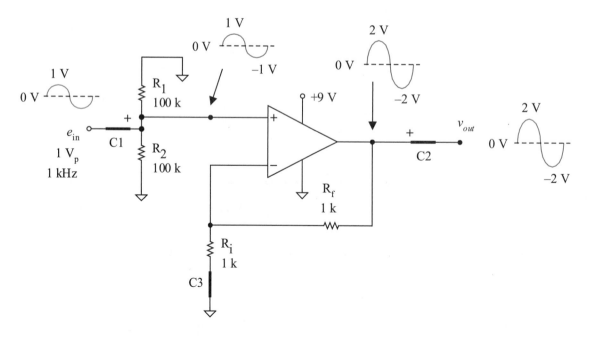

Figure 13-31(e) The ac model of circuit–*capacitors ideal short*

Power Supply with Filter Capacitor

The circuit of Figure 13-32 demonstrates the use of a filter capacitor in a dc power supply circuit. The half-wave rectifier of Figure 13-32(a) produces the positive half cycles at the output. The *filter capacitor* of Figure 13-32(b) charges up to the peak voltage and then holds a decaying dc voltage across the load. It is called a **filter capacitor** because it is filtering out the ac signal, bypassing it around the resistor R_{load}. The capacitor acts like an open to dc allowing a dc voltage drop across R_{load}.

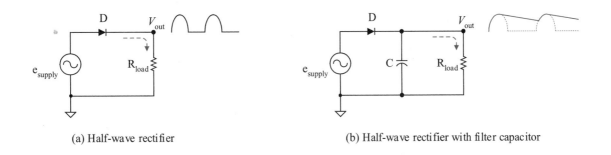

(a) Half-wave rectifier   (b) Half-wave rectifier with filter capacitor

Figure 13-32 Power supply filter capacitor

Summary

Capacitance is created when two conductors are separated by an insulator. A capacitor is a component specifically designed to have two conductors separated by an insulator. The insulating material separating the two plates is also called a dielectric. Dielectrics include air, vacuum, Teflon, paper, mica, Bakelite, and glass.

The capacitance value in terms of electrical units is defined to be

$$C = \frac{Q}{V}$$

where
C capacitance in units of farads (F)
Q charge in units of coulombs (C)
V voltage in units of volts (V)

Inserting an insulator between the plates of a capacitor increases its capacity to store charge per unit volt dropped across the capacitor. In terms of basic geometry, capacitance can be defined as

$$C = \varepsilon \frac{A}{d}$$

where

   C capacitance in units of farads (F)
   ε dielectric constant of insulator in farads per meter (F/m)
   A area of one of the plates in units of meters squared (m²)
   d distance between the plates in units of meters (m)

A few key characteristics that describe capacitance are that capacitance

- stores charge in the form of a voltage $V = Q/C$.

- retains voltage if disconnected from the circuit.

- returns charge (voltage) to a circuit like a voltage supply E.

- resists instantaneous change in voltage.

The instantaneous relationship of the voltage and current in a capacitor is

$$i_c(t) = C \frac{d v_c(t)}{dt}$$

Capacitance exhibits opposition to ac current. This reaction to ac current is called reactance. Specifically, it is called capacitive reactance for capacitance.

$$X_C = \frac{1}{2\pi f C}$$

where

   X_C capacitive reactance with units of ohms (Ω)
   f frequency of the ac sinusoidal signal with units of hertz (Hz)
   C capacitance with units of farads (F)

Capacitive reactance is in units of ohms like resistance, but the reactance magnitude cannot be directly added to resistance magnitude. A capacitor acts as an open to dc or low frequency sinusoids and as a short to high frequency sinusoids. Thus, dc and low frequency signals are *blocked* by a capacitor, while high frequency sinusoids are *coupled* or *bypassed* by a the capacitor. Capacitive reactance in series adds like resistance in series. The total capacitive reactance of three capacitors in series is

$$X_{C\text{ total}} = X_{C1} + X_{C2} + X_{C3}$$

Substituting the reactance for each capacitor

$$X_C = \frac{1}{2\pi f C}$$

leads to the formula for the equivalent total capacitance in series.

$$C_{total} = \left(\frac{1}{C1} + \frac{1}{C2} + \frac{1}{C3} \right)^{-1}$$

Capacitive reactance in parallel combines like resistance in parallel. The total reactance of three parallel capacitors is

$$\frac{1}{X_{C\,total}} = \frac{1}{X_{C1}} + \frac{1}{X_{C2}} + \frac{1}{X_{C3}}$$

Substituting individual reactances, the total capacitance of three parallel capacitances is

$$C_{total} = C1 + C2 + C3$$

The capacitor has many applications in electronic circuits. In an amplifier system, a coupling capacitor is used to block dc while coupling ac signals, and bypass capacitors pass ac signals around a parallel resistance while allowing a dc voltage drop. In a dc power supply, a filter capacitor shorts out unwanted ac signals while developing a dc voltage across the load.

Problems

Capacitors

13-1   The following voltages are applied to a 1 μF capacitor. Find the corresponding capacitor currents.

    a. $10\,V_{dc}$

    b. A $10\,V_p$ sine wave with a 100 Hz signal

    c. A $10\,V_p$ triangle wave with a 100 Hz signal

13-2   The following voltages are applied to a 1 nF capacitor. Find the corresponding capacitor currents.

    a. $-6\,V_{dc}$

    b. A $6\,V_p$ sine wave with a 10 kHz signal

    c. A $6\,V_p$ triangle wave with a 10 kHz signal

Capacitive Reactance

13-3   **a.** Find the reactance of a 100 µF capacitor at the frequencies of 0 Hz, 10 Hz, 100 Hz, 1 kHz, 10 kHz, and 100 kHz.

       b. Based upon your results would you model a capacitor with a signal of 0 Hz frequency as an *open* or a *short*?

       c. Based upon your results would you model a capacitor with a signal at very high frequencies as an *open* or a *short*?

13-4   Repeat Problem 13-3 for a 0.1 µF capacitor.

13-5   For the circuit of Figure 13-33, as frequency f *increases*, what happens to the capacitor reactance, the voltage across the capacitor, the voltage across the resistor, the current through the resistor, and the induced capacitor current?

13-6   For the circuit of Figure 13-33, as frequency f *decreases*, what happens to the capacitor reactance, the voltage across the capacitor, the voltage across the resistor, the current through the resistor, and the induced capacitor current?

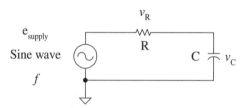

Figure 13-33

13-7   Use the circuit of Figure 13-34(a).

       a. If the frequency f is 10 Hz, find the reactance of the capacitor. Model the capacitor appropriately for this circuit as an open or a short, and find v_{out}.

       b. If the frequency f is 1 MHz, find the reactance of the capacitor. Model the capacitor appropriately for this circuit as an open or a short, and find v_{out}.

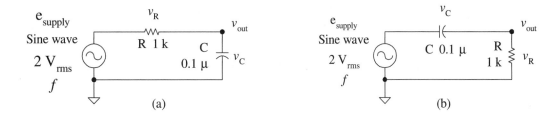

Figure 13-34

13-8   Repeat Problem 13-7 for the circuit of Figure 13-34(b).

Series and Parallel Capacitance

13-9   Three 100 μF capacitors are placed in series. Find the equivalent total series capacitance.

13-10   Three capacitors are placed in series: 100 μF, 47 μF, and 33 μF. Find the equivalent total series capacitance.

13-11   Three 100 μF capacitors are placed in parallel. Find the equivalent total parallel capacitance.

13-12   Three capacitors are placed in parallel: 100 μF, 47 μF, and 33 μF. Find the equivalent total parallel capacitance.

13-13   10 V is applied to the series combination of 100 μF, 47 μF, and 33 μF.
   a.   Find the total charge in the circuit.
   b.   Find the charge on each capacitor.
   c.   Find the voltage drop across each capacitor.

13-14   10 V is applied to the parallel combination of 100 μF, 47 μF, and 33 μF.
   a.   Find the total charge in the circuit.
   b.   Find the voltage drop across each capacitor.
   c.   Find the charge on each capacitor.

Capacitive Coupler and Filter

13-15 For the circuit of Figure 13-35(a):

    a. Draw the dc equivalent circuit with the capacitor modeled.

    b. Then find the dc voltage drop across the load resistor

    c. Find the dc voltage drop across the capacitor.

    d. Draw the ac equivalent circuit (assume ideal capacitor model).

    e. Then find the ac voltage drop across the load resistor.

    f. Find the ac voltage drop across the capacitor.

    g. Is the load voltage ac or dc?

13-16 Repeat Problem 13-15 for the circuit of Figure 13-35(b).

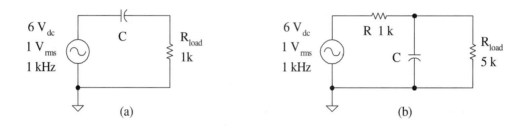

(a)   (b)

Figure 13-35

13-17 For the circuit of Figure 13-36:

    a. Find the dc voltage at the base, emitter, and collector of the BJT.

    b. Sketch the total voltage waveform at the base of the BJT.

    c. If the amplifier has a *voltage gain* of approximately R_C / R_E, find the peak-to-peak ac voltage at the collector.

    d. Draw the total waveform at the collector of the BJT.

    e. Draw the v_{out} waveform.

    f. What is the purpose of the two capacitors?

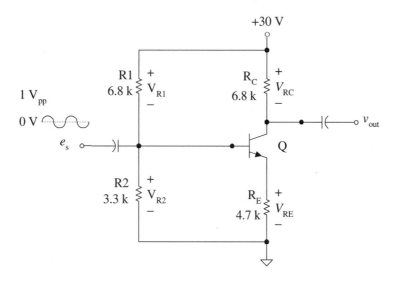

Figure 13-36

13-18 For the circuit of Figure 13-37:

   a. Draw the dc equivalent circuit.

   b. Then find the dc voltage at the op amp's output pin.

   c. Draw the ac equivalent circuit.

   d. Then find the ac voltage at the op amp's output pin.

   e. Draw the total waveform at the op amp's output pin.

   f. Draw the total waveform of the output voltage v_{out}.

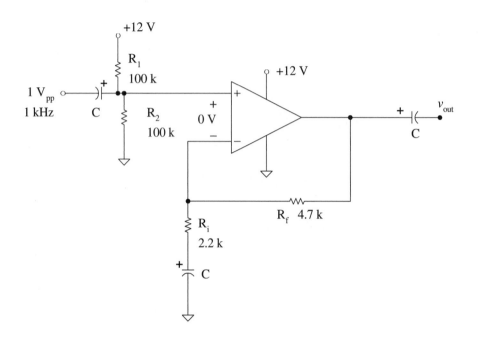

Figure 13-37

14

RC Switching Circuits

Introduction

When *RC* circuits are subjected to sudden changes in dc voltages or currents, exponential voltages and currents result throughout the *RC* circuit. A sudden change in dc voltage or current can occur from a switched circuit or from a rectangular supplied waveform.

Objectives

Upon completion of this chapter you will be able to:

- Model the capacitor based upon its initial conditions and its steady state condition.

- Find the initial and steady state values for a simple *RC* circuit.

- Find the *RC* circuit time constant tau τ.

- Calculate how long it takes to charge or discharge a capacitor in an *RC* circuit.

- Write the exponential general equations for an *RC* circuit subjected to a sudden change in dc voltage based upon its circuit time constant, its initial circuit values, and its steady state circuit values.

- Evaluate the instantaneous exponential general expression for a given time.

- Find the inverse solution of the exponential general expression for a given voltage or current.

- Define and calculate the time constant τ from an exponential rise curve or fall curve.

- Define and calculate the rise and fall time of a transient waveform.

- Sketch the voltage and current waveforms of an *RC* circuit subjected to a square wave.

14.1 Qualitative *RC* Transient

Figure 14-1 shows a simple *RC* switched circuit. This is a simple switched series circuit.

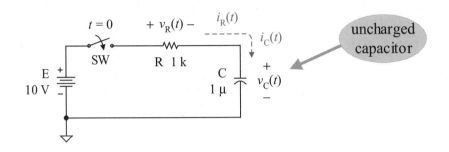

Figure 14-1 Simple *RC* transient dc circuit

Let's step through the thought process of what occurs before the switch is closed ($t < 0$), when the switch is closed ($t = 0$), and after the switch has been closed for a long time.

1. Before the switch is closed (t < 0), assume that the switch has been open for a long time and that the capacitor has been discharged. That is, the capacitor is initially uncharged. Since the capacitor is initially uncharged, the voltage across the capacitor is

$$V_C = \frac{Q}{C} = \frac{0 \text{ C}}{1 \text{ μF}} = 0 \text{ V} \qquad (t < 0)$$

    The switch is open, therefore no current is flowing through the circuit and the resistor has no voltage drop.

$$I_C = I_R = 0 \text{ mA} \qquad (t < 0)$$

$$V_R = 0 \text{ V} \qquad (t < 0)$$

2. Now assume the switch has just closed at t = 0 (or t = 0⁺) as shown in Figure 14-2. The capacitor resists any initial change in voltage but current starts rushing through the circuit to deposit charge on the plates of the capacitor. Electrons rush from the top capacitor plate to the (+) supply terminal and electrons rush from the (−) supply termi-

nal to the bottom capacitor plate. This is the induced current in the circuit due to charge rushing to and from the plates of the capacitor. There is no current in the capacitor itself; that would be equivalent to a lightning strike inside the capacitor.

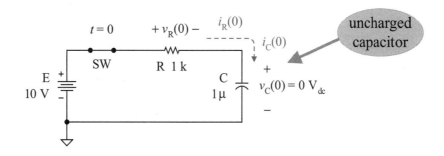

Figure 14-2 Switch has just closed at $t = 0$ (*initial circuit*)

The switch is closed at $t = 0$. Before the switch is closed, the capacitor voltage drop is 0 V (from step 1 above). *The capacitor resists a sudden change in voltage.* Since it is 0 V just before the switch is closed ($t = 0^-$), it is 0 V just after the switch is closed ($t = 0$). So,

$$V_C = 0 \text{ V} \qquad\qquad (t = 0)$$

Since the capacitor voltage drop is 0 V at $t = 0$, the resistor drops all of the applied voltage (Kirchhoff's voltage law).

$$V_R = 10 \text{ V} \qquad\qquad (t = 0)$$

Using Ohm's law for resistance, the resistor current is

$$I_R = \frac{10 \text{ V}}{1 \text{ k}\Omega} = 10 \text{ mA} \qquad (t = 0)$$

Since this is a series circuit,

$$I_C = I_R = 10 \text{ mA} \qquad\qquad (t = 0)$$

Summarizing, the *initial values* of this circuit at $t = 0$ are

$$V_C = 0 \text{ V}$$

$$V_R = 10 \text{ V}$$

$$I_C = I_R = 10 \text{ mA}$$

- The capacitor voltage has not changed instantly; it has remained at 0 V at the instant the switch is closed. *The uncharged capacitor is initially acting like a short.*

- The capacitor current has changed instantly from 0 mA to 10 mA (a *current spike*) at the instant the switch was closed.

Now assume the switch has been closed a very long time, that is $t \rightarrow \infty$ as shown in Figure 14-3. This is called the *steady state* condition since all the transients have died out. The voltages and currents are no longer changing; they have reached steady values.

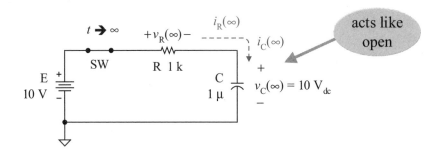

Figure 14-3 Switch closed for a very, very long time (*steady state circuit*)

3. The capacitor has fully charged to the applied voltage of 10 V.

$$V_C = 10 \text{ V} \qquad\qquad (t \rightarrow \infty)$$

Since the *capacitor voltage is no longer changing*, the induced current is 0 mA. Recall that a changing voltage across the capacitor is required to produce an induced capacitor current. This is equivalent to considering the capacitor to be an *open* in steady state.

$$I_C = 0 \text{ mA} \qquad\qquad (t \rightarrow \infty)$$

Since this is a series circuit,

$$I_R = I_C = 0 \text{ mA} \qquad\qquad (t \rightarrow \infty)$$

By Ohm's law, the resistor voltage drop must be 0 V.

$$V_R = 1 \text{ k}\Omega \times 0 \text{ mA} = 0 \text{ V} \qquad\qquad (t \rightarrow \infty)$$

Summarizing, the *steady state values* of this circuit as $t \rightarrow \infty$,

$$V_C = 10 \text{ V}$$

$$V_R = 0 \text{ V}$$

$$I_C = I_R = 0 \text{ mA}$$

- The capacitor has increased from an initial value of 0 V to a steady state, fully charged voltage of 10 V (the applied voltage).

- The capacitor current has decreased from an initial spike of 10 mA to a steady state value of 0 mA. *The fully charged capacitor is acting like an open.*

RC Time Constant τ

The time constant τ (the lowercase, Greek letter tau) of an *RC* circuit is defined to be its resistance value multiplied by its capacitance value.

$$\tau = RC$$

For the circuit of Figure 14-1 the time constant is

$$\tau = RC = 1 \text{ k}\Omega \times 1 \text{ μF} = 1 \text{ ms}$$

It takes approximately five (5) time constants to charge (or discharge) a capacitor to 99.3% of its steady state value. Thus, for the circuit of Figure 14-1, the capacitor is essentially fully charged in 5 ms.

$$t_{\text{full charge}} = 5\tau = 5 \times 1 \text{ ms} = 5 \text{ ms}$$

14.2 Switched *RC* Transient

Capacitor Models in a Switched DC *RC* Circuit

The following are models of a capacitor used in switched dc circuit analysis of an *RC* circuit. Initially the capacitor resists a change in voltage. If a capacitor voltage is 0 V just before the circuit is dc switched, the capacitor tends to stay at 0 V immediately after the switch is made. If the capacitor has a charge (that is, a voltage drop), it initially retains that voltage immediately after the switch is made. Thus the capacitors can be modeled initially as shown in Figure 14-4.

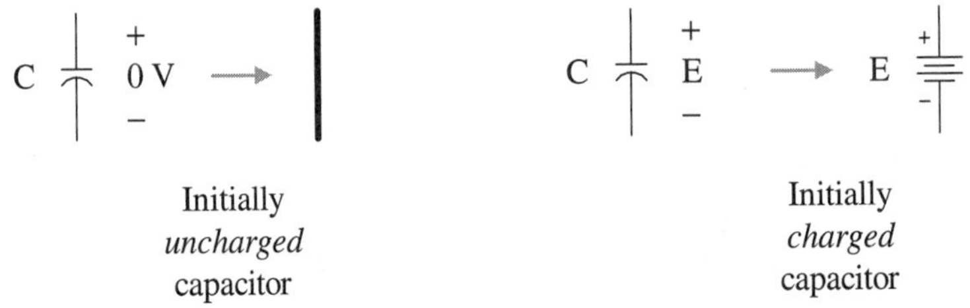

Figure 14-4 Possible initial dc models of a capacitor

In a steady state dc switched *RC* circuit, the capacitor is modeled as an ideal dc open as shown in Figure 14-5.

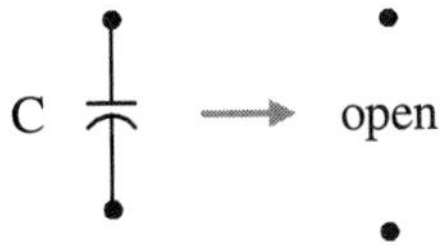

Transient Expression–Classical Calculus

Figure 14-5
Steady state dc
model

There are three distinct activities or phases in a dc transient: (1) initial circuit with its initial values, (2) the steady state circuit and its steady state values, and (3) the transient state as the voltages and currents go from their initial values to their steady state values. Initial and steady state circuits are relatively straightforward to analyze. Calculus provides the solution for the transient expressions of voltages and currents. Using the simple *RC* circuit of Figure 14-6, let's perform basic circuit analysis.

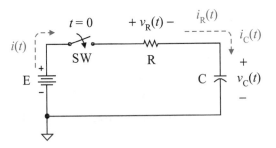

Figure 14-6 Basic *RC* transient circuit

This is a series circuit; therefore the currents in each component are the same at any instant in time.

$$i(t) = i_R(t) = i_C(t)$$

From Ohm's law,

$$v_R(t) = R\ i_R(t) = R\ i(t)$$

The voltage and current relationship of the capacitor

$$i_c(t) = C\frac{d\,v_c(t)}{dt}$$

can be rewritten as

$$v_c(t) = \frac{1}{C}\int_0^t i_c(t)\ dt$$

From Kirchhoff's voltage law,

$$E = v_R(t) + v_C(t)$$

Substituting the voltage expressions for $v_R(t)$ and $v_C(t)$, the KVL expression becomes

$$E = R\ i(t) + \frac{1}{C}\int_0^t i(t)\ dt$$

Using calculus (refer to a basic calculus text if so inclined), the solution to this integral-differential equation is

$$i(t) = i_R(t) = i_C(t) = \left(\frac{E}{R}\right)e^{-\frac{t}{RC}}$$

where e is the natural number forming an exponential decay function $e^{-t/RC}$, and RC is the time constant τ.

Using Ohm's law, the voltage drop across the resistor is

$$v_R(t) = E\ e^{-\frac{t}{RC}}$$

Using Kirchhoff's voltage law, the voltage drop across the capacitor is

$$v_C(t) = E - E\,e^{-\frac{t}{RC}} = E\left(1 - e^{-\frac{t}{RC}}\right)$$

Notice that the time constant RC is a natural outcome of this analysis. When $t = \tau$, the decay function is equal to e^{-1} (or, 0.37). That is equivalent to a decay of 63% from its starting value of 1.00.

General Transient Expression

The above process can be followed to find a general expression of all *RC* transient circuits. This approach requires that only three quantities be known and a generic expression can be written for any *RC* dc transient expression. The general solution for resistor or capacitor voltage is

$$v_x(t) = V_{ss} + (V_{init} - V_{ss})e^{-\frac{t}{\tau}}$$

where

$v_x(t)$	instantaneous voltage across a resistor or capacitor
V_{ss}	steady state voltage across the resistor or capacitor
V_{init}	initial voltage across the resistor or capacitor
τ	time constant *tau* or *RC* for resistance-capacitance circuits
$e^{-t/\tau}$	exponential decay function

The general solution for resistor or capacitor current is

$$i_x(t) = I_{ss} + (I_{init} - I_{ss})e^{-\frac{t}{\tau}}$$

where

$i_x(t)$	instantaneous current for the resistor or capacitor
I_{ss}	steady state current for the resistor or capacitor
I_{init}	initial current for the resistor or capacitor
τ	time constant *tau* or *RC* for resistance-capacitance circuits

These general expressions always work for any dc transient simple *RC* circuit. Later, this same general expression is used for *RL* (resistor-inductor) dc transient circuits.

Example 14-1

For the circuit of Figure 14-1, write the transient expression for the capacitor voltage and the capacitor current. Find the value of the capacitor voltage and current at $t = 0$ ms, 1 ms, 2 ms, and 5 ms.

Solution

Based upon preceding analysis, we know the following values:

$$\tau = RC = 1 \text{ ms}$$

$$V_{C \text{ init}} = 0 \text{ V}$$

$$V_{C \text{ ss}} = 10 \text{ V}$$

Substituting into the general expression of

$$v_x(t) = V_{ss} + (V_{init} - V_{ss}) \ e^{-\frac{t}{\tau}}$$

$$v_C(t) = 10 \text{ V} + (0 \text{ V} - 10 \text{ V}) e^{-\frac{t}{1 \text{ ms}}}$$

Simplifying the coefficient of the exponential term

$$v_C(t) = 10 \text{ V} - 10 \text{ V} \ e^{-\frac{t}{1 \text{ ms}}}$$

$$v_C(0 \text{ ms}) = 10 \text{ V} - 10 \text{ V} \ e^{-\frac{0 \text{ ms}}{1 \text{ ms}}} = 10 \text{ V} - 10 \text{ V} = 0 \text{ V}$$

$$v_C(1 \text{ ms}) = 10 \text{ V} - 10 \text{ V} \ e^{-\frac{1 \text{ ms}}{1 \text{ ms}}} = 10 \text{ V} - 10 \text{ V} \ e^{-1} = 6.32 \text{ V}$$

$$v_C(2 \text{ ms}) = 10 \text{ V} - 10 \text{ V} \ e^{-\frac{2 \text{ ms}}{1 \text{ ms}}} = 10 \text{ V} - 10 \text{ V} \ e^{-2} = 8.65 \text{ V}$$

$$v_C(5 \text{ ms}) = 10 \text{ V} - 10 \text{ V} \ e^{-\frac{5 \text{ ms}}{1 \text{ ms}}} = 10 \text{ V} - 10 \text{ V} \ e^{-5} = 9.93 \text{ V}$$

Notice that at $t = 5$ ms (that is, 5τ), the capacitor voltage is 9.93 V or 99.3% of its steady state value.

$$I_{C \text{ init}} = 10 \text{ mA}$$

$$I_{C \text{ ss}} = 0 \text{ mA}$$

Substituting into the general expression of

$$i_x(t) = i_{ss} + (i_{init} - i_{ss}) e^{-\frac{t}{\tau}}$$

$$i_C(t) = 0 \text{ mA} + (10 \text{ mA} - 0 \text{ mA}) e^{-\frac{t}{1 \text{ ms}}}$$

Simplifying the coefficient of the expression

$$i_C(t) = 10 \text{ mA } e^{-\frac{t}{1 \text{ ms}}}$$

$$i_C(0 \text{ ms}) = 10 \text{ mA } e^{-\frac{0 \text{ ms}}{1 \text{ ms}}} = 10 \text{ mA } e^{-0} = 10 \text{ mA}$$

$$i_C(1 \text{ ms}) = 10 \text{ mA } e^{-\frac{1 \text{ ms}}{1 \text{ ms}}} = 10 \text{ mA } e^{-1} = 3.68 \text{ mA}$$

$$i_C(2 \text{ ms}) = 10 \text{ mA } e^{-\frac{2 \text{ ms}}{1 \text{ ms}}} = 10 \text{ mA } e^{-2} = 1.35 \text{ mA}$$

$$i_C(5 \text{ ms}) = 10 \text{ mA } e^{-\frac{5 \text{ ms}}{1 \text{ ms}}} = 10 \text{ mA } e^{-5} = 0.07 \text{ mA}$$

Practice

For the circuit of Figure 14-1, write the transient expression for the resistor voltage and the resistor current. Find the value of the resistor voltage and current at $t = 0$ ms, 1 ms, 2 ms, and 5 ms.

Answer:

$$i_R(t) = 10 \text{ mA } e^{-\frac{t}{1 \text{ ms}}} \text{ , 10 mA, 3.68 mA, 1.35 mA, 0.07 mA}$$

$$v_R(t) = 10 \text{ V } e^{-\frac{t}{1 \text{ ms}}} \text{ , 10 V, 3.68 V, 1.35 V, 0.07 V}$$

Note that in the last practice problem, a quick solution would have been to equate the capacitor current and the resistor current from the example values, and then use Ohm's law to find the resistor voltage.

Unity Exponential Decay and Growth Functions

Figure 14-7 is a graph of the unity exponential decay curve (falling curve) and exponential growth curve (rising curve). The curves have been marked at arguments of 0, 1, 2, 3, 4, and 5. This is the basic shape of all exponential curves. The unity exponential decay curve starts at a value of 1 and falls to a value of 0 exponentially. The unity exponential growth curve starts at a value of 0 and rises to a value of 1 exponentially.

x	e^{-x}
0	1.000
1	0.368
2	0.135
3	0.050
4	0.018
5	0.007

x	$1 - e^{-x}$
0	0.000
1	0.632
2	0.865
3	0.950
4	0.982
5	0.993

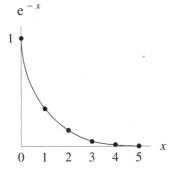

Exponential *fall* curve

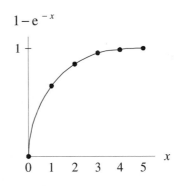

Exponential *rise* curve

Figure 14-7 Unity exponential *fall* and *rise* curves

The exponential function is the inverse function of the natural logarithm. Therefore, on the calculator, the e^x function key is typically the <2nd> function button of the <LN> key. For example, to find e^{-1} on the calculator, the key strokes are typically

2nd   LN   (−)   1

The (−) key is the unary negative key not the subtraction key.

A unique characteristic of the exponential curve is that if x is increased by one (1) the value of the function is always changed by a value of 63% toward its final destination. For the falling curve

$$e(1) = e(0) - 0.63\,(1-0) = 0.37$$

$$e(2) = e(1) - 0.63\,(0.37 - 0) = 0.14$$

$$e(3) = e(2) - 0.63\,(0.13 - 0) = 0.05$$

For the rising curve

$$e(1) = e(0) + 0.63\,(1-0) = 0.63$$

$$e(2) = e(1) + 0.63\,(1-0.63) = 0.86$$

$$e(3) = e(2) + 0.63\,(1-0.85) = 0.96$$

For quick sketching the exponential waveform, a simplifying assumption is made that if x is increased by one (1) the value of the function rises or falls by two-thirds (2/3) toward its final destination value.

Universal Exponential DC Transient Curves

Figure 14-8 is a graph of the universal exponential decay curve (falling curve) and a graph of the universal exponential growth curve (rising curve). These are the same as the unity exponential curves except the axes have been scaled indicating the initial value V_{init}, the steady state value V_{SS}, and the time scale in terms of the time constant τ. The curves have been marked at time t of 0, τ, 2τ, 3τ, 4τ, and 5τ. *These are the two basic waveshapes of all transient dc circuit responses.* The following relationships characterize all of these curves.

$$v(0) = V_{init}$$

$$v(5\tau) \approx V_{ss}$$

Figure 14-8 is a sketch of a dc voltage transient, but these curves are just as valid for *dc current transients*.

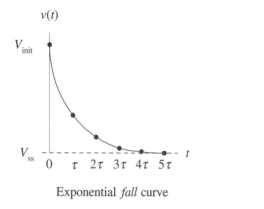

Exponential *fall* curve

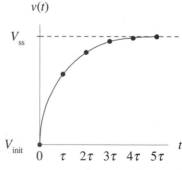

Exponential *rise* curve

Figure 14-8 Universal exponential dc fall and rise curves

Universal DC Transient Equation

Circuit analysis is performed and the initial, steady state, and time constants are found. Once these values are found, they are substituted into the universal dc transient equation

$$y(t) = Y_{ss} + \left(Y_{init} - Y_{ss}\right) e^{-\frac{t}{\tau}} \qquad \textbf{(14-1)}$$

where

$y(t)$	instantaneous R or C current or voltage
Y_{ss}	steady state R or C current or voltage
Y_{init}	initial R or C current or voltage
τ	circuit RC time constant
$e^{-t/\tau}$	exponential decay function

This universal dc transient equation works for every *RC* equivalent dc transient circuit whether the capacitor is charging or discharging from any initial value to any steady state value. The equation works for the transient capacitor voltage or current and for the transient resistor voltage or current.

Example 14-2

An *RC* transient dc circuit is analyzed. The initial capacitor voltage is −4 V, its steady state voltage is 10 V, and its time constant is 3 ms. Write the universal dc transient equation for this capacitor voltage. Verify the initial value at $t = 0$ and steady state value at $t = \infty$. Then evaluate the expression for $t = \tau, 2\tau, 3\tau, 4\tau,$ and 5τ. Sketch the waveform for five time constants.

Solution

Apply the universal dc transient equation for a capacitor voltage

$$v_C(t) = V_{C\ ss} + \left(V_{C\ init} - V_{C\ ss}\right) e^{-\frac{t}{\tau}}$$

$$V_{C\ init} = -4\ \text{V}$$

$$V_{C\ ss} = 10\ \text{V}$$

Substituting into the universal expression

$$v_C(t) = 10 \text{ V} + \left(-4 \text{ V} - (10 \text{ V})\right) e^{-\frac{t}{3 \text{ ms}}}$$

Simplifying the coefficient of the exponential term

$$v_C(t) = 10 \text{ V} - 14 \text{ V } e^{-\frac{t}{3 \text{ ms}}}$$

Guarantee the equation is correct by verifying the initial and steady state values.

$$v_C(0) = 10 \text{ V} - 14 \text{ V } e^{-0} = 10 \text{ V} - 14 \text{ V} = -4 \text{ V}$$

$$v_C(\infty) = 10 \text{ V} - 14 \text{ V } e^{-\infty} = 10 \text{ V}$$

The initial and steady state values have been verified, therefore this dc transient capacitor voltage equation is correct.

$$v_C(\tau) = v_C(3 \text{ ms}) = 10 \text{ V} - 14 \text{ V } e^{-1} = 4.85 \text{ V}$$

$$v_C(2\tau) = v_C(6 \text{ ms}) = 10 \text{ V} - 14 \text{ V } e^{-2} = 8.11 \text{ V}$$

$$v_C(3\tau) = v_C(9 \text{ ms}) = 10 \text{ V} - 14 \text{ V } e^{-3} = 9.30 \text{ V}$$

$$v_C(4\tau) = v_C(12 \text{ ms}) = 10 \text{ V} - 14 \text{ V } e^{-4} = 9.74 \text{ V}$$

$$v_C(5\tau) = v_C(15 \text{ ms}) = 10 \text{ V} - 14 \text{ V } e^{-5} = 9.91 \text{ V}$$

Sketch the equation. An accurate sketch would plot the points $(0, -4 \text{ V})$, $(1\tau, 4.85 \text{ V})$, $(2\tau, 8.11 \text{ V})$, $(3\tau, 9.30 \text{ V})$, $(4\tau, 9.74 \text{ V})$, and $(5\tau, 9.91 \text{ V})$. After 5τ, the curve has reached a steady state voltage of approximately 10 V. Figure 14-9 is a sketch of this dc voltage transient.

Let's look at a technique called a **quick sketch**. Draw and label the axes $v_C(V)$ and t. Scale the axes and put the values of V_{init}, V_{ss} and the time scale values in terms of time and τ. Then draw a dashed, horizontal line through the steady state point of 10 V. Since the initial value starts at −4 V and the steady state value is 10 V, this is a rising curve. First, plot the −4 V point. Start at $(0, -4 \text{ V})$, go over τ, and up 63% (about 2/3) of the way toward 10 V, and plot a dot. Continue this process for 2τ and 3τ. Now connect the dots and at 5τ draw the horizontal steady state line at 10 V. Try this technique on Figure 14-9. Your quick sketch technique should very closely match the accurate sketch. Unless great

graphing accuracy is required, the quick sketch technique is sufficient in sketching these curves.

t	t	$v_C(t)$
0	0 ms	-4.00 V
1τ	3 ms	4.85 V
2τ	6 ms	8.11 V
3τ	9 ms	9.30 V
4τ	12 ms	9.74 V
5τ	15 ms	9.91 V

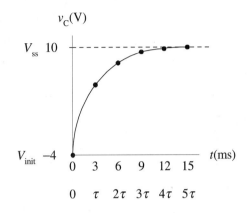

Exponential *rise* curve

Figure 14-9 Data table and sketch of $v_C(t) = 10$ V -14 V e$^{-\frac{t}{3\ \text{ms}}}$

Figure 14-10 demonstrates rising from 0 to τ, from τ to 2τ, and so
forth.

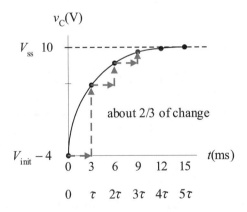

Exponential *rising* curve

Figure 14-10 Quick sketch technique

Practice

An *RC* transient dc circuit is analyzed. The initial capacitor voltage is 10 V, its steady state voltage is −4 V, and its time constant is 3 ms. Write the universal dc transient equation for this capacitor voltage. Verify the initial value at $t = 0$ and steady state value at $t = \infty$. Sketch the waveform for five time constants.

Answer: $v_C(t) = -4 \text{ V} + 14 \text{ V e}^{-\frac{t}{3 \text{ ms}}}$, $v_C(0) = 10 \text{ V}$, $v_C(\infty) = -4 \text{ V}$

See Figure 14-11, the quick sketch of the capacitor voltage.

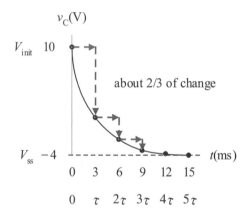

Figure 14-11 Quick sketch of practice problem

Universal Approach to Solve DC Transients

The following technique is used to solve all dc transient circuits with a simple *RC* circuit or Thévenin model equivalent circuit.

1. Find $V_C(0^-)$

    Find the capacitor voltage just before the switch is closed.

2. Find time constant τ

$$\tau = RC$$

If this is a complex circuit, other than a simple *RC* circuit, first find the Thévenin circuit using the capacitor as the load. Then use the Thévenin circuit for this process. If this is the Thévenin equivalent circuit, then the time constant is

$$\tau = R_{\text{Th}}C$$

3. Draw the modeled *initial* circuit (or Thévenin circuit). Properly model the capacitor as a *short* or *voltage supply*. If the capacitor is uncharged in step 1 above, then model the capacitor as a short. If the capacitor is charged in step 1 above, then model the capacitor as a voltage supply with the same polarity and voltage. See Figure 14-12. Find the desired initial voltages and currents.

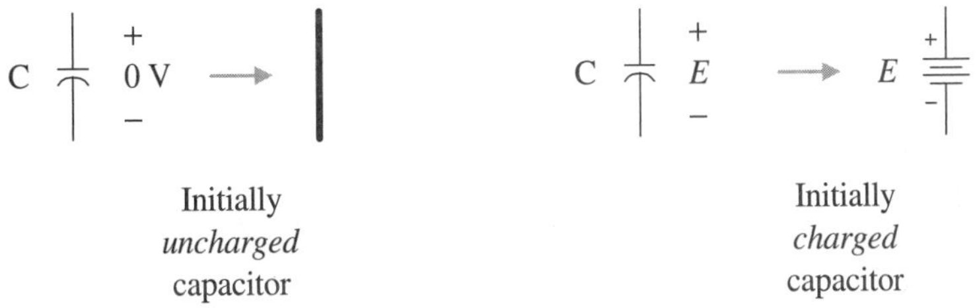

C ⊥ + 0 V → | C ⊥ + E → E

Initially
uncharged
capacitor

Initially
charged
capacitor

Figure 14-12 Possible initial models of capacitor

4. Draw the modeled *steady state* circuit or Thévenin circuit. Properly model the capacitor as an *open*. See Figure 14-13. Find the desired steady state voltages or currents.

5. If needed, write and apply the universal transient dc equation for the desired voltages and currents. Simplify equations.

$$y(t) = Y_{\text{ss}} + \left(Y_{\text{init}} - Y_{\text{ss}}\right)e^{-\frac{t}{\tau}}$$

6. If needed, sketch (or quick sketch) the desired voltages and currents.

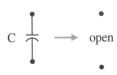

Figure 14-13
Steady state model

All simple *RC* circuits (or Thévenin modeled *RC* circuits) can be solved by this process. This textbook consistently follows this process for all transient dc circuit analysis.

Example 14-3

For the circuit of Figure 14-14(a), find the universal dc transient equation for the capacitor voltage and current. Sketch their exponential response for five time constants. Find the capacitor voltage and current at 3 ms. Assume that the switch has been open a long time and that the capacitor is completely discharged when the switch is closed at $t = 0$.

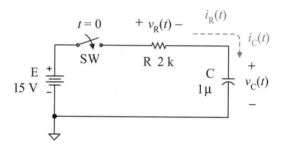

Figure 14-14(a) Example problem circuit

Solution

Follow the universal dc transient analysis process, step by step.

1. Find $v_C(0^-)$

    Since the capacitor is uncharged when the switch is closed at $t = 0$, then the capacitor voltage is 0 V just before the switch is closed.

    $$v_c(0^-) = 0 \text{ V}$$

2. Find the *time constant* τ. With the switch closed, this simple *RC* circuit is already in Thévenin model form.

    $$\tau = RC = 2 \text{ k}\Omega \times 1 \text{ }\mu\text{F} = 2 \text{ ms}$$

3. Model the *initial circuit* and find the initial capacitor voltage and current. To model the capacitor, go back to step 1. The capacitor is uncharged just before the switch is closed. The voltage across the capacitor is 0 V and therefore wants to stay at 0 V initially. Thus, the capacitor is modeled as a short. Draw the modeled circuit as shown in Figure 14-14(b). The switch is closed (short) and the capacitor is modeled as a *short*, dropping 0 V.

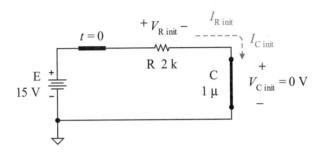

Figure 14-14(b) Initial circuit for example problem

$$V_{C\,init} = 0 \text{ V}$$

$$V_{R\,init} = 15 \text{ V}$$

$$I_{C\,init} = I_{R\,init} = \frac{15 \text{ V}}{2 \text{ k}\Omega} = 7.5 \text{ mA}$$

4. Model the *steady state circuit* and find the steady state capacitor voltage and current. The steady state capacitor is ideally an *open*. Draw the modeled circuit as shown in Figure 14-14(c). The switch is closed (short) and the capacitor is modeled as an *open*.

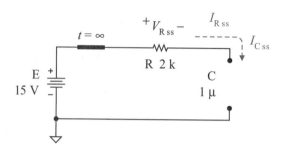

Figure 14-14(c) Steady state circuit for example problem

$$I_{C\,ss} = 0 \text{ mA}$$

$$V_{R\,ss} = 0 \text{ V}$$

$$V_{C\,ss} = 15 \text{ V}$$

5. Write and apply the universal transient dc equation for the desired voltages and currents. *Simplify the instantaneous equations.*

$$y(t) = Y_{ss} + (Y_{init} - Y_{ss}) e^{-\frac{t}{\tau}}$$

Write and simplify the capacitor transient voltage equation.

$$v_C(t) = 15 \text{ V} + (0 \text{ V} - (15 \text{ V})) e^{-\frac{t}{2 \text{ ms}}}$$

$$v_C(t) = 15 \text{ V} - 15 \text{ V } e^{-\frac{t}{2 \text{ ms}}}$$

Write and simplify the capacitor transient current equation.

$$i_C(t) = 0 \text{ mA} + (7.5 \text{ mA} - (0 \text{ mA})) e^{-\frac{t}{2 \text{ ms}}}$$

$$i_C(t) = 7.5 \text{ mA } e^{-\frac{t}{2 \text{ ms}}}$$

Find the capacitor voltage and current values at $t = 3$ ms.

$$v_C(3 \text{ ms}) = 15 \text{ V} - 15 \text{ V } e^{-\frac{3 \text{ ms}}{2 \text{ ms}}} = 11.65 \text{ V}$$

$$i_C(3 \text{ ms}) = 7.5 \text{ mA e}^{-\frac{3 \text{ ms}}{2 \text{ ms}}} = 1.67 \text{ mA}$$

6. Quick sketch the capacitor transient voltage and current. See
    Figure 14-14(d).

    The capacitor voltage and current values at $t = 3$ ms can be
    estimated. The above calculations of 11.65 V and 1.67 mA at
    3 ms match closely with the curves.

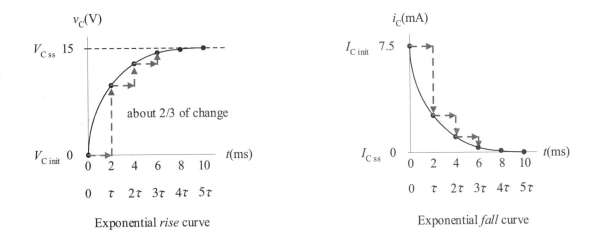

Figure 14-14(d) Quick sketch of example capacitor dc transient voltage and current

Practice

Use the circuit of the example problem of Figure 14-14(a) but
assume that the capacitor had been previously charged to 9 V and
that its voltage is 9 V just before the switch is closed. Find the
universal dc transient equation for the capacitor voltage and current. Sketch their exponential response for five time constants.

Hint: Since the capacitor has an initial charge, the model of the
capacitor in step 4 must be a 9 V voltage supply as shown in Figure 14-14(e). Thus the initial capacitor voltage and current values
change and their resulting equations change. Notice that the
curves remain the same except for the initial values.

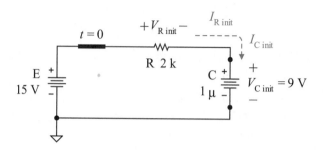

Figure 14-14(e) Initial model of practice problem

The capacitor transient dc voltage and current sketches are shown in Figure 14-14(f).

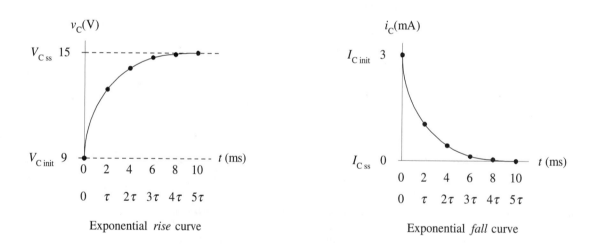

Exponential *rise* curve

Exponential *fall* curve

Figure 14-14(f) Practice problem capacitor transient voltage and current sketches

$$v_C(t) = 15 \text{ V} - 6 \text{ V e}^{-\frac{t}{2 \text{ ms}}}$$

$$i_C(t) = 3 \text{ mA e}^{-\frac{t}{2 \text{ ms}}}$$

Inverse Solution

In the last example, a time of 3 ms was specified and the capacitor voltage and current calculated with their respective dc transient equations and/or estimated from their sketches. Figure 14-15 demonstrates this technique of finding voltage and current graphically given the time.

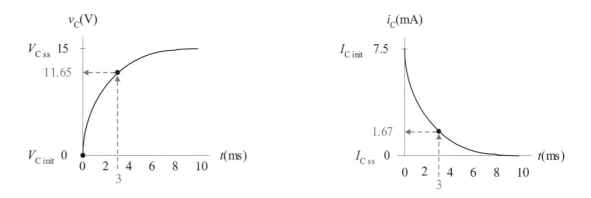

Figure 14-15 Given a time ($t = 3$ ms) find the voltage and current values

Suppose the voltage or current is given and the time must be found to achieve those values during the transient.

Example 14-4

Use the circuit of Example 14-3, Figure 14-14(a). Find the time required for the capacitor to charge to 8 V.

Solution

Figure 14-16 demonstrates the example problem. Start at 8 V, draw a line across until the curve is intersected, and draw a line down until time axis is intersected. That is the time that is needed. It appears to be approximately 1.5 ms. Is there a more accurate way to find this time for the capacitor to charge to 8 V?

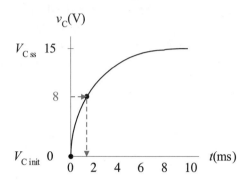

Figure 14-16 Example problem finding time

The capacitor dc transient voltage equation can be used to find this time more accurately. From Example 14-3, the capacitor dc transient voltage equation is

$$v_C(t) = 15 \text{ V} - 15 \text{ V e}^{-\frac{t}{2 \text{ ms}}}$$

Substitute the capacitor voltage value into the equation and solve the equation for time t. The intermediate goal is solve for the exponential term first.

$$8 \text{ V} = 15 \text{ V} - 15 \text{ V e}^{-\frac{t}{2 \text{ ms}}}$$

Subtract 15 V from both sides of the equation.

$$-7 \text{ V} = -15 \text{ V e}^{-\frac{t}{2 \text{ ms}}}$$

Divide both sides by –15 V.

$$\frac{-7 \text{ V}}{-15 \text{ V}} = e^{-\frac{t}{2 \text{ ms}}}$$

or

$$e^{-\frac{t}{2 \text{ ms}}} = 0.4666$$

The exponential term is now isolated. Our goal is still to solve for the time *t*. Time *t* is buried in the argument of the exponent. The only way to extract a function argument is to perform the inverse function of the function. The inverse function of the exponential function is the natural logarithm (ln). Take the natural log of both sides of the equation.

$$\ln\left(e^{-\frac{t}{2\text{ ms}}}\right) = \ln\left(0.4666\right)$$

The inverse function of a function extracts the argument. In this case, the argument is the entire exponent, namely, −*t*/2 ms. Perform the natural log (ln) operation on both sides of the equation.

$$-\frac{t}{2\text{ ms}} = -0.762$$

The exponent pops out of the inverse expression. Now simply use algebra to solve for time *t*.

$$t = 0.762 \times 2\text{ ms} = 1.524\text{ ms}$$

This result is very close to our graphical prediction of 1.5 ms, but the equation is much more accurate in forecasting the value.

Practice

Use the circuit of Example 14-3, Figure 14-14(a). Find the time required for the capacitor current to reach 2.5 mA.

Answer: 2.198 ms

Multiple *RC* Switched Circuit

The following example combines several concepts: a somewhat complex dc switching *RC* circuit with multiple resistors and capacitors, a partial charge with the switch closed, and opening the circuit producing a discharge circuit.

Example 14-5

The capacitors in the circuit of Figure 14-17(a) are uncharged before the switch is closed at $t = 0$. The switch is closed for 4 ms and then the switch is opened again. Find and sketch the C_{total} capacitor voltage and current waveforms for a total elapsed time of 34 ms. Be sure to show proper axes labels and proper scale values. Use the universal dc transient approach and equations for each transient.

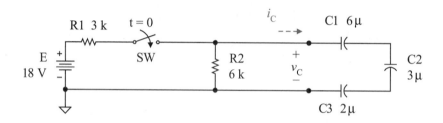

Figure 14-17(a) Example switched *RC* circuit

Solution

This solution requires several major steps consisting of several smaller steps. Let's plan the overall strategy before calculating any numbers.

Major step 1

Evaluate the circuit just before the switch is closed and determine the capacitive voltage $v_C(0^-)$.

Major step 2

For the charging cycle, produce the Thévenin model of the circuit with the switch closed. This will be needed for the first transient circuit analysis during the charging of capacitors.

Major step 3

For the charging cycle, use the Thévenin model of the circuit and perform universal dc transient circuit analysis on the charging cycle of the capacitance C_{total}. Evaluate the capacitance voltage and current response for 4 ms.

Major step 4

For the discharging cycle, produce the Thévenin model of the circuit when the switch is open. This will be needed for the second transient circuit analysis during the discharging of the capacitors.

Major step 5

Perform universal dc transient circuit analysis on the discharging cycle of the capacitance C_{total} after the switch is opened at $t = 4$ ms.

Major step 6

Sketch the voltage and current responses for capacitance C_{total} for a total elapsed time of 34 ms.

Start at step 1 and proceed with each step.

Major step 1

Evaluate $v_C(0^-)$. This is a relatively straightforward step. Since the capacitors are uncharged before the switch is closed and there is no supply voltage to the capacitors with the switch open, the voltage across the capacitance before the switch is closed is 0 V.

$$v_C(0^-) = 0 \text{ V}$$

Major step 2

Find the Thévenin model of the resistive circuit with switch closed and the capacitive load removed. Find the Thévenin model of the circuit of Figure 14-17(b).

Charging cycle before switch closed $v_C(0^-)$

Charging cycle Thévenin model

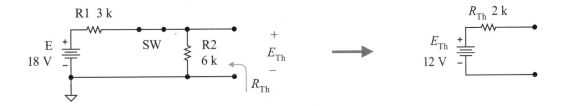

Figure 14-17(b) Charging Thévenin model circuit

The open circuit voltage is

$$E_{\text{Th}} = \left(\frac{6 \text{ k}\Omega}{9 \text{ k}\Omega}\right) 18 \text{ V} = 12 \text{ V}$$

Looking back into the load terminals and zeroing the voltage supply (replacing it with a short), the Thévenin resistance is

$$R_{\text{Th}} = \left(\frac{1}{3 \text{ k}\Omega} + \frac{1}{6 \text{ k}\Omega}\right)^{-1} = 2 \text{ k}\Omega$$

Major step 3

Perform universal dc transient analysis of the charging cycle.

It has been established that the voltage across capacitance C_{total} is 0 V before the switch is thrown.

$$v_C(0^-) = 0 \text{ V}$$

Find the charging circuit time constant τ_{charge}. The Thévenin resistance R_{Th} and total capacitance C_{total} must be used to determine the circuit time constant. See Figure 14-17(c) for the equivalent charging circuit. The total capacitance is

$$C_{\text{total}} = \left(\frac{1}{6 \text{ μF}} + \frac{1}{3 \text{ μF}} + \frac{1}{2 \text{ μF}}\right)^{-1} = 1 \text{ μF}$$

$$\tau = R_{\text{Th}} \, C_{\text{total}} = 2 \text{ k}\Omega \times 1 \text{ μF} = 2 \text{ ms}$$

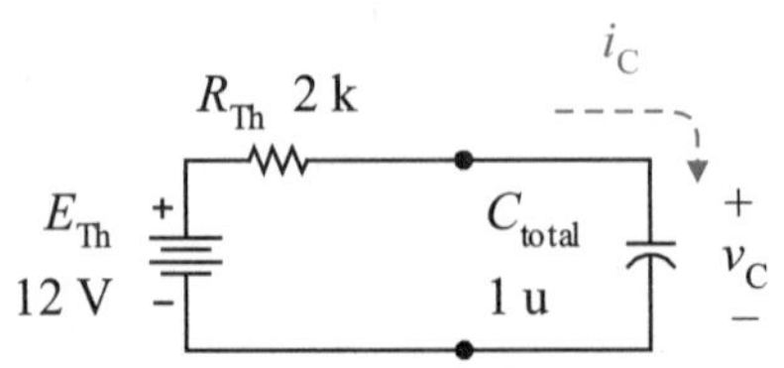

Figure 14-17(c) Charging Thévenin model with load

Draw the initial model of the charging circuit and find the initial voltage and current of C_{total}. Since the capacitors are uncharged just before the switch is thrown, they are modeled as shorts as shown in Figure 14-17(d).

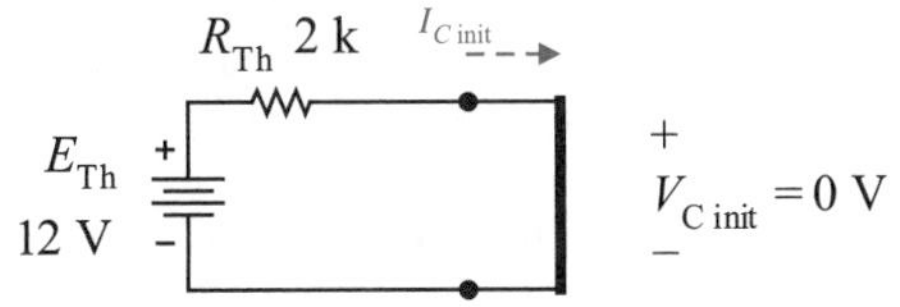

Figure 14-17(d) Charging initial circuit analysis

$$V_{C\ init} = v_C(0^-) = 0 \text{ V}$$

$$I_{C\ init} = \frac{12 \text{ V}}{2 \text{ k}\Omega} = 6 \text{ mA}$$

Draw the steady state model of the charging circuit and find the steady state voltage and current of the capacitance. A steady state capacitor is modeled as an open, thus the capacitance C_{total} is modeled as an open. The steady state circuit is shown in Figure 14-17(e). Since the capacitance is acting as an open, the circuit current is 0 mA and the drop across R_{Th} is 0 V. The capacitance combination C_{total} must therefore drop 12 V.

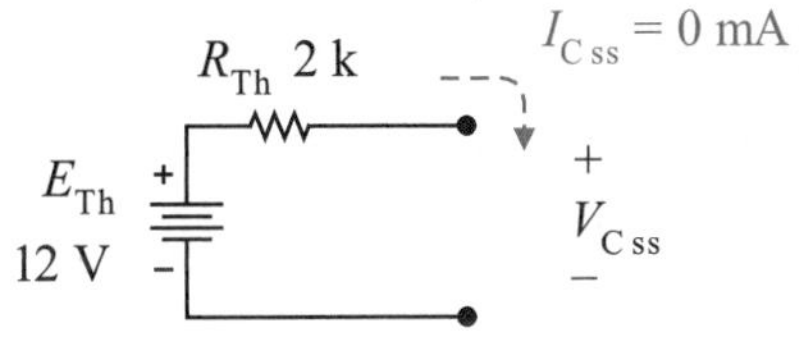

Figure 14-17(e) Charging steady state circuit analysis

$$I_{C\,ss} = 0 \text{ mA}$$

$$V_{C\,ss} = 12 \text{ V}$$

Write the universal dc transient *RC* equations and simplify them. Evaluate the expression for $t = 4$ ms when the switch is opened again to determine the capacitance voltage and current at that time.

The capacitance charging voltage expressions are

$$v_C(t) = 12 \text{ V} + \left(0 \text{ V} - (12 \text{ V})\right) e^{-\frac{t}{2 \text{ ms}}}$$

$$v_C(t) = 12 \text{ V} - 12 \text{ V } e^{-\frac{t}{2 \text{ ms}}}$$

$$v_C(4 \text{ ms}) = 12 \text{ V} - 12 \text{ V } e^{-\frac{4 \text{ ms}}{2 \text{ ms}}} = 10.4 \text{ V}$$

The capacitance charging current expressions are

$$i_C(t) = 0 \text{ mA} + \left(6 \text{ mA} - 0 \text{ mA}\right) e^{-\frac{t}{2 \text{ ms}}}$$

$$i_C(t) = 6 \text{ mA } e^{-\frac{t}{2 \text{ ms}}}$$

$$i_C(4 \text{ ms}) = 6 \text{ mA } e^{-\frac{4 \text{ ms}}{2 \text{ ms}}} = 0.812 \text{ mA}$$

Major step 4

Find the Thévenin model of the resistive circuit with the switch open and the capacitive load removed. This is simply the resistor R2's resistance as shown in Figure 14-17(f).

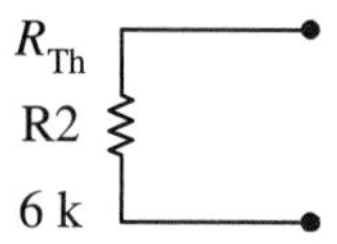

The open circuit voltage is

$$E_{Th} = 0 \text{ V}$$

$$R_{Th} = R2 = 6 \text{ k}\Omega$$

R_{Th}
R2
6 k

Figure 14-17(f)
Discharging cycle
Thévenin model

Major step 5

Perform universal dc transient circuit analysis on the discharging cycle of the capacitor C_{total} after the switch is opened at $t = 4$ ms.

Find the voltage just before the switch is opened. It has been established during the charge cycle that the voltage across capacitor C_{total} at 4 ms just before the switch is opened is 10.4 V. Thus, for the charging cycle

$$v_C(0^-) = 10.4 \text{ V}$$

Find the discharging circuit time constant $\tau_{discharge}$. The Thévenin resistance R_{Th} and total capacitance C_{total} must be used to determine the circuit time constant. See Figure 14-17(g) for the equivalent charging circuit. The total capacitance is the same as in the charging cycle.

$$C_{total} = 1 \, \mu F$$

$$\tau_{discharge} = R_{Th} \ C_{total} = 6 \text{ k}\Omega \times 1 \, \mu F = 6 \text{ ms}$$

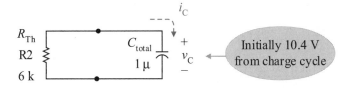

Figure 14-17(g) Example discharging circuit

Draw the initial model of the discharging circuit and find the initial voltage and current of C_{total}. Since capacitors are charged just before the switch is thrown, they are modeled as a voltage supply as shown in Figure 14-17(h).

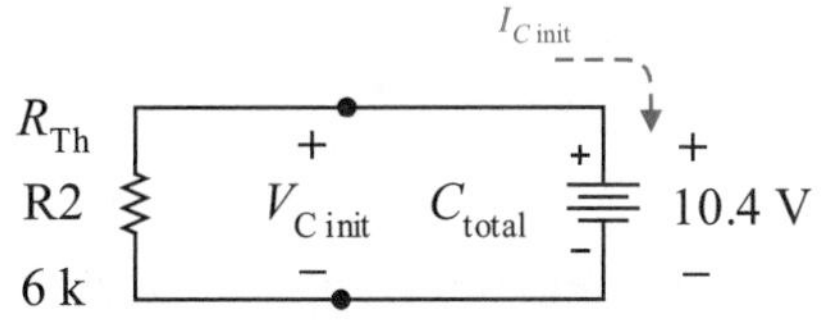

Figure 14-17(h) Discharging initial circuit analysis

$$V_{C\text{ init}} = v_C(0^-) = 10.4 \text{ V}$$

$$I_{C\text{ init}} = -\frac{10.4 \text{ V}}{6 \text{ k}\Omega} = -1.73 \text{ mA}$$

Draw the steady state model of the discharging circuit and find the steady state voltage and current of *C*. See Figure 14-17(i).

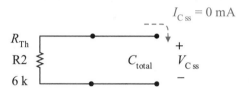

Figure 14-17(i) Discharging circuit steady analysis

$$I_{C\text{ ss}} = 0 \text{ mA}$$

$$V_{C\text{ ss}} = 0 \text{ V}$$

Write the *discharging* universal dc transient *RC* equations and simplify them. The C_{total} *discharging voltage* expressions are

$$v_C(t) = 0 \text{ V} + (10.4 \text{ V} - (0 \text{ V}))e^{-\frac{t}{6 \text{ ms}}}$$

$$v_C(t) = 10.4 \text{ V } e^{-\frac{t}{6 \text{ ms}}}$$

The C_{total} *discharging current* expressions are

$$i_{C2}(t) = 0 \text{ mA} + (-1.73 \text{ mA} - 0 \text{ mA})e^{-\frac{t}{6 \text{ ms}}}$$

$$i_{C2}(t) = -1.73 \text{ mA } e^{-\frac{t}{6 \text{ ms}}}$$

Major step 6
Figure 14-17(j) is sketch of the capacitance voltage and current response curves $v_C(t)$ and $i_C(t)$.

The charging cycle lasts for two time constants (2τ or 4 ms). The capacitor voltage starts at 0 V and climbs to 10.4 V at 4 ms. The switch is opened and the capacitor voltage decays with a time constant of 6 ms down to 0 V in 30 ms (34 ms total elapsed time).

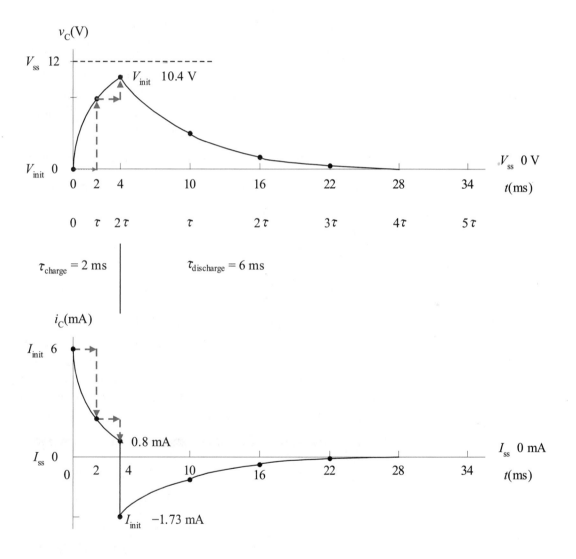

Figure 14-17(j) Capacitance C_{total} charging and discharging curves

The capacitance current response curve $i_C(t)$ charging cycle lasts for two time constants (2τ or 4 ms). The capacitor current spikes to 6 mA and falls to 0.8 mA at 4 ms. The switch is then opened and the capacitor current changes direction and magnitude. It spikes from 0.8 mA down to −1.73 mA and then rises to 0 mA with a time constant of 6 ms.

Note the change in capacitor current direction. During the charging cycle, current is flowing into the top of the capacitor, the same direction as the reference current. When the switch is thrown open at $t = 4$ ms, the capacitor discharges, creating a current up and out of the top capacitor plate. This current direction is opposite the reference and thus has a negative (−) value.

Practice
Find and sketch the voltage and current waveforms of the C2 capacitor.

Answer: Same current curve (series circuit), v_{C2} is one-third v_C voltage.

14.3 *RC* Square Wave Response

The square wave supply produces two dc voltages that the *RC* circuit responds to, V_{max} and V_{min}. It is equivalent to an *RC* switched circuit switching between two dc supply voltages. During the first half cycle the capacitor charges towards V_{max} and on the second half cycle the capacitor charges toward V_{min}. This is the key concept of viewing the action of the *RC* square wave response circuit.

Example 14-6
Sketch the capacitor and resistor voltage and current waveforms of the *RC* circuit of Figure 14-18(a).

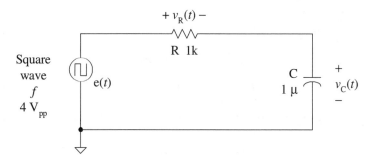

Figure 14-18(a) Example *RC* circuit square wave response

Solution

First find the maximum and minimum voltages of the supplied square wave voltage. The supply is a pure ac signal, therefore

$$E_{max} = \frac{4 \text{ V}_{pp}}{2} = 2 \text{ V}_p$$

$$E_{min} = -2 \text{ V}_{min}$$

If the capacitor is allowed to fully charge on each half cycle by the supplied voltage signal, that is, if

$$\frac{T}{2} \geq 5\tau$$

or

$$T \geq 10\tau$$

then there is sufficient time to fully charge the capacitor such that

$$V_{max \ C} = 2 \text{ V}_p$$

$$V_{min \ C} = -2 \text{ V}_{min}$$

Ideally, if the $T = 10\tau$, then the capacitor has just enough time to fully charge to E_{max} on the first half cycle. And it has just enough time to charge to E_{min} on the negative half cycle. Basically the capacitor is exponentially rising from -2 V to $+2$ V on the first half cycle, and falling exponentially from $+2$ V to -2 V on the second half cycle. The capacitor voltage is the key, so it needs to be the first sketch. For the circuit of Figure 14-18(a), this ideal period is

$$T = 10\tau$$

$$\tau = RC = 1 \text{ k}\Omega \times 1 \text{ } \mu\text{F} = 1 \text{ ms}$$

$$T = 10 \text{ } \tau = 10 \times 1 \text{ k}\Omega \times 1 \text{ } \mu\text{F} = 10 \text{ ms}$$

This requires that the frequency of the voltage supply be set to

$$f = \frac{1}{10 \text{ ms}} = 100 \text{ Hz}$$

to produce this ideally just charged capacitor waveform each half cycle.

Figure 14-18(b) shows a sketch of the voltage and current responses of this *RC* circuit. The capacitor at the end of a half cycle is fully charged, acts like an open, and drops all of the applied voltage (either +2 V or –2 V). Notice that the $v_C(t)$ curve is exponentially charging toward +2 V and then alternately to –2 V.

The $v_R(t)$ waveform can be found by Kirchhoff's voltage law

$$v_R(t) = e(t) - v_C(t)$$

Perform a point-by-point subtraction and the resistor voltage curve is generated. For example, at $t = 0^+$, just after the supply has changed to +2 V and the capacitor is still at –2 V,

$$v_R(0^+) = e(0^+) - v_C(0^+) = 2 \text{ V} - (-2 \text{ V}) = 4 \text{ V}$$

Use Ohm's law to find the current waveform.

$$i_R(t) = \frac{v_R(t)}{R} = \frac{v_R(t)}{1 \text{ k}\Omega}$$

$$I_{\text{max R}} = \frac{4 \text{ V}_p}{1 \text{ k}\Omega} = 4 \text{ mA}_p$$

$$I_{\text{min R}} = -\frac{4 \text{ V}_{\text{min}}}{1 \text{ k}\Omega} = -4 \text{ mA}_{\text{min}}$$

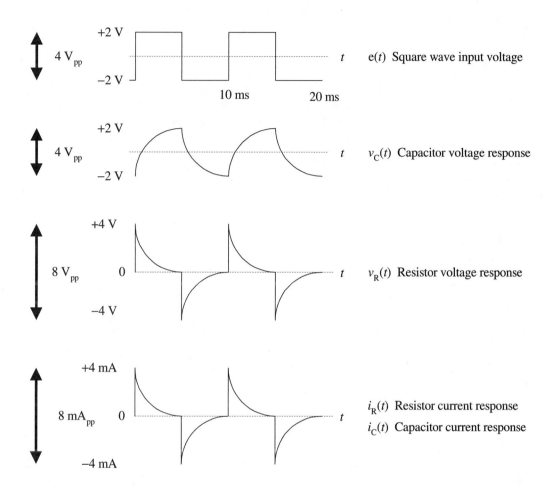

Figure 14-18(b) *RC* circuit square wave response waveforms

The shape of the resistor voltage and current waveforms must be identical since they are proportional (related by a constant).

Since this is a series circuit, the capacitor current must be the same as the resistor current. Therefore, it has the same shape and same values.

$$i_C(t) = i_R(t)$$

Practice

The supply in the example problem has a 2 V_{dc} offset added to the signal. Sketch the output capacitor and resistor voltage and current waveforms.

Answer: See Figure 14-18(c). Note that the capacitor voltage is offset also by the same 2 V_{dc}; however, since the capacitor current depends only upon the change in voltage (that is, its peak-to-peak value), the capacitor current, resistor current, and thus the resistor voltage waveforms are unchanged by the dc offset.

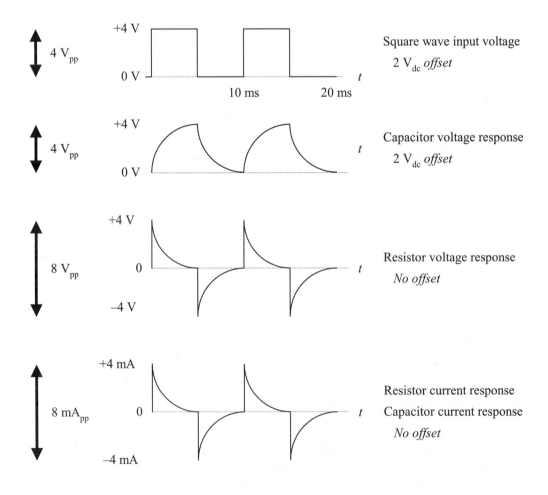

Figure 14-18(c) Practice solution with 2 V_{dc} offset in supplied voltage

Exploration

Perform a MultiSIM simulation of the circuit of Figure 14-18(a). Observe the output at a frequency of 100 Hz. See Figure 14-18(d) for the simulated output (use *auto* triggering). Then change the frequency to 10 Hz and to 100 Hz and observe the effect that a longer period and a shorter period have on the signal.

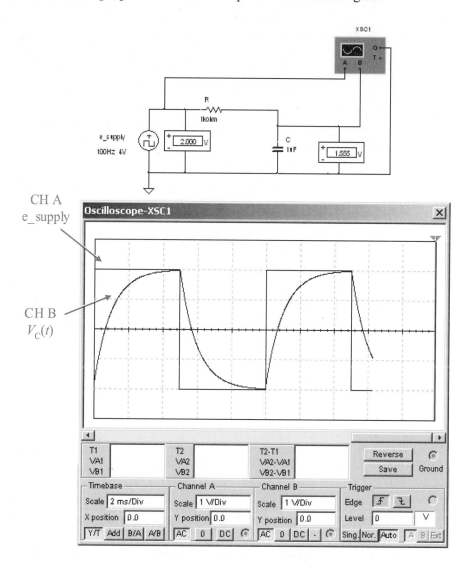

Figure 14-18(d) Exploration MultiSIM simulation of *RC* circuit driven by a square wave

RC Square Wave Frequency Effects

As noted in the last example, if $T = 10\tau$ then the frequency of the square wave allows just enough time during each half cycle to just charge the capacitor to the applied voltage. Figure 14-19 demonstrates the principle of the signal's period relative to the circuit time constant. Compare these wave shapes with those of the previous exploration problem.

If $T \gg 10\tau$, that is the frequency is relatively low, the capacitor charges to its maximum voltage and holds that voltage until the next half cycle drives the capacitor exponentially to its minimum voltage.

If $T \ll 10\tau$, the capacitor does not have time to fully charge and only a partial charge occurs each half cycle. If the period is too short, that is the frequency is too high, the capacitor charges only slightly towards its maximum and slightly towards its minimum and appears to be a linear triangle wave.

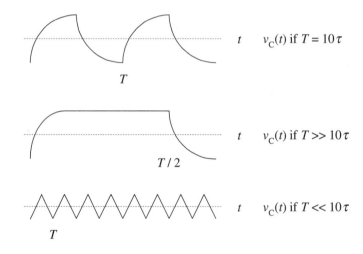

$t \qquad v_{\mathrm{C}}(t)$ if $T = 10\tau$

$t \qquad v_{\mathrm{C}}(t)$ if $T \gg 10\tau$

$t \qquad v_{\mathrm{C}}(t)$ if $T \ll 10\tau$

Figure 14-19 Relative comparison of signal period and 10τ

Measuring τ with an Oscilloscope

To measure the time constant with the scope, the transient portion of a square wave response is used. Figure 14-20 demonstrates the technique. First select a convenient transient portion of the curve to measure. The time constant is the same for all of the transient portions of the curve; therefore, choose whichever is the most convenient and accurate to measure.

1. Select the transient curve to use.

2. Expand the signal to fill as much of the screen as possible.

3. Measure the number of divisions between the steady state and the initial values (shown as Δ in Figure 14-20).

4. Calculate 63% of Δ.

5. Start at the initial value point and go up (or down) 63% of Δ.

6. Go across the screen and count the number of divisions until the transient curve is intersected.

7. Using the oscilloscope's time scale setting, translate the number of horizontal division into time. That is the time constant τ.

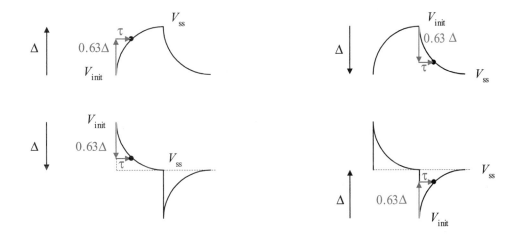

Figure 14-20 Measuring the time constant on an oscilloscope

14.4 Rise Time–Fall Time

It is often difficult to discern exactly when a transient signal starts or reaches its final value. The terms **rise time** and **fall time** are a standard that define how long it takes a signal to rise or fall from 10% of change to

90% of change. The 10% and 90% points are typically clearly discernable as opposed to the actual starting and ending points of the transient wave-form.

Rise Time

For a rising curve, the rise time t_r of a transient curve is defined to be the time required for the signal to go from 10% of its signal change to 90% of its signal change. Figure 14-21 demonstrates this signal parameter.

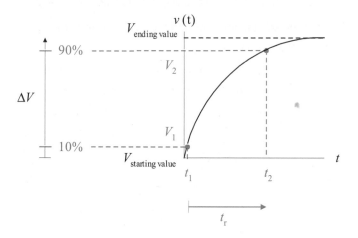

Figure 14-21 Rise time for a transient curve

The total change in the transient signal is

$$\Delta V = V_{\text{ending value}} - V_{\text{starting value}}$$

The time t_1 corresponds to the time for the signal to rise to 10% of its change:

$$v(t_1) = V_{\text{starting value}} + 0.1\,\Delta V$$

The time t_2 corresponds to the time for the signal to rise to 90% of its change:

$$v(t_2) = V_{\text{starting value}} + 0.9\,\Delta V$$

The rise time t_r is the difference of these times

$$t_r = t_2 - t_1$$

Example 14-7

Find the rise time t_r for the *RC* transient signal of Figure 14-22(a) plotted from a spreadsheet. A spreadsheet is a relatively quick and easy way to sketch waveforms. Find the time constant τ. For an *RC* transient circuit, there is a fixed relationship between signal rise time and the circuit *RC* time constant. Find the rise time relative to the time constant by dividing the rise time by the circuit time constant.

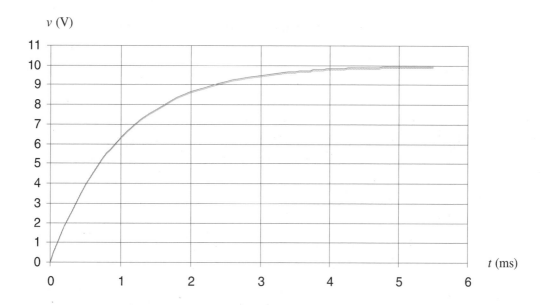

Figure 14-22(a) Example to determine rise time

Solution

Find the total signal voltage change of the transient.

$$\Delta V = 10 \text{ V} - (-0 \text{ V}) = 10 \text{ V}$$

The voltage V_1 corresponding to the time t_1 for the signal to rise to 10% of its change is

$$V_1 = 0 \text{ V} + 0.1 \,(10 \text{ V}) = 1 \text{ V}$$

The voltage V_2 corresponding to the time t_2 for the signal to rise to 90% of its change is

$$V_2 = 0 \text{ V} + 0.9 \,(10 \text{ V}) = 9 \text{ V}$$

Estimate t_1 corresponding to 1 V.

$$t_1 \approx 0.1 \text{ ms}$$

Estimate t_2 corresponding to 9 V.

$$t_2 \approx 2.3 \text{ ms}$$

See Figure 14-22(b) for the above data points.

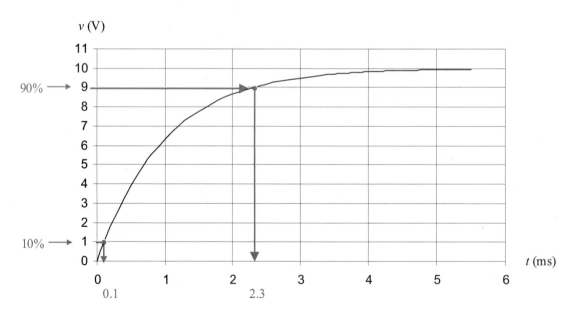

Figure 14-22(b) Example to determine rise time with data points

The rise time is

$$t_r = t_2 - t_1$$

$$t_r = 2.3 \text{ ms} - 0.1 \text{ ms} = 2.2 \text{ ms}$$

To find the *RC* time constant τ, use the fact that the *RC* curve reaches 63% of its change after charging for one time constant τ. The voltage across the capacitor after charging for 1 time constant is

$$v(\tau) = 0 \text{ V} + 0.63 \, (10 \text{ V}) = 6.3 \text{ V}$$

Estimate τ corresponding to 6.3 V from the graph as shown in Figure 14-22(c).

$$\tau \approx 1 \text{ ms}$$

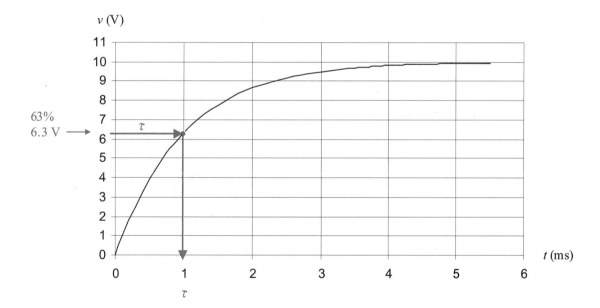

Figure 14-22(c) Example to determine time constant τ

The ratio of the rise time t_r to the time constant τ is

$$\frac{t_r}{\tau} = \frac{2.2 \text{ ms}}{1 \text{ ms}} = 2.2$$

$$t_r = 2.2\tau$$

$$t_r = 2.2RC$$

This is a fixed relationship for an exponential curve that is proven later.

Practice

An *RC* circuit has an initial capacitor voltage of −5 V, a steady state voltage of +5 V, and a time constant of 10 ms. (a) Find the capacitor voltage when the capacitor is 10% into its transient cycle and its corresponding time t_1. (b) Find the capacitor voltage when the capacitor is 90% into its transient cycle and its corresponding time t_2. Use the inverse relationship to accurately find t_1 and t_2. (c) Find the rise time from t_1 and t_2. (d) Find the rise time from the relationship that the rise time is 2.2*RC*. (e) Do parts *c* and *d* agree with each other?

Answer: (a) −4 V, 1.054 ms   (b) +4 V, 23.03 ms   (c) 21.98 ms (d) 22 ms (e) Parts *c* and *d* confirm each other: $t_r = 2.2\tau = 2.2RC$

Exploration

For a rising curve, pick any other initial value, steady state value, and time constant and repeat the practice problem. Prove to yourself that the relationship of $t_r = 2.2\tau = 2.2RC$ is valid.

Fall Time

For a falling curve, the fall time t_f of a transient curve is the time required for a signal to fall from 10% of its signal change to 90% of its signal change. Figure 14-23 demonstrates this signal parameter. For a falling curve the total change in the transient signal is a negative value since the starting value is greater than the ending value for a falling curve.

$$\Delta V = V_{\text{ending value}} - V_{\text{starting value}}$$

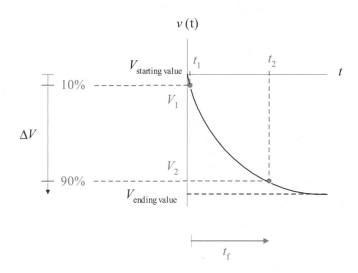

Figure 14-23 Fall time for a transient curve

The analysis for the fall time is the same as it is for the rise time. In the case of an *RC* circuit driven by a square wave, the rise time and the fall time should be the same.

Example 14-8

An *RC* circuit with a resistance of 100 Ω and a capacitance of 0.1 µF is driven by a 20 V$_{pp}$ square wave with a frequency of 10 kHz. Find the circuit time constant τ. Is the frequency sufficiently low to allow the capacitor to fully charge each half cycle? Write the universal *RC* transient equation for the falling curve. Find the time required for the signal to fall 10% and to fall 90% of its change. From those two times find the fall time for the waveform. Then find the fall time based upon its 2.2*RC* value. Are the two calculated fall times the same?

Solution

The *RC* circuit time constant τ is

$$\tau = 100\ \Omega \times 0.1\ \mu F = 10\ \mu s$$

To allow sufficient time for the capacitor to charge each half cycle, the signal period must be greater than or equal to 10τ.

$$T = \frac{1}{10 \text{ kHz}} = 100 \text{ μs}$$

$$10 \ \tau = 10 \times 10 \text{ μs} = 100 \text{ μs}$$

Yes, the period of the signal is sufficiently long enough to just charge the capacitor on each half cycle.

For the falling cycle, the initial value is 10 V, the steady state value is −10 V, and the time constant is 10 μs. Therefore, the falling curve's universal dc transient expression is

$$v(t) = -10 \text{ V} + \left(10 \text{ V} - (-10 \text{V})\right) e^{-\frac{t}{10 \text{ μs}}}$$

which simplifies to

$$v(t) = -10 \text{ V} + 20 \text{ V } e^{-\frac{t}{10 \text{ μs}}}$$

The total transient change in voltage is 20 V. Thus, 10% change into the transient is equivalent to a 2 V fall. Since the falling half cycle starts at +10 V, a fall of 2 V results in a capacitor voltage of 8 V. Substitute this value into the capacitor voltage transient expression and inversely solve for t_1.

$$8 \text{ V} = -10 \text{ V} + 20 \text{ V } e^{-\frac{t_1}{10 \text{ μs}}}$$

$$t_1 = 1.054 \text{ μs}$$

Again, the total transient change in voltage is 20 V. Thus, a 90% change into the transient is equivalent to an 18 V fall. Since the falling half cycle starts at +10 V, a fall of 18 V results in a capacitor voltage of −8 V. Substitute this value into the capacitor voltage transient expression and inversely solve for t_2.

$$-8 \text{ V} = -10 \text{ V} + 20 \text{ V } e^{-\frac{t_2}{10 \text{ μs}}}$$

$$t_2 = 23.03 \text{ μs}$$

The fall time is

$$t_f = t_2 - t_1$$

$$t_f = 23.03\ \mu s - 1.05\ \mu s = 22.0\ \mu s$$

Using the 2.2τ approach

$$t_f = 2.2 \times 10\ \mu s = 22.0\ \mu s$$

Consistently it has been demonstrated that the rise time and fall time of an RC circuit can be found as $2.2RC$.

Practice

An RC circuit has an initial capacitor voltage of 25 V, a steady state voltage of 5 V, and a time constant of 2 ms. Is this a rising or a falling curve? (b) Write the universal dc transient equation and simplify it. (c) Find the capacitor voltage when the capacitor is 10% into its transient cycle and its corresponding time t_1. (d) Find the capacitor voltage when the capacitor is 90% into its transient cycle and its corresponding time t_2. Use the inverse relationship to accurately find t_1 and t_2. (e) Find the rise time or fall time from t_1 and t_2. (f) Find the rise time or fall time from the relationship that the rise time is $2.2RC$. (g) Do parts d and e agree with each other?

Answer: (a) Falling (b) $v(t) = 5\ \text{V} + 20\ \text{V}\ e^{-\frac{t}{2\ ms}}$ (c) 23 V, 0.211 ms (d) +7 V, 4.605 ms (e) 4.39 ms (f) 4.4 ms (g) Parts e and f confirm each other: $t_f = 2.2\tau = 2.2RC$

Prove Rise Time t_r is 2.2τ

To prove that the rise time of an RC circuit response is 2.2τ, first express t_1 and t_2 in terms of its capacitor voltage. That requires an inverse solution of the universal dc transient equation corresponding to the 10% change and the 90% change in the universal dc transient expression. It can be readily shown that the rise time for an RC circuit is in fact 2.2 times its time constant.

First, find the time t_1 required for the signal to change 10% of its transient value. The left side of the equation represents the universal dc transient equation at $t = t_1$. The right side of the equation represents the value of the signal after it has risen for 10% of its transient change above the initial value (that is, the starting value). These two expressions must be the same.

$$Vss + \left(V_{\text{init}} - V_{\text{ss}}\right)e^{-t_1/\tau} = V_{\text{init}} + 0.1\left(V_{\text{ss}} - V_{\text{init}}\right)$$

Subtract V_{ss} from both sides of the expression.

$$\left(V_{\text{init}} - V_{\text{ss}}\right)\ e^{-t_1/\tau} = \left(V_{\text{init}} - V_{\text{ss}}\right) + 0.1\ \left(V_{\text{ss}} - V_{\text{Iinit}}\right)$$

Simplifying the right side of the equation algebraically gives

$$\left(V_{\text{init}} - V_{\text{ss}}\right)e^{-t_1/\tau} = 0.9\left(V_{\text{init}} - V_{\text{ss}}\right)$$

Divide both sides of the equation by the common term $(V_{\text{init}} - V_{\text{ss}})$.

$$e^{-t_1/\tau} = 0.9$$

Take the natural logarithm of both sides of the equation and then solve for t_1.

$$t_1 = -\ln(0.9)\tau$$

$$t_1 = 0.1054\tau$$

Second, find the time t_2 required for the signal to change 90% of its transient value. The left side of the equation represents the universal dc transient equation at $t = t_2$. The right side of the equation represents the value of the signal after it has risen for 90% of its transient change above the initial value (that is, the starting value). These two expressions must be the same.

$$V_{\text{ss}} + \left(V_{\text{init}} - V_{\text{ss}}\right)e^{-t_2/\tau} = V_{\text{init}} + 0.9\left(V_{\text{ss}} - V_{\text{init}}\right)$$

Follow the same process as was followed to find t_1. The result is

$$t_2 = -\ln(0.1)\tau$$

$$t_2 = 2.303\tau$$

The rise time t_r is the difference between these two times.

$$t_r = t_2 - t_1$$

$$t_r = 2.303\tau - 0.105\tau = 2.20\tau$$

As expected, the rise time for an RC transient curve is always 2.2 times its time constant. A similar proof demonstrates that this is also true for the fall time of an RC transient curve.

Summary

When RC circuits are subjected to sudden changes in dc voltages or currents, transient exponential voltages and currents result throughout the RC circuit. A sudden change in dc voltage or current can occur from a switched circuit or from a rectangular supplied waveform. The circuit voltages and currents go through a transition from their *initial values* when the sudden change occurs to their *steady state values* after the transition.

A key to analyzing these circuits is that a *capacitor stores charge (equivalent to a voltage), resists a sudden change in voltage, and returns energy back to the circuit in the form of voltage supply.* An uncharged capacitor drops 0 V. When subjected to a sudden dc change, the capacitor initially resists a voltage change and retains a 0 V drop. Therefore, an uncharged capacitor subjected to a sudden change can be initially modeled as a short.

A charged capacitor has a voltage drop. If subjected to a sudden change, the capacitor initially retains this voltage drop. The charged capacitor can be initially modeled as a voltage supply with the same voltage and polarity as the capacitor dropped before the sudden signal change.

The time required for the transient depends upon the circuit's time constant. The circuit's time constant τ is defined to be the product of the circuit's capacitance and the equivalent resistance (R_{Th}) seen by the capacitance.

$$\tau = R_{Th}C$$

The transient expression for the capacitor voltage is

$$v_C(t) = V_{ss} + \left(V_{init} - V_{ss}\right)e^{-\frac{t}{\tau}}$$

where V_{ss} is the steady state capacitor voltage after the transient has died out, V_{init} is the initial capacitor voltage at the start of the transient, and τ is the RC time constant during the transient. The transient exponential decay term dies out in approximately five time constants (5τ).

This universal exponential expression also applies to the capacitor current.

$$i_C(t) = I_{ss} + \left(I_{init} - I_{ss}\right)e^{-\frac{t}{\tau}}$$

This universal exponential expression also applies to circuit resistance voltages and currents.

A plot of the universal dc transient expression leads naturally to an exponentially increasing or an exponentially decreasing curve. You can plot a very precise graph based upon the evaluation of several data points of the universal dc transient expression. Or you can do a quick sketch using the exponential curve characteristic that an increase in time τ produces the next data point that is a 63% increase or decrease towards its steady state value.

The following step-by-step process consistently produces the three key parameters of the time constant, initial values, and steady state values for the universal dc transient equation:

1. At $t = 0^-$ (just before the switch is thrown or transient begins), find the voltage across the capacitance $V_C(0^-)$. This will also be the initial capacitor voltage once the transient begins.

2. After the switch is thrown (during the transient), find the *RC* time constant τ. If it is not a simple *RC* circuit, then the Thévenin resistance and/or equivalent capacitance must be found before calculating the *RC* time constant.

3. At $t = 0$ or $t = 0^+$, that is just after the switch is thrown or at the beginning of the transient, model the uncharged capacitor as a short or the charged capacitor as a voltage supply based upon step 1 of this process. Then evaluate the modeled circuit and find the desired initial voltages and currents.

4. At $t = \infty$, that is after the transient has died out and the circuit is in a steady state, model the capacitor as an open. Then evaluate the modeled circuit and find the desired steady state voltages and currents.

5. Substitute the *RC* time constant τ, the initial value, and steady state value into the universal dc transient equation and simplify the equation.

6. If appropriate, quick sketch or precisely sketch the transient response curve.

A square wave into an RC circuit is equivalent to switching between two dc supply voltages E_{max} and E_{min} to the RC circuit. If the square wave signal period T is greater than 10τ then the capacitor has sufficient time to essentially fully charge to its E_{max} each half cycle. The capacitor voltage exponentially and smoothly rises and falls to the voltage values of E_{max} and E_{min}. On the positive half cycle, the capacitor current spikes to a value of

$$I_{Cmax} = \frac{E_{max} - E_{min}}{R}$$

and falls to a steady state of 0. On the negative half cycle, the capacitor current spikes to $-I_{Cmax}$ and rises to 0. If $T \gg 10\tau$, then the capacitor charges to its steady state value and holds that value until the next half cycle starts. If $T \ll 10\tau$ then the capacitor does not have enough time to fully charge each half cycle and the output waveform is a reduced exponential waveform. If extreme, the output waveform reduces to a triangle waveform.

For a rising curve, the rise time t_r of a transient curve is defined to be the time required for the signal to go from 10% of its transient signal change to 90% of its transient signal change. In terms of an RC dc transient circuit, the rise time t_r is equal to the value of 2.2τ or $2.2RC$. The fall time for a falling transient signal is similarly defined. The fall time and rise time for RC circuit voltages and currents responding to a square wave is the same for all components in the circuit.

Problems

Switched RC Transient

14-1 For the circuit of Figure 14-24, the capacitor has an initial charge of 0 V.

    a. Find the circuit time constant τ.

    b. Find the initial capacitor voltage and current,

    c. Find the steady state capacitor voltage and current,

    d. Find the universal expression for capacitor voltage and current (with values and simplified).

    e. Find the capacitor voltage and current after two time constants.

f. Sketch the capacitor voltage and current waveforms for five time constants.

14-2 Repeat Problem 14-1 for the resistor's voltage and current.

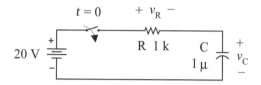

Figure 14-24

14-3 For the circuit of Figure 14-24, the capacitor has an initial charge of 8 V.

a. Find the circuit time constant τ.

b. Find the initial capacitor voltage and current.

c. Find the steady state capacitor voltage and current.

d. Find the universal expression for capacitor voltage and current (with values and simplified).

e. Find the capacitor voltage and current at two time constants.

f. Sketch the capacitor voltage and current waveforms for five time constants.

14-4 Repeat Problem 14-3 for the resistor's voltage and current.

14-5 In the circuit of Figure 14-25, the switch has been in position 1 for a long time. The switch is switched to position 2 at $t = 0$ s.

a. Find the capacitor voltage just before the switch is switched to position 2.

b. Find the circuit time constant τ.

c. Find the initial capacitor voltage and current.

d. Find the steady state capacitor voltage and current.

e. Find the universal expression for capacitor voltage and current (with values and simplified).

f. Find the capacitor voltage and current at two time constants.

g. Sketch the capacitor voltage and current waveforms for five time constants.

14-6 Repeat Problem 14-5 for the voltage and current of resistor R1.

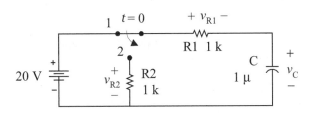

Figure 14-25

14-7 In an *RC* transient circuit with a series resistor of 1 kΩ and a capacitor of 4.7 μF, the initial capacitor voltage is 0 V and its steady state voltage is 12 V.

a. Find the circuit time constant τ.

b. Write the universal equation for the capacitor voltage (with values and simplified).

c. Find the voltage drop across the capacitor after charging for one, two, three, four, and five time constants.

d. Find the voltage dropped after 6 ms.

e. Find the time required for the capacitor to charge to 9 V.

f. Sketch the transient curve for ten time constants.

14-8 In an *RC* transient circuit with a series resistor of 100 Ω and a capacitor of 10 μF, the initial capacitor voltage is –4 V and its steady state voltage is 8 V.

a. Find the circuit time constant τ.

b. Write the universal equation for the capacitor voltage (with values and simplified).

c. Find the voltage drop across the capacitor after charging for one, two, three, four, and five time constants.

d. Find the voltage dropped after 1.5 ms.

e. Find the time required for the capacitor to charge to 6 V.

      f.   Sketch the transient curve for ten time constants.

14-9   In an *RC* transient circuit with a series resistor of 10 kΩ and a capacitor of 0.1 μF, the initial capacitor current is +6 mA and its steady state current is 0 mA.

      a.   Find the circuit time constant τ.

      b.   Write the universal equation for the capacitor current (with values and simplified).

      c.   Find the current induced in the capacitor after charging for one, two, three, four, and five time constants.

      d.   Find the current induced after 500 μs.

      e.   Find the time required for the current to reach 1 mA.

      f.   Sketch the transient curve for ten time constants.

14-10   In an *RC* transient circuit with a series resistor of 100 kΩ and a capacitor of 0.01 μF, the initial resistor voltage is 10 V and its steady state voltage is 0 V.

      a.   Find the circuit time constant τ.

      b.   Write the universal equation for the resistor voltage (with values and simplified).

      c.   Find the voltage drop across the resistor after the capacitor has charged for one, two, three, four, and five time constants.

      d.   Find the voltage dropped after 2.5 ms.

      e.   Find the time required for the resistor to drop to 9 V.

      f.   Sketch the transient curve for ten time constants.

RC Square Wave Response

14-11   For the circuit of Figure 14-26(a):

      a.   Find the time constant τ for this circuit.

      b.   Prove that the period *T* is long enough for the capacitor to fully charge each half cycle.

      c.   Sketch the waveform of the capacitor voltage.

      d.   Sketch the waveform of the resistor voltage.

      e.   Sketch the waveform of the capacitor current.

f.   How long does it take for the voltage across the capacitor to reach 2 V, assuming that the capacitor is initially –4 V at the beginning of its half cycle? Hint: Start with the general expression with its initial value, steady state value, and time constant τ.

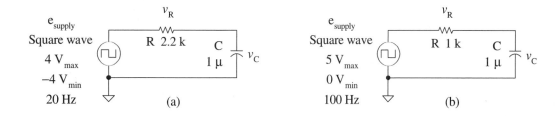

Figure 14-26

14-12   Use the circuit of Figure 14-26(b).

a.   Find the time constant τ for this circuit.

b.   Prove that the period T is long enough for the capacitor to fully charge each half cycle.

c.   Sketch the waveform of the capacitor voltage.

d.   Sketch the waveform of the resistor voltage.

e.   Sketch the waveform of the capacitor current.

f.   How long does it take for the voltage across the resistor to reach 0.8 V (a digital low), assuming that the resistor is initially 5 V at the beginning of its half cycle? Hint: Start with the general expression with its initial value, steady state value, and time constant τ.

14-13   For the circuit of Figure 14-26(a), find the rise time and fall time of the capacitor voltage, capacitor current, resistor voltage, and resistor current.

14-14   For the circuit of Figure 14-26(b), find the rise time and fall time of the capacitor voltage, capacitor current, resistor voltage, and resistor current.

Wave Shaping and Generation

Introduction

Waveshapes can be generated or converted using resistors, capacitors, diodes, active devices, and specially designed IC chips in combination. Several practical RC circuits are examined in this chapter. The nature of the RC circuit in charging and discharging exponentially in a time directly related to its RC time constant is exploited in signal waveshape conversion and signal generation. The RC universal dc transient expression is used to determine charge and discharge time which in turn can be used to determine such characteristics as pulse width, signal period, and signal frequency.

Objectives

Upon completion of this chapter you will be able to analyze the following RC application circuits:

- Multivibrators and the 555 timer
- Multivibrator circuits
- Op amp RC integrator circuit and wave shaping
- Op amp RC differentiator circuit and wave shaping
- Op amp RC waveform generators
- Natural hidden capacitance
- Capacitor leakage
- Clippers, clampers, and detectors
- Switched capacitor circuit

15.1 Multivibrators Using 555 Timers

Since the capacitor charges at a predictable rate in an *RC* circuit, the *RC* circuit is used for **timing circuits**. A timing circuit can control such things as the period and therefore the frequency of a signal that is being generated, the time between events to trigger other electronic circuit action, and so forth.

Multivibrators

A primary source of timing pulses and waveforms is the **multivibrator**. Multibrators have two possible outputs: a voltage high or a voltage low. Multivibrators are classified as follows.

Astable or free-running	*not stable, no stable state*
Monostable or one shot	*one stable state*
Bistable	*two stable states*

The **astable multivibrator** produces a periodic rectangular waveform as shown in Figure 15-1 where

V_H High voltage output signal (unstable)

V_L Low voltage output signal (unstable)

W Pulse width (same as T_H)

T_H Time the signal is high during its period

T_L Time the signal is low during its period

T Period of the repeating waveform

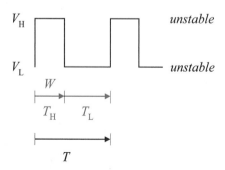

Figure 15-1 Astable multivibrator output waveform

The **duty cycle** of a periodic rectangular waveform is defined to be

$$D = \frac{W}{T} \times 100\%$$

A square wave has a duty cycle of 50%.

Example 15-1

A rectangular signal is high for 2 ms and low for 6 ms. Find the signal's period, pulse width, duty cycle, and frequency.

Solution

The signal period is

$$T = T_H + T_L$$

$$T = 2 \text{ ms} + 6 \text{ ms} = 8 \text{ ms}$$

The signal pulse width is the time that the signal is high during a period.

$$W = T_H = 2 \text{ ms}$$

The signal duty cycle is

$$D = \frac{W}{T} \times 100\%$$

$$D = \frac{2 \text{ ms}}{8 \text{ ms}} \times 100\% = 25\%$$

The signal frequency, also called its **repetition rate**, is

$$f = \frac{1}{8 \text{ ms}} = 125 \text{ Hz}$$

Practice

A rectangular waveform has a frequency of 100 Hz and a duty cycle of 20%. Find the signal's repetition rate, period, and pulse width.

Answer: 100 repetitions per second, 10 ms, 2 ms

The **monostable multivibrator** produces a well-defined pulse out for an input signal that may not be well defined (a narrow pulse or a spike) as shown in Figure 15-2 where

V_H High voltage output signal (unstable)

V_L Low voltage output signal (stable)

T_1 Start time of input signal which sends the output signal high

W Pulse width (when the pulse is high)

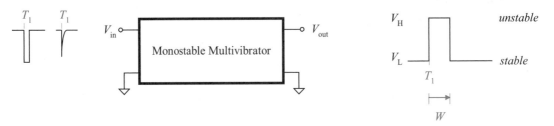

Figure 15-2 Monostable multivibrator output waveform

The **bistable multivibrator,** also known as a *Flip-Flop* or FF, has two stable states and utilizes one input pulse to take the output high and another input pulse to take the output low as shown in Figure 15-3 where

V_H High voltage output signal (stable)

V_L Low voltage output signal (stable)

T_1 Start time of input signal which starts the high output signal

T_2 Start time of input signal which starts the low output signal

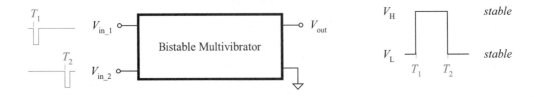

Figure 15-3 Bistable multivibrator output waveform

The input pulses to the monostable and bistable multivibrators need not be well defined, they must only reach a specific level to trigger the output change. The output pulse is well defined in terms of output voltage and pulse width even though the inputs may be somewhat distorted.

555 Timer

The 555 timer is a very versatile IC used in many applications. The basic function blocks of the 555 timer are shown in Figure 15-4, which has been set up for digital operations by connecting to a +5 V dc power supply. By connecting a few external components (resistors and capacitors) to the 555 timer chip, a wide variety of circuits can be created. The 555 timer has become one of the more popular components in electronics.

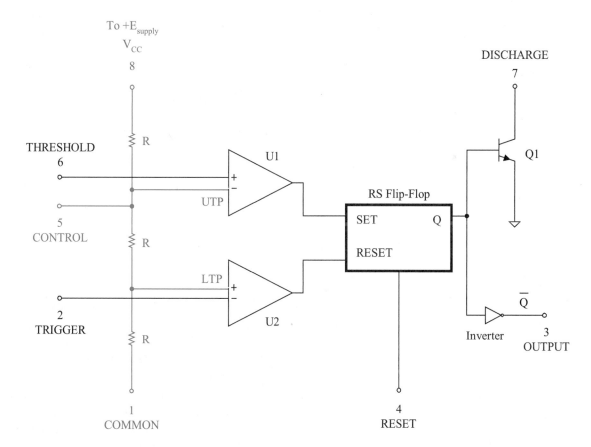

Figure 15-4 555 timer basic block diagram and signaling setup for digital operation

The key features of the 555 timer are:

- Q   FF output–high or low output
- $\overline{Q}$ Complement (opposite) of Q–low or high output
- SET   FF input–high signal sets Q high and $\overline{Q}$ low
- RESET FF input–high signal in resets Q low and $\overline{Q}$ high
- U1   Noninverting comparator driving the FF *set input*
- U2   Inverting comparator driving the FF *reset input*
- UTP   Upper trip point for U1 set by 2R/3R of E_{supply}
- LTP   Lower trip point for U2 set by R/3R of E_{supply}
- Q1   BJT used to discharge external capacitor when Q high
- Pin 7   Open or tied to common through BJT switched on

Once the flip-flop is set, a reset input is required to reset the flip-flop; or, once the flip-flop is reset, a set input is required to set the flip-flop.

For the circuit of Figure 15-4, assume pin 1 is connected to common and pin 8 is connected to a +5 V dc supply (digital system). Find the UTP and LTP voltages using the voltage divider rule on the essentially series three-resistor circuit combination. R = 5 kΩ

$$V_{UTP} = \left(\frac{10 \text{ k}\Omega}{15 \text{ k}\Omega} \right) 5 \text{ V} = 3.33 \text{ V}_{dc} \qquad\qquad (2/3 \text{ V}_{CC})$$

$$V_{LTP} = \left(\frac{5 \text{ k}\Omega}{15 \text{ k}\Omega} \right) 5 \text{ V} = 1.67 \text{ V}_{dc} \qquad\qquad (1/3 \text{ V}_{CC})$$

555 Timer Monostable Multivibrator

The circuit of Figure 15-5 is a 555 timer connected in a *monostable mul-tivibrator* configuration. There is only one stable state: the FF is set, the BJT is turned on acting like a short from pin 7 to common, the top of the capacitor is tied to common through the BJT, the capacitor drops 0 V, the input (pin 2) is high at 5 V, and the output (pin 3) is low at 0 V. If the input is suddenly changed from 5 V to below 1.67 V, the FF is reset, the BJT turns off, pin 7 is no longer shorted to common, the capacitor charges toward the supply voltage, and the output is driven high to 5 V. When the capacitor voltage (tied to pin 6) reaches the UTP, the FF is set again, the BJT turns on, the capacitor is shorted to common through pin 7 and the turned-on BJT, and the output returns to 0 V (its stable state). The follow-ing is a step-by-step thought process through the basic phases of this cir-cuit.

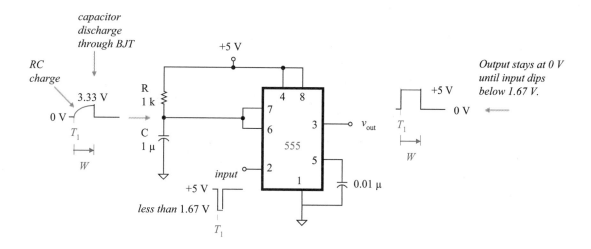

Figure 15-5 555 timer connected as a monostable multivibrator

First assume that the 555 timer is in its stable state with the FF set. The output (pin 3) is stable at 0 V as long as the input remains above 1.67 V (pin 2). Refer to Figure 15-4 and Figure 15-5 while stepping through this thought process.

- Output (pin 3) low at 0 V

- $\overline{Q}$ low

- Q high

- BJT turned on (ideally acts like a short)

- Capacitor (pin 7) ideally shorted to common through the BJT

- Capacitor drops 0 V (tied to common)

- U1 input (pin 6) at 0 V therefore U1 output low to FF set lead

- U2 input (pin 2) at 5 V therefore U2 output low to FF reset lead

- FF remains set since both FF inputs are low

Next assume that input drops to less than 1.67 V (pin 2) at time T_1. Again refer to Figure 15-4 and Figure 15-5 while stepping through this thought process.

- Input voltage (pin 2) dips below 1.67 V and returns to 5 V

- U2 input (pin 2) goes below 1.67 V (less than 1/3 V_{CC})

- U2 output goes high *resetting* the FF

- $\overline{Q}$ goes high sending the output (pin 3) high to 5 V

- Q goes low turning off the BJT and creating an open at pin 7

- Capacitor starts charging through *RC* circuit to voltage supply

- Capacitor charges based upon its *RC* time constant τ

- Capacitor charges from 1.67 V to 3.33 V (or 1/3 V_{CC})

- Capacitor voltage reaches 3.33 V (2/3 V_{CC})

- U1 input (pin 6) reaches 3.33 V and goes high setting the FF

- Q goes high, BJT turns on, and shorts out the capacitor

- Capacitor voltage drops to and stays at 0 V

- $\overline{Q}$ goes low taking the output (pin 3) low to 0 V

- The circuit is now stable again with a 0 V output and a 5 V input

Example 15-2

What is the width W of the pulse in the output of the circuit in Figure 15-5?

Solution

The width W of the pulse depends upon the *RC* time constant τ and the time it takes to charge the capacitor from 0 V to 3.33 V when the circuit transitions back to a stable state. The capacitor must follow the universal dc transient expression

$$v_C(t) = V_{ss} + (V_{init} - V_{ss})\,e^{-t/\tau}$$

The capacitor starts with a value of 0 V (initial value) and attempts to climb to 5 V (the steady state value). The *RC* time constant is

$$\tau = RC = 1\ k\Omega \times 1\ \mu F = 1\ ms$$

Substitute these values into the transient expression and simplify it.

$$v_C(t) = 5 \text{ V} + (0 \text{ V} - 5 \text{ V}) e^{-t/\tau}$$

$$v_C(t) = 5 \text{ V} - 5 \text{ V } e^{-t/1 \text{ ms}}$$

The capacitor charges to 3.33 V (or 2/3 of V_{CC}) when it transitions back to its stable state. Set the capacitor voltage expression equal to 3.33 V and solve the expression inversely for the time t required to charge the capacitor to 3.33 V.

$$3.33 \text{ V} = 5 \text{ V} - 5 \text{ V } e^{-t/1 \text{ ms}}$$

Use the inverse solution technique and find time t.

$$t = 1.10 \text{ ms}$$

The pulse width W is the time required to charge the capacitor; therefore,

$$W = 1.10 \text{ ms}$$

Practice

Use the generic term of V_{CC} and write the RC universal dc transient expression in terms of V_{CC} and time constant τ. Then inversely solve that expression to find the time of the pulse width W in terms of the RC time constant τ and the supply voltage V_{CC}. Verify your results using the example values.

Answer: $v_C(t) = V_{CC} - V_{CC} \, e^{-t/\tau}$; setup: $\dfrac{2 \, V_{CC}}{3} = V_{CC} - V_{CC} \, e^{-t/\tau}$;

$W = 1.10 \; \tau$; $W = 1.10$ ms (verified)

555 Timer Astable Multivibrator

The circuit of Figure 15-6 is a 555 timer connected as an **astable multivibrator**. This is also called a **free-running multivibrator,** which creates an output periodic rectangular waveform without any external input signal.

The sawtooth waveform (at pins 2 and 6) is created from charging and discharging the capacitor. While the BJT is turned off (pin 7), the capacitor charges through R1 and R2 toward the supply voltage. When the capacitor voltage at pin 6 reaches the UTP, the FF is set driving Q high, turning on the BJT, shorting the top of R2 to common through the BJT, and the capacitor starts its discharge cycle through R2. The capacitor then discharges through pin 7 until the capacitor voltage (pin 2) reaches the LTP. When the capacitor voltage falls to the LTP, then the FF is reset and Q goes low and the BJT turns off (opens). The capacitor now charges through R1 and R2 toward the supply voltage again, and the cycle continues. While the input is a self-generated exponential waveform created across capacitor C, a rectangular waveform is being generated at the output (pin 3). The following is a step-by-step thought process through the basic phases of this circuit.

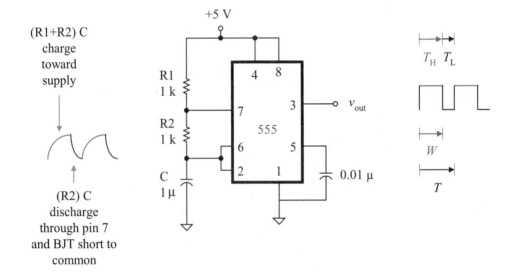

Figure 15-6 555 timer connected as an astable multivibrator

First, since you are working with a memory element (the flip-flop FF inside the 555 timer), you must assume the current state of the memory element before proceeding. Assume that the output is currently at 0 V just before it switches to 5 V. What is the status of the circuit? Refer to Figure 15-4 and Figure 15-6 as you step through this thought process.

The status of the circuit just before the output switches from a low of 0 V to a high of 5 V is:

- Output (pin 3) low at 0 V
- $\overline{Q}$ low
- Q high
- FF in the *set* position
- BJT turned on (ideally acts like a short) to common
- Pin 7 tied to common through turned-on BJT
- Capacitor discharging through R2 tied to common through pin 7
- Capacitor discharging down toward a steady state voltage of 0 V
- Capacitor can only discharge down to 1.67 V (1/3 V_{CC})
- U1 and U2 inputs following the capacitor voltage
- U1 and U2 inputs approaching 1.67 V
- U1 and U2 outputs are low to the FF
- FF stays in the *set* state

Next assume that input drops to 1.67 V (pin 2). Follow what happens in Figure 15-4 and Figure 15-6.

- Capacitor voltage (pin 2) has decreased to 1.67 V
- U2 input goes below 1.67 V (or, 1/3 V_{CC})
- U2 output goes high sending the FF *reset* input high
- FF resets
- $\overline{Q}$ goes high sending the output (pin 3) high to 5 V
- Q goes low turning off the BJT and creating an open at pin 7
- Capacitor charges through R1 and R2 toward the voltage supply
- Charging time constant τ_H is (R1 + R2) C
- Capacitor charges from 1.67 V to 3.33 V
- Output voltage remains high at 5 V while the capacitor charges

Next assume that the charging capacitor voltage reaches 3.33 V (or, 2/3 V_{CC}). Again, follow what happens in Figure 15-4 and Figure 15-6.

- Capacitor voltage (pin 6) increases until it reaches 3.33 V
- U1 input goes above 3.33 V (or, 2/3 V_{CC})
- U1 output goes high sending the FF *set* input high

- FF sets
- $\overline{Q}$ goes low sending the output (pin 3) low to 0 V
- Q goes high turning on the BJT and pin 7 is tied to common
- Capacitor discharges through R2 with pin 7 tied to common
- Discharging time constant τ_L is (R2) C
- Capacitor discharges from 3.33 V to 1.67 V
- Output voltage remains low at 0 V while the capacitor discharges

Once the capacitor voltage reaches 1.67 V, one complete cycle of a periodic rectangular waveform is completed and the next cycle begins. A self-generated rectangular signal is created. This circuit can then be used as a signal generator.

The next example calculates the pulse width W (or, T_H) of the output signal of an astable multivibrator.

Example 15-3

What is the width W of the pulse in the output of the circuit of Figure 15-6?

Solution

The pulse width W depends upon the (R1 + R2) C time constant τ_H and the capacitor charging from 1.67 V to 3.33 V while the output is in the high (or 5 V) state. The capacitor must follow the universal dc transient expression

$$v_C(t) = V_{ss} + (V_{init} - V_{ss}) e^{-t/\tau}$$

The capacitor starts with a value of 1.67 V (initial value) and attempts to climb to 5 V (the steady state value). The RC time constant is

$$\tau_H = (R1 + R2)C = 2\ k\Omega \times 1\ \mu F = 2\ ms$$

Substitute these values into the transient expression and simplify it.

$$v_C(t) = 5\ V + (1.67\ V - 5\ V) e^{-t/2\ ms}$$

$$v_C(t) = 5\ V - 3.33\ V\ e^{-t/2\ ms}$$

The capacitor charges to 3.33 V (or 2/3 of V_{CC}) when it transitions back to its stable state. Set the capacitor voltage expression equal to 3.33 V and solve the expression inversely for the time t required to charge the capacitor to 3.33 V.

$$3.33 \text{ V} = 5 \text{ V} - 3.33 \text{ V } e^{-t/2 \text{ ms}}$$

Use the inverse solution technique and find time t.

$$t = 1.386 \text{ ms}$$

The pulse width W is the time required to charge the capacitor, so

$$W = 1.386 \text{ ms}$$

Practice

Assume that the output is in its high (5 V) state. Write the universal dc transient capacitor voltage expression using the terms V_{CC} and time constant τ_H. Then inversely solve that expression to find the time that the pulse is in its high state T_H in terms of its time constant τ_H and the supply voltage V_{CC}. T_H is also the pulse width. Verify your results using the example values.

Answer $v_C(t) = V_{CC} - \dfrac{2}{3}V_{CC}\, e^{-t/\tau_H}$; set up: $\dfrac{2}{3}V_{CC} = V_{CC} - \dfrac{2}{3}V_{CC}\, e^{-t/\tau_H}$;

$W = 0.693\, \tau_H$ where $\tau_H = (R1+R2)\, C$; $W = T_H = 1.386 \text{ ms}$ (verified)

In preparation for finding the period of the signal, the next example finds the length of time during the period that the output is low (0 V).

Example 15-4

What is the time T_L during the period when the output signal is low in the circuit of Figure 15-6?

Solution

The time T_L of the pulse depends upon the (R2) C time constant τ_L and discharging the capacitor from 3.33 V to 1.67 V while the output is in the low (0 V) state. The capacitor must follow the universal dc transient expression

$$v_C(t) = V_{ss} + (V_{init} - V_{ss})\,e^{-t/\tau}$$

The capacitor starts with a value of 3.33 V (initial value) and attempts to fall to 0 V (the steady state value). The *RC* time constant is

$$\tau_L = (R2)C = 1\,k\Omega \times 1\,\mu F = 1\,ms$$

Substitute these values into the transient expression and simplify it.

$$v_C(t) = 0\,V + (3.33\,V - 0\,V)\,e^{-t/1\,ms}$$

$$v_C(t) = 3.33\,V\ e^{-t/1\,ms}$$

The capacitor discharges to 1.67 V (or 1/3 of V_{CC}), when it transitions back to its stable state. Set the capacitor voltage expression equal to 1.67 V and solve the expression inversely for the time t required to discharge the capacitor to 1.67 V.

$$1.67\,V = 3.33\,V\ e^{-t/1\,ms}$$

Use the inverse solution technique and find time t.

$$t = 0.693\,ms$$

The time that the output remains low during its cycle is

$$T_L = 0.693\,ms$$

Practice

Assume that the output is in its low (0 V) state. Write the universal dc transient capacitor voltage expression using the terms V_{CC} and time constant τ_L. Then inversely solve that expression to find the time that the pulse is in its low state T_L in terms of its time constant τ_L and the supply voltage V_{CC}. Verify your results using the example values.

Answer: $v_C(t) = \dfrac{2}{3}V_{CC}\ e^{-t/\tau_H}$; set up: $\dfrac{1}{3}V_{CC} = \dfrac{2}{3}V_{CC}\ e^{-t/\tau_H}$;

$T_L = 0.693\,\tau_L$ where $\tau_L = (R2)\,C$; $T_L = 0.693\,ms$ (verified)

The next example combines the results of the last two examples to find the 555 timer asynchronous output period, duty cycle, and frequency.

Example 15-5

Based upon the results of the last two examples, find the period, duty cycle, and frequency of the output signal of the circuit of Figure 15-6.

Solution

The signal period is

$$T = T_H + T_L$$

$$T = 1.386 \text{ ms} + 0.693 \text{ ms} = 2.079 \text{ ms}$$

The signal duty cycle is

$$D = \frac{W}{T} \times 100\%$$

$$D = \frac{1.386 \text{ ms}}{2.079 \text{ ms}} \times 100\% = 66.7\%$$

The signal frequency is

$$f = \frac{1}{T} = \frac{1}{2.079 \text{ ms}} = 481 \text{ Hz}$$

Practice

Based upon the results of the last two practice problems, write an equation for the period, duty cycle, and frequency of the output signal of the circuit of Figure 15-6.

Answer: $T = 0.693(R1 + 2(R2))\,C$, $D = \dfrac{R1 + R2}{R1 + 2(R2)} \times 100\%$,

$f = \dfrac{1.44}{(R1 + 2(R2))\,C}$

15.2 Op Amp *RC* Integrator Circuit

The op amp (operational amplifier) was originally created to perform mathematical operations, for example, addition, subtraction (inverter), multiplication, and so forth. The op amp can also be used in *RC* circuits to mathematically create the calculus integrator and differentiator operators. These mathematical operators form the building blocks of analog computer systems. With the advance of digital computers, analog computer usage has significantly declined. Analog computers are still in use but focus on specialized applications such as wave motion studies.

The **integrator** circuit utilizes a negative feedback amplifier and an *RC* feedback circuit as shown in Figure 15-7(a).

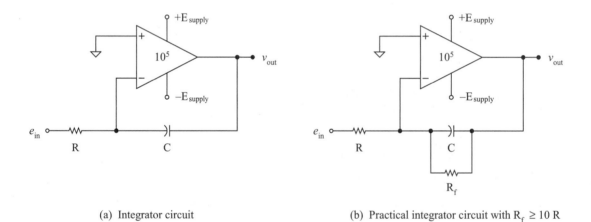

(a) Integrator circuit

(b) Practical integrator circuit with $R_f \geq 10\ R$

Figure 15-7(a,b) Integrator circuit (*C not polarized*)

Input offset currents

A more practical circuit is shown in Figure 15-7(b). A resistor with a resistance of about ten times the input resistance R is placed in parallel to the capacitor to prevent the op amp output from drifting to a rail voltage due to the internal differences in the op amp. Even with both inputs connected to common and no input signal currents, a very small unwanted difference in op amp input dc currents (called **input offset currents**) causes an op amp input voltage difference to be amplified by the dc open loop gain of the op amp (10^5). This in turn drives the op amp to a rail voltage. The feedback resistor R_f closes the loop for this dc offset signal, producing negative feedback with a fixed gain for this unwanted signal. The integration of the input signal is unaffected.

Referring to Figure 15-7(a), the input is being fed to the inverting op amp terminal; therefore the output voltage is inverted. The output voltage is being fed back to the inverting op amp terminal, therefore this is negative feedback and the op amp error voltage is approximately 0 V as long as the op amp is not hitting a rail voltage. The current into the op amp is assumed to be ideally 0 A.

Figure 15-7(c) shows the schematic prepared for circuit analysis. Note that a **virtual common** is formed between the resistor and feedback capacitor. This node is not really at common but is so close to 0 V that it is called a *virtual common*.

Virtual common

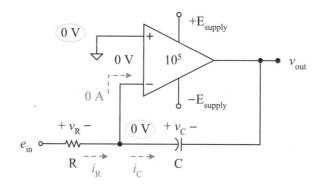

Figure 15-7(c) Integrator (or analog accumulator) circuit

Apply Ohm's law and KVL to the input resistor R.

$$i_R = \frac{v_R}{R} = \frac{e_{in}}{R}$$

Apply the current and voltage relationship and KVL to the capacitor C.

$$i_C = C\frac{d\,v_C}{dt} = -C\frac{d\,v_{out}}{dt}$$

Applying KCL at the *virtual common* node, $i_R = i_C$. Thus, the two above expressions are equal to each other.

$$-C\frac{d\,v_{\text{out}}}{dt} = \frac{e_{\text{in}}}{R}$$

Solve this differential equation for v_{out}, using calculus. Or, if you wish to ignore the calculus development, just trust the outcome.

$$v_{\text{out}} = -\frac{1}{RC}\int_0^t e_{\text{in}}\ dt + V_{C\,\text{init}}$$

Thus, this circuit is really a mathematical (electronic) integrator circuit. The input signal is integrated and multiplied by a constant to produce the output signal, assuming the output is not hitting the upper or lower rail. Note that the output signal is inverted as expected.

A special case is when the input voltage is a fixed voltage, that is, a dc voltage E_{in} and $V_{C\,\text{init}}$ of 0 V. Then the output voltage reduces to

$$v_{\text{out}} = -\frac{E_{\text{in}}}{RC}t$$

where t is the length of time the capacitor is charging. Note t and RC are in units of seconds (s) and thus cancel out each other's units. Thus, the remaining unit is volts (V).

Note that this is a linear function of time with a constant coefficient of $-E_{\text{in}}/RC$. If a positive dc voltage is applied continuously, the output eventually reaches the lower rail voltage. And, if a negative dc voltage is applied continuously, the output eventually reaches the upper rail voltage.

Another way to look at the circuit of Figure 15-7(c) with a dc input is to view the input as a *constant current generator*. The input current is fixed at

$$I_C = I_R = \frac{E_{\text{in}}}{R}$$

The basic definition of current is

$$I_C = \frac{Q}{t}$$

thus

$$\frac{Q}{t} = \frac{E_{\text{in}}}{R}$$

This *constant current* is driving the capacitor and storing charge on the plates of the capacitor. This accumulating charge produces a corresponding voltage drop across the capacitor. Substituting charge Q from the basic definition of capacitance into the preceding equation

$$Q = CV_C$$

produces the result that

$$\frac{CV_C}{t} = \frac{E_{in}}{R}$$

From KVL, $V_C = -V_{out}$ thus

$$\frac{C(-V_{out})}{t} = \frac{E_{in}}{R}$$

Solve for V_{out}.

$$V_{out} = -\frac{E_{in}}{RC}t \qquad \text{(for } V_{C\,init} \text{ of 0 V)}$$

This algebraic process produced the same result as the calculus approach assuming a fixed dc input voltage and using basic laws. Again, *a constant current into the capacitor produces a linearly increasing charge on the capacitor, which produces a linearly increasing voltage drop across the capacitor, which creates a voltage ramp across the capacitor, which creates a voltage ramp output.*

Example 15-6

For the circuit of Figure 15-8, what is the slope of the output ramp (linear) if the input is 1 V_{dc}? Find the output voltage at 0 s, 1 ms, 2 ms, and 3 ms. How much time does it take to reach a rail voltage of −13 V? What is a reasonable minimum value for R_f?

Solution

The output voltage for a fixed dc input is

$$V_{out}(t) = -\frac{E_{in}}{RC}t$$

The slope of the ramp is the coefficient of time variable t.

$$-\frac{E_{in}}{RC} = -\frac{1\text{ V}}{1\text{ k}\Omega \times 1\text{ }\mu\text{F}} = -1\,\frac{\text{V}}{\text{ms}}$$

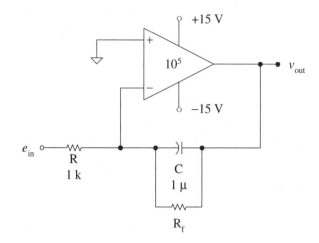

Figure 15-8 Example integrator circuit

For individual time t, the output voltages are:

$$V_{out}(0 \text{ s}) = \left(-\frac{1 \text{ V}}{\text{ms}}\right) \times 0 \text{ ms} = 0 \text{ V}$$

$$V_{out}(1 \text{ s}) = \left(-\frac{1 \text{ V}}{\text{ms}}\right) \times 1 \text{ ms} = -1 \text{ V}$$

$$V_{out}(2 \text{ s}) = \left(-\frac{1 \text{ V}}{\text{ms}}\right) \times 2 \text{ ms} = -2 \text{ V}$$

$$V_{out}(3 \text{ s}) = \left(-\frac{1 \text{ V}}{\text{ms}}\right) \times 3 \text{ ms} = -3 \text{ V}$$

Notice that the output is ramping down at 1 V/ms. To find the length of time it takes for the output voltage to hit the lower rail voltage, rearrange the output voltage expression to solve for the time t variable

$$t = \frac{V_{out}}{\left(-\dfrac{E_{in}}{RC}\right)}$$

and substitute the output rail voltage of −13 V.

$$t = \frac{-13 \text{ V}}{-1 \text{ V/ms}} = 13 \text{ ms}$$

The output falls at 1 V/ms until it hits the lower rail of −13 V in 13 ms.

Design R_f to be about 10 times R.

$$R_f = 10 \times 1 \text{ k}\Omega = 10 \text{ k}\Omega$$

Practice

Find the output of the circuit of Figure 15-8 if its input signal is a 600 µs rectangular pulse of 5 V.

Answer: Negative-going ramp. See input signal of Figure 15-9(a) and its corresponding output voltage of Figure 15-9(b).

Exploration

Design a system that receives a voltage spike (5 V down to 0 V) for 10 µs and outputs a ramp from 0 V up to +3 V over 600 µs. Think in terms of three separate circuits (or stages). *Stage 1*, a 10 µs spike must be converted to a 600 µs signal (think rectangular pulse). What circuit do you know that can do that? *Stage 2*, the 600 µs pulse must be converted to a ramp. However, the output ramp would be negative. *Stage 3*, convert the negative ramp output to a positive ramp. Draw the schematic of each stage with its corresponding values.

Answer: *Stage 1*: monostable 555 timer with E_{supply} of +5 V and pulse width W of 600 µs (you must select an R and a C combination such that 0.693RC is 600 µs); *stage 2*: *RC* integrator circuit (same as practice problem) or one where R and C are selected such that $RC = 1$ ms; *stage 3*: an inverting voltage amplifier with $R_i = R_f$, which is a voltage gain of −1 to invert the negative ramp to a positive ramp).

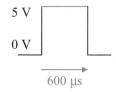

5 V

0 V

600 µs

(a) Input

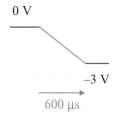

0 V

−3 V

600 µs

(b) output

Figure 15-9
Practice signals

An *RC* integrator circuit can also be used to change a signal's wave shape. A fairly common example is changing a square wave into a triangle wave. The *positive* half cycle of the input square wave creates a *negative* ramp output signal. The negative half signal of the input square wave generates a positive ramp output signal. Combining the positive and negative half cycles, a square wave input creates alternating negative and positive ramps, essentially creating a triangle wave output that stabilizes with a 0 V average. Since $V_{C\,init}$ is not 0 V,

$$\Delta v(t) = -\left(\frac{E_{in}}{RC}\right)t$$

5 V

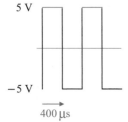

−5 V

400 μs

(a) Input

1 V

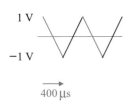

−1 V

400 μs

(b) Output

Figure 15-10
Example waveforms
of an integrator circuit

Example 15-7

A 10 V_{pp} square wave with a frequency of 1.25 kHz is fed into the integrator circuit of Figure 15-8. Find and sketch the output voltage waveform.

Solution

A 10 V_{pp} square wave translates to a fixed +5 V signal on the positive half cycle and a fixed −5 V signal on the negative half cycle as shown in Figure 15-10(a). The period of the input signal is

$$T = \frac{1}{1.25\ \text{Hz}} = 800\ \mu s$$

therefore each half cycle is 400 μs long. After 400 μs, the positive half square wave produces a negative going ramp voltage with a value of

$$\Delta v_{out}(t) = -\frac{E_{in}}{RC}t$$

$$\Delta v_{out}(400\ \mu s) = -\frac{5\ \text{V}}{(1\ \text{k}\Omega)(1\ \mu F)}(400\ \mu s) = -2\ V_{pp}$$

where the − value indicates a negative ramp.

The resulting output voltage waveform is an inverted 2 V_{pp} triangle wave with the same frequency as the original signal of 1.25 kHz as shown in Figure 15-10(b).

Practice

The capacitor in the integrator circuit of Figure 15-8 is changed to a 10 μF capacitor. Find and draw the output voltage waveform if the input is a 10 V_{pp} square wave at 1.25 kHz.

Answer: 200 mV_{pp} triangle wave

15.3 Op Amp *RC* Differentiator Circuit

The **differentiator** circuit utilizes a negative feedback amplifier and an *RC* feedback circuit as shown in Figure 15-11(a). The capacitor is the input element of the circuit instead of being the feedback element. There must be a changing input voltage to create an induced capacitor current, which in turn creates a voltage drop across the feedback resistor. A fixed dc voltage on the input does not create an input current and therefore does not create an output voltage.

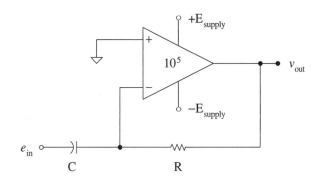

Figure 15-11(a) Differentiator circuit (*C not polarized*)

Referring to Figure 15-11(a), the input is being fed to the inverting op amp terminal; therefore the output voltage is inverted. The output voltage is being fed back to the inverting op amp terminal, therefore this is negative feedback and the op amp error voltage is approximately 0 V as long as the op amp is not hitting a rail voltage. The current into the op amp is assumed to be ideally 0 A.

Figure 15-11(b) shows the schematic prepared for circuit analysis. Note that a virtual common is formed between the capacitor and feedback resistor.

Apply Ohm's law and KVL to the feedback resistor R.

Virtual common

$$i_R = \frac{v_R}{R} = -\frac{v_{out}}{R}$$

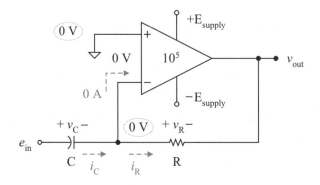

Figure 15-11(b) Differentiator circuit with references

Apply the current and voltage relationship and KVL to the capacitor C.

$$i_C = C \frac{d\,v_C}{dt} = C \frac{d\,e_{in}}{dt}$$

Applying KCL at the *virtual common* node, $i_R = i_C$. Thus, the two above expressions are equal to each other.

$$-\frac{v_{out}}{R} = C \frac{d\,e_{in}}{dt}$$

Algebraically for v_{out}.

$$v_{out} = -RC \frac{d\,e_{in}}{dt}$$

Thus, this circuit is really a mathematical (electronic) differentiator circuit. The input signal is differentiated and multiplied by a constant to produce the output signal, assuming the output is not hitting the upper or lower rail. Note that the output signal is inverted as expected.

Another way to think of the derivative (de_{in}/dt) is in terms of a change in voltage divided by a change in time ($\Delta e_{in}/\Delta t$).

$$v_{out} = -RC \frac{\Delta e_{in}}{\Delta t}$$

As noted above, if e_{in} is a constant the v_{out} is 0 V. Another special case is when the input voltage e_{in} is a ramp voltage; then the rate of change of the input voltage is a constant value (the slope of the ramp of e_{in}). The derivative reduces to a constant in units of V/s, and v_{out} is a fixed dc value.

Example 15-8

For the circuit of Figure 15-12, find and draw the output voltage waveform.

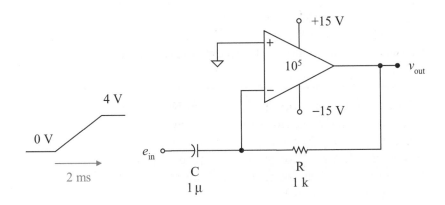

Figure 15-12 Example differentiator circuit

Solution

With the input at 0 V, the rate of change of the input is 0 V and therefore the output is at 0 V.

$$v_{out} = -RC\frac{\Delta e_{in}}{\Delta t} = -RC(0 \text{ V}) = 0 \text{ V}$$

With the input ramping up at 4 V per 2 ms, the rate of change of the input is a constant and therefore the output voltage is a fixed voltage.

$$v_{out} = -(1 \text{ μF})(1 \text{ k}\Omega)\left(\frac{4 \text{ V}}{2 \text{ ms}}\right) = -2 \text{ V}_{min}$$

With the input fixed at 4 V, the rate of change of the input is 0 and therefore the output is at 0 V.

$$v_{out} = -RC\left(0\ \frac{V}{s}\right) = 0\ V$$

The resulting output voltage waveform is a 2 ms inverted pulse as shown in Figure 15-13.

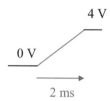

4 V

0 V

2 ms

(a) Input

Practice
Find the output of the circuit of Figure 15-12 if the capacitor is changed to 0.1 μF.

Answer: Same as Figure 15-13 except the pulse has a minimum voltage of −0.2 V.

An *RC* differentiator circuit can also be used to change a signal's waveshape. A fairly common example is converting a triangle wave into a square wave. The *positive slope ramp* of the input triangle wave creates the *negative half cycle* of the square wave output signal. The negative slope ramp of the input triangle wave creates the positive half cycle of the square wave output signal.

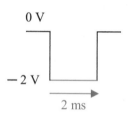

0 V

− 2 V

2 ms

(b) Output

Figure 15-13
Example wave-
forms of a differen-
tiator circuit

Example 15-9
A 2 V$_{pp}$ triangle wave with a frequency of 1.25 kHz is fed into the differentiator circuit of Figure 15-12. Find and sketch the output voltage waveform.

Solution

The period of the input triangle wave signal is

$$T = \frac{1}{1.25\ \text{kHz}} = 800\ \mu s$$

Therefore a half cycle is 400 μs. The positive slope ramp would have a constant slope of

$$\frac{\Delta e_{in}}{\Delta t} = \frac{2\ V}{400\ \mu s} = 5\frac{V}{ms}$$

The output voltage is a fixed voltage of

$$v_{out} = -RC\frac{\Delta e_{in}}{\Delta t}$$

$$v_{out} = -(1\ \mu F)(1\ k\Omega)\left(5\frac{V}{ms}\right) = -5\ V_{min}$$

A 2 V_{pp} input triangle wave with a frequency of 1.25 kHz into the example *RC* differentiator circuit of Figure 15-12 converts into an inverted 10 V_{pp} square wave signal output as shown in Figure 15-14.

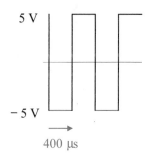

2 V

−2 V

→
400 μs

(a) Input

Practice

The capacitor in the integrator circuit of Figure 15-12 is changed to a 10 μF capacitor. Find and draw the output voltage waveform if the input is still a 40 mV$_{pp}$ triangle wave with a frequency of 1.25 kHz.

Answer: 2 V$_{pp}$ square wave

What happens when the differentiator circuit is driven with a square wave? The next example examines this question.

5 V

−5 V

→
400 μs

(b) Output

Figure 15-14
Example of a triangle wave input into a differentiator circuit

---

Example 15-10

A 10 V$_{pp}$ square wave with a frequency of 500 Hz is fed into the differentiator circuit of 15-12. The input waveform is shown in Figure 15-15(a). Find and sketch the output voltage waveform.

Solution

The period of the input square wave signal is

$$T = \frac{1}{500\ Hz} = 2\ ms$$

Therefore a half cycle is 1 ms. At time T_1, the beginning of the positive half cycle of the input square wave, the rate of change of the input signal is a +10 V change divided by a 0 ms change in time.

$$\frac{\Delta e_{in}}{\Delta t} = \frac{5\ V - (-5\ V)}{0\ ms} = +\infty\ \frac{V}{ms}$$

Mathematically this leads to the output voltage being infinite.

$$v_{out} = -RC\frac{\Delta e_{in}}{\Delta t}$$

$$v_{out}(T_1) = -(1\ \mu F)(1\ k\Omega)\left(\infty\frac{V}{ms}\right) = -\infty\ V_{min}$$

But that is impossible. Why? The output voltage of the op amp is limited to its rail voltages. Assume that the op amp requires 2 V headroom to operate; that creates rail voltages of ±13 V. Therefore the negative spike of voltage is limited to –13 V as shown in Figure 15-15(b).

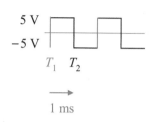

(a) Input

$$v_{out}(T_1) \approx -13\ V_{min}$$

The input square wave change at time T_2 is a sudden decrease from 5 V down to –5 V, or a sudden change of –10 V over 1 ms.

$$\frac{\Delta e_{in}}{\Delta t} = \frac{-5\ V - (5\ V)}{0\ ms} = -\infty\ \frac{V}{ms}$$

This sudden decrease in input voltage generates a +13 V output voltage spike at time T_2

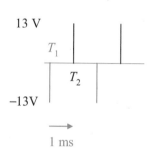

(b) Output

$$v_{out}(T_2) = -(1\ \mu F)(1\ k\Omega)\left(-\infty\frac{V}{ms}\right) = +\infty\ V_{max}$$

$$v_{out}(T_2) \approx +13\ V_{max}$$

The output is a nearly ideal spike voltage generator limited to the rail voltages of the op amp.

Figure 15-15
Example of a square wave input into a differentiator circuit

Practice

The supply voltages of the example are changed to ±12 V. What is the output voltage waveform?

Answer: Voltage spikes on the half cycle alternating between a –10 V spike and a +10 V spike.

15.4 Op Amp *RC* Waveform Generators

A circuit configured to generate an ac output signal without any external signal input is called a **waveform generator** or an **oscillator**. The asynchronous 555 timer circuit is an example of a circuit producing an output voltage waveform without an obvious input signal.

Positive feedback is used to create waveform-generating circuits. Positive feedback systems use very low voltage noise to start the process. For example, very weak noise signals at an op input are amplified and fed back to the input, adding to the original input signal. That net larger input signal is amplified producing a larger feedback signal, which increases the net input signal, and so forth. Eventually the system reaches a designed or natural circuit limit or a component limit.

Waveform generators create various waveshapes, for example, sine waves, square waves, exponential waves, and voltage spikes. This section covers waveform generators that are created using basic op amp circuits in combination with resistive and *RC* circuits.

Relaxation Oscillator

The relaxation oscillator of Figure 15-16(a) generates a square wave output at the output pin of the op amp. The op amp output is slamming its upper rail (maximum output voltage) or its lower rail (minimum output voltage).

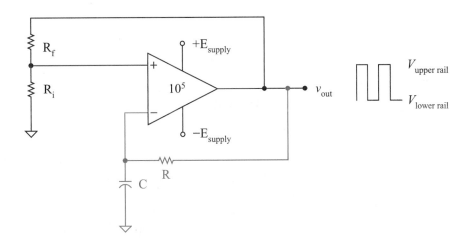

Figure 15-16(a) Relaxation oscillator

This circuit is primarily a *noninverting positive feedback comparator circuit* through the essentially series R_i-R_f circuit. This positive feedback circuit drives the op amp to its upper rail voltage or to its lower rail voltage. The R_i-R_f resistor circuit voltage divides the output signal and establishes the upper trip point (UTP) and the lower trip point (LTP) at the op amp's noninverting input pin as shown in Figure 15-16(b).

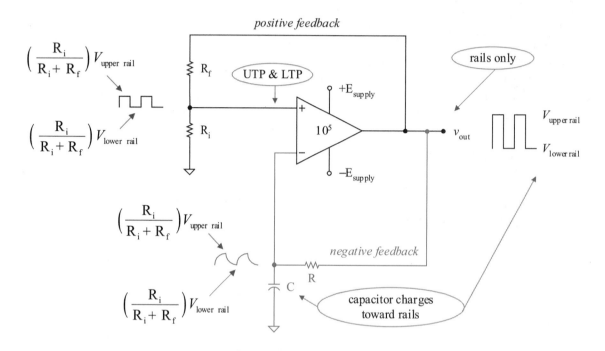

Figure 15-16(b) Relaxation oscillator

When the output is at the op amp's upper rail, the noninverting op amp pin voltage is at the UTP voltage.

$$V_{max\ NI} = V_{UTP} = \left(\frac{R_i}{R_i + R_f} \right) V_{upper\ rail} \qquad (UTP)$$

The output voltage is also across the essentially series *RC* circuit. The capacitor is always charging towards the op amp's output voltage. If the op amp's output voltage is at its upper rail, the capacitor is charging toward $V_{upper\ rail}$. When the capacitor voltage climbs to slightly above the value of the UTP, the op amp's input voltage goes negative and the op amp then slams its lower rail. The R_i-R_f positive feedback circuit now produces the LTP voltage at the op amp's noninverting pin.

$$V_{\text{min NI}} = V_{\text{LTP}} = \left(\frac{R_i}{R_i + R_f}\right) V_{\text{lower rail}} \qquad \text{(LTP)}$$

Since the op amp output is now at its lower rail voltage, the capacitor now discharges towards that lower rail output voltage and continues to discharge until it reaches the LTP voltage. When the capacitor voltage reaches the LTP voltage, the op amp input goes positive and the op amp switches back to its upper rail output voltage. This in turn changes the trip point voltage back to the UTP voltage. That completes one full cycle and the process continues into the next cycle.

The voltage divider of R_i and R_f forms the feedback voltage divider circuit, and the fraction of voltage being fed back to the input is given the special name of the *feedback factor* and is typically given the symbol B. B represents the fraction of the output signal fed back to the input.

$$B = \frac{R_i}{R_i + R_f}$$

It is often more convenient to use the feedback factor B in circuit calculations rather than using the actual feedback resistor symbols or values. For example,

$$V_{\text{UTP}} = B\, V_{\text{upper rail}}$$

$$V_{\text{LTP}} = B\, V_{\text{lower rail}}$$

Example 15-11

Assume that the op amp of the circuit in Figure 15-17(a) requires 2 V of headroom. Is there any signal input to this circuit? Find and sketch the output voltage waveform. Find the feedback factor B. Find the UTP and LTP voltages. Sketch the waveform at the op amp's noninverting input terminal. Sketch the waveform at the op amp's inverting input. Find the period of the output signal. Find the frequency of the output signal.

Solution

There is no apparent input signal to this circuit; however, very low noise voltage is always present. Positive feedback of this very low noise voltage eventually drives the op amp output to its rail voltages.

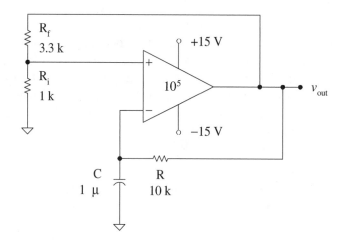

Figure 15-17(a) Example circuit of a relaxation oscillator

The output voltage is a square wave with a maximum value of

$$V_{out\ max} = V_{upper\ rail} = 15\ V - 2\ V = 13\ V$$

and a minimum value of

$$V_{out\ min} = V_{lower\ rail} = -15\ V + 2\ V = -13\ V$$

The positive feedback factor B is

$$B = \frac{1\ k\Omega}{1\ k\Omega + 3.3\ k\Omega} = 0.2326$$

which indicates 23.26% of the output signal is being fed back to the input.

The upper trip point (UTP) is determined by the positive feedback R_i-R_f voltage divider circuit.

$$V_{\text{max NI}} = V_{\text{UTP}} = \left(\frac{1 \text{ k}\Omega}{1 \text{ k}\Omega + 3.3 \text{ k}\Omega}\right) 13 \text{ V} = 3.023 \text{ V}$$

Or, using the feedback factor *B*

$$V_{\text{UTP}} = (0.2326)13 \text{ V} = 3.023 \text{ V}$$

The lower trip point (LTP) is

$$V_{\text{min NI}} = V_{\text{LTP}} = \left(\frac{1 \text{ k}\Omega}{1 \text{ k}\Omega + 3.3 \text{ k}\Omega}\right)(-13 \text{ V}) = -3.023 \text{ V}$$

Or, using the positive feedback factor *B*

$$V_{\text{LTP}} = -(0.2326)13 \text{ V} = -3.023 \text{ V}$$

The voltage at the op amp's noninverting input pin is a square wave with a maximum of +3.023 V and a minimum of −3.023 V. The voltage at the op amp's inverting input pin is an exponential waveform with a maximum of +3.023 V and a minimum of −3.023 V. Figure 15-17(b) shows the example relaxation oscillator circuit with its waveforms.

How can you find the period of the output voltage square wave? First note that all of the waveforms in Figure 15-17(b) have the same frequency *f* and therefore the same signal period *T*. Which waveform is the best candidate for solving for the period *T*? The *RC* charge-discharge curve must follow the *RC* universal dc transient expression, which includes the variable time *t* and the *RC* time constant τ.

The *RC* time constant τ is

$$\tau = 10 \text{ k}\Omega \times 1 \text{ μF} = 10 \text{ ms}$$

During a charge cycle, the capacitor voltage initially starts at the LTP of −3.023 V and climbs toward a steady state voltage of 13 V, but trips at the upper trip point voltage of 3.023 V. Substituting into the *RC* universal dc transient expression

$$v_C(t) = V_{\text{ss}} + (V_{\text{init}} - V_{\text{ss}}) \, e^{-t/\tau}$$

the expression becomes

$$3.023 \text{ V} = 13 \text{ V} + (-3.023 \text{ V} - 13 \text{ V}) \, e^{-t/10 \text{ ms}}$$

and simplifies to

$$3.023 \text{ V} = 13 \text{ V} - 16.023 \text{ V} \, e^{-t/10 \text{ ms}}$$

Inversely solve this expression to find the value of t required to charge the capacitor from the LTP to the UTP.

$$t = 4.737 \text{ ms}$$

The same amount of time is required to discharge the capacitor from the UTP to the LTP. A complete charge and discharge cycle creates one cycle of the waveform. Therefore,

$$T = 2 \times 4.737 \text{ ms} = 9.474 \text{ ms}$$

The signal frequency is the reciprocal of its period.

$$f = \frac{1}{9.474 \text{ ms}} = 105.6 \text{ Hz}$$

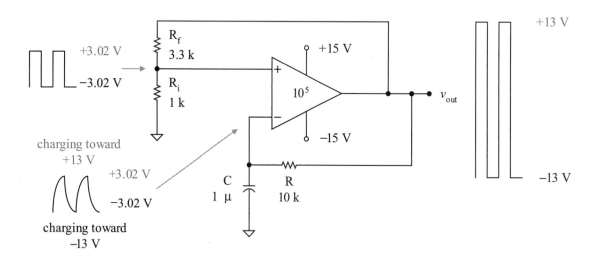

Figure 15-17(b) Example relaxation oscillator with waveform solutions

Practice

Use the generic terms of Figure 15-16(a). Write the expression for the charging of the capacitor from its LTP to its UTP. Solve this expression inversely and find the expression for half a cycle of the waveform, then write the expression for the period. Hint: use the same steps as in the example but with generic terms. Also, use the feedback factor B where appropriate to simplify the expressions. Verify the period of the example problem using your generic expression result.

Answer: Expression:

$$v_C(t) = V_{\text{upper rail}} + (V_{\text{LTP}} - V_{\text{upper rail}})\, e^{-t/RC}$$

Solving for t set up:

$$V_{\text{UTP}} = V_{\text{upper rail}} + (V_{\text{LTP}} - V_{\text{upper rail}})\, e^{-t/RC}$$

Simplifies to:

$$B = 1 + (-B - 1)\, e^{-t/RC}$$

Inverse solution for t:

$$t = RC \, \ln\!\left(\frac{1 + B}{1 - B}\right)$$

Resulting period:

$$T = 2\,RC \, \ln\!\left(\frac{1 + B}{1 - B}\right)$$

From example:

$$T = 2(10\text{ k}\Omega)(1\text{ }\mu\text{F}) \ln\!\left(\frac{1 + 0.2326}{1 - 0.2326}\right) = 9.47\text{ ms}$$

Triangle Wave Generator

The square wave output of a relaxation oscillator can be used to drive an op amp integrator circuit to produce a triangle wave output as shown in the circuit of Figure 15-18. This is a two-stage system with two distinct circuits. These circuits are said to be **cascaded** because the signal output of the first stage serves as the input signal for the next stage in a tandem like fashion. The output of the *first stage relaxation oscillator* is a square wave with:

Maximum voltage

$$V_{\text{max out}\_1} = V_{\text{upper rail}}$$

Minimum voltage

$$V_{\text{min out}\_1} = V_{\text{lower rail}}$$

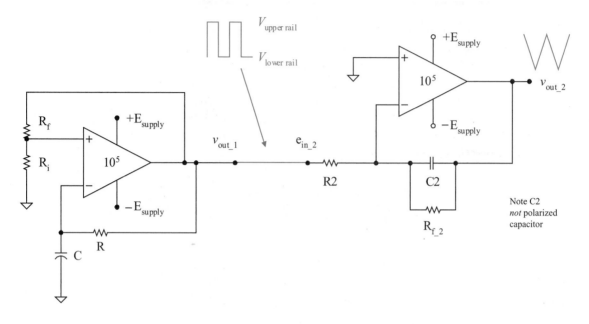

Figure 15-18 Triangle wave generator

Positive feedback factor:

$$B = \frac{R_i}{R_i + R_f}$$

Period:

$$T = 2\,RC \ln \frac{1 + B}{1 - B}$$

The positive half cycle of the square wave from stage 1 produces a negative ramp out of the integrator and reaches the minimum voltage of:

Stage 2 minimum output

$$V_{\text{min out\_2}} = -\frac{1}{2} \times \frac{V_{\text{upper rail}}}{R2 \times C2} \times \frac{T}{2}$$

The negative half cycle of the square wave from stage 1 produces a positive ramp out of the integrator and reaches the maximum voltage of:

Stage 2 maximum output

$$V_{\text{max out\_2}} = -\frac{1}{2} \times \frac{V_{\text{lower rail}}}{R2 \times C2} \times \frac{T}{2}$$

Example 15-12

For the circuit of Figure 15-19, find and sketch the signal output. Assume the op amp rail voltages are ±13 V. Note: the first stage is the same circuit as in the previous relaxation oscillator example.

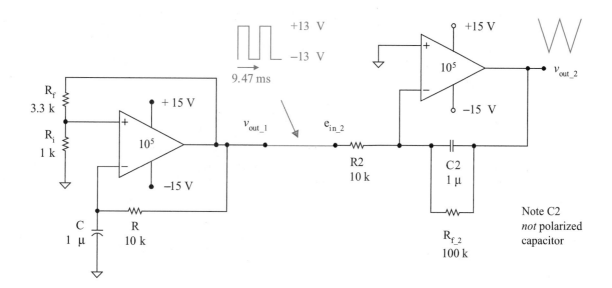

Figure 15-19 Triangle wave generator

Solution

The minimum output is

$$V_{\text{min out\_2}} = -\frac{1}{2} \times \frac{+13 \text{ V}}{10 \text{ k}\Omega \times 1 \text{ μF}} \times \frac{9.474 \text{ ms}}{2} = -3.08 \text{ V}$$

Then maximum output is

$$V_{\text{max out\_2}} = -\frac{1}{2} \times \frac{-13 \text{ V}}{10 \text{ k}\Omega \times 1 \text{ }\mu\text{F}} \times \frac{9.474 \text{ ms}}{2} = +3.08 \text{ V}$$

The output is a triangle wave with a maximum of +3.08 V, a minimum of −3.08 V, and a period from the relaxation oscillator of 9.474 ms producing a frequency of 105.6 Hz.

Practice

Change the resistor R2 to produce a 5 V_{pp} triangle wave output.

Answer: 1.23 kΩ

15.5 Natural Hidden Capacitance

Capacitors are components specifically designed with two metallic surfaces separated by an insulator. As noted earlier in this textbook, capacitance also occurs naturally with electronic components and between conductors in a circuit. This capacitance is not obvious or intended and is therefore called hidden or stray capacitance. Figure 15-20 shows a few examples of hidden capacitances:

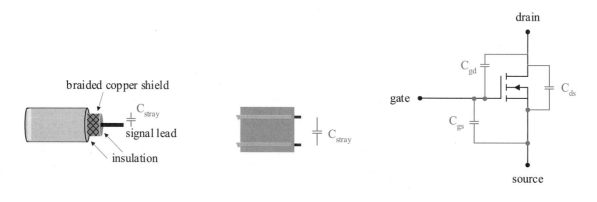

(a) 50 Ω coaxial cable   (b) 300 Ω twin lead   (c) n-channel MOSFET

Figure 15-20 Hidden capacitance

a. Coaxial cable (oscilloscope and high frequency test cable)

b. Twin lead cable for high frequency signals

c. Between terminals of a device (e.g., MOSFET gate-to-source capacitance C_{gs})

Hidden capacitance subject to continuous dc or low frequency continuous ac signals is not typically a problem. Recall that the formula for capacitive reactance is

$$X_C = \frac{1}{2\pi f C}$$

If the frequency f is low or 0 Hz (dc), the capacitive reactance is very high (approaching an *open*). A parallel open has no effect on the signal. However, at high frequencies, the capacitive reactance is very low (approaching a *short*). Imagine if each capacitance in Figure 15-20 were replaced with a *short* for *high frequencies*. What is the result? The high frequency signals are shorted between conductors. In Figure 15-20(a), a high frequency signal is shorted to the conductive mesh of the test cable. In Figure 15-20(c), a high frequency signal is shorted from the gate to source and the signal is not processed by the MOSFET. The signals are totally lost or significantly attenuated.

A square wave is equivalent to a very high frequency at its leading and trailing edges. A sudden change in voltage of the steady state voltage less its initial voltage ($V_{ss} - V_{init}$) with 0 change in time is equivalent to a very high frequency. Mathematically,

$$f = \frac{1}{T} = \frac{1}{0 \text{ s}} = \infty \text{ Hz}$$

Thus, the leading and lagging edges of a square or rectangular wave have the effect of creating what initially appear to be shorted capacitors. The capacitance combined with the effective resistance seen by that capacitance forms an RC time constant τ. After five time constants (5τ or $5RC$), the signal has reached steady state with a fixed voltage V_{ss} and the capacitance now appears to be an open in steady state. Thus, rectangular waves into a hidden RC circuit respond with each leading and each lagging edge as an RC switching circuit and follow the RC universal dc transient expression of

$$v_C(t) = V_{ss} + (V_{init} - V_{ss})\, e^{-t/\tau}$$

A square wave into a hidden capacitance (*RC*) circuit produces a voltage exponential response as shown in Figure 15-21 just like a circuit designed with *RC* components. Another way to think of this response is that the capacitor voltage resists a sudden change in voltage and therefore changes exponentially. Since the stray capacitance initially acts like a short, the stray capacitance induces a **surge current** spike in the circuit with each leading and lagging edge of the rectangular wave.

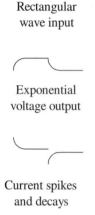

Rectangular
wave input

Exponential
voltage output

Current spikes
and decays

Figure 15-21
Rectangular signal
responses in an RC
circuit

Example 15-13

For the circuit of Figure 15-22, find the capacitive reactance of a 15 Hz signal (lower end of the audio range or human hearing range) to 15 kHz (the upper end of the human audio range). Would this serve as a good audio cable? Also find the time constant for this circuit. How long does it take for the circuit to respond to a square wave input? What is the rise time of the responding signal?

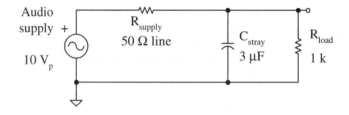

Figure 15-22 Capacitive audio cable

Solution

The capacitive reactance at 15 Hz is 3.54 kΩ. This reactance is greater than the parallel load resistance of the circuit but only by a factor of about 3. Therefore, the stray capacitance causes very little loss with a 15 Hz sine wave.

$$X_C = \frac{1}{2\pi \times 15 \text{ Hz} \times 3 \text{ μF}} = 3.54 \text{ k}\Omega$$

The capacitive reactance at 15 kHz is 3.54 Ω. This reactance is much lower than its parallel load resistance. Significant signal

current is being lost through the stray capacitance of this cable at
the high frequency end of the human audio spectrum.

$$X_C = \frac{1}{2\pi \times 15 \text{ kHz} \times 30 \text{ μF}} = 3.54 \text{ Ω}$$

This cable is not suitable for the audio range. A more appropriate
cable must be selected.

For the rectangular wave response, first find the equivalent resistance (namely the Thévenin resistance) seen by the stray capacitance. This equivalent resistance is the 50 Ω line in parallel with
the load resistance of 1 kΩ. The parallel combination is 47.62 Ω.
The hidden circuit time constant τ is

$$\tau = R_{Th}C$$

$$\tau = 47.62 \text{ Ω} \times 3 \text{ μF} = 142.9 \text{ μs}$$

After five time constants, the output is very close to its steady
state value. Therefore, it takes the circuit 714 μs to respond to the
leading and trailing edge of a rectangular waveform.

$$5\tau = 5 \times 142.9 \text{ ms} = 714 \text{ μs}$$

The rise time of the responding exponential signal to travel from
10% of its change to 90% of its change is

$$t_r = 2.2 \ \tau = 2.2 \times 142.9 \text{ μs} = 314 \text{ μs}$$

Practice

As will be shown in a later textbook in this series, a **half-power
point** occurs in an *RC* circuit when the capacitance reactance
equals the Thévenin resistance (R_{Th} as seen from the capacitance
perspective).

$$R_{Th} = X_C$$

$$R_{Th} = \frac{1}{2\pi \ f_{\text{half-power}} \ C}$$

At this frequency, the signal output from the *RC* circuit is at *half
of its original power*. Therefore it is called the *half-power point*.
If the frequency goes higher than this, there is significant loss in

output power of that signal. Find the half-power point frequency of the example RC circuit. Does this confirm the example problem result that demonstrated that this is not a good audio cable?

Answer: 1.11 kHz (all signals higher than 1.11 kHz have significant power loss; this is a very bad audio cable)

Exploration

A 2 kΩ microphone is connected through an 8 foot microphone cable to the input of a very high resistance amplifier. If the microphone cable has a capacitance of 3 pF per inch, prove that this cable easily passes the highest human audio frequency of 15 kHz by (a) comparing the capacitive reactance to the circuit Thévenin resistance and (b) finding the half-power point frequency.

Semiconductor devices with *pn* junctions exhibit capacitance since they have conductive material (*p* and *n* type material) with carriers separated by a *pn* junction area, which is depleted of carriers. The **varactor** (a reverse biased diode) takes advantage of this capacitive characteristic; it acts as a variable capacitor. As the reverse voltage is increased, the *pn* junction width increases. As the pn junction width increases, its capacitance decreases. A decrease in reverse voltage across a diode increases its capacitance.

The BJT and the MOSFET exhibit **input capacitance**, **output capacitance**, and **reverse transfer capacitance**. These capacitances are created by the internal junction capacitances of the semiconductor junctions in these devices. The next example demonstrates the effect of the inherent input capacitance of a MOSFET used in an audio amplifier.

Example 15-14

The circuit of Figure 15-23(a) is a push amplifier (part of a push-pull audio power amplifier). The MOSFET is rated to have an input capacitance C_{iss} of 670 pF. Draw the *ac model* of this circuit including C_{iss}. Find the equivalent resistance R_{Th} seen by the C_{iss} capacitance. Find the reactance of C_{iss} at 15 kHz. Find the half-power point frequency of this input RC circuit. Is there any significant loss of power for audio signals in this input MOSFET circuit?

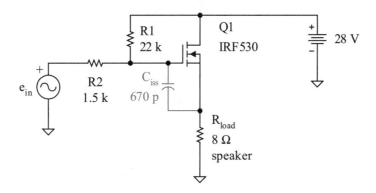

Figure 15-23(a) Push amplifier with MOSFET input capacitance

Solution

For ac analysis, model the dc power supply as a short. Then redraw the ac modeled circuit for ac signal calculations of the input *RC* circuit. See Figure 15-23(b) for the ac model to be used.

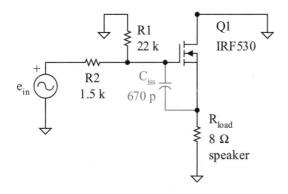

Figure 15-23(b) The ac modeled circuit

The input resistance into the MOSFET is very, very high (gigaohms), therefore the gate circuit is essentially an open and can be removed from the input circuit analysis. The circuit of Fig-

ure 15-23(c) is a simplified ac model of the input circuit to analyze the effect of the MOSFET's input capacitance.

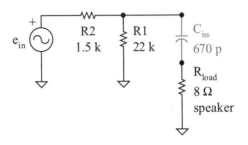

Figure 15-23(c) Simplified ac model for C_{iss} impact analysis

The equivalent resistance as viewed from the capacitance C_{iss} is
R_{load} in series with the parallel combination of R1 and R2. Therefore, the equivalent resistance R_{Th} is

$$R_{Th} = R_{load} + \left(R1^{-1} + R2^{-1} \right)^{-1}$$

$$R_{Th} = 8\ \Omega + \left((22\ k\Omega)^{-1} + 1.5\ k\Omega)^{-1} \right)^{-1} = 1.41\ k\Omega$$

The reactance of C_{iss} at 15 kHz is 15.8 kΩ, which is significantly
higher than the R_{Th} circuit resistance.

$$X_{Ciss} = \frac{1}{2\pi \times 15\ \text{kHz} \times 670\ \text{pF}} = 15.8\ k\Omega$$

The half-power point frequency for this input *RC* circuit is

$$f_{half\text{-}power} = \frac{1}{2\pi\ R_{Th} C_{iss}}$$

$$f_{half\text{-}power} = \frac{1}{2\pi\ (1.41\ k\Omega)(670\ \text{pF})} = 168\ \text{kHz}$$

Significant signal power loss starts occurring at 168 kHz, well
above the audio range. This frequency supports the above claim

that this input *RC* circuit does not negatively affect the audio range signals.

Practice

If a square wave were fed into the example circuit, find the input circuit time constant τ. Find the rise time of the signal responding to this input *RC* circuit.

Answer: 947 ns, 2.08 μs

The next example demonstrates the effect of test equipment frequency loading effects.

Example 15-15

The circuit of Figure 15-24 represents the input circuit to a digital voltmeter (DVM). Would this digital voltmeter give an accurate voltage measurement of a 1 kHz sinusoidal signal? Would a 100 kHz signal voltage measurement be accurate?

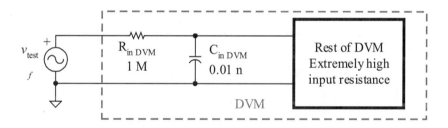

Figure 15-24 Equivalent input circuit of a digital voltmeter

Solution

The half-power point frequency for sinusoidal inputs is

$$f_{half\text{-}power} = \frac{1}{2\pi \left(1\,M\Omega\right)\left(0.01\,nF\right)} = 15.9 \ kHz$$

At this frequency the signal has actually been reduced to half power before it is delivered to the rest of the digital voltmeter circuit for measurement and display. Signal losses from the *RC* circuit become noticeable about a decade in frequency sooner than the half-power point frequency. Thus, a reasonable upper frequency for testing sinusoidal signals with this meter is

$$f_{\text{half-power}} = \frac{15.9 \text{ kHz}}{10} = 1.6 \text{ kHz}$$

The voltage of the 1 kHz signal is measured accurately while the voltage of the 100 kHz signal is not. A later textbook in this series examines frequency responses in detail.

Practice

The input specification of an oscilloscope is 10 MΩ in parallel with a 20 pF capacitance. Assume that a 50 Ω audio signal generator is driving the oscilloscope input. What is the time constant of this combined RC input circuit? If a square wave is measured with this scope, what is the resulting signal rise time? If a sine wave is generated and measured by this oscilloscope, what is the half-power point frequency of this combined RC input circuit? What is the maximum frequency that can be used for an accurate measurement?

Answer: 1 ns, 2.2 ns, 159 MHz, 15.9 MHz

15.6 Capacitor Leakage

An ideal capacitance holds charge forever, but in reality capacitance loses charge naturally without any external resistance connection. Capacitor leakage is a characteristic of capacitor types. For example, electrolytic capacitors are very leaky (lose charge quickly) while ceramic capacitors are not. Electrolytic capacitors do not hold a charge as long as a ceramic capacitor.

A leaky capacitor is modeled as shown in Figure 15-25 with a parallel resistance noted as R_{leakage}. This represents the leakage current path, while the ideal capacitor remains modeled as a steady state open. This leakage path exists on charge and discharge of the capacitor.

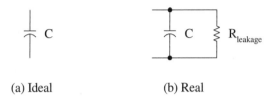

(a) Ideal   (b) Real

Figure 15-25 Capacitor leakage

The switching circuit of Figure 15-26(a) serves as a good example of how the leakage resistance affects an *RC* circuit. After the switch has been closed for at least five time constants, the fully charged capacitor achieves its maximum voltage of E_{Th} (as viewed from the capacitor C).

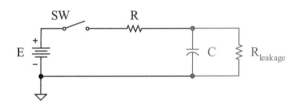

Figure 15-26(a) Switched RC circuit with lossy capacitor

Figure 15-26(b) shows the circuit used to find E_{Th}. The charge time *RC* time constant depends upon the resistance seen by the capacitance C, which is equivalent to the Thévenin resistance. Therefore, to solve this charging *RC* circuit you must first prepare the Thévenin model.

The circuit of Figure 15-26(b) is simply a voltage dividing series circuit. Apply the voltage divider rule.

$$E_{Th} = \left(\frac{R_{leakage}}{R + R_{leakage}} \right) E$$

Thus, the leakage resistance prevents the capacitor from fully charging to the applied voltage E.

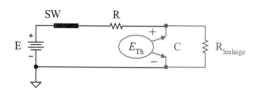

(b) Circuit to find Thévenin voltage

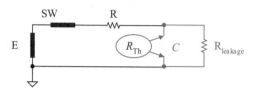

(c) Circuit to find Thévenin resistance

Figure 15-26(b,c) Circuits prepared for E_{Th} and R_{Th} calculation

To find the *RC* time constant, first you must find the Thévenin resistance. Using the prepared circuit of Figure 15-26(c), the Thévenin resistance is seen to be the parallel combination of R and $R_{leakage}$. Thus,

$$R_{Th} = R \; // \; R_{leakage}$$

The *RC* circuit time constant with the switch closed is therefore

$$\tau_{charge} = R_{Th} \; C$$

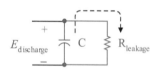

Figure 15-26(d)
Discharge through capacitor leakage resistance

where the R_{Th} parallel combination of R and $R_{leakage}$ is less than R, and the overall time constant has been reduced. If $R_{leakage}$ is comparable in value with R, the circuit time constant is significantly affected. On the other hand, if $R_{leakage} \gg R$, then $R_{leakage}$ can be ignored in all circuit calculations since it is not consequential to either the time constant or the steady state voltage value.

With the switch open, the capacitor discharges only through its leakage resistance as shown in Figure 15-26(d). The ideal capacitor would hold its voltage charge indefinitely, but in reality the capacitor charge leaks internally based upon its *RC* time constant of

$$\tau_{discharge} = R_{leakage} \; C$$

If $R_{leakage}$ is very high as it is in ceramic capacitors, the capacitor takes a long time to discharge. If $R_{leakage}$ is only moderately high as it is in electrolytic capacitors, the capacitor discharges relatively quickly.

Example 15-16

The ceramic capacitor of Figure 15-27 has a leakage resistance of 100 MΩ. (a) With the switch closed find the capacitor voltage when it is fully charged, its circuit time constant, and how long the capacitor takes to fully charge (five time constants). Assume that the capacitor fully charges and the switch opens. (b) Find the discharge time constant and how long it takes for the capacitor to fully discharge (five time constants).

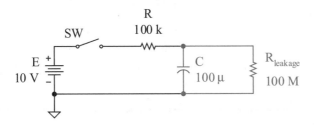

Figure 15-27 Example problem with ceramic capacitor

Solution

a. With the switch closed, find the Thévenin voltage as viewed from the capacitor C. That is, find the maximum voltage the capacitor can charge to. Note the result is very close to the ideal value (without a leakage resistance) of 10 V.

$$E_{Th} = V_{C \text{ fully charged}}$$

$$E_{Th} = \left(\frac{100 \text{ M}\Omega}{100 \text{ k}\Omega + 100 \text{ M}\Omega} \right) 10 \text{ V} = 9.990 \text{ V}$$

Find the Thévenin resistance as viewed from the capacitor. Note the result is very close to the ideal value of 100 kΩ.

$$R_{Th} = \left((100 \text{ k}\Omega)^{-1} + (100 \text{ M}\Omega)^{-1} \right)^{-1} = 99.90 \text{ k}\Omega$$

Find the circuit time constant. Note this is very close to the ideal value of 10 s.

$$\tau = R_{Th} \, C = 99.90 \text{ k}\Omega \times 100 \text{ }\mu\text{F} = 9.99 \text{ s}$$

Find the time to fully charge the capacitor (that is, 5τ).

$$t_{\text{full charge}} = 5\times9.99 \text{ s} = 49.95 \text{ s}$$

b. When the switch just opens, the capacitor begins to discharge through its leakage resistance R_{leakage}. The discharge time constant is

$$\tau = R_{\text{leakage}} \ C = 100 \text{ M}\Omega\times100 \text{ µF} = 10,000 \text{ s}$$

Find the time to fully discharge the capacitor.

$$t_{\text{full charge}} = 5\times10,000 \text{ s} = 50,000 \text{ s}$$

Since this is a low leakage capacitor, its discharge time constant is very long and the time to discharge the capacitor is reasonably long. The capacitor can charge much faster than it discharges.

Practice

Repeat the example with a capacitor with an R_{leakage} of 1 MΩ.

Answer: (a) 9.09 V, 90.9 kΩ, 9.09 s, 45.5 s (b) 100 s, 500 s

15.7 Clippers, Clampers, and Detectors

The diode and the diode in conjunction with capacitance are used to alter signal wave shapes. The clipper clips off part of a waveform using diodes. A limiter limits the maximum and/or minimum signal value using diodes, varistors, and other voltage limiting devices. A clamper adds a dc offset to an ac signal using diodes and capacitors.

Positive Clipper

A **clipper** clips off a portion of a signal's waveform. The clipper circuit uses a diode *in parallel* with the load. The circuit of Figure 15-28(a) is a **positive clipper**, since the positive half cycles of the output signal are missing (clipped off).

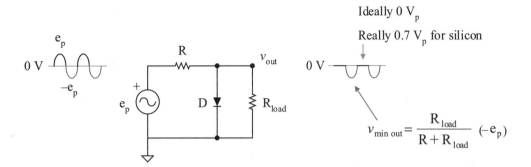

Figure 15-28(a) Positive clipper–parallel ideal diode shorts out positive signals

During the positive half cycle of the input, the diode is forward biased and ideally acts like a short as modeled in Figure 15-28(b). The *on* ideal diode drops 0 V, which in turn drops 0 V across its parallel load. Realistically, if this is a silicon diode, 0.7 V of forward voltage is required to turn the diode on and then the diode maintains the 0.7 V drop while turned on. For germanium, it is 0.2 V.

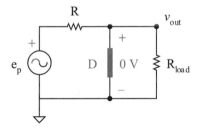

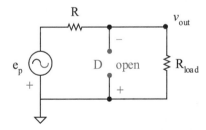

(b) Positive half cycle:
 forward biased *on ideal* diode

$$v_{out} = 0 \text{ V}$$

(c) Negative half cycle:
 reverse biased *off ideal* diode

$$v_{min\ out} = \frac{R_{load}}{R + R_{load}} (-e_p)$$

Figure 15-28(b,c) Positive clipper half cycle ideal models

Is it necessary to have a resistance R in this circuit? Looking at Figure 15-28(b), what is the problem with this circuit if there were not resistance R? The problem is that all of the supply voltage would drop across the diode, drawing excessive current and potentially destroying the diode and/or the supply. R is called a **current limiting resistor** since it limits the current through the diode.

During the negative half cycle, the reverse biased diode acts ideally like an *open* as modeled in Figure 15-28(c). Since the reverse biased diode acts like a *parallel open*, the diode plays no part in the negative half cycle analysis. What remains is a simple, essentially series circuit with two series resistors of R and R_{load} voltage dividing the supply voltage.

Negative Clipper

The circuit of Figure 15-29 is a **negative clipper**, since the negative half cycles of the output signal are missing (clipped off). During the input positive half cycles the diode is reverse biased *off* (*open*) and the supply voltage is divided between the two essentially series resistors. During the negative half cycle, the diode acts ideally like a *short* dropping 0 V across the load. In reality, a silicon diode drops about 0.7 V when forward biased *on* and a germanium diode 0.2 V.

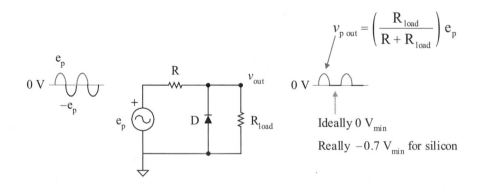

Figure 15-29 Negative clipper–*parallel* ideal diode shorts out negative signals

Example 15-17

Name the circuit in Figure 15-30(a). Sketch the voltage output waveforms of the circuit of Figure 15-30(a) assuming the diode is an ideal, a silicon, and a germanium.

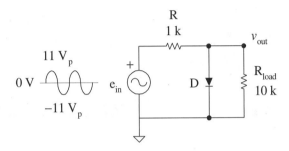

Figure 15-30(a) Example circuit

Solution

During the positive half cycles, the diode is forward biased on and develops a forward voltage drop of 0 V for an ideal diode, 0.7 V for a silicon diode, and 0.2 V for germanium. During the negative cycles, the diode is reverse biased and acts like an open. This is a *positive clipper circuit* based upon the essentially series, voltage divider circuit. See Figure 15-30(b,c,d) for the respective output voltage waveforms.

$$V_{\text{min out}} = \frac{10\ \text{k}\Omega}{1\ \text{k}\Omega + 10\ \text{k}\Omega}\left(-11\ V_p\right) = -10\ V_{\text{min}}$$

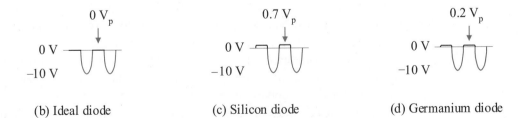

Figure 15-30(b,c,d) Example problem–*positive clipper circuit*

Practice

Repeat the example problem with the diode reversed.

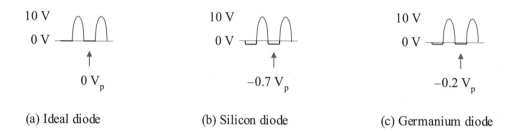

(a) Ideal diode   (b) Silicon diode   (c) Germanium diode

Figure 15-31 Practice problem–negative clipper circuit

Answer: Negative clipper. See Figure 15-31 for output waveforms.

Rectifier and Filtered Rectifier

A **rectifier** is a *series connection* with the load that eliminates the positive or negative part of an ac signal. It is a similar effect to the clipper, but the rectifier is connected in series with the load while the clipper is connected in parallel with the load. You have seen the rectifier circuit several times previously in this text. The circuit of Figure 15-32(a) is a review of a circuit previously studied.

Figure 15-32(a) is a *rectifier circuit* that allows current to flow in only one direction through the load. When the diode is forward biased from the input signal's positive half cycle, the diode is forward biased *on* and the series circuit is completed, and current flows through the load. The load current direction establishes the load voltage polarity that establishes this as a positive or negative voltage supply system. Note that the ideal diode drops 0 V and a silicon diode drops 0.7 V when forward biased. Note that the output peak voltage is reduced by the diode drop and that the diode is on less than 50% of the time (not a true half wave). When the diode is reverse biased during the input signal's negative half cycle, the diode is reverse biased *off* and is an *open*. No current flows through the load, therefore 0 V is dropped across the load.

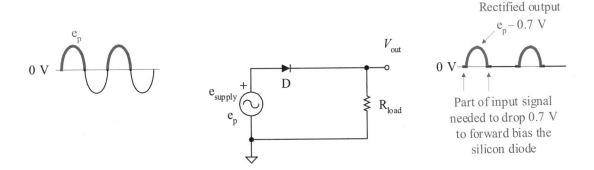

Figure 15-32(a) Half-wave rectifier circuit

Figure 15-32(b) is the same *rectifier circuit* with a capacitor placed across the load. The capacitor charges to the peak voltage out of the rectifier circuit and ideally holds that voltage across the load. When the diode is *on*, the capacitor charges through the diode with *a short time constant* of

$$\tau_{charge} = R_{diode} \ C$$

When the diode is *off*, the capacitor discharges through the load with a long time constant. The capacitor *charges quickly* with a short time constant τ_{charge} and *discharges very slowly* with a long time constant $\tau_{discharge}$.

$$\tau_{discharge} = R_{load} \ C$$

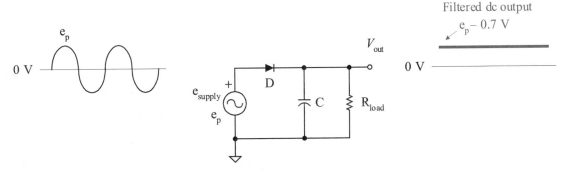

Figure 15-32(b) Half-wave rectifier circuit with filter capacitor

In reality, the capacitor discharge occurs during each cycle as shown in Figure 15-32(c).

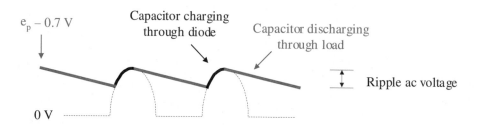

Figure 15-32(c) Half-wave rectifier output if significant capacitor discharge

The lower the capacitance value of filter capacitor C and/or the lower the value of the load resistance of R_{load}, the faster the discharge through the load and the deeper the discharge. This creates an ac ripple voltage, which is the difference between the maximum and minimum output voltages. Ideally or with a very long discharge time constant, the output voltage would be nearly a straight line as shown in Figure 15-32(b), but that requires very large, very expensive capacitance values.

As will be shown later, a regulator circuit can be placed between the filter capacitor and the load to produce a constant dc output as shown in Figure 15-32(d).

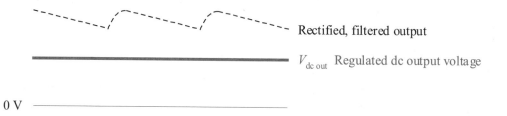

Figure 15-32(d) Regulated output voltage using a zener or IC regulator

This allows the circuit designer to use a lower valued, less expensive capacitor and compensate for the excessive ripple created by using an inexpensive voltage regulator such as a zener diode or an IC regulator.

The minimum capacitor voltage must be slightly greater than the desired regulated dc output voltage so that the capacitor voltage does not dip below the regulated output value.

Example 15-18

Perform a MultiSIM simulation run of the circuit of Figure 15-33(a). Measure the output with an oscilloscope. Use the T1 and the T2 vertical timer values to find the maximum, minimum, and ripple output voltages. Use CHA for the supplied signal and CHB for the output voltage signal to compare them.

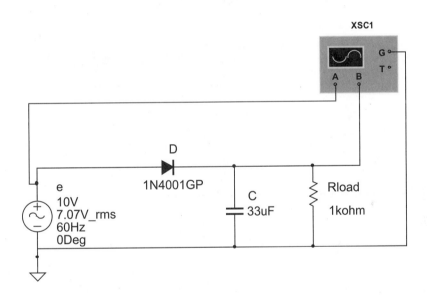

Figure 15-33(a) Example circuit of filtered half-wave rectifier

Solution

See Figure 15-33(b) for the simulated oscilloscope output voltage measurements. The peak output voltage, which is also the peak capacitor voltage, is VB2 with a value of 9.4 V_p. The minimum output voltage, which is also the minimum capacitor voltage, is

VB1 with a value of 6.2 V_{min}. The ac ripple voltage is the difference reading of VB2 − VB1, which has a value of 3.2 V_{pp}.

Also, use the oscilloscope voltage scales to be sure that you can properly interpret voltage values based upon the voltage scale setting in units of V per division.

This simulated graph of the output waveform clearly compares the input sine wave with the output charging and discharging *RC* curve. The diode is on a relatively short time to quickly recharge the capacitor before the input signal reaches its maximum again and turns off the diode. The capacitor then begins its relatively long discharge to complete a full cycle.

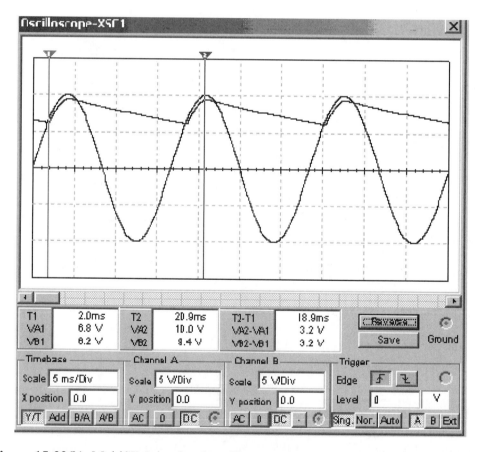

Figure 15-33(b) MultiSIM simulated oscilloscope display to determine output voltages

Practice

Reset the T1 and the T2 timer lines to measure the period of the ripple voltage. Based upon your measured value, calculate the frequency of the ripple.

Answer: See Figure 15-33(c). The T1 and T2 vertical lines have been placed at the beginning of a charge cycle. The T2–T1 measurement is a period reading of 16.6 ms, which is equivalent to 60 Hz (the same as the supply frequency).

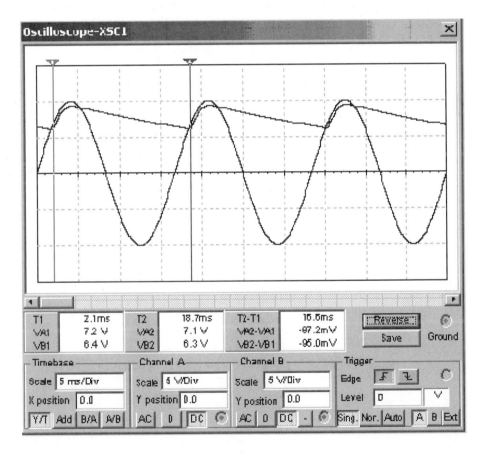

Figure 15-33(c) MultiSIM simulated oscilloscope display to calculate the output period

Exploration

Repeat the example and practice with the diode reversed.

Peak Detector

The circuit of Figure 15-32(b) can also be used as a **peak detector** since it in essence outputs a fixed voltage equal to the peak voltage of an input signal. For small signal detection, overcoming the forward voltage drop of the diode is an impossibility. For example, suppose you need to detect a 500 mV$_p$ signal. To eliminate the diode drop and produce a more accurate output, an active diode can be used within a *voltage follower amplifier* as shown in the first stage of the circuit in Figure 15-34. For a silicon diode, the *voltage follower amplifier* turns on when the input reaches

$$v_{\text{in op amp}} = \frac{0.7 \text{ V}}{A_{\text{op amp}}}$$

Only microvolts are needed to turn on the diode with a high gain op amp, and the output peak detector becomes a very accurate peak detector. The second stage is a voltage follower, which isolates the load resistance from the discharging capacitor. The discharging capacitor is looking into gigaohms of resistance in the second stage and the discharging time constant becomes very, very long (minutes instead of seconds). The leakage current of the capacitor becomes the limiting factor.

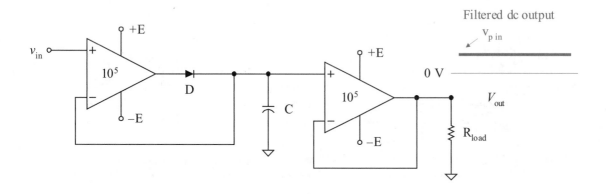

Figure 15-34 Active ideal signal peak detector

Positive Clamper

A **clamper** circuit provides a positive dc offset voltage to a signal, typically clamping the bottom of the signal to the 0 V line as shown in Figure 15-35. Figure 15-35(a) demonstrates a pure sine wave being positively clamped to 0 V. The peak-to-peak voltage remains the same; only a positive offset has been added to raise the total signal to the desired level. Figure 15-35(b) generally demonstrates the clamping of a general ac waveform from a minimum value of V_{min} to a new clamped value of $V_{min} + V_{dc}$. One way to add an offset to a signal is to insert a dc voltage supply into the circuit, but this is usually an expensive and inconvenient solution. A better, much less costly way to passively introduce dc voltage is to use the natural ability of the capacitor to store voltage.

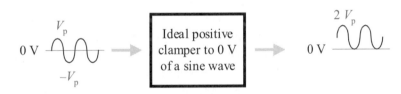

(a) Ideal positive clamper of an ac sine wave

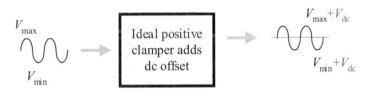

(b) Ideal positive clamper of adding a +dc offset

Figure 15-35 Positive clamper

The circuit in Figure 15-36 is a positive clamper circuit that utilizes a simple combination of a diode D with a capacitor C to clamp the ac signal to 0 V ideally (realistically, −0.7 V for a silicon diode).

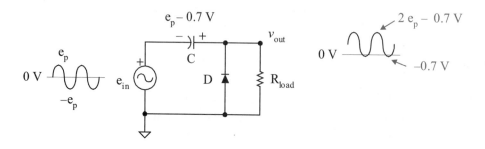

Figure 15-36 Positive clamper circuit

During the negative half cycle, the diode is forward biased after achieving a 0.7 V drop, turns on and the capacitor charges to the peak input voltage less the diode drop. The *RC* charge time constant is very short since the diode resistance is very low. Once the input signal starts climbing up from its minimum input voltage, the diode turns off and the capacitor discharges through supply e_{in} and the load R_{load}. This is designed to be long *RC* time constant relative to the signal's period so that the capacitor holds its charge. The capacitor then holds its charge while the diode is off and essentially introduces a dc supply in series with the ac signal.

$$V_{out} = e_{in} + V_{dc}$$

where

$$V_{dc} = e_{p\,in} - V_{diode}$$

For a silicon diode,

$$V_{p\,out} = 2e_p - 0.7 \text{ V}$$

Example 15-19

For the circuit of Figure 15-37(a) draw the equivalent modeled circuit for the minimum input signal (negative half cycle) when the diode is forward biased and the capacitor is charging. Use that modeled circuit to find the maximum voltage dropped across the capacitor. Assuming that the capacitor charges to and holds that maximum voltage, model the circuit with the capacitor as a voltage supply. Then find and sketch the output voltage.

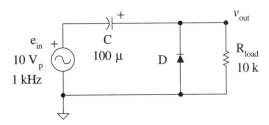

Figure 15-37(a) Example positive clamper circuit

Calculate the circuit charging time constant and the discharging time constant. Assume that the diode's on resistance is 1 Ω. How does the discharging time constant compare to the charging time constant? How does the discharging time constant compare to the period of the supply signal? Is this a good clamper circuit?

Solution

Assume that the input is at its minimum value, which produces the maximum forward bias voltage for the diode and the maximum voltage drop across the capacitor as demonstrated in the modeled circuit of Figure 15-37(b). The supply is labeled with the appropriate voltage and its polarity. The silicon diode has been replaced with its forward biased second approximation model of a 0.7 V voltage supply.

During the negative half cycle when the input is at its minimum of −10 V, find the output voltage. Starting at common and falling 0.7 V across the diode to the output node, the output voltage is

$$V_{\text{min out}} = 0 \text{ V} - 0.7 \text{ V} = -0.7 \text{ V}$$

To find the voltage across the capacitor when the diode is forward biased *on*, perform a KVL walkabout from $-v_{\text{C}}$ to $+v_{\text{C}}$. The capacitor charges to and holds this voltage when the diode is turned off.

$$V_{\text{C}} = -0.7 \text{ V} + 10 \text{ V} = 9.3 \text{ V}$$

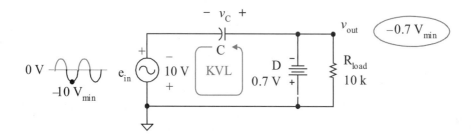

Figure 15-37(b) Modeled circuit with diode on

The capacitor fully charges to 9.3 V when the input signal is at its minimum of −10 V_{min}. Then the input signal starts climbing upward and the diode immediately *turns off* since it has less than the 0.7 V needed to keep it turned on. The modeled circuit now looks like that of Figure 15-37(c) with the diode turned *off* and acting like an *open*. The supply is shown at its peak value for maximum signal analysis. Assuming the capacitor holds its charge, find the maximum output voltage.

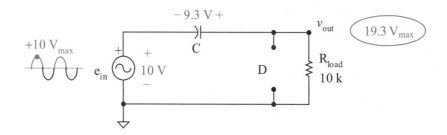

Figure 15-37(c) Modeled circuit with diode off

Starting at common, the supply creates a rise of 10 V and the capacitor a rise of 9.3 V for a total of 19.3 V.

$$V_{max\ out} = 10\ V + 9.3\ V = 19.3\ V_p$$

The output signal is the same 20 V_{pp} sine wave as the input signal but is now offset by 19.3 V_{dc} as shown in Figure 15-37(d).

To calculate the charging and discharging time constants, the equivalent Thévenin resistance as viewed from the capacitor must be determined. The capacitor sees the supply resistance in series with the parallel combination of the diode resistance and the load.

$$R_{Th} = R_{supply} + \left((R_{diode})^{-1} + (R_{load})^{-1} \right)^{-1}$$

Assuming an ideal voltage supply with an internal supply resistance of 0 Ω, this expression reduces to just the parallel combination of the diode's resistance and the load resistance.

$$R_{Th} = \left((R_{diode})^{-1} + (R_{load})^{-1} \right)^{-1}$$

During the charging cycle, the diode's forward resistance is very, very low and the parallel resistance combination is approximately equal to the diode's *on forward resistance*.

$$R_{Th\ charging} \approx R_{on\ diode}$$

Therefore the charging time constant is 100 μs and the capacitor is fully charged in 500 μs.

$$\tau_{charge} \approx 1\ \Omega \times 100\ \mu F = 100\ \mu s$$

When the diode is off (*open*) during the discharging of the capacitor, the parallel resistance combination is equivalent to just the load resistance.

$$R_{Th\ discharge} \approx R_{load}$$

Therefore the discharging time constant is 1 s and the capacitor would take 5 s to discharge.

$$\tau_{discharge} \approx 10\ k\Omega \times 100\ \mu F = 1\ s$$

The discharge time constant is very, very long compared to the charging time constant as is required for a good clamping circuit. The capacitor charges very quickly through the diode compared to the long discharge time through the load.

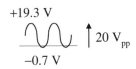

+19.3 V

−0.7 V

20 V_{pp}

Figure 15-37(d)
Positive clamper output

The other critical factor is that the capacitor must hold its charge during nearly a full cycle of the input signal. The period of the input signal is

$$T = \frac{1}{1 \text{ kHz}} = 1 \text{ ms}$$

$\tau_{\text{discharge}} \gg T$, therefore the capacitor holds the charge between input cycles and this is an excellent clamper circuit.

Practice

The silicon diode in the circuit of Figure 15-37 is reversed. What are the minimum and maximum voltages? What kind of circuit is this?

Answer: $+0.7 \text{ V}_{\text{max}}$, $-19.3 \text{ V}_{\text{min}}$, negative clamper

---

Peak-to-Peak Voltage Detector

Figure 15-38 shows the block diagram of an ideal **peak-to-peak voltage detector**.

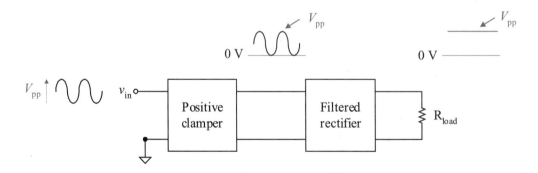

Figure 15-38 Ideal peak-to-peak voltage detector block diagram

This circuit is a *cascaded* two-stage circuit. The first stage is a *positive clamper* and the second stage is a *peak detector* (filtered rectifier). The positive clamper clamps the input signal to 0 V and ideally produces an output of

$$V_{\text{p clamper out}} \approx V_{\text{pp in}}$$

The second stage, which is a peak detector, detects the peak of its input signal and produces a fixed dc output equal to its input peak voltage.

$$V_{\text{dc out}} \approx V_{\text{p clamper out}}$$

Thus this cascaded two-stage circuit produces a dc voltage output ideally equal to the original signal's peak-to-peak input voltage.

$$v_{\text{out}} = V_{\text{dc out}} \approx V_{\text{pp in}}$$

The following example utilizes simulation to examine a practical clamper circuit, which includes the voltage drops of the diodes.

Example 15-20

Create a Cadence PSpice simulation of the peak-to-peak detector of Figure 15-39(a) using 100 μF capacitors, 1N4002 diodes, and a load resistance of 10 kΩ. Use the probe feature to observe the input, positive clamper output, and filtered rectifier output voltage waveforms. Compare the results with the ideal model of Figure 15-38 and speculate on the differences. Also, from probe, *delete all traces* and *add* a trace of the average *avg()* function on the output voltage for 1 second and determine its data point near 1 s. This is the stable average (dc) output voltage.

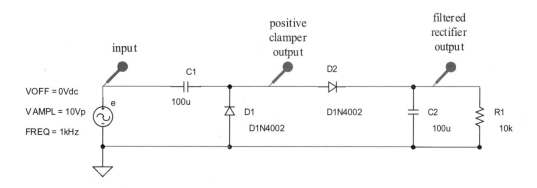

Figure 15-39(a) Example peak-to-peak voltage detector circuit

Solution

Figure 15-39(b) is the Cadence PSpice simulation output with the three required traces and two significant data points. Note, the text has been augmented to improve readability.

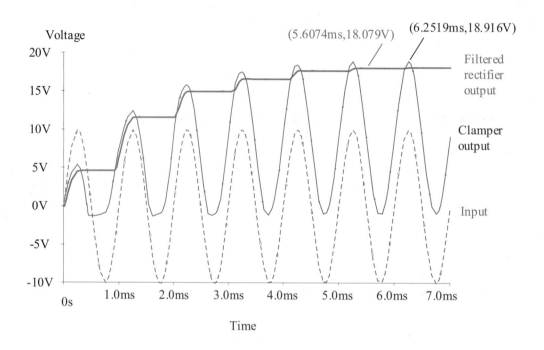

Figure 15-39(b) Cadence PSpice probe waveforms of peak-to-peak voltage detector circuit

The input signal is a 20 V_{pp} sine wave. The ideal clamper circuit should be 20 V_p, but its simulated output is 18.916 V_p. The simulator is including the clamper diode drop of 0.7 V, which would more realistically produce a clamper circuit output of

$$V_{p \text{ clamper out}} \approx V_{pp \text{ in}} - 0.7 \text{ V}$$

$$V_{p\ clamper\ out} \approx 20\ V_{pp} - 0.7\ V = 19.3\ V_p$$

This value is still higher than the simulated output of 18.916 V_p, indicating that the capacitor is not fully charging.

The simulated input to the second stage peak detector is 18.916 V_p, therefore its output ideally would be 18.916 V_p. However, the peak detector silicon diode realistically drops about 0.7 V. Therefore the simulated output voltage of the second stage is now expected to be

$$V_{p\ out} \approx 18.916\ V_p - 0.7\ V \approx 18.2\ V_p$$

This is very close to the observed simulated output voltage of 18.079 V_p. Note that there is some ripple voltage and that the output is not truly a fixed dc 18.1 V. The ripple voltage is so small with these R and C values, however, that the output voltage can be considered to be approximately a dc voltage.

$$v_{out} \approx V_{p\ out} \approx 18.1\ V_{dc}$$

Using the ideal diode approximation model produces about a 10% error compared to simulated outcomes. Using the second approximation model of the silicon diode, which includes its 0.7 V drop, produces about a 3% error compared to the simulation. The only real test is to build the circuit, take measurements, and compare with the calculated and simulated values. Even then you do not have a true comprehensive picture. Components vary slightly in their values based upon things like production batch, manufacturer, construction, temperature and so forth. The same circuit can produce slightly different outputs based upon its components, construction, and its temperature, pressure, and humidity environment.

Careful observation of the simulated output trace shows that the signal has not truly stabilized (not reached steady state) after only 7 cycles and that more *run time* is needed to obtain a true steady state result. The simulated output of the *average* of the output voltage using the avg() function in probe is demonstrated in Figure 15-39(c) over a 1 second run time. You must allow enough time for the average (dc) value to reach its steady state value to obtain a reasonably accurate measurement.

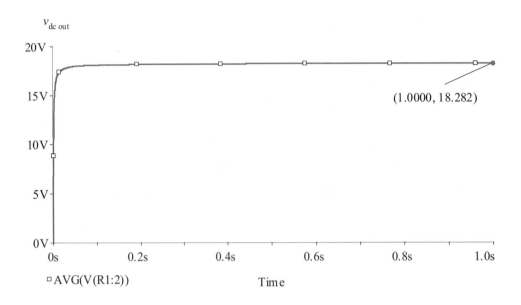

Figure 15-39(c) Probe plot of the average (dc) output voltage

The plot shows the simulated output data point to be 18.282 V_{dc} at 1 s. This is the average value after 1000 cycles, which is not much different than it is after 7 cycles. The number of cycles that must be run for reasonable accuracy depends upon the R and C values and how quickly the signal reaches steady state operation.

Practice

Run the simulation again with the following changes and note the effect on the simulated signals. In each case, also use the average function of the transient trace to find the dc voltage output after 1000 cycles. Each case is separate; not cumulative. Make each change based upon the original circuit. (a) Change the input signal to have an offset of 10 V_{dc}. (b) Change the capacitors to 10 μF and the load resistance to 1 kΩ. (c) Change the capacitors to 10,000 μF and the load resistance to 100 kΩ.

Answer: (a) Effectively there is no change in the clamper output signal; therefore there is no significant change in the peak-to-peak detector output (1 s, 18.286 V_{dc}). (b) The output has picked up a noticeable ripple voltage of 1.308 V_{pp} ripple voltage (16.944 V_{max}, 15.636 V_{min}). The average voltage must fall between the ripple maximum and minimum voltages and does (1 s, 16.577 V_{dc}). (c) Requires longer for the output to climb to steady state (1 s, 18.211 V_{dc}). Note, the signal is very close to steady state but if you run the simulations again and determine the average value at 10 s, all the above cases produce slightly higher dc voltages out. However, realize that this is only a simulation and carrying simulations out to more than 3 significant figures is really a futile exercise. Device and environmental variations cause much greater variations.

Voltage Doubler (Multipliers)

The circuit of Figure 15-40 is a **multiplier** circuit or more specifically a **voltage doubler** circuit. Ideally, it produces an output dc voltage equal to twice its peak input signal. This *doubler circuit* is exactly the same circuit as Figure 15-38. It is really *a peak-to-peak voltage detector* with the only additional constraint that the input signal be a true ac signal with no dc offset. Then,

$$V_{pp\ in} = 2\ V_{p\ in}$$

The output of the *voltage doubler circuit* is a dc voltage that is ideally (assuming ideal diodes) equal to two times the input peak voltage.

$$v_{out} \approx V_{dc\ out} \approx 2\ V_{p\ in}$$

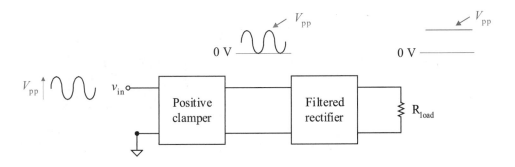

Figure 15-40 Voltage doubler circuit

The previous example is actually a peak-to-peak detector without an offset; therefore the last example also serves as a voltage doubler example.

15.8 Switched Capacitor Converter

A **switched capacitor** circuit demonstrates a technique of switching capacitors on and off in such a way as to charge capacitors in one switch position and deliver the charged capacitor voltage in a different switched position. Switched capacitor circuits have several applications. The application examined in this section is the application of a switched capacitor to convert a dc positive voltage to dc negative voltage. One of the ICs specifically designed for this function is the MAX1044/ICL7660 chip. It is a +dc voltage to −dc voltage converter.

For the switched capacitor circuit of Figure 15-41, switches S1 and S3 are **ganged** together, as indicated by the dashed line. The S1-S3 combination closes and opens together under the control of the square wave into the inverter component. A positive square wave pulse results in the closing of the S1-S3 switches, while a negative pulse results in opening the switches. Ganged switches S2 and S4 operate in the same fashion.

The circuit of Figure 15-41 is called a **charge-pump voltage converter**. C1 is a **bucket capacitor**; it accumulates charge and delivers it to C2. C2 is a **reservoir capacitor** and serves as a reservoir of charge to the load.

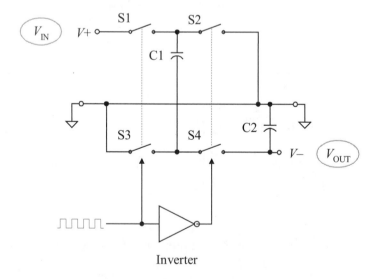

Figure 15-41 Switched capacitor circuit

Example 15-21

For the circuit of Figure 15-42, the input voltage V_{IN} is +5 V_{dc} and the square wave frequency driving the inverter component is 10 kHz. Find the output voltage. Also refer to Figure 15-41 and describe the switched capacitor action.

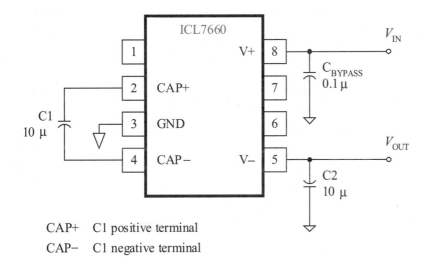

CAP+   C1 positive terminal
CAP−   C1 negative terminal

Figure 15-42 Switching capacitor +dc to −dc converter

Solution

During the positive half cycle of the square wave into the inverter S1-S3 switches are closed and S2-S4 switches are open. The input voltage is applied to the capacitor C1 and it charges up to 5 V. It may take a few cycles to fully charge, but once fully charged the bucket capacitor C1 is charged up to the input voltage of +5 V_{dc}.

During the negative half cycle of the square wave into the inverter, the output of the inverter is a positive half cycle. S1-S3 switches now open and S2-S4 switches close. The C1 bucket capacitor delivers its 5 V charge to the reservoir capacitor C2. The reservoir capacitor C2 eventually charges to 5 V and holds

5 V. But notice that the output V_{OUT} is −5 V relative to common. The +5 V_{dc} input creates an output voltage V_{OUT} of −5 V_{dc}.

Practice

The input voltage of Figure 15-42 is set to +4.5 V_{dc}. Find the output voltage.

Answer: −4.5 V_{dc}

Exploration

Using data books or the Internet, find the data sheets and application notes for the MAX1044, LMC7760, or ICL7660 IC chips. Read the specifications especially for the different models of the chip available with different packaging, somewhat different specifications, and especially their maximum values (values not to be exceeded during device operation). Read the *typical application* notes for understanding and potential application circuit ideas including voltage divider, voltage multiplier, the output supply resistance calculation, and output resistance reduction techniques.

Summary

Wave shapes can be generated or converted using resistors, capacitors, diodes, active devices, and specially designed IC chips. A multivibrator is a rectangular wave generator and wave shape converter that typically utilizes *RC* timing circuits to determine pulse widths and frequencies of generated signal waves. Multivibrators are *astable* (not stable or free running oscillators), *monostable* (one stable state to create a pulse with a designed pulse width), or *bistable* (two stable states to create a pulse with a designed starting time and ending time). The duty cycle of a periodic rectangular wave defines what percent of the time the rectangular wave is high.

$$D = \frac{W}{T} \times 100\%$$

where W is the time of the pulse width within the period T. The reciprocal of the period is the *frequency,* or the *repetition rate*, of the rectangular signal.

The 555 timer is a very versatile IC chip that utilizes two comparators testing for 1/3 and 2/3 of the supplied dc voltage to set and reset a flip flop (FF). The complemented output of the FF is fed to the output pin; therefore, the output is either a high or a low and forms rectangular pulses or a periodic rectangular wave. Capacitors and resistors are connected externally to the 555 timer to create a variety of electronic circuits that convert or generate waveforms. The *RC* universal dc transient expression is used to determine charge and discharge, which in turn can be used to determine such characteristics as pulse width, signal period, and signal frequency.

An inverting voltage amplifier with a capacitor in its feedback loop creates an integrator circuit. A voltage applied to the input resistor creates an amplifier input current. That current is fed through the feedback capacitor. Capacitor voltage is related to capacitor current by the expression

$$i_C = C \frac{d v_C}{dt}$$

Performing circuit analysis and applying calculus leads to the output voltage as a function of the integral of the input voltage with respect to time.

$$v_{out} = -\frac{1}{RC} \int_0^t e_{in} \, dt + V_{C\,init}$$

The integrator circuit is mathematically an integrator circuit. If the output of the integrator is fed to an inverting voltage amplifier with a voltage gain of $-RC$, the output of the two-stage amplifier is truly a mathematical integrator.

$$v_{out} = \int_0^t e_{in} \, dt$$

This circuit is also used as a wave shaping circuit. For example, a fixed dc voltage is converted to an inverted ramp, and a square wave input is converted to an inverted triangle wave output.

An inverting voltage amplifier with a capacitor as its input component creates a differentiator circuit. Performing circuit analysis and applying calculus leads to the output voltage as a function of the derivative of the input voltage.

$$v_{out} = -RC \frac{dv_{in}}{dt}$$

The differentiator circuit is mathematically a differentiator multiplied by a constant. If the output of the differentiator is fed to an inverting voltage amplifier with a voltage gain of $-1/RC$, the output of the two-stage amplifier is truly a mathematical differentiator.

$$v_{out} = \frac{dv_{in}}{dt}$$

The differentiator circuit is also used as a wave shaping circuit. For example, a pulse input signal is converted to an output voltage with a negative spike from the leading pulse edge and a positive spike from the lagging pulse edge. A ramp is converted to a pulse. A triangle wave is converted to a square wave.

Integrators, differentiators, positive feedback comparators, and positive feedback systems are used in various combinations to create wave-generating circuits. Typical waveforms for these circuit combinations include exponential waves, triangle waves, and square waves.

Hidden capacitance that naturally occurs within components, in test equipment input circuits, test leads, and in circuit construction significantly impacts wave shapes and signal amplitudes at high frequencies. Significant loss and wave shape distortion of a signal occur at very high frequencies due to capacitive reactance, which is inversely proportional to frequency.

$$X_C = \frac{1}{2\pi f C}$$

As frequency f increases, X_C is reduced. Capacitance has a very low reactance at very high frequencies, appearing as unwanted shorts in a circuit. The *half-power frequency* is a specific frequency in an RC circuit for which half of the power is dissipated in R_{Th} and the other half of the power is stored in C. This occurs when the capacitive reactance equals the equivalent circuit resistance R_{Th} as viewed from the capacitance. Setting X_C equal to R_{Th}, the half power frequency can be found as

$$f_{half\text{-}power} = \frac{1}{2\pi R_{Th} C}$$

Significant signal power loss and distortion occur above this frequency.

A rectangular wave's leading and trailing edges cause high frequency effect. The leading edges and trailing edges of a rectangular wave create transient RC effects whether the capacitance is embodied in a capacitor component or is in the form of natural hidden capacitance.

An ideal capacitance holds charge forever, but real capacitance loses charge (*discharges*) naturally without any external resistive connection. Capacitor leakage is a characteristic of capacitor types. For example, electrolytic capacitors are very leaky (lose charge quickly) while ceramic capacitors are not. Electrolytic capacitors do not hold a charge as long as a ceramic capacitor. The capacitance and capacitor's leaky resistance form an *RC* circuit; the leakage resistance effectively combines in parallel with the external circuit resistance to form the total equivalent resistance R_{Th} seen by the capacitance. This effective parallel combination reduces the total effective resistance of the *RC* circuit; therefore the capacitor charges and discharges faster and can significantly impact the circuit's effect on its high frequency and rectangular signals.

The diode and the diode in conjunction with capacitance are used to alter signal waveshapes. A *clipper* circuit clips off part of a waveform using diodes. A limiter restricts the maximum and/or minimum signal value using diodes. A *clamper* circuit combines the use of the diode and the capacitor to insert a dc offset into a signal.

The *switched capacitor* circuit switches between charging and discharging capacitors. The example selected was a +dc to −dc converter. During the charge cycle, the bucket capacitor collected charge. Then the charging circuit was turned off and the bucket capacitor charge was deposited into a reservoir capacitor. The reservoir capacitor is across the load such that it delivers a negative dc voltage with a positive voltage input.

Problems

Multivibrators Using 555 Timers

15-1   A rectangular signal is high for 1 μs and low for 9 μs. Find the signal's period, pulse width, duty cycle, and frequency.

15-2   A rectangular signal has a repetition rate of 100 pulses per second and a pulse width of 3 ms. Find its period and duty cycle.

15-3   The capacitor of Figure 15-5 is changed to 10 μF. Using the *RC* universal dc equation approach, find the pulse width of the output voltage. Verify that the pulse width is 1.1τ.

15-4   The capacitor of Figure 15-6 is changed to 10 μF. Using the *RC* universal dc equation approach, find the pulse width *W* of the output and the time the signal is low. Then find the output signal's period, duty cycle, and frequency.

15-5 Verify that in the previous problem $W = 0.693\ \tau_H$ where $\tau_H = (R1+R2)\ C$, and $T_L = 0.693\ \tau_L$ where $\tau_L = (R2)\ C$.

Op Amp *RC* Integrator Circuit

15-6 For the circuit of Figure 15-43(a):

  a. Is this a negative feedback circuit?

  b. Is it an integrator circuit? Why?

  c. What is a reasonable minimum value for R_f?

  d. The input signal e_{in} is a 100 mV$_{pp}$ rectangular pulse with a width of 100 μs. Find and sketch the output signal.

  e. The input signal e_{in} is a 1 V$_{pp}$ square wave with a period of 100 μs. Find and sketch the output signal.

15-7 Repeat Problem 15-6 if the input resistance is changed to 10 kΩ.

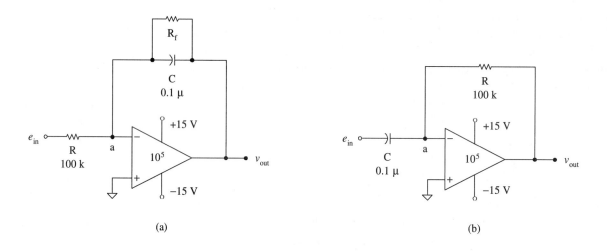

(a)   (b)

Figure 15-43

Op Amp *RC* Differentiator Circuit

15-8 For the circuit of Figure 15-43(b):

  a. Is this a negative feedback circuit?

  b. Is it a differentiator circuit? Why?

c. The input e_{in} is a square wave. What is the output wave shape?

d. The input signal e_{in} is a 2 V_{pp} triangle wave with a period of 100 ms. Find and sketch the output signal.

15-9   Repeat Problem 15-8 if the input capacitor is changed to 0.01 µF.

15-10   If the output signal of the circuit in Figure 15-43(a) is fed as the input signal to the circuit of Figure 15-43(b), what is the resulting output signal?

15-11   If the output signal of the circuit in Figure 15-43(b) is fed as the input signal to the circuit of Figure 15-43(a), what is the resulting output signal?

Op Amp *RC* Waveform Generators

15-12   Assume that the op amp of the circuit in Figure 15-44 requires 2 V of headroom.

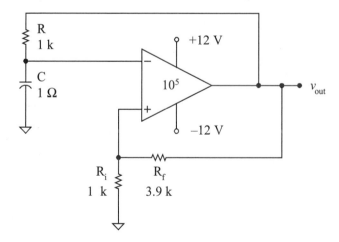

Figure 15-44

a. What are the components in the positive feedback circuit?

b. What are the components in the negative feedback circuit?

c. Find the positive feedback factor *B*.

d. Find the upper and lower trip point voltages.

e. Find the period and the frequency of the waveform.

f. Sketch the output, op amp NI terminal, and the op amp INV terminal waveforms.

15-13 For the circuit of Figure 15-44, find the output frequency generated if R_f is changed to 10 kΩ. Assume the op amp requires 2 V of headroom.

15-14 For the circuit of Figure 15-45 assume that the op amp rail voltages are ±13 V.

a. What is the name of the first stage circuit?

b. What is the output wave shape of the first stage?

c. Find the first stage output signal's peak voltage and frequency.

d. What is the name of the second stage circuit?

e. What is the output wave shape of the second stage?

f. Find the second stage output signal's peak voltage and frequency.

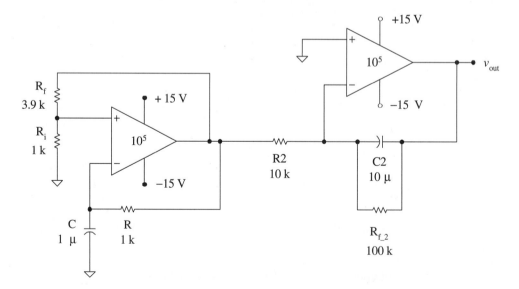

Figure 15-45

15-15 For the circuit of Figure 15-45:

    a. Name at least two different changes that could be made to increase the signal's output frequency

    b. Name at least two different changes that could be made to increase the signal's output amplitude without changing its frequency.

Natural Hidden Capacitance

15-16 For the circuit of Figure 15-46:

    a. Find the capacitive reactance of a 15 Hz signal.

    b. Find the capacitive reactance of a 15 kHz signal.

    c. Find the half-power point frequency. Is this above the audio range? Would this stray capacitance cause a problem for audio signals?

    d. Also find the RC time constant for this circuit.

    e. How long does it take for the circuit to respond to a square wave input?

    f. What is the rise time of the output signal to a square wave input signal?

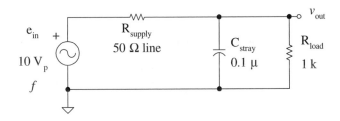

Figure 15-46

15-17 For the circuit of Figure 15-46, what value of stray capacitance causes 15 kHz to be the half-power point frequency?

15-18 For the digital voltmeter of Figure 15-47:

    a. Find the capacitive reactance of a 1 kHz signal.

    b. Find the capacitive reactance of a 10 kHz signal.

c. Find the half-power point frequency for this input test measurement circuit. Would this cause a measurement error for either the 1 kHz or the 10 kHz signal voltage?

d. Also find the RC time constant for this circuit.

e. How long does it take for the circuit to respond to a square wave input?

f. What is the rise time of the output signal to a square wave input signal?

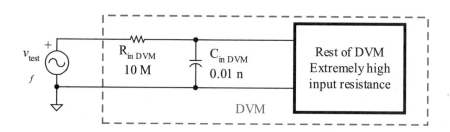

Figure 15-47

Capacitor Leakage

15-19 The leakage resistance of the capacitor in the circuit of Figure 15-48 is 10 MΩ.

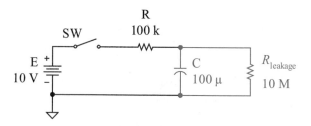

Figure 15-48

a. With the switch closed find the capacitor voltage when it is fully charged, its circuit time constant, and how long the capacitor takes to fully charge (five time constants). What is its rise time?

b. Assume that the capacitor fully charges and the switch opens. Find the discharge time constant and how long it takes for the capacitor to fully discharge. What is its fall time?

15-20   Repeat Problem 15-19 for a circuit resistance of 1 kΩ.

Clippers, Clampers, and Detectors

15-21   For the circuit of Figure 15-49(a), the input signal is a 5 V$_p$ sine wave.

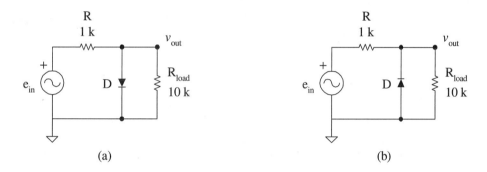

(a)   (b)

Figure 15-49

a. Name the circuit.

b. Describe and sketch the output voltage waveform (including maximum and minimum values) assuming an *ideal diode*.

c. Sketch the output voltage waveform assuming a *germanium diode*.

d. Sketch the output voltage waveform assuming a *silicon diode*.

e. Find the peak forward diode current for a silicon diode.

f. Sketch the load current waveform assuming a silicon diode.

g. Find the peak reverse bias voltage across the diode.

15-22 Repeat Problem 15-21 for the circuit of Figure 15-49(b).

15-23 For the circuit of Figure 15-50(a), the input signal is a 5 V_p sine wave.

    a. Name the circuit.

    b. Describe and sketch the output voltage waveform (including maximum and minimum values) assuming an *ideal diode*.

    c. Sketch the output voltage waveform assuming a *germanium diode*.

    d. Sketch the output voltage waveform assuming a *silicon diode*.

    e. Find the peak forward diode current for a silicon diode.

    f. Sketch the load current waveform assuming a silicon diode.

    g. Find the peak reverse bias voltage across the diode.

15-24 Repeat Problem 15-23 (parts a, b, c, and d) for the circuit of Figure 15-50(b).

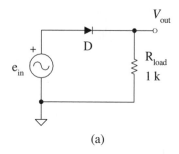

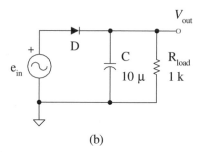

(a)   (b)

Figure 15-50

Switched Capacitor Circuit

15-25 The input voltage of Figure 15-42 is set to +4 V_{dc}. Find the output voltage.

16

Inductance and *RL* Circuits

Introduction

This chapter introduces inductance, inductor components, inductive reactance, frequency effects of inductance, combined inductive effects, switching transient analysis of *RL* circuits, and *RL* circuit ac signal responses.

Objectives

Upon completion of this chapter you will be able to:

- List the key characteristics of inductance.

- Define and apply the inductive reactance formula.

- Find the total inductance of a multiple inductor circuit.

- Find the *RL* circuit time constant tau (τ) and the time it takes to energize or de-energize an inductor.

- Write the *RL* exponential universal dc transient equations for a switched dc circuit based upon its circuit time constant, initial circuit values, and steady state circuit values.

- Evaluate the instantaneous exponential general expression for voltage and current in an *RL* circuit.

- Completely evaluate a multiple resistor-inductor switched *RL* circuit and sketch the inductor voltage and current waveforms.

- Determine and sketch the waveforms in an *RL* circuit with a square wave supplied signal.

- Determine the time constant of an *RL* circuit from its transient square response curves.

- Determine and sketch the waveforms in an *RL* circuit with a sinusoidal supplied signal.

16.1 Inductors

Electromagnet

Flowing charge (current) creates an **electromagnetic (em) field**. There-fore, current flow through a wire creates an electromagnetic field around the wire as shown in Figure 16-1(a). The electromagnetic field consists of electromagnetic flux lines. The direction of the *em* field, or flux lines, is predicted by the right-hand rule of electromagnetics. Using your right hand, place your thumb in the direction of the current; then wrap your fin-gers around the wire (physically or mentally). Your curled fingers point in the direction of the electromagnetic field. Since the *magnetic field* is cre-ated by *electricity* (a current), it is called an *electromagnetic* field. Remove the current and the electromagnetic field collapses.

If a wire is coiled as shown in Figure 16-1(b), the *em* field effect is significantly enhanced as shown in Figure 16-1(c). The coils of wire must not touch each other. This would create a short defeating the coiling effect. The wire must be insulated such that the conductors do not touch each other. This insulation is typically a chemical coating that insulates the windings from each other. The *em* field around each wire is rein-forced, creating a larger magnetic effect. An electromagnet is created with a north pole and a south pole. Again, remove the current through the wire, and the electromagnetic field collapses.

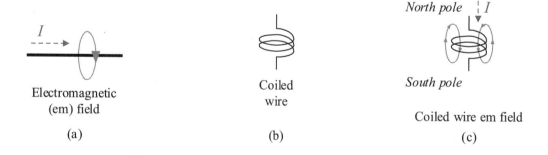

Electromagnetic
(em) field

(a)

Coiled
wire

(b)

North pole   *I*

South pole

Coiled wire em field

(c)

Figure 16-1 Electromagnetic effect

Flux Density

Flux density is defined to be

$$B = \frac{\Phi}{A}$$

where
    B flux density in units of teslas (T)
    Φ number of flux lines in units of webers (Wb)
    A cross section area in units of meters squared (m^2)

The greater the flux density is, the stronger the electromagnetic field. If the field is strong enough, the electromagnet picks up materials that can be magnetized (such as soft iron filings).

Figure 16-2(a)
Soft iron core
electromagnet

Inductance and the Inductor

A change in current *I* creates an opposing force, resisting the change in current. **Self-inductance** is the characteristic subscribed to this opposition to current change in a conductor. Two conductors near each other whose electromagnetic fields interact with each other have a shared or **mutual inductance** as well. The term **inductance** is used to include any inductive effect (self or mutual). The greater the inductance is, the greater the opposition to a change in current in the conductor.

An **inductor** component is basically a coiled wire specifically designed to take advantage of this *em* effect and increasing its inductive effect as shown in Figure 16-1(c). This *em* effect is increased by inserting a material like a soft iron core inside the coiled wire as shown in Figure 16-2(a). The *flux density* is significantly increased in the soft iron core. This significantly increases the **inductive effect**, the opposition to a change in current. This also significantly increases the electromagnetic field strength.

Figure 16-2(b)
Relay or elec-
tromechanical
switch

While current flows through the inductor, the soft iron core forms a north-south pole magnet that can be used to create devices such as a **relay**, that is, an **electromechanical switch** as shown in Figure 16-2(b). This is a relay with **normally open contacts**. That is, the contacts are open when the inductor is not energized and their core has no magnetic effect. When current does flow, the north-south pole magnet is created, the top arm of the switch is attracted downward to the magnetized soft iron core, and the top arm makes contact with the bottom switch arm. This closes the mechanical switch. A **normally closed switch** is closed when the inductor is de-energized and opens when the inductor is energized with current.

The schematic symbol for the inductor is shown in Figure 16-3(a), where L is the symbol for inductance. Inductance is measured in units of henries (H).

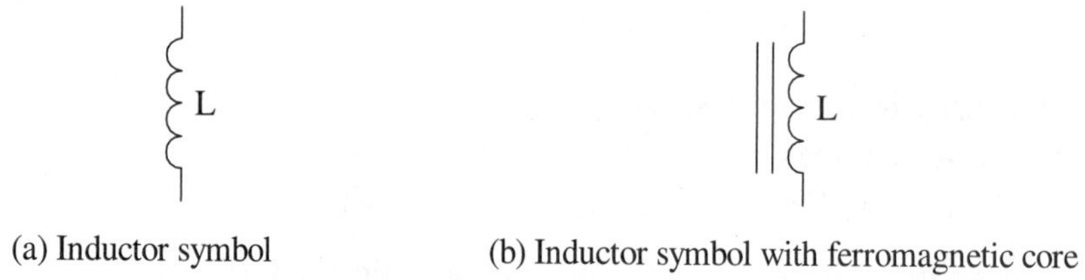

(a) Inductor symbol   (b) Inductor symbol with ferromagnetic core

Figure 16-3 Inductor symbols

The inductor schematic symbol is similar to the inductor's construction, a coiled wire. In fact, **coil** is another name used for an inductor because of its appearance as a coiled wire. **Solenoid** is another name used for an inductor, but solenoid is strictly interpreted to be a specific construction of a helix winding with small pitch. Another name for the inductor is the **choke**, since an inductor chokes or inhibits current changes.

A material, such as soft iron, that is easily magnetized is called **ferromagnetic** material. The presence of a ferromagnetic core inside the inductor winding is indicated by two parallel lines, as shown in Figure 16-3(b). Ferromagnetic material focuses the electromagnetic flux lines and creates a stronger electromagnetic effect and a much greater value of total inductance.

A *time varying current* in an inductor *induces* a voltage drop across the inductor as shown in Figure 16-4. The relationship of the ac voltage and current in an inductor is

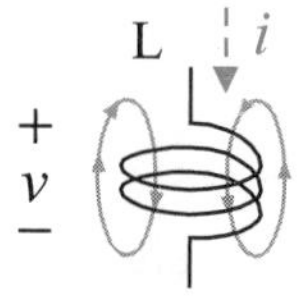

$$v = L\,\frac{di}{dt}$$

Figure 16-4
Lenz's law

where

v	voltage drop across the inductor in units of volts (V)
L	inductance in units of henries (H)
$\dfrac{di}{dt}$	inductor current time rate of change in units A/s

A time varying current is required to create an induced voltage. A fixed dc current does not produce an induced voltage across the inductance.

The voltage induced across the inductor creates an induced effect that opposes the original change in current. **Lenz's law** *is a general law that states that an induced effect tends to oppose the cause that created that effect in the first place.* Simply stated for the inductor, the inductor resists a change in current because the changing current induces a voltage that opposes that change.

Permeability

The type of material used for the core material impacts the effective strength of the inductance. Air, wood, glass, copper, and so forth have virtually no effect on the flux lines created. A soft iron core significantly increases the number of and concentrates the electromagnetic flux lines. Core materials that facilitate the creation of and the concentration of *em* flux lines are called **magnetic** materials. **Permeability**, represented by the Greek symbol mu (μ), is the characteristic that quantifies the magnetic property of materials. The permeability of free space is

$$\mu_0 = 4\pi \times 10^{-7} \frac{\text{Wb}}{\text{A} \cdot \text{m}}$$

A material that has a permeability significantly higher than free space, such as soft iron, cobalt, steel, and alloys, are called *ferromagnetic* materials. Relative permeability μ_r is measured relative to μ_0 such that

$$\mu = \mu_r \, \mu_0$$

A ferromagnetic (iron-like) material has a relative permeability greater than 100, that is $\mu_r > 100$. A ferromagnetic material is hundreds or thousands of times greater than that of free space.

Measuring Inductance

Inductance must be measured with an inductance meter, which is usually combined into an *LCR* (inductance, capacitance, and resistance) meter.

Inductance Calculations

Inductance of a simple conductor such as wire or a printed circuit board (PCB) trace is typically specified in terms of inductance per unit length.

Example 16-1

A 1 foot piece of 22 gauge wire exhibits inductance of 75 nH per inch. What is its total inductance?

Solution

First convert feet to inches. Then it is a simple multiplication.

$$L = (12 \text{ in})\left(75 \ \frac{\text{nH}}{\text{in}}\right) = 900 \text{ nH}$$

Practice

A 10 foot piece of wire is measured to be 9 μH. What is its inductance per inch?

Answer: 75 nH per inch (22 gauge wire)

Generally, calculations of inductance of more complex arrangements can be very challenging. For a simple solenoid with an air core, the formula for an *approximate* value of inductance is

$$L_0 = \frac{N^2 \mu_0 A}{l}$$

where

L_0 inductance with an air core (H)
N number of turns or loops of the coiled inductor
μ_0 permeability of free space or a vacuum (Wb/A·m)
A cross-section area of the core (m²)
l mean or average length of the core (m)

If a ferromagnetic core is used, the estimated value of inductance is

$$L = \mu_r L_0$$

where

μ_r relative permeability

Example 16-2

A 20-turn air core solenoid is 100 mm long and 2 mm in radius. Find its inductance.

Solution

First convert the length and diameter to meters of 0.1 m and 0.002 m, respectively. The cross-section area of the core is

$$A = \pi (0.002 \text{ m})^2 = 12.57 \times 10^{-6} \text{ m}^2$$

The air core inductance is approximately

$$L_0 = \frac{N^2 \mu_0 A}{l}$$

$$L_0 = \frac{(20^2) \left(4\pi \times 10^{-7} \dfrac{\text{Wb}}{\text{A} \cdot \text{m}}\right) (12.57 \times 10^{-6} \text{ m}^2)}{0.1 \quad \text{m}}$$

$$L_0 = 63.2 \text{ nH}$$

Practice

A soft iron core with a relative permeability of 1500 is inserted into the example solenoid. What is the solenoid's new inductance?

Answer: 94.8 μH

Inductance Key Characteristics

A few key characteristics that describe inductance are that inductance

- stores energy in the form of a current supply (*em* field)

- does not retain the *em* field if current is removed

- returns current (via a collapsing *em* field) to a circuit like a current supply I

- resists change in current

The instantaneous relationship of the inductor voltage and current is expressed in Equation 16-1. This is a *fundamental* component relationship.

$$v_L(t) = L\frac{d\,i_L(t)}{dt} \qquad \text{(16-1)}$$

Or, if working with linear changes,

$$v_L(t) = L\frac{\Delta i_L(t)}{\Delta t}$$

A changing inductor current is required to produce an induced inductor voltage. Inductor dc voltage drops are due to a dc current flowing through the series resistance of the inductor winding (Ohm's law of resistance, not Equation 16-1).

Current Ramp to Create Constant Voltage

Based upon the basic relationship of Equation 16-1, an inductor current ramp creates a constant voltage drop across the inductor.

$$\frac{\Delta i_L(t)}{\Delta t} = \text{constant}$$

$$v_L(t) = L \times \text{constant}$$

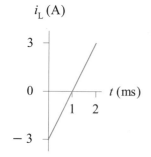

Figure 16-5(a)
Example inductor
current ramp

Example 16-3

The current ramp of Figure 16-5(a) is passed through a 10 mH inductor. Find and draw the corresponding inductor voltage drop.

Solution

The current is a linear ramp with a slope of

$$\frac{\Delta i_L(t)}{\Delta t} = \frac{3\text{ A} - (-3\text{ A})}{2\text{ ms} - 0\text{ s}} = 3\,\frac{\text{A}}{\text{ms}}$$

Therefore, the inductor has a constant voltage during the current ramp. See Figure 16-5(b).

$$v_L(t) = 10\text{ mH}\left(3\,\frac{\text{A}}{\text{ms}}\right) = 30\text{ V}$$

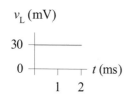

Figure 16-5(b)
Example inductor
voltage response to a
current ramp

Practice

The current ramp of Figure 16-6(a) is passed through a 10 mH inductor. Find and draw the corresponding inductor voltage drop.

Answer: See Figure 16-6(b)

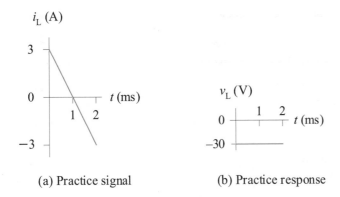

(a) Practice signal   (b) Practice response

Figure 16-6 Inductor voltage response to a negative current ramp

Voltage Response to Simple Current Waveforms

The following are example voltage responses to inductor signals.

Example 16-4

A 50 mA dc current flows through an ideal inductor of 10 mH. Find the inductor dc and ac voltage drops.

Solution

An ideal inductor has an internal series resistance of 0 Ω (equivalent to an ideal dc short). By Ohm's law,

$$V_L = R_S I_L = 0\ \Omega \times 50\ \text{mA} = 0\ V_{dc}$$

The only signal present in the inductor is a dc current; therefore, there is no ac signal present and $i_L(t)$ is 0.

$$\frac{\Delta i_L(t)}{\Delta t} = 0$$

therefore the inductor ac voltage drop is

$$v_L(t) = 10 \text{ mH} \left(0 \frac{\text{A}}{\text{s}} \right) = 0 \text{ V}$$

Practice

A 50 mA dc current flows through a 10 mH inductor with an internal series resistance of 20 Ω. Find the inductor dc and ac voltage drops.

Answer: 1 V_{dc}, 0 V of ac

The next example and practice examine the relationship of inductor voltage and current with sinusoidal, trapezoidal, triangle, and square wave current signals.

Example 16-5

Given the inductor current waveforms of Figure 16-7(a,b), find the corresponding inductor voltage waveforms for a 10 μH inductor.

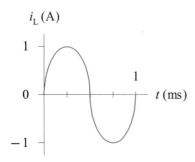

(a) Sinusoidal inductor current

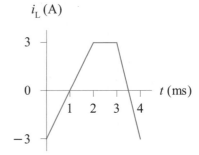

(b) Trapezoidal inductor current

Figure 16-7(a,b) Example inductor current waveforms

Solution

a. For the sine wave of Figure 16-7(a), substitute the current sine wave function of sin(2π*ft*) into the basic definition defining inductor voltage. The inductor voltage is

$$v_{\mathrm{L}}(t) = L\frac{d\,i_{\mathrm{L}}(t)}{dt}$$

$$v_{\mathrm{L}}(t) = L\frac{d\,\sin(2\pi ft)}{dt}$$

If the frequency is fixed at a constant value, the derivative of the sine function is the cosine function multiplied by the constant $2\pi f$. The inductor voltage becomes

$$v_{\mathrm{L}}(t) = L\,(2\pi f)\,\cos(2\pi ft)$$

simplifying to

$$v_{\mathrm{L}}(t) = 2\pi fL\cos(2\pi ft)$$

Thus, if a sinusoid current is applied to a inductor, the inductor voltage is a sinusoid. Substituting the inductance value of 10 μH and the frequency of 1 kHz, the inductor voltage expression is

$$v_{\mathrm{L}}(t) = 2\pi \times 1\ \text{kHz} \times 10\ \mu\text{H} \times \cos(2\pi ft)$$

and reduces to

$$v_{\mathrm{L}}(t) = 62.8\,\cos(2\pi ft)\,\text{mV}$$

as shown in Figure 16-7(c).

b. For the inductor signal current of Figure 16-7(b), separate the problem into three parts for each of the three distinct parts of the waveform.

For $0 \le t < 2$ ms, the rate of change is the slope of the line (up 6 A and over 2 ms).

$$v_{\mathrm{L}}(t) = L\frac{\Delta i_{\mathrm{L}}(t)}{\Delta t} = 10\ \mu\text{H}\left(\frac{6\ \text{A}}{2\ \text{ms}}\right) = 30\ \text{mV}$$

For 2 ms $\le t < 3$ ms, the rate of change of a constant inductor current of 6 A is zero (0).

$$v_{\mathrm{L}}(t) = L\frac{\Delta i_{\mathrm{L}}(t)}{\Delta t} = 10\ \mu\text{H}\left(\frac{0\ \text{A}}{1\ \text{ms}}\right) = 0\ \text{mV}$$

For $3 \leq t < 4$ ms, the rate of change is the slope of the line (down 6 V and over 1 ms).

$$v_L(t) = L \frac{\Delta i_L(t)}{\Delta t} = 10 \ \mu H \left(\frac{-6 \ A}{1 \ ms} \right) = -60 \ mV$$

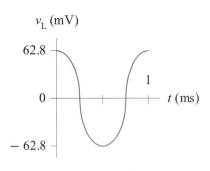

(c) Sinusoidal response   (d) Trapezoidal response

Figure 16-7(c,d) Example resulting inductor voltage waveforms

Practice

Given the inductor current waveforms of Figure 16-8(a,b), find the corresponding inductor voltage waveforms for a 1 mH inductor.

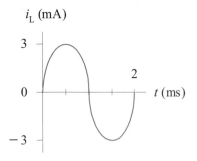

(a) Sinusoidal inductor current   (b) Triangle inductor current

Figure 16-8(a,b) Practice inductor current waveforms

Answer: See Figure 16-8(c,d) for inductor voltage response waveforms.

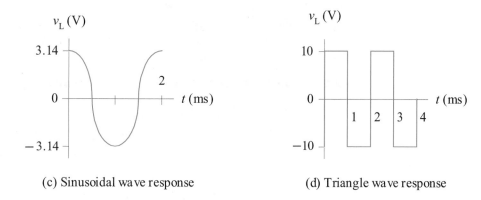

(c) Sinusoidal wave response (d) Triangle wave response

Figure 16-8(c,d) Practice response capacitor current waveforms

16.2 Inductive Reactance

Inductance exhibits opposition to ac current (changing current). This reaction to ac current is called **reactance**. Specifically, it is called **inductive reactance** for inductance. It is similar to *capacitive reactance*, but capacitive reactance and inductive reactance have an *opposite* effect on current and cancel each other out. For example, 1 kΩ of inductive reactance in series with 1 kΩ of capacitive reactance results in a net of 0 Ω reactance

This opposition to ac current is different from resistance. Resistance dissipates energy in the form of heat, light, and so forth. The inductor stores energy in the form of electromagnetic (*em*) energy that can be returned to the circuit in the form of a supplied current. Rather than dissipating the energy, energy is absorbed and returned by the inductor.

The formula for inductive reactance to an ac sinusoidal signal is

$$X_L = 2\pi fL \tag{16-2}$$

Inductive reactance

where

X_L inductive reactance with units of ohms (Ω)
f frequency of the ac sinusoidal signal with units of hertz (Hz)
L inductance with units of henries (H)

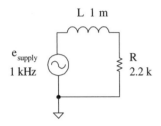

L 1 m

e_{supply}
1 kHz

R
2.2 k

Figure 16-9
Example circuit

Inductive reactance is in units of ohms like resistance, but as was noted for capacitive reactance the reactance magnitude cannot be directly added to resistance magnitude. They are fundamentally different in that resistance dissipates energy and reactance stores energy. A subsequent textbook in this textbook series explores the combination of resistance and reactance; it is a vector sum relationship.

Example 16-6

In the *RL* circuit of Figure 16-9, find the reactance of the inductor. Can the circuit resistance be added directly to the inductive reactance?

Solution

Apply Equation 16-2. The units of henries (H) are typically assumed on a schematic.

$$X_L = 2\pi f L = 2\pi (1\ kHz)(1\ mH) = 6.28\ \Omega$$

Reactance and resistance magnitudes cannot be directly added; they have a vector sum relationship.

Reactance ohms and
resistive ohms add
as a vector sum.

Practice

The inductance of the inductor in the circuit of Figure 16-9 is changed to 100 mH. Find its inductive reactance.

Answer: 628 Ω

$f \uparrow \quad X_L \uparrow$

$L \uparrow \quad X_L \uparrow$

Inductive reactance is proportional to inductance and frequency. As frequency or inductance is increased, the inductive reactance increases.

Example 16-7

In a series *RL* circuit, the inductive reactance of an inductor is 10 Ω. The inductor is replaced by an inductor 100 times its original value. What is the reactance of this inductor? Would you expect the ac voltage drop across the new inductor to increase or decrease?

Solution

The inductance of the inductor is *increased* by a factor of 100; therefore, its reactance is *increased* by the same factor.

$$X_L = 100 \, (10 \, \Omega) = 1 \, k\Omega$$

As reactance of an inductor is increased in a series circuit it drops more voltage.

Practice

In a series *RL* circuit, the inductive reactance of an inductor is 10 kΩ. The inductor is replaced by an inductor 100 times smaller than its original value. What is the reactance of this inductor? Would you expect the ac voltage drop across the new inductor to increase or decrease?

Answer: 100 Ω, drops less voltage

Quick-look Frequency Effects

If the input signal is a sinusoid, then the output signal is a sinusoid. Figure 16-10 demonstrates the responses of a sinusoid in an *RL* circuit to very low and very high frequencies.

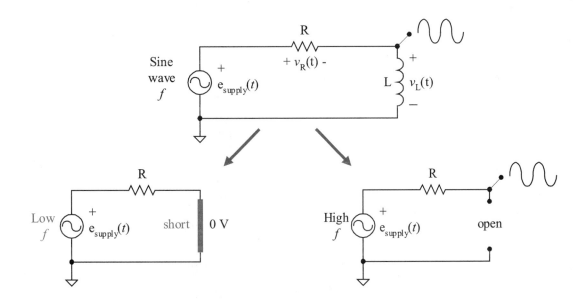

Figure 16-10 Extreme frequency responses of sinusoids in an *RL* circuit

Recall the reactance equation for an inductance of Equation 16-2.

$$X_L = 2\pi fL$$

If the signal frequency *f* is very low, then the inductor acts like an *ac short* and drops 0 V *ac* leaving most of the *ac* supply voltage to drop across series resistor R. If the signal frequency *f* is relatively high, then the inductor acts like an open and it drops the applied voltage. Therefore, this circuit arrangement delivers high frequencies to the output while low frequencies appear as a short across the load (the inductor).

Example 16-8

For the circuit of Figure 16-10, resistance R is 1 kΩ and inductor L is 1 H. Find the inductive reactance at 1 Hz and 1 MHz, and compare each resulting reactance with resistance R. What conclusions can you draw about the signal delivered to the load?

Solution

At a signal frequency of 1 Hz,

$$X_L = 2\pi fL = 2\pi \left(1 \text{ Hz}\right)\left(1 \text{ H}\right) = 6.28 \ \Omega$$

At a signal frequency of 1 MHz,

$$X_L = 2\pi \left(1 \text{ MHz}\right)\left(1 \text{ H}\right) = 6.28 \text{ M}\Omega$$

At a signal frequency of 1 Hz the inductive reactance is much, much less than 1 kΩ; therefore, essentially the entire 1 Hz low frequency signal is dropped across the resistor R. At 1 MHz the inductive reactance is much, much greater than the 1 kΩ; therefore, essentially the entire 1 MHz signal is dropped across the inductor. Since the output is across the inductor, the high frequency signal is delivered to the output and the low frequency signal is not.

Practice

Repeat the example with an inductor of 10 μH.

Answer: 62.8 mΩ, 62.8 Ω, both signals create a reactance much, much less than 1 kΩ and therefore appear as 0 V at the output.

The next circuit of Figure 16-11 is a more likely arrangement with the inductor in series with the load. In this configuration a *dc* and an *ac* signal are present. An ideal inductor appears as a short to *dc*; therefore the entire dc signal is dropped across the load. The inductive reactance is designed to be very high (ideally an *open*) for the input signal frequency *f*; in which case the entire *ac* signal is dropped across the inductor and none of the *ac* signal appears across the load. This circuit configuration is called a **choke filter** since it *chokes off ac changes to the load but passes the dc to the load*.

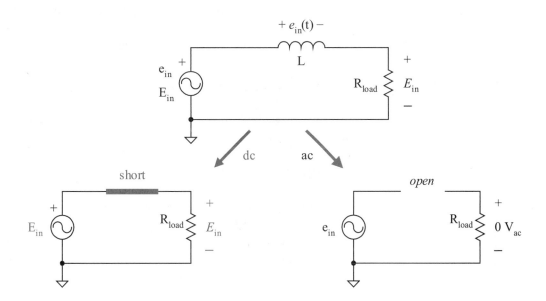

Figure 16-11 Choke filter circuit

Example 16-9

For the circuit of Figure 16-11, resistance R_{load} is 1 kΩ, the inductor L is an ideal inductor of 100 mH, the dc supply is 10 V_{dc} and the ac supply is 1 V_{rms} at 100 kHz. Find the dc output voltage. Find the inductive reactance and compare the resulting reactance with resistance R_{load}. What conclusions can you draw about the signal delivered to the load?

Solution

Since the inductor is an ideal short to dc, all of the dc voltage is delivered to the load.

$$V_{\text{load}} = 10 \ V_{\text{dc}}$$

At a signal frequency of 100 kHz,

$$X_L = 2\pi \left(100 \ \text{kHz}\right)\left(100 \, \text{mH}\right) = 62.8 \ \text{k}\Omega$$

The inductive reactance is much, much greater than the load resistance; therefore, the inductor appears essentially as an open to the ac signal. The output ac signal is essentially 0 V. This circuit delivers the dc input signal to the load and chokes (blocks) the ac signal to the load.

Practice

Repeat the example if the inductor has an internal series resistance of 100 Ω.

Answer: dc voltage divides, 9.09 V_{dc}; essentially no change in the ac signal

The circuit of Figure 16-12(a) is the half-wave capacitor input filter dc power supply that has been previously described. The supplied signal is an ac signal, the diode rectifies the signal, and the capacitor charges and holds a nearly dc voltage across the load.

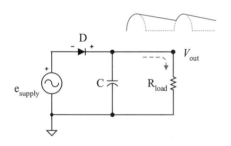

(a) Half-wave rectifier with capacitor input filter

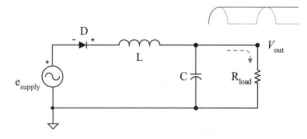

(b) Half-wave rectifier with choke input filter

Figure 16-12(a,b) Half-wave rectifier

The circuit of Figure 16-12(b) is called a **choke input filter** since the choke is the first input reactive element as viewed from the signal source. The capacitor is still placed across the load to charge and hold voltage.

   The ideal dc equivalent circuit is shown in Figure 16-12(c) with the inductor modeled as a dc short and the capacitor modeled as a dc open. The circuit is designed at the operating frequency to have a very high X_L and a very low X_C. The ideal ac equivalent circuit is shown in Figure 16-12(d) with the inductor modeled as an ac open and the capacitor modeled as an ac short. Ideally this delivers pure dc to the load while filtering out any unwanted ac ripple. In reality, such an arrangement does produce much less ripple with the dual filtering action of the inductor and the capacitor, but still there is some ac ripple voltage remaining.

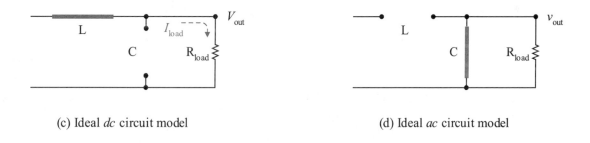

(c) Ideal *dc* circuit model   (d) Ideal *ac* circuit model

Figure 16-12(c,d) Equivalent filter ideal models of L and C

Example 16-10

   For the circuit of Figure 16-12(b), resistance R_{load} is 10 Ω, inductor L is an ideal inductor of 100 mH, the capacitor is 1000 μF, and the ac supply is 120 V_{rms} at 60 Hz. Find the value of X_C and X_L. Are these reasonable values for the choke input filter circuit? Do you have any concerns if you were mass producing this power supply?

Solution

   At a signal frequency of 60 Hz,

$$X_L = 2\pi \ (60 \ Hz)(100 \, mH) = 37.7 \ \Omega$$

$$X_C = \frac{1}{2\pi\,(60\ \text{Hz})\,(1000\ \mu\text{F})} = 2.56\ \Omega$$

This is a reasonable design. X_L is much greater than X_C and R_{load}; and X_C is much, much less than X_L. The major concern is the size and cost of the inductor. At 60 Hz, large inductances are required to keep the X_L value high enough to be effective. This is a very expensive, bulky circuit. The circuit can be improved by use of a voltage regulator (zener or IC) or the redesigning of the power supply to be a switched power supply. These circuits are covered later in more detail in this textbook.

Practice

Repeat the example if the inductor is changed to a value of 1 H. What is the problem with this design?

Answer: X_C unchanged; X_L changed to 377 Ω.

A switching power supply can be used to overcome the large, expensive inductor in the previous example by operating at much higher frequencies and thus requiring lower values of inductance and capacitance. The circuit of Figure 16-13(a) shows the foundation of a switching power supply filter system, specifically for a **buck regulated power supply**. The output of an input capacitor filtered power supply is switched on and off using a switching BJT or MOSFET and fed into the filter system of Figure 16-13(a). A **Schottky diode** is being used in this application. It is a fast switching diode that does not have the typical *pn* junction. Schottky diodes can drop significantly less than 0.7 V forward voltage; however, they also have lower reverse breakdown voltages.

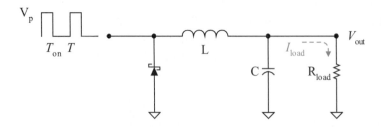

Figure 16-13(a) Switching power supply filter system

When the input goes high as demonstrated in Figure 16-13(b), the input to the filter system is V_p. The diode is off and the supply current flows through the inductor and splits between the load and the capacitor. The capacitor charges toward the applied voltage. The inductor resists a change in current and the capacitor resists a change in voltage.

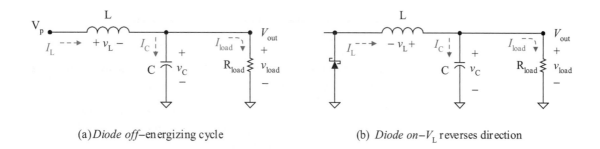

(a) *Diode off–energizing cycle*   (b) *Diode on–V_L reverses direction*

Figure 16-13 (b,c) Switching power supply energize and de-energize cycles

Figure 16-13(c) demonstrates the de-energize cycle, that is, when the input goes low (0 V). Since the inductor voltage is induced on the energizing cycle, its polarity reverses on the de-energize cycle. The diode becomes forward biased and turns on. The collapsing electromagnetic field of the inductor generates current to the capacitor and the load through the on diode.

The output voltage is an average voltage based upon the V_p voltage of the input and the switching wave's duty cycle.

$$V_{dc\ out} = D\ V_p$$

where

    D duty cycle

    V_p peak of the switching waveform

The duty cycle D as a fraction is

$$D = \frac{T_{on}}{T}$$

Example 16-11

For the circuit of Figure 16-13(a), the switching waveform out of the capacitor filtered power supply and into the filter system is a rectangular wave with a peak voltage of 10 V$_p$, a signal frequency of 10 kHz, and a duty cycle of 40%. Find the dc output voltage.

Solution

The output voltage is simply

$$V_{OUT} = 0.4 \times 10 \text{ V}_p = 4 \text{ V}_{dc}$$

Practice

Find the value of L needed to create a 10 kΩ inductive reactance and the value of C needed for a capacitive reactance of 10 Ω, at 10 kHz. What advantages become obvious?

Answer: 159 mH; 1.6 μF; much smaller and less expensive parts required for the higher frequency operation using relatively inexpensive parts of diode, BJT or MOSFET, and an additional capacitor.

A full description and complete circuit analysis of a switching power supply is given in a subsequent textbook in this series. Major advantages of such a supply are smaller, less expensive parts; less energy consumed; and the power supply runs cooler and costs less to operate.

16.3 Series and Parallel Inductance

Total Series Reactance and Inductance

Inductive reactance in series adds like resistance in series. The total inductive reactance of inductors in series is the sum of each individual reactance. Thus, the total inductive reactance of the three inductors in the series circuit of Figure 16-14 is

Total series inductive reactance

$$X_{L \text{ total}} = X_{L1} + X_{L2} + X_{L3} \qquad (16\text{-}3)$$

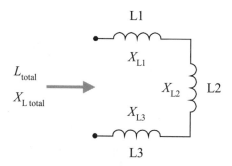

Figure 16-14 Three series inductors

The total equivalent series inductance can be found from substituting each individual inductive reactance form of Equation 16-2 into the series circuit inductive reactance equation of 16-3.

$$2\pi f\, L_{\text{total}} = 2\pi f\, L1 + 2\pi f\, L2 + 2\pi f\, L3$$

Total inductive reactance

The term $2\pi f$ is common in each term. Algebraically divide both sides of the equation and thus each term by $2\pi f$, canceling out the $2\pi f$ terms. The equation reduces to

$$L_{\text{total}} = L1 + L2 + L3 \qquad\qquad (16\text{-}4)$$

Total series inductive

This is the same as the form for series resistances. *A good memory trick is to think that inductance in series combines like resistance in series.*

Example 16-12

A 100 mH inductor, a 47 mH inductor, and a 10 mH inductor are in series. The circuit is supplied with a 1 kHz sinusoidal supply. Find the reactance of each inductor, the total circuit inductive reactance, and the combined total circuit inductance.

Solution

The individual inductive reactances are

$$X_{L1} = 2\pi f\, L1 = 2\pi\,(1\text{ kHz})(100\text{ mH}) = 628\ \Omega$$

$$X_{L2} = 2\pi f\, L2 = 2\pi\,(1\text{ kHz})(47\text{ mH}) = 295\ \Omega$$

$$X_{L3} = 2\pi f\, L3 = 2\pi\,(1\text{ kHz})(10\text{ mH}) = 62.8\ \Omega$$

The total circuit inductive reactance is the sum of the individual series reactances.

$$X_{L\ total} = 628\ \Omega + 295\ \Omega + 62.8\ \Omega = 986\ \Omega$$

Rewriting Equation 16-2 to solve for inductance

$$L_{total} = \frac{X_{L\ total}}{2\pi f} = \frac{986\ \Omega}{2\pi(1\ kHz)} = 157\ mH$$

This answer may be confirmed by using the total series inductance form of Equation 16-4.

$$L_{total} = 100\ mH + 47\ mH + 10\ mH = 157\ mH$$

Practice

Repeat the example with a sinusoidal frequency of 100 kHz. You can repeat the above solution process or use a shortcut. If the frequency is increased by a factor of 100, what is the impact on each inductive reactance and the total inductance?

Answer: 62.8 kΩ, 29.5 kΩ, 6.28 kΩ, 98.6 kΩ, 157 mH (unchanged)

Total Parallel Reactance and Inductance

Inductive reactance in parallel combines like resistance in parallel. Thus, the total inductive reactance of the three inductors in the parallel circuit of Figure 16-15 is

Total parallel inductive reactance

$$\frac{1}{X_{L\ total}} = \frac{1}{X_{L1}} + \frac{1}{X_{L2}} + \frac{1}{X_{L3}} \qquad \textbf{(16-5)}$$

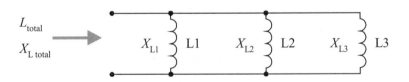

Figure 16-15 Three parallel inductors

The total equivalent parallel inductance can be found from substituting each individual inductive reactance form of Equation 16-2 into the parallel circuit inductive reactance Equation of 16-5.

$$\frac{1}{2\pi f\,L_{\text{total}}} = \frac{1}{2\pi f\,L1} + \frac{1}{2\pi f\,L2} + \frac{1}{2\pi f\,L3}$$

Algebraically multiplying each term by the common factor of $2\pi f$, this equation simplifies to

$$\frac{1}{L_{\text{total}}} = \frac{1}{L1} + \frac{1}{L2} + \frac{1}{L3} \qquad\qquad \textbf{(16-6)}$$

or

$$L_{\text{total}} = \left(\frac{1}{L1} + \frac{1}{L2} + \frac{1}{L3}\right)^{-1}$$

Total parallel inductance is the reciprocal of the sum of the reciprocals of the individual inductances. This is the same as the form for parallel resistances. *A good memory trick is to think that inductance in parallel combines like resistance in parallel.*

Example 16-13

A 100 mH inductor, a 47 mH inductor, and a 10 mH inductor are in parallel. Find the total circuit inductance and its reactance to a 1 kHz sinusoid signal.

Solution

The total circuit inductance is

$$L_{\text{total}} = \left(\frac{1}{100\ \text{mH}} + \frac{1}{47\ \text{mH}} + \frac{1}{10\ \text{mH}}\right)^{-1} = 7.62\ \text{mH}$$

The total circuit inductive reactance is

$$X_{\text{L total}} = 2\pi f\,L_{\text{total}} = 2\pi\,(1\ \text{kHz})(7.62\ \text{mH}) = 47.9\ \Omega$$

Practice

The total circuit inductance of three equally valued inductors in parallel is 300 mH. Find the inductance of each inductor. Find the total inductive reactance at a frequency of 60 Hz.

Answer: 900 mH each, 113 Ω

Series-parallel Inductance Reduction

For series-parallel inductor circuits, reduce simple series inductance combinations and simple parallel inductance combinations. These must be pure inductance combinations. Other components such as resistors cannot be simply combined with inductors. Resistance and reactance require vector addition.

Example 16-14

Reduce the circuit of Figure 16-16(a) to a single inductance.

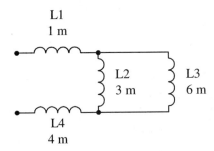

Figure 16-16(a) Circuit of example problem

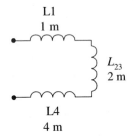

Figure 16-16(b)
Example circuit
reduced

Solution

The simplest combination is the parallel combination of L2 and L3, that is, L2 // L3. Combine the parallel capacitances.

$$L_{23} = \left(\frac{1}{6 \text{ mH}} + \frac{1}{3 \text{ mH}} \right)^{-1} = 2 \text{ mH}$$

Combining L2 and L3, the circuit is reduced to the equivalent series circuit of L1, L_{23}, and L4 of Figure 16-16(b). Combine these purely series inductances to a single total inductance value.

$$L_{\text{total}} = 1 \text{ mH} + 2 \text{ mH} + 4 \text{ mH} = 7 \text{ mH}$$

Practice

Find the total inductance for the circuit of Figure 16-17.

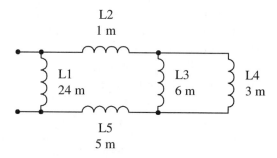

Figure 16-17 Practice circuit

Answer: 6 mH

16.4 Qualitative *RL* Transient

Figure 16-18(a) shows a simple *RL* switched circuit. This is a simple switched series circuit; however, practically, this circuit has an inherent switch arcing problem, which is addressed later in this section.

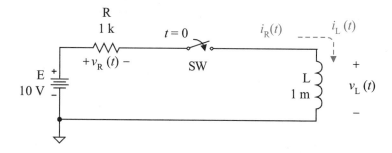

Figure 16-18(a) Simple (or Thévenin model) *RL* switched dc circuit

Energizing Cycle

During the energizing cycle, the switch closes and the current builds from 0 to its maximum value when the inductor is fully energized.

Let's step through the thought process of what occurs before the switch is closed ($t < 0$), when the switch is closed ($t = 0$), and after the switch has been closed for a long time.

Before the switch is closed ($t < 0$), assume that the switch has been open for a long time and that the inductor has been de-energized. That is, the inductor is initially unfluxed. Since the inductor is de-energized and the switch is open, the values just before the switch is closed are

$$I_L = I_R = 0 \text{ mA}$$

$$V_R = V_L = 0 \text{ V}$$

Now assume the switch has just closed at $t = 0$ (or $t = 0^+$) as shown in Figure 16-18(b). The inductor resists any initial change in current but induced voltage can change instantly and does.

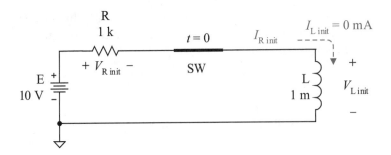

Figure 16-18(b) Switch has just closed at $t = 0$ (*initial circuit*)

1. The switch is closed at $t = 0$. Just before the switch is closed at $t = 0^-$, the inductor current is 0 mA (from above). *The inductor resists a change in current.* Since it is 0 mA just before the switch is closed ($t = 0^-$), it is still 0 mA just after the switch is closed ($t = 0$). Therefore,

$$I_{L \text{ init}} = 0 \text{ mA}$$

2. Since the inductor current is 0 mA at $t = 0$, the resistor current is 0 mA also since this is a series circuit.

$$I_{R \text{ init}} = 0 \text{ mA}$$

3. Since the resistor current is 0 mA at $t = 0$, the resistor drops 0 V (Ohm's law).

$$V_{R \text{ init}} = 0 \text{ V}$$

4. The inductor is acting like an open and all of the applied voltage (or E_{Th}) is dropped across the inductor.

$$V_{L \text{ init}} = 10 \text{ V}$$

Summarizing, the *initial values* of this circuit at $t = 0$ are

$$I_{L \text{ init}} = I_{R \text{ init}} = 0 \text{ mA}$$

$$V_{R \text{ init}} = 0 \text{ V}$$

$$V_{L \text{ init}} = 10 \text{ V}$$

- The inductor current has not changed instantly; it has remained at 0 mA at the instant the switch is closed. *The unenergized inductor is initially acting like an open.*

- The inductor voltage has changed instantly from 0 V to 10 V (*a voltage spike*) at the instant the switch was closed.

Now assume the switch has been closed a very long time, that is $t \rightarrow \infty$ as shown in Figure 16-18(c). This is called the *steady state* condition since all the transients have died out. The voltages and currents are no longer changing; they have reached steady values.

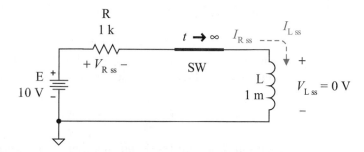

Figure 16-18(c) Switch closed for a very, very long time (*steady state circuit*)

1. The inductor has fully energized, current is no longer changing, and therefore the inductor drops 0 V.

$$V_{L \, ss} = 0 \text{ V}$$

2. The applied voltage must be dropped across the resistor (Kirchhoff's voltage law).

$$V_{R \, ss} = 10 \text{ V}$$

3. By Ohm's law, the resistor current must be 10 mA.

$$I_{R \, ss} = \frac{10 \text{ V}}{1 \text{ k}\Omega} = 10 \text{ mA}$$

4. The inductor current must be the same as the resistor current since this is a series circuit.

$$I_{L \, ss} = I_{R \, ss} = 10 \text{ mA}$$

5. Summarizing, the *steady state values* of this circuit as $t \rightarrow \infty$,

$$V_{L \, ss} = 0 \text{ V}$$

$$V_{R \, ss} = 10 \text{ V}$$

$$I_{L \, ss} = I_{R \, ss} = 10 \text{ mA}$$

- The inductor has increased from an initial current of 0 mA (an *open*) to a steady state, fully energized current of 10 mA.

- The inductor voltage has decreased from an initial spike of 10 V to a steady state value of 0 V. *The fully energized, ideal inductor is acting like a short.*

RL Time Constant τ

The time constant τ (the lowercase, Greek letter tau) of an *RL* circuit is defined to be its inductance value divided by the equivalent circuit resistance value as viewed from the inductor.

$$\tau = \frac{L}{R}$$

For the circuit of Figure 16-18 the energizing time constant is

$$\tau_{\text{energize}} = \frac{1 \text{ mH}}{1 \text{ k}\Omega} = 1 \text{ μs}$$

It takes approximately five (5) time constants to energize (or de-energize) an inductor to 99.3% of its steady state value. Thus, for the circuit of Figure 16-18, the inductor is essentially fully energized in 5 μs.

$$t_{\text{fully energized}} = 5\tau = 5 \times 1 \text{ μs} = 5 \text{ μs}$$

De-energizing Cycle

If the switch suddenly opens as shown in Figure 16-19, the electromagnetic field collapses and the inductor becomes a current generator with the same value of current and current direction the inductor had just before the switch is opened. Do you see a problem with this initial *RL* de-energizing circuit?

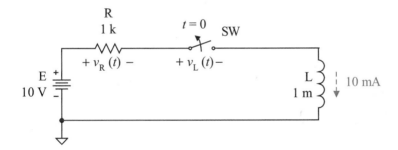

Figure 16-19 *RL* de-energizing cycle–initial circuit

If the switch just opened and the inductor's collapsing *em* field is creating a 10 mA circuit current, what is the voltage drop across the just opened switch? Theoretically, a 10 mA current is driving an open. By Ohm's law that is an infinite voltage drop for the inductor

$$v_{\text{L}}(0) = 10 \text{ mA} \times \infty \text{ } \Omega = \infty \text{ V}$$

That is impossible; however, a *voltage spike* does appear across the switch just as it is opened. That voltage spike creates a very high electric field between the switch contacts, which causes arcing between the switch contacts. This is very similar to a miniature lightning strike between the switch contacts. Eventually this arcing causes scoring, burning, and pitting of the contacts. The contacts must be occasionally **burnished** to clean and smooth the contacts. Eventually this process destroys the switch contacts.

To prevent arcing and eventual destruction of the switch contacts, a diode is placed in parallel with the inductor as shown in Figure 16-20. The diode D has the schematic symbol for a **Schottky diode**, which is a fast switching diode used for such purposes if fast switching is needed. During the energizing cycle the diode is reverse biased *off* (open) and plays no part in the energizing cycle. During the de-energizing cycle, the inductor voltage polarity is reversed, forward biasing the diode and providing a low resistance path around the entire circuit. This type of diode function is called a **compensating** or **compensation diode**, since this compensates for the inductive kick effect from the inductor. This type of diode function is also called a **flyback diode**; this terminology stems from the flyback circuit of a television in which a diode is used to provide the rapid return of the beam in a **cathode ray tube** (CRT).

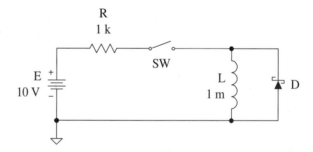

Figure 16-20 *RL* circuit with *compensating diode*

Recall that the inductor had been energized and had a 10 mA current flowing down through the inductor. When the switch is opened, the inductor's electromagnetic field collapses and initially generates that same 10 mA as a current supply. The diode is forward biased and the inductor current flows through the diode and not through the circuit. Eventually the inductor's electromagnetic field totally collapses and all energy is expended, resulting in a steady state current of 0 and a steady state voltage of 0.

First let's examine the *initial circuit* just after the switch has opened at $t = 0$ of the de-energizing circuit. Refer to the initial circuit conditions of Figure 16-21.

- SW is opened at $t = 0$ of the de-energizing cycle.

- The inductor has an initial current of 10 mA downward which was its condition at the end of the energizing cycle.

- The inductor voltage polarity reverses and builds up to 0.7 V, at which point the diode is forward biased and turns on.

- Also, at that point the current no longer flows through the switched circuit but only flows up through the turned-on diode.

$$I_{\text{L init}} = I_{\text{D init}} = 10 \text{ mA}$$

$$V_{\text{L init}} = -0.7 \text{ V} \qquad \text{(once diode turns on)}$$

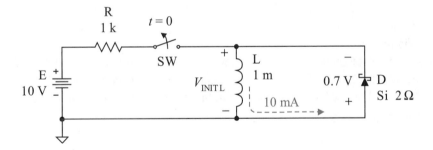

Figure 16-21 Initial de-energizing circuit at $t = 0$

After the inductor *em* field has totally collapsed, the inductor current and voltage are 0.

$$I_{\text{L ss}} = 0 \text{ mA}$$

$$V_{\text{L ss}} = 0 \text{ V}$$

For the de-energizing circuit of Figure 16-20 the time constant is based upon the L of the inductor and the resistance of the diode D. If the inductor is not ideal, then its internal series resistance must be included in the time constant calculation.

$$\tau_{\text{de-energize}} = \frac{L}{R} = \frac{1 \text{ mH}}{2 \text{ }\Omega} = 500 \text{ }\mu s$$

To fully de-energize the inductor requires approximately five time constants.

$$t_{\text{de-energize}} = 5 \times 500 \text{ }\mu s = 2.5 \text{ ms}$$

16.5 Switched *RL* Transient

Inductor Models in a Switched DC *RL* Circuit

The following is a review of the dc transient models of an inductor. Initially the inductor resists a change in current. If an inductor current is 0 A just before the circuit is dc switched, the inductor tends to stay at 0 A at that instant of time. It can be initially modeled as an open. An inductor with a current tends to initially retain that current at the instant the circuit is switched and can therefore be initially modeled as a current supply. See Figure 16-22(a).

In a steady state dc switched *RL* circuit, the inductor is modeled as an ideal wire (short). Or, if the wire resistance is significant in the circuit, the steady state model is a resistor with the dc resistance of the inductor wire. See Figure 16-22(b).

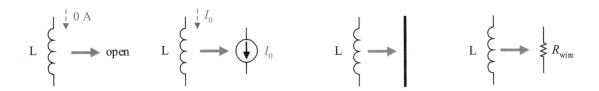

(a) Initially switched inductor models   (b) Steady state inductor models

Figure 16-22 Initial and steady state dc models of a switched *RL* circuit

Transient Expression–Classical Calculus

There are three distinct activities or phases in a dc transient: (1) initial circuit with its initial values, (2) the steady state circuit and its steady state values, and (3) the transient state as the voltages and currents go from their initial values to their steady state values. Initial and steady state circuits are relatively straightforward to analyze. Calculus provides the solution for the transient expressions of voltages and currents. Using the simple *RL* circuit of Figure 16-23, perform basic circuit analysis.

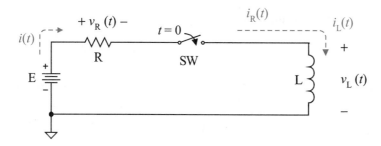

Figure 16-23 Basic *RL* transient circuit

This is a series circuit; therefore the currents in each component are the same at any instant in time.

$$i(t) = i_R(t) = i_L(t)$$

From Ohm's law,

$$v_R(t) = R\, i_R(t) = R\, i(t)$$

The voltage and current relationship of the inductor is

$$v_L(t) = L\frac{d\,i_L(t)}{dt}$$

From Kirchhoff's voltage law,

$$E = v_R(t) + v_L(t)$$

Substituting the voltage expressions for $v_R(t)$ and $v_L(t)$, the KVL expression becomes

$$E = R\, i(t) + L\frac{d\,i(t)}{dt}$$

Using calculus (refer to a basic calculus text if so inclined), the solution to this differential equation is

$$i(t) = i_R(t) = i_L(t) = \left(\frac{E}{R}\right) - \left(\frac{E}{R}\right)e^{-\frac{t}{L/R}}$$

where e is the natural number forming an exponential decay function $e^{-t/\tau}$, where τ is the circuit time constant L/R.

Using Ohm's law, the voltage drop across the resistor is

$$v_R(t) = E - E\,e^{-\frac{t}{L/R}}$$

Using Kirchhoff's voltage law, the voltage drop across the inductor is

$$v_L(t) = E\,e^{-\frac{t}{L/R}}$$

Notice that the time constant L/R is a natural outcome of this analysis. When $t = \tau$, the decay function is equal to e^{-1} (or, 0.37). That is equivalent to a decay of 63% from its starting value of 1.00.

RL Universal DC Transient Expression

The universal dc transient process used for *RC* circuits can be applied to *RL* circuits. The inductor is modeled differently for the initial and steady state values but the universal dc transient expression still applies.

$$y_X(t) = Y_{ss} + \left(Y_{init} - Y_{ss}\right)e^{-\frac{t}{\tau}}$$

where

$y_X(t)$ instantaneous voltage or current of the resistor or inductor

Y_{ss} steady state voltage or current of the resistor or inductor

Y_{init} initial voltage or current of the resistor or inductor

τ time constant *tau* where $\tau = L/R$

$e^{-t/\tau}$ exponential decay function

This universal dc transient expression always works for any dc transient simple *RL* circuit.

Example 16-15

For the circuit of Figure 16-18, write the transient expression for the inductor voltage and the inductor current. Find the value of the inductor voltage and current at $t = 0$ μs, 1 μs, 2 μs, and 5 μs.

Solution

Based upon preceding analysis, we know the following values:

$$\tau = \frac{1\ \text{mH}}{1\ \text{k}\Omega} = 1\ \mu\text{s}$$

$$I_{\text{L init}} = 0\ \text{mA}$$

$$I_{\text{L ss}} = 10\ \text{mA}$$

Substituting into the general expression of

$$i_X(t) = i_{ss} + \left(i_{init} - i_{ss}\right)e^{-\frac{t}{\tau}}$$

$$i_L(t) = 10\ \text{mA} + \left(0\ \text{mA} - 10\ \text{mA}\right)e^{-\frac{t}{1\ \mu s}}$$

Simplifying the coefficient of the expression

$$i_L(t) = 10\ \text{mA} - 10\ \text{mA}\ e^{-\frac{t}{1\ \mu s}}$$

Evaluate the transient current at $t = 0$ μs, 1 μs, 2 μs, and 5 μs.

$$i_L(0\ \mu\text{s}) = 10\ \text{mA} - 10\ \text{mA}\ e^{-\frac{0\ \mu s}{1\ \mu s}} = 0\ \text{mA}$$

$$i_L(1\ \mu\text{s}) = 10\ \text{mA} - 10\ \text{mA}\ e^{-\frac{1\ \mu s}{1\ \mu s}} = 6.32\ \text{mA}$$

$$i_L(2\ \mu\text{s}) = 10\ \text{mA} - 10\ \text{mA}\ e^{-\frac{2\ \mu s}{1\ \mu s}} = 8.65\ \text{mA}$$

$$i_L(5\ \mu\text{s}) = 10\ \text{mA} - 10\ \text{mA}\ e^{-\frac{5\ \mu s}{1\ \mu s}} = 9.93\ \text{mA}$$

The key inductor voltage parameters are

$$\tau = 1 \ \mu s$$

$$V_{\text{L init}} = 10 \ \text{V}$$

$$V_{\text{L ss}} = 0 \ \text{V}$$

Substituting into the general expression of

$$v_X(t) = V_{\text{ss}} + \left(V_{\text{init}} - V_{\text{ss}}\right) e^{-\frac{t}{\tau}}$$

$$v_L(t) = 0 \ \text{V} + \left(10 \ \text{V} - 0 \ \text{V}\right) e^{-\frac{t}{1\ \mu s}}$$

Simplifying the coefficient of the exponential term

$$v_L(t) = 10 \ \text{V} \ e^{-\frac{t}{1\ \mu s}}$$

Evaluate the transient voltage at $t = 0$ μs, 1 μs, 2 μs, and 5 μs.

$$v_L(0 \ \mu s) = 10 \ \text{V} \ e^{-\frac{0\ \mu s}{1\ \mu s}} = 10 \ \text{V}$$

$$v_L(1 \ \mu s) = 10 \ \text{V} \ e^{-\frac{1\ \mu s}{1\ \mu s}} = 3.68 \ \text{V}$$

$$v_L(2 \ \mu s) = 10 \ \text{V} \ e^{-\frac{2\ \mu s}{1\ \mu s}} = 1.35 \ \text{V}$$

$$v_L(5 \ \mu s) = 10 \ \text{V} \ e^{-\frac{5\ \mu s}{1\ \mu s}} = 0.07 \ \text{V}$$

Practice

For the circuit of Figure 16-18, write the transient expression for the resistor voltage and the resistor current. Find the value of the resistor voltage and current and evaluate the transient current at $t = 0$ μs, 1 μs, 2 μs, and 5 μs.

Answer:

$$i_R(t) = 10 \ \text{mA} - 10 \ \text{mA} \ e^{-\frac{t}{1\ \mu s}} \ ; \quad 0 \ \text{mA}, \ 6.32 \ \text{mA}, \ 8.65 \ \text{mA}, \ 9.93 \ \text{mA}$$

$$v_R(t) = 10 \ \text{V} - 10 \ \text{V} \ e^{-\frac{t}{1\ \mu s}} \ ; \quad 0 \ \text{V}, \ 6.32 \ \text{V}, \ 8.65 \ \text{V}, \ 9.93 \ \text{V}$$

Note that in the last practice problem, a quick solution would have been to equate the inductor current and the resistor current from the example values, and then use Ohm's law to find the resistor voltage.

Universal *RL* Approach to Solve DC Transients

The following technique is used to solve all dc transient circuits with a simple *RL* circuit or Thévenin model equivalent circuit.

1. Find $i_L(0^-)$

Find the inductor current just before the switch is closed.

2. Find the time constant τ

$$\tau = \frac{L}{R}$$

If this is a complex circuit, other than a simple *RL* circuit, first find the Thévenin circuit using the inductor as the load. Then use the Thévenin circuit for this process. If this is the Thévenin equivalent circuit, then the time constant is

$$\tau = \frac{L}{R_{Th}}$$

If the internal series resistance of the inductor wire R_{wire} is meaningful, then its resistance must be included in the calculation of the time constant

$$\tau = \frac{L}{R_{Th} + R_{wire}}$$

3. Draw the modeled *initial* circuit (or Thévenin circuit). Properly model the inductor as an *open (unenergized)* or a *current supply (energized)*. Refer back to Figure 16-22(a). Find the desired initial voltages and currents.

4. Draw the modeled *steady state* circuit or Thévenin circuit. Properly model the inductor as an *ideal wire (short)* or as a *real wire (its dc resistance)*. Refer to Figure 16-22(b). Find the desired steady state voltages or currents.

5. If needed, write and apply the universal transient dc equation for the desired voltages and currents. Simplify equations.

$$y(t) = Y_{ss} + (Y_{init} - Y_{ss})e^{-\frac{t}{\tau}}$$

6. If needed, sketch (or quick sketch) the desired voltages and currents.

All simple *RL* circuits (or Thévenin modeled *RL* circuits) can be solved by this process. This textbook consistently follows this process for all transient dc circuit analysis.

Example 16-16

For the circuit of Figure 16-24(a), find the universal dc transient equation for the inductor voltage and current. Sketch their exponential response for five time constants. Find the inductor voltage and current at 3 µs. Assume that at the instant before the switch is closed, an external circuit (not shown in the schematic) was producing an inductor current of 3 mA.

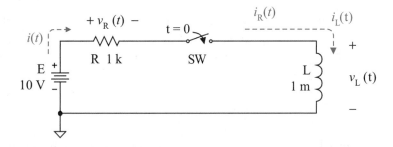

Figure 16-24(a) Example circuit

Solution

Follow the universal dc transient analysis process, step by step.

1. Find $i_L(0^-)$

It is given that just before the switch is closed, an external circuit is producing an inductor current of 3 mA.

$$i_L(0^-) = 3 \text{ mA}$$

2. Find the *time constant* τ. With the switch closed, this simple *RL* circuit is already in Thévenin model form.

$$\tau = \frac{L}{R} = \frac{1\text{ mH}}{1\text{ k}\Omega} = 1\ \mu s$$

3. Model the *initial circuit* and find the initial inductor current and voltage drop. To model the inductor, go back to step 1. The inductor is energized and has an equivalent current of 3 mA flowing just before the switch is closed. The current in the inductor must initially be 3 mA, since the inductor resists a change in current. Thus, the inductor is modeled as a current supply. Draw the modeled circuit as shown in Figure 16-24(b). The switch is closed (short) and the inductor is modeled as a 3 mA current supply.

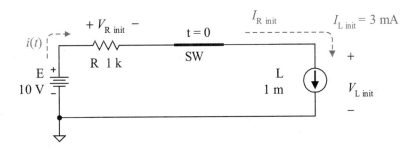

Figure 16-24(b) Initial circuit for example

$$I_{L\text{ init}} = I_{R\text{ init}} = 3\text{ mA}$$

$$V_{R\text{ init}} = 3\text{ mA} \times 1\text{ k}\Omega = 3\text{ V}$$

$$V_{L\text{ init}} = 10\text{ V} - 3\text{ V} = 7\text{ V}$$

4. Model the *steady state circuit* and find the steady state inductor voltage and current. The steady state inductor is ideally a *short*. Draw the modeled circuit as shown in Figure 16-24(c).

The switch is closed (short) and the inductor is modeled as a *short*.

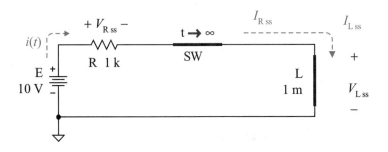

Figure 16-24(c) Steady state circuit for example

$$V_{L\,ss} = 0 \text{ V}$$

$$V_{R\,ss} = 10 \text{ V}$$

$$I_{L\,ss} = I_{R\,ss} = \frac{10 \text{ V}}{1 \text{ k}\Omega} = 10 \text{ mA}$$

5. Write and apply the universal transient dc equation for the desired voltages and currents. *Simplify the instantaneous equations.*

$$y(t) = Y_{ss} + (Y_{init} - Y_{ss})e^{-\frac{t}{\tau}}$$

Write and simplify the inductor dc transient current equation.

$$i_L(t) = 10 \text{ mA} + (3 \text{ mA} - (10 \text{ mA}))e^{-\frac{t}{1\,\mu s}}$$

$$i_L(t) = 10 \text{ mA} - 7 \text{ mA}\, e^{-\frac{t}{1\,\mu s}}$$

Write and simplify the inductor dc transient voltage equation.

$$V_L(t) = 0\text{ V} + \left(7\text{ V} - (0\text{ V})\right)e^{-\frac{t}{1\,\mu s}}$$

$$V_L(t) = 7\text{ V }e^{-\frac{t}{1\,\mu s}}$$

Find the inductor current and voltage values at $t = 3$ μs.

$$i_L(3\text{ μs}) = 10\text{ mA} - 7\text{ mA }e^{-\frac{3\,\mu s}{1\,\mu s}} = 9.65\text{ mA}$$

$$V_L(3\text{ μs}) = 7\text{ V }e^{-\frac{3\,\mu s}{1\,\mu s}} = 349\text{ mV}$$

6. Quick sketch the inductor transient current and voltage. See Figure 16-24(d). Use the rule of thumb of going over τ and up/down 63% (roughly 2/3) the distance toward the steady state value.

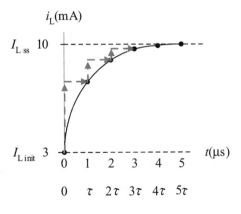

Exponential *rising* curve

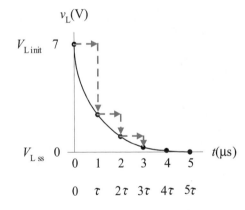

Exponential *falling* curve

Figure 16-24(d) Quick sketch of example inductor dc transient current and voltage

The inductor current and voltage values at $t = 3$ μs can be estimated. The above calculations of 9.65 mA and 394 mV at 3 μs match closely with the curves.

Practice

Use the circuit of the example problem of Figure 16-24(a) but assume that the inductor has an internal resistance of 1 kΩ. Assume that the inductor still has a current of 3 mA just before the switch is closed. Find the universal dc transient equation for the inductor current and voltage. Sketch their exponential response for five time constants.

Answer: The inductor transient dc current and voltage sketches are shown in Figure 16-24(e).

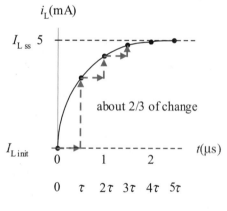

Exponential *rising* curve

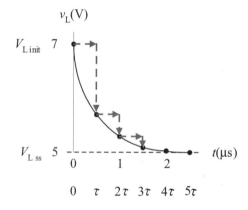

Exponential *falling* curve

Figure 16-24(e) Practice inductor transient current and voltage sketches

$$i_L(t) = 5 \text{ mA} - 2 \text{ mA } e^{-\frac{t}{0.5 \, \mu s}}, \quad v_L(t) = 5 \text{ V} + 2 \text{ V } e^{-\frac{t}{0.5 \, \mu s}}$$

Multiple Inductors—Currents and Voltages

Let's look at a multiple inductor circuit and determine the initial and steady values relevant to the circuit and their dc transient expressions.

Example 16-17

For the circuit of Figure 16-25(a), assume that the inductors are all de-energized before the switch is closed at $t = 0$. Find the voltage drop across and the current through L1 just before the switch is closed, just after the switch is closed, and in steady state. Find the time constant of the circuit and determine how long before the circuit reaches steady state. Finally, write the transient dc equation for the L1 inductor voltage. Note that resistance and inductor units have been included for clarity in this schematic.

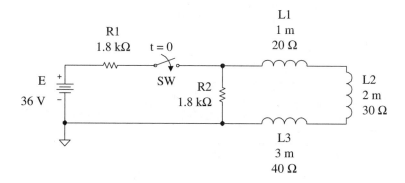

Figure 16-25(a) Example circuit

Solution

At $t = 0^-$, just before the switch is closed, the inductors are de-energized and there is no applied voltage or current to the circuit.

$$i_{L1}(0^-) = 0 \text{ mA}$$

$$v_{L1}(0^-) = 0 \text{ V}$$

Since this is a series parallel resistance circuit, it is necessary to produce the Thévenin circuit model of the circuit with the inductor circuit removed. The resistor circuit is simply a voltage divider circuit when the switch is closed.

$$E_{Th} = \left(\frac{1.8 \text{ k}\Omega}{3.6 \text{ k}\Omega}\right) 36 \text{ V} = 18 \text{ V}$$

Find the Thévenin resistance. Zero the voltage supply (replace it with its model of a short), look back into the load terminals (from the viewpoint of the inductors), and find the equivalent resistance. The equivalent resistance is R1 in parallel with R2.

$$R_{Th} = \frac{1.8 \text{ k}\Omega}{2} = 900 \text{ }\Omega$$

The dc transient Thévenin equivalent circuit with the equivalent total inductance L_{total} attached is shown in Figure 16-25(b).

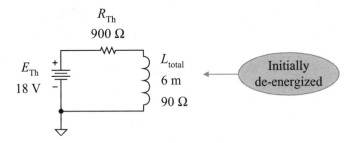

Figure 16-25(b) Equivalent energizing dc transient circuit

$$L_{total} = 1 \text{ mH} + 2 \text{ mH} + 3 \text{ mH} = 6 \text{ mH}$$

$$R_{total \text{ wire}} = 20 \text{ }\Omega + 30 \text{ }\Omega + 40 \text{ }\Omega = 90 \text{ }\Omega$$

At $t = 0$ (*initial circuit conditions*), the switch closes and 18 V is applied to the *RL* circuit. Initially the de-energized inductors ideally act like opens but in fact *their very large reactances dominate initially*. The circuit supplied Thévenin voltage has jumped from 0 V to 18 V essentially instantaneously. Initially that is a frequency equivalent to infinity ($f = \infty$ Hz). Therefore, the initial inductive reactance acts like an open, which is consistent with our model of Figure 16-22(a).

$$X_{\mathrm{L}} = 2\pi f\, L = 2\pi \left(\infty\ \mathrm{Hz}\right) L = \infty\ \Omega \qquad \text{(open)}$$

Since the inductors are acting like opens, the initial circuit current is 0 mA.

$$i_{\mathrm{L1}}(0) = I_{\text{init L1}} = 0\ \mathrm{mA}$$

Since the inductors are acting ideally as opens, the Thévenin supplied voltage is dropped across the total inductance.

$$v_{\mathrm{L\,total}}(0) = V_{\text{init L total}} = E_{\mathrm{Th}} = 18\ \mathrm{V}$$

The Thévenin supplied voltage is then initially voltage divided among the three inductors based upon each inductor's relative inductive reactance. See Figure 16-25(c).

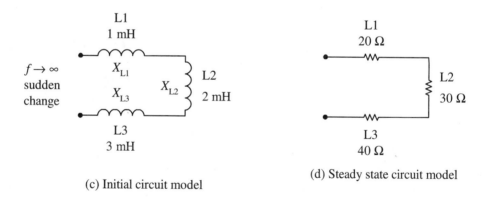

(c) Initial circuit model

(d) Steady state circuit model

Figure 16-25(c,d) Initial and steady state models of series inductor example circuit

Apply the voltage divider law to the three series inductive reactances to find the drop across the L1 inductor.

$$v_{\mathrm{L1}}(0) = V_{\text{init L1}} = \left(\frac{X_{\mathrm{L1}}}{X_{\mathrm{L\,total}}}\right) E_{\mathrm{Th}}$$

The voltage divider fraction is simplified by canceling out the common term of $2\pi f$.

$$\frac{X_{L1}}{X_{L\,total}} = \frac{2\pi f\ L1}{2\pi f\ L_{total}} = \frac{L1}{L_{total}}$$

Thus the initial voltage across L1 is

$$v_{L1}(0) = V_{init\ L1} = \left(\frac{1\ mH}{1\ mH + 2\ mH + 3\ mH}\right) 18\ V = 3\ V$$

As t → ∞ (*steady state circuit conditions*), the inductors reach steady state and are modeled as ideal shorts or by their wire resistances. In this case, the wire resistances add up to be 90 Ω, that is, 10 percent of the external 900 Ω resistance. Ignoring the inductor internal wire resistance causes a 10% error. This example includes the inductor wire resistance for increased accuracy as shown in Figure 16-25(d). Apply the voltage divider rule to this series resistive circuit of Figure 16-25(e) with four components, and find the steady state voltage drop across L1.

$$v_{L1}(\infty) = V_{ss\ L1} = \left(\frac{20\ \Omega}{900\ \Omega + 20\ \Omega + 30\ \Omega + 40\ \Omega}\right) 18\ V = 364\ mV$$

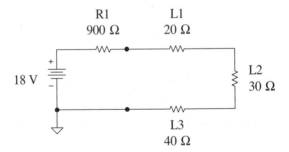

Figure 16-25(e) Steady state circuit model of example circuit

The steady state current in this series circuit and inductor L1 is

$$i_{L1}(\infty) = I_{ss\ L1} = \frac{364\ mV}{20\ \Omega} = 18.2\ mA$$

The time constant for the R_{Th} and each L in this circuit is

$$\tau_{circuit} = \frac{L_{total}}{R_{Th} + R_{wire}} = \frac{6 \text{ mH}}{990 \text{ } \Omega} = 6.06 \text{ } \mu s$$

The total time to energize this circuit is about five time constants.

$$t_{energize} = 5\tau_{circuit} = 5 \times 6.06 \text{ } \mu s = 30.3 \text{ } \mu s$$

The L1 inductor current dc transient expression is

$$i_{L1}(t) = I_{ss\,L1} + (I_{init\,L1} - I_{ss\,L1}) \, e^{-\frac{t}{\tau}}$$

$$i_{L1}(t) = 18.2 \text{ mA} + (0 \text{ mA} - 18.2 \text{ mA}) \, e^{-\frac{t}{6.06 \text{ } \mu s}}$$

The L1 inductor voltage dc transient expression is

$$v_{L1}(t) = V_{ss\,L1} + (V_{init\,L1} - V_{ss\,L1}) \, e^{-\frac{t}{\tau}}$$

$$v_{L1}(t) = 0.364 \text{ V} + (3 \text{ V} - 0.364 \text{ V}) e^{-\frac{t}{6.06 \text{ } \mu s}}$$

$$v_{L1}(t) = 0.364 \text{ V} + 2.636 \text{ V} e^{-\frac{t}{6.06 \text{ } \mu s}}$$

Practice

For the circuit of Figure 16-25(a), assume that the inductors are all de-energized before the switch is closed at $t = 0$. Find the voltage drop across and current through L2 just before the switch is closed, just after the switch is closed, and in steady state. Find the time constant of the circuit and determine how long before the circuit reaches steady state. Finally, write the transient dc equation for the L2 inductor voltage.

Answer: 0 mA, 0 V; 0 mA, 6 V, 18.2 mA, 545 mV; 6.06 µs, 30.3 µs;

$$i_{L2}(t) = 18.2 \text{ mA} - 18.2 \text{ mA} \, e^{-\frac{t}{6.06 \text{ } \mu s}}; \quad v_{L2}(t) = 0.545 \text{ V} + 5.455 \text{ V} \, e^{-\frac{t}{6.06 \text{ } \mu s}}$$

16.6 *RL* Square Wave Response

The square wave supply produces two dc voltages that the *RL* circuit responds to, E_{max} and E_{min}. It is equivalent to an *RL* switched circuit switching between two dc supply voltages. During the first half cycle the inductor energizes towards $I_{L\,max}$ and on the second half cycle the inductor energizes towards $I_{L\,min}$. This is the key concept of viewing the action of this the *RL* square wave response circuit.

Problem 16-18

Sketch the inductor and resistor voltage and current waveforms of the *RL* circuit of Figure 16-26(a).

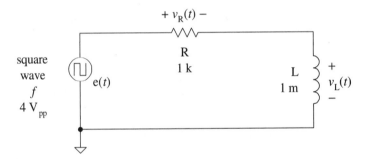

Figure 16-26(a) Example *RL* circuit square wave response

Solution

First find the maximum and minimum voltages of the supplied square wave voltage. The supply is a pure ac signal, therefore

$$E_{max} = \frac{4\;V_{pp}}{2} = 2\;V_p$$

$$E_{min} = -2\;V_{min}$$

If the inductor is allowed to fully energize on each half cycle by the supplied voltage signal, that is, if

$$\frac{T}{2} \geq 5\tau$$

or

$$T \geq 10\tau$$

then there is sufficient time to fully energize the inductor such that the steady state inductor voltage is 0 V and that of the resistor is

$$V_{\text{max R}} = 2\ V_{\text{p}}$$

$$V_{\text{min R}} = -2\ V_{\text{min}}$$

Ideally, if $T = 10\tau$, then the inductor has just enough time to fully energize such that the resistor voltage reaches E_{max} on the first half cycle, and just enough time to reach E_{min} on the negative half cycle. Basically the *resistor voltage drop* is exponentially rising from -2 V to $+2$ V on the first half cycle, and falling exponentially from $+2$ V to -2 V on the second half cycle.

For the circuit of Figure 16-26(a), this ideal period is

$$T = 10\tau$$

$$\tau = \frac{L}{R} = \frac{1\ \text{mH}}{1\ \text{k}\Omega} = 1\ \mu\text{s}$$

$$T = 10\tau = 10 \times 1\ \mu\text{s} = 10\ \mu\text{s}$$

This requires that the frequency of the voltage supply be set to

$$f = \frac{1}{10\ \mu\text{s}} = 100\ \text{kHz}$$

to produce this ideally just energized inductor waveform each half cycle.

Figure 16-26(b) shows a sketch of the voltage and current responses of this *RL* circuit. The inductor at the end of a half cycle is fully energized, acts like an ideal short, drops 0 V, leaving all of the applied voltage to drop across the resistor (either $+2$ V or -2 V). Notice that the $v_R(t)$ curve is exponentially rising toward $+2$ V and then alternately falling toward -2 V.

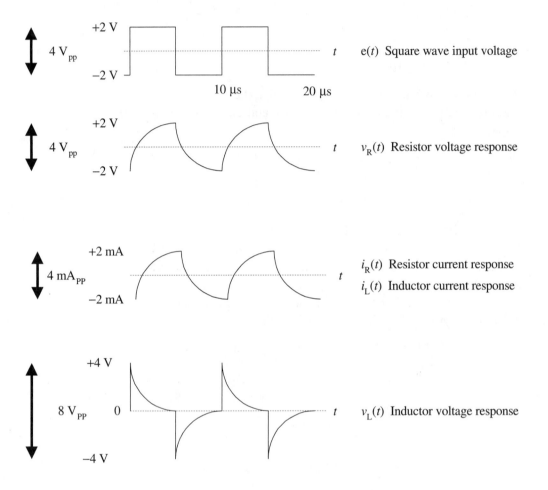

Figure 16-26(b) *RL* circuit square wave response waveforms

Use Ohm's law to find the resistor current waveform maximum and minimum values.

$$i_R(t) = \frac{v_R(t)}{R} = \frac{v_R(t)}{1\text{ k}\Omega}$$

$$I_{\max R} = \frac{2\text{ V}_p}{1\text{ k}\Omega} = 2\text{ mA}_p$$

$$I_{\min R} = -\frac{2\,V_{\min}}{1\ k\Omega} = -2\ mA_{\min}$$

Because of Ohm's law for resistance, the shape of the resistor voltage and current waveforms must be identical since they are proportional (related by a constant).

Since this is a series circuit, the inductor current must be the same as the resistor current. Therefore, it has the same shape and same values.

$$i_L(t) = i_R(t)$$

The $v_L(t)$ waveform can be found by Kirchhoff's voltage law

$$v_L(t) = e(t) - v_R(t)$$

Perform a point-by-point subtraction and the resistor voltage curve is generated. For example, at $t = 0^+$, just after the supply has changed to +2 V and the inductor is still at –2 V,

$$v_L(0^+) = e(0^+) - v_R(0^+) = 2\ V - (-2\ V) = 4\ V$$

Practice

The supply in the example problem has a 2 V_{dc} offset added to the signal. Sketch the output inductor and resistor voltage and current waveforms.

Answer: See Figure 16-26(c). Note that the resistor voltage, resistor current and inductor current are also offset. However, the inductor voltage depends only upon its *change* Δe_{in} (that is, its peak-to-peak value).

RL Square Wave Frequency Effects

As noted in the last example, if $T = 10\tau$ then the frequency of the square wave allows just enough time during each half cycle to just energize the inductor to its steady state current. Figure 16-27 demonstrates the principle of the signal's period relative to the circuit time constant.

If $T \gg 10\tau$, that is, the frequency is relatively low, the inductor energizes to its steady state maximum current and holds that current until the

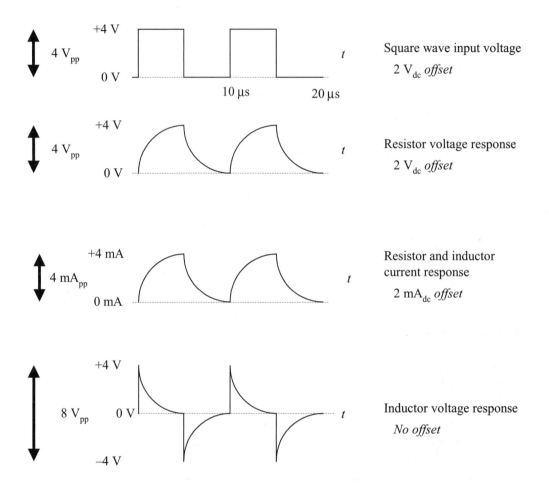

Figure 16-26(c) Practice solution with 2 V_{dc} offset in supplied voltage

next half cycle drives the inductor exponentially to its minimum steady state current.

If $T \ll 10\tau$, the inductor does not have time to fully energize and only partially energizes each half cycle. If the period is too short, that is the frequency is too high, the inductor energizes only slightly towards its maximum current and slightly towards its minimum current, and appears to be a linear triangle wave.

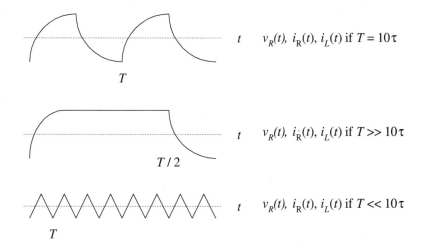

t $v_R(t)$, $i_R(t)$, $i_L(t)$ if $T = 10\tau$

t $v_R(t)$, $i_R(t)$, $i_L(t)$ if $T \gg 10\tau$

t $v_R(t)$, $i_R(t)$, $i_L(t)$ if $T \ll 10\tau$

Figure 16-27 Relative comparison of signal period and 10τ

Measuring τ with an Oscilloscope

To measure the time constant with the scope, the transient portion of an exponential waveform response is used. The procedure to measure the time constant τ is the same for *RL* circuits and for *RC* circuits. Figure 16-28 reviews this technique. First select a convenient transient portion of the exponential curve to measure. The time constant is the same for all of the transient portions of the curve, therefore, whichever is the most convenient and accurate to measure, choose it.

1. Select the transient exponential curve to use

2. Set up the time scale such that the signal fills up as much of the display as possible to obtain the most accurate measurement

3. Measure the number of divisions between the steady state and the initial values (shown as Δ in Figure 16-28)

4. Calculate 63% of Δ

5. Start at the initial value point and go up (or down) 63% of Δ

6. Go across the screen and count the number of divisions until the transient curve is intersected

7. Using the scope's time scale setting, translate the number of horizontal divisions into time. That is the time constant τ.

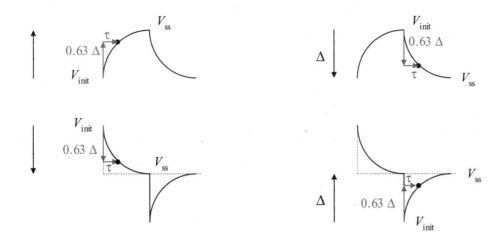

Figure 16-28 Measuring the time constant on an oscilloscope

16.7 *RLC* Effects and Compensation

Compensating Diodes

As demonstrated in the last section, energy storage devices can create special situations such as the inductive reverse voltage spike. Another example is that of the circuit in Figure 16-29. This is a MultiSIM simulation of a digitally controlled electrical circuit. Digitally switching Q1 with very low current controls the k1 relay, which controls a high current electrical circuit. Note, the spacebar is used to switch J1 so that you may coordinate the switching action of J1 with the lighting of lamp X1.

J1   Digitally switch between low (0 V) and high (5 V)

U1A   Digital NOT gate that requires **pull-up resistor** R1
        R1 is connected to the output BJT collector of U1A
        and completes the output circuit of U1A

Q1   Medium current BJT switch (on-off device)

K1   Relay with normally closed contacts
        Normally closed contacts 2 and 4

D1   Compensating diode for K1's inductor

X1   120 V lamp requires 120 V_{rms} to operate

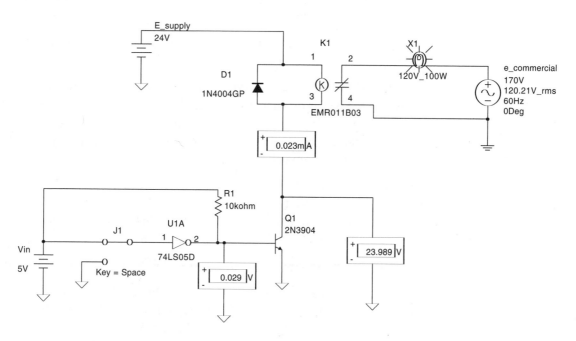

Figure 16-29 Digitally controlled electrical circuit

Example 16-19

Construct the simulation of Figure 16-29 and verify the circuit's operation for the switch J1 connecting a high (5 V) to the NOT gate. Also, walk through the logic of the circuit to confirm its logic. Connect a current meter to measure Q1 I_C and voltmeters to measure V_B and V_C to determine if the BJT is on (short) or off (open) during this state.

Solution

Let's follow the action of the circuit in its present state.

- J1 connects high (5 V) to the input of U1A

- U1A NOTs the high (5 V) into a low (0 V) output through its pull-up resistor

- Q1 base is low, therefore the BJT is off (open)

- Q1 is open, therefore no current flows through K1

- K1 relay is de-energized and the normally closed contacts are closed

- K1 contacts are closed completing the circuit to light the lamp

For Q1, I_C is 23 μA (virtually 0 current), V_B is 29 mV indicating the BJT is off (not dropping 0.7 V from base to emitter), and V_C is approximately 24 V indicating that the BJT is acting as an open and dropping all of the 24 V supply across it. This indicates that relay current is 0 mA and not energized.

Note: if faster switching of the compensating diode D1 is needed, it could be replaced with a Schottky diode. But select a Schottky diode with a sufficient maximum reverse voltage rating for the application.

Practice

Press the spacebar and note the action of the relay contacts and the lamp. Again, verify the action of the J1 switch and walk through the logic of the entire circuit. What are V_B and V_C of Q1 when the lamp is off? What is the purpose of D1? If you disconnect D1 from the circuit, is the simulation changed?

Answer: With J1 connected to common the relay contacts open and the lamp turns off. I_C = 48 mA, V_B = 0.78 V_{dc}, and V_C = 0.24 V_{dc}, indicating that the BJT is turned on and acting as a short. This completes the circuit for the relay to energize with a 48 mA current and open its contacts. The purpose of D1 is to provide a path for the de-energizing current when the inductor *em* field collapses. The simulator does not discern the difference with D1 in or out of the circuit. Warning: simulators are not real circuits and do not simulate all conditions.

Figure 16-30
Compensating Schottky diode used with switching MOSFET

High current switching MOSFETs are used with inductive devices like lamps, relays, motors, and switching power supplies. A compensating diode can be used in conjunction with a high current switching MOSFET by placing a compensating diode across the MOSFET as shown in Figure 16-30 to prevent inductive reverse voltage spike effects. This, in effect, is placing the compensating diode across the switch instead of across the inductive load. The polarity shown is normal operation. When the inductive reverse spike would occur, the diode becomes forward biased and provides a low resistive de-energizing path.

RLC Effects of Components and Circuits

All components and circuit configurations have inherent *RLC* effects. For example, wire has resistance but it also has inductance. An inductor has conductor windings separated by an insulator. That is the definition of capacitance. Figure 16-31(a) demonstrates this stray capacitance effect between inductor windings.

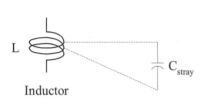

(a) Inductor stray capacitance affects

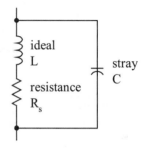

(b) Inductor model with *RLC* effects

Figure 16-31(a,b) *RLC* effects inherent in an inductor

A more realistic model of an inductor is shown in Figure 16-31(b). If inductor internal series resistance R_S is very small compared to its external circuit resistance, then it is negligible and can be removed and replaced by a short in the model. If the circuit frequency is sufficiently low so that X_C is sufficiently high as to be nonconsequential, then the stray capacitance C_{stray} can be modeled as an open and ignored. However, at very high frequencies, X_C exhibits very low reactance that approaches a short around the inductance L.

Even a simple resistor can exhibit serious frequency effects at very high frequencies. Figure 16-31(c) is a schematic model of a resistor including its high frequency effects. The resistor leads become reactive opens at very high frequencies and the parallel capacitance becomes a reactive short at very high frequencies.

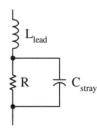

Figure 16-31(c) Resistor frequency model

Similarly, a capacitor has *RLC* elements as shown in Figure 16-31(d). Its leads are also inductive and it has a very large leakage resistance as described previously in this textbook. There are also stray and leakage capacitance that could be included in the model but Figure 16-31(d) includes the major elements.

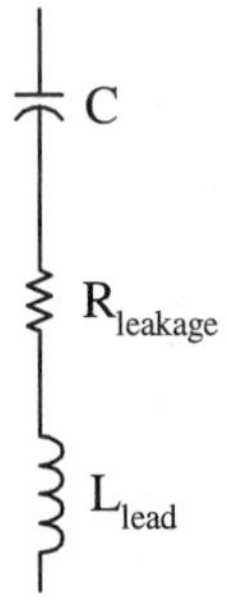

C

$R_{leakage}$

L_{lead}

Figure 16-31(d)
Capacitor model

Compensating Capacitors

Another problem area is providing instantaneous supply current for active devices such as op amps in a circuit. Although the lead length may appear relatively short, electronically it is not, especially at high frequencies or for dc transient circuits. The wire and PCB traces are conductors and resist change in current. For example, suppose there is a sudden voltage change that an op amp and its dc supply system must respond to immediately. The dc supply is a relatively long way away from the op amp. To compensate for this inductive effect compensating capacitors can be placed near the op amp dc supply pins to provide immediate charge as shown in Figure 16-32. Also, capacitors with capacitances on the order of 10 μF or greater can be placed at the beginning of a long dc supply trace to create a pool of charge that can feed the individual circuits. The more circuits that require an immediate charge pool, the higher this trace support capacitance should be.

Another way to think of these compensating capacitors is to view them as ac signal shorts to common. This prevents ac signals from using the dc supply system as an unwanted positive feedback path creating unwanted frequencies in the system.

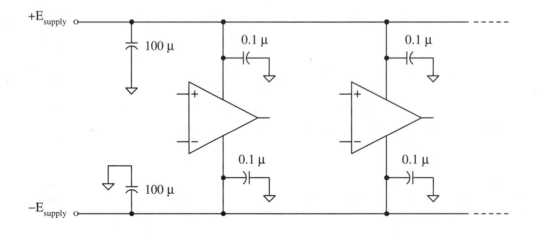

Figure 16-32 Compensating capacitors

Summary

Inductance is created by current, which creates an electromagnetic (*em*) field surrounding the current (flowing charge). *Inductors* are components that take advantage of this effect of storing *em* energy, which can be returned to the circuit. Coiling the conductor (wire) enhances the inductive affect and the ability to store more energy. Inserting a ferromagnetic material inside the inductor windings also significantly increases its inductance. Ferromagnetic material such as soft iron is easily magnetized and forms an electromagnet. The electromagnetic effect only exists when current exists in the inductor. If the current is turned off to the inductor, the *em* field collapses and returns its energy to the circuit.

The practical calculation of inductance based upon geometry is relatively complex, but it can be readily discerned from its instantaneous current and voltage relationship.

$$v = L\frac{di}{dt}$$

where L is the symbol for inductance measured in units of henries (H).

A few key characteristics that describe inductance are that inductance

- stores energy in its *em* field proportional to its current

- does not retain its *em* energy field if current is removed

- returns energy to a circuit like a current supply I

- resists change in current

An inductor exhibits opposition to ac signals. This reaction to ac signals is called its *reactance*. Specifically, it is called *inductive reactance* for inductance.

$$X_L = 2\pi f L$$

where

X_L	inductive reactance with units of ohms (Ω)
f	frequency of the ac sinusoidal signal with units of hertz (Hz)
L	inductance with units of henries (H)

Inductive reactance is in units of ohms like resistance, but the reactance magnitude cannot be directly added to resistance magnitude. An inductor acts as a short to dc or low-frequency sinusoids and as an open to high-frequency sinusoids. Thus, dc and low-frequency signals are *passed* by an inductor, while high-frequency sinusoids are *blocked* by an inductor.

Inductive reactance in series adds like resistance in series. The total inductive reactance of three inductors in series is

$$X_{L\,total} = X_{L1} + X_{L2} + X_{L3}$$

Substituting the reactance for each inductor

$$X_L = 2\pi f L$$

leads to the formula for the equivalent total inductance of three series inductances.

$$L_{total} = L1 + L2 + L3$$

Inductive reactance in parallel combines like resistance in parallel. The total reactance of three parallel inductances is

$$\frac{1}{X_{L\,total}} = \frac{1}{X_{L1}} + \frac{1}{X_{L2}} + \frac{1}{X_{L3}}$$

Substituting individual reactances, the total inductance of three parallel inductances is

$$L_{total} = \left(\frac{1}{L1} + \frac{1}{L2} + \frac{1}{L3}\right)^{-1}$$

The analysis of simple *RL* dc transient circuits is very similar to that of *RC* circuits. A simple *RL* combination is where the circuit can be reduced to a single equivalent Thévenin resistance R_{Th} and a single equivalent *L*. The time required for the transient depends upon the circuit's time constant. The *RL* circuit's time constant τ is defined to be

$$\tau = \frac{L}{R_{Th} + R_{wire}}$$

where R_{wire} is the internal inductor wire resistance.

The dc transient expression forms are the same for the inductor *RL* circuits as they are for the capacitor *RC* circuits

$$i_L(t) = I_{ss} + \left(I_{init} - I_{ss}\right)e^{-\frac{t}{\tau}}$$

$$v_L(t) = V_{ss} + \left(V_{init} - V_{ss}\right)e^{-\frac{t}{\tau}}$$

where the ss subscript represents the *steady state* value and the init subscript represents the *initial* value. The transient exponential decay term dies out in approximately five time constants (5τ). The only difference in analyzing *RL* dc transient circuits compared to *RC* transient circuits is the modeling of the inductor for initial and steady state values and the determination of the time constant.

- Initially de-energized inductor is modeled as an open
- Initially energized inductor is modeled as a current supply determined by its previous state before the beginning of the transient
- Steady state ideal inductor is modeled as a conductor or a short
- Steady state inductor is modeled as resistance R_S if meaningful

DC transient analysis applies to both switched *RL* circuits and to *RL* circuits subjected to rectangular waveform signals.

Depending upon the circuit configuration, a compensating diode may be needed to prevent an inductor's collapsing *em* field from creating huge voltage spikes in circuit. The compensating diode bypasses the inductor's current generated by a collapsing field around the inductor or switching components in the circuit (mechanical switches, BJT switch, MOSFET switch, and so forth).

Every component and circuit contains hidden *RLC* effects. These hidden effects may or may not be significant depending on circuit conditions and frequencies of operation.

Compensating capacitors are required in circuits, subcircuits, and components that need immediate charge fed from the dc power supply. Compensating capacitors also short out unwanted ac signals and oscillations to common and prevent the dc power supply from being used as a positive feedback network for unwanted ac signals.

Problems

Inductors

16-1   The following currents are applied to a 1 μH inductor. Find the corresponding inductor voltages.

    a. 10 A_{dc}
    b. A 10 A_p sine wave with a 100 Hz signal
    c. A 10 A_p triangle wave with a 100 Hz signal

16-2   The following currents are applied to a 1 nH inductor. Find the corresponding inductor voltages.

a. $6\,A_{dc}$

b. A $6\,A_p$ sine wave with a 10 kHz signal

c. A $6\,A_p$ triangle wave with a 10 kHz signal

Inductive Reactance

16-3   Given a 22 mH inductor:

   a. Find its reactance at signal frequencies of 0 Hz, 10 Hz, 100 Hz, 1 kHz, 10 kHz, and 100 kHz.

   b. Based upon your results would you model an inductor with a 0 Hz frequency as an open or a short?

   c. Based upon your results would you model an inductor operating at very high frequencies as an open or a short?

16-4   Repeat Problem 16-3 for a 0.1 mH inductor.

16-5   For the circuit of Figure 16-33, as frequency *increases*, what happens to the inductor reactance, the induced voltage across the inductor, the voltage across the resistor, the current through the resistor, and the current through the inductor?

16-6   For the circuit of Figure 16-33, as frequency *decreases*, what happens to the inductor reactance, the induced voltage across the inductor, the voltage across the resistor, the current through the resistor, and the current through the inductor?

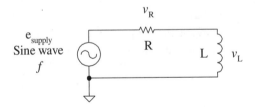

Figure 16-33

16-7   Use the circuit of Figure 16-34(a).

   a. If the frequency *f* is 1 Hz, find the reactance of the inductor, model the inductor appropriately for this circuit as an open or a short, and find v_{out}.

b. If the frequency f is 10 MHz, find the reactance of the inductor, model the inductor appropriately for this circuit as an open or a short, and find v_{out}.

16-8 Repeat Problem 16-7 for the circuit of Figure 16-34(b).

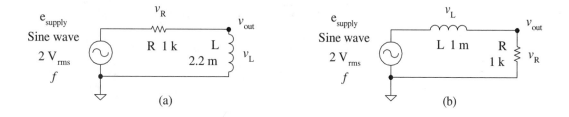

Figure 16-34

Series and Parallel Inductance

16-9 Three 3 mH inductors are in series.

   a. Find the total series inductance.

   b. Find the series circuit total inductive reactance at a frequency of 1 kHz.

16-10 Three 3 mH inductors are in parallel.

   a. Find the total parallel inductance.

   b. Find the parallel circuit total inductive reactance at a frequency of 1 kHz.

Switched *RL* Transient

16-11 For the circuit of Figure 16-35, the inductor is initially not energized.

   a. Find the circuit time constant τ.

   b. Find the initial inductor voltage and current.

   c. Find the steady state inductor voltage and current.

   d. Find the universal expression for inductor voltage and current (with values and simplified).

   e. Find the inductor voltage and current after two time constants.

    f. Sketch the inductor voltage and current waveforms for five time constants.

16-12 Repeat Problem 16-11 for the resistor's voltage and current.

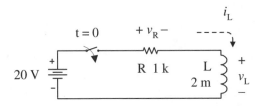

Figure 16-35

16-13 For the circuit of Figure 16-35, assume that the inductor has an initial current of 2 mA.

    a. Find the circuit time constant τ.

    b. Find the initial inductor voltage and current.

    c. Find the steady state inductor voltage and current.

    d. Find the universal expression for inductor voltage and current (with values and simplified).

    e. Find the inductor voltage and current after 3 μs.

    f. Sketch the inductor voltage and current waveforms for five time constants.

16-14 Repeat Problem 16-13 for the resistor's voltage and current.

RL Square Wave Response

16-15 Use the circuit of Figure 16-36(a).

    a. Find the time constant τ for this circuit.

    b. Prove that the period T is long enough for the inductor to fully energize each half cycle.

    c. Sketch the waveform of the resistor voltage.

    d. Sketch the waveform of the resistor current (the inductor current).

e. Sketch the waveform of the inductor voltage.

f. How long does it take for the current through the inductor to reach 1 mA, assuming that the supply is starting a positive half cycle? Hint: Start with the general universal expression with its initial value, steady state value, and time constant τ.

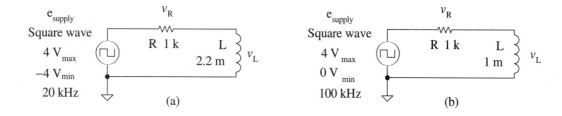

Figure 16-36

16-16 Use the circuit of Figure 16-36(b).

a. Find the time constant τ for this circuit.

b. Prove that the period T is long enough for the inductor to fully energize each half cycle.

c. Sketch the waveform of the resistor voltage.

d. Sketch the waveform of the resistor current (the inductor current).

e. Sketch the waveform of the inductor voltage.

f. How long does it take for the current through the inductor to reach 2 mA, assuming that the supply is starting a positive half cycle? Hint: Start with the general universal expression with its initial value, steady state value, and time constant τ.

17

Transformers

Introduction

A transformer is a device consisting of coiled wire windings (inductors) that couples an ac voltage from a supplied winding (the primary) to another winding or windings (the secondary). The coupled voltage can be increased or decreased in magnitude by increasing or decreasing the relative number of loops in the windings.

The ideal transformer with its primary connected to its supply circuit can be modeled by looking into the secondary. The Thévenin model looking into the secondary is a Thévenin voltage supply in series with a Thévenin resistance that represents the entire transformer and supply circuit combination. That Thévenin model can then be used to determine load voltages and load currents for any attached load.

Objectives

Upon completion of this chapter you will be able to:

- Describe the basic elements of a transformer.

- Identify the schematic symbols of air core and iron core transformers.

- Identify the primary and secondary winds of a transformer in a circuit.

- Analyze the primary and secondary voltages, currents, average powers, and ac referred resistances in an ideal transformer circuit with and without a supply resistance.

- Find the transformer referred ac resistance of an open and a short.

- Produce and apply the Thévenin model of a transformer circuit from the viewpoint of the secondary.

17.1 Ideal Transformer

A transformer is a device that raises or lowers an ac voltage through the use of coiled wires (or inductors). The wire loops of each inductor must be insulated from each other. The schematic symbol for a simple transformer with two windings is shown in Figure 17-1. The coiled wires or **windings** are essentially individual inductors. Like the inductor, a transformer can have an air core, or it can have an iron core to concentrate its electromagnetic flux lines. The vertical lines in Figure 17-1(b) indicate the use of an iron core, while the absence of these lines in Figure 17-1(a) indicates an air core transformer.

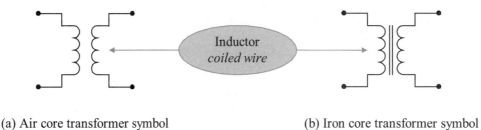

(a) Air core transformer symbol   (b) Iron core transformer symbol

Figure 17-1 Transformer schematic symbols

An ac supply signal is fed to one of the windings, called the **primary winding** as shown in Figure 17-2. An ac signal is electromagnetically coupled from the primary winding to its secondary winding or windings. A varying primary current creates a varying electromagnetic field that cuts the coils of a secondary winding. This in turn creates an induced voltage on the secondary winding. An **ideal transformer** couples 100% of its energy and power to the secondary; that is, there is no loss of energy or power in the secondary signal. Each turn on the winding of the primary induces the same voltage on a turn in the secondary winding. $N_1:N_2$ of Figure 17-2 is called the transformer turns ratio. For example, a ratio of one to five (1:5) indicates that the secondary has five times as many turns as the primary. For this case, since the secondary has five times as many turns as the primary, the secondary develops five times as much voltage. This is called a **step-up voltage transformer** since the secondary has more turns than the primary and produces a greater secondary voltage. If the secondary has fewer turns, then the transformer is a **step-down volt-**

age transformer. In this configuration, there is no direct physical contact of the primary with the secondary; therefore, the transformer **isolates** the attached load from the supply in terms of a direct connection to the supply.

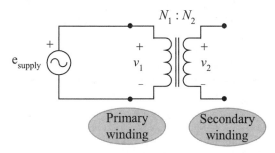

Figure 17-2 Transformer primary and secondary windings

For the circuit of Figure 17-2,

$$v_2 = \frac{N_2}{N_1} v_1 \qquad \text{(17-1)}$$

where
v_2	secondary voltage
N_2	number of secondary turns
N_1	number of primary turns
v_1	primary voltage

If $N_2 > N_1$, then the transformer is a *step-up voltage transformer*. If $N_2 < N_1$, then the transformer is a *step-down voltage transformer*.

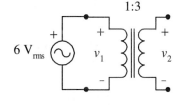

Figure 17-3
Example circuit

Example 17-1

For the circuit of Figure 17-3, what is the transformer turns ratio? Is this a step-up or step-down voltage transformer? Find the voltage induced on the secondary.

Solution

The turns ratio is one to three (1:3). This is a step-up voltage transformer. Apply Equation 17-1 to find the induced secondary voltage.

$$v_2 = \frac{N_2}{N_1} v_1 = \left(\frac{3}{1}\right) 6 \, V_{rms} = 18 \, V_{rms}$$

Practice

The transformer in the circuit of Figure 17-3 is replaced with a transformer with a turns ratio of two to one (2:1). Is this a step-up or step-down transformer? Find the transformer secondary voltage.

Answer: Step-down transformer, 3 V_{rms}

Transformer RMS Secondary Rating

A transformer to be connected to the electrical energy system of 120 V_{rms} is typically rated in terms of its secondary rms voltage. It is assumed that the primary voltage is 120 V_{rms}. For example, a transformer rated at 12.6 V_{rms} at 1 A_{rms} indicates that the secondary voltage of the transformer is 12.6 V_{rms} when the secondary is loaded at 1 A_{rms}. If the loaded transformer current is less than or greater than 1 A_{rms}, the actual secondary voltage may be different from the rated value. For example, if this transformer is lightly loaded with only 10 mA_{rms} of current, the output voltage might be 15 V_{rms}. Do not assume the secondary voltage is the rated output voltage with an underloaded transformer; check the specifications and test the transformer under normal load conditions.

Example 17-2

Find the load voltage, load current, and load power for the circuit of Figure 17-4.

Solution

Assuming the transformer primary is connected to 120 V_{rms}, the secondary voltage is 12.6 V_{rms} if loaded at 1 A_{rms}. Assuming the rated loading of the transformer, the secondary voltage is assumed to be 12.6 V_{rms}. This secondary voltage is delivered to the load resistance and thus the load voltage is 12.6 V_{rms} under rated load conditions.

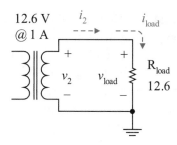

Figure 17-4 Example circuit

$$v_2 = v_{load} = 12.6 \text{ V}_{rms} \qquad \text{(rated loading)}$$

The secondary circuit is a series circuit with the transformer secondary being the voltage supply and R_{load} being the secondary load. From Ohm's law, the load current and thus the secondary current is

$$i_2 = i_{load} = \frac{12.6 \text{ V}_{rms}}{12.6 \text{ }\Omega} = 1 \text{ A}_{rms}$$

The transformer is at rated load conditions and therefore the secondary voltage should be 12.6 V$_{rms}$ and the secondary current is 1 A$_{rms}$. *Average power is found using rms voltage and current values.* The power dissipated by the load resistance and thus the power delivered by the transformer is

$$P_{average \text{ } load} = I_{rms}V_{rms} = 1 \text{ A}_{rms} \times 12.6 \text{ V}_{rms} = 12.6 \text{ W}$$

Practice

The load resistance in the circuit of Figure 17-4 is changed to a 1 kΩ resistance. If lightly loaded, the transformer secondary voltage delivers 15 V$_{rms}$ to the secondary. Find the load voltage, load current, and average load power.

Answer: The secondary is lightly loaded, 15 V$_{rms}$, 15 mA$_{rms}$, 225 mW

Ideal Transformer Referred AC Resistance

The **ac referred primary resistance** of a transformer is the ac resistance seen looking into the primary winding of a circuit. Referred resistance is also known as *reflected resistance*. The ideal transformer with a turns ratio of N_1 to N_2 in the circuit of Figure 17-5 has an attached load resistance R_{load}. The turns ratio and the secondary load resistance combine to produce a referred ac resistance of the primary winding r_1.

The ideal dc resistance of the primary winding is 0 Ω (a short); therefore, a dc voltage supply must not be directly applied to the transformer.

An ideal transformer is 100% efficient. No power is lost in the coupling of the signal to the secondary winding. The power of the secondary signal is equal to the power of the primary signal. The next example utilizes this characteristic of an ideal transformer (delivers 100% of the input signal power to the secondary) to find circuit currents, average power, and referred resistance.

Ideal transformer
$P_{primary} = P_{secondary}$

Example 17-3

The ideal transformer in the circuit of Figure 17-5(a) has a 10:1 turns ratio. Find the secondary voltage v_2, load voltage v_{load}, load current i_{load}, average load power, average primary power, the primary current i_1, and the ac referred primary resistance r_1. Sketch one cycle of the load voltage and the load current.

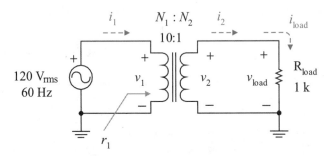

Figure 17-5(a) Example circuit

Solution

The primary circuit is a series circuit with a 120 V_{rms}, 60 Hz sinusoidal supply. Therefore, the primary voltage is

$$v_1 = e_{supply} = 120 \ V_{rms}$$

The transformer is a 10:1 *step-down transformer*, therefore the secondary voltage is

$$v_2 = \left(\frac{1}{10}\right) 120 \ V_{rms} = 12 \ V_{rms}$$

The secondary is a series circuit and all 12 V_{rms} is delivered to the load.

$$v_{load} = v_2 = 12 \ V_{rms}$$

The secondary series circuit current is found by applying Ohm's law to the load resistance.

$$i_2 = i_{load} = \frac{12 \ V_{rms}}{1 \ k\Omega} = 12 \ mA_{rms}$$

The average load power delivered by the secondary is found by finding the power dissipated by the secondary load. Use *rms* voltage and current to find *average* power.

$$P_{2 \ average} = P_{load \ average} = 12 \ mA_{rms} \times 12 \ V_{rms} = 144 \ mW$$

Since this is an ideal transformer, the primary average power equals the secondary average power.

$$P_{1 \ average} = P_{2 \ average} = 144 \ mW$$

Apply the power rule for resistance to the primary winding to find the primary current.

$$i_1 = \frac{P_1}{v_1} = \frac{144 \ mW}{120 \ V_{rms}} = 1.2 \ mA_{rms}$$

Apply Ohm's law to the primary winding to find the primary winding referred ac resistance.

$R_1 = 0 \ \Omega$

$r_1 = 100 \ k\Omega$

$$r_1 = \frac{v_1}{i_1} = \frac{120 \ V_{rms}}{1.2 \ mA_{rms}} = 100 \ k\Omega$$

The primary winding *dc resistance* is 0 Ω, but the primary winding *ac referred resistance* is a significant 100 kΩ.

The rms load voltage and current must be converted to peak values before sketching their waveforms.

$$v_{\text{p load}} = \sqrt{2} \times 12 \ V_{\text{rms}} = 17 \ V_{\text{p}}$$

$$i_{\text{p load}} = \sqrt{2} \times 12 \ mA_{\text{rms}} = 17 \ mA_{\text{p}}$$

The frequency and period of the voltages and currents in the circuit are the same as the supplied frequency.

$$f_{\text{load}} = f_2 = f_1 = f_{\text{supply}} = 60 \ \text{Hz}$$

$$T = \frac{1}{f} = \frac{1}{60 \ \text{Hz}} = 16.7 \ \text{ms}$$

Figures 17-5(b) and (c) show sketches of the load voltage and load current, respectively.

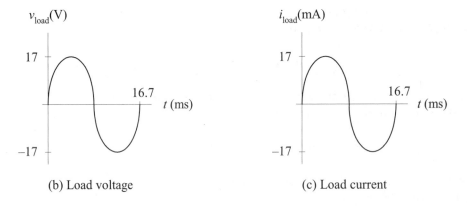

(b) Load voltage   (c) Load current

Figure 17-5(b,c) Load voltage and current waveforms

Practice

The ideal transformer in the circuit of Figure 17-5(a) is changed to a 1:2 turns ratio. Find the secondary voltage v_2, secondary current i_2, average load and primary winding power, the primary current i_1, and the ac referred primary resistance r_1.

Answer: 240 V_{rms}, 240 mA_{rms}, 57.6 W, 480 mA_{rms}, 250 Ω

In the previous example problem, notice that the *voltage is stepped down* by a factor of 10 but the *current is stepped up* by a factor of 10. This makes sense for an ideal transformer since the power of the primary must be equal to the power of the secondary.

$$P_1 = P_2$$

$$i_1 v_1 = i_2 v_2$$

Rearranging the terms and solving for the secondary current i_2,

$$\frac{i_2}{i_1} = \frac{v_1}{v_2} = \frac{v_1}{\left(\dfrac{N_2}{N_1}\right)v_1} = \left(\frac{N_1}{N_2}\right)$$

$$i_2 = \left(\frac{N_1}{N_2}\right)i_1 \qquad\qquad\qquad (17\text{-}2)$$

The secondary current has the reciprocal turns ratio relationship of the voltage. If the voltage steps up, the current must step down. If the voltage steps down, the current must step up.

Let's develop an expression for the referred primary winding ac resistance based upon the transformer turns ratio and the secondary load resistance. Based upon the above example and practice problems, both of these values are factors in determining this ac referred resistance. Use the current, voltage, and resistance designations of the circuit of Figure 17-6 to find the relationship between the ac primary referred resistance r_1 and the turns ratio and the secondary load resistance.

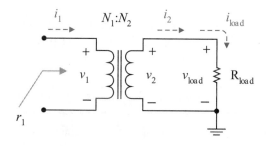

Figure 17-6 Ideal transformer circuit

$$P_1 = P_2 \qquad\qquad\qquad \text{(ideal transformer)}$$

$$i_1 v_1 = (i_2)^2 R_{load} \qquad\qquad\qquad \text{(power rule)}$$

$$i_1 v_1 = \left(\frac{N_1}{N_2}\right)^2 (i_1)^2 R_{load} \qquad\qquad \text{(substitute Equation 17-2)}$$

$$\frac{v_1}{i_1} = \left(\frac{N_1}{N_2}\right)^2 R_{load} \qquad\qquad \text{(divide by } i_1{}^2)$$

Applying Ohm's law, the referred ac primary resistance r_1 is

$$r_1 = \left(\frac{N_1}{N_2}\right)^2 R_{load} \qquad\qquad\qquad \textbf{(17-3)}$$

Example 17-4

    For the circuit of Figure 17-5, find the primary ac referred resistance r_1 directly.

Solution

    Apply Equation 17-3.

$$r_1 = \left(\frac{10}{1}\right)^2 1 \text{ k}\Omega = 100 \text{ k}\Omega$$

This result agrees with the previous example's result.

Practice

    For the circuit of Figure 17-5, find the primary ac referred resistance r_1 directly if the ideal transformer turns ratio is changed to one to two (1:2).

Answer: 250 Ω (agrees with the last practice problem)

In the next example the primary ac referred resistance must be found first so that the primary voltage drop can be determined.

Example 17-5

For the circuit of Figure 17-7, find the primary resistance r_1, the primary load current i_1, the primary voltage v_1, the secondary load voltage v_{load}, and the secondary load current i_{load}.

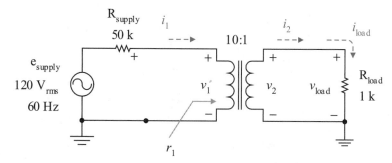

Figure 17-7 Example problem circuit

Solution

In this case, the primary ac referred resistance must be found first to determine the primary voltage drop. Use Equation 17-3.

$$r_1 = \left(\frac{10}{1}\right)^2 1 \text{ k}\Omega = 100 \text{ k}\Omega$$

$$i_1 = \frac{e_{supply}}{R_{supply} + r_1} = \frac{120 \text{ V}_{rms}}{50 \text{ k}\Omega + 100 \text{ k}\Omega} = 800 \text{ }\mu\text{A}_{rms}$$

$$v_1 = r_1 i_1 = 100 \text{ k}\Omega \times 800 \text{ }\mu\text{A}_{rms} = 80 \text{ V}_{rms}$$

Notice that 40 V_{rms} is lost across the supply resistance and that only 80 V_{rms} is applied to the primary. That primary voltage is then stepped down by a factor of 10.

$$v_{load} = v_2 = \left(\frac{1}{10}\right) 80 \text{ V}_{rms} = 8 \text{ V}_{rms}$$

$$i_{load} = i_2 = \frac{8 \text{ V}_{rms}}{1 \text{ k}\Omega} = 8 \text{ mA}_{rms}$$

Practice

For the circuit of Figure 17-8, find the primary resistance r_1, the primary load current i_1, the primary voltage v_1, the secondary load voltage v_{load}, and the secondary load current i_{load}.

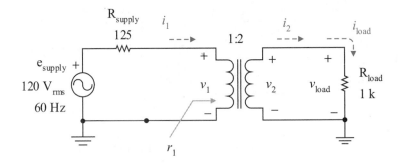

Figure 17-8 Practice circuit

Answer: 250 Ω, 320 mA$_{rms}$, 80 V$_{rms}$, 160 V$_{rms}$, 160 mA$_{rms}$

Electrical Energy Distribution

An electrical system transmits voltages on the order of 100 kV over high voltage transmission lines. Figure 17-9 is a simplified view of an electrical distribution system from a high voltage supply to a house (local drop). This distribution system has been greatly simplified to focus on the use of the transformer. At a main electrical distribution station, 138 kV$_{rms}$ is stepped down to 12.5 kV$_{rms}$ and transmitted to local (neighborhood) distribution transformers. That 12.5 kV$_{rms}$ is stepped down to 120 V$_{rms}$ for local drops into houses. Note that the **neutral** center line is connected to earth common, creating two lines of 120 V$_{rms}$ relative to earth common. The voltage difference between these two outside lines forms the 240 V$_{rms}$ which is used for major electrical appliances like electric furnaces.

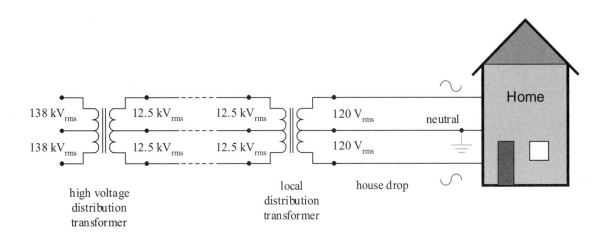

Figure 17-9 Simplified electrical distribution system to a home

Converters and Inverters

Other very common applications of transformers are simple **converters** used to

- Step commercial ac voltage (120 V_{rms}) down to low voltage ac (for example, door bell ringers)

- Step commercial ac voltage down to rectified, filtered dc voltage (for example, an inexpensive battery charger)

- Step commercial ac voltage down to a rectified, filtered, and regulated dc voltage (for example, convert ac voltage to dc voltage to power a laptop computer or charge a cell phone)

Such devices plug into an ac wall socket with a connecting cable coming out of them. They are sometimes referred to as wall warts since they are typically small units that plug into an ac wall socket. The output of a converter is a stepped down ac, a filtered dc, or a regulated dc voltage for a specific application. Figure 17-10 shows the schematic of a simple converter that consists of a transformer, a diode, a filter capacitor, and a regulator zener diode.

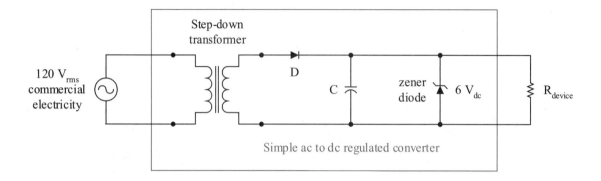

Figure 17-10 A 120 V_{rms} to 6 V_{dc} converter

A transformer can also be used to match resistance as in the next example. Note the term **inverter** is used for a device that converts a dc signal into an ac signal. These are popular in automobiles where the inverter transforms the electrical outlet dc voltage into an ac voltage to power laptop computers, portable televisions, and so forth. Laptops in turn have converters or internal dc power supplies to create a regulated dc output for their operation.

Example 17-6

There are electronic applications where matching resistances is important. Suppose you needed to match a 500 kΩ transducer to a 5 kΩ load. Find the turns ratio of the transformer in Figure 17-11(a) required to match the transducer's 500 kΩ resistance to a 5 kΩ load. Find the load voltage, current, and average power dissipation.

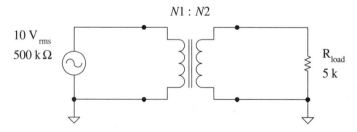

Figure 17-11(a) Example resistance matching circuit

Solution

The referred (reflected) resistance looking into the primary winding is

$$r_1 = \left(\frac{N_1}{N_2}\right)^2 R_{\text{load}}$$

Substitute the supply and load resistance values.

$$500 \text{ k}\Omega = \left(\frac{N_1}{N_2}\right)^2 5 \text{ k}\Omega$$

Solve for the turns ratio term.

$$\left(\frac{N_1}{N_2}\right)^2 = \frac{500 \text{ k}\Omega}{5 \text{ k}\Omega}$$

Simplify by taking the square root of both sides of the equation.

$$\frac{N_1}{N_2} = \sqrt{\frac{500 \text{ k}}{5 \text{ k}}} = 10$$

The turns ratio must be 10 to 1 for the supply resistance to match the referred resistance looking into the primary.

This is a step down transformer. The load voltage, which is the same as the transformer secondary voltage, is

$$V_{\text{load}} = \left(\frac{1}{10}\right) 10 \text{ V} = 1 \text{ V}_{\text{rms}}$$

Use Ohm's law to find the load current.

$$I_{\text{load}} = \frac{1 \text{ V}}{5 \text{ k}\Omega} = 200 \text{ } \mu\text{A}_{\text{rms}}$$

Use the *rms* voltage and/or current to find the average load voltage.

$$P_{\text{average}} = 1 \text{ V}_{\text{rms}} \times 200 \text{ } \mu\text{A}_{\text{rms}} = 200 \text{ } \mu\text{W}$$

multiSIM

Practice

Verify the load voltage, current, and average power with a Multi-SIM simulation.

Answer: See Figures 17-11(b) and Figure 17-11(c). Note the secondary ratio has been set up in MultiSIM to be 0.1 since the simulator treats the primary turns N_1 to be 1. Note: a wattmeter is a combined volt-ammeter.

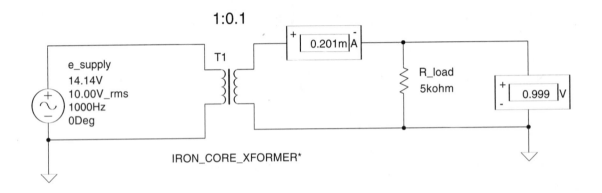

Figure 17-11(b) MultiSIM simulation current and voltage measurements

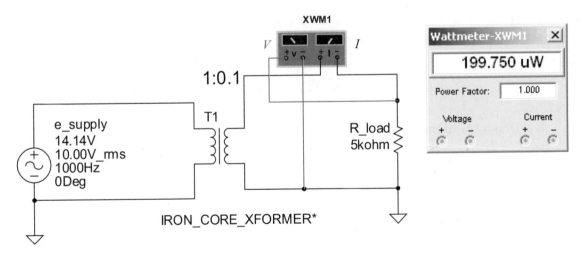

Figure 17-11(c) MultiSIM simulation power measurements

17.2 Transformer Thévenin Model

A transformer circuit can be modeled as a Thévenin model looking into the secondary. First let's examine the transformer referred ac resistance of an *open* and a *short*. Then a transformer circuit without supply resistance and a transformer circuit with a supply resistance are modeled as Thévenin models. The Thévenin models are then used to analyze the load voltage and current.

Referred Open

Figure 17-12 demonstrates looking into a transformer winding with the other winding being open.

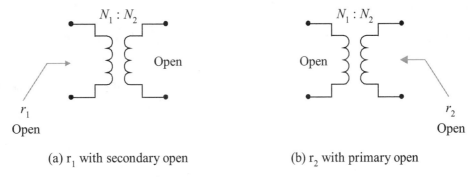

(a) r₁ with secondary open (b) r₂ with primary open

Figure 17-12 Referred ac resistance of an open

Figure 17-12(a) demonstrates looking into a primary winding with the secondary winding unloaded (open). Using the primary referred ac resistance formula of Equation 17-3,

$$r_1 = \left(\frac{N_1}{N_2}\right)^2 R_{load} = \left(\frac{N_1}{N_2}\right)^2 \infty \, \Omega = \infty \, \Omega \qquad \text{(open)}$$

A secondary open is referred back to the primary as an ac open. The primary is a dc short, but the primary referred ac resistance is an open. That is, *the primary winding is a dc short but an ac open.*

Figure 17-12(b) demonstrates looking into a secondary winding with the primary winding open. First, express the equivalent form of Equation 17-3 to find the secondary ac referred resistance.

$$r_2 = \left(\frac{N_2}{N_1}\right)^2 R_{\text{supply}}$$

Note the turns ratio factor. It is inverted since we are looking from the secondary towards the supply. A simple way to visualize this formula is to start by looking into the secondary and move to the left. The first term encountered is the N_2 term, then the N_1, term and then the primary's supply side resistance. Or, the referred resistance is the first turn encountered divided by the second turn encountered quantity squared times the resistance looking into the remainder of the corresponding circuit. Whether looking into the primary or the secondary, this visualization works to help create the appropriate referred ac resistance formula. Now substitute the resistance of an open ($\infty\ \Omega$) for the supply resistance. Again, an open is referred.

$$r_2 = \left(\frac{N_2}{N_1}\right)^2 R_{\text{supply}} = \left(\frac{N_2}{N_1}\right)^2 \infty\ \Omega = \infty\ \Omega \qquad \text{(open)}$$

The result is the same. Whether looking into the primary or into the secondary, *an open is referred as an open.*

Referred Short

Figure 17-13 demonstrates looking into a transformer winding with the other winding being shorted.

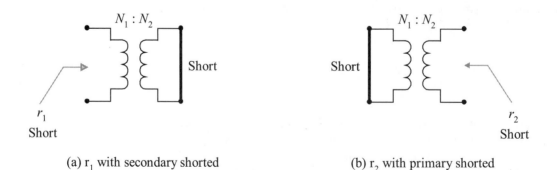

(a) r_1 with secondary shorted   (b) r_2 with primary shorted

Figure 17-13 Referred ac resistance of a short

Figure 17-13(a) demonstrates looking into a primary winding with the secondary winding shorted (0 Ω). Using the primary referred ac resistance formula

$$r_1 = \left(\frac{N_1}{N_2}\right)^2 R_{\text{load}} = \left(\frac{N_1}{N_2}\right)^2 0\,\Omega = 0\,\Omega \qquad \text{(short)}$$

A secondary short is referred back to the primary as an ac short.

Figure 17-13(b) demonstrates looking into a secondary winding with the primary winding without a supply resistance (short). Using the secondary referred ac resistance formula

$$r_2 = \left(\frac{N_2}{N_1}\right)^2 R_{\text{supply}} = \left(\frac{N_2}{N_1}\right)^2 0\,\Omega = 0\,\Omega \qquad \text{(short)}$$

This too looks like a short. The result is the same. Whether looking into the primary or into the secondary, *a short is referred as a short.*

Thévenin Model–Without Supply Resistance

The next example is a transformer circuit with an ideal voltage supply; that is, there is no supply resistance.

Example 17-7

Find the Thévenin model of the ideal transformer circuit of Figure 17-14(a). Then use the Thévenin model to find the load voltage and load current.

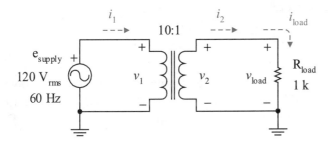

Figure 17-14(a) Example circuit

Solution

Remove the load and find the open circuit voltage v_{oc} (same as the Thévenin voltage e_{Th}) at the load terminals as illustrated in the circuit of Figure 17-14(b).

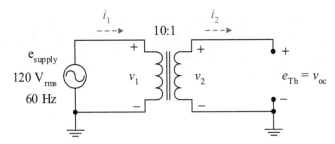

Figure 17-14(b) Find the Thévenin voltage e_{Th}

The primary circuit is a series circuit with a 120 V_{rms} sinusoidal supply driving the primary. Therefore, the primary voltage is

$$v_1 = e_{supply} = 120 \ V_{rms}$$

The transformer is a 10:1 *step-down transformer*; therefore, the secondary voltage is

$$v_2 = \left(\frac{1}{10}\right)120 \ V_{rms} = 12 \ V_{rms}$$

This is the same voltage that appears across the open load terminals; therefore, the open circuit (Thévenin) voltage is

$$e_{Th} = v_{oc} = v_2 = 12 \ V_{rms}$$

To find the Thévenin resistance, zero the voltage supply (replace it with a short) and look back into the secondary as demonstrated by the circuit of Figure 17-14(c).

Figure 17-14(c) Find the Thévenin resistance r_{Th}

$$r_{\text{Th}} = r_2 = \left(\frac{N_2}{N_1}\right)^2 R_{\text{supply}} = \left(\frac{N_2}{N_1}\right)^2 0\ \Omega = 0\ \Omega \qquad \textit{(short)}$$

The circuit of Figure 17-14(d) is the Thévenin model of the original transformer circuit with the load attached. Note that r_{Th} is modeled as a short and therefore does not appear in the Thévenin model. The load voltage and load current are easily determined from this simplified modeled circuit.

$$v_{\text{load}} = e_{\text{Th}} = 12\ V_{\text{rms}}$$

$$i_{\text{load}} = \frac{12\ V_{\text{rms}}}{1\ k\Omega} = 12\ mA_{\text{rms}}$$

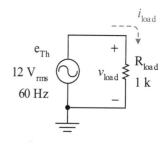

Figure 17-14(d)
Thévenin model with load

Practice

The transformer turns ratio of the circuit of Figure 17-14 is changed to one to two (1:2). Find the Thévenin voltage and Thévenin resistance. Use the Thévenin model to find the load voltage and load current.

Answer: Thévenin model (240V_{rms}, 0 Ω); 240 V_{rms}, 240 mA_{rms}

Thévenin Model–With Supply Resistance

The next example is a transformer circuit that has supply resistance.

Example 17-8

Find the Thévenin model of the ideal transformer circuit of Figure 17-15(a). Then use the Thévenin model to find the load voltage and load current.

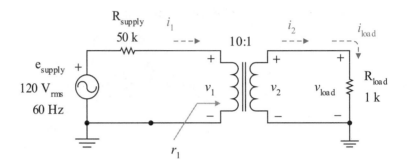

Figure 17-15(a) Example circuit

Solution

Remove the load and find the open circuit voltage v_{oc} at the load terminals as illustrated in the circuit of Figure 17-15(b).

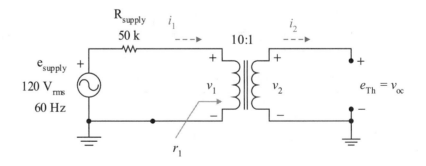

Figure 17-15(b) Find the Thévenin voltage e_{Th}

The primary circuit is a series circuit with a 120 V_{rms} sinusoidal supply, a 50 kΩ supply resistor, and the referred ac primary resistance. An *open* is referred to the primary; therefore, the supply voltage is dropped across the open primary.

$$v_1 = e_{supply} = 120 \text{ V}_{rms}$$

The transformer is a 10:1 *step-down transformer*; therefore, the secondary voltage is

$$v_2 = \left(\frac{1}{10}\right) 120 \text{ V}_{rms} = 12 \text{ V}_{rms}$$

This is the same voltage that appears across the open load terminals; therefore, the open circuit (Thévenin) voltage is

$$e_{Th} = v_{oc} = v_2 = 12 \text{ V}_{rms}$$

To find the Thévenin resistance, zero the voltage supply (replace it with a short) and look back into the secondary as demonstrated by the circuit of Figure 17-15(c).

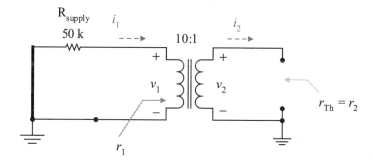

Figure 17-15(c) Find the Thévenin resistance r_{Th}

The supply resistance as viewed from the primary is 50 kΩ. The supply resistance as viewed from the secondary is 500 Ω.

$$r_{Th} = r_2 = \left(\frac{N_2}{N_1}\right)^2 R_{supply} = \left(\frac{1}{10}\right)^2 50 \text{ k}\Omega = 500 \text{ }\Omega$$

The circuit of Figure 17-15(d) is the Thévenin model of the original transformer circuit with the load attached. The load voltage and load current are easily determined from this simplified modeled series circuit.

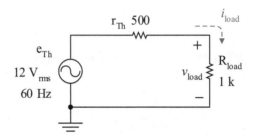

Figure 17-15(d) Thévenin model with load attached

$$i_{load} = \frac{12 \text{ V}_{rms}}{500 \text{ } \Omega + 1 \text{ k}\Omega} = 8 \text{ mA}_{rms}$$

$$v_{load} = 8 \text{ mA}_{rms} \times 1 \text{ k}\Omega = 8 \text{ V}_{rms}$$

Practice

Find the Thévenin model of the ideal transformer circuit of Figure 17-16. Then use the Thévenin model to find the load voltage and load current.

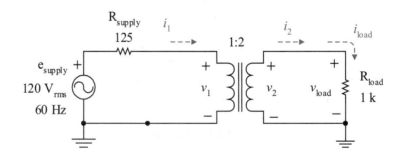

Figure 17-16 Practice circuit

Answer: Thévenin model (240 V$_{rms}$, 500 Ω), 160 mA$_{rms}$, 160 V$_{rms}$

The next example is the black box approach to determining the Thévenin model with a live circuit. You do not know the supply voltage, the turns ratio, nor the supply resistance but must use laboratory measurements with a live circuit to determine the Thévenin model. You need to load the transformer within its operating range, and thus measuring short circuit current is not an option (that would be dead short and perhaps smoke the transformer).

Example 17-9

A transformer's normal operating range is 0 mA to 10 mA. Its open circuit voltage is 10 V_{rms}. A test load of 4 kΩ produces a test load voltage of 8 V_{rms}. Find and draw the Thévenin model of the transformer. Then use this model to find the expected load voltage when an 9 kΩ load is attached.

Solution

Use the *what's in the box* modeling technique, also called *black box modeling*. Visualize a Thévenin model within the black box, which represents the transformer. You take two sets of measurements within the load requirements of 0 to 10 mA. The first is the open circuit voltage measurement V_{oc} of Figure 17-17(a). Since an open circuit produces 0 current through the first test circuit, the voltage drop across r_{Th} is 0 V. Apply Kirchhoff's voltage law.

$$e_{Th} = v_{oc} = 10 \ V_{rms}$$

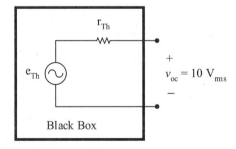 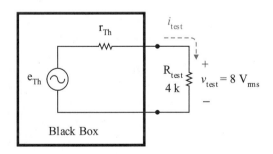

(a) Open circuit measurement   (b) Test circuit measurement

Figure 17-17(a,b) Example black box circuit measurements

Use the result of the open circuit measurement. Set e_{Th} of the loaded circuit of Figure 17-17(b) equal to 10 V_{rms} and use circuit analysis techniques to find r_{Th}. The voltage drop across r_{Th} is

$$v_{rTh} = 10\ V - 8\ V = 2\ V_{rms}$$

The current through r_{Th} is the same as its series test load current.

$$i_{rTh} = i_{test} = \frac{8\ V_{rms}}{4\ k\Omega} = 2\ mA_{rms}$$

Use Ohm's law to find the Thévenin resistance r_{Th}.

$$r_{Th} = \frac{v_{rTh}}{i_{rTh}} = \frac{2\ V_{rms}}{2\ mA_{rms}} = 1\ k\Omega$$

Now draw the Thévenin model as shown in Figure 17-17(c). Recall that the Thévenin model is a circuit model and never includes a load.

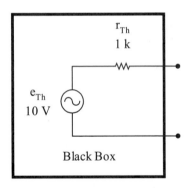

(c) Thévenin model

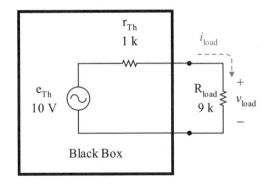

(d) Loaded model

Figure 17-17(c,d) Transformer Thévenin model

Figure 17-17(d) loads the modeled circuit with the requested load resistance of 9 kΩ. This is a simple series circuit. Apply the voltage divider rule to find the load voltage in one algebraic step.

$$V_{load} = \left(\frac{9\ k\Omega}{10\ k\Omega}\right) 10\ V = 9\ V_{rms}$$

Were the test circuits and the loaded circuit within the current specifications of the transformer? The test currents were 0 mA and 2 mA. Use Ohm's law and verify that the load current also meets specifications.

$$I_{\text{load}} = \left(\frac{9 \text{ V}}{9 \text{ k}\Omega} \right) = 1 \text{ mA}_{\text{rms}}$$

Practice

The transformer in the example problem is loaded with a 250 Ω resistor. Find the load current, load voltage, and average power.

Answer: 2 V_{rms}, 8 mA_{rms}, 16 mW

Summary

An ac signal on the primary of the transformer induces a voltage on the secondary based upon the turns ratio of the primary and secondary windings. Assuming an ideal transformer, that is, one in which 100% of the energy is transferred from the primary to the secondary,

$$v_2 = \frac{N_2}{N_1} v_1$$

where

v_2	secondary voltage
N_2	number of secondary turns
N_1	number of primary turns
v_1	primary voltage

If the secondary has more windings than the primary it is a step-up transformer. If the secondary has fewer turns, it is a step-down transformer. If the voltage steps up, the current in the secondary must step down by the same factor.

The transformer also isolates the secondary circuit from being directly connected to the supply.

Referred ac resistance looking into the primary is the ac resistance reflected to the primary based upon the transformer's turns ratio and the load resistance.

$$r_1 = \left(\frac{N_1}{N_2}\right)^2 R_{\text{load}}$$

Likewise, looking into the secondary, the referred ac resistance is

$$r_2 = \left(\frac{N_2}{N_1}\right)^2 R_{\text{supply}}$$

The transformer with a signal on the primary may be modeled looking into the secondary as a Thévenin model with an ac voltage supply in series with a series resistance. Such models may be created theoretically from circuit analysis or determined experimentally with some calculations. However, beware not to load a circuit when measuring to determine Thévenin values, otherwise you will obtain false data and create an invalid Thévenin model.

Problems

Ideal Transformer

17-1   For the transformer circuit in Figure 17-18:

   a. Find primary voltage v_1

   b. Find secondary voltage v_2

   c. Find secondary current i_2

   d. Find secondary average power p_2

   e. Find primary average power p_1

   f. Find primary current i_1

   g. Find the ac primary referred resistance r_1 using Ohm's law

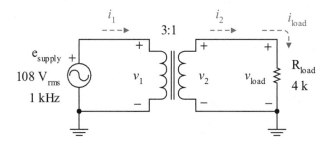

Figure 17-18

17-2   For the transformer circuit in Figure 17-19:

    a. First, find the ac primary referred resistance r_1

    b. Find primary current i_1

    c. Find primary voltage v_1

    d. Find secondary voltage v_2

    e. Find secondary current i_2

    f. Find secondary average power p_2

    g. Find primary average power p_1

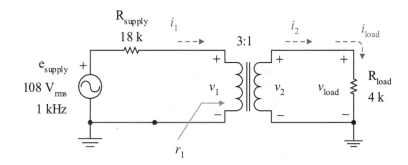

Figure 17-19

Transformer Thévenin Model

17-3   For the transformer circuit in Figure 17-18:

    a. Find the Thévenin voltage e_{Th}

    b. Find the Thévenin resistance r_{Th}

    c. Draw the Thévenin model

    d. Draw the Thévenin model with the load attached and find the load voltage

    e. Use the loaded Thévenin model above and find the load current

17-4   For the transformer circuit in Figure 17-19:

    a. Find the Thévenin voltage e_{Th}

    b. Find the Thévenin resistance r_{Th}

    c. Draw the Thévenin model

    d. Draw the Thévenin model with the load attached and find the load voltage

    e. Use the loaded Thévenin model above and find the load current

17-5   Repeat Problem 17-1 with a turns ratio of 1:2.

17-6   Repeat Problem 17-2 with a turns ratio of 1:2.

17-7   Repeat Problem 17-3 with a turns ratio of 1:2.

17-8   Repeat Problem 17-4 with a turns ratio of 1:2.

17-9   A transformer's normal operating range is 0 mA to 20 mA. Its open circuit voltage is 20 V_{rms}. A test load of 8 kΩ produces a test load voltage of 16 V_{rms}. Find and draw the Thévenin model of the transformer. Then use this model to find the expected load voltage when an 18 kΩ load is attached.

17-10   A transformer is rated at 15 V_{rms} at 1 A_{rms} of current. This assumes that there is a 120 V_{rms} input voltage. If a 15 Ω resistance is attached, find its peak and peak-to-peak load voltage.

18

Power Supply Applications

Introduction

An alternate title for this chapter is "AC to DC: Pulling the Pieces Together." You have the fundamental circuit analysis understanding and many of the major electronic pieces (components) to pull together into a major application: dc power supply circuits. DC power supplies are needed in most electronic circuit applications to supply and bias components such as op amps, BJTs, and MOSFETs. An ac signal from commercial electrical energy or an ac generator must be converted into a clean dc voltage that is fixed and regulated to maintain a fixed output voltage under varying source and load conditions.

Objectives

Upon completion of this chapter you will be able to:

- Identify and state the basic building blocks of a dc power supply.

- Identify and analyze the three basic rectifier circuits (half-wave, full-wave center-tapped, and full-wave bridge) and sketch their output voltage and current waveforms.

- Identify, apply, and analyze the input capacitor filter circuit.

- Analyze lightly loaded to heavily loaded filtered but unregulated dc power supplies.

- Apply and analyze the zener voltage regulator, loaded and unloaded.

- Analyze and apply the pass transistor circuit and the current limiting circuits in the output stages of a dc power supply.

- Incorporate the pass transistor into the negative feedback loop of an operational amplifier.

- Analyze and apply three-terminal IC regulators to a dc power supply.

- Determine the heating effects on the maximum power dissipation of a regulating device and the potential need for heat sinking.

18.1 The DC Power Supply Overview

The dc voltage power supply converts an ac voltage electrical signal into a dc voltage that can then be applied to circuits requiring dc operating voltages. The basic building blocks of a dc power supply using a **half-wave rectifier** are shown in the block diagram of Figure 18-1. Each block represents a subcircuit of the dc power supply.

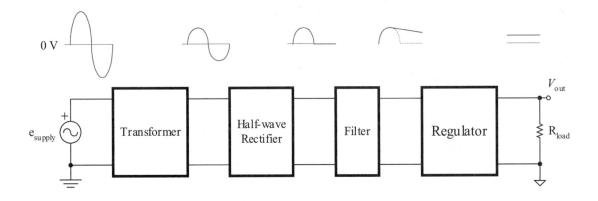

Figure 18-1 Block diagram of a basic dc power supply with a *half-wave rectifier*

- The *ac signal* e_{supply} (usually commercial ac electrical voltage) is supplied to the transformer circuit.

- The *transformer* can step up or step down the voltage and electrically isolates the secondary from the primary supply. The output is a similarly shaped waveform, usually with a different peak voltage but the same frequency. Note that the secondary may or may not be connected to earth common.

- The *half-wave rectifier* produces an output voltage during its positive (or negative) half cycle only. During the other half cycle, the output is 0 V. The output voltage peak is the same as its input peak voltage less its forward biased diode voltage drops. The half-wave output signal has the same frequency as its input signal. Each cycle in produces a single cycle out.

- The *filter* reduces (filters out) much of the ac signal and can produce a signal that approaches a dc voltage. This can be a nearly ideal dc voltage or it can be a voltage that still contains a significant *ac ripple voltage* as shown in Figure 18-1.

• The *voltage regulator* circuit regulates that output voltage and produces a nearly ideal *dc output voltage* for the load. As its name implies, it regulates the output voltage to a fixed dc value even with varying source or load conditions.

Figure 18-2 shows the block diagram of a **full-wave rectifier** circuit. The major difference between this and the half-wave rectifier is that each cycle into the full-wave rectifier produces two output cycles. This rectified signal has twice the frequency and thus has half the period of the input voltage. Also, the filter output has less ripple voltage since its output has less time to decay before the next half cycle refreshes it. The choice of full-wave or half-wave is a design consideration. In fact, it may be a better, more economical design to have a large ripple voltage out of the filter and let the regulator significantly reduce the ripple voltage while regulating the output. The regulator performs both functions.

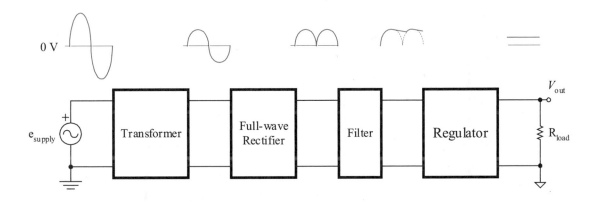

Figure 18-2 Block diagram of a basic dc power supply with a *full-wave rectifier*

The regulator circuit produces a fixed voltage output and can operate at higher currents and produce significant power dissipation in some of the components. Thermal (temperature) considerations must be taken into account. It may be necessary to **heat sink** components (attach cooling metal) to aid in the component's cooling, carrying away destructive heat from the component. Also, if maximum temperature ratings are reached,

thermal shut down protection circuitry can be provided to turn off the output voltage supply.

    Current limiting of the output can be employed to prevent the output current from becoming excessive in the case of a significantly reduced load resistance. This protects the power supply and potentially the load from a destructively high current. The **short circuit current** is the current produced in a short placed across the load terminals. For example, a bench power supply rated 0–18 V at 1 A can produce a dc voltage between 0 V and 18 V, but it is typically limited to a maximum of 1 A by design. Even with a shorted load, the output current is limited to 1 A_{dc}.

    DC power supplies can produce *positive* dc voltage, *negative* dc voltage, or *both positive and negative* dc voltage called a **dual supply** as shown in Figure 18-3. If the negative supply tracks the positive supply so that as the positive dc supply is changed the negative voltage supply follows it, the supply is called a **tracking dual power supply**.

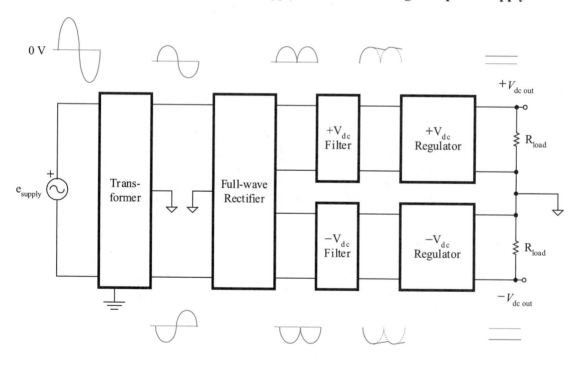

Figure 18-3 Block diagram of a *dual power supply*

You have studied and already have command of many of these sub-circuits: commercial signal supply, transformer, rectifier circuit, filtered rectifier circuit, and a zener regulator. You will now use that basic knowledge and build upon with some new devices, circuits, and concepts.

18.2 Rectifier Circuits

The three basic rectifier circuits used in dc power supplies are shown in Figure 18-4, where V_{p2} is the peak voltage of the full secondary voltage, which is a sine wave; and V_D is the forward voltage drop across a diode. Before analyzing each rectifier circuit in detail, first look at the basic configurations and the resulting output wave shapes and peak output voltage values.

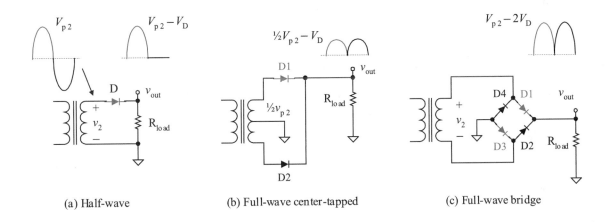

(a) Half-wave (b) Full-wave center-tapped (c) Full-wave bridge

Figure 18-4 Basic rectifier circuits

a. The **half-wave rectifier** circuit of Figure 18-4(a) produces *one cycle* of output for each cycle of input. On the *positive half cycle*, the diode D is forward biased on and the output voltage is the *full* secondary voltage v_2 less the diode D voltage drop. The output signal is always 0 V or greater, therefore the average or dc voltage is positive.

b. The **full-wave center-tapped rectifier** circuit shown in Figure 18-4(b) produces *two cycles* of output for each cycle of input. On the *positive half cycle*, the diode D1 is forward biased on and the output voltage is *half* of the secondary voltage ($\frac{1}{2}$ v_2) less the D1 forward voltage drop. On the *negative half cycle*, the diode D2 is forward biased on and the output voltage is *half* of the secondary voltage ($\frac{1}{2}$ v_2) less the D2 voltage drop. The output voltage is 0 V or greater, therefore the average dc voltage is positive.

c. The **full-wave bridge rectifier** circuit of Figure 18-4(c) produces *two cycles* of output for each cycle of input. On the *positive half cycle*, the D1 and D3 diodes are forward biased on and the output voltage is the *full* secondary voltage v_2 less the D1 and D3 forward voltage drops. On the *negative half cycle*, the D2 and D4 diodes are forward biased on and the output voltage is the *full* secondary voltage v_2 less the D2 and D4 voltage drops. The output voltage is 0 V or greater, therefore the average or dc voltage is positive.

Half-wave Rectifier

The *half-wave rectifier* circuit of Figure 18-5(a) has a single diode in series with the load.

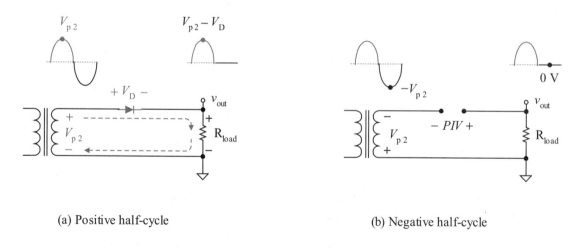

(a) Positive half-cycle   (b) Negative half-cycle

Figure 18-5 Half-wave rectifier

As shown in Figure 18-5(a), during the input positive half-cycle, when the diode is forward biased on, the current flows from the top of the transformer secondary (+) through the diode, down through the load resistor R_{load} dropping a + to − voltage, and to the bottom of the transformer secondary (−). This creates a positive output signal. When the diode is forward biased on, the voltage delivered to the load is the secondary voltage less the voltage drop across diode D. Therefore, the maximum or peak voltage dropped across the load for the half-wave rectifier is

$$V_{p\ out} = V_{p\ 2} - V_D$$

where

$V_{p\ out}$ peak or maximum output voltage
$V_{p\ 2}$ transformer full secondary peak voltage
V_D forward biased diode voltage drop

The **nonrepetitive peak forward surge current** rating (symbol I_{FSM}) of a diode specifies a single time event current surge maximum rating of the diode. For example, the 1N400x series of rectifier diodes has a rating of 30 A for one cycle. That is, the diode can handle a 30 A surge current as long as it does not repeat. If this maximum current rating is exceeded, the diode could self-destruct. When the diode is forward biased

$$v_{out} = v_{load} = v_2 - V_D$$

The forward current though the diode, which is the same as its series current through the load, is determined by the applied voltage and the current limiting load resistor. When the diode is forward biased

$$i_D = i_{load} = \frac{v_2 - V_D}{R_{load}}$$

The diode current is at its maximum forward current when the transformer secondary voltage is at its maximum.

$$I_{p\ load} = \frac{V_{p\ 2} - V_D}{R_{load}}$$

Another forward biased current constraint is the **average rectified forward current**, which has a symbol of $I_{F(AV)}$. This is based on a repetitive rectified current rating over time. The 1N4001 diode has an average forward current rating of 1 A. Thus, this diode could have an initial surge current of 30 A, but settle down to an average current of 1 A, and the diode is within specifications and safe.

From calculus, the average (dc) value of an ideally half-wave rectified sine wave is

$$Y_{average} = \frac{Y_{peak}}{\pi}$$

Therefore, the average output voltage and load currents are

$$I_{dc} = I_{average} = \frac{I_p}{\pi}$$

$$V_{dc} = V_{average} = \frac{V_p}{\pi}$$

From calculus, the rms value of an ideal half-wave sine wave is

$$I_{rms} = \frac{I_p}{2}$$

$$V_{rms} = \frac{V_p}{2}$$

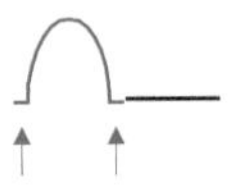

Figure 18-6
Overcoming diode *pn* junction barrier potential

The diode is reverse biased off (open) during the negative half-cycle; therefore, the output voltage is 0 V during the negative half cycle. Also note the output voltage is not a complete half-cycle. At the beginning of the positive half-cycle, the diode barrier voltage must be overcome before the diode turns on and conducts. Likewise, at the end of the positive half cycle the diode turns off when the input voltage falls below its *pn* junction barrier potential voltage. This effect is demonstrated in Figure 18-6. Thus the output positive wave is slightly less than a complete half cycle.

The **peak inverse voltage** (*PIV*) is the maximum reverse voltage that appears across the diode in an operating circuit as demonstrated in Figure 18-5(b). Even though this is a reverse voltage, the *PIV* value is considered a positive voltage since it is a reverse voltage measurement. The diode is acting like an open, therefore for a half-wave rectifier there is no current through the load while the diode is off and the load drops 0 V. When the secondary transformer signal is at its minimum, the voltage appearing across the diode is $-V_{p\,2}$. Thus, the *PIV* voltage appearing across the off diode is

$$PIV = \left| -V_{p\,2} \right|$$

The **peak repetitive reverse voltage rating** (symbol V_{RRM}) of the diode is the maximum repeating reverse voltage that the diode can safely tolerate. This diode rating must be greater than the *PIV* voltage the diode experiences in an operating circuit to ensure that the diode does not break down and start conducting in the reverse direction. For example, the 1N4001 diode has a maximum reverse voltage rating of 50 V. As long as the *PIV* across a 1N4001 diode is less than 50 V, the diode operates properly and does not break down. If the *PIV* exceeds 50 V, then the diode breaks down and the diode conducts in the reverse direction causing the distorted signal of Figure 18-7. In this case you should select the 1N4002, which has a breakdown voltage rating of 100 V.

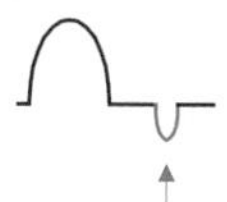

The voltage experienced across the diode is shown in Figure 18-8. Note that once it is forward biased on, it drops its forward voltage of V_D (0.2 V for germanium and 0.7 V for silicon). When not forward biased, the diode drops the applied voltage from the secondary since the resistor drops 0 V while the diode is turned off.

For each sine wave cycle into the half-wave rectifier there is one rectified half-wave output cycle. Therefore, the output frequency is equal to the input frequency, and the output period is equal to the input period.

$$f_{out} = f_{in}$$

$$T_{out} = T_{in}$$

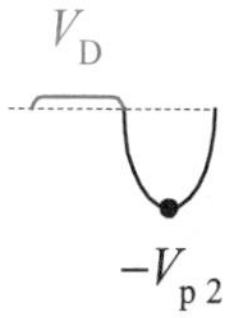

Figure 18-7
Diode reverse voltage breakdown

Figure 18-8
Diode voltage waveform

Example 18-1

For the circuit of Figure 18-9(a), find the transformer secondary peak voltage, the peak output voltage, and the signal frequencies and periods, and draw the circuit voltage waveforms. Find the peak load and diode currents, the average (dc) output voltage, and the average diode and load currents. Sketch two cycles of the output voltage waveform and of the load current. Find the peak inverse voltage *PIV* experienced by the diode. Where appropriate, confirm that the diode is operating within its specifications.

Solution

First you must find and use the *peak voltage* of the primary.

$$V_{p1} = \sqrt{2}\,(120\ \text{V}) = 170\ V_p$$

The transformer is a 17 to 2 step-down transformer, therefore the transformer *secondary peak voltage* is

$$V_{p\,2} = \frac{2}{17}\left(170\;V_p\right) = 20\;V_p$$

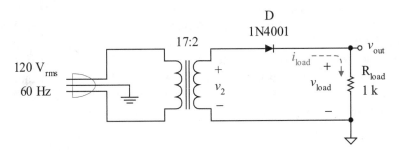

Figure 18-9(a) Example half-wave rectifier circuit

The forward voltage drop of the silicon diode is 0.7 V, therefore the peak output voltage is

$$V_{p\,out} = 20\;V_p - 0.7\;V = 19.3\;V_p$$

The output frequency is the same as the input frequency since for each full sine wave cycle input there is one half-wave signal output.

$$f_{out} = f_{in} = 60\;Hz$$

$$T_{out} = \frac{1}{60\;Hz} = 16.7\;ms$$

The circuit and its respective voltage waveforms are shown in Figure 18-9(b). The 17:2 voltage step-down transformer steps down the sine wave voltage from 170 V_p to 20 V_p. The diode rectifies the voltage to a half-wave voltage with a peak voltage of 19.3 V_p since the forward biased silicon diode drops 0.7 V.

The output voltage is across the load resistance; therefore,

$$V_{load} = V_{out} = 19.3\;V_p$$

By Ohm's law for the load resistance, the peak forward current in this series circuit is

$$I_{p\ load} = \frac{19.3\ V_p}{1\ k\Omega} = 19.3\ mA_p$$

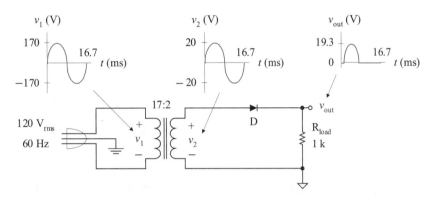

Figure 18-9(b) Half-wave rectifier voltage waveforms

The diode and the load resistance are in series; they therefore experience the same current.

$$I_{p\ D} = I_{p\ load} = 19.3\ mA_p$$

This peak forward current is significantly less than the 1N001 diode's specification of the *non-repetitive peak forward surge current* I_{FSM} of 30 A. This rating is not exceeded.

The output waveform is not an ideal positive half-wave signal since the diode junction barrier voltage must be overcome during the positive half cycle. But it is nearly ideal, thus the approximate average (dc) output voltage is

$$V_{dc\ out} \approx \frac{19.3\ V_p}{\pi} = 6.14\ V_{dc}$$

Likewise, the dc load current is

$$I_{dc\ load} \approx \frac{19.3\ mA_p}{\pi} = 6.14\ mA_{dc}$$

or by Ohm's law it is the dc load voltage divided by the load resistance

$$I_{dc\ load} \approx \frac{6.14\ V_{dc}}{1\ k\Omega} = 6.14\ mA_{dc}$$

Since this is a series circuit, the average current of the diode and load are the same.

$$I_{dc\ D} = I_{dc\ load} \approx 6.14\ mA_{dc}$$

The diode's *average rectified forward current specification* $I_{F(AV)}$ must be greater than 6.14 mA_{dc} or else the diode overheats and self-destructs. The $I_{F(AV)}$ rating of the 1N4001 diode is 1 A_{dc}, therefore this diode operates substantially below this maximum rating.

Two cycles of the output voltage waveform and load current waveform are shown Figure 18-9(c) and Figure 18-9(d), respectively.

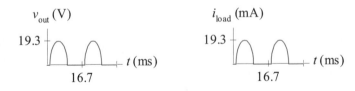

(c) Output voltage   (d) Load current

Figure 18-9(c,d) Output voltage and load current waveforms

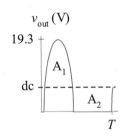

Figure 18-9(e)
Visualize the dc value (about one third of the peak value)

Let's stop and visualize why the dc (or average) values of the half-wave rectifier are reasonable. Visually draw a line though the output voltage waveform so that the area above your line is equal to the area below the line as shown in Figure 18-9(e). The area A_1 above the line is about the same as the area A_2 below the line. This line represents the waveform's average or dc value. If you visualized correctly, your line should be drawn at about one third of the peak voltage value, in this case about 6.4 V_{dc}. This is reasonably close to the calculated value of 6.14 V_{dc} and reaffirms the mathematical calculation for the average or dc value of a half-wave rectified waveform.

The diode's forward specifications have been satisfied. Now examine the diode's reverse voltage maximum rating as shown in Figure 18-9(f).

Performing a KVL counterclockwise walkabout from (–) *PIV* to (+) *PIV* references,

$$PIV = +20 \text{ V} - 0 \text{ V} = 20 \text{ V}_p$$

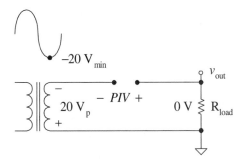

Figure 18-9(f) Half-wave rectifier negative half cycle

The *peak repetitive reverse voltage rating* of the 1N4001 diode is 50 V; therefore this specification is satisfied. This indicates that the diode *does not break down* in the reverse direction and does not conduct in the reverse direction.

Practice

Repeat the example problem with the transformer turns ratio changed to 17:1.

Answer: $V_{p\,2}$ (10 V_p), $V_{p\,out}$ and $V_{p\,load}$ (9.3 V_p), $I_{p\,load}$ and $I_{p\,D}$ (9.3 mA$_p$), f_{out} and f_{in} (60 Hz), T_{out} and T_{in} (16.7 ms), $V_{dc\,out}$ and $V_{dc\,load}$ (2.96 V_{dc}), $I_{dc\,load}$ and $I_{dc\,D}$ (2.96 mA$_{dc}$), *PIV* (10 V), diode specifications satisfied

---

Full-wave Center-tapped Rectifier

The *full-wave center-tapped rectifier* circuit of Figure 18-10(a) has two diodes. The center tap CT of the transformer secondary splits the secondary voltage into two secondary ac supplies. The top half supplies the D1 circuit and the bottom half supplies the D2 circuit. Essentially this circuit has two half-wave rectifier circuits; one operates during the positive half cycle and the other operates during the negative half cycle. The output voltage is always positive since current can only flow down through the load, dropping a positive (+) to negative (−) voltage polarity. To help aid in thinking through the circuit action of this rectifier the circuit is examined separately during its positive and negative half cycles.

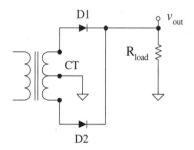

Figure 18-10(a) Full-wave center-tapped rectifier

The circuit of Figure 18-10(b) demonstrates the circuit action during the *positive half cycle* when diode D1 is biased *on* and diode D2 is biased *off*. Since diode D2 is biased *off*, D2 is essentially an *open* and current does not flow through D2 during the positive half cycle. D1 is forward biased *on* allowing a *series circuit current* to flow from the (+) top of the transformer, through D1, down through the load (dropping + to − voltage), and back to the (−) center tap CT of the transformer. Only half of the secondary voltage ($\frac{1}{2} v_2$) is applied to the rectifier circuit since the transformer is center-tapped.

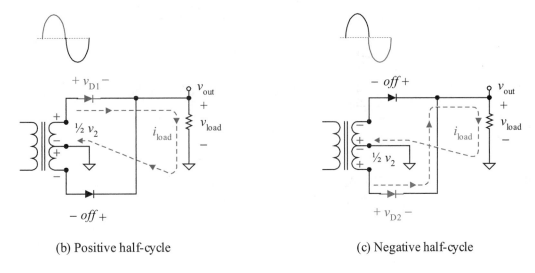

(b) Positive half-cycle   (c) Negative half-cycle

Figure 18-10(b,c) Full-wave center-tapped rectifier half-cycle circuit action

Therefore, the maximum or peak voltage dropped across the load for the positive half cycle is

$$V_{p\,out} = \tfrac{1}{2}V_{p\,2} - V_D$$

where

$V_{p\,out}$	peak or maximum output voltage
$V_{p\,2}$	transformer full secondary peak voltage
V_D	forward biased diode voltage drop

The circuit of Figure 18-10(c) demonstrates the circuit action during the *negative half cycle* when diode D2 is biased *on* and diode D1 is biased *off*. Since diode D1 is biased *off*, D1 is essentially an *open* and current does not flow through D1 during the negative half cycle. D2 is forward biased *on* allowing a *series circuit current* to flow from the (+) center tap CT of the transformer, through D2, down through the load (dropping + to − voltage), and back to the (−) center tap CT of the transformer. Again, only half of the secondary voltage (½ v_2) is applied to the rectifier circuit since the transformer is center-tapped. Therefore, the maximum or peak voltage dropped across the load for the negative half cycle is the same as it is for the positive half cycle.

$$V_{p\,out} = \tfrac{1}{2}V_{p\,2} - V_D$$

Since the diode junction barrier voltage must be overcome on each half cycle, the output voltage waveform is not truly a fully rectified sine wave as shown in Figure 18-11. The output voltage is at 0 V during the dead time when both diodes are off.

The turned on diode forms a series circuit with the load and therefore the diode experiences the same current as the load. When the diode is forward biased

$$i_D = i_{load} = \frac{\frac{1}{2}v_2 - V_D}{R_{load}}$$

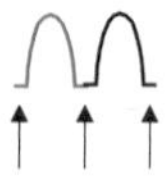

Figure 18-11
Dead time when
both diodes are off

The diode current is at its maximum forward current when the transformer secondary voltage is at its maximum.

$$I_{pD} = I_{p\,load} = \frac{\frac{1}{2}V_{p2} - V_D}{R_{load}}$$

From calculus, the average value of an ideally full-wave rectified-sine wave is

$$Y_{average} = \frac{2Y_{peak}}{\pi}$$

This makes sense; this is twice the average value of a half-wave rectifier where half of the input cycle is wasted, while the full-wave rectifier produces two output half cycles for each input cycle. However, since the center-tapped transformer secondary supplies only half of the full secondary voltage to the rectifier circuit, the average output voltage is

$$V_{dc\,out} = \frac{2\left(\frac{1}{2}V_{p2} - V_D\right)}{\pi}$$

The output voltage is across the load, therefore

$$V_{dc\,load} = V_{dc\,out}$$

By Ohm's law for resistance, the average (dc) load current is

$$I_{dc\,load} = \frac{V_{dc\,load}}{R_{load}}$$

What about the average diode current? The load experiences current during both half cycles, but *each diode is on half of the time* and experiences half of the current. Therefore, the average diode current is only half of the average load current.

$$I_{\text{average diode}} = I_{\text{dc D}} = \tfrac{1}{2} I_{\text{dc load}}$$

From calculus, the rms value of an ideally full-wave rectified sine wave is the same as it is for the sine wave. This makes sense since the first step in calculating rms is squaring the function, and the sine wave squared is the same waveshape as the fully rectified sine wave squared.

$$V_{\text{rms}} = \frac{V_{\text{p}}}{\sqrt{2}}$$

$$I_{\text{rms}} = \frac{I_{\text{p}}}{\sqrt{2}}$$

To determine the *peak inverse voltage* (*PIV*) experienced by the diodes in the full-wave center-tapped rectifier circuit, examine the circuit of Figure 18-12 representing the positive half cycle. The top diode D1 is on and dropping its forward voltage V_{D} (0.2 V for germanium and 0.7 V for silicon). The bottom diode D2 is off. Its peak inverse voltage *PIV* occurs when the secondary voltage is at its peak or maximum. Performing a KVL walkabout from the (−) *PIV* to the (+) *PIV* references,

$$PIV = +V_{\text{p 2}} - V_{\text{D}}$$

This result is valid for both the D1 and the D2 diodes.

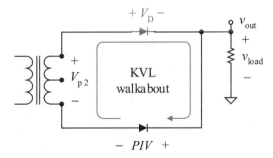

Figure 18-12 Schematic setup to find the peak inverse voltage

For each sine wave cycle into the full-wave rectifier circuit there are two rectified output cycles as shown in Figure 18-13. Therefore, the output frequency is equal to twice the input frequency, and the output period is equal to half of the input period.

$$f_{out} = 2f_{in}$$

$$T_{out} = \tfrac{1}{2}T_{in}$$

Input

Output

Figure 18-13
Full-wave rectifier has two output cycles for each input cycle

Example 18-2

For the circuit of Figure 18-14(a), find the transformer secondary voltage drop $V_{2\,p}$, the peak secondary voltage relative to the center tap CT, the peak output voltage, and the signal frequencies and periods, and draw the circuit voltage waveforms. Find the peak load and diode currents, the average (dc) output voltage, and the average diode and load currents. Find the peak inverse voltage *PIV* experienced by each diode.

Solution

The peak of the primary voltage is

$$V_{p1} = \sqrt{2}\ (120\ V) = 170\ V_p$$

The peak of the full secondary voltage is

$$V_{p2} = \frac{2}{17}(170\ V_p) = 20\ V_p$$

The center-tapped peak voltage, that is, the peak of half of the secondary voltage, is

$$\tfrac{1}{2}V_{p2} = \tfrac{1}{2}(20\ V_p) = 10\ V_p$$

The forward voltage drop of the silicon diode is 0.7 V, therefore the peak output voltage for each half cycle is

$$V_{p\,out} = 10\ V_p - 0.7\ V = 9.3\ V_p$$

The output frequency is twice the input frequency since for each full sine wave cycle input the output has two positive half waves.

$$f_{out} = 2 \times 60\ Hz = 120\ Hz$$

$$T_{out} = \frac{1}{120\ Hz} = 8.33\ ms$$

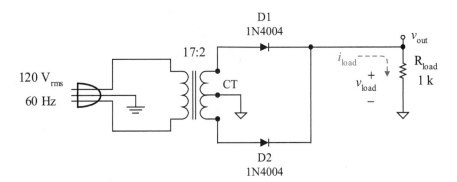

Figure 18-14(a) Example full-wave center-tapped rectifier circuit

The circuit and its respective voltage waveforms are shown in Figure 18-14(b). The 17:2 voltage step-down transformer steps down the sine wave voltage from 170 V_p to 20 V_p. The center-tapped secondary produces a 10 V_p sine wave driving the rectifier circuits. Each diode is a rectifier and produces a half-wave rectified voltage output with a peak voltage of 9.3 V_p since the forward biased silicon diode drops 0.7 V.

The output voltage across the load resistance is a fully rectified sine wave, each diode producing half of the output.

$$V_{load} = V_{out} = 9.3 \ V_p$$

By Ohm's law for the load resistance, the peak forward current in this series circuit is

$$I_{p\,load} = \frac{9.3 \ V_p}{1 \ k\Omega} = 9.3 \ mA_p$$

The *on* diode and the load resistance are in series; they therefore experience the same peak current.

$$I_{p\,D} = I_{p\,load} = 9.3 \ mA_p$$

The output waveform is not an ideal positive half-wave signal since the diode junction barrier voltage must be overcome during the positive half cycle. But it is nearly ideal, thus the approximate average (dc) output voltage is

$$V_{dc\ out} \approx \frac{2\left(9.3\ V_p\right)}{\pi} = 5.92\ V_{dc}$$

Likewise, the dc load current is

$$I_{dc\ load} \approx \frac{2\left(9.3\ mA_p\right)}{\pi} = 5.92\ mA_{dc}$$

or by Ohm's law it is the dc load voltage divided by the load resistance

$$I_{dc\ load} \approx \frac{12.3\ V_{dc}}{1\ k\Omega} = 5.92\ mA_{dc}$$

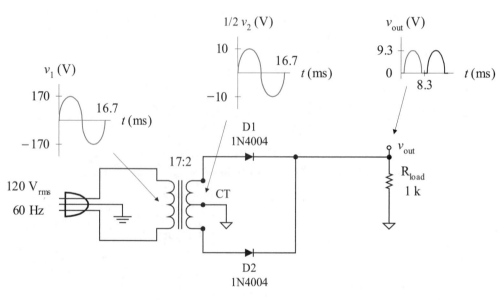

Figure 18-14(b) Full-wave center-tapped rectifier voltage waveforms

Each diode is on half the time and thus produces half the average load current.

$$I_{dc\ D} = \frac{1}{2} I_{dc\ load}$$

$$I_{dc\ D} = \tfrac{1}{2}\left(5.92 \text{ mA}_{dc}\right) = 2.96 \text{ mA}_{dc}$$

Check with the 1N4004 diode specifications and verify that the diode's forward specifications have been satisfied. Now examine the diode's maximum reverse voltage with the circuit shown in Figure 18-14(c). Performing a KVL clockwise walkabout from the (−) *PIV* to the (+) *PIV* references,

$$PIV = +20 \text{ V} - 0.7 \text{ V} = 19.3 \text{ V}_p$$

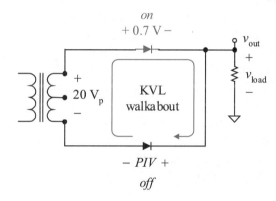

Figure 18-14(c) Full-wave center-tapped rectifier *PIV* calculation

The *peak repetitive reverse voltage rating* of the 1N4001 diode is 400 V; therefore this specification is satisfied. This indicates that the diode *does not break down* in the reverse direction.

Practice

Repeat the example problem with the transformer turns ratio changed to 17:1.

Answer: V_{p2} (10 V_p), $\tfrac{1}{2}V_{p2}$ (5 V_p), $V_{p\ out}$ and $V_{p\ load}$ (4.3 V_p), $I_{p\ load}$ and $I_{p\ D}$ (4.3 mA$_p$), f_{out} (120 Hz), T_{out} (8.3 ms), $V_{dc\ out}$ and $V_{dc\ load}$ (2.74 V_{dc}), $I_{dc\ load}$ (2.74 mA$_{dc}$), $I_{dc\ D}$ (1.37 mA$_{dc}$), *PIV* (19.3 V), diode specifications satisfied. Note the diode drop of 0.7 V is becoming more significant relative to the peak output voltage, the fully rectified signal is becoming less ideal, and therefore a higher percentage of error is creeping into the dc calculations.

multiSIM

Exploration

Simulate the practice problem with MultiSIM and measure the output peak and dc voltages. Compare your result with the practice problem result.

Full-wave Bridge Rectifier

The *full-wave bridge rectifier* circuit of Figure 18-15(a) has four diodes. Opposite pairs D1-D3 and D2-D4 work together. While one pair is forward biased *on*, the other pair is reverse biased *off* or *open*. The full secondary voltage is applied to the bridge rectifier circuit. When a pair of forward biased diodes is *on*, they complete a series circuit to the load from the full transformer secondary.

Essentially this circuit has two half-wave rectifier circuits; D1-D3 operates during the positive half cycle, and D2-D4 operates during the negative half cycle. The output voltage is always positive since current can only flow *down through the load*, dropping a positive (+) to negative (−) voltage polarity. To help aid in thinking through the circuit action of this rectifier, the circuit is examined separately during its positive and negative half cycles.

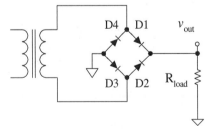

Figure 18-15(a) Full-wave bridge rectifier

The circuit of Figure 18-15(b) demonstrates the circuit action during the *positive half cycle* when diodes D1-D3 are biased *on* (red) and diodes

D2-D4 (black) are biased *off*. Since diodes D2 and D4 are biased *off*, D2 and D4 are essentially *open* and current does not flow through either diode during the positive half cycle. D1 and D3 are forward biased *on* allowing a *series circuit current* to flow from the (+) top of the transformer, through D1, down through the load (dropping + to − voltage), down through common at the bottom of the load to common on the bridge, down through D3, and back to the (−) bottom of the transformer secondary.

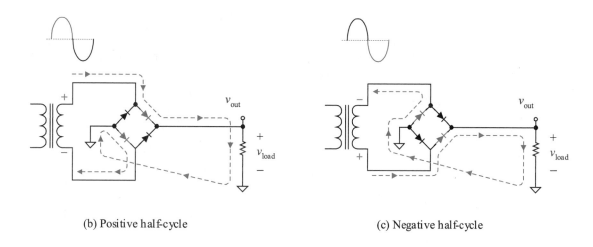

(b) Positive half-cycle   (c) Negative half-cycle

Figure 18-15(b,c) Full-wave bridge rectifier half-cycle circuit action

Therefore, the maximum or peak voltage dropped across the load for the positive half cycle is

$$V_{p\,out} = V_{p\,2} - 2V_D$$

where

$V_{p\,out}$	peak or maximum output voltage
$V_{p\,2}$	transformer full secondary peak voltage
V_D	forward biased diode voltage drop

The circuit of Figure 18-15(c) demonstrates the circuit action during the *negative half cycle* when diodes D2-D4 are biased *on* (red) and diodes D1-D3 (black) are biased *off*. Since diodes D1 and D3 are biased *off*, D1 and D3 are essentially *open* and current does not flow through either diode during the negative half cycle. D2 and D4 are forward biased *on* allowing a *series circuit current* to flow from the (+) bottom of the transformer, up through D2, down through the load (dropping + to − voltage), down through common at the bottom of the load to common on the bridge, up through D4, and back to the (−) top of the transformer secondary.

The turned on diodes form a series circuit with the load, and therefore the diodes experience the same current as the load. When the diode is forward biased

$$i_D = i_{load} = \frac{v_2 - 2V_D}{R_{load}}$$

The diode current is at its maximum forward current when the transformer secondary voltage is at its maximum.

$$I_{pD} = I_{p\,load} = \frac{V_{p2} - 2V_D}{R_{load}}$$

Again, the average (dc) value of a fully rectified sine wave is

$$Y_{average} = \frac{2Y_{peak}}{\pi}$$

therefore

$$V_{dc\ out} = 2\left(\frac{V_{p2} - 2V_D}{\pi}\right)$$

The output voltage is across the load, therefore

$$V_{dc\ load} = V_{dc\ out}$$

By Ohm's law for resistance, the average (dc) load current is

$$I_{dc\ load} = \frac{V_{dc\ load}}{R_{load}}$$

Since this is a full-wave rectifier, the average diode current is half of the load current.

$$I_{average\ diode} = I_{dc\ D} = \tfrac{1}{2}I_{dc\ load}$$

To determine the *peak inverse voltage (PIV)* experienced by the diodes in the full-wave bridge rectifier circuit, examine the circuit of Figure 18-15(d) representing the positive half cycle. Diode D3 is *on* (red) and drops a forward diode voltage V_D (0.2 V for germanium and 0.7 V for silicon). Diode D4 is off. Its peak inverse voltage *PIV* occurs when the secondary voltage is at its peak or maximum. Performing a KVL walkabout from the (–) *PIV* to the (+) *PIV* references,

$$PIV = -V_D + V_{p2}$$

The same approach with different loops and reversing the input voltage polarity produce the same result for all of the diodes.

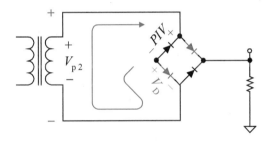

Figure 18-15(d) Schematic setup to find the peak inverse voltage

For full-wave rectifiers there are two rectified output cycles for each input cycle. Therefore, the output frequency is equal to twice the input frequency, and the output period is equal to half of the input period.

$$f_{out} = 2f_{in}$$

$$T_{out} = \tfrac{1}{2}T_{in}$$

Example 18-3

For the circuit of Figure 18-16(a), find the transformer secondary peak voltage, the peak output voltage, and the signal frequencies and periods, and draw the circuit voltage waveforms. Find the peak load and diode currents, the average (dc) output voltage, and the average diode and load currents. Find the peak inverse voltage

PIV experienced by each diode. Note, the diodes are 1N4004 rectifier diodes.

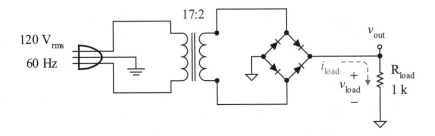

Figure 18-16(a) Example full-wave bridge rectifier circuit

Solution

The peak of the primary voltage is

$$V_{p1} = \sqrt{2} \ (120 \text{ V}) = 170 \text{ V}_p$$

The peak of the full secondary voltage is

$$V_{p2} = \frac{2}{17}(170 \text{ V}_p) = 20 \text{ V}_p$$

The forward voltage drop of the silicon diode is 0.7 V, therefore the peak output voltage for each half cycle is

$$V_{p \text{ out}} = 20 \text{ V}_p - 2(0.7 \text{ V}) = 18.6 \text{ V}_p$$

The output frequency is twice the input frequency since for each full sine wave cycle input the output has two positive half waves.

$$f_{\text{out}} = 2 \times 60 \text{ Hz} = 120 \text{ Hz}$$

$$T_{\text{out}} = \frac{1}{120 \text{ Hz}} = 8.33 \text{ ms}$$

The circuit and its respective voltage waveforms are shown in Figure 18-16(b). The 17:2 voltage step-down transformer steps down the sine wave voltage from 170 V_p to 20 V_p. Each diode pair rectifies the sine wave and produces a fully rectified voltage output with a peak voltage of 18.6 V_p since the forward biased silicon diode pair drops 1.4 V.

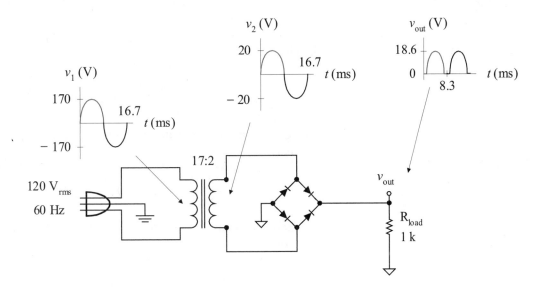

Figure 18-16(b) Full-wave bridge rectifier voltage waveforms

The output voltage is across the load resistance; therefore,

$$V_{load} = V_{out} = 18.6 \ V_p$$

By Ohm's law for the load resistance, the peak forward current in this series circuit is

$$I_{p \ load} = \frac{18.6 \ V_p}{1 \ k\Omega} = 18.6 \ mA_p$$

The *on* diodes and the load resistance are in series; they therefore experience the same peak current.

$$I_{p \ D} = I_{p \ load} = 18.6 \ mA_p$$

The output waveform is not an ideal positive half-wave signal since the diode junction barrier voltage must be overcome during the positive half cycle. But it is nearly ideal, thus the approximate average (dc) output voltage is

$$V_{dc\ out} \approx \frac{2\ (18.6\ V_p)}{\pi} = 11.8\ V_{dc}$$

Likewise, the dc load current is

$$I_{dc\ load} \approx \frac{2(18.6\ mA_p)}{\pi} = 11.8\ mA_{dc}$$

or by Ohm's law it is the dc load voltage divided by the load resistance

$$I_{dc\ load} \approx \frac{11.8\ V_{dc}}{1\ k\Omega} = 11.8\ mA_{dc}$$

Each diode is on half the time and thus produces half the average load current.

$$I_{dc\ D} = \tfrac{1}{2} I_{dc\ load}$$

$$I_{dc\ D} = \tfrac{1}{2} (11.8\ mA_{dc}) = 5.9\ mA_{dc}$$

Check with the 1N4004 diode specifications and verify that the diode's forward specifications have been satisfied. Now examine the diode's reverse voltage maximum rating as shown in Figure 18-16(c). Performing a KVL clockwise walkabout from the (–) *PIV* to the (+) *PIV* references.

$$PIV = -0.7\ V + 20\ V = 19.3\ V_p$$

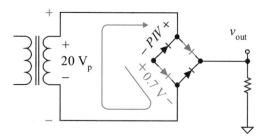

Figure 18-16(c) Full-wave bridge rectifier *PIV* calculation setup

The *peak repetitive reverse voltage rating* of the 1N4001 diode is 400 V; therefore this specification is satisfied. This indicates that the diode *does not break down* in the reverse direction.

Practice

Repeat the example problem with the transformer turns ratio changed to 17:1.

Answer: V_{p2} (10 V$_p$), $V_{p \text{ out}}= V_{p \text{ load}}$ (8.6 V$_p$), $I_{p \text{ load}} = I_{p \text{ D}}$ (8.6 mA$_p$), f_{out} (120 Hz), T_{out} (8.3 ms), $V_{\text{dc out}}$ and $V_{\text{dc load}}$ (5.47 V$_{\text{dc}}$), $I_{\text{dc load}}$ (5.47 mA$_{\text{dc}}$), $I_{\text{dc D}}$ (2.74 mA$_{\text{dc}}$), PIV (19.3 V), diode specifications satisfied. Note the diode pair drop of 1.4 V is becoming more significant relative to the peak output voltage, the fully rectified signal is becoming less ideal, and therefore a higher percentage of error is creeping into the dc calculations.

Exploration

Simulate the practice problem with MultiSIM and measure the output peak and dc voltages. Compare your result with the practice problem result.

↗
multi**SIM**

Reversing the diodes of the previous rectifier circuits creates negative dc power supplies. Figure 18-17 demonstrates this principle.

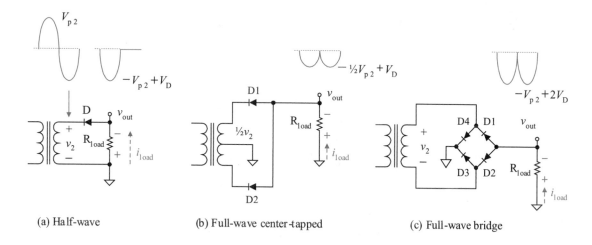

(a) Half-wave   (b) Full-wave center-tapped   (c) Full-wave bridge

Figure 18-17 Negative output rectifier circuits

The load current now flows up through common dropping positive (+) to negative (−) voltage across the load. The output is now at a lower potential than common, thus creating a negative rectified output.

The next example demonstrates the use of a bridge rectifier circuit combined with a center-tapped transformer to create a **split dc power supply system** with a positive (+) dc voltage output and a negative (−) dc voltage output. Such supplies are needed to power op amps.

Example 18-4

For the split dc supply circuit of Figure 18-18, verify the output voltage waveforms.

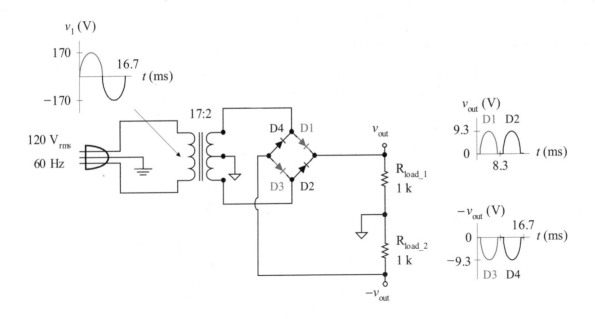

Figure 18-18 Split power supply using a center-tapped transformer and a bridge rectifier

Solution

D1 and D2 create the positive dc supply. D3 and D4 create the negative dc supply. D1 and D2 form a full-wave center-tapped rectifier for the positive supply. D3 and D4 form a full-wave center-tapped rectifier for the negative supply. During the positive half cycle of the input, the diodes shown in red (D1 and D3) are turned on, drop 0.7 V, and provide a current path to their respective loads.

As in previous examples, the center-tapped secondary feeds half of the full secondary voltage to each rectifier circuit. Thus, the rectifier supply signal from the transformer is

$$\tfrac{1}{2}\,V_{p\,2} = \tfrac{1}{2}\left(20\ \mathrm{V_p}\right) = 10\ \mathrm{V_p}$$

The forward biased diodes drop 0.7 V, therefore the peak output voltage of the positive supply is

$$V_{p\,\mathrm{out}} = 10\ \mathrm{V_p} - 0.7\ \mathrm{V} = 9.3\ \mathrm{V_p}$$

Likewise the negative supply output is $-9.3\ \mathrm{V_p}$.

Since it is a full-wave rectifier, the output frequency is twice the input frequency.

$$f_{\mathrm{out}} = 2 \times 60\ \mathrm{Hz} = 120\ \mathrm{Hz}$$

$$T_{\mathrm{out}} = \frac{1}{120\ \mathrm{Hz}} = 8.33\ \mathrm{ms}$$

Practice

Use MultiSIM or Cadence PSpice to verify the example results.

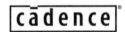

how big can you dream?™

18.3 Capacitive Input Filter

A rectifier circuit converts a pure ac input signal ($0\ \mathrm{V_{dc}}$) to a half-wave or a full-wave rectified signal that has either positive or negative average (dc) voltage. The rectified output signal, however, is far from being a clean fixed dc voltage that can be used to supply electronic circuits. A filter is needed to eliminate (or filter out) much of the ac ripple voltage in the output voltage while passing the dc voltage through to the output as shown in Figure 18-19. The filtered output may still have some ac ripple voltage in it, but the ac ripple voltage has been significantly reduced while maintaining the dc output voltage.

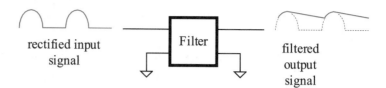

Figure 18-19 Filter to increase dc voltage and reduce ac voltage

Capacitor and Inductor Filter Elements

As previously presented in this textbook, the capacitor and inductor are components that are sensitive to frequency and can be used to pass dc voltage and eliminate (short out or block) or significantly reduce the ac voltage in a signal. Figure 18-20 shows several types of filters that could be used to pass dc and block or short out ac.

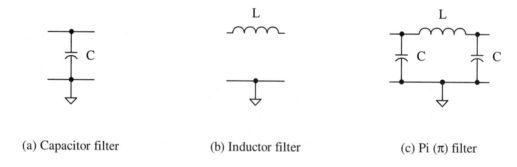

(a) Capacitor filter   (b) Inductor filter   (c) Pi (π) filter

Figure 18-20 Various types of filters

Recall that the capacitive reactance X_C formula is

$$X_C = \frac{1}{2\pi f C}$$

The capacitance C is an ideal *open to dc* while it is an *ideal short to ac* if the frequency f is high enough for the selected values of C and the load resistance.

The inductive reactance X_L formula is

$$X_L = 2\pi f L$$

The inductance L is an ideal *short to dc*, while it is an ideal *open to ac* if the frequency f is high enough for the selected values of L and the load resistance.

Figure 18-21 shows the corresponding *ideal dc models* for the three sample filters of Figure 18-20. Note that dc has an ideal conducting path to the output, while the path to common is a dc open. Ideally all of the dc is transferred to the output without loss to common.

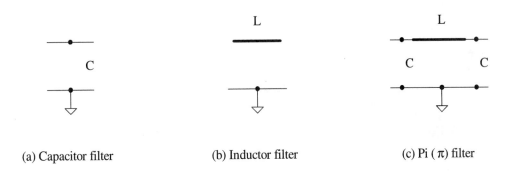

(a) Capacitor filter   (b) Inductor filter   (c) Pi (π) filter

Figure 18-21 *Ideal dc models* of the filters in Figure 18-20 (dc delivered to the load)

Figure 18-22 shows the corresponding *ideal ac models* for the three sample filters of Figure 18-20. Note that the capacitors provide an ideal short to common for ac signals and that the series inductor acts as an open. The ac signal is ideally shorted to common through a capacitor or has an ideal open path to the load. Ideally, the ac ripple voltage is eliminated and pure dc is delivered to the load.

At the input commercial frequency of 60 Hz, an inductor must be very large to produce a sufficiently high reactance at 60 Hz for a half-wave rectifier and 120 Hz for a full-wave rectifier to be effective. Since the operating frequencies are so low, the inductance must be relatively large. For example, the inductive reactance of a 10 H inductor for a fully rectified sine wave is only 7.54 kΩ.

$$X_L = 2\pi (120 \text{ Hz})(10 \text{ H}) = 7.54 \text{ k}\Omega$$

The essentially series inductive reactance must be much greater than the resistive load resistance to drop a significant ac signal. Thus, a 10 H inductor would be reasonable for a 1 kΩ load resistance. But a 10 H inductor is relatively expensive and bulky, whereas a capacitor that is reasonably inexpensive, lighter, and requires less space can produce the same or a better result by shorting out the ac signal instead of blocking it. The *capacitor filter* of Figure 18-20(a) is therefore a very popular choice for a simple, inexpensive power supply filter.

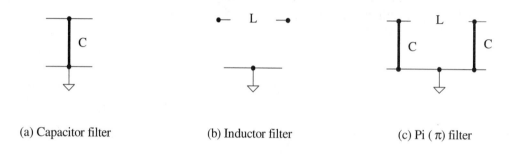

(a) Capacitor filter   (b) Inductor filter   (c) Pi (π) filter

Figure 18-22 *Ideal ac models* of the filters in Figure 18-20 (ac filtered out)

Capacitive Input Filter

A power supply filter where the capacitor is the first filtering element of the filter is known as a **capacitive input filter**. We shall focus our attention on the single input capacitor filter of Figure 18-20(a). As demonstrated later in this chapter, a simple input filter capacitor is usually sufficient when the filtered output is fed to a voltage or current regulator where significant reduction of the ac ripple naturally occurs before the dc voltage is delivered to the load.

The circuit of Figure 18-23(a) demonstrates the use of a capacitance input filter to smooth out the output voltage waveform (filter out the ac and deliver the dc). You have seen this circuit in a previous chapter when you were studying capacitors and *RC* circuits. This is an ideal case in which the capacitor charges quickly through the low forward resistance of the *on diode* and discharges very, very slowly through the load when the *diode is biased off*. Once charged, the capacitor ideally holds this voltage during the discharge cycle across the load. This requires that the $R_{load}C$ time constant be very long compared to the period of the output.

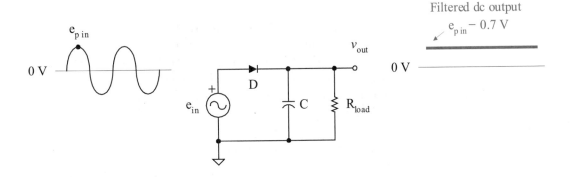

Figure 18-23(a) *Ideal* half-wave capacitive input filtered dc power supply

Figure 18-23(b) is a more realistic output voltage from this circuit. A high capacitance valued capacitor is needed for the ideal case, but the ideal case is not typically needed or preferred. High capacitance values require relatively larger and more costly capacitors than really needed. Lower valued capacitors usually suffice even though a larger ac ripple voltage results as displayed in Figure 18-23(b).

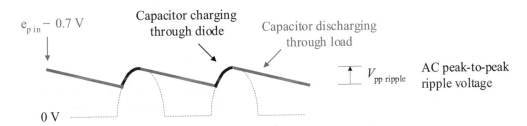

Figure 18-23(b) Realistic half-wave capacitive input filtered dc power supply output voltage

The dashed line represents the half-wave rectified output without filtering. The black line is the output voltage during the recharging of the capacitor through the on diode, while the red line represents the output voltage during the discharge of the capacitor through the load when the diode is biased off. The ac peak-to-peak ripple voltage $V_{pp\ ripple}$ is defined by

$$V_{pp\ ripple} = V_{max} - V_{min}$$

Light loading occurs if the capacitor discharges no more than 10% of its voltage; the ripple wave shape is then essentially a triangle wave. This is called a **light discharge** of the capacitor. The smaller the ripple, the more closely the ripple approaches a triangle wave. The discharge becomes naturally an RC exponential discharge decay curve if the capacitor is allowed to discharge significantly. A significant capacitor discharge is called a **deep discharge** of the capacitor.

Figure 18-23(c) represents the ac ripple voltage of a lightly discharged capacitor filter. The waveform is essentially a linear triangle wave shape with a very short rising ramp and a relatively long falling ramp. Recall that the rise curve is really part of a sine wave and the falling curve is really part of an exponential RC discharge wave.

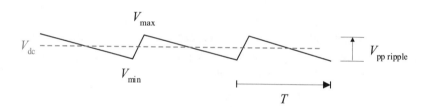

Figure 18-23(c) AC ripple voltage from a light discharge

As noted earlier, the period T is equal to the input signal period if it is a half-wave rectifier or it is half of the input period if it is a full-wave rectifier. For a lightly loaded system, a formula can be developed to estimate the output voltages.

As noted in the previous section, V_{max} (the peak or maximum voltage out of a rectifier) is determined by the transformer secondary peak voltage $V_{p\,2}$, possibly split in half by a center-tapped transformer secondary, less any diode forward voltage drops V_D in the conducting path. The following is a summary of the maximum output voltages that rectifiers produce for a rectified positive dc supply.

$$V_{max} = V_{p\,2} - 2V_D \qquad \text{(Half-wave)}$$

$$V_{max} = \tfrac{1}{2}V_{p\,2} - V_D \qquad \text{(Full-wave center-tapped)}$$

$$V_{max} = V_{p\,2} - 2V_D \qquad \text{(Full-wave bridge)}$$

$$V_{max} = \tfrac{1}{2}V_{p\,2} - V_D \qquad \text{(Split supply full-wave bridge)}$$

The peak-to-peak ripple voltage for a lightly loaded system can be estimated from the RC discharging expression for the capacitor.

$$v_C(t) = V_{ss} + \left(V_{init} - V_{ss}\right)e^{-t/RC}$$

where

$v_C(t)$	capacitor voltage at time t of the discharge cycle
V_{init}	initial voltage (V_{max} for each discharge cycle)
V_{ss}	steady state (0 V if capacitor fully discharges)
R	resistance of the load as viewed from the capacitor
C	capacitance value of the capacitor

The capacitor voltage discharge expression reduces to

$$v_C(t) = V_{max}\, e^{-t/RC}$$

In a lightly discharged system, the RC time constant is much greater than the period T and the exponent of e must be reasonably small. For a lightly discharged system to produce less than 10% error

$$\frac{T}{RC} \le 0.4$$

For this case, from calculus (using the McClaurin series expansion),

$$e^{-t/RC} \approx 1 - \frac{t}{RC}$$

Try your calculator and compare your results with Table 18-1. The first table estimate has a percent error of less than 1%. The last table estimate has a percent error of about 10%.

The capacitor voltage expression can now be approximated as

$$v_C(t) \approx V_{max}\left(1 - \frac{t}{RC}\right)$$

For a light discharge system, the capacitor discharge time is approximately equal to the period T of the ripple. The capacitor voltage at the end of the discharge cycle V_{min} can be estimated as

Table 18-1
Approximating $e^{-t/RC}$

	Accurate	Estimate
$e^{-0.1}$	0.905	0.900
$e^{-0.2}$	0.819	0.800
$e^{-0.3}$	0.741	0.700
$e^{-0.4}$	0.670	0.600

$$V_{min} = v_C(T) \approx V_{max}\left(1 - \frac{T}{RC}\right)$$

The ac peak-to-peak ripple can now be estimated as

$$V_{pp\ ripple} = V_{max} - V_{min}$$

Substituting for V_{min},

$$V_{pp\ ripple} \approx V_{max} - V_{max}\left(1 - \frac{T}{RC}\right)$$

which reduces to

$$V_{pp\ ripple} \approx V_{max}\frac{T}{RC}$$

This can be written in terms of the ripple frequency f, which is the reciprocal of the ripple period T, as

$$V_{pp\ ripple} \approx \frac{V_{max}}{RfC}$$

Ohm's law can be applied and the term V_{max} divided by R can be written as the maximum current I_{max} delivered by the capacitor toward the load.

$$V_{pp\ ripple} \approx \frac{I_{max}}{fC}$$

Again, this approximation only applies to lightly loaded capacitive input filters where

- the capacitor charge time is much less than its discharge time.

- the ripple period T is much less than the $R_{load}C$ time constant.

- The load resistance R_{load} is sufficiently high to produce a sufficiently long $R_{load}C$ time constant that in turn produces modest load currents.

The average (dc) voltage across the capacitor, and thus across the load, can be calculated from any of the basic wave shape voltage values.

$$V_{dc} = \tfrac{1}{2}\left(V_{max} + V_{min}\right)$$

$$V_{dc} = V_{max} - \tfrac{1}{2}V_{pp\ ripple}$$

$$V_{dc} = V_{min} + \tfrac{1}{2}V_{pp\ ripple}$$

The **percent ripple** quantifies the ripple content relative to the output dc voltage. Ideally, a pure dc voltage has 0% ripple.

$$\% \, ripple = \frac{V_{\text{rms ripple output}}}{V_{\text{dc output}}} \times 100\%$$

A lightly loaded capacitor input filter circuit has an ac ripple voltage that approximates a triangle wave. Recall that the rms value of an ac triangle wave is

$$V_{\text{rms}} \approx \frac{V_{\text{p}}}{\sqrt{3}} = \frac{V_{\text{pp}}}{2\sqrt{3}}$$

Example 18-5

For the circuit of Figure 18-24(a), find the ac ripple frequency and period, the peak output voltage, the peak load current, the peak-to-peak output voltage, the minimum output voltage, and the average dc output voltage. Find the percent ripple. Find the rms voltage of the ripple voltage and the peak inverse voltage experienced by the diode. Sketch the output waveform.

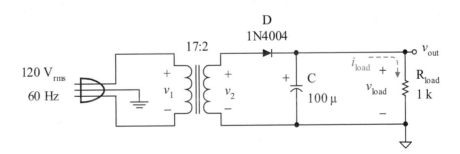

Figure 18-24(a) Example capacitor input filter of a half-wave rectifier

Solution

This is the same half-wave rectifier circuit as shown in the previous example except that an input filter capacitor has been inserted to reduce the ac ripple and raise the dc voltage level. Since this is a half-wave rectifier, the output frequency and period are the same as the input frequency and period.

$$f_{ripple} = f_{in} = 60 \text{ Hz}$$

$$T_{ripple} = \frac{1}{60 \text{ Hz}} = 16.7 \text{ ms}$$

The transformer secondary voltage is

$$V_{p2} = \frac{2}{17}\left(170 \text{ V}_p\right) = 20 \text{ V}_p$$

The diode drop of 0.7 V across the diode creates a peak capacitor voltage of

$$V_{max \text{ out}} = V_{max \text{ C}} = 20 \text{ V}_p - 0.7 \text{ V} = 19.3 \text{ V}_p$$

By Ohm's law the maximum or peak load current is

$$I_{max \text{ load}} = \frac{19.3 \text{ V}_p}{1 \text{ k}\Omega} = 19.3 \text{ mA}$$

Is this a light load condition? Is the period sufficiently less than the $R_{load}C$ time constant to apply the approximation of a lightly loaded condition?

$$\frac{T_{ripple}}{R_{load}C} \leq 0.2 \text{ ?}$$

$$\frac{16.7 \text{ ms}}{\left(1 \text{ k}\Omega\right)\left(100 \text{ } \mu\text{F}\right)} = 0.17$$

Yes, the condition is just barely met. An expected error of close to 10% is expected. The lightly loaded approximation can be used.

$$V_{pp \text{ ripple}} \approx \frac{I_{max}}{fC}$$

$$V_{pp \text{ ripple}} \approx \frac{19.3 \text{ mA}_p}{\left(60 \text{ Hz}\right)\left(100 \text{ } \mu\text{F}\right)} = 3.22 \text{ V}_{pp}$$

The estimated minimum output voltage is

$$V_{min \text{ out}} \approx 19.3 \text{ V}_p - 3.22 \text{ V}_{pp} = 16.1 \text{ V}_{min}$$

The average (dc) output voltage is

$$V_{dc\ out} \approx 19.3\ V_p - \tfrac{1}{2}(3.22\ V_{pp}) = 17.7\ V_{dc}$$

The rms voltage of the ac ripple output voltage is

$$V_{rms\ ripple\ output} = \frac{3.22\ V_{pp}}{2\sqrt{3}} = 930\ mV_{rms}$$

The power supply percent ripple is fairly high at approximately

$$\%\ ripple = \frac{930\ mV_{rms}}{17.7\ V_{dc\ output}} \times 100\% = 5.3\%$$

The predicted output voltage waveform ripple is shown in Figure 18-24(b).

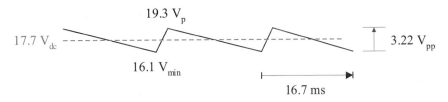

17.7 V$_{dc}$ 19.3 V$_p$ 16.1 V$_{min}$ 3.22 V$_{pp}$ 16.7 ms

Figure 18-24(b) Example ripple output voltage of capacitor input filter

To find the peak inverse voltage (*PIV*) experienced by the diode, redraw the circuit as shown in Figure 18-24(c) with the input at its minimum value, treat the diode as an open, and find its reverse voltage drop. The peak inverse voltage (*PIV*) can be found by setting up a closed loop KVL equation and solving for *PIV*; or perform a KVL walkabout from (−) *PIV* to (+) *PIV*.

$$PIV = V_{p\,2} + V_{p\,C}$$

$$PIV = 20\ V + 19.3\ V = 39.3\ V$$

This is a very unusual case. In general, except for half-wave rectifier input filtered capacitor circuits, the *PIV* is approximately equal to the peak voltage of the secondary $V_{p\,2}$.

$$PIV \approx V_{p\,2} \qquad \text{(all but half-wave capacitor filtered)}$$

However, the half-wave input capacitive filtered rectifier circuit produces a peak inverse voltage across the diode that is approximately twice the secondary peak voltage.

$$PIV \approx 2 \times V_{p\,2} \qquad \text{(half-wave capacitor filtered)}$$

$$PIV \approx 2 \times 20\ V_p = 40\ V$$

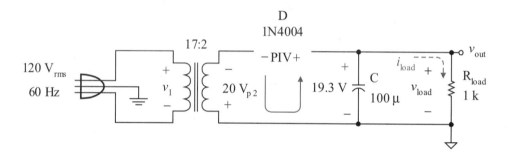

Figure 18-24(c) Peak inverse voltage (*PIV*) experienced by diode

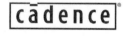

how big can you dream?™

Practice

Use MultiSIM or Cadence PSpice and verify the waveform, the ripple period, the output dc voltage, and the output rms voltage.

Answer: MultiSIM results: half-wave filtered waveform had an initial surge current creating a maximum output voltage of 31.7 V_p and did not stabilize for 70 ms before settling down to the expected waveform with a period of 16.7 ms, 18.3 V_{dc}, and 0.77 V_{rms}. If you place a sensor resistor in series with the diode, and use the oscilloscope to measure the test resistor's peak voltage drop (channel *A* less channel *B* mode), you can determine that there is an initial surge current of 0.67 A_p with a repeating diode forward current of about 0.15 A_p (the diode and the load experience different peak and average currents).

Exploration

Raise and lower the value of the capacitance *C* and resistance value *R* and note the effect on the output voltage wave shape, dc voltage, and rms voltage. As R and C are decreased, what is the effect on the ac ripple and the dc voltage? Use both oscilloscope measurements and dc/ac digital multimeter measurements to make your observations.

The simulator can be a very valuable tool in circuit analysis conditions like this. A simulation lets you very quickly analyze circuit values and wave shapes, and visualize a circuit's simulated responses. You can also change component values quickly and rerun the simulation to look at the effects of the change. This can be very helpful in circuit design in trying out different component values. You can use an oscilloscope to view the response including the initial surge current response to look at maximum voltage and current conditions. And you can use a simulated digital multimeter to quickly measure the simulated results of an ac or dc voltage or current.

18.4 Filtered Unregulated— Light to Heavy Load

The circuits of the previous section are called **unregulated** filtered rectifier circuits (or power supplies). Typically a *voltage regulator circuit* follows the filter circuit to regulate the output voltage to the load to maintain a fixed dc output voltage even with variations in the supply signal and/or the load current. This section examines the loading effects on unregulated filtered rectified voltages, from light loading to heavy loading, which cause deep discharge of the capacitor. Under light loading conditions with relatively modest capacitor discharge, the output ac ripple voltage can be estimated as a triangle wave and the voltage values reasonably estimated. However, the charge curve is really part of a sine wave, and the discharge wave is an exponentially decaying voltage. Even with modest discharge of the capacitor, the lightly loaded model begins causing significant error. For moderate to heavy loading, experimental data or simulated results are needed to accurately determine the actual output voltage. Moderate to heavy loading is examined using simulated results. Realize that the real goal is to produce a filtered output voltage such that its minimum voltage is not too low, so that it can feed a voltage regulator circuit. *Significant ripple may be reasonable.*

Example 18-6

Use MultiSIM to simulate the full-wave input capacitor filtered bridge rectifier of the circuit of Figure 18-25(a) for

a. C of 47 μF and R_{load} of 1 kΩ

b. C of 47 μF and R_{load} of 100 Ω

From your simulated results, find the output period and frequency, the peak output voltage, the minimum output voltage, and the peak-to-peak output voltage. Use a DMM voltmeter and current meter to find the dc output voltage, dc load current, ac output ripple rms voltage, and ac output ripple rms current. Calculate the % ripple.

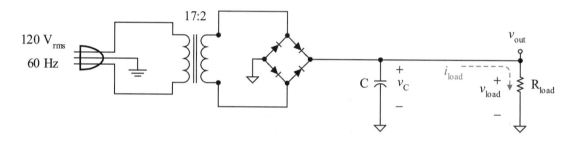

Figure 18-25(a) Input capacitive filter full-wave bridge rectifier example

Solution

a. Figure 18-25(b) shows the MultiSIM simulation with dc DMM meters, the XSC1 oscilloscope across the secondary in the difference mode, and the XSC2 oscilloscope across the load. The time for one output cycle of the ripple voltage is measured to be 8.4 ms, verifying that the full-wave rectifier has a frequency of 120 Hz. This is a lightly loaded filter with a percent ripple of 4%.

$$\% \ ripple = \frac{730 \ \mu V_{rms}}{18.0 \ V_{dc}} \times 100\% = 4\%$$

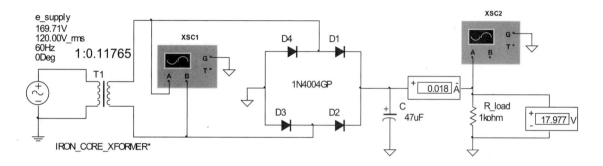

Meters shown in dc mode

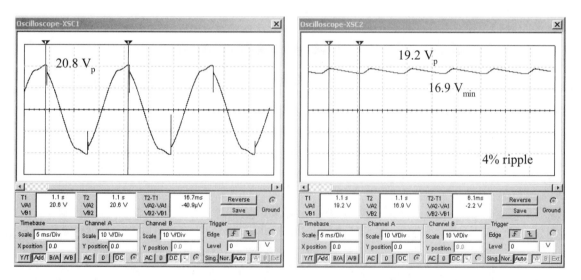

XSC1 v_2 (dc mode: Channel A—Channel B)   XSC2 v_{out} (dc mode: Channel A)

		DMM
V_{out}	dc	18.0 V_{dc}
I_{load}	dc	18 mA_{dc}
V_{out}	ac	730 mV_{rms}
I_{load}	ac	730 μA_{rms}

Period	T	8.4 ms
XSC1	$V_{p\,2}$	20.8 V_p
XSC2	$V_{p\,out}$	19.2 V_p
XSC2	$V_{min\,out}$	16.9 V_{min}
XSC2	$V_{pp\,out}$	2.2 V_{pp}

Figure 18-25(b) Example with C = 47 μF and R_{load} = 1 kΩ

b. Figure 18-25(c) shows the simulation with a 100 Ω load. The reduced *RC* time constant creates much more ripple.

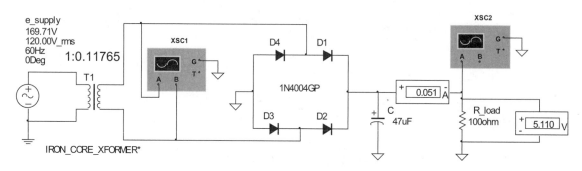

Meters shown in ac mode

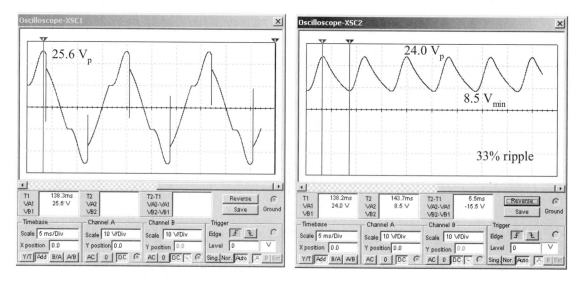

XSC1 v_2 (dc mode: Channel A—Channel B)   XSC2 v_{out} (dc mode: Channel A)

	DMM	
V_{out}	dc	18.0 V_{dc}
I_{load}	dc	18 mA_{dc}
V_{out}	ac	730 mV_{rms}
I_{load}	ac	730 μA_{rms}

Period	T	8.4 ms
XSC1	$V_{p\,2}$	20.8 V_p
XSC2	$V_{p\,out}$	19.2 V_p
XSC2	$V_{min\,out}$	16.9 V_{min}
XSC2	$V_{pp\,out}$	2.2 V_{pp}

Figure 18-25(c) Example with C = 47 μF and R_{load} = 100 Ω

Practice

For the example problem, place the output oscilloscope in the ac mode and adjust the sensitivity to more accurately display the ac ripple voltage for each case.

Answer: See Figure 18-25(d). *The voltage sensitivity for each case is different.* Note the triangular shape of the lightly loaded case with a load of 1 kΩ as compared to the sinusoidal rise and exponential decay curves of the heavily loaded case with a load of 100 Ω. The transformer secondary voltage has increased based upon the increased loading (increased load current). Transformer secondary voltages may increase, decrease, or stay the same with increasing load currents; or may be rated to operate at a specific secondary voltage with a specified load current (for example, 20 V at 1 A). Light to heavy loading can create significant variation in the secondary transformer voltages. *Be sure to match component specifications to your load conditions.*

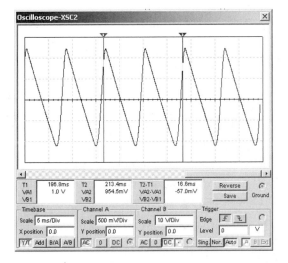

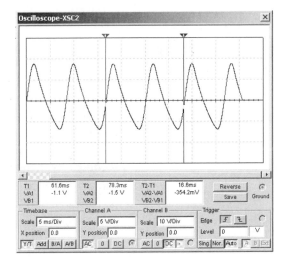

Ripple v_{out} (ac mode 500 mV/Div) with R $_{load}$ = 1k Ω   Ripple v_{out} (ac mode 5 V/Div) with R $_{load}$ = 100 Ω

Figure 18-25(d) Example ac ripple voltage oscilloscope displays

Exploration

Try various values of C and R_{load} in the example circuit and observe the capacitor filtered output wave shapes and dc and ac values.

Surge Current

Recall that an uncharged capacitor initially acts like a short when subjected to a suddenly applied voltage, creating a surge of current or what is called a **surge current**. To visualize this instantaneous effect, visualize the capacitor in each of the rectifier circuits being a short circuit at the instant that voltage is first applied to the circuit. The instantaneous voltage of the transformer secondary is applied to the rectifier circuit, which is driving a short circuit (initial model of the uncharged capacitor). The maximum surge current possible is when the initial instantaneously applied voltage is at its maximum (peak) or at its minimum. In that case, surge current through the diodes is

$$I_{surge} = \frac{V_{p\,2}}{R_{wire} + R_{secondary\ winding} + R_{diode} + R_{ESR\ C}}$$

which includes the circuit wire resistance, the transformer secondary resistance, the forward-biased turned-on diode resistance, and the **equivalent series resistance** (ESR) of the capacitor (primarily the contact resistance of its internal connectors). In our previous examples, the peak secondary voltage is about 20 V. Suppose the initial secondary circuit has a total resistance of about 10 Ω. Then the nonrepetitive initial current, that is, the surge current, of the secondary is very high.

$$I_{surge} = \frac{20\,V_p}{10\ \Omega} = 2\ A_p$$

The rectifier diodes must survive this initially high current; therefore the surge current must be taken into account when selecting the rectifier diodes such that its maximum nonrepetitive forward current rating I_{FSM} is greater than I_{surge} of the circuit. A simulator can be used effectively to measure the initial currents and voltages by the use of the oscilloscope during the first few cycles. To observe current waveforms on the oscilloscope use a series test resistor, measure the test resistor's peak voltage drop, and then apply Ohm's law to find the peak current.

Regulation

Once the filter becomes moderately to heavily loaded with significant discharge of the capacitor, you must examine the output waveform to see if the filtered output can serve the requirements of the system. The output waveform of Figure 18-25(c) appears to have too much ripple in it. However, if this signal is being used to drive a voltage regulator circuit that only requires a 6 V input then this signal is more than adequate. Figure 18-26 demonstrates this principle.

$$V_{\text{min C filter}} > V_{\text{min into regulator specification}} > V_{\text{dc out of regulator}}$$

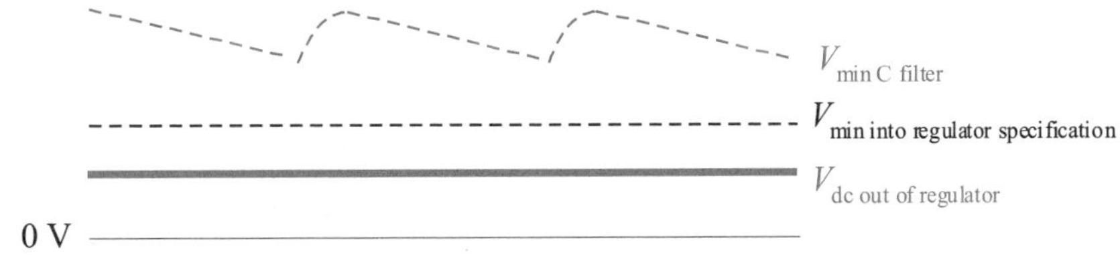

Figure 18-26 Regulated fixed dc output voltage

The next section utilizes the zener diode as a voltage regulator to produce a quality fixed dc output voltage for light current loads. If heavy load currents are expected, then an IC voltage regulator is needed as described later in this chapter.

18.5 Zener Voltage Regulator

The characteristic curve of the zener diode and its normal operating region are shown in Figure 18-27(a). V_Z is called the *zener voltage* and is the voltage maintained across the zener diode once the zener voltage is achieved. Specifications associated with the zener diode are stated as positive values with the understanding that the diode is actually reverse biased. Therefore, V_Z is stated as a positive number (for example, 6.2 V) even though it is a negative value on its characteristic curve. V_F is the forward biased voltage drop across the diode and is approximately 0.7 V to 0.9 V for a silicon diode.

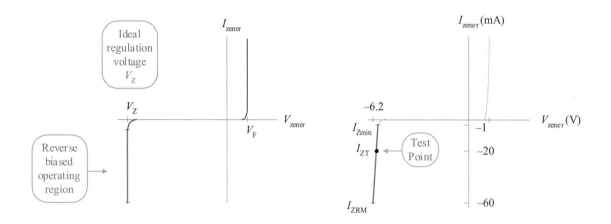

(a) Zener diode characteristic curve   (b) 1N753 zener diode characteristic curve

Figure 18-27 The zener diode characteristic curve

Figure 18-27(b) shows the key specification values associated with the zener diode and specific ratings for the 1N753 zener diode.

V_Z	6.2 V	Zener voltage (regulation voltage)
I_{ZT}	20 mA	Test current at the rated V_Z voltage
I_{Zmin}	1 mA	Minimum zener current needed to hold regulation
I_{ZRM}	60 mA	Zener reverse maximum current

Another key specification is the reverse resistance rating Z_Z. For the 1N753 zener diode, the reverse resistance rating Z_Z is 7.0 Ω at 20 mA of test current. The slope of the reverse zener regulation curve is the reciprocal of Z_Z. The voltage drop across the zener diode D increases with increasing zener current through it due to the zener diode's internal resistance Z_Z. Generally, as the current through the zener increases, its voltage drop increases slightly across Z_z, which in turn increases the total zener diode voltage drop and the load voltage drop since $V_{dc\ out} \approx V_D$.

The basic circuit configuration of a zener voltage regulating circuit is shown in Figure 18-28. A filtered rectified signal is fed to the series resistance R_S and the zener diode Z. The load is placed across or in parallel with the zener. That is, the output dc voltage across the load is the same as the zener voltage V_Z since they are in parallel.

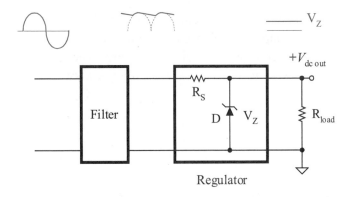

Figure 18-28 Zener regulator circuit

What is the purpose of R_S? Imagine that R_S were not there but instead the filtered output voltage is fed directly across the zener diode. Suppose the filtered input signal to the regulator were 14 V. Then 14 V would appear across the zener diode with a resistance of 7 Ω producing a zener current of 2 A. Recall that the maximum zener current rating for the 1N753 is 60 mA. Do you see the problem? The current through the zener diode must be limited by a series resistance; therefore, R_S is required to limit the current to the zener diode to prevent the destruction of the zener diode. The current limiting resistor R_S must be large enough to prevent the zener from exceeding I_{ZRM} but not so large that it reduces the zener current below its regulating range. Also, R_S can potentially dissipate (waste) relatively significant power, and the power rating of R_S must be taken into consideration.

Also, apply the rule of thumb that the minimum voltage into the zener regulator should be at least 2 V greater than the zener voltage to ensure that the zener does not drop out of regulation. Figure 18-29 demonstrates the result of a zener dropping out of regulation because the input signal from the filter falls below the zener voltage rating.

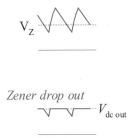

Figure 18-29
Zener dropping out of
regulation

Example 18-7

For the zener regulating circuit of Figure 18-30(a), find the ideal output dc voltage. Find the maximum zener current without the load connected. Is the zener safe? With the *load attached*, find the *maximum* values of V_{RS}, I_{RS}, instantaneous power dissipation P_{RS}, and I_Z. With the *load attached*, find the *minimum* values of V_{RS}, I_{RS}, instantaneous power dissipation P_{RS}, and I_Z. Find the rms voltage, current, and average power dissipation of R_S. What power dissipation rating should you select for R_S to ensure its safety? Solve this a step at a time using basic circuit knowledge and analysis. It is not magic; it is circuit analysis basics.

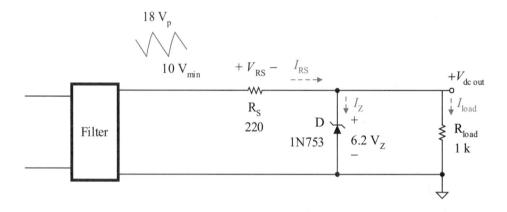

Figure 18-30(a) Example zener regulated circuit

Solution

The *ideal output voltage* is the zener voltage, which is across the load.

$$V_{dc\ out} = V_D = V_Z = 6.2\ V_{dc}$$

Always find the maximum zener current without the load attached as shown in Figure 18-30(b). This is a conservative estimate and includes the possibility that the load could be detached. The maximum zener current occurs when R_S drops its maximum voltage, and that occurs when the input to the regulator is at its maximum. The maximum voltage drop across R_S is

$$V_{max\ RS} = V_{max\ filter} - V_{dc\ out}$$

$$V_{max\ RS} = 18\ V_p - 6.2\ V_{dc} = 11.8\ V_p$$

By Ohm's law, the maximum R_S current is

$$I_{max\ RS} = \frac{11.8\ V}{220\ \Omega} = 53.6\ mA_p$$

Since this is now a series circuit with the load removed

$$I_{max\ Z\ without\ load} = I_{max\ RS} = 53.6\ mA_p$$

and the zener's instantaneous maximum power dissipation without a load is

$$P_{max\ Z\ without\ load} = 53.6\ mA_p \times 11.8\ V_p = 633\ mW_p$$

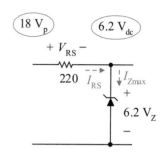

Figure 18-30(b)
Circuit to find maximum zener current without a load

The 1N753 diode is safe since the maximum current through the zener is 50.9 mA and its I_{ZRM} rating is 60 mA. The power rating for this zener diode is a half-watt, so why is it safe? The power rating is the average power sustained by the diode, not its instantaneous maximum. We shall calculate the average power dissipation later in this solution.

Now assume that the *load is attached* and the regulator is at its *maximum input voltage* of 18 V_p as shown in Figure 18-30(c). This produces *maximum* voltages, currents, and power dissipations in the regulator circuit *with the load attached*.

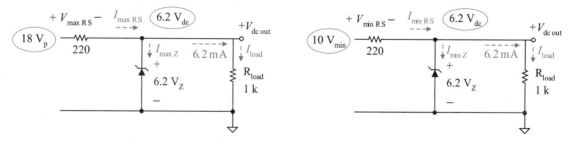

(c) Maximum input voltage circuit values   (d) Minimum input voltage circuit values

Figure 18-30(c,d) Regulator circuit analysis with maximum and minimum input voltages

Applying basic laws and the power rule, the maximum values for the regulator circuit are as follows.

$$V_{max\ RS} = 18\ V - 6.2\ V = 11.8\ V_p$$

$$I_{max\ RS} = \frac{11.8\ V}{220\ \Omega} = 53.6\ mA_p$$

$$P_{max\ RS} = 53.6\ mA \times 11.8\ V = 633\ mW_p$$

Warning! The peak instantaneous power dissipation of this current limiting resistor R_S is 633 mW. This is a potential problem. We need to examine the average power dissipation later in this solution to ensure that R_S has the appropriate power dissipation rating.

By Kirchhoff's current law at the output node,

$$I_{max\ Z} = 53.6\ mA - 6.2\ mA = 47.4\ mA_p$$

$$P_{max\ Z} = 47.4\ mA \times 6.2\ V = 294\ mW_p$$

Under loaded conditions, the zener is well within its half-watt average power rating.

Now assume that the *load is still attached* and that the regulator is operating at its *minimum input voltage* of 10 V_{min} as shown in Figure 18-30(d). This produces *minimum* voltages, currents, and power dissipations in the regulator circuit *with the load attached*. Applying basic laws and the power rule, the minimum values for the regulator circuit are as follows.

$$V_{min\ RS} = 10\ V - 6.2\ V = 3.8\ V_{min}$$

$$I_{min\ RS} = \frac{3.8\ V}{220\ \Omega} = 17.3\ mA_{min}$$

$$I_{min\ Z} = 17.3\ mA - 6.2\ mA = 11.1\ mA_{min}$$

This minimum zener current is well within the regulation region of this zener. Based upon the 1N753 data sheets, this zener diode is specified by its characteristic curves as operating down to a minimum of 1 mA and up to maximum of 60 mA.

The minimum instantaneous power dissipation of the series resistor is

$$P_{\text{min RS}} = 17.3 \text{ mA} \times 6.2 \text{ V} = 107 \text{ mW}_{\text{min}}$$

The average power dissipated by the series resistor R_S in the regulator is somewhere between its maximum instantaneous power dissipation of 633 mW and its minimum of 107 mW. To be economical, you want to design the resistor for the lowest reasonable power rating that is safe. A higher power rated resistor costs more and if your company makes millions of these regulators that becomes a major expense. Production volume does make a difference. So what is the average power dissipated by this regulator resistor? Before you can answer that question, you need to find the rms value of the voltage across R_S. The signal appearing across R_S is shown in Figure 30(e). Recall that average power is based upon the rms values of voltage and current.

$$V_{\text{pp RS}} = 11.8 \text{ V} - 3.8 \text{ V} = 8 \text{ V}_{\text{pp}}$$

The rms voltage of the triangular shaped ac ripple voltage is

$$V_{\text{rms ac RS}} = \frac{8 \text{ V}}{2\sqrt{3}} = 2.31 \text{ V}_{\text{rms}}$$

The average (dc) value of the total signal appearing across R_S is

$$V_{\text{dc RS}} = \tfrac{1}{2}\left(11.8 \text{ V} + 3.8 \text{ V}\right) = 7.8 \text{ V}_{\text{dc}}$$

The total signal rms voltage appearing across the R_S resistor is

$$V_{\text{rms total signal RS}} = \sqrt{(2.31 \text{ V})^2 + (7.8 \text{ V})^2} = 8.13 \text{ V}_{\text{rms}}$$

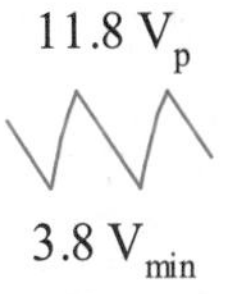

Using the power rule with the total signal rms voltage, calculate the average power dissipated by the R_S resistance.

$$P_{\text{average RS}} = \frac{V_{\text{rms}}^2}{R_S} = \frac{(8.31 \text{ V})^2}{220 \ \Omega} = 301 \text{ mW}_{\text{average}}$$

The result shows that a half-watt resistor is sufficient to dissipate the *average power* required in this regulator circuit.

11.8 V$_{\text{p}}$

3.8 V$_{\text{min}}$

Figure 18-30(e)
V_{RS} signal voltage

Practice

Use the MultiSIM simulator and verify the wave shapes and the values calculated in the example problem. Specifically, compare your calculated results with your simulated results for $V_{dc\ out}$, $V_{dc\ RS}$, $V_{rms\ ac\ RS}$, $P_{average\ RS}$. Why might there be differences between our manual calculations and the simulated results?

Answer: See Figure 18-30(f). Meters are in the dc mode. To display rms values, change the meters to the ac mode.

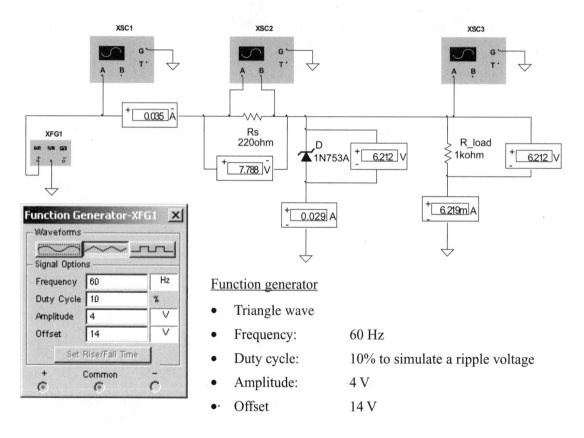

Function generator

- Triangle wave

- Frequency: 60 Hz

- Duty cycle: 10% to simulate a ripple voltage

- Amplitude: 4 V

-· Offset   14 V

Figure 18-30(f) MultiSIM simulation with dc meter readings displayed

The simulated results compare very well with the calculated values, showing less than 1% difference: simulated $V_{dc\ out}$ of 6.212 V_{dc} (6.2 V_{dc} calculated); $V_{dc\ RS}$ of 7.79 V_{dc} (7.8 V_{dc} calculated); $V_{rms\ ac\ RS}$ of 2.29 V_{rms} (2.31 V_{rms} calculated); average P_{RS} of 300 mW (301 mW calculated). The manual calculations used the ideal model of the zener dropping 6.2 V with an internal resistance of 0 Ω, while the simulator dynamically represented the zener and therefore produced a zener voltage slightly greater than the rated value of 6.2 V. Also, an ideal input triangle waveform was assumed in the calculations, but with a transformer supplied voltage and an input filter capacitor, the output voltage of the filter is not a pure triangle wave and its peak voltage could vary from the transformer based upon the transformer's secondary loading.

Exploration

Simulate the circuit of Figure 18-25(b) with a zener voltage regulator circuit inserted as shown in Figure 18-31. Use a regulator current limiting resistance of 220 Ω and a 1N753 zener diode. Do not recreate the entire circuit, but copy the previous example circuit simulation file to a new file and insert the zener circuit. Place the oscilloscope Channel A at the input of the regulator and Channel B at the output of the regulator. Reduce the value of load resistance R_{load} until output distortion occurs and notice the effect on the dc and ac output voltage. Speculate on why the zener starts dropping out when the load resistance is too low.

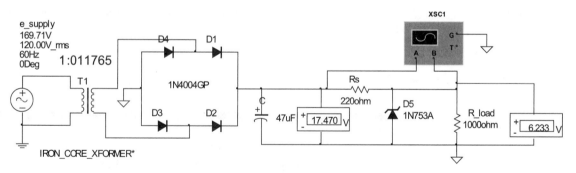

Figure 18-31 Exploration circuit

If you performed the above exploration, you should have noticed that the zener started dropping out below 180 Ω. Why? Channel A of the oscilloscope XSC1 shows the minimum input voltage to the regulator to be about 14.8 V. That voltage is voltage divided between R_S and the load R_{load}. Assume that the zener is taken out of the circuit for this calculation; then the output voltage is

$$V_{min\ out} = \left(\frac{180\ \Omega}{180\ \Omega + 220\ \Omega} \right) 14.8\ V_{min} = 6.66\ V_{min}$$

Do you see the problem yet? Try 150 Ω.

$$V_{min\ out} = \left(\frac{150\ \Omega}{150\ \Omega + 220\ \Omega} \right) 14.8\ V_{min} = 6.00\ V_{min}$$

If the load is reduced below 180 Ω, the voltage drop across R_{load} decreases until it is below the 6.2 V zener voltage. When the load voltage dips below 6.2 V, the zener drops out of regulation and zener drop out occurs resulting in a distorted dc output and a reduced dc output voltage. Thus there are two major conditions that must occur to prevent zener dropout. The minimum input voltage from the filter must be high enough to drive the regulator circuit. For the zener voltage regulator, the rule of thumb is that the input regulator voltage must be at least 2 V above the zener voltage. This design criterion is satisfied with a 14.8 V_{min} input.

The other criterion is that there must be enough voltage developed across the load to sustain zener regulation.

$$\left(\frac{R_{load}}{R_{load} + R_S} \right) V_{min\ into\ regulator} > V_Z$$

Example 18-8

For the simulated circuit of Figure 18-31, find the minimum value of R_{load} to prevent zener drop out. Find the minimum input voltage to the regulator to prevent zener drop out. For that given minimum input voltage, find the new minimum load resistance required to prevent zener drop out.

Solution

To maintain sufficient voltage across the zener, the minimum value of R_{load} must satisfy the condition

$$\left(\frac{R_{load}}{R_{load} + 220\ \Omega} \right) 14.8\ V_{min} = 6.2\ V$$

Solve this equation for R_{load}. The minimum value of R_{load} required to maintain zener regulation is

$$R_{\text{load}} = 159 \ \Omega$$

The next available standard resistor value is 180 Ω, which confirms the simulation results from the previous exploration problem.

The minimum input voltage to the regulator from the filter circuit must be at least 2 V higher than the zener voltage rating. For the 1N753 zener diode with a zener voltage of 6.2 V, the minimum input voltage to the regulator circuit is 8.2 V. Now the new minimum load resistance must satisfy the equation

$$\left(\frac{R_{\text{load}}}{R_{\text{load}} + 220 \ \Omega} \right) 8.2 \ V_{\text{min}} = 6.2 \ V$$

resulting in a new value of R_{load} of 221 Ω. You need to select the next highest standard resistance above 221 Ω to ensure that zener drop out does not occur.

Practice

Given the circuit of Figure 18-31 and that the load resistance must be 100 Ω, what must you change in the circuit to allow the regulator to work properly?

Answer: Change the regulator series resistance R_S to less than 138 Ω, but *recheck the power dissipation* of the regulator series resistor.

multiSIM

Next let's quantify the effect of the regulator circuit on the ripple voltage. We want to compare the output ripple voltage to the input ripple voltage. An ideal dc supply has no output ripple voltage, and a quality dc supply has very little ac ripple voltage compared to its dc voltage.

One way to compare the output ripple to the input ripple is to express regulator ripple voltage gain, which is the output ripple voltage divided by the input ripple voltage (a pure number without units).

$$A_{\text{v}} = \frac{v_{\text{out}}}{v_{\text{in}}}$$

For an amplifier this would typically be a voltage gain, where A_{v} is greater than or equal to one (1). But the output ripple of a voltage regulator is much, much less than the input ripple voltage, so this is not a gain but is a voltage loss. The output ripple voltage is less than the input ripple. This is an **attenuation** or loss of a signal. Attenuation is the reciprocal of gain; therefore, the voltage attenuation can be defined as

$$voltage\ attenuation = \frac{v_{in}}{v_{out}}$$

In electronics, gains and attenuations may be very large numbers; therefore base 10 logarithms (log) are used to describe large gains and attenuations with more reasonable numbers. We shall discuss this in greater detail in the next chapter of this textbook, but for now we shall simply define the **logarithm voltage gain** or **dB voltage gain** as

$$A_v\,(dB) = 20\ \log\left(\frac{v_{out}}{v_{in}}\right)$$

and the **dB voltage attenuation** as

$$voltage\ attenuation\ (dB) = 20\ \log\left(\frac{v_{in}}{v_{out}}\right)$$

where the abbreviation **dB** represents the pure unit **decibel**. The term dB is not a unit in the usual sense. For example, if the input voltage is 2 mV and the output voltage is 200 mV, the voltage gain A_v is 100, a pure number without units.

$$A_v = \frac{200\ mV}{2\ mV} = 100$$

The dB voltage gain is

$$A_v\,(dB) = 20\ \log(100) = 40\ dB$$

The dB indicates that this is the logarithmic gain and not the pure voltage gain form.

Example 18-9

Perform the simulation of Figure 18-30(f) measuring the ac ripple voltage into and out of the regulator circuit. Use your measured results to calculate the regulator's voltage gain and attenuation. Also express the ripple voltage attenuation in decibels.

Solution

The simulated regulator's input rms voltage as measured with an ac meter is 2.31 V_{rms} while the simulated regulator output voltage is 16 mV_{rms}. The regulator has an ac voltage gain A_v of

$$A_v = \frac{16 \text{ mV}_{rms}}{2.31 \text{ V}_{rms}} = 0.00693$$

which is really an attenuation since the voltage gain is less than unity. The ac ripple attenuation is

$$voltage \ attenuation = \frac{2.31 \text{ V}_{rms}}{16 \text{ mV}_{rms}} = 144$$

The dB or logarithmic attenuation is

$$voltage \ attenuation \ (dB) = 20 \ \log(144) = 43.2 \text{ dB}$$

Practice

A voltage regulator has a dB attenuation rating of 60 dB. Find its actual voltage attenuation ratio. If the input ripple voltage to this voltage regulator is 5 V_{rms}, find its output ripple voltage.

Answer: 1000, 5 mV_{rms}. Hint: substitute 60 dB into the dB attenuation gain formula and solve for the attenuation inversely. Then divide the input ripple by the actual attenuation to find the output ripple.

18.6 Pass Transistor

The zener regulated circuit has limited load current capacity. As the load resistance is lowered to create higher load currents, the voltage divider of R_S and R_{load} eventually reduces the load voltage to the point of zener drop out. Zener regulators typically operate with load currents into the hundreds of milliamps range. To operate at higher load currents, a zener regulated circuit load current capacity can be increased with the use of a high current capacity BJT. The circuit of Figure 18-32 shows the schematic of the **pass transistor** (BJT) working with a zener regulator circuit. It is called a pass transistor since the input current to the regulator essentially passes through the collector-to-emitter of the BJT to the load ($I_C \approx I_E$). The output voltage is

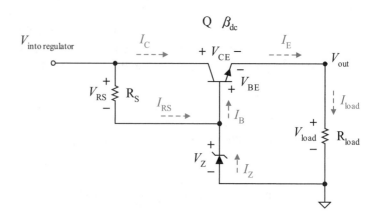

Figure 18-32 Pass transistor regulator basic circuit configuration

$$V_{out} = V_Z - V_{BE}$$

The silicon BJT V_{BE} is typically 0.8 V to 0.9 V for heavy loads.

Note that BJT current gain beta is being designated as the dc beta β_{dc}. The dc beta β_{dc} (that is, I_C / I_B) can be different from its ac beta β_{ac} (that is, $\Delta I_C / \Delta I_B$) especially for high current power transistors that are used in pass transistor circuits. The next example demonstrates several characteristics of this circuit as compared to the previous zener regulated circuit without the use of the pass transistor. One thing that you must be mindful of is the power dissipated by this high current capacity BJT. The power dissipated by the BJT ($P_{D\ BJT}$) is essentially the power dissipated in the collector section of the BJT and is found by applying the power rule to the BJT.

$$P_{D\ BJT} = I_C\ V_{CE}$$

Example 18-10

The zener regulated circuit is enhanced with the inclusion of a pass transistor as shown in the circuit of Figure 18-33(a). Find the output dc voltage. Find the maximum and the minimum currents and voltages in the circuit. Ensure that the zener diode is always operating in its zener voltage regulation range. Find the power dissipated by the current limiting resistor for the zener diode and the pass transistor. Use the current and voltage references established in the circuit of Figure 18-32.

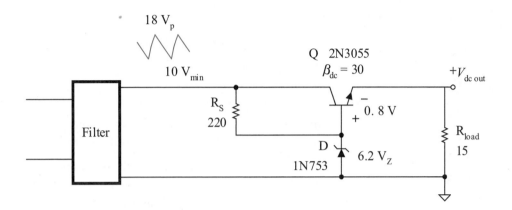

Figure 18-33(a) Example using a pass transistor

Solution

The output voltage and load current are approximately

$$V_{dc\ out} \approx 6.2\ V - 0.8\ V = 5.4\ V_{dc}$$

$$I_{dc\ load} \approx \frac{5.4\ V}{15\ \Omega} = 360\ mA_{dc}$$

$$P_{dc\ load} = 5.4\ V \times 360\ mA = 1.94\ W$$

The approximate dc currents of the BJT are

$$I_C \approx I_E = I_{load} \approx 360\ mA_{dc}$$

$$I_B \approx \frac{360\ mA}{30} = 12.0\ mA_{dc}$$

Now analyze the circuit values with the maximum (peak) input voltage of $18\ V_p$ and show that the zener is safely under its maximum rating of 60 mA.

$$V_{p\ CE} = 18\ V - 5.4\ V = 12.6\ V_p$$

$$V_{p\ RS} = 18\ V - 6.2\ V = 11.8\ V_p$$

$$I_{p\,RS} = \frac{11.8\ \text{V}}{220\ \Omega} = 53.6\ \text{mA}_p$$

$$I_{p\,Z} \approx 53.6\ \text{mA} - 12.0\ \text{mA} = 41.6\ \text{mA}_p$$

The zener current without a load is

$$I_{p\,Z} \approx 53.6\ \text{mA}_p$$

Zener is safely below its maximum current rating.

An unloaded circuit produces the maximum possible zener current, and it is within its 60 mA maximum specification.

Now analyze the circuit values with the minimum input voltage of 10 V_{min}. and show that the zener minimum current is still sufficient to maintain regulation (above 1 mA).

$$V_{min\,CE} = 10\ \text{V} - 5.4\ \text{V} = 4.6\ \text{V}_{min}$$

$$V_{min\,RS} = 10\ \text{V} - 6.2\ \text{V} = 3.8\ \text{V}_{min}$$

$$I_{min\,RS} = \frac{3.8\ \text{V}}{220\ \Omega} = 17.3\ \text{mA}_{min}$$

Minimum zener current sufficient to hold regulation

$$I_{min\,Z} \approx 17.3\ \text{mA} - 12.0\ \text{mA} = 5.3\ \text{mA}_{min}$$

The zener diode experiences a minimum current of 5.3 mA_{min} and an unloaded maximum current of 53.6 mA_p; therefore, *the zener diode is operating well within its zener voltage regulation region*. The average zener current in the loaded circuit is

$$I_{dc\,Z} = \tfrac{1}{2}\left(41.6\ \text{mA}_p + 5.3\ \text{mA}_{min}\right) = 23.5\ \text{mA}_{dc}$$

However, the zener current does depend on the BJT β_{dc} current gain, and a sufficiently high β_{dc} must be maintained to keep the base current sufficiently low to maintain minimum zener current for regulation.

Now find the BJT collector rms current, the collector-to-emitter rms voltage drop, and average power dissipated. The load current and thus the collector current have been assumed to basically be a dc current; therefore, the dc current and the rms current are the same.

$$I_C \approx 360\ \text{mA}_{rms}$$

The rms of the ac only signal is

$$V_{CE \text{ rms ac}} = \frac{12.6 \text{ V}_p - 4.6 \text{ V}_{min}}{2\sqrt{3}} = \frac{8 \text{ V}_{pp}}{2\sqrt{3}} = 2.31 \text{ V}_{rms}$$

Notice that ripple voltage across the BJT is approximately the same as the input ripple voltage to the regulator. The collector-to-emitter dc voltage is

$$V_{CE \text{ dc}} = \frac{1}{2}(12.6 \text{ V}_p + 4.6 \text{ V}_{min}) = 8.6 \text{ V}_{dc}$$

The collector-to-emitter total signal rms is

$$V_{CE \text{ rms total}} = \sqrt{(8.6 \text{ V}_{dc})^2 + (2.31 \text{ V}_{rms})^2} = 8.90 \text{ V}_{rms}$$

Be aware of the BJT *average power dissipation*. If the average power dissipation exceeds the BJT rating, then a cooling heat sink will be needed to conduct heat away from the BJT faster to keep the BJT cooler. Or else pick a BJT with a higher power rating. Heat sinking is covered later in this chapter and can be the better, cheaper solution than a more expensive power BJT.

$$P_{D \text{ average BJT}} = 8.90 \text{ V}_{rms} \times 360 \text{ mA}_{rms} = 3.20 \text{ W}_{average}$$

Since the signal energy is primarily dc, it is common practice to calculate $P_{D \text{ dc BJT}}$ by using the BJT's dc values. For this example, the rms voltage drop across the BJT is approximately its dc drop.

$$V_{CE \text{ rms total}} = \sqrt{(8.6 \text{ V}_{dc})^2 + (2.31 \text{ V}_{rms})^2} \approx \sqrt{(8.6 \text{ V}_{dc})^2}$$

$$V_{CE \text{ rms total}} \approx V_{CE \text{ dc}} = 8.60 \text{ V}_{dc}$$

Then the BJT dc power dissipation can be calculated as

$$P_{D \text{ dc BJT}} \approx V_{CE \text{ dc}} I_{C \text{ dc}}$$

$$P_{D \text{ dc BJT}} \approx (14 \text{ V}_{dc} - 5.4 \text{ V}_{dc}) \times 360 \text{ mA}_{dc}$$

$$P_{D \text{ dc BJT}} \approx (8.6 \text{ V}_{dc}) \times 360 \text{ mA}_{dc} = 3.10 \text{ W}_{dc}$$

which is a 3% difference, even with a significant ripple. You can estimate the average BJT power dissipation using dc values.

$$P_{D \text{ average BJT}} \approx P_{D \text{ dc BJT}} = V_{CE \text{ dc}} I_{C \text{ dc}}$$

multi**SIM**

Practice

Verify the values of the example problem using a simulator. Also, measure the simulated ripple rms input and output voltages and calculate the percent ripple, ripple voltage attenuation, and dB ripple voltage attenuation.

Answer: See Figure 18-33(b) for the simulated circuit.

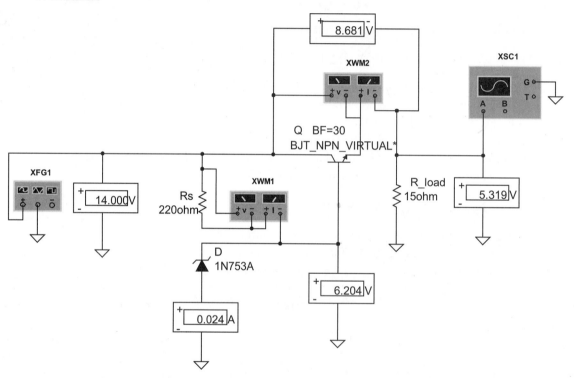

Figure 18-33(b) Simulated circuit of Figure 18-33(a) with DMM dc values

The simulated values are within 5% difference of the manual calculations. The simulated circuit does show that the BJT V_{BE} is closer to 0.9 V instead of the assumed 0.8 V for manual calculations, causing a slightly higher percent difference throughout the calculations. $V_{rms\ input\ ripple}$ (2.132 V_{rms}); $V_{rms\ output\ ripple}$ (26 mV_{rms}); output percent ripple (0.5%); ripple voltage attenuation (82); ripple voltage dB attenuation (38.3 dB). The percent ripple is very low and the ripple voltage attenuation is still good, even under heavy load.

Pass Transistor Inside Negative Feedback Loop

The circuit of Figure 18-34 demonstrates placing the pass transistor inside the negative feedback loop of an op amp circuit. The major advantage of this circuit is that the open loop gain of the op amp turns on the BJTs nearly instantly, overcoming the 0.7 V base-emitter voltage with microvolts of input. The output voltage is now essentially the zener voltage multiplied by the gain of the negative feedback amplifier circuit. The negative feedback closed loop gain $A_{\text{closed loop}}$ is

$$A_{\text{closed loop}} = \frac{R_f + R_i}{R_i}$$

$$V_{\text{out}} = A_{\text{closed loop}} V_Z$$

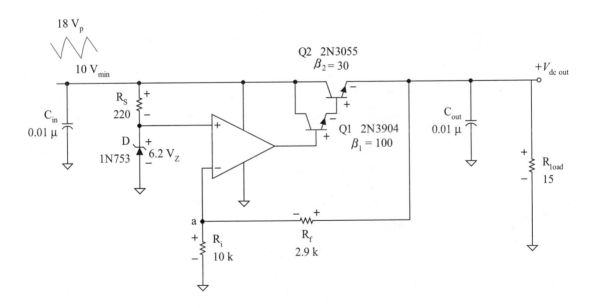

Figure 18-34 Pass transistor Darlington pair in op amp negative feedback loop

If a voltage follower amplifier is used, the closed loop gain is unity (1) and the output voltage is the same as the zener voltage.

$$V_{\text{out}} = V_Z$$

A **Darlington pair** arrangement of transistors is used to significantly reduce the op amp output current by the product of the two BJT beta factors.

$$I_{\text{out op amp}} = \frac{I_{E\,Q2}}{(\beta_1 + 1)(\beta_2 + 1)}$$

$$I_{\text{out op amp}} \approx \frac{I_{E\,Q2}}{\beta_1 \beta_2}$$

Q1 is typically a low current (small signal) BJT with a high β factor, while Q2 is a high current (large signal) BJT with a low β factor. For example, suppose the emitter current of Q2 were 120 mA, β_1 were 30, and β_2 were 100, then the op amp current would be approximately

$$I_{\text{out op amp}} \approx \frac{120 \text{ mA}}{(30)(100)} = 40 \text{ μA}$$

A heavy load current is reduced to a very low op amp output current, keeping the op amp within its output current range.

The compensating capacitors at the input and output of the regulator are used to prevent parasitic oscillations through the dc power supply.

Example 18-11

For the circuit of Figure 18-34, find the maximum and minimum zener diode current, the dc output voltage, the dc load current, the Q2 base current, the op amp output current, and the power dissipated by the Q2 large signal transistor.

Solution

The zener diode and R_S form an essentially series circuit since the current into the op amp is approximately 0. Thus, the zener current has essentially the same current as the limiting resistor R_S.

$$I_{\text{min Z}} = I_{\text{min RS}} = \frac{10 \text{ V} - 6.2 \text{ V}}{220 \text{ Ω}} = 17.3 \text{ mA}_{\text{min}}$$

$$I_{\text{max Z}} = I_{\text{max RS}} = \frac{18 \text{ V} - 6.2 \text{ V}}{220 \text{ }\Omega} = 53.6 \text{ mA}_{\text{min}}$$

The zener diode is operating near its maximum reverse current rating. As a circuit designer, you may want to consider increasing the R_S value to a slightly higher resistance value to reduce the zener operating currents.

Since this is a negative feedback amplifier system and assuming that the op amp is not hitting a rail, the input voltage drop across the op amp is essentially 0 V. Therefore, the inverting input is at the same node voltage as the noninverting input; therefore, the voltage at node a is the same as the drop across the zener.

$$V_{\text{dc a}} \approx V_Z = 6.2 \text{ V}_{\text{dc}}$$

$$V_{\text{dc Ri}} = V_a = 6.2 \text{ V}_{\text{dc}}$$

The remaining calculations are the same as the basic negative feedback op amp amplifier circuit.

$$I_{\text{dc Ri}} = \frac{6.2 \text{ V}}{10 \text{ k}\Omega} = 0.62 \text{ mA}_{\text{dc}}$$

$$I_{\text{dc Rf}} = I_{\text{dc Ri}} = 0.62 \text{ mA}_{\text{dc}}$$

$$V_{\text{dc Rf}} = 0.62 \text{ mA}_{\text{dc}} \times 2.9 \text{ k}\Omega = 1.80 \text{ V}_{\text{dc}}$$

$$V_{\text{dc out}} = 6.2 \text{ V} + 1.8 \text{ V} = 8.0 \text{ V}_{\text{dc}}$$

$$I_{\text{dc load}} = \frac{8.0 \text{ V}}{15 \text{ }\Omega} = 533 \text{ mA}_{\text{dc}}$$

$$P_{\text{dc load}} = 8.0 \text{ V} \times 533 \text{ mA} = 4.26 \text{ W}$$

The Q2 emitter current is the load current less the feedback current.

$$I_{\text{E dc}} = 533 \text{ mA} - 0.62 \text{ mA} \approx 532 \text{ mA}$$

The Q2 base current is

$$I_{\text{B dc Q2}} = \frac{532 \text{ mA}}{31} = 17.2 \text{ mA}_{\text{dc}}$$

The op amp output current feeds the base of the Q1 BJT and thus

$$I_{\text{out op amp}} = I_{\text{B dc Q1}} = \frac{17.2 \text{ mA}}{101} = 172 \text{ } \mu\text{A}_{\text{dc}}$$

The output current for the op amp is well below the mA range.

Also, check the output voltage of the op amp, which is the output voltage *plus* the two base-emitter voltage drops, to be sure that it falls within the op amp's rail voltage level.

$$V_{\text{out op amp}} \approx 8.0 \text{ V} + 0.8 \text{ V} + 0.7 \text{ V} = 9.5 \text{ V}_{\text{dc}}$$

There is a problem. The minimum voltage to the op amp positive supply is 10 V_{dc} from the input ripple voltage. Normally at this point you should increase the filter capacitance to reduce the ripple to provide at least 2 V of headroom for the op amp. The minimum op amp input voltage should be at least 12 V to satisfy the op amp. The op amp is likely to be hitting its upper rail voltage. *Continue the manual calculations but realize that this is a potential problem.*

The ripple voltage across Q2 is the same as the input ripple voltage since the load voltage is approximately a fixed dc voltage. Therefore the rms of the ac ripple voltage is

$$V_{\text{CE rms ac}} = \frac{8 \text{ V}_{\text{pp}}}{2\sqrt{3}} = 2.31 \text{ V}_{\text{rms}}$$

Next find the dc voltage drop across Q2 and its power dissipation.

$$V_{\text{CE max Q2}} = 18 \text{ V}_{\text{p}} - 8.0 \text{ V}_{\text{dc}} = 10.0 \text{ V}_{\text{p}}$$

$$V_{\text{CE min Q2}} = 10 \text{ V}_{\text{min}} - 8.0 \text{ V}_{\text{dc}} = 2.0 \text{ V}_{\text{min}}$$

$$V_{\text{CE dc}} = \frac{1}{2}\left(10.0 \text{ V}_{\text{p}} + 2.0 \text{ V}_{\text{min}}\right) = 6.0 \text{ V}_{\text{dc}}$$

$$V_{\text{CE rms total Q2}} = \sqrt{\left(6.0 \text{ V}_{\text{dc}}\right)^2 + \left(2.31 \text{ V}_{\text{rms}}\right)^2} = 6.43 \text{ V}_{\text{rms}}$$

$$P_{\text{D average Q2}} = 6.43 \text{ V}_{\text{rms}} \times 533 \text{ mA}_{\text{rms}} = 3.4 \text{ W}_{\text{average}}$$

Note this could be approximated as the dc power with only a slightly lower result even with a large ripple.

$$P_{\text{D dc Q2}} = 6.0 \text{ V}_{\text{dc}} \times 533 \text{ mA}_{\text{dc}} = 3.2 \text{ W}_{\text{dc}}$$

Therefore, a shortcut to finding the power transistor's power dissipation is to use the dc values, but realize this value is a little less than the average power dissipation.

$$P_{\text{D BJT}} \approx I_{\text{C dc}} \times V_{\text{CE dc}}$$

Practice
Verify the values of the example problem using a simulator. Use an LM324 op amp.

Answer: The simulated values are very close to the manual calculations. The output dc voltage is simulated to be 8.012 V_{dc}, compared to a calculated value of 8.0 V_{dc}.

multi**SIM**

Current Limiting

To prevent producing excessive load current if a short circuit load condition occurs, a simple BJT-resistor circuit combination can be used to limit the output current. A basic *current limiting* circuit is shown in Figure 18-35(a) and consists of BJT Q3 and current limiting resistor R_{limit}.

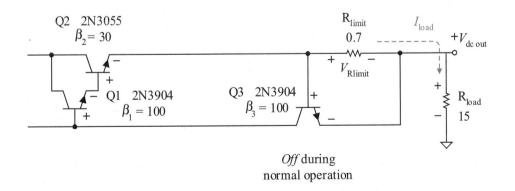

Figure 18-35(a) Current limiting circuit to limit output current

The voltage drop across R_{limit} establishes the voltage drop across the base-emitter voltage drop across the Q3.

$$V_{BE\ Q3} = V_{Rlimit} = R_{limit} \times I_{load}$$

For a silicon transistor, Q3 is turned off (open) until 0.7 V is developed across R_{limit}, which in turn applies 0.7 V to the base-emitter junction of Q3. If the load current is below the maximum established by the current limiting circuit, Q3 remains open and plays no part in the power supply circuit.

For the circuit of Figure 18-35(a), Q3 turns on when the voltage drop across the 0.7 Ω R_{limit} reaches 0.7 V. Thus, the maximum load current possible for this circuit is

$$I_{maximum\ load} = \frac{0.7\ V}{0.7\ \Omega} = 1\ A$$

If the load resistance is reduced further, the load current remains fixed at 1 A, the output load voltage goes down, and excess current is bypassed through Q3. Figure 18-35(b) demonstrates this current limiting action. Even if a short is placed across the output, the load current is 1 A. Therefore this is also called the **short circuit current**.

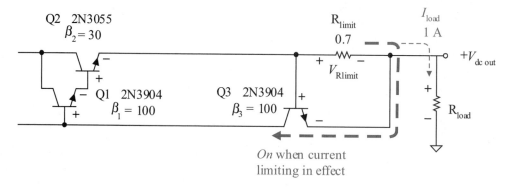

Figure 18-35(b) Current limiting action

Example 18-12

For the circuit of Figure 18-35(b), the load resistance is suddenly reduced to 1 Ω. Find the output current and output voltage.

Solution

The power supply is into current limiting producing a 1 A_{dc} load current. Therefore, the load current is fixed at 1 A_{dc} and the output voltage is found with Ohm's law.

$$I_{load} = 1 \text{ A}_{dc}$$

$$V_{dc\ out} = 1 \text{ A} \times 1 \text{ }\Omega = 1 \text{ V}_{dc}$$

Practice

Revise the power supply to produce 6.2 V_{dc} output and current limit at 2 A.

Answer: Replace the op amplifier circuit with an op amp voltage follower with a voltage gain of unity (1) as shown in Figure 18-35(c). The current limiting resistance is 0.35 Ω.

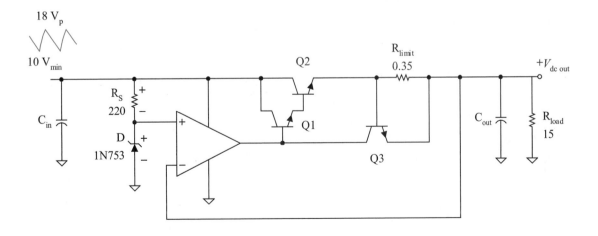

Figure 18-35(c) Practice circuit solution

Tracking Dual Regulated Power Supply

The circuit of Figure 18-36 is the schematic for a very basic tracking dual regulator dc power supply. This is a very basic circuit that gives you the basic concepts of a tracking dual power supply circuit.

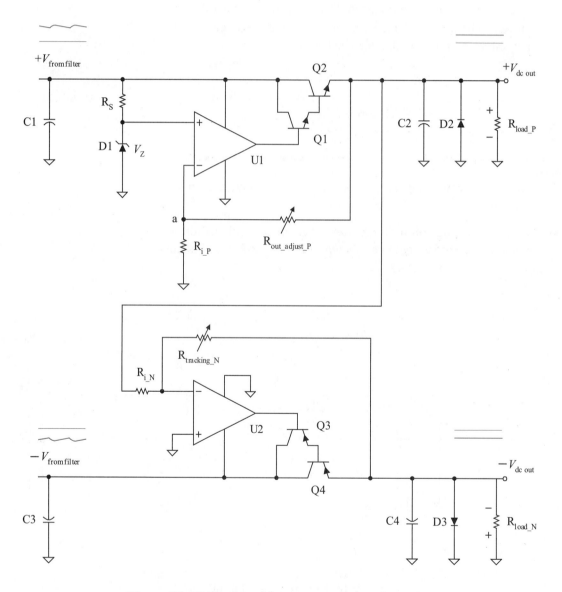

Figure 18-36 Dual tracking power supply schematic

The top circuit produces a *positive* dc supply output voltage $+V_{dc\ out}$ from a positive filtered input voltage with ripple ($+V_{from\ filter}$). The negative feedback loop produces an op amp input V_{error} voltage of about 0 V. Thus, the voltage at node a is approximately equal to V_Z.

$$V_a = V_Z$$

The variable resistor $R_{out\_adjust\_P}$ allows the positive dc output voltage to be adjusted from a minimum value of V_Z to a higher output voltage. The voltage gain A^+ of the U1 positive supply amplifier circuit is

$$A^+ = \frac{R_{out\_adjust\_P} + R_{i\_P}}{R_{i\_P}}$$

$$A^+ \geq 1$$

The positive output supply voltage is

$$+V_{dc\ out} = A^+ V_Z$$

Placing the Q1 and Q2 transistors in the negative feedback loop eliminates their 0.7 V drops from affecting the output voltage. But do check the op amp output voltage to ensure that it is within the rail voltages.

The bottom circuit produces a *negative* dc supply output voltage ($-V_{dc\ out}$) from a negative filtered input voltage with ripple ($-V_{from\ filter}$). The output of the positive supply is fed down and into the inverting voltage amplifier circuit of U2. Thus the negative supply *tracks* the positive supply.

$$V_a = +V_{dc\ out}$$

The voltage gain A^- of the inverting voltage amplifier circuit of U2 is

$$A^- = -\frac{R_{tracking\_N}}{R_{i\_N}}$$

and the negative output supply voltage is

$$-V_{dc\ out} = A^- (+V_{dc\ out})$$

If the $R_{tracking\_N}$ resistance is set equal to the R_{i_N} resistance, the output of the negative supply is the same magnitude as the positive dc supply.

$$\left| +V_{dc\ out} \right| = \left| -V_{dc\ out} \right|$$

If $R_{tracking\_N}$ is adjusted to be less than R_{i_N}, then the negative supply produces less voltage (a fraction of the positive supply voltage).

Q1 and Q2 are *npn* transistors to set up the positive power supply. Q3 and Q4 must be *pnp* transistors to set up a negative supply system. To determine the type of transistor, look at the emitter arrow. The emitter arrow always points at the *n*-type material. For an *npn* BJT, the arrow points out to the emitter. For a *pnp* BJT, the emitter arrow points into the base. Also, the emitter arrow indicates the direction of flowing conventional current. For *npn* transistors, current flows out of the emitter. For *pnp* transistors the current flows into the emitter. The *pnp* transistor is basically the complement of an *npn* transistor. The *pnp* voltage polarities and current directions are the opposite of the *npn* transistors, but the analysis is exactly the same process. Fundamental circuit analysis laws and techniques are applied the same for *pnp* transistor circuits as they are for *npn* transistor circuits.

The D2 and D3 diodes serve as protection diodes during the first cycle of powering up the supplies. To prevent reverse biasing of semiconductor components and potential damage to the components, the protective diodes provide a low resistance reverse path until the supplies become stable with the appropriate polarities.

Example 18-13

The circuit of Figure 18-36 uses a 5.1 V zener, R_{i_P} of 10 kΩ, $R_{out\_adjust\_P}$ of 6.13 kΩ, R_{i_N} of 10 kΩ, and $R_{tracking\_N}$ of 6 kΩ. Find the positive supply amplifier voltage gain and output voltage. Find the negative supply amplifier voltage gain and output voltage.

Solution

The voltage gain A^+ of the positive supply amplifier is

$$A^+ = \frac{6.13 \text{ k}\Omega + 10 \text{ k}\Omega}{10 \text{ k}\Omega} = 1.613$$

and the positive dc output supply voltage is

$$+V_{dc\ out} = 1.613 \times 5.1 \text{ V} = 10 \text{ V}_{dc}$$

The voltage gain A^- of the negative supply amplifier is

$$A^- = -\frac{6\ \text{k}\Omega}{10\ \text{k}\Omega} = -0.6$$

and the negative dc output supply voltage is

$$-V_{\text{dc out}} = -0.6 \times 10\ \text{V} = -6\ \text{V}_{\text{dc}}$$

Practice

Given the zener voltage, R_{i_P} and R_{i_N} values of the example circuit, what values must $R_{\text{out\_adjust\_P}}$ and $R_{\text{tracking\_N}}$ be to produce a split supply system of $\pm 8\ V_{\text{dc}}$?

Answer: 2.90 kΩ; 10 kΩ

18.7 Three-Terminal IC Regulators

Three-terminal IC regulators incorporate the previously noted features of fixed or adjustable voltage regulation, current limiting, and reduced ripple in a single compact component. Additionally they offer additional features of higher output currents to the load and **thermal shutdown** if the regulator chip becomes too hot. An IC regulator schematic symbol is shown in Figure 18-37 along with its associated signal voltages.

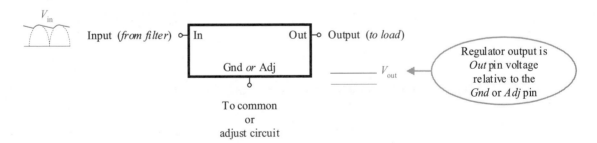

Figure 18-37 Three-terminal IC regulator schematic symbol

Although internally complex, the three-terminal IC regulator is an easy-to-use three-terminal component with an *input pin* (*In*), an *output pin* (*Out*), and a *ground pin* (*Gnd*) *or adjust pin* (*Adj*). A rectified, filtered voltage with ripple is fed to the regulator input pin (*In*). The voltage regulator output is a fixed dc voltage at its output pin (*Out*) relative to its ground pin (*Gnd*). An **adjustable regulator**, with an adjust pin (*Adj*), is specifically designed to support an external output circuit that can change the dc voltage delivered to the external circuit. The regulator output dc voltage at the *Out* pin relative to the *Gnd* pin (or *Adj* pin) is always fixed, even for adjustable regulators.

$$V_{\text{regulator output}} = V_{\text{out reg}} - V_{\text{gnd reg}} \qquad \text{(Fixed)}$$

$$V_{\text{regulator output}} = V_{\text{out reg}} - V_{\text{adj reg}} \qquad \text{(Adjustable)}$$

The *Gnd* or *Adj* pin must be connected to common or to an adjust circuit that is connected to common to complete the connection of the regulator.

Figure 18-38 demonstrates two ways to deliver regulated voltage to a system. A centralized regulator system distributes regulated dc voltage to an entire system. Or, since regulators are simple and relatively inexpensive, they can be distributed throughout a system such that each printed circuit board (PCB) or subsystem can have its own regulator.

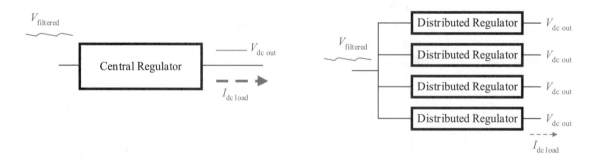

Figure 18-38 Central regulator versus distributed regulators

A centralized system requires a relatively more expensive IC to support the total current to the whole system, whereas a distributed system requires several lower current, less expensive IC chips. A centralized regulator would produce a higher current, greater power dissipation, and potentially the need for a heat sink to dissipate excessive heat. A centralized system with a higher single supply current would create greater line power losses (called I^2R losses) in long runs of wire. A disadvantage of a distributed regulator system is that the filtered signal with a significant 60 Hz or 120 Hz ripple is distributed throughout the system, possibly requiring additional protection from ripple frequency interference.

To protect or enhance the operation of an IC regulator, external components can be connected to it as shown in Figure 18-39. A distributed system would require regulator input capacitors. If a regulator is connected more than a few inches from its filter capacitor, C1 is needed to provide a pool of charge close to the regulator so that it is readily available for sudden changes in current experienced by the regulator input. Output capacitor C2 may or may not be needed for stability and may be used to improve transient response. If C2 or an equivalent output load capacitance is large enough (see specifications), then a protection diode is needed for regulator input short circuit protection. If the regulator input were shorted, the regulator input drops toward 0 V causing a charged C2 capacitor to discharge back through the regulator. This could destroy the IC regulator. If this occurs, the protection diode becomes forward biased, turns on, and much of C2's discharge current discharges through the protection diode D. That significantly reduces the reverse current through the IC to a safe level.

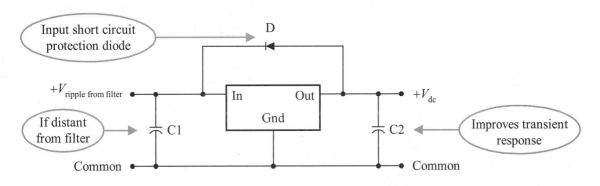

Figure 18-39 Positive IC regulator supporting components

Fixed Positive DC Voltage Regulator

A **fixed positive voltage regulator** produces a fixed positive dc output voltage at its *Out* pin relative to its *Gnd* pin even with varying input and load current conditions. The LM340-xx is such a voltage regulator series. The LM340-5, LM340-12, LM340-15 are 5 V, 12 V, and 15 V regulators, respectively. The schematic symbol for the LM340-xx component is shown in Figure 18-40. See data sheets for specifications.

The LM340-5 three-terminal IC fixed regulator is used for several examples. The key specifications utilized in these examples are:

Typical output voltage at 25°C	$V_{out\ dc} = 5.0$ V
Maximum output current without heat sinking	$I_{out\ dc} = 1$ A
Short circuit current	$I_{SC} = 2.1$ A
Quiescent current (*Gnd* pin)	$I_Q = 6$ mA
Minimum input voltage	$V_{in\ min} = 7.5$ V
Maximum input voltage	$V_{in\ max} = 20$ V
Dropout voltage	$V_{drop\ out} = 2$ V
Output resistance	$R_O = 8$ mΩ
Ripple rejection (dB attenuation)	$\dfrac{\Delta V_{IN}}{\Delta V_{OUT}} = 80$ dB

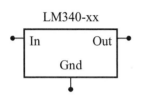

Figure 18-40
IC *positive* dc voltage regulator LM340-xx series

The chip has internal thermal shutdown and current limiting protection to automatically maintain a safe level of power dissipation; therefore, no maximum power dissipation rating is specified. This concept of thermal protection and heat dissipation is discussed later in this chapter.

From the specifications, the input ripple voltage from the filter must satisfy the following requirement to produce a regulated +5 V_{dc} output voltage.

$$7.5\ V_{min} \le V_{in\ ripple} \le 20\ V_p$$

The IC regulator output voltage of 5 V_{dc} is at the *Out* pin relative to the *Gnd* pin.

$$V_{out\ reg} - V_{gnd\ reg} = 5\ V_{dc}$$

The first example demonstrates the simplest use of the regulator with the *Gnd* pin connected to common and the output to the load.

Example 18-14
        For the circuit of Figure 18-41, are the required input specifications of the LM340-5 chip met? Find the dc output voltage, the

load current, and output ac ripple voltage. Find the power dissipated by the chip.

Solution

The input voltage meets the requirements of the LM340-5 voltage regulator chip. The 18 V_p input is less than the 20 V maximum chip specification, and the 10 V_{min} input is greater than the 7.5 V chip minimum specification.

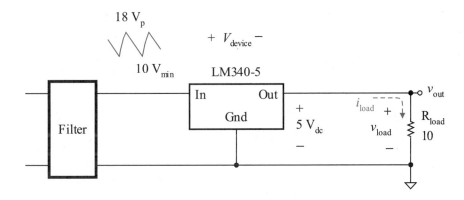

Figure 18-41 IC dc voltage regulator

The output of the regulator is 5 V and appears across the load.

$$V_{out} = V_{out\ reg} - V_{gnd\ reg}$$

$$V_{out} = 5\ V_{dc}$$

Use Ohm's law to find the load current and ensure that is within chip specification of a maximum current of 1 A.

$$I_{dc\ load} = \frac{5\ V_{dc}}{10\ \Omega} = 500\ mA_{dc}$$

Use the ripple rejection (attenuation) specification to calculate the output ripple voltage. Typically the output ripple from an IC regulator is so small that it is no longer a major concern.

$$db\ attenuation = 80\ dB$$

$$20\ \log\left(\frac{\Delta V_{\text{IN}}}{\Delta V_{\text{OUT}}}\right) = 80\ dB$$

$$\log\left(\frac{\Delta V_{\text{IN}}}{\Delta V_{\text{OUT}}}\right) = 4$$

$$attenuation = \frac{\Delta V_{\text{IN}}}{\Delta V_{\text{OUT}}} = 10^4$$

$$V_{\text{pp out ripple}} = \frac{V_{\text{pp in ripple}}}{attenuation}$$

$$V_{\text{pp out ripple}} = \frac{18\ V_p - 10\ V_{\min}}{10^4} = 0.8\ \text{mV}_{\text{pp}}$$

As noted earlier when examining the pass transistor, we can estimate the average power dissipation with less than 10% error by using the dc values and finding the power dissipation due to dc. The regulator input voltage with a peak of 18 V_p and a minimum of 10 $V_{\min}$ is 14 V_{dc}. Therefore, the *device* dc voltage drop V_{device} *across* the IC regulator is

$$V_{\text{device}} = V_{\text{in reg}} - V_{\text{out reg}}$$
$$V_{\text{dc device}} = 14\ V_{\text{dc}} - 5\ V_{\text{dc}} = 9\ V_{\text{dc}}$$

The dc current passing through the IC device is the same as the dc load current.

$$I_{\text{dc device}} = I_{\text{dc load}} = 500\ \text{mA}_{\text{dc}}$$

The average power dissipated by the IC device is calculated from its rms voltage and rms current; however, in this case, average power can be *estimated* from its dc values with less than 10% error.

$$P_{\text{average device}} \approx P_{\text{dc device}}$$

$$P_{\text{dc device}} = I_{\text{dc device}} \times V_{\text{dc device}}$$

$$P_{\text{dc device}} = 500\ \text{mA}_{\text{dc}} \times 9\ V_{\text{dc}} = 4.5\ \text{W}$$

The average power dissipated by the device is therefore less than 5 W. However, these linear IC regulators are designed with internal protection. The IC power dissipation is not allowed to become excessive. The IC has current limiting, so that the IC output current is restricted to a maximum value, which limits the power dissipation. The IC also has thermal detection and protection to prevent excessive temperatures.

Practice

Repeat the example if the regulator input voltage of the example is changed to have a minimum of 14 V_{min}.

Answer: 5 V_{dc}, 500 mA_{dc}, 0.4 mV_{pp}, 11 V_{dc}, 5.5 W_{dc}

Fixed DC Current Regulator

The voltage regulator IC can also be arranged to provide a constant current to a load as demonstrated by the circuit of Figure 18-42.

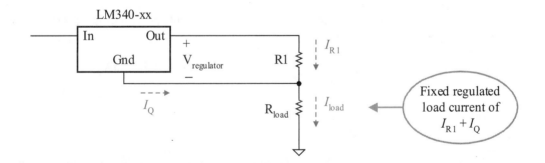

Figure 18-42 Fixed current regulator

The current I_{R1} is fixed by the regulator voltage and the R1 resistor.

$$I_{R1} = \frac{V_{regulator}}{R1}$$

Then by Kirchhoff's current law, the load current is fixed by the R1 current I_{R1} plus the quiescent current I_Q of the IC.

$$I_{\text{load}} = I_{R1} + I_Q$$

Typically the R1 current I_{R1} is much greater than the quiescent current I_Q, reducing the load current to that fixed by the IC regulator output voltage across the R1 resistor.

$$I_{\text{load}} \approx I_{R1}$$

Example 18-15

For the circuit of Figure 18-42, an LM340-5 IC is used with R1 equal to 10 Ω and R_{load} equal to 5 Ω. Find the load current. Check the IC *Out* pin node voltage to ensure it is still within specifications.

Solution

The R1 current is

$$I_{R1} = \frac{5 \text{ V}_{\text{dc}}}{10 \text{ }\Omega} = 500 \text{ mA}_{\text{dc}}$$

The LM340 quiescent current specification I_Q is 6 mA, therefore the load current is

$$I_{\text{dc load}} = 500 \text{ mA} + 6 \text{ mA} = 506 \text{ mA}_{\text{dc}}$$

Note this could have been approximated as

$$I_{\text{dc load}} \approx I_{R1} = 500 \text{ mA}_{\text{dc}}$$

The IC *Out* pin voltage is the sum of the R1 and R_{load} voltages.

$$V_{\text{dc out reg}} = 500 \text{ mA} \times 10 \text{ }\Omega + 506 \text{ mA} \times 5 \text{ }\Omega = 7.53 \text{ V}_{\text{dc}}$$

$$V_{\text{dc out reg}} \approx 500 \text{ mA} \times 15 \text{ }\Omega = 7.5 \text{ V}_{\text{dc}}$$

The IC drop out voltage is 2 V. Thus the input voltage must be greater than 9.5 V_{min} to prevent drop out, and it is.

Practice

The load resistor is changed to 7 Ω. Find the load current.

Answer: The load current remains *fixed* at 506 mA$_{dc}$ even with a change in load resistance. The IC *Out* pin voltage is 8.54 V$_{dc}$, so there is no drop out problem.

Fixed Negative DC Voltage Regulator

A **fixed negative voltage regulator** produces a fixed negative dc output voltage at its *Out* pin relative to its *Gnd* pin even with varying input and load current conditions. The LM320-xx is such a negative voltage regulator series. The LM320-5, LM320-12, LM320-15 are −5 V, −12 V, and −15 V regulators, respectively. The schematic symbol for the LM320-xx component is shown in Figure 18-43. See data sheets for specifications.

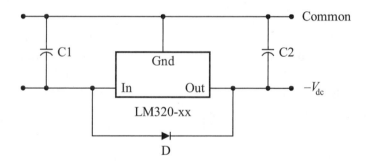

Figure 18-43 LM320-xx IC *negative* dc voltage regulator series

Dual (Split) Fixed DC Voltage Regulated Supply

A dual (or split) regulated fixed dc voltage supply produces a fixed positive dc regulated supply and a fixed negative dc regulated supply as shown in Figure 18-44. Such a system can be used, for example, to supply dc voltage to a dual-supplied op amp.

The positive supply requires filtered positive input voltage and the negative supply requires a filtered negative input voltage. A center-tapped bridge rectifier circuit can supply the rectified positive and negative voltages, with each voltage fed to its respective filter capacitor. The positive and negative IC voltage regulator outputs are nearly ideal dc output voltages even with significant ripple input voltage.

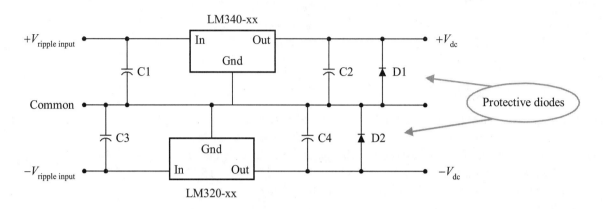

Figure 18-44 Dual regulated dc supply

The C1, C2, C3, and C4 capacitors may or may not be needed; check the IC regulator specifications. Either supply could start charging first; it depends upon when the circuit is switched on relative to the input ac voltage. Protective diodes D1 and D2 are required during signal start up to prevent possible reverse voltages in the IC regulators.

Adjustable Voltage Regulators

An **adjustable IC voltage regulator** is designed to support an externally adjustable circuit that can produce a variable regulated output voltage. Figure 18-45(a) is a schematic symbol of an adjustable LM317 IC voltage regulator. The term *adjustable* is a bit misleading. The output of the IC at the *Out* pin relative to the adjust (*Adj*) pin is a fixed 1.25 V_{dc}. As demonstrated by the circuit of Figure 18-45(b), the output voltage of the circuit $V_{dc\ out}$ is adjustable primarily based upon the values of R1 and R2.

The R1 dc current is fixed.

$$I_{R1} = \frac{1.25\ V}{R1}$$

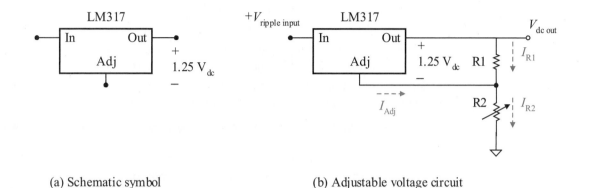

(a) Schematic symbol   (b) Adjustable voltage circuit

Figure 18-45 Adjustable IC voltage regulator

Use R1 to establish the minimum regulator output current.

$$I_{R1} = \frac{1.25 \text{ V}}{R1} > I_{\text{min regulator output}}$$

Typically the minimum output current established as I_{R1} is much, much greater than the current out of the adjust (Adj) pin I_{adj}.

$$I_{R1} \gg I_{\text{adj}}$$

This creates an essentially series circuit of R1 and R2, so that

$$I_{R2} \approx I_{R1} = \frac{1.25 \text{ V}}{R1}$$

The sum of the R1 current and the load current establishes the maximum output load current.

Assuming I_{Adj} is much, much less than I_{R1}, the ideal output voltage of this adjustable regulator circuit can be determined directly from circuit analysis.

$$V_{\text{dc out}} = V_{R1} + V_{R2}$$

$$V_{\text{dc out}} = 1.25 \text{ V} + R2\left(I_{R2}\right)$$

$$V_{\text{dc out}} \approx 1.25 \text{ V} + R2\left(\frac{1.25 \text{ V}}{R1}\right)$$

$$V_{dc\ out} \approx 1.25\ V\left(1+\frac{R2}{R1}\right)$$

The minimum output voltage is the reference output voltage, or typically 1.25 V_{dc} for the LM317. Check device specifications to determine minimum and maximum device parameters. Also, see associated application notes for several applications of IC regulators and device protection.

Example 18-16

For the circuit of Figure 18-45(b), find the value of R1 to establish a minimum regulator output current of 5 mA.

Solution

The reference voltage across R1 is 1.25 V_{dc}. Use Ohm's law and solve for the maximum R1 resistance needed to establish the minimum regulator current.

$$R1_{max} < \frac{1.25\ V}{5\ mA} = 250\ \Omega$$

The nearest standard reference value from Appendix C is 240 Ω.

Practice

If the minimum regulator output current is 2.5 mA, find the maximum standard value of R1 to satisfy this requirement.

Answer: 470 Ω

The next example calculates the range of output voltage of a regulator circuit.

Example 18-17

For the circuit of Figure 18-45(b), an LM317 IC is used with an R1 value of 240 Ω and an R2 potentiometer value of 1 kΩ. Calculate the range of the regulated output voltage using basic circuit law analysis. Verify these results using the R1 and R2 expression to calculate the range of output voltages. Find the regulated output voltage more accurately by including the adjustment current. If the specified maximum regulator output current is specified to be 200 mA, what is the maximum allowable load current and minimum allowable load resistance?

Solution

The minimum load voltage is the fixed output voltage of the IC and occurs with the potentiometer set to 0 Ω.

$$V_{\text{dc output}} = 1.25 \ V_{\text{dc}}$$

The maximum output voltage occurs with the potentiometer set to 1 kΩ.

$$I_{R2} \approx I_{R1} = \frac{1.25 \ V}{240 \ \Omega} = 5.208 \ \text{mA}$$

$$V_{\text{dc out}} = 1.25 \ V + 1 \ k\Omega \left(5.208 \ \text{mA}\right) = 6.46 \ V_{\text{dc}}$$

Therefore, the range of the regulated output voltage is $1.25 \ V_{\text{dc}}$ to $6.46 \ V_{\text{dc}}$.

Using the R1 and R2 expression, the minimum output voltage with R2 set to 0 Ω is

$$V_{\text{dc output min}} \approx 1.25 \ V \left(1 + \frac{0 \ \Omega}{240 \ \Omega}\right) = 1.25 \ V_{\text{dc}}$$

The maximum output voltage with R2 set to 1 kΩ is

$$V_{\text{dc output max}} \approx 1.25 \ V \left(1 + \frac{1 \ k\Omega}{240 \ \Omega}\right) = 6.46 \ V_{\text{dc}}$$

If the adjust current I_{adj} is 0.1 mA, there is no change with R2 set to 0 Ω. But if R2 has resistance, an additional voltage drop is developed across R2. Set R2 to its maximum value of 1 kΩ. By Kirchhoff's current law

$$I_{R2} = I_{R1} + I_{\text{adj}}$$

$$I_{R2} = 5.208 \ \text{mA} + 0.1 \ \text{mA} = 5.308 \ \text{mA}$$

The maximum output voltage becomes

$$V_{\text{dc output}} = 1.25 \ V + 1 \ k\Omega \left(5.308 \ \text{mA}\right) = 6.56 \ V_{\text{dc}}$$

This is a 0.1 V increase, a 1.5% error, compared to the ideally calculated value.

The maximum allowable load current is the regulator maximum current specification less the R1 current.

$$I_{\text{max load}} = I_{\text{max regulator}} - I_{\text{R1}}$$

$$I_{\text{max load}} = 200 \text{ mA} - 5.21 \text{ mA} \approx 195 \text{ mA}$$

Therefore the minimum allowable load resistance is

$$R_{\text{load min}} > \frac{6.56 \text{ V}}{195 \text{ mA}} = 34 \text{ }\Omega$$

Practice

Design the regulator circuit of Figure 18-45(b) to produce a regulated output voltage of 10 V_{dc}, assuming R1 is 240 Ω.

Answer: R2 resistance of 1.68 kΩ

IC negative dc voltage regulators such as the LM337 driving their adjustment circuit produce an adjustable negative dc regulated voltage.

18.8 Heat Sinking

Devices that carry current and have real resistance normally dissipate power in the form of heat. This is of special interest for semiconductor devices, which are very heat sensitive. A semiconductor IC device is typically designed to operate at temperatures up to 125°C or 150°C. Warning, this is hot enough to boil water or burn skin.

Heat flows from hotter media to cooler media. An analogy is current flowing from a higher to a lower voltage potential. The semiconductor device creates the heat. The heat dissipates from the device outward through its cooler package, and then from the package into a cooler surrounding environment (normally air) as pictured in Figure 18-46.

Thermal Resistance and Junction Temperature

Different types of materials conduct heat better than others. For example, notice the difference between stepping onto a rug or stepping onto a tiled floor in your bare feet. The rug and tile are both at room temperature but the tile feels colder because it is conducting heat from your feet faster

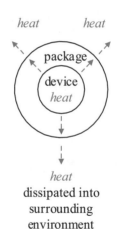

Figure 18-46
Device heat dissipation

than the rug. The rug has greater resistance to heat flow. **Thermal resistance** quantifies how quickly heat is dissipated through a medium and is measured in units of degrees per watt. For example,

$$\Theta = 50°C/W$$

where Θ is the symbol representing thermal resistance.

Standard terminology is used in describing this electrical/thermal activity. **Wafer**, **semiconductor**, or **die** refers to the actual semiconductor device (not the entire component). **Package** is that part of the component that houses the semiconductor and has internal connectors to external connecting terminals to the outside world. **Junction** is that area of the semiconductor immediately next to the package, connecting the device to the package. The **junction temperature** T_J is the temperature at that junction. **Ambient temperature** T_A is the temperature in the immediately surrounding environment of the whole component (at the surface of the package). The surface interfaces and the standard terminology used to describe this thermal activity are show in Figure 18-47. The term Θ_{JA} represents the thermal resistance of the package from junction-to-ambient. The relationship between power dissipation, temperature, and thermal resistance is given by the following formula.

$$T_J = T_A + \Theta_{JA} P_D$$

where

T_J	Junction temperature in units of °C
T_A	Ambient temperature in units of °C
P_D	Device power dissipation in units of watts (W)
Θ_{JA}	Junction-to-ambient thermal resistance in units of °C/W

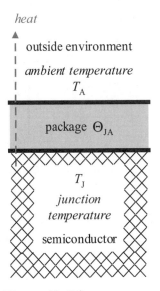

Figure 18-47
Thermal terminology

Semiconductor components have a specified maximum temperature, for example, the LM317 junction temperature is rated at 125°C. If this temperature is exceeded the LM317 has internal thermal protection that shuts down the IC to prevent component damage. Its thermal resistance depends upon the package on which it is mounted (for example, TO-3).

Example 18-18

In a previous example using the circuit of Figure 18-48, the power dissipation of the LM340-5 was found to be 4.5 W based upon the LM340 dc voltage drop and its dc load current. Verify this previous calculation. Assuming an ambient temperature of room temperature (that is about 25°C, which is 72°F), find the IC's junction temperature if a TO-3 package is used. Is this a reasonable temperature for the LM340? If not, what can you do? What is the maximum possible dc load current in this situation?

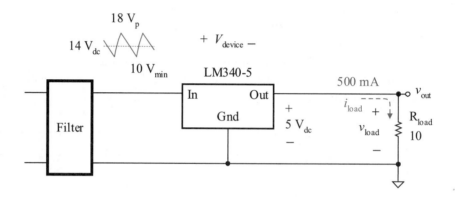

Figure 18-48 Thermal effects example

Solution

Estimate the device power dissipation P_D based upon its dc values. The dc voltage dropped from the input to the output is

$$V_{\text{dc across regulator device}} = V_{\text{dc in}} - V_{\text{dc out}}$$

$$V_{\text{dc across regulator device}} = 14 \text{ V}_{\text{dc}} - 5 \text{ V}_{\text{dc}} = 9 \text{ V}_{\text{dc}}$$

Use the power rule and estimate the IC regulator's power dissipation.

$$P_D \approx 9 \text{ V}_{\text{dc}} \times 500 \text{ mA}_{\text{dc}} = 4.5 \text{ W}$$

Go to the specification sheets for the LM340 and find its maximum rated junction temperature and its thermal resistance rating for the TO-3 package. The thermal resistance rating of the TO-3 package is a little difficult to find but it is there. The thermal information is found in the numbered notes following the table specifications. The thermal resistance Θ_{JA} of the TO-3 package is specified as 39°C/W. Substitute the known values into the equation

$$T_J = T_A + \Theta_{JA} P_D$$

$$T_J \approx 25°C + (39°C/W)\ 4.5\ W = 200\ °C$$

Warning! Warning! The LM340 junction temperature rating is listed to be 125°C. The IC would be cooking at 200°C, but the IC does have thermal protection and shuts down when a junction temperature of 125°C is reached. Actually the situation is worse than ideally calculated. The actual average power dissipation based upon rms values is closer to 5 W. Therefore, more accurately, the junction temperature is closer to 220°C. The IC is not cooling fast enough. How can you cool it faster? You could operate it in a freezer; however, that would reduce the ambient temperature to only 0°C and the junction temperature would still be too high at 175°C ideally. The answer is to improve the cooling of the package by attaching the appropriate heat sink (metal) to the IC to cool it faster.

Ideally, what is the maximum load (output regulator) current possible without heat sinking? Reorganize the junction temperature equation by solving for the maximum power dissipation P_D without additional cooling.

$$P_{D\ max} = \frac{T_{J\ max} - T_A}{\Theta_{JA}}$$

$$P_{D\ max} = \frac{125°C - 25°C}{39°\ C/W} = 2.56\ W$$

Using the ideal dc voltages and current, the maximum regulator output current, and thus load current, is

$$I_{dc\ out\ max} = \frac{2.56\ W}{9\ V_{dc}} = 285\ mA_{dc}$$

Again, realize that this is about 10% high, so more realistically the maximum load current without cooling is about 250 mA_{dc}.

Practice

Repeat the example for a TO-220 package. Does this help?

Answer: Θ_{JA} of 54°C/W, ideal T_J of 268°C, ideal maximum 206 mA$_{dc}$; this case type makes the temperature problem worse

---

Let's make a few observations. With an IC regulator, it is to your advantage to select a lower valued filter capacitor. A greater ripple is created but this creates a lower ripple dc input voltage, which creates a lower voltage drop across the IC regulator. This in turn creates less power dissipation by the regulator. A larger ripple in is typically not a problem since the ripple attenuation by an IC regulator is excellent. If greater power dissipation is needed by the regulator, then a properly mounted and matching heat sink can significantly increase the output current capability of the IC regulator.

Heat Sinks

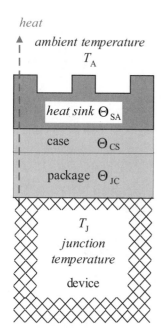

Figure 18-49
Thermal terminology with heat sinking

If a regulator is to operate at realistic load currents, we must find a way to reduce the device's thermal resistance and remove heat faster from its package. Fans cool the circuit by blowing hot air away from the surface of the component so the air temperature near a device is cooler. The circuits could be placed in a subfreezing environment, but operating circuits at −100°C is not a reasonable solution physically or economically.

If you were to gently grip the device with a pair of heavy pliers, the device would cool much faster. The more surface area you gripped or the thicker the metal, the faster the device would cool, and it would operate at a lower temperature. An analogy is placing a metal spoon in a cup of hot chocolate to cool it faster. The bigger the metal spoon, with more surface area, the faster the hot chocolate would cool. Hiring someone to grip each device with a pair of metal pliers is not a practical solution.

Heat sinks are metal devices designed to fit over specific package types and make maximum surface contact with the devices. By placing a heat sink over a device, that device's thermal resistance is substantially reduced and the junction temperature significantly reduced. The package case to heat sink thermal resistance is reduced by the use of an electrically insulated **mica wafer**. If an insulator is not needed, this surface contact is significantly improved by using **heat sink compound** or **heat sink grease**. Figure 18-49 demonstrates the thermal effects and terminology with heat sinking, where

T_J Junction temperature in units of °C

T_A Ambient temperature in units of °C

Θ_{JC} Junction-to-case thermal resistance in units of °C/W

Θ_{CS} Case-to-sink thermal resistance in units of °C/W

Θ_{SA} Sink-to-ambient thermal resistance in units of °C/W

The total thermal resistance from junction-to-ambient is just the sum of the series thermal resistances.

$$\Theta_{JA} = \Theta_{JC} + \Theta_{CS} + \Theta_{SA}$$

The net result is that the junction-to-ambient thermal resistance Θ_{JA} is reduced substantially using a heat sink and even further by using heat sink grease or a mica wafer. The lower the total thermal resistance Θ_{JA}, the less heat generated per watt of operation; therefore, the device can dissipate more power and operate with higher load currents.

Example 18-19

A heat sink is used in the circuit of Figure 18-48 of the last example. The junction-to-case thermal resistance becomes 4°C/W. The heat sink is mounted with heat sink grease creating a case-to-sink thermal resistance Θ_{CS} of 0.2°C/W. The heat sink was selected to have a sink-to-ambient thermal resistance Θ_{SA} of 5°C/W. Find the junction-to-ambient thermal resistance Θ_{JA}. Find the junction temperature. Is the LM340 operating inside its temperature range? What is the maximum output current the LM340 can produce with this heat sink?

Solution

The total junction-to-ambient thermal resistance is

$$\Theta_{JA} = \Theta_{JC} + \Theta_{CS} + \Theta_{SA}$$

$$\Theta_{JA} = 4°C/W + 0.2°C/W + 5°C/W = 9.2°C/W$$

Substitute the ideal power dissipation of 4.5 W and thermal resistance values into the equation junction temperature equation.

$$T_J = T_A + \Theta_{JA} P_D$$

$$T_J \approx 25°C + (9.2°C/W)4.5\ W = 66°C$$

The LM340 is running very cool, within specifications.

Ideally, what is the maximum load (output regulator) current possible? Again use the reorganized junction temperature equation by solving for the maximum power dissipation P_D.

$$P_{D\,max} = \frac{T_{J\,max} - T_A}{\Theta_{JA}}$$

$$P_{D\,max} = \frac{125°C - 25°C}{9.2°C/W} = 10.9\ W$$

Using the ideal dc voltages and current, the maximum regulator output current, and thus load current, is

$$I_{dc\,out\,max} = \frac{10.9\ W}{9\ V_{dc}} = 1.2\ A_{dc}$$

Compare this result with the specifications for the power and current ratings of the LM340 with and without heat sinking. Based upon specifications, the LM340 is tested at up to 15 W (with heat sinking) and a maximum current of 5 A.

Practice

Repeat the example for a TO-220 package. Is the IC still operating within its junction temperature specification?

Answer: Θ_{JA} is the same of 9.2°C/W, ideal T_J of 66°C, ideal maximum 1.21 A_{dc}, same result for TO-3 and TO-220 packages.

An IC with **thermal protection** turns off as its junction temperature rating is exceeded. Its output voltage and current drop toward 0. Once the IC is sufficiently cooled, it turns back on and operates properly until once again its rated junction temperature is exceeded. This cycle could continue indefinitely. Note that this is different from current limiting. If the component is within temperature ratings but the current output becomes excessive, then the current is limited to its specified value. Its output voltage drops to maintain a fixed maximum output current.

Component Power Dissipation Derating

The maximum device power dissipation rating is the maximum average power that the device can dissipate and remain safe from destruction. For example, at room temperature 25°C, the rating for the 2N3904 BJT is 625 mW. If operating above room temperature (for example, in a closed cabinet without significant ventilation), the power rating must be de-rated at a value of 5.0 mW/°C. When operating in an environment above 25°C, the adjusted derated power dissipation rating is

$$P_{\text{D derated}} = P_{\text{D max}} - (derating\ factor)\ (T - 25°C)$$

For example, if a 2N3904 BJT were operating at a temperature of 50°C, then the derated maximum power is

$$P_{\text{D derated}} = 625\ \text{MW} - (5.0\ \text{mW/°C})(50°C - 25°C) = 500\ \text{mW}$$

If this is not tolerable, then a heat sink could be attached to the 2N3904 and the maximum power that the BJT could dissipate would be significantly increased.

Summary

This chapter pulled many basic circuit concepts together and incorporated them into the basic concepts of the circuitry of a dc power supply. A transformer is used to step up or down commercial ac voltage. That transformed voltage is fed to a rectifier circuit to produce an average positive and/or negative voltage supply or supplies. A filter component or circuit is then used to significantly reduce the ac ripple voltage. Typically an input filter capacitor is used. If a large filter capacitor is used, the ac ripple is significantly reduced. This rectified, filtered voltage may be sufficient for many simple applications such as simple plug-in wall supplies like chargers for cell phones. A more sophisticated dc supply that requires voltage or current regulation as well uses a regulator device and/or circuit. A simple regulator for low current applications is the simple zener diode with a series limiting resistor. The zener is placed in parallel with the load so that it is reverse biased and operates at its zener voltage. The output voltage is fixed to the same value as the zener voltage. If greater load currents are required or if the zener current needs to be desensitized from significant load current changes, a pass transistor can be used. In this case the output voltage is fixed to the zener voltage less the base-emitter drop of the pass transistor.

The BJT can be placed within the negative feedback loop of an operational amplifier, boosting the current from the op amp to the load and allowing dc output voltage variation based upon the gain of the negative feedback loop. A current sensor can be added to the pass transistor. When a maximum allowable current is reached, any increased demand for current by the load is rerouted through an internal circuit. This maintains a fixed maximum load current, and any increased demand merely reduces the load voltage.

For higher load currents, improved ripple rejection, and improved voltage regulation, three-terminal IC voltage regulators can be used. These can be fixed positive or fixed negative voltage regulators, or can be adjustable regulators designed to work with an external adjustment circuit that can produce a range of regulated dc voltage outputs. The voltage regulator component can also be used in a circuit to produce a current regulator.

The power dissipation of the pass transistor and IC voltage regulators can be significant. The power of the regulator device is approximately determined by its dc load current times its dc voltage drop (namely, its input dc voltage less its output dc voltage). Increased power dissipation inherently leads to an increase in the semiconductor's temperature (called the junction temperature). The semiconductor is encased in a package, which radiates the heat generated by the semiconductor. The faster this heat is conducted away from the semiconductor, the cooler the semiconductor is. The opposition to heat transfer from the semiconductor to the outside environment is called thermal resistance. Of primary interest is the thermal resistance from the junction (surface of the semiconductor) to the outside environment, called the junction-to-ambient thermal resistance. The lower this value, the faster heat is transferred away from the semiconductor. A critical regulator IC rating is its maximum junction temperature rating. If the semiconductor reaches this maximum temperature rating, the IC goes into thermal shutdown, reducing the output current and voltage to 0. Once cooled, the IC turns back on and produces its normal outputs. If the IC needs to operate at high currents and thus high temperatures, its heat dissipation rate can be increased by heat sinking it. A heat sink is a metal attachment, some with fins to increase cooling, which significantly reduces the junction-to-ambient thermal resistance. Since the heat dissipates much more rapidly, the IC semiconductor operates at a cooler temperature. This allows ICs that would normally operate at a few hundred milliamps to operate in the ampere range.

Problems

Rectifier Circuits
18-1 The switch SW in the circuit of Figure 18-50 is open.

a. Find the peak secondary voltage, peak output voltage, and peak load current.

b. Find the ripple voltage frequency and period.

c. Sketch the output voltage and load current waveforms.

d. Find the average load current and diode current. Is the diode safe?

e. Find the peak inverse voltage experienced by the diode. Does it break down during the negative half cycle?

f. Use a simulator and verify your output voltage calculations.

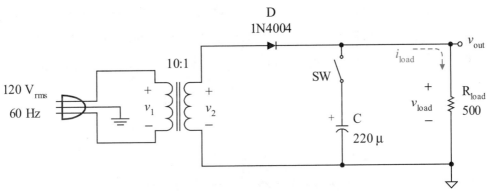

Figure 18-50 Half-wave rectifier circuit

18-2 Repeat Problem 18-1 with a 1 kΩ load resistance.

18-3 The switch SW in the circuit of Figure 18-50 is closed.

a. Find the peak secondary voltage, the stable peak output voltage, and the stable peak load current.

b. Find the ripple voltage frequency and period.

c. Prove that the capacitor is lightly discharged. Estimate the peak-to-peak ripple voltage and its rms value.

d. Find the approximate minimum output voltage and approximate dc output voltage.

    e.   Sketch the output voltage.

    f.   Find the percent ripple in the output voltage.

    g.   Find the peak inverse voltage experienced by the diode. Does the diode break down during the negative half cycle?

    h.   Use a simulator and verify your output voltage calculations.

    i.   Use a simulator and estimate the diode surge current.

18-4   Repeat Problem 18-3 with a 1 kΩ load resistance.

18-5   Repeat Problem 18-1 with the circuit of Figure 18-51.

18-6   Repeat Problem 18-2 with the circuit of Figure 18-51.

18-7   Repeat Problem 18-3 with the circuit of Figure 18-51.

18-8   Repeat Problem 18-4 with the circuit of Figure 18-51.

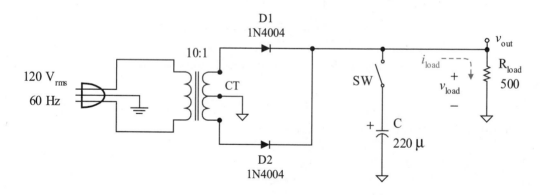

Figure 18-51 Full-wave center-tapped rectifier circuit

18-9   Repeat Problem 18-1 with the circuit of Figure 18-52.

18-10   Repeat Problem 18-2 with the circuit of Figure 18-52.

18-11   Repeat Problem 18-3 with the circuit of Figure 18-52.

18-12 Repeat Problem 18-4 with the circuit of Figure 18-52.

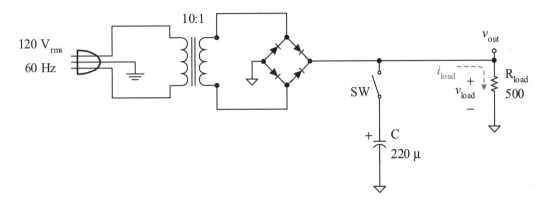

Figure 18-52 Full-wave bridge rectifier circuit with 1N4004 rectifier diodes

Zener Voltage Regulator

18-13 A zener regulator circuit is inserted into Figure 18-52 to produce the circuit of Figure 18-53.

    a. What is the approximate regulated dc output voltage and dc load current?

    b. Select the minimum standard value of R_S to meet the zener diode's maximum specification.

    c. With the load connected, find the *maximum* values of the regulator circuit input voltage, the series resistor voltage drop and current, the zener current, and instantaneous power dissipated by R_S.

    d. With the load connected, find the *minimum* values of the regulator circuit input voltage, the series resistor voltage drop and current, the zener current, and instantaneous power dissipated by R_S. Is the minimum current specification of the zener diode met? Assume that the peak-to-peak ripple is

$$V_{pp\ ripple} \approx \frac{I_{p\ into\ regulator}}{fC}$$

e. With the load connected, find the estimated *dc* (approximate average) values of the regulator circuit input voltage, the series resistor voltage drop and current, the zener current, and power dissipated by R_S. What is the wattage rating required by the series current limiting resistor R_S?

f. Use a simulator to measure the regulator input and output ripple rms and dc voltages. Calculate the percent ripple into the regulator and the percent ripple out of the regulator based upon your measured values. What is the regulator's attenuation and dB attenuation?

18-14 Repeat Problem 18-13 with a 1 kΩ load resistance.

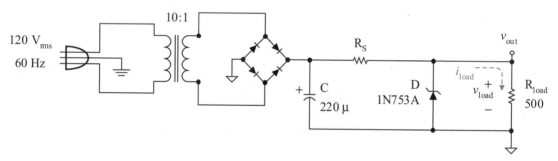

Figure 18-53 Zener regulated full-wave filtered rectifier

Pass Transistor

18-15 A pass transistor is inserted into the circuit of Figure 18-53 to produce a higher load current for a lower load resistance as shown in the circuit of Figure 18-54. Use the same R_S value as the preceding problem. Assume that the BJT base-emitter drop is $0.8\ V_{dc}$.

a. Find the approximate dc output voltage, dc load current, and dc load power.

b. Find the approximate BJT dc base current.

c. Assuming the minimum regulator input, is there sufficient zener current to operate in the zener region?

   d. Use the estimated dc circuit values to estimate the power dissipated by the pass transistor. Is the power dissipation of the pass transistor within specifications? If not, what can be done?

   e. Simulate the circuit of Figure 18-54 and verify your calculations.

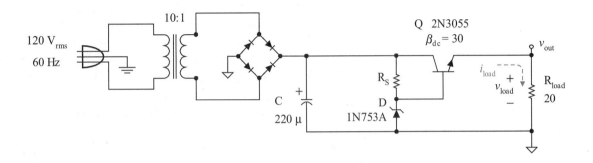

Figure 18-54 Pass transistor regulated output

18-16 Repeat Problem 18-15 with a 10 Ω load resistance.

18-17 For the circuit of Figure 18-34 (page 1013), the feedback resistor is changed to 1 kΩ. Find the amplifier voltage gain, the output voltage, the op amp output voltage, the load current, and the op amp output current. Is $V_{out\,op\,amp}$ within the op amp rail voltages if the supply headroom required is 1 V?

18-18 For the circuit of Figure 18-34, find the value of R_f required to produce a 6.6 V_{dc} output voltage.

18-19 For the circuit of Figure 18-35(a), find the current limiting resistance required to limit the load current to 500 mA.

18-20 Repeat Problem 18-19 to limit current to 250 mA.

18-21   The circuit of Figure 18-36 uses a 6.2 V zener, R_{i_P} of 100 kΩ, $R_{out\_adjust\_P}$ of 49 kΩ, R_{i_N} of 100 kΩ, and $R_{tracking\_N}$ of 100 kΩ. Find the positive supply amplifier voltage gain and output voltage. Find the negative supply amplifier voltage gain and output voltage.

18-22   Repeat Problem 18-21 with $R_{out\_adjust\_P}$ changed to 33 kΩ and $R_{tracking\_N}$ changed to 80 kΩ.

IC Regulators and Heat Sinking

18-23   An LM340-12 IC regulator in a TO-3 package has a filtered input ripple voltage of 22 V_{max} and 18 V_{min} with a load resistance of 25 Ω. It is operating in an environment with an ambient temperature of 30°C.

   a.   Find the approximate dc output voltage, dc load current, and dc load power.

   b.   Find the power dissipation of the IC using dc values and the IC semiconductor's junction temperature.

   c.   Is the device operating within temperature specifications? If not, find the heat sink thermal resistance required (assuming heat sinking grease is used) to operate at 100°C.

   d.   Find the output ripple voltage and percent voltage regulation.

18-24   Repeat Problem 18-23 with a load resistance of 35 Ω.

18-25   The LM317 IC regulator circuit of Figure 18-45 (page 1033) has an input of 15 V_{dc}. Assume that R1 is 240 Ω and the ambient temperature is 35°C. The IC is mounted in a TO-92 package with thermal resistance of 160°C/W.

   a.   Find the value of R2 needed to produce an output of 5 V_{dc}.

   b.   A 100 Ω load is connected; find the IC output current.

   c.   Find the IC power dissipation and junction temperature.

   d.   Is the IC operating within specifications? If not, what can you do?

18-26   Repeat Problem 18-25 for an output voltage of 10 V_{dc}.

19

Dependent Sources

Introduction

An active device inherently contains one or more dependent sources. The dependent source acts like a voltage source or a current source and can be controlled by another voltage or current in the circuit. You must be able to visualize how a circuit or subsystem is functioning by modeling and visualizing it in terms of characteristics such as input impedance, output impedance, current or voltage gain, and so forth. This chapter provides the process to examine circuits and systems in terms of their major characteristics.

Objectives

Upon completion of this chapter you will be able to:

- Identify and create modeled circuits of basic active devices and circuits based upon their primary characteristics.

- Analyze and design circuits to transfer maximum voltage, maximum current, or maximum power.

- Produce and use the models of BJTs, MOSFETs, and op amps in circuit analysis.

- Identify, model, and apply basic active circuit models in circuit analysis with voltage or current controlled inputs and voltage or current sourced outputs.

- Identify and analyze fundamental circuits such as ideal voltage amplifiers, current-to-voltage converters, voltage-to-current converters, and current amplifiers.

- Identify and analyze fundamental circuits such as the noninverting and inverting op amp voltage amplifiers, common drain and common source MOSFET amplifiers, common collector and common emitter BJT amplifiers, and the MOSFET and BJT difference amplifiers.

1051

19.1 Ideal Dependent Source Models

A **dependent source** depends on another variable. The dependent source can be a *voltage source* or a *current source*. The controlling variable, called the **independent variable**, can be a controlling voltage or a controlling current. For example, the output voltage of an op amp depends upon the controlling input voltage of the op amp. The symbols of Figure 19-1 represent the four basic ideal **dependent source models**:

VCVS	voltage controlled (VC) voltage source (VS)
CCVS	current controlled (CC) voltage source (VS)
VCCS	voltage controlled (VC) current source (CS)
CCCS	current controlled (CC) current source (CS)

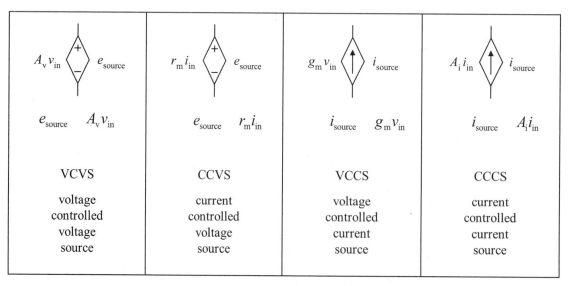

Figure 19-1 Ideal dependent source models

The *diamond* schematic symbol shape represents a *dependent voltage source* or a *dependent current source* that depends upon another variable (a controlling voltage or current). For example, the voltage controlled voltage source (VCVS) has a dependent voltage source e_{source} that depends upon the value of the *controlling voltage* v_{in}.

The first two letters of the four-letter abbreviation indicate the controlling input, that is, **VC** for **voltage controlled** input and **CC** for **cur-**

rent controlled input. The last two letters of the abbreviation indicate the type of dependent source, that is, **VS** for a dependent **voltage source** or **CS** for a dependent **current source**. Thus, VCVS is a voltage controlled (VC) voltage source (VS). The corresponding dependent relationship is shown with each schematic symbol in Figure 19-1 where the constant multipliers of the dependent sources are

A_v voltage gain

r_m **mutual resistance** or **transresistance**

g_m **mutual conductance** or **transconductance**

A_i current gain

The voltage and current variables are all shown as lowercase letters representing ac signals, but corresponding models are also possible with uppercase letters to indicate dc signal modeling.

Example 19-1

The input voltage to a VCVS circuit is 10 mV$_{rms}$ and its voltage gain is 100. Find the voltage generated by the *dependent voltage source*.

Solution

The dependent source voltage generated is

$$e_{source} = A_v v_{in} = 100 \times 10 \text{ mV}_{rms} = 1 \text{ V}_{rms}$$

Practice

The input current to a CCVS circuit is 10 mA$_{rms}$ and its mutual resistance (or transresistance) is 100 Ω. Find the voltage generated by the *dependent voltage source*.

Answer: 1 V$_{rms}$

The next example and practice examine the dependent current source.

Example 19-2

The input voltage to a VCCS circuit is 10 mV$_{rms}$ and its mutual conductance (or transconductance) is 100 mS. Find the current generated by the *dependent current source*.

Solution

The dependent current voltage generated is

$$i_{\text{source}} = g_m v_{\text{in}} = 100 \text{ mS} \times 10 \text{ mV}_{\text{rms}} = 1 \text{ mA}_{\text{rms}}$$

Practice

The input current to a CCCS circuit is 10 mA$_{\text{rms}}$. Its current gain is 100. Find the current generated by the *dependent current source*.

Answer: 1 A$_{\text{rms}}$

19.2 Active Devices

An **active device** can increase signal power, for example, a BJT, a MOS-FET, or an op amp, whereas a passive device such as a resistor, capacitor, inductor, diode, or transformer cannot. The model of an active device always includes a dependent source where the generated source output is dependent upon its input signal. The BJT is an example of a current controlled current source (CCCS), where the much larger collector current i_c is controlled by the much smaller base current i_b as shown in Figure 19-2(a).

Figure 19-2(a) BJT–CCCS device

A MOSFET is an example of a voltage controlled current source (VCCS) as shown in Figure 19-2(b). A vacuum tube triode is also an example of a voltage controlled current source.

Figure 19-2(b) MOSFET–VCCS device where v_{gs} controls i_d

The inherent characteristic of a general-purpose op amp is that it is a voltage controlled voltage source (VCVS). The op amp's voltage output depends upon its input voltage signal (v_{error}) as shown in Figure 19-2(c). The op amp typically consists of hundreds of parts such as BJTs, resistors, and diodes. The circuitry within the op amp creates a voltage amplifying device even though the internal active components are current controlled BJT devices. The overall characteristic of the circuit is *determined primarily by the circuit configuration.*

Figure 19-2(c) Ideal general-purpose op amp–VCVS device

19.3 Maximum Signal Transfer

What is the purpose of an amplifier circuit? Is it to deliver a voltage signal or a current signal to a load? Or must the circuit deliver maximum power to the load without significant power loss? Let's use the Thévenin model of Figure 19-3 to examine these three possibilities. Given that the source circuit is fixed with values of e_{source} and R_{source}, what is the relationship of R_{load} to R_{source} to transfer maximum voltage, maximum current, or maximum power to the load? Note that e_{source} is equivalent to the source circuit's Thévenin voltage e_{Th} value, and R_{source} is equivalent to the source circuit's Thévenin resistance R_{Th} value.

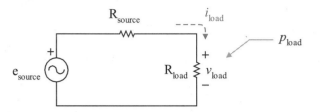

Figure 19-3 Thévenin model to examine maximum signal transfer

Maximum Voltage Transfer

The circuit of Figure 19-3 is a series circuit and therefore forms a voltage divider. The load voltage can be expressed in terms of the voltage divider rule.

$$v_{load} = \left(\frac{R_{load}}{R_{source} + R_{load}} \right) e_{source}$$

If $R_{load} \gg R_{source}$, then R_{load} dominates the denominator and the expression reduces to

$$v_{load} \approx e_{source}$$

and the output voltage is at its maximum possible value of e_{source}. Ideally all of the source signal voltage would be delivered to the load if the load resistance R_{load} were an open ($\infty\ \Omega$) or if the source resistance R_{source} is a short ($0\ \Omega$). In either case, if $R_{load} \gg R_{source}$ then essentially the source signal is delivered to the load. Ideally,

$$v_{load} = e_{source}$$

Another way to characterize voltage transfer is in terms of the circuit's *voltage gain*. For the series voltage dividing circuit, the voltage gain is

$$A_v = \frac{v_{load}}{e_{source}} = \frac{R_{load}}{R_{source} + R_{load}}$$

so, in this case,

$$0 \le A_v \le 1$$

Maximum voltage transfer occurs in this passive circuit when the circuit or system voltage gain is equal to unity.

$$A_v = 1$$

This indicates that 100% of the signal voltage is delivered to the load.

Note that if the load is essentially an open, the circuit would produce a maximum transfer of voltage, but the load current would be ideally zero and the load power would be ideally zero. Therefore, it is possible to deliver maximum voltage to the load but produce no output current or output power.

Example 19-3

For the circuit of Figure 19-3, the source resistance is 10 Ω and the load resistance 1 kΩ. Find the load voltage. Prove that this is a good voltage transfer circuit by comparing the source and load voltages, the source and load resistances, and the circuit voltage gain. If the load resistance is doubled, find the new values of load voltage and load current. What are the effects of the change on the load voltage and load current and the implications of the effect?

Solution

The load voltage is

$$v_{load} = \left(\frac{1 \text{ k}\Omega}{10 \text{ }\Omega + 1 \text{ k}\Omega} \right) 10 \text{ V} = 9.90 \text{ V}$$

The load current is

$$i_{load} = \frac{9.90 \text{ V}}{1 \text{ k}\Omega} = 9.90 \text{ mA}$$

The load voltage of 9.9 V is very close to the value of the source voltage of 10 V. There is very little loss of voltage across the source resistance. Therefore, this circuit is considered to be a good voltage transfer circuit.

A higher load resistance relative to the source resistance transfers more voltage to the load. Compare the source and load resistances.

$$R_{load} \gg R_{source} \quad ?$$

$$1 \text{ k}\Omega \gg 10 \text{ }\Omega$$

Yes, this is essentially a voltage transfer circuit. Verify this by calculating the circuit voltage gain, which is very close to unity.

$$A_v = \frac{1 \text{ k}\Omega}{10 \text{ }\Omega + 1 \text{ k}\Omega} = 0.99$$

If the load resistance is doubled to 2 kΩ, then the load voltage and load current become, respectively,

$$v_{\text{load}} = \left(\frac{2 \text{ k}\Omega}{10 \text{ }\Omega + 2 \text{ k}\Omega} \right) 10 \text{ V} = 9.95 \text{ V}$$

$$i_{\text{load}} = \frac{9.95 \text{ V}}{2 \text{ k}\Omega} = 4.98 \text{ mA}$$

The load voltage has changed very little but the load current was cut in half. This is an indicator that this is a *stable voltage source* and *cannot* be considered a robust current source.

Practice

The load resistance of the example circuit is changed to an open. Find the load voltage, load current, load power dissipation, and circuit voltage gain.

Answer: 10 V, 0 mA, 0 W, voltage gain of 1.0 (an ideal case)

Maximum Current Transfer

The circuit of Figure 19-3 is a series circuit; therefore, the circuit current is found from the Ohm's law expression.

$$i_{\text{load}} = \frac{e_{\text{source}}}{R_{\text{source}} + R_{\text{load}}}$$

The maximum load current is created when the load resistance is 0 Ω, or a short. To be characterized as a current transfer circuit, load resistance must be much, much less than the source resistance.

$$R_{\text{load}} \ll R_{\text{source}} \qquad \text{(maximum current transfer)}$$

Use the current divider rule with the Norton model of the circuit of Figure 19-4 to express the output current in terms of the input current.

$$i_{load} = \left(\frac{R_{source}}{R_{source} + R_{load}} \right) i_{source}$$

i_{source} is equivalent to the source circuit's Norton current i_N value, and R_{source} is equivalent to the source circuit's Norton resistance R_N value. Thévenin resistance and Norton resistance are the same resistance value.

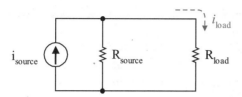

Figure 19-4 Norton model to examine maximum current transfer

The maximum current is transferred to the load if the load resistance is much, much less than the source resistance, $R_{load} \ll R_{source}$. Then R_{source} dominates the denominator of the current divider rule and the expression reduces to

$$i_{load} \approx i_{source}$$

The output current is at its maximum possible value. Ideally, all the source signal current would be delivered to the load if the load resistance R_{load} is a short (0 Ω), or if the source resistance R_{source} is an open (∞ Ω). In either case, if $R_{load} \ll R_{source}$ then essentially all of the source signal is delivered to the load.

$$i_{load} = i_{source}$$

Another way to characterize current transfer of the passive current divider circuit is in terms of the circuit's current gain.

$$A_i = \frac{i_{load}}{i_{source}} = \frac{R_{source}}{R_{source} + R_{load}}$$

where

$$0 \le A_i \le 1$$

Maximum current transfer occurs when the circuit or system current gain is equal to unity.

$$A_i = 1 \qquad \text{(maximum current transfer)}$$

This indicates that 100% of the signal current is delivered to the load.

If the load is essentially a *short*, the circuit would produce a maximum transfer of current, but the load voltage and load power would be ideally zero. Therefore, it is possible to deliver maximum current to the load but produce no output voltage or output power.

Example 19-4

For the Norton circuit of Figure 19-4, the source resistance is 100 kΩ and the load resistance 1 kΩ. Find the load current and load voltage. Prove that this is a good current transfer circuit by comparing the source and load currents, the source and load resistances, and the circuit current gain. If the load resistance is doubled, find the new values of load current and load voltage. What is the effect on the load current and load voltage and the implications of the effect?

Solution

The load current is

$$i_{load} = \left(\frac{100 \text{ k}\Omega}{100 \text{ k}\Omega + 1 \text{ k}\Omega} \right) 10 \text{ mA} = 9.90 \text{ mA}$$

The load voltage is

$$v_{load} = 9.90 \text{ mA} \times 1 \text{ k}\Omega = 9.90 \text{ V}$$

The load current of 9.90 mA is very close to the value of the source current of 10 mA. There is very little loss of current through the relatively high internal source resistance. Therefore, this circuit is considered to be a robust current transfer circuit.

A relatively lower load resistance generates more current to the load. Compare the load resistance to the source resistance.

$$R_{load} \ll R_{source} \qquad ?$$

$$1 \text{ k}\Omega \ll 100 \text{ k}\Omega$$

Yes, this is essentially a current transfer circuit. Verify this by calculating the circuit current gain, which is very close to unity.

$$A_\mathrm{i} = \frac{100\ \mathrm{k\Omega}}{100\ \mathrm{k\Omega} + 1\ \mathrm{k\Omega}} = 0.99$$

If the load resistance is doubled to 2 kΩ, then the load voltage and load current become

$$i_\mathrm{load} = \left(\frac{100\ \mathrm{k\Omega}}{2\ \mathrm{k\Omega} + 100\ \mathrm{k\Omega}}\right) 10\ \mathrm{mA} = 9.80\ \mathrm{mA}$$

$$v_\mathrm{load} = 9.80\ \mathrm{mA} \times 2\ \mathrm{k\Omega} = 18.6\ \mathrm{V}$$

The load current has changed very little but the load voltage was doubled. This is an indicator that this is a stable, robust current source but *cannot* be considered a practical voltage source.

Practice

The example load resistance is changed to a short. Find the load current, voltage, power dissipation, and the circuit current gain.

Answer: 10 mA, 0 V, 0 W, current gain of 1.0 (an ideal case)

Note that a practical current supply or source most efficiently drives a relatively low resistive load, while a practical voltage supply or source most effectively drives a relatively high resistive load.

Maximum Power Transfer

What about **maximum power transfer** to the load with a given source or supply circuit? As we have seen, maximum transfer of voltage or current does not necessarily lead to a maximum transfer of power. We shall show that for a given supply or source circuit the *maximum transfer of power occurs when the load resistance is equal to the source's resistance.*

$$R_\mathrm{load} = R_\mathrm{source}$$

that is when

$$R_\mathrm{load} = R_\mathrm{Th} = R_\mathrm{N} \qquad \text{(maximum power transfer)}$$

Either the Thévenin model of the circuit of Figure 19-3 or the Norton model of the circuit in Figure 19-4 can be used to prove this relationship. Let's use the Thévenin model of Figure 19-3. The power dissipated in the load is based upon the power rule.

$$p_{load} = i_{load} \, v_{load}$$

Substituting

$$p_{load} = \frac{R_{load}}{\left(R_{source} + R_{load}\right)^2} \, e^2_{source}$$

Calculus can be used to find when the power is a *maximum* relative to R_{load}, treating R_{source} and e_{source} as constants. A simple approach is demonstrated in the next example to deduce the maximum power transfer rule.

Example 19-5

For the circuit of Figure 19-5, use a spreadsheet to calculate the load power for a range of load resistance R_{load} values from 0 Ω to 100 Ω in steps of 1 Ω. What can you conclude from your data?

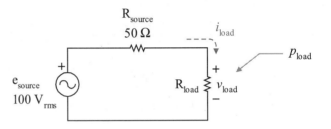

Figure 19-5 Example circuit to find maximum power transfer

Solution

Table 19-1 shows an abbreviated version of the spreadsheet in increments of 10 Ω for the load resistance. As the load resistance is increased from 0 Ω to 100 Ω,

- load current starts at 2 A and decreases toward 0 A
- load voltage starts at 0 V and increases toward 100 V
- load power increases from 0 W to a maximum of 50 W at R_{load} equals R_{source}
- load power decreases from a maximum of 50 W toward 0 W as the load resistance value increases above the source resistance value

For a given source circuit, maximum power transfer to the load occurs when the load resistance R_{load} is equal in value to the source circuit's source or Thévenin resistance R_{Th}.

Table 19-1 Maximum power spreadsheet data

e_{source}	R_{source}	R_{load}	i_{load}	v_{load}	p_{load}
(V)	(Ω)	(Ω)	(A)	(V)	(W)
100	**50**	0	2.00	0.0	0.0
		10	1.67	16.7	27.8
		20	1.43	28.6	40.8
		30	1.25	37.5	46.9
		40	1.11	44.4	49.4
$R_{load} = R_{source}$ ⟶		**50**	**1.00**	**50.0**	**50.0**
		60	0.91	54.5	49.6
		70	0.83	58.3	48.6
		80	0.77	61.5	47.3
		90	0.71	64.3	45.9
		100	0.67	66.7	44.4

Practice

Change the source resistance in your spreadsheet to 100 Ω and repeat the example's approach to find maximum power transfer.

Answer: Verified, maximum power output when R_{load} is 100 Ω

Maximum power transfer is also referred to as **matching power** or a power matching condition, critical in high frequency communications.

Example 19-6

The output impedance of an electronics communication amplifier must drive a 75 Ω twin-pair lead to an antenna. What must the output resistance of the amplifier be?

Solution

75 Ω. The resistances must match to transfer the maximum power and prevent signal reflection losses.

Practice

An electronic communications amplifier circuit has a source resistance of 300 Ω and is driving a 300 Ω antenna. Is this a maximum voltage, current, or power transfer circuit?

Answer: Matching resistance, maximum power transfer circuit

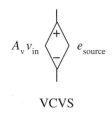

VCVS

Figure 19-6
Voltage controlled
voltage source

19.4 Voltage Controlled Voltage Source

Again, the symbol of Figure 19-6 represents an *ideal* voltage controlled (VC) voltage source (VS). The dependent voltage source's output e_{source} depends upon the voltage v_{in}. *Think voltage in, voltage out.*

$$e_{\text{source}} = A_v v_{\text{in}}$$

where A_v is the voltage gain of the dependent voltage source. A model block diagram of a VCVS circuit that includes input and output impedances is shown in Figure 19-7 along with the ideal VCVS model with Z_{in} an *open* and Z_{out} a *short*.

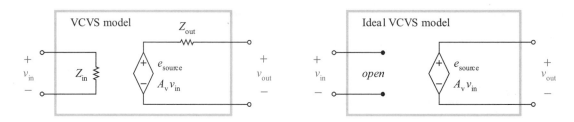

Figure 19-7 Signal models of the voltage controlled voltage source

Z_{in} is the circuit's input resistance, generally called **input impedance** to allow for the possibility of reactance as well as resistance. Z is the symbol of impedance, which could be purely resistive, purely reactive, or a combination of resistance and reactance. An external voltage supply circuit or signal drives the input circuit to create the input voltage v_{in}.

The voltage source e_{source} is *dependent* on the input voltage v_{in}, which is developed across the input impedance. The input voltage v_{in} *controls* the generated voltage. The input voltage v_{in} is multiplied by the voltage gain A_v to create the dependent source voltage e_{source}.

This generated voltage e_{source} then feeds the loaded output circuit of Z_{out} in series with the load. The term Z_{out} is the circuit's output resistance, generally called **output impedance** to allow for the possibility of reactance as well as resistance.

The *ideal* VCVS model circuit of Figure 19-7 is also called an ideal *voltage amplifier*. The input circuit and the output circuit are both *ideal voltage transfer circuits* without loss of any signal voltage. If input impedance Z_{in} is extremely high (ideally an *open*) the input voltage supply system delivers all of its supply signal voltage to the input circuit. If the output impedance Z_{out} is extremely low (ideally a *short*) the VCVS generator e_{source} delivers all of its generated source voltage to the load. The characteristics of a nearly ideal VCVS circuit are:

- Input impedance Z_{in} is very high (ideally an *open*)
- Output impedance Z_{out} is very low (ideally a *short*)
- Output voltage is related to and directly controlled by v_{in}

Example 19-7

A voltage supply with 1 V_{rms} and an internal resistance of 1 kΩ drives a VCVS circuit with an input resistance of 9 kΩ, output resistance of 2 kΩ, and a dependent source voltage gain of 10. The load is an 8 kΩ resistance. Draw the circuit schematic with the VCVS circuit model. Find the unloaded output voltage and the loaded output voltage. Find the system voltage $A_{v\ system}$ gain from e_{supply} to v_{out} with the load connected.

Solution

The loaded modeled circuit is shown in Figure 19-8.

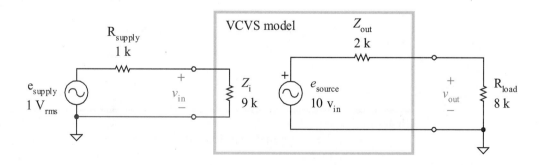

Figure 19-8 Example voltage controlled voltage source circuit

Apply the voltage divider rule to the input circuit.

$$V_{\text{rms in}} = \left(\frac{9\ \text{k}\Omega}{10\ \text{k}\Omega}\right) 1\ \text{V} = 900\ \text{mV}_{\text{rms}}$$

Find the VCVS generated source voltage.

$$e_{\text{rms source}} = 10\ (900\ \text{mV}) = 9\ \text{V}_{\text{rms}}$$

This is also the *open circuit* or *no load* voltage.

$$V_{\text{rms out unloaded}} = e_{\text{source}} = 9\ \text{V}_{\text{rms}}$$

With the load connected.

$$V_{\text{rms out loaded}} = \left(\frac{8\ \text{k}\Omega}{10\ \text{k}\Omega}\right) 9\ \text{V} = 7.2\ \text{V}_{\text{rms}}$$

The system voltage gain from the supply voltage to the load is

$$A_{\text{v system}} = \frac{7.2\ \text{V}_{\text{rms}}}{1\ \text{V}_{\text{rms}}} = 7.2$$

The system voltage gain can be divided into three distinct voltage gains: (1) the input voltage divider, (2) the VCVS voltage gain, and (3) the output circuit voltage divider.

$$A_{\text{v input}} = \frac{9\ \text{k}\Omega}{10\ \text{k}\Omega} = 0.9$$

$$A_{\text{v VCVS}} = 10$$

$$A_{\text{v output}} = \frac{8\ \text{k}\Omega}{10\ \text{k}\Omega} = 0.8$$

The total system voltage gain is the product of the individual voltage gains.

$$A_{\text{v system}} = 0.9 \times 10 \times 0.8 = 7.2$$

Practice

Find the example circuit loaded output voltage if Z_{in} is changed to 10 MΩ and Z_{out} to 1 Ω.

Answer: 10 V_{rms}, essentially an ideal voltage amplifier

Non-ideal Op Amp Characteristics

The symbol of the op amp is shown in Figure 19-9. If the op amp's output voltage is within its rail voltages, the op amp is a voltage controlled voltage source (VCVS) device. The input signal difference voltage (also called v_{error}) is multiplied by the voltage gain A of the op amp to produce its output voltage. As previously noted, the voltage gain A decreases for signal frequencies above the op amp's critical frequency.

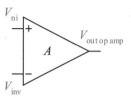

Figure 19-9
Op amp symbol

$$V_{\text{out op amp}} = A\left(V_{\text{ni}} - V_{\text{i nv}}\right) \qquad \text{(ideal)}$$

Figure 19-10(a) is the ideal signal model of the op amp where the op amp voltage gain A is equivalent to the VCVS voltage gain A_v. The input circuit is modeled as an ideal open that draws ideally no current. The input signal difference across the op amp ($V_{\text{ni}} - V_{\text{inv}}$) is multiplied by the op amp voltage gain A (that is, A_v) and delivered to the output of the op amp without loss. Using the ideal model, the input circuit and the output circuit produce no loading effects. The input signal voltage supply delivers all of its voltage to the input of the VCVS open, and the op amp delivers its entire generated signal to the load.

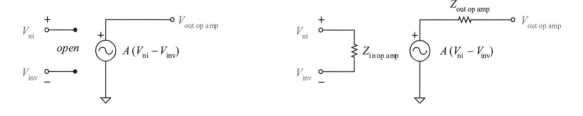

(a) Ideal signal model   (b) Second approximation signal model

Figure 19-10 VCVS signal models of the operational amplifier

Figure 19-10(b) is the second approximation signal model of the op amp and incorporates input impedance Z_{in} and output impedance Z_{out} as well as the VCVS voltage source generator with the op amp voltage gain A. Typical operational amplifiers have very high input impedances (in the low megohms range) and very low output impedances (in the tens of ohms range). Since the input impedance is very high and the output impedance is very low, the op amp is nearly an ideal voltage amplifier. However, the op amp voltage gain A is inconsistent from op amp to op amp even within the same op amp model and manufacturing batch. It is also sensitive to temperature changes. The op amp voltage gain is typically so high (for example, 10^5), that it is not practical. Noise or an imbalanced input biased op amp circuit can drive the op amp to its rail voltages. Negative feedback produces a stable controllable circuit voltage gain and enhances the input and output impedances as well.

Ideal Voltage Amplifier–Closed Loop System

The noninverting voltage amplifier is a nearly ideal voltage amplifier with very high input impedance, very low output impedance, and a controllable voltage gain, which is dependent on circuit resistor values. Figure 19-11(a) is the schematic of a noninverting voltage amplifier. Figure 19-11(b) highlights the feedback circuit and its *feedback factor B*. The system signal input is v_{in}, the input to the op amp is v_{error}, the system output voltage is v_{out}, and the feedback voltage out of the feedback circuit is identified as v_f.

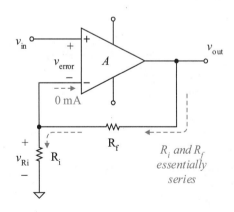

(a) Schematic   (b) Identifying the feedback factor B

Figure 19-11 Noninverting voltage amplifier

From the input circuit and Kirchhoff's voltage law

$$v_{in} = v_{error} + v_f$$

Substitute into this v_{in} expression the following:

$$v_{error} = \frac{v_{out}}{A} \qquad \text{(op amp gain)}$$

$$v_f = B v_{out} \qquad \text{(feedback circut gain)}$$

where B is called the *feedback factor* (that is, the gain of the feedback circuit) and is formed from the voltage divider circuit of R_i and R_f.

$$B = \frac{R_i}{R_i + R_f}$$

Therefore

$$v_{in} = \frac{v_{out}}{A} + B v_{out}$$

Solve for the output signal voltage in terms of the input signal voltage.

$$v_{out} = \frac{A}{1 + AB} v_{in}$$

The closed loop voltage gain A_{v_CL} (that is, the system voltage gain) is

$$A_{v\_CL} = \frac{A}{1 + AB}$$

The $(1+AB)$ term is so prevalent in the calculations of the closed loop characteristics of voltage gain, input impedance, output impedance, and so forth that this term has been named the **sacrifice factor**.

$$S = 1 + AB$$

It is called a *sacrifice factor* because the **open loop voltage gain** A is reduced (sacrificed) by the factor of $(1 + AB)$. An *open loop* configuration is the circuit configuration with the feedback loop removed and the remaining circuit properly grounded or connected to common. The feedback factor B is a voltage divider fraction.

$$0 \le B \le 1$$

A feedback factor B of unity (1) indicates 100% feedback. This is equivalent to a *voltage follower circuit*. If the op amp gain A is a very large number in a circuit with 100% feedback,

$$A_{v\_CL} = \frac{A}{1 + A(1)} \approx 1$$

At the other extreme, a feedback factor B of zero (0) indicates 0% feedback. This is equivalent to no feedback or what is called the *open loop* circuit configuration. The *open loop voltage gain* A_{v_OL} of the system would be equivalent to the op amp voltage gain A.

$$A_{v\_OL} = \frac{A}{1 + A(0)} = A$$

For negative feedback (closed loop) operation, if the denominator term $AB \gg 1$, then the closed loop voltage gain reduces to

$$A_{v\_CL} = \frac{A}{1 + AB} \approx \frac{A}{AB}$$

This reduces to

$$A_{v\_CL} \approx \frac{1}{B}$$

Substituting R_i and R_f for the feedback factor B, the closed loop voltage gain is approximately

$$A_{v\_CL} \approx \frac{R_i + R_f}{R_i}$$

or

$$A_{v\_CL} \approx 1 + \frac{R_f}{R_i}$$

This is the same result we determined and applied in earlier chapters by assuming that v_{error} is approximately zero (0 V). Therefore the VCVS *voltage gain is fixed and controllable* by the selection of the values of R_i and R_f as long as $AB \gg 1$.

Next look at the *input impedance* (or *resistance*) of this circuit.

$$Z_{in\_CL} = \frac{v_{in}}{i_{in}}$$

Substitute for v_{in} the output voltage v_{out} divided by the system voltage gain A_{v_CL} in terms of A and B.

$$Z_{\text{in\_CL}} = \frac{v_{\text{out}}\left(\dfrac{A}{1+AB}\right)}{i_{\text{in}}}$$

Invert, multiply and simplify.

$$Z_{\text{in\_CL}} = \frac{\dfrac{v_{\text{out}}}{A}\left(1+AB\right)}{i_{\text{in}}}$$

The term v_{out} divided by A is just the op amp input voltage v_{error}.

$$Z_{\text{in\_CL}} = \frac{v_{\text{error}}\left(1+AB\right)}{i_{\text{in}}}$$

v_{error} divided by i_{in} is the input impedance of the op amp $Z_{\text{in op amp}}$.

$$Z_{\text{in\_CL}} = \left(1+AB\right)Z_{\text{in op amp}}$$

Thus the input resistance to the system $Z_{\text{in\_CL}}$ is the op amp's input resistance $Z_{\text{in op amp}}$ multiplied by the sacrifice factor $(1 + AB)$. Assuming the sacrifice factor is a large number to produce a stable voltage gain, the input impedance (or resistance) is significantly larger than an already very large op amp input resistance. The closed loop input resistance is much higher than the op amp's input resistance.

A similar exercise to find the system or closed loop output resistance looking back into the system results in

$$Z_{\text{out\_CL}} = \frac{Z_{\text{out op amp}}}{1+AB}$$

An already low output resistance of the op amp is significantly reduced further by a large denominator. A VCVS model of this amplifier circuit is shown in Figure 19-12 where

$$B = \frac{R_i}{R_i + R_f}$$

$$A_{v\_CL} = \frac{A}{1+AB}$$

$$Z_{in\_CL} = (1 + AB)Z_{in\ op\ amp}$$

$$Z_{out\_CL} = \frac{Z_{out\ op\ amp}}{1 + AB}$$

Negative feedback enhances each of the major characteristics of the VCVS voltage amplifier circuit:

- A significant increase in an already high input impedance to prevent loss of the input voltage signal
- A significant decrease in an already low output impedance to prevent loss of the output voltage signal
- A reduced but *circuit controlled* output signal voltage gain directly proportional to the input voltage signal

The voltage gain is reduced by the *sacrifice factor* $(1+AB)$, the input impedance is increased by the sacrifice factor, and the output impedance is reduced by the sacrifice factor.

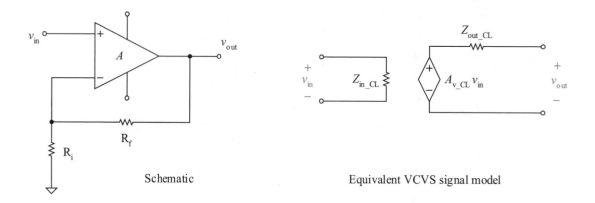

Schematic   Equivalent VCVS signal model

Figure 19-12 Noninverting voltage amplifier and its system VCVS model

Example 19-8

The op amp of Figure 19-12 has a voltage gain of 10^5, an input resistance of 2 MΩ, and an output resistance of 75 Ω. The feedback resistor values are 1 kΩ for R_i and 9 kΩ for R_f. Find the closed loop feedback factor, sacrifice factor, voltage gain, input resistance, and output resistance. Draw the VCVS equivalent circuit model. Explain why this is an ideal voltage amplifier circuit.

Solution

The feedback factor is

$$B = \frac{1 \text{ k}\Omega}{1 \text{ k}\Omega + 9 \text{ k}\Omega} = 0.1$$

$$S = 1 + (0.1)(10^5) = 10{,}001 \approx 10^4$$

$$A_{v\_CL} = \frac{10^5}{1 + (0.1)(10^5)} = 9.9990 \approx 10$$

The closed loop or system input resistance (impedance) is significantly increased over the open loop input resistance.

$$Z_{in\_CL} = 2 \text{ M}\Omega \left(1 + (0.1)\ (10^5)\right) = 20 \text{ G}\Omega$$

The closed loop or system output resistance (impedance) is significantly reduced over the open loop resistance.

$$Z_{out\_CL} = \frac{75 \ \Omega}{1 + (0.1)(10^5)} = 0.0075 \ \Omega$$

The VCVS circuit model is shown in Figure 19-13 and is essentially an ideal voltage amplifier.

Z_{in_CL}	very, very high (20 GΩ)
Z_{out_CL}	very, very low (0.0075 Ω)
v_{out}	controllable and directly related to v_{in}

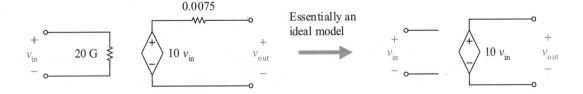

Figure 19-13 Example noninverting voltage amplifier VCVS circuit model

Practice

Repeat the example problem with a 99 kΩ feedback resistor.

Answer: 0.01, 10^3, 100, 2 GΩ, 0.075 Ω, 100, very high Z_{in_CL}, very low Z_{out_CL}, fixed voltage gain related directly to v_{in}, same as Figure 19-13 with new parameter values found.

Inverting Voltage Amplifier–Not Ideal

The inverting voltage amplifier circuit of Figure 19-14 is a voltage controlled voltage source VCVS, but is *not* an ideal voltage amplifier. Since the right side of R_i is connected to virtual common, the input signal essentially sees R_i of resistance looking into the amplifier. An ideal voltage amplifier would see an open looking into the amplifier.

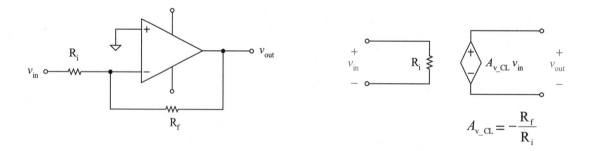

Figure 19-14 Op amp inverting voltage amplifier and its VCVS signal model

As shown in previous work, the voltage gain of the inverting voltage amplifier is

$$A_{v\_CL} = -\frac{R_f}{R_i}$$

The output voltage is controllable and dependent upon the input voltage v_{in}. The closed loop voltage gain is negative indicating that the output signal is inverted relative to the input signal.

Like the output circuit of the noninverting voltage amplifier, the output resistance looking back into the parallel connection at the amplifier output is nearly ideal. That is, the output resistance of the voltage amplifier is very, very low and for all practical purposes is ideally 0 Ω.

Example 19-9

Convert the inverting voltage amplifier circuit of Figure 19-15(a) to its VCVS model and redraw the schematic with this VCVS model. Use the modeled circuit and find the voltage drop across the input of the inverting voltage amplifier, the voltage generated by the VCVS dependent voltage source, and the output voltage.

Solution

The input impedance to the inverting voltage amplifier is

$$Z_{in\_CL} = R_i = 2 \text{ k}\Omega$$

The voltage gain of the inverting voltage amplifier circuit is

$$A_{v\_CL} = -\frac{10 \text{ k}\Omega}{2 \text{ k}\Omega} = -5$$

The output impedance of the inverting voltage amplifier is extremely low, in the milliohms. Effectively,

$$Z_{out\_CL} = 0 \text{ }\Omega$$

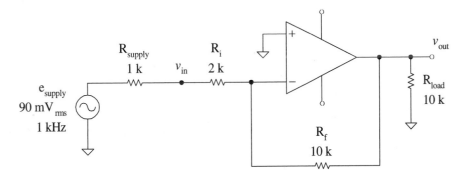

Figure 19-15(a) Example inverting voltage amplifier circuit

The modeled VCVS circuit model of the inverting voltage amplifier circuit is shown in Figure 19-15(b). Find the input signal voltage from the voltage divider input circuit.

$$v_{in} = \left(\frac{2 \text{ k}\Omega}{2 \text{ k}\Omega + 1 \text{ k}\Omega} \right) 90 \text{ mV}_{rms} = 60 \text{ mV}_{rms}$$

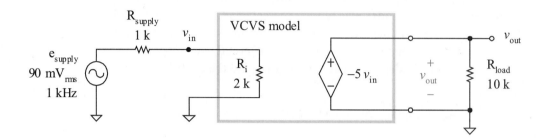

Figure 19-15(b) Example circuit with VCVS amplifier model

The VCVS dependent voltage generated is

$$e_{source} = -5 \left(60 \text{ mV}_{rms} \right) = -300 \text{ mV}_{rms}$$

Note that rms voltage is *always positive*, therefore the *negative sign* indicates that the output voltage is *inverted relative to the input voltage*. Since the amplifier out resistance is 0 Ω, the dependent voltage source is behaving like an ideal voltage source and delivering all of its voltage to the load without loss.

$$v_{out} = e_{source} = -300 \text{ mV}_{rms}$$

Practice

For the example circuit, find the voltage gain of the input circuit, the voltage gain of the inverting voltage amplifier circuit, and the voltage gain of the output circuit. Find the total voltage gain of the system from the supply voltage e_{supply} to the output. Use the overall system gain to find the output voltage.

Answer: 0.667, −5, 1, −3.33, −300 mV$_{rms}$

19.5 Current Controlled Voltage Source

The symbol of Figure 19-16 represents an ideal current controlled (CC) voltage source (VS). The dependent voltage source's output e_{source} depends upon the current i_{in}. *Think current in, voltage out.*

$$e_{source} = r_m i_{in}$$

where r_m is the *mutual resistance* or *transresistance* of the circuit. A model block diagram of a CCVS circuit that includes input and output impedances is shown in Figure 19-17 along with the ideal CCVS model with Z_{in} a short and Z_{out} a short.

An external supply circuit or signal drives the input circuit to create the input current i_{in}. The voltage source e_{source} is *dependent* on the input current i_{in}. The input current *controls* the generated voltage.

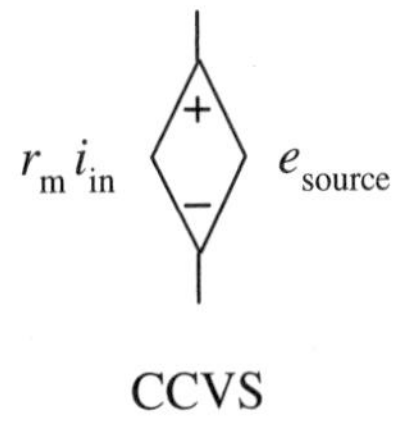

Figure 19-16
Current controlled voltage source

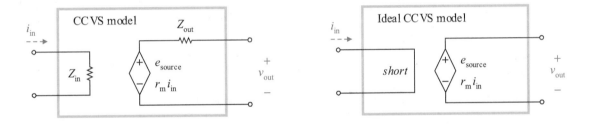

Figure 19-17 Signal models of the current controlled voltage source

The *ideal* CCVS circuit is also called an ideal *current-to-voltage converter*. The input circuit is an *ideal current transfer circuit*. The output circuit is an *ideal voltage transfer circuit*. The characteristics of a nearly ideal CCVS circuit are:
- Input impedance Z_{in} is very low (ideally a *short*)
- Output impedance Z_{out} is very low (ideally a *short*)
- Output voltage is related to and controlled by the input current

Ideal Current-to-Voltage Converter

The ideal **current-to-voltage converter** circuit using an op amp with negative feedback is shown in Figure 19-18.

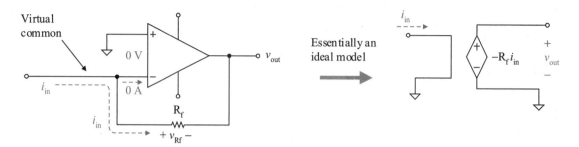

Figure 19-18 Ideal current-to-voltage converter

Since the input current to the op amp is 0 A, the input current i_{in} flows into the circuit then up and through the feedback resistor R_f dropping voltage + to − across R_f. Negative feedback produces ideally a 0 V drop across the op input terminals producing a *virtual common* at the op inverting input pin. Starting at the virtual common with a node voltage of approximately 0 V, a falling voltage across R_f produces a negative output voltage with a positive input current.

$$i_{Rf} = i_{in}$$

$$v_{Rf} = R_f \, i_{in}$$

$$v_{out} = 0 \text{ V} - v_{Rf} = -v_{Rf}$$

$$v_{out} = -R_f \, i_{in}$$

The ideal current controlled voltage source (CCVS) model of Figure 19-18 represents this ideal current-to-voltage converting circuit. Negative feedback enhances each of the major characteristics of the CCVS circuit:

- Converts a high input resistance op amp into a very low resistance (virtually a *short*) input circuit resistance
- A significant decrease in an already low output impedance to prevent loss of the output voltage signal
- A controlled output signal voltage directly proportional to the input current signal

Example 19-10

Design a circuit that converts a 4–20 mA input current loop to a 1–5 V output. That is, 4 mA converts to 1 V, 8 mA converts to 2 V, and so forth.

Solution

The two-stage amplifier of Figure 19-19(a) is a possible circuit design that would satisfy this requirement for low current load applications. The first stage is an inverting current-to-voltage converter. The second stage is a voltage inverter with unity gain to compensate for the inversion in the first stage.

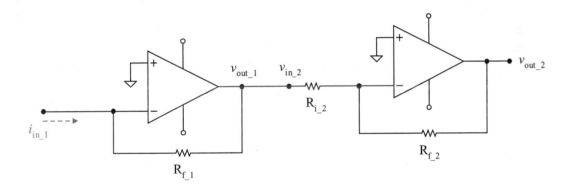

Figure 19-19(a) 4–20 mA converter to 1–5 V

The first stage CCVS circuit produces an ideal current-to-voltage converter. The feedback resistance R_{f_1} must satisfy Ohm's law of converting 1 V to 4 mA.

$$R_{f\_1} = \frac{v_{out\_1}}{i_{in\_1}}$$

$$R_{f\_1} = \frac{1 \text{ V}}{4 \text{ mA}} = 250 \ \Omega$$

Other combinations such as 5 V and 20 mA would also produce the same feedback resistance value. This current-to-voltage converter circuit produces an inverted output voltage range of -1 V to -5 V.

A second stage inverting voltage amplifier with a voltage gain A_{v_2} of -1 is needed to produce a positive output voltage range of 1 V to 5 V. This requires an inverting voltage amplifier with an input resistance R_{i_2} equal to the feedback resistance R_{f_2}. The value of R_{i_2} and R_{f_2} should be reasonably large to produce an output op amp current that is within the op amp's output current specifications and not load the output of the op amp. For example, 100 kΩ produces a maximum feedback current of 50 μA with an input voltage of -5 V.

The two-stage model for this circuit is shown in Figure 19-19(b). Note the *compatible interfaces* between the stages. For example, the output voltage source (VS) of the first stage drives the input voltage control (VC) circuit of the second stage for maximum transfer of voltage. The input current source i_{in_1} should be from a high resistance source to transfer maximum current to the first stage. The output of the second stage v_{out_2} should drive a load with sufficient resistance to transfer maximum voltage to the load.

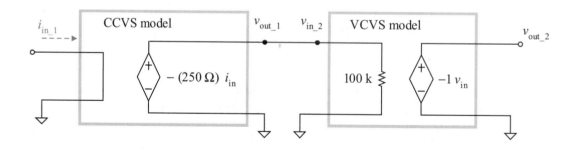

Figure 19-19(b) Example problem two-stage model

Practice

Use the model circuit of Figure 19-19(b) with an input control current of 8 mA and calculate the first stage dependent voltage source output, the first stage output voltage, the second stage input voltage, the second stage dependent voltage source output, and the second stage circuit output voltage.

Answer: $e_{source\_1} = v_{out\_1} = v_{in\_2} = -2$ V, $e_{source\_2} = v_{out\_2} = 2$ V

Photosensitive Circuit

Photovoltaic cells, photoresistors, photodiodes, and phototransistors are examples of sensor devices that convert light energy changes into voltage, current, or a resistance changes. The next example utilizes a photovoltaic diode. The input is a light signal and the output of the photovoltaic diode is a proportional current. That current is then fed into a current-to-voltage converter circuit to produce an output voltage that is proportional to the surrounding light intensity.

Example 19-11

Determine the current-to-voltage conversion factor for the circuit of Figure 19-20. Note R_{comp} is a compensation resistance to help offset slight difference in input op amp bias currents. Slight current differences, although very small, can create an erroneous v_{error} signal that is amplified by the op amp voltage gain.

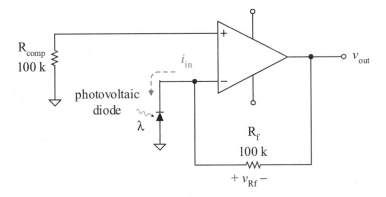

Figure 19-20 Example light-to-voltage converter circuit

Solution

Note that the photovoltaic diode produces a reverse current so that the output voltage v_{out} is a positive voltage. The transresistance of the circuit is R_f with a value of 100 kΩ.

$$\frac{v_{out}}{i_{in}} = 100 \text{ k}\Omega = 0.1 \frac{\text{V}}{\mu\text{A}}$$

Practice

For the circuit of Figure 19-20, find the value of the feedback resistance required to create 1 V per microampere output.

Answer: 1 MΩ

Inverting Voltage Amplifier–Inherently a CCVS

The *inverting voltage amplifier* uses the *ideal current-to-voltage converter* (a CCVS circuit) as its fundamental building block circuit as demonstrated in Figure 19-21. Because of the virtual common, the input signal v_{in} appears across the input resistance R_i to form the input current i_{in}.

$$i_{in} \approx \frac{v_{in}}{R_i}$$

This i_{in} current is then fed to the *ideal current-to-voltage circuit* to create an inverted voltage output.

$$v_{out} \approx -R_f i_{in}$$

$$v_{out} \approx -R_f \frac{v_{in}}{R_i}$$

$$v_{out} \approx -\left(\frac{R_f}{R_i}\right) v_{in}$$

The complete system is an *inverting voltage amplifier* but is built around the *ideal current-to-voltage converter*. But as previously noted, this voltage amplifier is not ideal. The inverting voltage amplifier has the disadvantage of an input resistance of R_i. An ideal voltage amplifier has infinite input impedance or resistance.

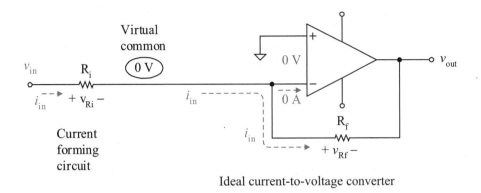

Figure 19-21 Inverting voltage amplifier formed with CCVS circuit

19.6 Voltage Controlled Current Source

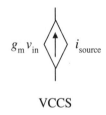

VCCS

Figure 19-22
Voltage controlled
current source

The symbol of Figure 19-22 represents an ideal voltage controlled (VC) current source (CS). The dependent current source's output i_{source} depends upon the voltage v_{in}. *Think voltage in, current out.*

$$i_{source} = g_m v_{in}$$

Where g_m is the *mutual conductance* or *transconductance* of the circuit. A model schematic of a VCCS circuit that includes input and output impedances is shown in Figure 19-23 along with the ideal VCCS model with Z_{in} an open and Z_{out} an open.

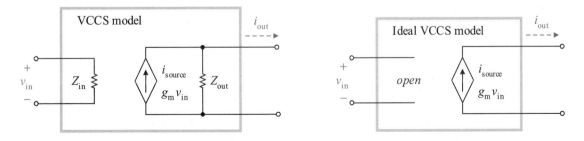

Figure 19-23 Signal models of the voltage controlled current source

The *ideal* VCCS circuit is also called an ideal *voltage-to-current converter*. The input circuit is an *ideal voltage transfer circuit*. The characteristics of a nearly ideal VCCS circuit are:

- Input impedance Z_{in} is very high (ideally an *open*)
- Output impedance Z_{out} is very high (ideally an *open*)
- Output voltage is related to and controlled by the input current

Ideal Voltage-to-Current Converter

The ideal **voltage-to-current converter** circuit using an op amp with negative feedback is shown in Figure 19-24.

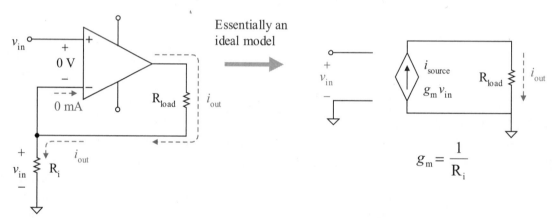

Figure 19-24 Ideal voltage-to-current converter

Essentially all of the input voltage v_{in} is dropped across R_i. The output current i_{out} flows out of the op amp in series with the load resistance R_{load} and essentially in series with R_i. Therefore, the current through R_i is approximately equal to i_{out}. Apply Ohm's law to the resistance R_i.

$$i_{out} = \frac{1}{R_i} v_{in}$$

The ideal voltage controlled current source (VCCS) model of Figure 19-24 represents the ideal voltage-to-current converting circuit. The input circuit is an ideal voltage transfer circuit. The output circuit is an ideal current transfer circuit. A load resistance is needed to complete the current path of the dependent current source. Negative feedback enhances each of the major characteristics of the voltage-to-current converter circuit:

- Increases an already high input resistance op amp into a much higher (virtually an *open*) input circuit resistance
- Converts a very low op amp output resistance into a very high (virtually an *open*) output source resistance
- Input voltage signal controls a current output signal

Example 19-12

The circuit of Figure 19-24 has a resistance R_i value of 1 kΩ. Find the output current in the load resistance.

Solution

The output current is

$$i_{out} = \left(\frac{1}{1 \text{ k}\Omega} \right) 2 \text{ V} = 2 \text{ mA}$$

Practice

Design a voltage-to-current converter to convert a 1–5 V control loop voltage to a 4–20 mA current control loop.

Answer: Circuit of Figure 19-24 with R_i resistance of 250 Ω, requires an op amp that can handle a 20 mA output current or a BJT current booster

Enhancement Mode MOSFET

The enhancement mode MOSFET (metal oxide semiconductor field effect transistor) was introduced earlier as an example active device in the study of basic circuit laws and analysis. Other types of FETs include the junction field effect transistor (JFET) and the depletion mode MOSFET. This text focuses on the enhancement mode MOSFET since it is the most extensively used MOSFET. The basic construction of an enhancement mode MOSFET (abbreviated E-MOSFET) semiconductor is shown in Figure 19-25(a); in this case it is an *n*-channel E-MOSFET. The silicon dioxide is an insulator and thus I_G is always 0. The schematic symbol for an *n*-channel E-MOSFET is shown in Figure 9-25(b). When biased off, as shown in Figure 19-25(a), there is no path for current through the MOS-FET from drain to source; it is *off* or *open*. If sufficient gate-to-source voltage V_{GS} is applied, the MOSFET's threshold voltage $V_{GS(th)}$ is reached, an **n-channel** forms at the insulated gate semiconductor surface, and drain current I_D can occur through the n-channel formed, as shown in Figure 19-26(c).

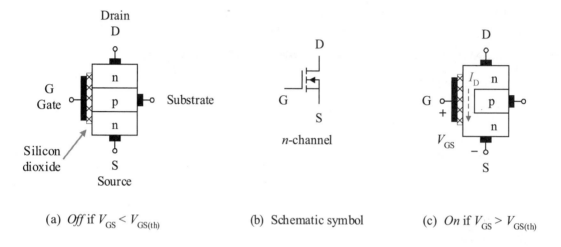

(a) *Off* if $V_{GS} < V_{GS(th)}$ (b) Schematic symbol   (c) *On* if $V_{GS} > V_{GS(th)}$

Figure 19-25 Construction, symbol, and operation of the n-channel E-MOSFET

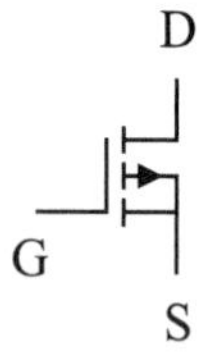

Figure 19-26
P-channel
E-MOSFET

Figure 19-26 shows the schematic symbol for the p-channel E-MOS-FET. The *arrow always points* to the n-type semiconductor material. Thus, the n-channel MOSFET's arrow points *into* the channel, while the p-channel MOSFET's arrow points *out,* away from the channel. In the schematic symbol, the gate G is shown insulated from the rest of the device representing the physical reality. The symbol also shows a dashed line for the channel. This is its natural open state if the MOSFET is biased off. Once the MOSFET turns on by sufficient gate-to-source voltage, visualize the *dashed channel* as a solid line and current flowing through the channel as demonstrated in Figure 19-25(c). The p-type channel requires a reverse voltage polarity to turn on its MOSFET and form a channel, therefore, $V_{GS(th)}$ must be a negative voltage.

The E-MOSFET's silicon dioxide layer is *especially sensitive.* MOS-FETs should not be removed from or inserted into a live circuit. Some MOSFETs are so sensitive that grounding strap systems must be used when handling MOSFETs to prevent a charge buildup and a destructive discharge. The silicon dioxide SiO_2 layer is sufficiently sensitive that a maximum gate-to-source voltage $V_{GS(max)}$ is specified for the E-MOS-FETs. The E-MOSFET may be protected from excessive voltage with a zener diode as shown in Figure 19-27. Some models of E-MOSFETs are constructed with zener diode protection built in.

The *characteristic curves* for an *n*-channel E-MOSFET are shown in Figure 19-28. Figure 19-28(a) shows the transconductance curve of I_D as a function of its input gate-to-source voltage V_{GS}. Once the MOSFET turns on, it is in its **active region**. As the gate-to-source voltage increases the current drain current increases. Note that the relationship is nonlinear when operating over a wide range of V_{GS}. In the active region, the drain current is proportional to the square of the gate-to-source voltage. For small-signal operation, the drain current is approximately linearly proportional to the gate-to-source voltage and the MOSFET is behaving like a voltage controlled (VC) current source (CS) device. The gate-to-source voltage V_{GS} controls the drain current I_D in the active region.

Figure 19-27
Zener protection of an *n*-channel E-MOSFET

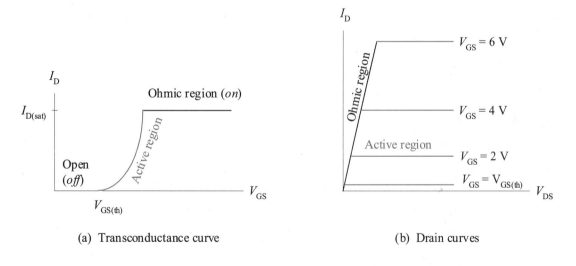

(a) Transconductance curve

(b) Drain curves

Figure 19-28 Characteristic curves of an *n*-channel E-MOSFET

Once the MOSFET reaches its maximum circuit current $I_{D(sat)}$ based upon the external circuit, the drain current remains fixed at $I_{D(sat)}$ and the MOSFET acts like a fixed resistance. Therefore this is called the **ohmic region** of operation and the MOSFET exhibits a very low resistance (on the order of tenths of an ohm to a few ohms). The **ohmic region resistance** is defined to be the static resistance at a test point in the ohmic region as demonstrated in Figure 19-29.

$$R_{DS(on)} = \frac{V_{DS(on)}}{I_{D(on)}}$$

Figure 19-28(b) shows the drain characteristic curves of the drain current I_D versus the drain-to-source voltage V_{DS}. The gate-to-source voltage V_{GS} is fixed for each curve. Note that the V_{GS} curves are not equally spaced due to the nonlinear relationship of V_{GS} with I_D. Compare the *active* and *ohmic* regions between the transconductance curve and the drain curves.

The MOSFET is approximately linear for small-signal operation where it behaves like a voltage controlled (VC) current source (CS) as shown in Figure 19-30. The MOSFET has the distinct advantage of being less noisy than a BJT and therefore can be used in small-signal operation where it is approximately linear. The MOSFET also has the advantage of a virtually open input, optimum for maximum voltage transfer to the input of the MOSFET. However, the BJT does have the advantage of being more linear in its active region of operation for signal amplification and can therefore be used in large-signal amplification.

Keep distinct:
dependent source
from
MOSFET source terminal

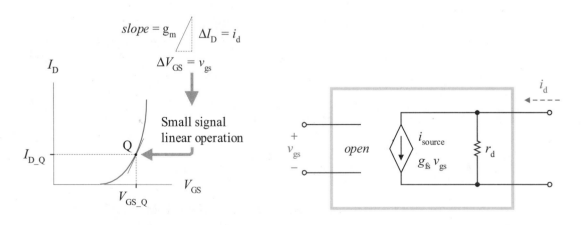

(a) Small-signal transconductance region   (b) MOSFET small-signal model

Figure 19-30 MOSFET small-signal operation at quiescent point Q

Figure 19-30(a) demonstrates the nearly linear operation for a small-signal operation at a **quiescent operating point** Q. **Quiescence** is the condition in which the *ac signal is zero* (0) and only the dc biasing values are present in the system, that is, when the ac signal is quiet and only the dc values are present. The transconductance g_m is the slope of the line tangent to the active curve at the operating **Q point**. For the MOSFET, transconductance g_m is the specification g_{fs}, which is the **forward transconductance**. The value of g_{fs} varies with the Q point on the curve. A change in the Q point moves the tangent line, which changes the slope of the tangent line (g_{fs}).

Figure 19-30(b) shows the corresponding small-signal model for the MOSFET. As demonstrated in Figure 19-30(a), the forward transconductance is defined to be the change in drain current divided by the change in gate-to-source voltage at a Q point with drain-to-source voltage held to a constant value.

$$g_{fs} = g_m = \frac{\Delta I_D}{\Delta V_{GS}} \bigg|_{V_{DS} = \text{constant}}$$

The MOSFET forward transconductance is very sensitive to its Q point location on the transconductance characteristic curve.

The MOSFET output resistance r_d is defined to be the change in drain-to-source voltage divided by the change in drain current with the gate-to-source voltage held constant.

$$r_d = \frac{\Delta V_{DS}}{\Delta I_D} \bigg|_{V_{GS} = \text{constant}}$$

The output resistance r_d is very sensitive to the Q point of operation but is typically so high (for example, 50 kΩ or higher) that it can be practically ignored for reasonable resistive loads. This is especially true since the MOSFET output circuit is designed to be a *maximum current transfer circuit* with a low value of parallel load resistance. So, with $r_d \gg R_{load}$, the VCCS model of the E-MOSFET simplifies to the *ideal* and *practical* model of Figure 19-31.

The specifications of the E-MOSFET tend to be ranges of values rather than specific values. For example, an E-MOSFET threshold voltage rating could be a minimum of 2 V and a maximum of 4 V without a typical threshold voltage rating. And the forward transconductance could be a minimum of 5 S (at a specific operating point) without a maximum or typical rating. You must consider this in your design of circuits. Either these are not crucial values or your external circuit design is such that the overall circuit performance is not dependent on these specifications. If you need to know the specific values of your MOSFET, you must measure them. But be careful, if you replace this MOSFET or are manufacturing circuits using thousands of MOSFETS, you cannot count on the values of each MOSFET. You can count on the shape of the characteristic curves, but expect a wide range of characteristic values.

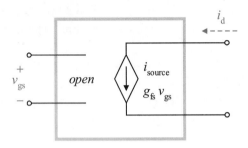

Figure 19-31 Ideal VCCS small-signal model of the E-MOSFET

Common Drain Amplifier–Source Follower

The ac **common drain amplifier** circuit of Figure 19-32(a) delivers a significant current gain and therefore a significant power gain to low impedance devices such as motor circuits and speakers. The signal voltage drives the gate terminal and the output voltage is taken off the source terminal. *The drain is at ac common potential.* The resistance r_S represents the *ac resistance to common as viewed from the source terminal* of the MOSFET. The ideal ac small-signal MOSFET model of this circuit is shown in Figure 19-32(b). Use this ideal circuit model and basic circuit analysis techniques to find the circuit voltage gain.

$$v_{in} = v_{gs} + v_{out} \hspace{4cm} \text{(KVL)}$$

$$v_{in} = \frac{i_{out}}{g_m} + v_{out} \hspace{3cm} \text{(current source substitution)}$$

$$v_{in} = \frac{v_{out}/r_S}{g_m} + v_{out} \hspace{3cm} \text{(Ohm's law for } r_s)$$

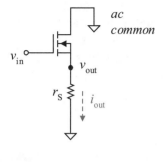

(a) AC common drain amplifier

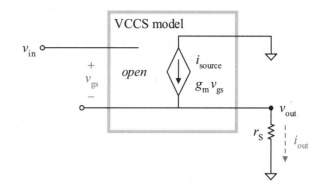

(b) Modeled ac small signal MOSFET

Figure 19-32 AC common drain (or source follower) model

$$v_{in} = \left(1 + \frac{1}{g_m r_S}\right) v_{out}$$
(simplify the expression)

$$A_v = \frac{v_{out}}{v_{in}} = \frac{g_m r_S}{1 + g_m r_S}$$
(solve for voltage gain)

Note that even though the MOSFET is a current output device, it is being used as an active device in a circuit to create an output voltage. This is essentially a voltage amplifier circuit. However, the voltage gain ranges from 0 to 1, so the input voltage is not really amplified but the current is.

$$0 < A_v < 1$$

If $g_m r_S \gg 1$, then the $A_v \approx 1$ and the circuit becomes a **source follower** (that is, a *voltage follower*) since the output ac signal has approximately the *same voltage* value and is *not inverted*.

So what is the advantage of this circuit? Ideally the input current to the MOSFET is 0 A with an output current equal to $g_m v_{gs}$. This implies that the current gain A_i is ideally infinite which would create an infinite power gain. The reality is that a dc bias circuit is needed to properly bias the MOSFET and that the ac current into the resistive bias network creates an input current to the amplifier stage.

Also this circuit is typically used to drive low resistance devices, such as audio speakers, with little loss of signal voltage and significant increase in signal power. The signal power gain is defined to be the *signal power out* divided by the *signal power in* but this simplifies to the product of the voltage and the current gain.

$$A_p = \frac{p_{out}}{p_{in}} = \frac{v_{out} i_{out}}{v_{in} i_{in}} = \frac{v_{out}}{v_{in}} \frac{i_{out}}{i_{in}} = A_v A_i$$

multi**SIM**

Example 19-13

Simulate with MultiSIM the circuit of Figure 19-33(a), which is an ac common drain amplifier. The drain is connected to ac common through the dc power supply E_supply.

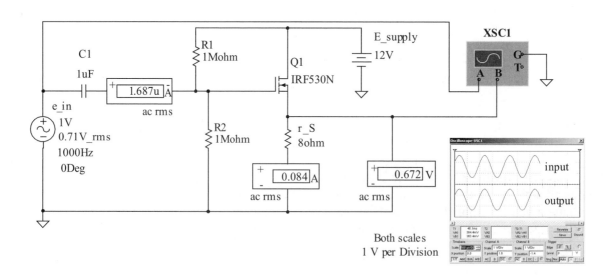

Figure 19-33(a) Example common drain circuit

Calculate the capacitive reactance of C1 and verify that virtually no ac signal is lost across the capacitor. Use the simulated measurements to calculate the amplifier input resistance, the simulated MOSFET input resistance, the circuit voltage gain, the circuit current gain, the circuit power gain, and the MOSFET's transconductance at this circuit's operating point. Use the oscilloscope to verify the output is not inverted. Warning: beware of the voltmeter and oscilloscope internal instruments resistance and their potential impact on this high resistance circuit. Change the voltmeter internal resistances to the $T\Omega$ range.

Solution

The input capacitor C1 capacitive reactance is

$$X_{C1} = \frac{1}{2\pi(1 \text{ kHz})(1 \text{ }\mu\text{F})} = 159 \text{ }\Omega$$

The top of R1 and the bottom of R2 are at *ac common potential*. The ac input circuit impedance (resistance) looking into the bias circuit is $1 \text{ M}\Omega$ // $1 \text{ M}\Omega$ // $Z_{\text{in\_MOSFET}}$. Since $Z_{\text{in\_MOSFET}}$ is ideally an open, the input amplifier resistance is ideally 500 kΩ. The capacitor and the input impedance of the amplifier are in series. Since 500 kΩ >> 159 Ω, the ac signal is dropped entirely across the amplifier circuit with insignificant ac signal loss across C1. Therefore, the full ac input signal of 0.707 V_{rms} is applied to the amplifier. *Based upon the simulation*, the amplifier circuit input resistance is

$$r_{\text{in\_amplifier}} \approx \frac{0.707 \text{ V}_{\text{rms}}}{1.687 \text{ }\mu\text{A}_{\text{rms}}} = 419 \text{ k}\Omega$$

Caution: ideally this should be 500 kΩ, but the simulation is showing a MOSFET input resistance of 3.67 MΩ.

$$\frac{1}{419 \text{ k}\Omega} = \frac{1}{R_{\text{in\_MOSFET}}} + \frac{1}{1 \text{ M}\Omega} + \frac{1}{1 \text{ M}\Omega}$$

$$R_{\text{in\_MOSFET}} = 2.59 \text{ M}\Omega$$

The amplifier circuit voltage, current, and power gains follow.

$$A_v = \frac{0.672 \text{ V}_{\text{rms}}}{0.707 \text{ V}_{\text{rms}}} = 0.9505$$

$$A_i = \frac{0.084 \text{ A}_{\text{rms}}}{1.687 \text{ }\mu\text{A}_{\text{rms}}} = 49,800$$

$$p_{\text{out}} = 0.084 \text{ A}_{\text{rms}} \times 0.672 \text{ V}_{\text{rms}} = 56.45 \text{ mW}$$

$$p_{in} = 1.687 \ \mu A_{rms} \times 0.707 \ V_{rms} = 1.193 \ \mu W$$

$$A_p = \frac{56.45 \ mW}{1.193 \ \mu W} = 47{,}300$$

The power gain based upon the voltage and current gains confirms this result.

$$A_p = 0.9505 \times 49{,}800 = 47{,}300$$

Find the forward transconductance g_m from the voltage gain.

$$A_v = \frac{g_m \times 8 \ \Omega}{1 + g_m \times 8 \ \Omega} = 0.9505$$

Solve for g_m.

$$g_m = \frac{0.9505}{8 \ \Omega \ (1 - 0.9505)}$$

$$g_m = g_{fs} = 2.40 \ S$$

Practice

Draw the VCCS model of the MOSFET and the VCVS model of the complete amplifier circuit of the example problem.

Answer: See Figure 19-33(b).

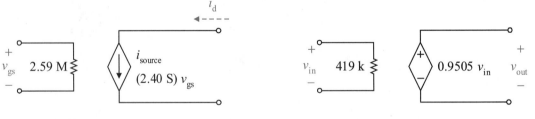

E-MOSFET model   Voltage amplifier circuit model

Figure 19-33(b) MOSFET and voltage amplifier models of example problem

Common Source Amplifier

The ac **common source amplifier** circuit of Figure 19-34(a) delivers a significant current gain and possibly a voltage gain. Therefore, this ac circuit delivers a significant power gain to an ac load connected to the MOSFET drain terminal. *The MOSFET source is at ac common potential.* The resistance r_D represents the *ac load to common as viewed from the drain terminal* of the MOSFET. The ideal ac small-signal MOSFET model of this circuit is shown in Figure 19-34(b).

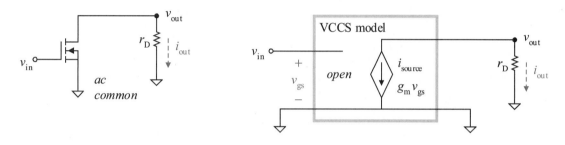

(a) AC common source amplifier   (b) Modeled ac small-signal MOSFET

Figure 19-34 AC common source model

Use the circuit model of Figure 19-34(b) and basic circuit analysis techniques to find the circuit voltage gain.

$$v_{out} = i_{out} \times r_D \qquad\qquad \text{(Ohm's law)}$$

$$v_{out} = -g_m v_{gs} \times r_D \qquad \text{(current source substitution)}$$

$$v_{in} = v_{gs} \qquad\qquad \text{(KVL)}$$

$$v_{out} = -g_m r_D v_{in} \qquad\qquad (v_{in} \text{ substitution)}$$

$$A_v = \frac{v_{out}}{v_{in}} = -g_m r_D \qquad\qquad \text{(solve for voltage gain)}$$

Again, it is assumed ideally that the output resistance of the MOSFET r_d is very high and significantly greater than the parallel load resistance.

Note, even though the MOSFET is a current output device, it is being used as an active device in a circuit to create a voltage amplifier. If $g_m r_D$ is greater than unity, there is a voltage gain. The *negative* voltage gain indicates that a gate driven, drain output signal is *inverted*. A single-stage, ac common source MOSFET small-signal amplifier is shown in Figure 19-35(a) with an ac input signal source and a load.

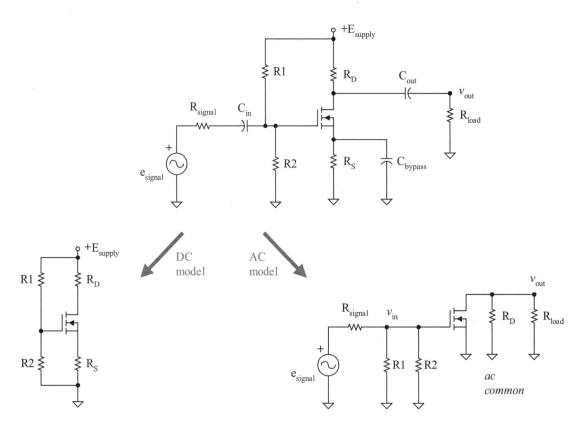

Figure 19-35(a) Single-stage, ac common source small-signal amplifier

First analyze the dc circuit with capacitors modeled as *opens. Secondly* analyze the ac circuit with the capacitors modeled as *shorts* and the dc voltage supply modeled as a *short.* The ac model can be further simplified to the circuit of Figure 19-35(b) by modeling the MOSFET then analyzing (1) the input circuit, (2) the dependent current source value, and (3) the signal delivered to the output.

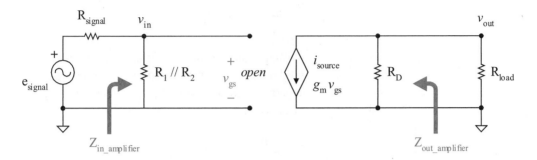

Figure 19-35(b) Modeled small-signal, ac common drain amplifier

The Norton dependent source circuit can be converted to the Thévenin equivalent model of the circuit in Figure 19-35(c), converting this model to a voltage amplifier model. Note the polarity of the dependent voltage source e_{source}. The dependent source e_{source} can also be modeled with reversed polarity and the negative value of $-g_m R_D v_{gs}$ to show inversion.

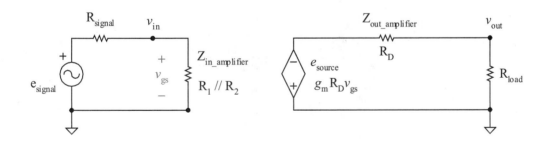

Figure 19-35(c) AC model converted to Thévenin dependent source

$$v_{in} = \frac{Z_{in\_amplifier}}{R_{signal} + Z_{in\_amplifier}} e_{signal} \qquad \text{(VDR)}$$

$$v_{in} = \frac{R1 // R2}{R_{signal} + R1 // R2} e_{signal}$$

$$e_{source} = g_m R_D v_{in} \qquad \text{(dependent source)}$$

$$v_{out} = -\frac{R_{load}}{R_D + R_{load}}\left(g_m R_D v_{in}\right) \qquad\qquad \text{(VDR)}$$

The total signal voltage gain of this circuit is the product of the input circuit voltage gain, dependent source voltage gain, and output circuit voltage gain. The negative sign indicates that the output voltage is inverted relative to the input voltage.

$$A_v = -\left(\frac{R1//R2}{R_{signal} + R1//R2}\right)\left(g_m R_D\right)\left(\frac{R_{load}}{R_D + R_{load}}\right)$$

Example 19-14

A common drain amplifier circuit is shown in Figure 19-36(a) along with the MOSFET's transconductance curve and selected Q point of operation. Completely analyze the dc and ac circuits. Find the value of R2 needed to properly bias the MOSFET.

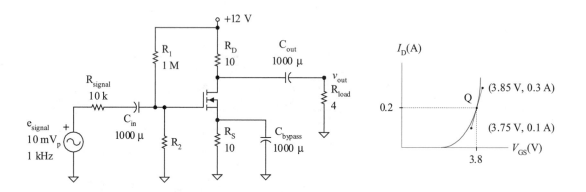

Figure 19-36(a) Example ac common drain amplifier and MOSFET transconductance curve

Solution

The dc model of the amplifier stage (the circuit between the input and output capacitors) is shown in Figure 19-36(b). From the transconductance curve, the Q point is (3.8 V, 2.4 mA), therefore at quiescence the following dc values can be calculated.

$$I_D = I_{D\_Q} = 0.2 \text{ A}_{dc}$$

$$V_{GS} = V_{GS\_Q} = 3.8 \ V_{dc}$$

$$V_S = 0.2 \ A \times 10 \ \Omega = 2 \ V_{dc}$$

$$V_G = 3.6 \ V + 2.4 \ V = 6.0 \ V_{dc}$$

Find the value of R2 needed to create this gate voltage to properly bias the FET. The bias resistors of R1 and R2 form an essentially series circuit since the gate current is essentially 0 A.

$$V_G = \left(\frac{R2}{1 \ M\Omega + R2} \right) 12 \ V = 5.8 \ V$$

$$R2 = 0.94 \ M\Omega$$

Find the node voltage at the drain by subtracting the drain resistor voltage drop V_{RD} from the supply voltage.

$$V_D = 12 \ V - 0.2 \ A \times 10 \ \Omega = 10.0 \ V_{dc}$$

The drain-to-source voltage drop is the drain node voltage less the source node voltage.

$$V_{DS} = 10.0 \ V - 2.0 \ V = 8.0 \ V_{dc}$$

The capacitor views of the circuit are shown in Figure 19-36(c).

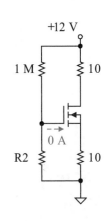

+12 V

Figure 19-36(b)
DC model

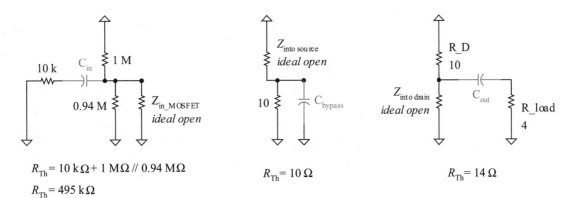

Figure 19-36(c) Capacitor views of the circuit (r_{Th} is the ac Thévenin resistance)

The capacitive reactance of each capacitor is

$$X_C = \frac{1}{2\pi(1\text{ kHz})(1000\ \mu\text{F})} = 0.159\ \Omega$$

The equivalent Thévenin resistance seen by each capacitor is modeled in the circuits of Figure 19-36(c). Recall that the resistance looking into the source and drain terminals of the MOSFET component operating in the active region are ideally ac opens since they are looking into an ac current source (ideally an open). In each case the capacitive reactance is much, much less than the capacitor's external Thévenin resistance; therefore, the capacitors can be modeled as ideal shorts for the ac signal analysis.

The circuit of Figure 19-36(d) shows the ac equivalent circuit and the corresponding VCCS MOSFET modeled circuit.

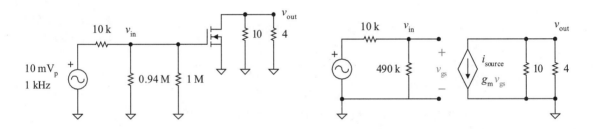

Figure 19-36(d) Equivalent ac circuit and its corresponding VCCS MOSFET model

The forward transconductance g_m of the E-MOSFET (that is, its g_{fs}) can be found from its transconductance curve.

$$g_m = \frac{\Delta I_D}{\Delta V_{GS}} = \frac{0.3\text{ A} - 0.1\text{ mA}}{3.85\text{ V} - 3.75\text{ V}} = 2\text{ S}$$

The Thévenin equivalent model converts this to a VCVS approach as shown in Figure 19-36(e). The dependent voltage source e_{source} (essentially e_{Th}) is

$$e_{source} = g_m R_D v_{gs} = (2\text{ S} \times 10\ \Omega)v_{gs} = 20\ v_{gs}$$

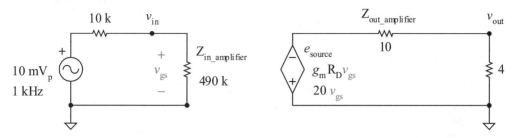

Figure 19-36(e) Circuit with VCVS model of amplifier stage

The Thévenin model is now used for the remaining ac signal analysis; however, the Norton model could also be used.

$$v_{in} = v_{gs} = v_g = \left(\frac{490 \text{ k}\Omega}{10 \text{ k}\Omega + 490 \text{ k}\Omega}\right) 10 \text{ mV}_p = 9.80 \text{ mV}_p$$

$$e_{source} = 20 \times 9.80 \text{ mV} = 196 \text{ mV}_p$$

Dependent source voltage

$$v_s = 0 \text{ V}$$

MOSFET source terminal at ac common potential

$$v_{out} = v_d = \left(\frac{4 \text{ }\Omega}{10 \text{ }\Omega + 4 \text{ }\Omega}\right) 196 \text{ mV} = 56 \text{ mV}_p$$

Practice
Find the amplifier voltage gain, current gain, and power gain.
Answer: 5.6, 750, 4200

CMOS Difference Amplifier

The input of an operational amplifier (op amp) is a difference amplifier that has two input terminals, a noninverting (+) and an inverting (−). This difference amplifier can be created with BJTs only, a mixture of BJTs and FETs (called a BiFET or BiMOS circuit), or completely of MOSFETs. The circuit of Figure 19-37 is composed of complementary *p*-channel and *n*-channel E-MOSFETs. **Complementary MOSFETs** are called **CMOS** and have various applications in digital and analog electronics.

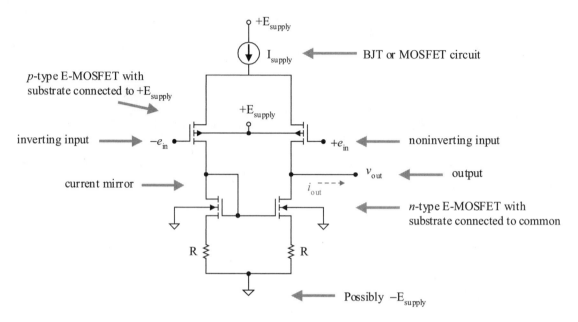

Figure 19-37 CMOS difference amplifier

Like the op amp input terminals, each CMOS difference amplifier's input terminal can be connected to a voltage signal source to produce a dual input, or one of the input terminals can be connected to common to produce a single-ended input. FETs and particularly MOSFETs have significantly higher input impedance and much lower noise distortion than BJTs.

Figure 19-38 is a simplified schematic of a CMOS pair used in a digital inverter circuit. The top MOSFET Q1 is a p-channel and the bottom MOSFET Q2 is an n-channel. In this case, the MOSFETs are acting as electronic switches, either on (ideally a short) or off (open). The supply voltage is set to a digital high voltage (for example, 5 V) and the input is either a digital high or digital low (0 V). When the input e_{in} goes *high*, Q1 is biased *off*, Q2 is biased *on*, and the output is connected to common (low) through Q2. When the input e_{in} goes *low*, Q1 is biased *on*, Q2 is biased *off*, and the output is connected to $+E_{supply}$ (high) through Q1. A high input creates a low output; and a low input creates a high output. Therefore, this is a digital inverter circuit.

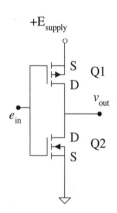

Figure 19-38
Simplified CMOS pair digital inverter circuit

19.7 Current Controlled Current Source

The symbol of Figure 19-39 represents an ideal current controlled (CC) current source (CS). The dependent current source's output i_{source} depends upon the current i_{in}. *Think current in, current out.*

$$i_{source} = A_i i_{in}$$

where A_i is the *current gain*. A model schematic of a CCCS circuit that includes input and output impedances is shown in Figure 19-40 along with the ideal CCCS model with Z_{in} a short and Z_{out} a short.

An external supply circuit or signal drives the input circuit to create the input current i_{in}. The current source i_{source} is *dependent* on the input current i_{in}. The input current *controls* the generated output current.

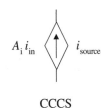

CCCS

Figure 19-39
Current controlled
current source

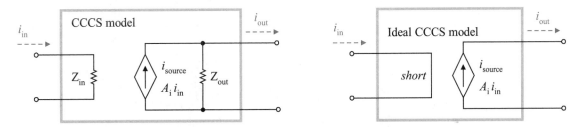

Figure 19-40 Signal models of the current controlled current source

An *ideal* CCCS circuit is also called an ideal *current amplifier*. The input circuit is an *ideal current transfer circuit*. The output circuit is an *ideal current transfer circuit* with $\infty \ \Omega$ of internal resistance. The characteristics of a nearly ideal CCCS circuit are:
- Input impedance Z_{in} is very low (ideally a *short*)
- Output impedance Z_{out} is very high (ideally an *open*)
- Output current is related to and controlled by the input current

Ideal Current Amplifier

The ideal **current amplifier** circuit using an op amp with negative feedback is shown in Figure 19-41. Since the input current to the op amp is ideally 0 A, the input current i_{in} flows into the circuit then down and through the feedback resistor R_f dropping voltage + to – across R_f. Negative feedback produces ideally a 0 V drop across the op amp input terminals producing a *virtual common* at the op amp noninverting input pin. Starting at the virtual common with a node voltage of approximately 0 V, the current i_{in} creates a falling voltage across R_f.

$$i_{Rf} = i_{in}$$

$$v_{Rf} = R_f \, i_{in}$$

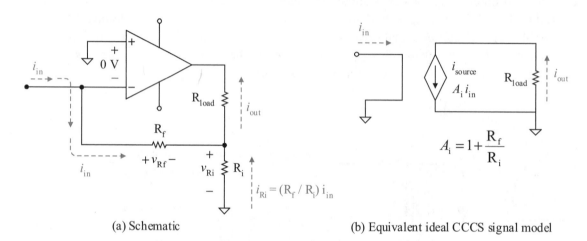

(a) Schematic

(b) Equivalent ideal CCCS signal model

Figure 19-41 Ideal current amplifier using an op amp

This same voltage is impressed across R_i (note the polarity of the drop).

$$v_{Ri} = v_{Rf}$$

$$R_i \, i_{Ri} = R_f \, i_{in}$$

Solve for i_{Ri}.

$$i_{Ri} = \frac{R_f}{R_i} i_{in}$$

Use Kirchhoff's current law, substitute currents, and simplify to find the output current relationship to the input current.

$$i_{out} = i_{Rf} + i_{Ri}$$

$$i_{out} = i_{in} + \frac{R_f}{R_i} i_{in}$$

$$i_{out} = \left(1 + \frac{R_f}{R_i}\right) i_{in}$$

$$A_i = \frac{i_{out}}{i_{in}} = 1 + \frac{R_f}{R_i}$$

Negative feedback enhances each of the major characteristics of the CCCS voltage amplifier circuit:

- Converts a high input resistance op amp into a very low resistance (virtually a *short*) input circuit resistance
- Converts a low output resistance op amp into a very high resistance (virtually an *open*) output circuit resistance
- A controlled output signal current directly proportional to the input signal current

Example 19-15

For the circuit of Figure 19-42(a), use basic laws and circuit analysis techniques to find the amplifier input current and amplifier output current. Find amplifier current gain and verify the relationship of the current gain based upon the input and feedback resistor values. Confirm that the op amp is operating within its specifications (assume 2 V of headroom is needed for the rails).

Solution

The noninverting pin of the op amp is at *virtual common*, thus

$$i_{in} = \frac{10\ V}{10\ k\Omega} = 1\ mA_p$$

The current i_{in} flows through R_f, dropping voltage across R_f and thus across R_i.

$$v_{Rf} = 1\ mA \times 2.2\ k\Omega = 2.2\ V_p$$

$$v_a = 0\ V - 2.2\ V = -2.2\ V_p$$

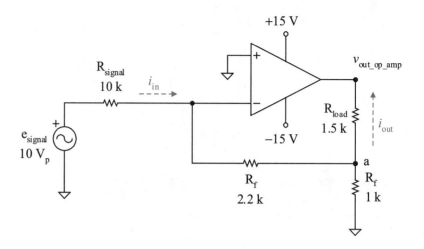

Figure 19-42(a) Example current amplifier

The current flow is *up* through R_i.

$$i_{Ri} = \frac{2.2 \text{ V}}{1 \text{ k}\Omega} = 2.2 \text{ mA}_p$$

$$i_{out} = 1 \text{ mA} + 2.2 \text{ mA} = 3.2 \text{ mA}_p \qquad \text{(KCL)}$$

The amplifier current gain is

$$A_i = \frac{i_{out}}{i_{in}} = \frac{3.2 \text{ mA}}{1 \text{ mA}} = 3.2$$

which is confirmed using the current gain based upon the input and feedback resistance values.

$$A_i = 1 + \frac{2.2 \text{ k}\Omega}{1 \text{ k}\Omega} = 3.2$$

The op amp output current of 3.2 mA_p is within the specifications of the LM741 and the LM324. The voltage at the output of the op amp is equal to the node voltage v_a less the fall across the load resistor. Yes, it is within the rail voltage of ± 13 V.

$$v_{out\_op\_amp} = -2.2 \text{ V} - 2.2 \text{ mA} \times 1.5 \text{ k}\Omega = -5.5 \text{ V}_p$$

Practice
Draw the example circuit in its CCCS signal model form and use the model to confirm the ac load current value.

Answer: See Figure 19-42(b).

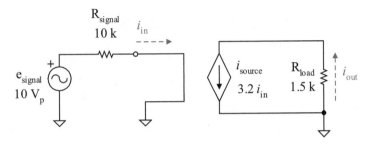

Figure 19-42(b) Practice solution of example circuit modeled

BJT–Bipolar Junction Transistor Operation

The bipolar junction transistor BJT was introduced previously as a current amplifying device and has been used in several application circuits. Basic BJT construction and symbols are shown in Figure 19-43.

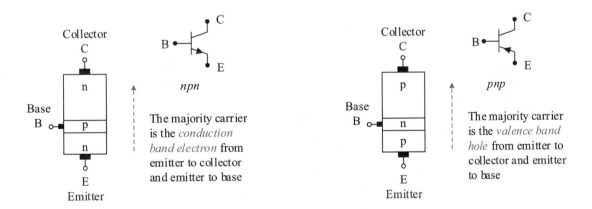

Figure 19-43 Basic BJT construction and schematic symbols

What follows is a *very brief description* of semiconductor physics. Extensive information is available on this subject area elsewhere. The primary focus of this text is on the principles and applications of semiconductor components. **Intrinsic** (pure) silicon material with a valence of four (+4) outer band electrons bonds with other neighbor atoms to form **covalent bond pairs** (sharing electron pairs with neighboring atoms). Each silicon atom then appears to be stable with 8 valence band electrons in its outer band. Intrinsic semiconductor material can be used as a resistance. Its resistance is affected by its size, shape, and temperature. As its temperature is increased, valence band electrons obtain enough thermal energy to escape to the conduction band, leaving a **hole** (lack of an electron) in the valence band. If subjected to an electric field, the conduction band electrons (called **electron carriers**) drift toward the positive external charge. The valence band electrons move from hole to hole, drifting toward the external positive charge. As a valence band electron moves from one atom to the next, the hole (or lack of electron) appears to move in the other direction toward the external negative charge. These virtual holes are called **hole carriers** and appear to move toward the external negative charge. This theory also applies to the germanium semiconductors.

If the intrinsic semiconductor is **doped** (mixed with an impurity) with a +5 valence atom, a conduction band *electron* is created for each electron **donor atom**. If the intrinsic semiconductor is doped with a +3 valence atom, a valence band *hole* is created for each **acceptor atom**. The term *acceptor atom* has been adapted since that atom is missing a valence band electron, a hole is created, and that atom can accept electrons from neighboring atoms.

Material heavily doped with donor atoms is called *n*-**type** material because it contains excessive conduction band electron carriers. Therefore the electron carrier is termed the **majority carrier**. Some **electron-hole pairs** are thermally generated and aid in conduction, but this is a very small contribution. Thus *n*-type material has relatively few hole carriers and the hole is termed the **minority carrier** in *n*-type material. The opposite is true for *p*-type material. The hole is the majority carrier and the electron carrier is the minority carrier in *p*-type material.

When *n*-type and *p*-type materials are joined during manufacturing to create diodes, transistors, and so forth, a barrier forms between the *p* and *n* junctions (called the ***pn* junction**) as shown in Figure 19-44.

At the junction, the *n*-type material gives up electrons to the *p*-type material and the *p*-type material accepts these electrons. The *n*-type side loses electrons and therefore forms positive ions at the *pn* junction.

The *p*-type side accepts electrons and therefore forms negative ions at the *pn* junction. This *pn* junction is devoid of carriers, forming a **barrier potential** of 0.7 V for silicon and 0.2 V for germanium. This barrier potential voltage must be overcome in a circuit before the device conducts. If a *pn* junction is forward biased sufficiently, the width of the *pn* junction is reduced, electron and hole carriers flow, and the semiconductor allows conduction. If reverse voltage is applied, the *pn* junction widens, and conduction does not occur (the junction is effectively an *open*). If excessive reverse voltage is applied, **avalanche breakdown** or **zener breakdown** occurs and the semiconductor conducts in the reverse direction (for example, the zener regulator diode).

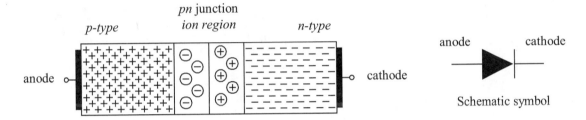

Figure 19-44 Construction and symbol of a diode or *pn* junction

The BJT has two *pn* junctions. The base-to-emitter junction performs as a controlling diode for the BJT. If this diode is forward biased and overcomes the *pn* junction barrier voltage (0.2 V for germanium and 0.7 V for silicon), carriers are emitted from the emitter. If the collector-base junction is reverse biased, the collector is biased such that it collects the emitted carriers. This action creates a relatively heavy emitter-to-collector current and a relatively small emitter-to-base current. A small change in the base current creates a relatively linear, magnified change in the collector current. Thus, the BJT is considered a current amplifier with a relatively small base current controlling a relatively larger collector current. The current gain from base to collector of a BJT is called its *beta* specification, β_{dc} for dc beta and β for ac beta (no subscript). Sample linear approximations of the forward characteristic curves for a silicon BJT are shown in Figure 19-45.

Figure 19-45(a) is a forward biased second approximation curve of the diode effect of the base-emitter junction. Once V_{BE} exceeds 0.7 V, the base-emitter diode turns on, drops approximately 0.7 V, and base current exists. Figure 19-45(b) is the BJT output collector characteristic curves.

Since the BJT depends upon three variables, the input variable I_B is held constant for a given collector current versus its collector-to-emitter voltage drop. The collector currents are ideally shown as constant for a given base current but more accurately there is a slight slope. Refer to BJT data sheets to determine responses more accurately if needed. Look at the collector curve for a base I_B current of 60 µA. The corresponding ideal collector current I_C is 6 mA. Thus the dc beta of this BJT is ideally

$$\beta_{dc} = \frac{6 \text{ mA}}{100 \text{ µA}} = 100$$

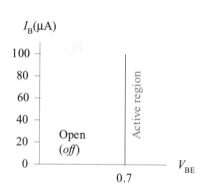

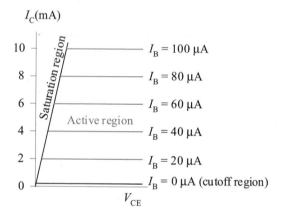

(a) Input characteristic curve

(b) Collector characteristic curves

Figure 19-45 Silicon BJT sample forward characteristic curves–second approximation

With the base-emitter junction forward biased *on* and the base-collector junction appropriately *reverse biased*, the BJT operates in the **active region** and produces a collector current proportional to the base current. In the active region, the BJT is a current amplifier.

$$I_C = \beta_{dc} I_B \qquad \qquad \text{(dc)}$$

$$i_c = \beta i_b \qquad \qquad \text{(ac)}$$

The static dc beta is usually close to but not necessarily the same value as the dynamic ac beta. Beta is normally kept constant to simplify circuit analysis, but beta is temperature sensitive. The beta factor can vary widely with manufacturing and with circuit temperature conditions. Circuits can be designed to rely on a substantial BJT current gain, but still be **beta independent** with varying values of beta.

The *second approximation dc model* of the silicon BJT is shown in Figure 19-46(a), incorporating the 0.7 V drop across the turned-on base-emitter diode junction and the dependent collector current source. This model can be substituted in the dc equivalent circuit to perform dc circuit analysis. Reverse the voltage polarity and current direction for a *pnp* BJT.

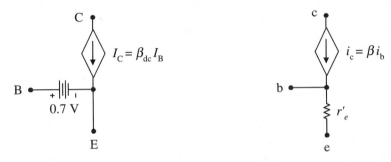

(a) DC model of *npn* BJT   (b) AC model of *npn* and *pnp* BJT

Figure 19-46 Silicon BJT dc and ac models

The *second approximation ac model* of the silicon BJT is shown in Figure 19-46(b), incorporating the BJT emitter resistance r'_e and the dependent collector current source. The r' parameter model is one of the ways to model the BJT. Another popular parameter system in modeling the BJT is the h parameter model. In data sheets, the forward current parameter h_{FE} and h_{fe} are approximately equal to β_{dc} and β, respectively. The internal BJT **ac emitter resistance** r'_e is estimated by

$$r'_e = \frac{26 \text{ mV}}{I_E} + 0.5 \ \Omega$$

In large-signal analysis, the 0.5 Ω resistance must be included as it provides a significantly relevant voltage drop. The BJT ac emitter resistance r'_e is inversely proportional to the dc emitter current I_E. *DC circuit analysis must be performed first* to determine I_E before the ac emitter resistance r'_e can be determined and used in the ac circuit analysis.

Common Collector Amplifier–Emitter Follower

The **common collector circuit**, also called the **emitter follower** circuit, is shown in the circuit of Figure 19-47(a).

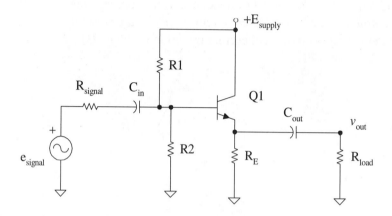

19-47(a) Common collector or emitter follower amplifier

The capacitors couple the ac signal and block the dc, therefore no dc voltage or currents exist in the signal circuit or the load circuit. The circuit of Figure 19-47(b) is the dc equivalent circuit to be analyzed.

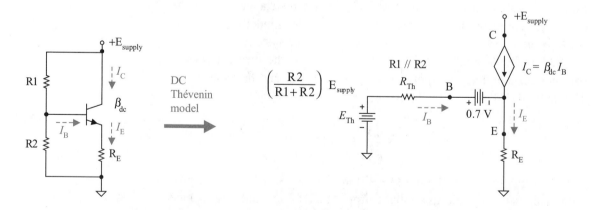

Figure 19-47(b) DC equivalent circuit and model

$$E_{Th} - R_{Th}I_B - 0.7 \text{ V} - R_E I_E = 0 \qquad \text{(KVL B-E loop)}$$

$$I_E = I_B + I_C \qquad \text{(KCL)}$$

$$I_E = I_B + \beta_{dc} I_B \qquad \text{(substitution)}$$

$$I_B = \frac{I_E}{1 + \beta_{dc}} \qquad \text{(substitution)}$$

$$E_{Th} - R_{Th}\left(\frac{I_E}{1 + \beta_{dc}}\right) - 0.7 \text{ V} - R_E I_E = 0 \qquad \text{(substitution)}$$

$$I_E = \frac{E_{Th} - 0.7 \text{ V}}{R_E + \dfrac{R_{Th}}{1 + \beta_{dc}}} \qquad \text{(algebra)}$$

This form of I_E is very accurate and includes the β_{dc} of the BJT. If the following condition holds, then the circuit is *dc beta independent*.

$$\text{If} \quad R_E \gg \frac{R_{Th}}{1 + \beta_{dc}} \qquad \text{then} \quad I_E \approx \frac{E_{Th} - 0.7 \text{ V}}{R_E} \; .$$

This ensures that the base current must be much less than the bias currents of R1 and R2, and R1 and R2 form an essentially series circuit.

Once you have solved for I_E, you can then solve for all the other dc values in the circuit. Also, you can now solve for the BJT emitter ac resistance r'_e and begin the ac signal analysis of the amplifier.

The ac equivalent circuit and its equivalent BJT ac modeled circuit are shown in Figure 19-47(c). The input circuit has been modeled as an ac Thévenin model with the BJT base-emitter junction considered the load. The ac resistance r_E represents the ac resistance as viewed from the BJT emitter terminal to ac common, which in this case is the parallel combination of R_E and R_{load}. Again, the key to solving this circuit is finding the emitter current i_e.

$$r'_e = \frac{26 \text{ mV}}{I_E} + 0.5 \text{ }\Omega$$

$$e_{Th} - r_{Th}i_b - r'_e \, i_e - r_E i_e = 0 \qquad \text{(KVL)}$$

$$i_e = i_b + i_c \qquad \text{(KCL)}$$

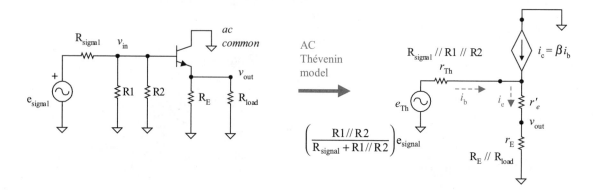

Figure 19-47(c) AC equivalent circuit and model

$$i_e = i_b + \beta\, i_b \qquad \text{(substitution)}$$

$$i_b = \frac{i_e}{1+\beta} \qquad \text{(substitution)}$$

$$e_{Th} - r_{Th}\left(\frac{i_e}{1+\beta}\right) - r'_e\, i_e - r_E i_e = 0 \qquad \text{(substitution)}$$

$$i_e = \frac{e_{Th}}{r'_e + r_E + \dfrac{r_{Th}}{1+\beta}} \qquad \text{(algebra)}$$

If the following condition holds, then the circuit is *ac beta independent*.

$$\text{If} \quad r'_e + r_E \gg \frac{r_{Th}}{1+\beta} \quad \text{then} \quad i_e \approx \frac{e_{Th}}{r'_e + r_E}\,.$$

The ac output voltage then becomes

$$v_{out} = r_E\, i_e$$

$$v_{out} \approx \frac{r_E}{r'_e + r_E}\, e_{Th}$$

The input voltage is essentially voltage divided by r'_e and r_E to form the output voltage v_{out}. If $r_E \gg r'_e$, then this circuit becomes a voltage follower circuit with the output voltage following the input in value and polarity.

$$v_{out} \approx e_{Th}$$

Example 19-16

Completely analyze the circuit of Figure 19-48 assuming the capacitors are functioning ideally.

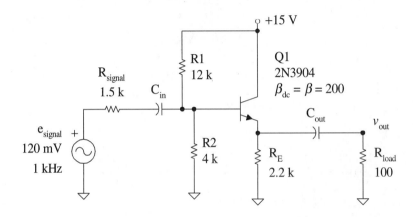

Figure 19-48 Example common collector amplifier

Solution

Model the dc circuit as shown in Figure 19-47(b) and solve.

$$E_{Th} = \left(\frac{4 \ k\Omega}{12 \ k\Omega + 4 \ k\Omega}\right) 12 \ V = 3 \ V$$

$$R_{Th} = \frac{4 \ k\Omega \times 12 \ k\Omega}{4 \ k\Omega + 12 \ k\Omega} = 3 \ k\Omega$$

$$I_E = \frac{3 \ V - 0.7 \ V}{2.2 \ k\Omega + \dfrac{3 \ k\Omega}{1 + 200}} = 1.038 \ mA$$

$$I_B = \frac{1.038 \text{ mA}}{1+200} = 5.166 \ \mu A$$

$$I_C = 200 \times 5.166 \ \mu A = 1.033 \text{ mA}$$

$$V_C = 12 \text{ V} \qquad\qquad\qquad \text{(dc supply)}$$

$$V_E = 0 \text{ V} + 2.2 \text{ k}\Omega \times 1.038 \text{ mA} = 2.283 \text{ V}$$

$$V_B = 2.283 \text{ V} + 0.7 \text{ V} = 2.983 \text{ V}$$

$$V_{CE} = 12 \text{ V} - 2.983 \text{ V} = 10.02 \text{ V}$$

$$Z_{in\_base} = \frac{2.983 \text{ V}}{5.166 \ \mu A} = 577 \text{ k}\Omega$$

Model the ac circuit as shown in Figure 19-47(c) and solve. Note lowercase literals are used to indicate ac variables.

$$r'_e = \frac{26 \text{ mV}}{1.038 \text{ mA}} + 0.5 \ \Omega = 24.84 \ \Omega$$

$$e_{Th} = \left(\frac{3 \text{ k}\Omega}{1.5 \text{ k}\Omega + 3 \text{ k}\Omega} \right) 120 \text{ mV} = 80 \text{ mV}$$

$$\frac{1}{r_{Th}} = \frac{1}{1.5 \text{ k}\Omega} + \frac{1}{4 \text{ k}\Omega} + \frac{1}{12 \text{ k}\Omega}$$

$$r_{Th} = 1 \text{ k}\Omega$$

$$r_E = \frac{100 \ \Omega \times 2.2 \text{ k}\Omega}{100 \ \Omega + 2.2 \text{ k}\Omega} = 95.65 \ \Omega$$

$$i_e = \frac{80 \text{ mV}}{24.84 \ \Omega + 95.65 \ \Omega + \dfrac{1 \text{ k}\Omega}{1+200}} = 637.6 \ \mu A$$

$$i_b = \frac{637.6 \ \mu A}{1+200} = 3.172 \ \mu A$$

$$i_c = 200 \times 3.172 \ \mu A = 634.4 \ \mu A$$

$$v_c = 0 \text{ V} \qquad\qquad\qquad \text{(ac common)}$$

$$v_{out} = v_e = 637.6 \ \mu A \times 95.65 \ \Omega = 60.99 \ mV$$

$$i_{load} = \frac{60.99 \ mV}{100 \ \Omega} = 609.9 \ \mu A$$

$$v_b = 637.6 \ \mu A \ (24.84 \ \Omega + 95.6 \ \Omega) = 76.77 \ mV$$

$$z_{in\_base} = \frac{76.77 \ mV}{3.172 \ \mu A} = 24.2 \ k\Omega$$

$$z_{in\_stage} = \frac{76.77 \ mV}{28.8 \ \mu A} = 2.66 \ k\Omega$$

$$i_{signal} = \frac{120 \ mV - 76.77 \ mV}{1.5 \ k\Omega} = 28.8 \ \mu A$$

$$A_v = \frac{v_{out}}{e_{signal}} = \frac{60.99 \ mV}{120 \ mV} = 0.508$$

$$A_i = \frac{i_{out}}{i_{signal}} = \frac{609.9 \ \mu A}{28.8 \ \mu A} = 21.2$$

$$A_p = 0.508 \times 21.2 = 10.8$$

Practice

Verify the example voltages and currents by simulation. Assume the capacitors are 100 μF. Observe the input and output voltages with an oscilloscope.

Exploration

Increase the signal input and note wave shape effects. Note shape distortion and eventually clipping distortion.

Input and output impedances are key characteristics of a circuit. Referring to Figure 19-47(c), the signal input impedance to the amplifier stage is the parallel combination of R1 // R2 // z_{in_base}. Where z_{in_base} of the BJT is

$$z_{\text{in\_base}} = \frac{v_b}{i_b}$$

$$z_{\text{in\_base}} = \frac{(r'_e + r_E)i_e}{i_b}$$

$$z_{\text{in\_base}} = \frac{(r'_e + r_E)(1+\beta)i_b}{i_b}$$

$$z_{\text{in\_base}} = (1+\beta)(r'_e + r_E)$$

To analyze the output impedance looking back into the emitter circuit, use the test circuit of Figure 19-49 without the load connected. The signal voltage source and capacitors have been replaced with their equivalent impedances of a short. The collector current source has been replaced with its equivalent resistance of an open. The source is being driven with a test signal to determine the resistance viewed from the test signal source. A complete analysis generates the equivalent impedance model shown in Figure 19-49.

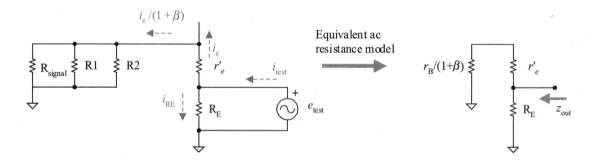

Figure 19-49 Output impedance test circuit and model

$$\frac{1}{z_{\text{out}}} = \frac{1}{R_E} + \frac{1}{r'_e + \dfrac{r_B}{1+\beta}}$$

where r_B is the ac parallel resistance looking from the BJT base pin toward ac circuit common, that is, R_{signal} // R1 //R2. In effect, it is the Thévenin resistance looking from the BJT base to external common.

When looking into the base, the impedance is a *multiplier* of $(1+\beta)$. When looking back into the emitter, the equivalent base-to-common resistance is *divided by* the factor of $(1+\beta)$. Both of these effects are based upon the relationship of the base current to the emitter current.

Example 19-17

For the circuit of Figure 19-48, find the amplifier stage ac input impedance and output impedance. Compare z_{in_stage} with the value found in the last example.

Solution

$$z_{in\_base} = (1+200)(24.84\ \Omega + 95.65\ \Omega) = 24.21\ k\Omega$$

$$\frac{1}{z_{in\_stage}} = \frac{1}{4\ k\Omega} + \frac{1}{12\ k\Omega} + \frac{1}{24.2\ k\Omega}$$

$$z_{in\_stage} = 2.67\ k\Omega \qquad \text{(confirm with previous example)}$$

$$\frac{1}{r_B} = \frac{1}{1.5\ k\Omega} + \frac{1}{4\ k\Omega} + \frac{1}{12\ k\Omega}$$

$$r_B = 1\ k\Omega$$

$$\frac{1}{z_{out\_stage}} = \frac{1}{2.2\ k\Omega} + \frac{1}{24.84\ \Omega + \dfrac{1\ k\Omega}{1+200}}$$

$$z_{out\_stage} = 29.6\ \Omega$$

Practice

Review the previous two examples and attempt to make some general observations. For example, the parallel $r'_e + r_B/(1+\beta)$ branch dominates z_{out_stage}, creating a very low output impedance. This is a reasonably good buffer amplifier. Why?

Common Emitter Amplifier

The circuit of Figure 19-50 is an ac **common emitter amplifier** since its BJT emitter terminal is connected to ac common. Figure 19-51 is a common emitter amplifier with **swamping resistor** R_{E1} to produce voltage gain stability.

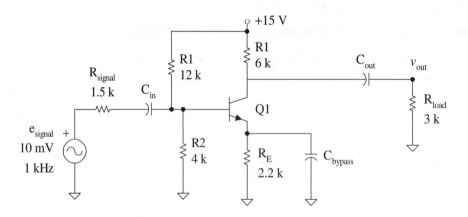

Figure 19-50 Common emitter amplifier

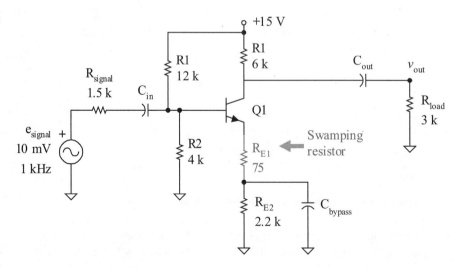

Figure 19-51 Common emitter amplifier with swamping resistor

Example 19-18

Use the circuit solution techniques you have developed to model and find all of the dc and ac voltages and currents in the circuit of Figure 19-50. Assume Q1 is a 2N3904 BJT with a current gain of 200. First develop all of your expressions using literals. Then substitute appropriate values into your literal expressions to find numerical values. Prove your results by simulation. Hint: the voltage gain from base to collector of this amplifier stage is

$$A_{v\ base\_to\_collector} = -\frac{r_C}{r'_e}$$

multi**SIM**

Practice

Repeat the example problem for the circuit of Figure 19-51. Hint: the voltage gain from base to collector of this amplifier stage is

$$A_{v\ base\_to\_collector} = -\frac{r_C}{r'_e + R_{E1}}$$

Note, the ac emitter resistance r'_e is susceptible to changes in the dc bias current I_E. The swamping resistance reduces the voltage gain but does improve the stability of the voltage gain.

BJT Difference Amplifier

Figure 19-52 is the schematic of the rudimentary elements of a **BJT difference amplifier** with its associated signal analysis model. This basic circuit is at the heart of the input to the op amp circuit. The input voltage terminals at v_{in_1} and v_{in_2} *must be tied to* a signal voltage source or to common. If both inputs have signal voltages, this is a **differential input**. If there is a single input and the other terminal is grounded (tied to common), then this is a **single-ended input**. If both terminals are tied to the same signal source, then this is called the **common-mode** configuration. The output can be single-ended or **differential output**. The differential output is $V_{C2} - V_{C1}$ for both dc and ac.

The dc analysis must be done first. In this case the **tail current** is created by a defined current source. In real circuits, this must be a current output device circuit with a BJT or FET. The input connections complete the base-emitter bias circuit. DC analysis shows that the tail current I_T splits approximately equally between the two BJTs so that I_E is approximately $I_T / 2$. The dc collector voltages are then at the same potential, so the dc difference between collectors is 0 V_{dc}. There is however an ac signal difference. As noted in the ac model of Figure 19-52, an input signal v_{in_1} produces an inverted amplified output at the collector of Q1 while that same input signal produces a noninverted amplified signal at the collector of Q2. The difference of the ac output signals terminals produces a difference output voltage twice that of either collector ac voltage. Common signals at the input terminals, such as noise, ideally produce 0 V differential output (very desirable). The signal voltage v_{in} is amplified while a common noise signal v_{in_CM} is ideally eliminated.

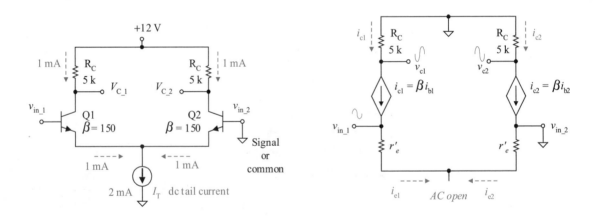

Figure 19-52 Basic BJT differential amplifier and corresponding ac model

Even though manufactured on the same IC (integrated circuit), slight differences between components exist creating slightly different voltages, currents, and voltage gains. Since this is the front end of an op amp with significant gain, manufacturer's specifications account for these basic differences. The difference in input base currents is specified as the **input offset current** I_{OS}.

$$I_{OS} = |I_{B1} - I_{B2}|$$

The average of input base currents is specified as the **input bias current** I_{BIAS}.

$$I_{BIAS} = \tfrac{1}{2}\left(I_{B1} + I_{B2}\right)$$

The base-emitter voltage drop difference is specified as the **input offset voltage** V_{OS}. Even a 2 mV input difference can have a tremendous impact when the op amp gain is 10^5 or a closed loop gain is significant.

$$V_{OS} = \left|V_{B1} - V_{B2}\right|$$

The voltage gain of a common-mode configuration is called the **common-mode voltage gain** A_{v_CM} and is ideally 0. However, due to circuit differences, there can be a relatively small common mode gain. To quantify this effect, op amp specifications include the common-mode rejection ratio **CMRR**, usually specified in decibels (dB).

$$CMRR(\text{dB}) = 20\log CMRR$$

$$CMRR = \frac{A_v}{A_{v\_CM}}$$

$$v_{\text{out\_CM}} = A_{v\_CM} v_{\text{in\_CM}}$$

This text has assumed ideal conditions to gain fundamental understanding of devices and the application of basic laws and circuit analysis techniques. Later textbooks in this series incorporate the nonideal characteristics into analysis as your skills and knowledge progress.

Example 19-19

For the circuit of Figure 19-52, prove the following using basic laws and circuit analysis techniques.

$$v_{c1} = -\frac{R_C}{2r'_e}\left(v_{b1} - v_{b2}\right)$$

$$v_{c2} = \frac{R_C}{2r'_e}\left(v_{b1} - v_{b2}\right)$$

$$v_{\text{out\_differential}} = v_{c2} - v_{c1} = \frac{R_C}{r'_e}\left(v_{b1} - v_{b2}\right)$$

$$z_{\text{in\_base}} = 2r'_e\left(1 + \beta\right) \qquad \text{(other base at common)}$$

$$z_{\text{out}} = R_C$$

multiSIM

Practice

Find the values of the variables expressed in the example based on the schematic values of Figure 19-52. Verify your results using simulation.

A more practical difference amplifier circuit is shown in Figure 19-53. The bias circuit now includes a current source circuit consisting of the dc supplies, R1, Q1 and Q2. Notice that the base and the collector of Q1 are tied together turning Q1 into an **active diode**. It acts like a silicon diode, dropping 0.7 V, but uses an active device to create the diode effect. Thus the drop across R1 is the difference in the supply voltages less the 0.7 V Q1 drop. This creates the 2 mA Q1 collector current. Q1 and Q2 form a **current mirror** circuit. The Q1 collector current is mirrored into the Q2 collector, creating the Q2 collector current. This can be demonstrated with basic circuit analysis techniques. The Q2 collector current forms the I_T tail current. Assuming that Darlington pair Q3-Q4 is identical to the Q5-Q6 pair, the tail current splits equally to create the two dc emitter bias currents of 1 mA_{dc} each. Since it is a single-ended output, the Q3-Q4 collector resistor R_C is not needed and is removed to save power.

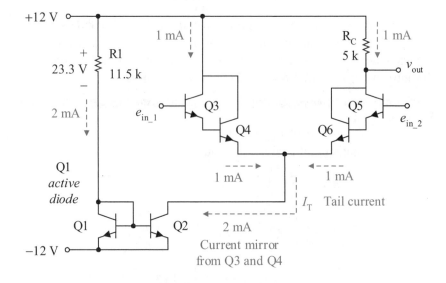

Figure 19-53 Single-ended output difference amplifier

Once I_E is determined and the BJT betas are known, the ac emitter resistances can be found for the Darlington-pair transistors. For example,

$$r'_{e\_Q4} = \frac{26 \text{ mV}}{I_{E\_Q4}} + 0.5 \text{ }\Omega$$

$$I_{E\_Q3} = \frac{I_{E\_Q4}}{1 + \beta_4}$$

$$r'_{e\_Q3} = \frac{26 \text{ mV}}{I_{E\_Q3}} + 0.5 \text{ }\Omega$$

Now, the signal model can be formed and the signal values and parameters determined; however, that approach is excessively and unnecessarily complex. There is an approach in engineering called "Keep it simple, stupid!", the **KISS** approach. A simple approach that retains the inherent, important characteristics has fewer operations or calculations and therefore less chance for error. The KISS approach for this signal analysis is to consider a Darlington pair such as Q3-Q4 as a single transistor with a single current gain beta and solve this problem like you did for the circuit of Figure 19-52. The Darlington pair current gain β_{DP} is

$$\beta_{DP} \approx \beta_{Q3} \times \beta_{Q4}$$

The effective Darlington pair ac emitter resistance r'_{e_DP} is

$$r'_{e\_DP} \approx 2\left(\frac{26 \text{ mV}}{I_{E\_Q4}} + 0.5 \text{ }\Omega\right)$$

A final note: the collector resistor R_C could be replaced by an active circuit, creating another current mirror circuit. Then the only passive resistance in this circuit would be R1.

Summary

Active devices and circuits can be modeled as voltage controlled voltage sources (VCVS), current controlled voltage sources (CCVS), voltage controlled current sources (VCCS), or current controlled current sources (CCCS). Fundamental active devices include the BJT, the FET, and the MOSFET. Other devices such as operational amplifiers are constructed of these fundamental components or combinations of these fundamental components. Input and output impedances can be incorporated into a model if these additional characteristics are significant.

An ideal voltage transfer circuit has a load resistance much, much greater than its source resistance. An ideal current transfer circuit has a load resistance much, much less than its source resistance. An ideal power transfer system has a load resistance matched to the source (or Thévenin) resistance. An op amp can be used with negative feedback to create ideal voltage amplifiers, current-to-voltage converters, voltage-to-current converters, and current amplifiers.

The ideal and second approximation characteristics of the active devices were utilized to keep the circuit analysis focused on the essentials of circuit analysis and circuit conceptualization. Later textbooks in this series add increasing component detail and characteristics to provide deeper understanding in a just-in-time approach to learning.

Problems

Ideal Dependent Source Models

19-1   For a VCVS, find the output if the input is 5 mV and voltage gain is 8. Draw the device model.

19-2   For a CCVS, find the output if the input is 5 mA and the transresistance is 10 Ω. Draw the model.

19-3   For a VCCS, find the output if the input is 5 mV and the transconductance is 10 mS. Draw the model.

19-4   For a CCCS, find the output if the input is 5 mA and the current gain is 10.

Maximum Signal Transfer

19-5   A voltage source of 20 V_{rms} with an internal resistance of 4 kΩ drives a load resistance. Find the value of the load resistance to transfer the maximum load voltage, load current, and load power and the corresponding value of load voltage, current, and power.

19-6   A current source of 20 mA_{rms} with an internal resistance of 4 kΩ drives a load resistance. Find the value of the load resistance to transfer the maximum load voltage, load current, and load power and the corresponding value of load voltage, current, and power.

19-7   Find the maximum and minimum load voltage, load current, and average load power for the circuit in Figure 19-54.

19-8   The wiper arm of the circuit in Figure 19-54 is placed in the middle of the linear potentiometer. Find the value of the load resistance to transfer the maximum load voltage, load current, and load power and the corresponding value of load voltage, current, and power.

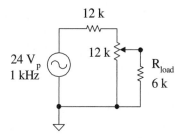

Figure 19-54

Voltage Controlled Voltage Source

19-9   For the circuit of Figure 19-55(a), find the output voltage and load current. Draw the VCVS model of this circuit.

19-10   Repeat Problem 19-9 for the circuit of Figure 19-55(b).

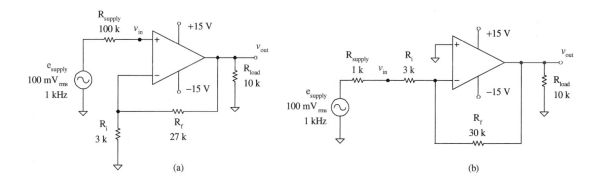

Figure 19-55

19-11 For the circuit of Figure 19-55(a) find the sacrifice factor, the input impedance, and the output impedance of the amplifier stage (v_{in} to v_{out}) if the op amp has a 2 MΩ input impedance, a 75 Ω output impedance, and an open loop voltage gain of 10^5.

19-12 Based upon the results of Problem 19-11, find and draw the VCVS circuit model of the amplifier and use it to find v_{out}.

Current Controlled Voltage Source

19-13 The resistors in the circuit of Figure 19-19(a) are each 10 kΩ and the input current is 1 μA$_p$. Find stage 1 and 2 output voltages.

19-14 Which subcircuit in the circuits of Figure 19-55 is inherently a CCVS? Redraw that circuit using the CCVS model of the amplifier and find the output voltage.

Voltage Controlled Current Source

19-15 Design a VCCS op amp circuit to convert 10 V to 2 mA.

19-16 The transconductance of a VCCS circuit is 20 mS and the output current is 10 mA. Find the input voltage.

19-17 The input signal e_{in} in the circuit of Figure 19-33(a) is changed to 500 mV$_{rms}$. Find the load voltage, current, and average power.

19-18 The load in the circuit of Figure 19-36(a) is changed to an 8Ω load. Find the load voltage, current, and average power.

Current Controlled Current Source

19-19 Design an op amp current amplifier for a current gain of 100.

19-20 For the circuit of Figure 19-48, the input signal is changed to 100 mV$_p$. Find the output signal voltage. Think! Use known information before charging off into lots of equations.

19-21 Bias resistance R1 is changed to 23 k? in the circuit of Figure 19-53. Find the dc tail current and the dc emitter currents for Q4. Each BJT β is 150. Find the current gain β_{DP} for the Q3-Q4 Darlington pair combination, the Darlington pair ac emitter resistance r'_{e_DP}, and the amplifier input impedance. What overall impact does the Darlington pair have on input impedance and voltage gain compared to a single BJT?

19-22 The voltage gain of a differential amplifier is 30 and its CMRR(dB) rating is 60 dB. If a signal of 100 mV$_p$ is applied and there is a common noise of 10 mV$_p$ on both input terminals, find the output signal, noise, and signal-to-noise ratio.

20

Special Analog Integrated Circuits

Introduction

Throughout this text you have seen a variety of electronic circuits using resistors, capacitors, diodes, transistors and operational amplifiers. Combine enough of these five building blocks correctly and any analog operation can be performed. However, the price of designing, manufacturing, testing, and supporting a piece of electronic equipment is dependent on the *number* of parts on the printed circuit board, *not* the total cost of those parts. Ordering, stocking, board space, part placement, testing, and repair are all more expensive as the *number* of parts goes up.

Lattice Semiconductor Corporation makes a line of *programmable* analog circuits. In a single package may be differential amplifiers with adjustable gains, integrators, comparators, and digital-to-analog converters. Gain, capacitance, and the interconnection of these blocks to each other and to the outside are set through a personal computer and a drag-and-drop schematic generation program, with no proto-boards, wiring, or soldering. If you want to change the design, just click.

Exar provides a variety of single IC signal generators. The XR-2206 provides a square wave, triangle wave, and sine wave. The frequency, pulse width, amplitude and dc level of each may be controlled by an external resistor, capacitor, or voltage.

Objectives

Upon completion of this chapter, you will be able to do the following for the Lattice ispPAC 10 and the Exar XR-2206:
- Define power supply requirements.
- Define input and output signal limitations.
- With the ispPAC 10, design
  - Single and differential input amplifiers
  - Fractional gain amplifiers
  - Summers
  - Comparators with a reference and hysteresis
  - Relaxation oscillator
- With the XR-2206, design a square or sine generator with fixed or variable duty cycle, adjustable frequency, and adjustable amplitude.

20.1 Programmable Analog Circuit

Usually an electronic circuit is developed in a series of steps. First you determine the design. You then enter the schematic into a simulator to verify its expected performance, and alter the design until the simulation indicates the results you need. Then you wire all the parts together on a breadboard. Again you test, redesign, resimulate, and rebuild the circuit until the breadboard performance is correct. Next, the schematic must be entered into a printed circuit board (PCB) layout package, and the PCB designed. The board design is sent to be built. Under the *best* of conditions, several days pass. The prototype PCB arrives, you install all of the parts, and test it. *If* it works correctly, you can then integrate that part of the system with others. If it does not, you must repeat the entire process again.

The process is *very* time-consuming, and filled with opportunities for minor mistakes to delay it. You may misread a resistor color code, or wire to the wrong hole during bread boarding, or create a solder bridge while installing parts on the PCB, or lift a pad off of the prototype PCB when you change a component for the fourth time. Once the board is working correctly, you have created yet *another* different item for the company to keep track of, keep in stock, maintain, and provide customer service on.

When it is time to create the "new-and-improved" version, you repeat the entire process. Then you must contact your customers who have the original model to convince them that they should upgrade their hardware. At best, you ship them a new board, and their technician replaces the old one. At worst, they have to throw their old widget away and buy your new one.

In the digital world, microprocessors and programmable logic devices (**PLDs**) are developed by altering their *instructions*. A very limited number of standard boards are kept by your company. Design, simulation, and testing are all done at a computer terminal. Change your mind, change the design, change an instruction, click-click-click. Download the code to the IC and it's time to test. Model upgrades may be done by downloading new code over the Internet. This is precisely why personal computers are so widespread and relatively inexpensive. I watch movies and talk to friends with my PC, you design a new rocket booster with yours. They are exactly the same hardware, we just use different instructions.

Lattice Semiconductor Corporation makes a series of in-system Programmable *Analog* Circuits. These ispPAC™s bring the advantages of programming rather than wiring to the analog portion of your project. Design, simulation and instruction downloading are done from a drag-and-drop schematic entry program that runs on your PC and connects to the analog IC through a parallel port cable. You can develop a single printed circuit board that is a truck scale in one project, and an answering machine in another. The *analog* hardware and printed circuit board are exactly the same. You just changed the instructions. Combine these programmable *analog* ICs with a microprocessor, a few PLDs and some electronically controlled switches and you have a fully amorphous system. Your company has only a few different printed circuit boards and a very limited number of parts to buy and keep track of. Hardware manufacturing for every product is the same. You just download one set of instructions to build a controller for a washing machine, and a different set for an mp3 player.

The PAC 10 is shown in Figure 20-1. It contains four analog blocks. Each block has two input instrumentation amplifiers (**IA**). The gain of each of these may be programmed and may be noninverting or inverting. The signal from each of the two instrumentation amplifiers is summed and sent to the output amplifier (**OA**). There is a selection of 255 capacitors that may by connected across the feedback resistor in the output amplifier. This allows you to remove high frequency noise, or to build a variety of filters. The feedback resistor in the output amplifier may be removed, allowing the amplifier to run open-loop. You can then use that element as a comparator, or with the correct capacitor, as an integrator. The outputs of each **OA** are hardwired to specific pins, but they may *also* be routed internally to the input of other blocks. You select which input or output pins are connected to the **IA** inputs of each block.

The PAC 20 is shown in Figure 20-2(a). It contains two of the amplifier blocks found in the PAC 10. It also has two elements optimized to work as comparators, a precision reference and a digital-to-analog converter. While the PAC 10 is a purely analog part, the PAC 20 provides the tools necessary to allow the analog world to interact with a microprocessor.

The PAC 30 offers more digital control options and more flexibility in how you interconnect the elements. Its pin out diagram is shown in Figure 20-2(b).

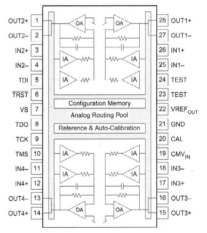

Figure 20-1
PAC 10 pin out diagram
(*courtesy of Lattice Semiconductor Corporation*)

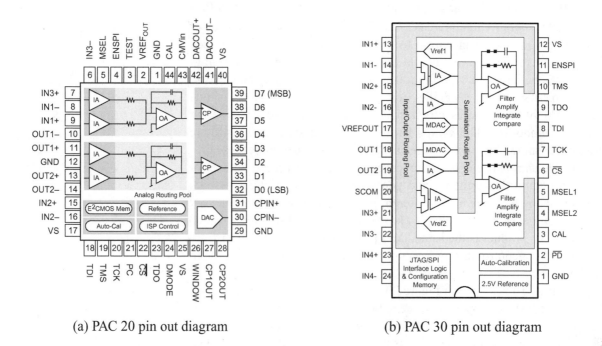

(a) PAC 20 pin out diagram   (b) PAC 30 pin out diagram

Figure 20-2 Programmable analog ICs with digital interfaces (*courtesy of Lattice Semiconductor Corporation*)

Introduction to the PAC 10

$$4.75\ V_{dc} \le E_{supply} \le 5.25\ V_{dc}$$
$$I_{supply} < 23\ mA_{dc}$$

The focus of this text is *analog* electronics. So, the remainder of this section will discuss the PAC 10. This is a *single* supply part. A positive supply voltage between 4.75 V_{dc} and 5.25 V_{dc} (nominally 5 V_{dc}) is connected to **VS** (pin 7). Circuit common is connected to **GND** (pin 21). The manufacturer guarantees that the IC will use no more than 23 mA_{dc} from the +5 V_{dc} power supply.

As with other analog integrated circuits, the input voltages must be within the rails of the IC's power supply. For this part, the condition is even more restrictive. You must assure that the input voltages are between 1 V and 4 V at all times. Similarly, the IC can output voltages between 1 V

$$1\ V \le E_{in} \le 4\ V$$
$$1V \le E_{out} \le 4\ V$$
$$I_{load} < 10\ mA_{dc}$$

and 4 V, and can source or sink at least 10 mA. So be sure to keep your
load resistances above 400 Ω.

Input signals that go both up and down, such as sine waves, must have
a dc offset voltage added to them to assure that they stay within the 1 V to
4 V restriction. Although adding any offset that moves the signals into the
1 V to 4 V range works, the PAC 10 provides a **VREF**$_{OUT}$ voltage of 2.500 V$_{dc}$
at pin 21. This signal is intended to shift your input's level. It can only
source 50 μA$_{dc}$ and sink 350 μA$_{dc}$. So be sure to limit the current into or
out of this pin by connecting only high impedance loads.

All input pins, internal signal lines and output pins are differential.
Even though the diagrams *may* indicate only one connection or one line,
in reality there is a pair. The IC is specifically designed to process two
signals simultaneously, one going up from V$_{ref}$ and the other going equally down from V$_{ref}$. The *difference* between these two signals, then, is
twice as large as you could obtain if only a single signal with respect to
V$_{ref}$ or common were used. So, there is an **IN1+** and an **IN1-** connection,
and an **OUT1+** and an **OUT1-**. At each instant, the signal at **OUT1+** is equally
as far above V$_{ref}$ as the signal at **OUT1-** is below V$_{ref}$.

You route signals from the input pins to the PAC blocks, set the gains,
feedback capacitor, enable the feedback resistor, and route each PAC
block's output through a drag-and-drop schematic capture program on
your personal computer. A typical screen is shown in Figure 20-3. To
connect the pair of signals at **IN1** to **IA1** of **PAC BLOCK 1**, just click on the
input to **IA1**, and drag the resulting wire over to the **IN1** pad. If you want to
remove a connection you have already made, double click on that wire,
and select **NONE** from the connection dialog box that appears.

The gain for each **IA** can also be set with a dialog box. Double click on
the **IA** symbol. A dialog box allows you to scroll through gains from −10
to +10, in steps of 1. Each **PAC BLOCK** has two **IA**s. The signals from each
are added and sent to the output amplifier, **OA**. Clicking on the link just
before the feedback resistor in the output amplifier opens it. This forces
that amplifier to run open-loop, driving its gain up very high. When you
want to use the output amplifier as a comparator or as an integrator, open
the feedback resistor's link. Otherwise be sure to keep the link closed so
that the **OA** acts as a linear amplifier.

There is a wide variety of capacitors available to be connected across
the output amplifier. These vary from 1 pF to 62 pF. This lets you reject
high frequencies, and design many different types of filters and integrators. Typically, the frequency limits are between 10 kHz and 100 kHz.

VREF$_{OUT}$ = 2.500 V$_{dc}$

Signals are *differential.*

IA gains:
   Selectable from −10 to
   +10 in steps 1.

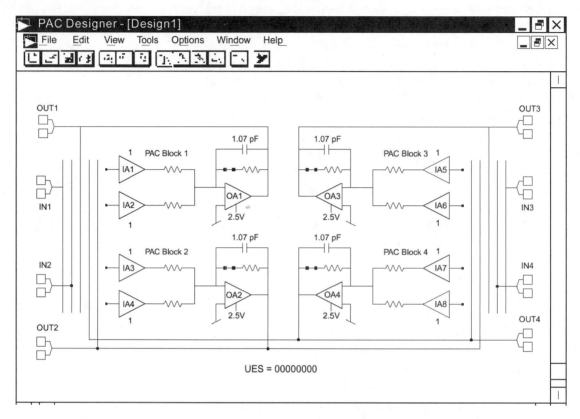

Figure 20-3 PAC 10 programming window (*courtesy of Lattice Semiconductor Corporation*)

Just click on the capacitor and select the value you want from the dialog box.

There is a cable and adapter that connect the parallel port of your personal computer to the PAC 10 IC. When you have completed the design in the **PAC Designer** window, connect the cable, apply power to the IC and click the far right button in the window. The program is downloaded to the IC, and takes effect immediately. You are ready to use the amplifier, comparator, integrator, or filter you have designed. The IC retains its configuration even when power is removed. Turn the IC on and off as many times as you like. Each time power is applied, it remembers the design you sent it. The personal computer and cable are only needed if you want to send the IC another configuration.

Basic Amplifiers with the PAC 10

Each PAC block is designed to use two differential input signals. These signals should be equal in amplitude, opposite in phase and each offset from common by the same dc level, preferably 2.5 V_{dc}. Look at Figure 20-4.

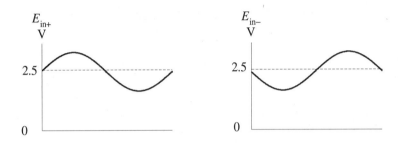

Figure 20-4 Differential signals used by a PAC block

The input amplifier (**IA**) processes the *difference* of the two input signals. Any voltage that is common to both E_{in+} and E_{in-} subtracts out. So, the 2.5 V_{dc} offset that is necessary to assure that the inputs are within the supply limits (0 V to 5 V) is removed. This subtraction also removes any common-mode noise. When small signals are transmitted, electrical interference, such as 60 Hz from the electrical power grid, infects both lines. This noise is injected exactly the same on both inputs. So it too subtracts out. Finally, by driving both input lines, the difference between the two is twice as large as it would be if you placed the signal on E_{in+} and tied E_{in-} to 2.5 V_{dc}.

Many small signals are inherently differential. Dynamic microphones, and guitar and other magnetic pickup provide two signals. Most commercial function generators are *not* referenced to earth. Usually, you connect their − output to circuit common. However, the signal floats, and may be 10's V away from earth. Look carefully at the front panel of your signal generator. Bridge circuits, such as those used in many weight and pressure sensors, are differential. Figure 20-5 shows several ways to connect differential and floating signals to a PAC block.

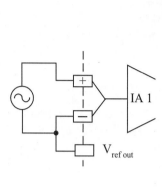

(a) Function generator

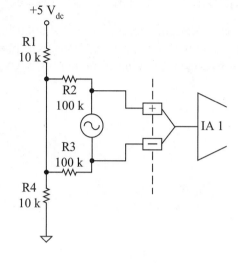

(b) Other float signals

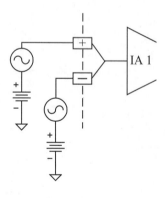

(c) Signals referenced to common

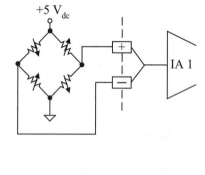

(d) Bridge sensors

Figure 20-5 Differential inputs to a PAC block

The simplest solution for a floating signal is shown in Figure 20-5(a). Connect the signal between E_{in+} and E_{in-}. Provide the 2.5 V_{dc} offset voltage from the $V_{ref\,out}$ pin. This works if the dc bias current needed by the signal generator is less than the 50 μA_{dc} available from $V_{\text{ref out}}$. For testing with a function generator, this technique works well.

Many differential sources, such as dynamic microphones and magnetic pick-ups require a bias current > 50 μA_{dc}. The four-resistor network shown in Figure 20-5(b) can be used. Resistors R1 and R4 divide the +5 V_{dc}, placing 2.5 V_{dc} at each input. Resistors R2 and R3 limit the bias current to the source, and allow each input to vary up and down around 2.5 V_{dc}, independently, in response to the ac signal from the source.

Occasionally, the source may provide two separate signals, opposite in phase and each referenced to common. This is shown in Figure 20-5(c). It is important that you also assure that the dc levels of the signals are equal, and of a value that ensures the signals will always be between 0 V and 5 V.

Figure 20-5(d) is a bridge circuit, typical of a force or pressure transducer. When at rest, all four resistors have exactly the same resistance, producing 2.5 V_{dc} at both $E_{\text{in}+}$ and $E_{\text{in}-}$. This is processed as a difference of 0 V by the PAC block's **IA**. When force is applied, the upper right and lower left resistances are decreased. The upper left and lower right resistances increase the same amount. This sends the voltage at $E_{\text{in}+}$ up above 2.5 V_{dc} and the voltage at $E_{\text{in}-}$ down the same amount.

Just as a PAC block has two input signals, one positive and one negative, it also has two output signals, $E_{\text{out}+}$ and $E_{\text{out}-}$. The two output signals each ride on a 2.5 V_{dc} level. They are equal in amplitude and opposite in phase. These are shown in Figure 20-6.

The gain for each PAC block is defined as

$$A = A_{\text{diff}} = \frac{E_{\text{out}+} - E_{\text{out}-}}{E_{\text{in}+} - E_{\text{in}-}}$$

It is assumed that the input is driven differentially, and the output is taken *between* the two output pins. If you use *either* $E_{\text{out}+}$ or $E_{\text{out}-}$ with respect to power supply common, the gain is one-half the programmed setting.

$$A_{\text{single}} = \frac{E_{\text{out}+}}{E_{\text{in}+} - E_{\text{in}-}} = -\frac{E_{\text{out}-}}{E_{\text{in}+} - E_{\text{in}-}} = \frac{1}{2} A_{\text{diff}}$$

You may set the gain of each **IA** to any integer between −10 and +10 using the PAC Designer software. Just click on the **IA**, and scroll through the dialog box. Then click the download button. The circuit in Figure 20-7

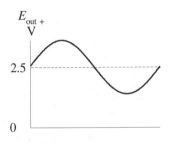

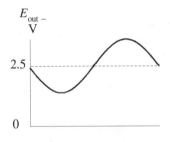

Figure 20-6
PAC differential outputs

has a differential gain of 4. The unused summer contributes nothing to the output. Gain accuracy is 3%, considerably better than building the same circuit with standard 5% resistors and an op amp.

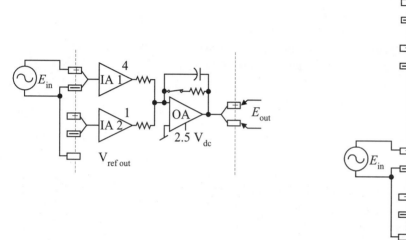

Figure 20-7 Simple differential gain   **Figure 20-8** Two digit gain up to 99

Two digits of gain resolution with values up to 99 can be set using the circuit in Figure 20–8. **PAC Block 2** provides a gain of 10. Its output, $10\,E_{in}$ is routed internally to **IA1** of **PAC Block 1**. The input is also routed internally to **IA2** of **PAC Block 1**. In **PAC Block 1 – IA1** the $10\,E_{in}$ is multiplied by 4, giving $40\,E_{in}$. E_{in} is multiplied by 7 in **PAC Block 1 – IA2**. The output amplifier, **OA**, of **PAC Block 1** combines these two signals to produce

$$E_{out} = 47\,E_{in}$$

To change the gain, simply alter the two digits in **PAC Block 1**. Remember, these signals are all *differential*. Also, the 3% error accumulates. Processing E_{in} by **PAC Block 2** then **PAC Block 1** may result in more than 3% gain error.

A similar technique can be used to provide a decimal gain such as 4.7 rather than 47. To understand how this can be done without an external resistor voltage divider, recall how an op amp with negative feedback works. Look at Figure 20-9. An amplifier with very high internal gain, and negative feedback (such as an op amp) alters its output to whatever voltage is necessary to drive the negative feedback voltage to equal the input voltage. To produce a *gain* of 10, you then connect a voltage *divider* to provide the negative feedback. For the negative feedback voltage to equal the input, the output is driven to 10 times the input.

The inverse is also true. If the negative feedback is given a *gain* of 10, then the output must be 0.1 E_{in}. Then, when the output is multiplied by 10 in the feedback block, it matches the input as V_{-fb}. This all seems ridiculous when building circuits with op amps and resistors. But with a PAC, this is programmed easily, and no additional, external components are needed. Look at Figure 20-10.

The feedback link in **PAC Block 2's OA** has been opened. That output is then fed back to the block's input with a gain of −10. The result is that **PAC Block 2** outputs 0.1 E_{in}. This is then scaled and combined with the scaled input in PAC Block 1. $E_{out} = 4.7\, E_{in}$.

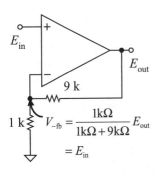

Figure 20-9
Attenuation in the negative feedback path produces amplification

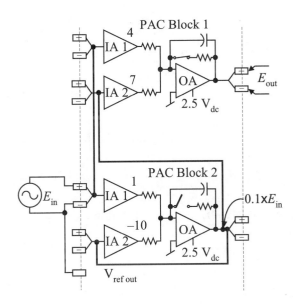

Figure 20-10 Decimal gain (4.7)

Simple Comparators

Opening the feedback link around the **OA** provides that amplifier with a very large gain, just like an op amp. This allows you to build a variety of comparator circuits. The **OA**'s reference level is 2.5 V_{dc}. So, whenever the signal going into the **OA** is greater than 2.5 V, the output is driven to +5 V. When the input to the **OA** falls below 2.5 V, the output is 0 V. The circuit configuration and signals are shown in Figure 20-11(a) and (b).

If the input signal is small, it is wise to provide gain in the input **IA**. This increases the size of the signal going into the **OA** and reduces errors. Figure 20-11(c) shows a gain of 3, but much higher gains for the **IA** are appropriate if the input signal is quite small.

An inverting comparator can be built by setting the **IA**'s gain to a negative value. Figure 20-11(d) illustrates the signals for an **IA** gain of −1. Smaller signals may warrant a much higher negative gain.

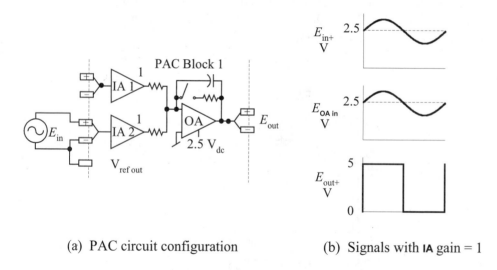

(a) PAC circuit configuration   (b) Signals with **IA** gain = 1

Figure 20-11(a,b) PAC 10 as a simple comparator

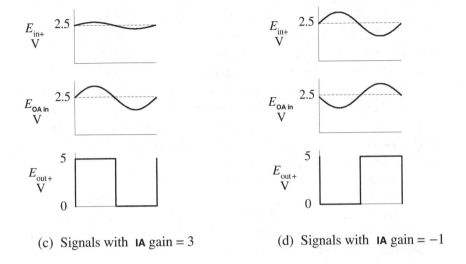

(c) Signals with **IA** gain = 3   (d) Signals with **IA** gain = −1

Figure 20-11(c,d) PAC 10 as a simple comparator

Referenced Comparator

The simple comparators in Figure 20-11 compare at 2.5 V. When the signal is above 2.5 V the output goes high. When it falls below 2.5 V the signal goes low. You can change the point on the input wave at which this comparison happens by adding a voltage to the input signal. For example, if E_{in+} has a 1.0 V_p amplitude, and the **IA2** has a gain of 1, then the signal at the input of the **OA** goes from 2.5 V up to 3.5 V, down to 1.5 V and then back to 2.5 V. In the circuit in Figure 20-11, comparisons are made at the mid-point, producing a 50% duty cycle output.

Input into **IA1** a 3.2 V_{dc} level. This level is 0.7 V above the reference. With a gain of 1, **IA1** outputs a shift of 0.7 V. This is added to the waveform from **IA2**. The signal into the **OA** now goes from 3.2 V up to 4.2 V, down to 2.2 V, and then back to 3.2 V. It has been shifted up by the offset introduced at the input of **IA2**.

With its feedback link open, the **OA** still compares at 2.5 V. However, the signal has been shifted up, so it crosses 2.5 V at a lower point. The duty cycle out of the **OA** is narrower. Look at Figure 20-12.

Remember, as the voltage at the input of **IA1** goes above 2.5 V_{dc}, the input to the **OA** is sent *up*. This means it crosses the 2.5 V internal reference level at a *lower* point on its wave. The duty cycle *decreases*. The overall effect of an input voltage above 2.5 V_{dc} is to cause the comparison to occur at a proportionally *lower* point on the input signal.

If this reversed effect is a problem, change the gain of **IA1** to −1. A 3.2 V_{dc} input into **IA1** now sends its input *down* 0.7 V_{dc}. This lowers the signal into the OA by 0.7 V. The 2.5 V internal reference is crossed at a higher point on the input signal. The dc into **IA1** now precisely defines the point at which the comparison occurs.

This technique is called **pulse width modulation**. The dc level into **IA1** varies how long the output signal is high (its pulse width). This high level may be used to drive a power switch *on*. When the output is low, the power switch goes *off*. Changing the dc into **IA1** directly alters how long the switch is on and how much power is applied to the load. This technique is the core of switching power supplies, many dc motor drivers (such as in your printer), and even the class D audio amplifier.

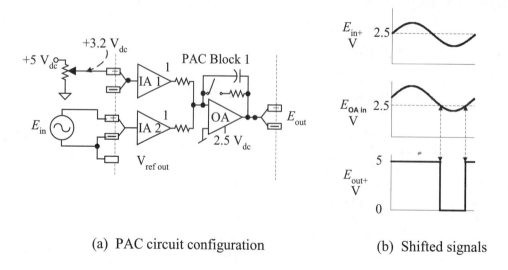

(a) PAC circuit configuration   (b) Shifted signals

Figure 20-12 Adjusting the duty cycle with a reference level

Comparator with Hysteresis

The comparators in the preceding sections have a single trip point. When the input goes above that level, the output goes to one extreme. When the input goes below that same level, the output goes to the other extreme. Generally, this is preferred. However, any noise in the system can easily make a small, slow input jiggle back and forth across the trip point repeatedly, causing several transitions at the output when there should only be one (see Section 9.3). Also, to build the relaxation oscillator and triangle wave generator you saw in Section 15.4, you need a comparator that switches at two different levels.

Two different trip points can be added to the comparator in Figure 20-12 with only a small change. Look at Figure 20-13. $E_{out\,+}$ is feedback through resistors R_f and R_i to the input of IA1. When the output is 5 V, V_{trip} is pulled up, above the 2.5 V_{dc} reference. This moves the first part of $E_{OA\,in}$ up, as shown in the first part of the waveforms. Eventually, that signal drops far enough to cross below 2.5 V_{dc}. The output, $E_{out\,+}$, steps down to 0 V.

With $V_{ref\,out}$ at one end of the R_f-R_i voltage divider, this drop at the other end causes V_{trip} to step down. That step down at the input of IA1 causes $E_{OA\,in}$ to step down by the same amount. This is shown in the center part of the waveforms in Figure 20-13. To switch again, that signal

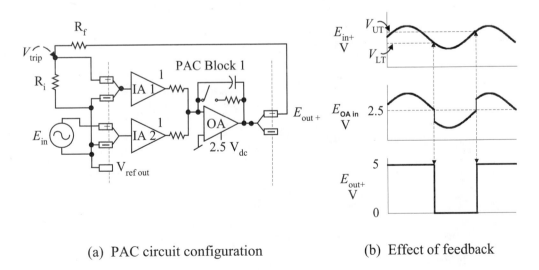

(a) PAC circuit configuration   (b) Effect of feedback

Figure 20-13 PAC block configured as a comparator with hysteresis

must overcome the step down caused by the positive feedback, and climb back up above 2.5 V_{dc}. When that finally happens, output $E_{out\,+}$ steps up again.

This step *up* at the output pulls V_{trip} up, sending $E_{OA\,in}$ up. The wave into the comparator (output amp) is offset positively, as shown in the last third of the middle waveform in Figure 20-13. This is right where we came in. The cycle repeats. When the output is high, the input to the **OA** is offset upward. It must come further down to cross 2.5 V_{dc} and switch. When the output is low, the input to the **OA** is offset downward. It must climb further up to cross 2.5 V_{dc} and switch. Two different switching thresholds, V_{LT} and V_{UT}, have been created by the positive feedback.

Resistors R_f and R_i set these two threshold voltages. Look at Figure 20-14. The output, $E_{out\,+}$ drives one end of the voltage divider between 5 V and 0 V. The other end is tied to V_{ref}, 2.5 V_{dc}. The voltage at the input to **IA1** is

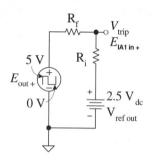

Figure 20-14
Trip point calculations

$$V_{trip} = 2.5\,\text{V} + I_{Ri}R_i$$

The current through R_i is the current through the series combination of $R_f + R_i$.

$$I_{Ri} = \frac{E_{out\,+} - 2.5\,\text{V}}{R_f + R_i}$$

When $E_{out\,+}$ is 5 V this current is

$$I_{Ri} = \frac{5\,\text{V} - 2.5\,\text{V}}{R_f + R_i} = \frac{2.5\,\text{V}}{R_f + R_i}$$

$$V_{UT} = 2.5\,\text{V} + 2.5\,\text{V}\frac{R_i}{R_f + R_i}$$

$$V_{UT} = 2.5\,\text{V}\left(1 + \frac{R_i}{R_f + R_i}\right)$$

When $E_{out\,+}$ is 0 V this current is

$$I_{Ri} = \frac{0\,\text{V} - 2.5\,\text{V}}{R_f + R_i} = -\frac{2.5\,\text{V}}{R_f + R_i}$$

$$V_{LT} = 2.5\,\text{V} - 2.5\,\text{V}\frac{R_i}{R_f + R_i}$$

$$V_{LT} = 2.5\,\text{V}\left(1 - \frac{R_i}{R_f + R_i}\right)$$

The upper and lower thresholds are symmetric around 2.5 V_{dc}. The total distance between them is

$$V_{UT} - V_{LT} = 5\,\text{V}\,\frac{R_i}{R_f + R_i}$$

When selecting these resistors, remember that the *total* current available into $V_{ref\,out}$ is 50 μA_p. Keep the resistors large enough to provide current to other loads that may be connected to $E_{out\,+}$ and still drive R_f and R_i. Very large resistors, however, are susceptible to noise. Values in the range of 100 kΩ should be reasonable.

Example 20-1
Select R_f and R_i for trip points of 1.5 V and 3.5 V.

Solution

$$V_{UT} - V_{LT} = 3.5\,\text{V} - 1.5\,\text{V} = 2\,\text{V}$$

$$V_{UT} - V_{LT} = 5\,\text{V}\,\frac{R_i}{R_f + R_i} = 2\,\text{V}$$

$$\frac{R_i}{R_f + R_i} = 0.4$$

$$R_i = 0.4R_f + 0.4R_i$$

$$0.6R_i = 0.4R_f$$

$$R_i = 0.667R_f$$

Look through the list of standard resistors.
$$R_f = 330\text{ kΩ}, \quad R_i = 220\text{ kΩ}$$

Practice   What is the effect of changing the gain of IA2 to 2?

Answer: Increasing the gain of IA2 means that the input signal may be smaller, but V_{LT} and V_{UT} must be proportionally adjusted to 2 V and 3 V. $R_f = 82$ kΩ, $R_i = 20$ kΩ

Relaxation Oscillator

In Chapter 15, the noninverting positive feedback op amp based comparator is combined with another resistor and a capacitor to produce a square wave generator. It is called a **relaxation oscillator** and is shown, again, in Figure 20-15.

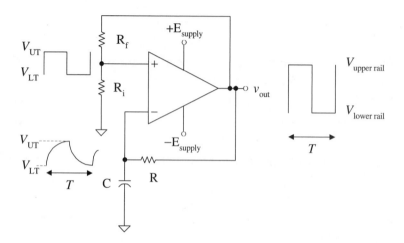

Figure 20-15 Op amp based relaxation oscillation and waveforms

When power is first applied, there is no charge on the capacitor. It holds the voltage at the op amp's noninverting input low. This forces the output to the upper rail voltage. Positive feedback resistors R_f and R_i divide that voltage, creating the upper threshold voltage, V_{UT}, at the op amp's non-inverting input. Since $V_{UT} > 0$ V, the output stays at $V_{upper\ rail}$.

The capacitor charges through R, and its voltage rises. Eventually, that charge exceeds V_{UT}. At that point, the voltage on the op amp's inverting input is above the voltage at its noninverting input. The output steps to the lower rail voltage, $V_{lower\ rail}$.

This also drops the positive feedback voltage to V_{LT}. The capacitor now discharges through R toward $V_{lower\ rail}$. This continues until the voltage across the capacitor has fallen below the voltage at the noninverting input, V_{LT}. At that point, the voltage at the inverting input is below the voltage at the op amp's noninverting input. The output steps up to $V_{\text{upper rail}}$, and the process repeats.

The upper and lower threshold voltages are set by the output rail voltages and by the positive feedback resistors. This also determines how far the capacitor must charge and discharge.

$$V_{\text{UT}} = \frac{R_i}{R_i + R_f} V_{\text{upper rail}}$$

$$V_{\text{LT}} = \frac{R_i}{R_i + R_f} V_{\text{lower rail}}$$

For convenience, define

$$B = \frac{R_i}{R_i + R_f}$$

The derivation for the time it takes the capacitor to charge, or discharge, between V_{UT} and V_{LT} is given in Section 15.4. The result is

$$t = RC \ln\left(\frac{1+B}{1-B}\right)$$

A full cycle requires that the capacitor both charge from V_{LT} to V_{UT}, then discharge from V_{UT} to V_{LT}. The *period* of the waveform is

$$T = 2RC \ln\left(\frac{1+B}{1-B}\right)$$

This same relaxation oscillator can be built with a PAC 10. Look at Figure 20-16. This is the same as the comparator with hysteresis, shown in Figure 20-13. The input source, E_{in}, has been replaced with a capacitor, C, connected to $V_{\text{ref out}}$ and a resistor, R, to the output.

$V_{\text{ref out}}$ must sink the current through RC and from R_f - R_i. So, keep these resistors in the 100 kΩ range to limit this current to about 50 μA. If that is not practical, move the connection from $V_{\text{ref out}}$ to the *output* of an unused PAC block. When the inputs of a PAC block are not used, its output is set to 2.5 V_{dc}. Each output is able to sink or source 10 mA. This allows you to lower R_f, R_i, and R into the kΩ range.

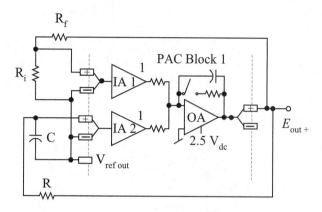

Figure 20-16 Relaxation oscillator built with a PAC 10

Example 20-2

Design a relaxation oscillator using a PAC 10, with f = 60 Hz.

Solution

The comparator with hysteresis from Example 20-1 provides trip points at 1.5 V and 3.5 V. These are reasonable levels, far enough apart to reduce the effect of noise. From that example,

$$R_f = 330 \text{ k}\Omega, \ R_i = 220 \text{ k}\Omega$$

$$B = \frac{R_i}{R_i + R_f} = \frac{220 \text{ k}\Omega}{220 \text{ k}\Omega + 330 \text{ k}\Omega} = 0.40$$

Also pick R = 470 kΩ, and solve the period equation of C.

$$C = \frac{T}{2R \ \ln\left(\frac{1+B}{1-B}\right)} = \frac{\dfrac{1}{60 \text{ Hz}}}{2 \times 470 \text{ k}\Omega \times \ln\left(\dfrac{1+0.4}{1-0.4}\right)}$$

$$C = 20.9 \text{ nF} \quad \text{Pick C} = 22 \text{ nF}$$

Practice   Change the components to set f = 20 kHz.

Answer: R_f = 330 kΩ, R_i = 220 kΩ, R = 820 kΩ, C = 36 pF

20.2 Function Generator IC

The XR-2206 function generator IC from Exar is a flexible and easy-to-use waveform generator. It outputs a square wave, and either a triangle or sine wave. Their frequency may vary down to 0.01 Hz to 1 MHz. This frequency may be adjusted by changing an external capacitor or resistor, by a logic level, or with an analog voltage. A single resistor controls the duty cycle. The amplitude of the sine and triangle waves also may be controlled by external resistors, or with an analog voltage. The dc level of the sine/triangle output is adjusted with an external voltage. The XR-2206's block diagram is shown in Figure 20-17.

Power Supply Requirements

Power supply voltages are applied between pins 4 and 12. The IC may be operated either single supply or from a dual supply. In either case, there must be at least 13 V_{dc} difference between the supply pins, but no more than 26 V_{dc}.

Single supply
(V– to common)
$$10\,V_{dc} \le V+ \le 26\,V_{dc}$$

Dual supply
$$\pm 5\,V_{dc} \le \pm V \le \pm 13\,V_{dc}$$

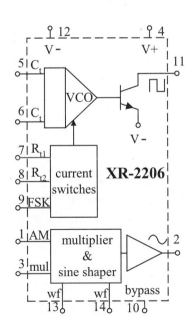

Figure 20-17 XR-2206 block diagram (*redrawn courtesy of Exar*)

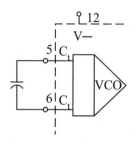

Figure 20-18
Timing capacitor

When powered from a single supply, you must provide a dc offset for the sine wave, biasing the output to about half of the single supply voltage. The control voltages are all reference to V−. So, if you plan to control the frequency or amplitude of the signals with external signals, power the IC from a single voltage of at least 10 V_{dc}. Unfortunately, a single +5 V_{dc} does *not* work.

If you want the output sine and triangle waves to ride evenly above and below common, then provide the IC with symmetric ± V_{dc} supply voltages. The output offset can then be set to common. However, remember all control voltages are referenced to the V− pin. Supply voltages of ± 5 V_{dc} means that external signals setting frequency and amplitude command their lowest values when they are − 5V_{dc}, not 0 V_{dc}.

Basic Timing

The fundamental waveform is a triangle wave. It is created by driving a constant current through an external capacitor. The capacitor is connected between the two C_t pins (pin 5 and pin 6). Look at Figure 20-18. Select capacitors in the range of

Timing capacitor

$$1 \text{ nF} \leq C_t \leq 100 \text{ µF}$$

Applying a constant current to a capacitor causes the voltage across it to *change* at a constant rate. A ramp has a constant slope.

$$i = C_t \frac{dv}{dt}$$

$$\frac{dv}{dt} = \frac{i}{C_t}$$

The derivative of the voltage across the capacitor is the slope of that voltage. When the current is a constant, I, then the derivative and the slope of the voltage are constant.

$$\frac{dv}{dt} = \frac{I}{C_t}$$

This current and capacitor set the frequency.

Frequency equation

$$f = 0.32 \frac{I}{C_t}$$

Limit $1 \text{ µA} \leq I \leq 3 \text{ mA}$

This current is set with a pair of external timing resistors, R_{t1} and R_{t2}, and a voltage. Look at Figure 20-19. The voltage at each timing resistor pin (pin 7 and pin 8) is held at a constant 3 V above V−. The current *out of* each of these pins sets the timing current. If you connect the timing resistors between their pin and V−, then there is a constant 3 V across each resistor. So the current through each resistor is a constant

$$I = \frac{3\,V}{R_t}$$

$$f = 0.32\frac{\dfrac{3V}{R_t}}{C_t} = \frac{0.96}{R_t C_t}$$

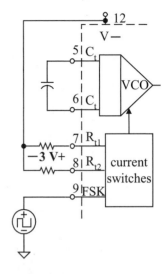

Figure 20-19
Timing resistor and FSK

Frequency shift keying, FSK, causes the frequency to change between two values depending on a logic level. A logic 0 causes a signal at a low frequency, while a logic 1 shifts the frequency to a higher value. This allows logic signals to be transmitted as tones.

The voltage at the FSK pin (pin 9) determines which timing resistor, and therefore what frequency, is used. Be careful. This voltage is measured with respect to V−. If you are powering the IC from a single supply voltage, with V− connected to common, then

$$V_{FSK} < 1\,V \rightarrow R_{t2}\ (\text{pin 8})$$

$$V_{FSK} > 2\,V \rightarrow R_{t1}\ (\text{pin 7})$$

FSK voltage requirements: single supply

This assures that a TTL signal may be used to select which timing resistor is used, and set the frequency accordingly.

However, if you are using a split supply, such as ± 5 V_{dc}, then these levels are to set with respect to −5 V. The result is

$$V_{FSK} < -4\,V \rightarrow R_{t2}\ (\text{pin 8})$$

$$V_{FSK} > -3\,V \rightarrow R_{t1}\ (\text{pin 7})$$

Split supply, ± 5 V_{dc}

These levels are *not* TTL compatible. So be sure to consider how you are going to control the FSK pin when you select the power supply voltages.

You do not have to use an external signal to select the timing resistor. Connecting the FSK pin (pin 9) to V− (pin 12) selects R_{t2} (pin 8). Wiring the FSK voltage directly to V+ (pin 4) selects R_{t1} (pin 7).

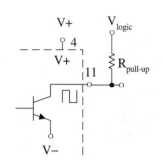

Figure 20-20
Square wave output

$V_{high} = V_{logic}$

$V_{low} = V-$

Square Wave Output

The square wave output (pin 11) is connected to the collector of a transistor switch. Refer to Figure 20-20. When that transistor switch is *off* pin 11 is open. To create a voltage level, you must connect an external pull-up resistor between the square wave output pin, and whatever voltage you want as the high level. When the transistor goes *off*, the voltage is pulled up to V_{logic}. This may be the same as V+, but may be some other voltage. For example, if you are running the IC single supply from +12 V_{dc}, and want the high level of the square wave to be TTL compatible, set V_{logic} to + 5 V_{dc}.

When the transistor turns *on*, pin 11 is shorted to V−. If you are running the IC single supply, then the low level out of pin 11 is 0.2 V. This is also a TTL compatible signal. However, if you have powered the IC from ± 5 V_{dc}, then the low level is about −4.8 V. This may damage many logic families. So when you are considering how to provide power to the IC, remember that the low level is equal to the negative supply.

When the transistor is *on*, current flows from V_{logic}, through $R_{pull-up}$ and through that transistor to V−. You must select $R_{pull-up}$ large enough to limit this current to 2 mA_p or less.

$$R_{pull-up} \geq \frac{V_{logic} - (V-)}{2\,mA_p}$$

When the transistor turns *off*, the voltage at pin 11 does not *immediately* jump up to V_{logic}. Any parasitic capacitance between that pin and common must be charged through $R_{pull-up}$. The larger $R_{pull-up}$, the longer it takes that capacitance to charge, and the more gradual the rise. For a nice sharp edge, make $R_{pull-up}$ as small as possible, just barely above the value needed to limit the transistor's current to 2 mA_p.

Example 20-3

Design a square wave generator that outputs TTL signals. The frequency is 2 kHz when the FSK pin is at a logic high, and is 300 Hz when the FSK pin is at a logic low.

Solution

Since the output is to be TTL compatible and the FSK pin is driven from logic level signals, V− must be connected to common. So, power the IC using a single supply voltage of 12 V_{dc}.

$$V- = \text{common} \qquad V+ = +12 \text{ V}_{dc}$$

The output is to be TTL compatible, so the logic high level from the square wave pin should be 5 V. Select

$$V_{logic} = +5 \text{ V}_{dc}$$

The pull-up resistor can now be calculated.

$$R_{pull-up} \geq \frac{V_{logic} - (V-)}{2 \text{ mA}_p}$$

$$R_{pull-up} \geq \frac{5 \text{ V} - 0 \text{ V}}{2 \text{ mA}_p} = 2.5 \text{ k}\Omega$$

Pick $\qquad R_{pull-up} = 2.7 \text{ k}\Omega$

When the

$$I_{Rt1} < 3 \text{ mA} = \frac{3 \text{ V}}{R_{t1}}$$

$$R_{t1} > \frac{3 \text{ V}}{3 \text{ mA}} = 1 \text{ k}\Omega$$

Using the relationship

$$f = \frac{0.96}{R_t C_t}$$

find a pair of standard R and C that produces 2 kHz.

Pick $\qquad R_{t1} = 2.2 \text{ k}\Omega$ and $C_t = 0.22 \text{ }\mu\text{F}$

When the voltage at the FSK pin goes low, timing resistor R_{t2} is used. The frequency is to shift to 300 Hz.

$$R_{t2} = \frac{0.96}{f \text{ } C_t} = \frac{0.96}{300 \text{ Hz} \times 0.22 \text{ }\mu\text{F}} = 14.5 \text{ k}\Omega$$

Pick $\qquad R_{t2} = 15 \text{ k}\Omega$

This timing resistor produces a current of

$$I = \frac{3 \text{ V}}{15 \text{ k}\Omega} = 200 \text{ }\mu\text{A}$$

This is well above the minimum timing current of 1 μA.

Practice Repeat the problem for 540 kHz and 600 kHz.

Answer: $C_t = 1$ nF, $R_{t1} = 1.6$ kΩ, $R_{t2} = 1.8$ kΩ

The basic XR-2206 circuit produces a 50% duty cycle. The time that the square wave output is high, t_{high}, is the same as the time that it is low, t_{low}. These two times may be set independently with the circuit in Figure 20-21. When the square wave output is high, the FSK pin is high, and R_{t1} sets the current that charges C_t. This is the high time, t_{high}. When the square wave goes low, FSK is low, and R_{t2} sets the current that discharges C_t, setting the low time.

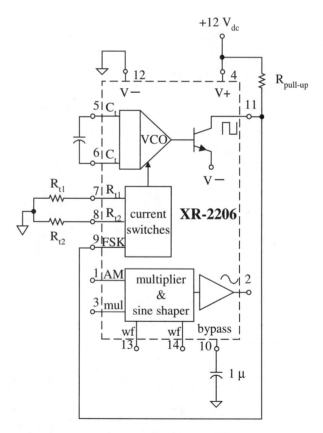

Figure 20-21 Variable duty cycle output

Timing resistor R_{t1} sets the time high and R_{t2} sets the time low.

$$f = \frac{0.96}{R_t C_t}$$

$$T = \frac{1}{f} = \frac{R_t C_t}{0.96}$$

This relationship was designed assuming a 50% duty cycle, and the same timing resistor used to both charge and discharge C_t.

$$t_{high} = \frac{1}{2} T = \frac{1}{2} \times \frac{R_{t1} C_t}{0.96}$$

$$t_{high} = \frac{R_{t1} C_t}{1.92}$$

Similarly $$t_{low} = \frac{R_{t2} C_t}{1.92}$$

Variable duty cycle
t_{high} and t_{low}

These adjustments are *independent*. You have separate control over the time high and the time low. However, you no longer have a single component that changes the frequency without affecting the duty cycle. A much more sophisticated feedback and control scheme is needed to provide duty cycle adjustment on one control and independent frequency adjustment on another.

Sine-Triangle Wave Output

The output from pin 2 is a triangle wave. Connecting a resistor between the waveform pins (pins 3 and 4) alters the output amplifier's characteristics, and shapes the triangle into a sine wave. Nominally set that resistor to 180 Ω. Adjustment around this value changes the quality (distortion) of the sine wave. Experiment to determine the exact value that works in your application.

The resistance between the multiplier pin (pin 3) and ac common sets the magnitude of both the triangle and the sine wave.

$$V_{p\ triangle} = \frac{160\ \mu V_p}{\Omega} \times R_{mult}$$

$$V_{p\ sine} = \frac{60\ \mu V_p}{\Omega} \times R_{mult}$$

The dc level of the triangle and sine (pin 2) is equal to the dc at the multiplier pin. Figure 20-22 shows proper connections when using the IC with dual supplies and with a single supply voltage.

For dual supply operation, connect the FSK pin and R_{t2} to the negative supply. Also set the AM pin to V–. This disables its effect. (More on that soon.) The multiply pin (pin 3) is connected through R_{mult} to common. The resistor R_{mult} sets the amplitude of the sine/triangle output. Connecting it to common allows this signal to go up and down evenly about 0 V.

When running from a single supply, resistor R_{t2}, the FSK pin, and the AM pin are still tied to the most negative voltage in the circuit. In Figure 20-22(b), this is common.

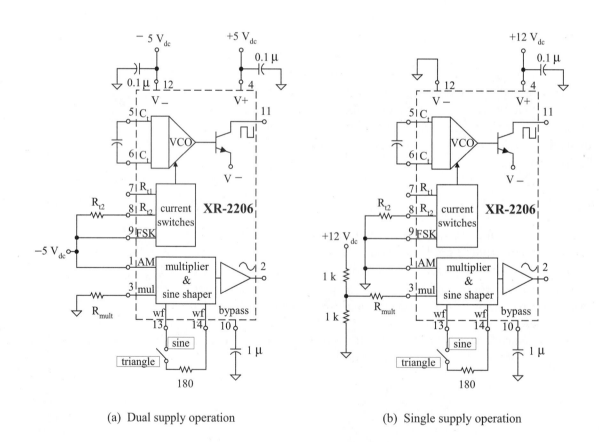

(a) Dual supply operation   (b) Single supply operation

Figure 20-22 Simple sine/triangle generators

Since the output voltage may only be positive, you must offset the dc voltage at pin 2 to allow the signal to go both up and down. The two 1 kΩ resistors set the voltage at their junction, at the multiply pin and at the output to 6 V$_{dc}$.

Example 20-4

Design a sine wave generator.

> Frequency adjustable from 60 Hz to 5 kHz
> Amplitude adjustable from 100 mV$_{rms}$ to 2 V$_{rms}$
> dc offset = 0 V$_{dc}$

Solution

Use the schematic in Figure 20 –22(a).
At the upper frequency, limit $I_t < 3$ mA.

$$R_{t2\ min} < \frac{3\ V}{3\ mA} = 1\ k\Omega$$

Configure R$_{t2}$ as a 1 kΩ fixed resistor in series with a potentiometer. At the upper frequency limit, 10 kHz, R$_{t2}$ is at the minimum.

$$f = \frac{0.96}{R_{t2}C_t}$$

$$C_t = \frac{0.96}{R_{t2}f} = \frac{0.96}{1\ k\Omega \times 5\ kHz} = 192\ nF$$

Pick $C_t = 180$ nF
To lower the frequency to 100 Hz, R$_{t2}$ must be increased.

$$R_{t2\ max} = \frac{0.96}{f\ C_t} = \frac{0.96}{60\ Hz \times 180\ nF} = 89\ k\Omega$$

Pick R$_{t2}$ as a 1 kΩ resistor in series with a 100 kΩ rheostat.
A minimum amplitude of 0.1 V$_{rms}$ means a peak of

$$V_{out\ p\ max} = 0.1\ V_{rms} \times \sqrt{2} = 141\ mV_p$$

This is established by a R$_{mult\ min}$ of

$$V_{p\,sine} = \frac{60\ \mu V_p}{\Omega} \times R_{mult}$$

$$R_{mult\,min} = \frac{0.141\ V_p}{\dfrac{60\ \mu V_p}{\Omega}} = 2.4\ k\Omega$$

To output 2 V_{rms}, set

$$V_{out\,p\,max} = 2\ V_{rms} \times \sqrt{2} = 2.8\ V_p$$

$$R_{mult\,max} = \frac{2.8\ V_p}{\dfrac{60\ \mu V_p}{\Omega}} = 47\ k\Omega$$

Pick R_{mult} as a 2.4 kΩ resistor in series with a 50 kΩ rheostat.

Practice Find f and V_{out} for $C_t = 1$ nF, $R_{t2} = 2.2$ kΩ, $R_{mult} = 33$ kΩ.

Answer: $f = 436$ kHz, $V_{out\,p} = 1.98\ V_p$

Voltage Control

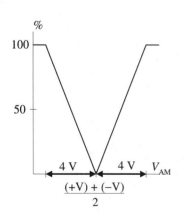

Figure 20-23
AM voltage control
(courtesy of Exar)

The **amplitude modulation** pin (AM pin 1) allows you to change the amplitude of the sine/triangle with a voltage. Look at Figure 20-23. This graph is a plot of the voltage on the AM pin on the x axis versus the amplitude (as a percent) on the y axis. When the AM pin voltage is at either extreme, the output amplitude is 100% of that calculated using the R_{mult} equations. So if you do *not* want to use the AM feature, just connect that pin to either supply voltage.

When the voltage at the AM pin is exactly half-way between the two supplies, the output amplitude is driven to zero. Increasing or decreasing the AM pin voltage from this mid-point causes the amplitude to increase linearly with voltage, 25%/V. At that slope, a 4 V change in the AM pin voltage sends the amplitude to its full value.

Increasing the voltage from the mid-point causes the amplitude to increase. Decreasing the voltage from the mid-point increases the amplitude exactly the same amount, but with the opposite phase. This allows the resulting output to match that needed for commercial AM communication systems.

Example 20-5

Given that R_{mult} is 47 kΩ, power is provided from ±5 V_{dc} and the voltage on the AM pin is 2.5 V, calculate the amplitude of an output sine wave.

Solution

The full amplitude is

$$V_{p\,sine} = \frac{60\ \mu V_p}{\Omega} \times R_{mult}$$

$$V_{p\,sine\,max} = \frac{60\ \mu V_p}{\Omega} \times 47\ k\Omega = 2.8\ V_p$$

Halfway between ±5 V_{dc} is 0 V. From 0 V the amplitude increases 25%/V. With $V_{AM} = 2.5$ V,

$$V_{p\,out} = \left(\frac{0.25}{V} \times 2.5\,V \right) \times 2.8\,V_p = 1.75\,V_p$$

Practice What is the effect if $R_{mult} = 68$ kΩ, and $V_{AM} = -1$ V?

Answer: $V_{p\,out} = 1.02\ V_p$. The phase is reversed.

The frequency can also be controlled with an externally applied voltage. This is called **frequency modulation**. It is done more effectively when powering the IC from a single supply. Look at Figure 20-24. Instead of connecting R_{t2} between pin 8 and V−, the left end of R_{t2} is connected to the output of an op amp. This point must be driven from a low impedance source (such as an op amp) since all resistance seen from pin 8 affects the frequency. If you use a potentiometer instead of an op amp, then the resistance of that potentiometer must be added to R_{t2}. And, as you move the potentiometer, both its voltage and its resistance affect the frequency, producing a nonlinear change.

The current *out* of pin 3 determines the frequency by

$$f = 0.32 \frac{I_{Rt}}{C_t}$$

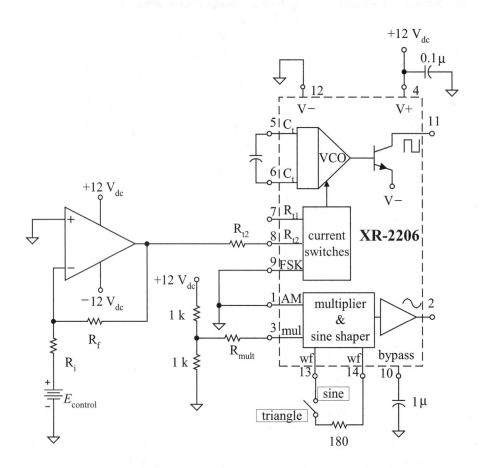

Figure 20-24 Voltage control of frequency

The voltage at pin 8 is kept 3 V above V−, +3 V_{dc} in Figure 20-24. When the op amp outputs 3 V, there is no difference in potential across R_{t2}, so there is no current, and no signal is produced, $f = 0$.

When the op amp's output goes *down* below +3 V, and then negative, current is drawn from pin 8. The IC begins to oscillate. The further below +3 V the output of the op amp, the more current is drawn, and the higher the frequency. Select R_{t2} so that at the most negative voltage from the op amp, the current from the XR-2206 timing pin is no more than 3 mA, and the maximum desired frequency is produced.

The gain of the op amp may be adjusted so that when $E_{control}$ swings to its most positive, the frequency goes to its maximum.

Example 20-6

For the circuit in Figure 20-24, given:

$$R_{t2} = 15 \text{ k}\Omega, \ C_t = 0.22 \ \mu\text{F}, R_f = 22 \text{ k}\Omega, R_i = 10 \text{ k}\Omega,$$

Calculate the output frequency for:

$$E_{control} = 0 \text{ V}_{dc}, \ 2.5 \text{ V}_{dc}, \text{ and } 5.0 \text{ V}_{dc}$$

Solution

$$E_{control} = 0 \text{ V}_{dc}$$

$$V_{\text{out op amp}} = 0 \text{ V}_{dc}$$

$$I_{Rt2} = \frac{3 \text{ V}_{dc} - 0 \text{ V}_{dc}}{15 \text{ k}\Omega} = 200 \ \mu\text{A}_{dc}$$

$$f = 0.32 \times \frac{200 \ \mu\text{A}_{dc}}{0.22 \ \mu\text{F}} = 291 \text{ Hz}$$

$$E_{control} = 2.5 \text{ V}_{dc}$$

$$V_{\text{out op amp}} = 2.5 \text{ V}_{dc} \times \left(-\frac{22 \text{ k}\Omega}{10 \ \Omega} \right) = -5.5 \text{ V}_{dc}$$

$$I_{Rt2} = \frac{3 \text{ V}_{dc} - (-5.5 \text{ V}_{dc})}{15 \text{ k}\Omega} = 567 \ \mu\text{A}_{dc}$$

$$f = 0.32 \times \frac{567 \ \mu\text{A}_{dc}}{0.22 \ \mu\text{F}} = 824 \text{ Hz}$$

$$E_{control} = 5.0 \text{ V}_{dc}$$

$$V_{\text{out op amp}} = -11 \text{ V}_{dc}$$

$$I_{Rt2} = 933 \ \mu\text{A}_{dc}$$

$$f = 1.35 \text{ kHz}$$

Practice Find R_{t2} and C_t so that f_{max} = 10 kHz. Pick R_f so that E_{in} = 5 V$_{dc}$ produces this maximum frequency. ($V_{\text{out op amp min}}$ = -14 V$_{dc}$)

Answer: R_{t2} = 5.6 kHz, C_t = 91 nF, R_f = 28 kΩ

Summary

Overall system cost is reduced when many analog functions can be combined within a single integrated circuit. Three different Programmable Analog Circuits from Lattice combine input and output amplifiers, comparators, and a digital-to-analog converter in a series of ICs. Their precise function, parameters, and interconnections can be programmed via a drag-and-drop window running on a personal computer, then downloaded into the IC. This significantly lowers design time, and increases product flexibility.

The PAC 10 is powered from a single +5 V_{dc} supply, and requires that all signals be referenced to +2.5 V_{dc}. Inputs and outputs are differential. Each of the input amplifiers has a programmable gain, adjustable in integer steps between ±10. Each output amplifier may be run closed or open loop. Properly configured, this open loop feature allows fractional (decimal) gain, and comparators. Simple comparators, as well as those with an external reference and with hysteresis, can be programmed. An additional external feedback path around a PAC 10 cell produces a relaxation oscillator, just as you built with op amps in Chapter 15.

The XR-2206 function generator IC from Exar produces square, triangle, and sinusoidal signals. It may be powered from a single supply or a dual supply, but requires at least 10 V_{dc}. There are two external time resistors, selected by the FSK pin. Current through the selected resistor linearly charges an external capacitor, producing a ramp. An internal comparator drives an open collector (square wave) output. This pin must be pulled-up to the desired logic high voltage with an external resistor. The logic low voltage is the negative supply. So be sure to consider these logic requirements when setting the power supply values. Feedback from the square wave output to the FSK produces a variable duty cycle.

The triangle wave may be converted into a sine by adding an external wave-shaping resistor. The external R_{mult} determines the amplitude of the sine and the triangle wave. The dc voltage at the mult pin sets the dc level of these two signals. For single supply operation, be sure to set this to about half of the supply to allow the waves to both rise and fall.

The amplitude may also be controlled with a voltage applied to the AM pin. Voltage control of frequency is possible by adding an op amp and driving the left end of the timing resistor. This alters the current through the resistor, the charge rate and therefore the frequency.

Problems

Programmable Analog Circuit

20-1   Locate the web site for the Lattice Programmable Analog Circuits. From there determine a vendor. At the vendor's web site determine the price of the PAC 10, PAC 20, and PAC 30.

20-2   From the Lattice Programmable Analog Circuits' web page, find the data sheets for the PAC 10. Locate the following specifications:

    a.   What is the input impedance?

    b.   What size capacitor should be added to the VREF pin?

    c.   How many programmable cycles are guaranteed?

    d.   What is the maximum power supply current required?

    e.   What is the typical highest frequency for a differential input of 6 V and a gain of 1?

20-3   The gain assumes that the input is driven differentially, and that you measure one output to common. What happens to the gain if you measure between the two outputs (differential output)?

20-4   What happens to the gain if you drive one input, referenced to common (single-ended), rather than driving the input differentially?

20-5   Draw the schematic of a circuit using an *RC* coupler that allows a single input (with respect to common) to be connected to the IC. This is called a single-ended input. Remember that the input must be referenced to 2.5 V_{dc}. (Look at the data sheets for hints.)

20-6   Alter your design for Problem 20-5 to replace the *RC* coupler with (an) op amp(s) that adds the 2.5 V_{dc} offset to the input signal.

20-7   Design an amplifier using a PAC 10 that has a gain of 82.

20-8   Design an amplifier using a PAC 10 that has a gain of 0.82.

20-9   Design a circuit using a PAC 10 that

    a.   amplifies an external, single-ended signal by 3.

    b.   adds a second external, single-ended signal to that.

    c.   compares the composite sum to 2.5 V_{dc}, going high when it is above 2.5 V_{dc} and low when it is below 2.5 V_{dc}.

20-10   Alter the design from Problem 20-9 to allow an external comparison level, instead of 2.5 V_{dc}.

20-11   Design a circuit using a PAC 10 as a comparator with trip points at 3.2 V and 1.8 V.

20-12   Design a circuit using a PAC 10 as a comparator with trip points at 2.8 V and 2.2 V.

20-13   Use the circuit from Problem 20-11 to design a relaxation oscillator that runs at 440 Hz.

20-14   Use the circuit from Problem 20-12 to design a relaxation oscillator that runs at 20 kHz.

Function Generator IC

20-15   Locate the web site for Exar Semiconductor. From there determine a vendor. At the vendor's web site determine the price of the XR-2206.

20-16   From the Exar web page, find the data sheets for the XR-2206. Locate the following specifications:

    a.   Maximum current required from the power supply.

    b.   Lowest practical period.

    c.   Maximum sine/triangle output amplitude.

    d.   How close to V− does the low level of the square wave output come?

20-17   Design a square wave generator with an output from 5 V to −5 V, and a frequency of 200 kHz.

20-18   Design a variable duty cycle square wave generator with an output from 5 V to common. The frequency should be 30 Hz when the duty cycle is 50%.

20-19   Design a simple sine wave generator with a 1 V_{rms} amplitude, riding above and below common, and a frequency of 100 Hz.

20-20 Design a triangle wave generator with a ramp that starts at 1 V and goes to 5 V in 3 ms.

20-21 Modify the design from Problem 20-19 to allow the amplitude to be adjusted to a maximum of 2 V_{rms} by an externally provided dc voltage. The sine wave's amplitude should increase proportionally as the externally voltage goes from 0 V (no output amplitude) to 5 V (full output amplitude).

20-22 Modify the design of Problem 20-20 to allow the ramp's rise time to be changed with an external voltage. The time should be a minimum of 100 μs when the external voltage is 5 V.

Answers to Odd-Numbered Problems

Chapter 1

1-1 Foot; meter; centimeter; meter

1-3 −1.33%

1-5 1, 3, 1, 4

1-7 Nearest millimeter

1-9 $10.5 \text{ m} \leq X < 11.5 \text{ m}$

1-11 20 (unit place)

1-13 1000 (1 significant figure)

1-15 $10^0, 10^2, 10^4, 10^{-1}, 10^{-3}, 10^{-5}$

1-17 $5\times10^1, 12\times10^4, 43\times10^{-3}, 7\times10^{-5}$

1-19 $10^5, 10^1, 10^3, 10^6, 10^6, 10^{-6}, -10^4, 10^{-4}$

1-21 $6\times10^2, 33\times10^2, 8\times10^6, 12\times10^4, 6\times10^3,$ $6\times10^6, 16\times10^6, 8\times10^{-6}$

1-23 $128, -72\times10^7$

1-25 $3.4\times10^3, 5.60\times10^{-3}$

1-27 $1.2\times10^3, 340\times10^3, 56.0\times10^{-3}, 780\times10^{-6}$

1-29 1.2 ks, 3.4 Ms, 56.0 ms, 780 µs

1-31 120 ks, 0.12 Ms, 0.00012 Gs, 0.03 ms, 30 µs, 30000 ns

1-33 120 Ms, 0.12 Gs, 30 µs, 30000 ns

1-35 $864,000,000

1-37 9144 cm

1-39 120 ks, 0.12 Ms, 30 µs, 30000 ns

1-41 120 Ms, 30 µs

1-43 390.6×10^3

Chapter 2

2-1 proton: +, neutron: neutral, electron: −

2-3 attract

2-5 $6.242 \times 10^{18}, 6.242 \times 10^{18}$

2-7 $F = \dfrac{kQ_1Q_2}{r^2}$ where k = 9.0×10^9Nm²/C²

2-9 7.2×10^9 N

2-11 **a.** 1st shell of 2 (2). 2nd shell of 8 (2+6), 3rd shell of 18 (2+6+10); 4th shell of 19 (2+6+10+1) **b.** conductor since the last subshell has only 1 electron

2-13 **a.** Easily create free electrons and conduct them **b.** Very hard to create free electrons and conduct them **c.** Its ability to create free electrons and conduct them falls between the conductor and insulator

2-15 An atom missing an electron

2-17 Yes

2-19 Electron current: electrons flow from the negative supply terminal to the positive supply terminal. Conventional current: current modeled as a positive charge flowing from the positive supply terminal to its negative supply terminal.

2-21 30 A

2-23 500 s

2-25 Yes, to insert the ammeter

2-27 Break circuit, insert current meter with red lead toward the positive supply terminal; supply must be off to insert the meter but then turned on to measure the current

2-29 The current is actually flowing the opposite direction of the current meter connection; that is, current is flowing into the black lead instead of the red lead

2-31 600,000 ft-lb, 813.8 MJ

2-33 5.085 kW

2-35 V (drop) or E (supply), volt, V

2-37 90 J

2-39 1.2 V

2-41 100 V

2-43 Rise

2-45 (b) Across

2-47 No (acts like an open that does not draw current)

2-49 Voltmeter put across the lamp with the red lead on the relatively positive terminal (top of the lamp)

2-51 Red lead on node a and black lead on node b.

2-53 See text

2-55 See text

2-57 **a.** 45 Ah, **b.** 162 kC, **c.** 2.041 MJ

2-59 **a.** out of **b.** clockwise **c.** see text **d.** positive **e.** 0 V, 5 V, 5 V **f.** 5 Ah **g.** 5 V

2-61 **a.** see text **b.** 0 V, 6 V, −6 V **c.** red lead on node a and black lead on common **d.** 12 V **e.** 4 V drops, + at top of each **f.** 8 V, −8 V **g.** red lead on node a and

black lead on node c **h.** 8 V, −8 V **i.** +6 V bubble at top, −6 V bubble at bottom, no lamps connected to common

2-63 **a.** 9 V, 9 V **b.** 0 V **c.** no

Chapter 3

3-1 5.211 Ω

3-3 256.4 μS

3-5 2 V, 100 Ω, 10 mS, 0 mA, −30 mA, 100 Ω, linear, positive

3-7 (b) 0 V

3-9 SW 1 down, SW 3 closed

3-11 reverse, off, no

3-13 **a.** 270 kΩ ± 5%, 283.5 kΩ, 256.5 kΩ **b.** 1.8 kΩ ± 10%, 1.98 kΩ, 1.62 kΩ **c.** 10 Ω ± 20%, 12 Ω, 8 Ω

3-15 No

3-17 100 Ω

Chapter 4

4-1 AWG 14, 2.08 mm^2; AWG 24, 0.205 mm^2, factor of 10.15

4-3 AWG 10

4-5 Slow blow allows for current spikes and must remain above rated current capacity for a few seconds.

4-7 15 kΩ

4-9 Wiper arm shorted to unused wiper arm to prevent unused terminal from acting like an antenna

4-11 272 Ω

Chapter 5

5-1 9 A

5-3 **a.** 20 mA, 11 mA, 9 mA, 4 mA, 11 mA, 20 mA **b.** circle cuts only I_2 & I_5 **c.** circle cuts only I_1 & I_6

5-5 **a.** 2 A, 3 A **b.** 3 A **c.** 3 A

5-7 **a.** 50 µA **b.** 10.05 mA **c.** 10.05 mA **d.** 0 mA, no

5-9 **a.** 5 mA **b.** 15 mA

Chapter 6

6-1 **a.** 2 V **b.** 2 V **c.** 18 V, 14 V, 12 V **d.** 2 V

6-3 **a.** 3 V **b.** 2 V **c.** 1 V **d.** 1 V

6-5 14 V, 2.4 V, 0.75 Ω or 750 mΩ

Chapter 7

7-1 5.455 mA

7-3 1.8 kΩ

7-5 **a.** See text **b.** −6 V, −5 mA **c.** See text **d.** line from (10 V, 8.33 mA) through origin to (−10 V, −8.33 mA)

7-7 20 mA

7-9 4 W, unsafe

7-11 100 mA, 100 V

7-13 37.3 mA

7-15 **a.** 5.3 V **b.** 53 µA **c.** 37 µW **d.** 16 V **e.** 160 mW **f.** 189

7-17 **a.** 0 V **b.** 1 V **c.** 1 mA **d.** 1mA **e.** 2.7 V **f.** −2.7 V **g.** −2.7 V **h.** −0.27 mA **i.** 1.27 mA into op amp

7-19 2.4 W, 72 cents

7-21 into positive terminal, sinking

Chapter 8

8-1 **a.** See Figure 8-40 **b.** 13 V **c.** 13 mA

8-3 **a.** E=I(R1+R2+R3) **b.** 10 mA **c.** 1.5 V, 2.7 V, 6.8 V **d.** 11 V, 9.5 V, 6.8 V, 0 V **e.** 4.2 V, −4.2 V **f.** 1.1 kΩ

8-5 **a.** 1.1 kΩ **b.** 10 mA

8-7 **a.** 650 Ω **b.** 100 Ω **c.** 260 mW **d.** 132 mW, yes

8-9 **a.** 500 mA **b.** 476 mA **c.** −4.76%, yes

8-11 **a.** 1.5 V, 2.7 V, 6.8 V **b.** 4.2 V **c.** 9.5 V **d.** −9.5 V

8-13 **a.** 4.4 V **b.** 3.7 V **c.** 3.7 mA **d.** 3.7 mA **e.** 16 V **f.** 12.3 V **g.** 15 µA **h.** yes, 2 mA >> 15 µA

8-15 4.5 kΩ

8-17 20.7 kΩ

8-19 **a.** 1.6 V **b.** 1.45 Ω **c.** 0.15 V **d.** 0.15 Ω **e.** −10.34%

8-21 **a.** 20 V **b.** 3 kΩ **c.** 20 V series with 3 kΩ **d.** 8 V, 4 mA

8-23 **a.** See text **b.** 3 V

8-25 **a.** 4 V, 33.3 mA, 5 V **b.** 3.75 V, 31.25 mA, 5.25 V

Chapter 9

9-1   **a.** 5 μV **b.** 7 μV **c.** 7 μV **d.** −7 μV

9-3   **a.** safe **b.** safe **c.** unsafe **d.** unsafe

9-5   **a.** −200 mV **b.** 700 mV **c.** 13 V

9-7   **a.** $-10.4 \text{ V} \leq V_{in} \leq +11 \text{ V}$ **b.** noninverting comparator, 0 V crossover (−13 V for $V_{error} < 0$ V, +13 V for $V_{error} > 0$ V)

9-9   2 kΩ, inverting

9-11 **a.** 1 V **b.** 1.67 kΩ **c.** red, 8 mA; green, 5.33 mA **d.** $T > 100°F$

9-13 **a.** 1 mA, 1 V, −4 V **b.** +1.44 V

9-15 **a.** 1.3 mA, 1.3 V, 1.7 V **b.** −1.3 V

9-17 Noninverting voltage amplifier (negative feedback), infinite ohms (open), 500 mV, 500 μA, 500 μA, 4.5 V, 5 V, 5 V, 1 mA, 1.5 mA. No op amp limitations reached.

9-19 16.14 mA. Still > 10 mA so needs a BJT with higher β for this to be reasonable

9-21 0 V, 2 mA, 2 mA, 18 V, −18 V, 18 V (hits upper rail at +13 V), op amp in saturation and is no longer acting as an amplifier so none of the calculated values are valid since $V_{error} \neq 0$ V.

9-23 −277 mV, 277 mV

Chapter 10

10-1 **a.** 12 V **b.** 12 A, 12 A, 24 A **c.** 2 Ω **d.** 0.5 Ω **e.** 2 S

10-3 **a.** 15 V **b.** 15 mA, 6.818 mA, 4.545 mA **c.** 26.36 mA **d.** 569 Ω **e.** 1.758 mS

10-5 **a.** 2 Ω **b.** 0.5 Ω **c.** 1 Ω **d.** 0.5 S **e.** 2 S **f.** 1 S

10-7 **a.** 569 Ω **b.** 26.36 mA **c.** 395.5 mW **d.** 225 mW, 102.3 mW, 68.2 mW **e.** 395.5 mW **f.** yes

10-9 202.8 mA, 116.6 mA, 80.6 mA, currents sum to 400 mA

10-11 **a.** 63.16 Ω **b.** 237.5 mA **c.** 150 mA

10-13 **a.** 1 mA **b.** 0 V **c.** 1 mV **d.** 1 V **e.** 1 kV **f.** 1 MV **g.** ∞V (not really) **h.** no, circuit limitation or component limitation will be reached (perhaps arcing, the breaking down of air like a lightning strike)

10-15 **a.** 20 mA **b.** 19.9998 mA **c.** 19.998 mA **d.** 19.98 mA **e.** 19.8 mA **f.** 18.8 mA **g.** 10 mA **h.** between 10 kΩ and 100 kΩ

10-17 **a.** 6 mA into top node **b.** 6 mA **c.** 6 V **d.** 36 mW

10-19 **a.** 10 mA current source with parallel 99 kΩ resistance **b.** 9.547 mA, 44.87 V

10-21 10 V in series with 10 kΩ

10-23 **a.** 36 V in series with 3 kΩ **b.** 3 mA, 18 V **c.** 9 mA, 27 V **d.** 27 V

10-25 **a.** 4 mA, 2 mA **b.** 6 mA **c.** 2 mA **d.** 12 V

Chapter 11

11-1 **a.** 1 and 2 **b.** 3 and 4 **c.** 1 with 2 with 3//4 combination with 5

11-3 5.5 kΩ

11-5 **a.** 2.4 kΩ **b.** 923 Ω **c.** 3 kΩ

11-7 **a.** 2 kΩ **b.** 0 Ω (short) **c.** 3 kΩ

11-9   **a.** 15 mA   **b.** 8 mA, 24 V
       c. 18 V, −6 V, −18 V
       d. 36 V, 24 V, 12 V   **e.** 21 mA   **f.** 3 mA

11-11   **a.** 12 mA   **b.** 4 mA, 16 V   **c.** 3 mA,
       8 mA   **d.** 9 V, 7 V, −7 V, −9 V
       e. 18 V, 2 V, 14 V

11-13   **a.** 5 kΩ   **b.** 9 mA   **c.** 1 mA, 16 V
       d. 45 mA; jeopardy: supply, 1 kΩ resistor

11-15   **a.** 60 V   **b.** 12 mA   **c.** 5 kΩ   **d.** 10 mA,
       10 V   **e.** 10 mA, 10 V   **f.** 10 mA, 10 V
       g. Thévenin model, Norton model and
       actual circuit produced the same load cur-
       rent and load voltage.

11-17   **a.** 80 μA, 40 μA, 20 μA, 10 μA
       b. 80 μA, 120 μA, 140 μA, 150 μA
       c. −150 mV

11-19   −100 mV, −90 mV, −70 mV

11-21   **a.** 5.00 V   **b.** 1.519 mA   **c.** 3.29 kΩ
       d. 495 μA, 3.37 V   **e.** 911 μA   **f.** 2.2 kΩ

Chapter 12

12-1 $3 V_p$, $-3 V_{min}$, $6 V_{pp}$,
       2 ms, 500 Hz, 2 cycles

12-3   **a.** 10 ms   **b.** 200 ns

12-5   **a.** 100 Hz   **b.** 30.3 MHz

12-7   **a.** $0 V_{dc}$ **b.** $6 V_p$, $0 V_{min}$, $6 V_{pp}$

12-9 $4 V_p$, $-2 V_{min}$, $6 V_{pp}$, $1 V_{dc}$

12-11   **a.** $30 V_p$ **b.** $12.27 V_p$ **c.** $0 V_p$

12-13   **a.** 75 mW   **b.** 12.55 mW   **c.** 0 mW

12-15   **a.** $21.21 V_p$ **b.** $20.51 V_p$ **c.** half wave,
       $20.51 V_p$, 16.7 ms   **d.** $11.39 mA_p$ **e.** half
       wave, $11.39 mA_p$, 16.7 ms   **f.** $-21.21 V_p$

12-17   **a.** negative feedback   **b.** noninverting
       c. $100 mV_{rms}$ **d.** $100 mV_{rms}$ **e.** $45.5 μA_{rms}$
       f. $214 mV_{rms}$ **g.** 314 mV noninverted
       h. 9.71 μW

12-19   **a.** $24 V_{dc}$ **b.** $6 V_{dc}$ **c.** $30 V_{dc}$ **d.** $5 mA_{dc}$
       e. 150 mW

12-21   **a.** $2 mA_{dc}$ **b.** $-1 mA_{dc}$ **c.** $2 mA_{dc}$
       d. $3 mA_{dc}$ **e.** $36 V_{dc}$

12-23   **a.** $0 mA_{dc}$ **b.** $4 mA_{dc}$ **c.** $4 mA_{dc}$
       d. $12 V_{dc}$

12-25   **a.** $0 mA_{rms}$ **b.** $10 mA_{rms}$ **c.** $10 mA_{rms}$
       d. $50 V_{rms}$ **e.** 500 mW   **f.** $70.7 V_p$
       g. $14.1 mA_p$

12-27   **a.** $-1 mA_{dc}$ **b.** $4 mA_{rms}$ **c.** $4.123 mA_{rms}$
       d. 340 mW
       e. sine wave, $4.66 mA_{max}$, $-6.66 mA_{min}$,
       period: 1 ms

12-29   **a.** $-900 mV_{dc}$ **b.** $+600 mV_{dc}$
       c. $-500 mV_{dc}$ **d.** $-800 mV_{dc}$

12-31   **a.** $1.767 V_{rms}$ **b.** $1.767 V_{rms}$
       c. $5.303 V_{rms}$

Chapter 13

13-1   **a.** 0 A   **b.** cosine wave, $628 μA_p$
       c. square wave, $4 mA_p$

13-3   **a.** ∞ Ω, 159 Ω, 15.9 Ω, 1.59 Ω, 0.16 Ω,
       0.016 Ω   **b.** open   **c.** short

13-5 $f\!\uparrow\ X_C\!\downarrow\ v_C\!\downarrow\ v_R\!\uparrow\ i_R\!\uparrow\ i_C\!\uparrow$

13-7   **a.** 159 kΩ, open, $2 V_{rms}$
       b. 1.59 Ω, short, $0 V_{rms}$

13-9   33.33 μF

13-11   300 μF

13-13 **a.** 162.4 μC **b.** 162.4 μC each
c. 1.624 V, 3.455 V, 4.921 V
note: capacitor voltages in series add

13-15 **a.** capacitor open **b.** 0 V_{dc} **c.** 6 V_{dc}
d. capacitor short **e.** 1 V_{rms} **f.** 0 V_{rms}
g. ac output

13-17 **a.** 9.8 V_{dc}, 9.1 V_{dc}, 16.8 V_{dc}
b. 9.8 V_{dc}, 1 V_{pp} sine wave
c. gain 1.45, 1.45 V_{pp}
d. 16.8 V_{dc}, 1.45 V_{pp} inverted sine wave
e. 0 V_{dc}, 1.45 V_{pp} sine wave
f. couple ac signals through (short to ac),
block dc (open to dc)

Chapter 14

14-1 **a.** 1 ms **b.** 0 V, 20 mA with C *modeled* as
a *short* **c.** 20 V, 0 mA with C *modeled* as
an *open* **d.** 20 V – 20 V·$e^{-t/1\,ms}$,
20 mA·$e^{-t/1\,ms}$ **e.** 17.3 V, 2.71 mA
f. exponential rise 0 V to 20 V in 5 ms,
exponential fall 20 mA to 0 mA in 5 ms

14-3 **a.** 1 ms **b.** 8 V, 12 mA with C *modeled* as
supply E of 8 V **c.** 20 V, 0 mA with C
modeled as an *open*
d. 20 V – 12 V·$e^{-t/1\,ms}$, 12 mA·$e^{-t/1\,ms}$
e. 18.4 V, 1.62 mA
f. exponential rise 8 V to 20 V in 5 ms,
exponential fall 12 mA to 0 mA in 5 ms

14-5 **a.** 20 V **b.** 2 ms **c.** 20 V, –10 mA (direction opposite reference) with C *modeled* as
E of 20 V **d.** 0 V, 0 mA
e. 20 V·$e^{-t/2\,ms}$, –10 mA·$e^{-t/2\,ms}$
f. 2.71 V, –1.35 mA **g.** exponential fall
20 V to 0 V in 10 ms, exponential rise
–10 mA to 0 mA in 10 ms

14-7 **a.** 4.7 ms **b.** 12 V – 12 V·$e^{-t/4.7\,ms}$
c. 7.59 V, 10.4 V, 11.4 V, 11.8 V,

11.9 V **d.** 8.65 V **e.** 6.52 ms
f. exponential rise 0 V to 12 V
in 23.5 ms and stays at 12 V

14-9 **a.** 1 ms **b.** 6 mA·$e^{-t/1\,ms}$ **c.** 2.21 mA,
812 μA, 299 μA, 110 μA, 40 μA
d. 3.64 mA **e.** 1.79 ms
f. exponential fall 6 mA to 0 mA in 5 ms
and stays at 0 mA

14-11 **a.** 2.2 ms **b.** $T \geq 10\tau$? 50 ms > 22 ms ?
yes.
c. 0-25 ms exponential rise
     –4 V to +4 V in 11 ms
     25-50 ms exponential fall
     +4 V to –4 V in 11 ms
d. 0-25 ms exponential fall
     +8 V to 0 V in 11 ms
     25-50 ms exponential rise
     –8 V to 0 V in 11 ms
e. Use Ohm's Law: $i_R(t) = v_R(t) / R$
     0-25 ms exponential fall
     +3.64 mA to 0 mA in 11 ms
     25-50 ms exponential rise
     –3.64 mA to 0 mA in 11 ms
f. 3.05 ms

14-13 4.83 ms

Chapter 15

15-1 10 μs, 1 μs, 10%, 100 kHz

15-3 11 ms

15-5 13.86 ms, 6.93 ms, 20.79 ms, 66.7%,
48.1 Hz

15-7 **a.** yes **b.** input R, feedback C **c.** 100 kΩ
d. ramp 0 V to –10 mV in 100 μs
e. inverted triangle wave, 50 mV_p, 10 kHz

15-9 **a.** yes **b.** input C, feedback R **c.** inverted
spikes based on leading and lagging
square wave edges and hitting rails of

±13 V **d.** inverted square wave, 40 mV$_p$, 10 Hz

15-11 the original input signal if the rails are not reached

15-13 2.74 kHz

15-15 **a.** reduce R, C, or B (i.e., reduce R$_i$ or raise R$_f$) **b.** decrease R2, decrease C2

15-17 223 nF

15-19 **a.** 9.90 V, 9.9 s, 49.5 s, 21.8 s
b. 1000 s, 5000 s, 2200 s

15-21 **a.** positive clipper
b. 0 V$_p$, −4.55 V$_{min}$ **c.** 0.2 V$_p$, −4.55 V$_{min}$
d. 0.7 V$_p$, −4.55 V$_{min}$ **e.** 4.3 mA$_p$
f. 70 μA$_p$, −455 μA$_{min}$ **g.** 4.55 V

15-23 **a.** rectifier circuit, positive half wave signal out
b. 5.0 V$_p$, 0 V$_{min}$ **c.** 4.8 V$_p$, 0 V$_{min}$
d. 4.3 V$_p$, 0 V$_{min}$ **e.** 4.3 mA$_p$ **f.** positive half wave, 4.3 mA$_p$, 0 V$_{min}$ **g.** 5 V

15-25 −4 V$_{dc}$

Chapter 16

16-1 **a.** 0 A **b.** cosine wave, 628 μV$_p$
c. square wave, 4 mV$_p$

16-3 **a.** 0 Ω, 1.38 Ω, 13.8 Ω, 138 Ω, 1.38 kΩ, 13.8 k Ω **b.** short **c.** open

16-5 f↑ X_L↑ v_L↑ V_R↓ i_R↓ i_L↓

16-7 **a.** 13.8 mΩ, short, 0 V$_{rms}$
b. 138 kΩ, open, 2 V$_{rms}$

16-9 **a.** 9 mH **b.** 56.5 Ω

16-11 **a.** 2 μs
b. 20 V, 0 mA with L *modeled* as *open*
c. 0 V, 20 mA with L *modeled* as *short*
d. 20 V·e$^{-t/2\,μs}$, 20 mA − 20 mA·e$^{-t/2\,μs}$
e. 2.71 V, 17.3 mA

f. spike 0 V to 20 V then exponential fall 20 V to 0 V in 10 μs, exponential rise from 0 mA to 20 mA in 10 μs

16-13 **a.** 2 μs **b.** 18 V, 2 mA with L *modeled* as *supply* I of 2 mA
c. 0 V, 20 mA with L *modeled* as a *short*
d. 18 V e$^{-t/2μs}$, 20 mA − 18 mA e$^{-t/2μs}$
e. 4.02 V, 16.0 mA
f. spike 0 V to 18 V then exponential fall 18 V to 0 V in 10 μs, exponential rise 2 mA to 20 mA in 10 μs

16-15 **a.** 2.2 μs
b. $T \geq 10\tau$? 50 μs > 22 μs ? yes.
c. 0-25 μs exponential rise
  −4 V to +4 V in 11 μs
  25-50 μs exponential fall
  +4 V to −4 V in 11 μs
d. Use Ohm's Law: $i_R(t) = v_R(t) / R$
  0-25 μs exponential rise
  −4 mA to +4 mA in 11 μs
  25-50 μs exponential fall
  +4 mA to −4 mA in 11 μs
e. 0-25 μs spike to 8 V then exponential fall +8 V to 0 V in 11 μs
  25-50 μs spike to −8 V then exponential rise −8 V to 0 V in 11 μs
f. 2.16 μs

Chapter 17

17-1 **a.** 108 V$_{rms}$ **b.** 36 V$_{rms}$ **c.** 9 mA$_{rms}$
d. 324 mW **e.** 324 mW **f.** 3 mA$_{rms}$
g. 36 kΩ

17-3 **a.** 36 V$_{rms}$ **b.** 0 Ω **c.** 36 V$_{rms}$ with 0 Ω
d. 36 V$_{rms}$ **e.** 9 mA$_{rms}$

17-5   **a.** 108 V_{rms} **b.** 216 V_{rms} **c.** 54 mA_{rms}
d. 11.66 W **e.** 11.66 W **f.** 108 mA_{rms}
g. 1 kΩ

17-7   **a.** 216 V_{rms} **b.** 0 Ω **c.** 216 V_{rms} with 0 Ω
d. 216 V_{rms} **e.** 54 mA_{rms}

17-9   Thévenin model: 20 V_{rms}, 2 kΩ;
load: 18 V_{rms}, 1 mA_{rms}

Chapter 18

18-1   **a.** 17 V_p, 16.3 V_p, 32.6 mA_p **b.** 60 Hz,
16.7 ms **c.** half-wave, 16.7 ms, 16.3 V_p;
half-wave, 16.7 ms, 32.6 mA_p
d. 10.4 mA_{dc}, 10.4 mA_{dc}, diode safe
e. 17 V_p, diode does not break down

18-3   **a.** 17 V_p, 16.3 V_p, 32.6 mA_p **b.** 60 Hz,
16.7 ms **c.** 16.7 ms << 110 ms,
2.47 V_{pp}, 713 mV_{rms} **d.** 13.8 V_{min},
15.0 V_{dc} **e.** triangle wave, 16.7 ms,
16.3 V_p, 13.8 V_{min} **f.** 4.77% **g.** 33.3 V_p,
diode does not break down

18-5   **a.** 17 V_p, 7.8 V_p, 15.6 mA_p **b.** 120 Hz,
8.33 ms **c.** full-wave, 8.33 ms, 7.8 V_p;
full-wave, 8.33 ms, 15.6 mA_p
d. 9.93 mA_{dc}, 4.97 mA_{dc}, diode safe
e. 16.3 V_p, diode does not break down

18-7   **a.** 17 V_p, 7.8 V_p, 15.6 mA_p **b.** 120 Hz,
8.33 ms **c.** 8.33 ms << 110 ms,
591 mV_{pp}, 171 mV_{rms} **d.** 7.2 V_{min},
7.5 V_{dc} **e.** triangle wave, 8.33 ms,
7.8 V_p, 7.2 V_{min} **f.** 2.3% **g.** 16.3 V_p, diode
does not break down

18-9   **a.** 17 V_p, 15.6 V_p, 31.2 mA_p **b.** 120 Hz,
8.33 ms **c.** full-wave, 8.33 ms, 15.6 V_p;
full-wave, 8.33 ms, 31.2 mA_p
d. 19.9 mA_{dc}, 9.93 mA_{dc}, diode safe
e. 16.3 V_p, diode does not break down

18-11 **a.** 17 V_p, 15.6 V_p, 31.2 mA_p **b.** 120 Hz,
8.33 ms **c.** 8.33 ms << 110 ms,
1.18 mV_{pp}, 341 mV_{rms} **d.** 14.4 V_{min},
15 V_{dc} **e.** triangle wave, 8.33 ms,
15.6 V_p, 14.4 V_{min} **f.** 2.3% **g.** 16.3 V_p,
diode does not break down

18-13 **a.** 6.2 V_{dc}, 12.4 mA_{dc} **b.** 270 Ω
c. 15.6 V_p, 9.4 V_p, 34.8 mA_p, 22.4 mA_p,
327 mW_p **d.** 14.3 V_{min}, 8.1 V_{min},
30 mA_{min}, 17.6 mA_{min} (I_{Zmin} satisfied)
243 mW_{min} **e.** 14.95 V_{dc}, 8.75 V_{dc},
32.4 mA_{dc}, 20 mA_{dc}, 284 mW_{dc}, select
half-watt R_s

18-15 **a.** 5.4 V_{dc}, 270 mA_{dc}, 1.46 W_{dc}
b. 9 mA_{dc}, **c.** 21 mA_{min} (I_{Zmin} satisfied)
d. 2.59 W_{dc}, power dissipation okay but
could heat sink to operate cooler

18-17 1.10, 6.82 V_{dc}, 8.32 V_{dc}, 455 mA_{dc},
152 μA_{dc}, rail voltage marginal

18-19 1.4 Ω

18-21 1.49, 9.24 V_{dc}, −1.0, −9.24 V_{dc}

18-23 **a.** 12.0 V_{dc}, 480 mA_{dc}, 5.76 W_{dc}
b. 3.84 W_{dc}, 180°C **c.** No, 18.2°C/W
d. 0.4 mV_{pp}, 0.3%

18-25 **a.** 720 Ω **b.** 55 mA_{dc} **c.** 550 mW_{dc}, 123°C
d. Yes but too close to specification, heat
sinking recommended

Chapter 19

19-1   40 mV voltage source

19-3   50 μA current source

19-5   20 V_{rms} with open, 5 mA_{rms} with short,
25 mW with matched load of 4 kΩ

19-7   Maximum values with wiper arm at top:
6 V_p, 1 mA_p, 3 mW; minimum values

with wiper arm at the bottom: 0 V, 0 A, 0 W

19-9 1.0 V_{rms}, 0.1 mA_{rms}, ideal VCVS with dependent voltage source of $10 \times v_{in}$

19-11 10001 (about 10^4), 20 GΩ, 7.5 mΩ

19-13 −10 mV_p, 10 mV_p

19-15 Ideal voltage to current converter op amp circuit with a 5 kΩ feedback resistor

19-17 443 V_{rms}, 55.6 mA_{rms}, 24.6 mW

19-19 Ideal current amplifier op amp circuit with $R_f = 99\ R_i$ (one design: an R_i value of 1 kΩ and an R_f value of 99 kΩ)

19-21 1.0 mA_{dc}, 0.5 mA_{dc} 22.5×10^3, 108Ω, 2.43 MΩ, single BJT input resistance is 8.1 kΩ therefore there is a300 factor increase (that is, 2β) producing an input impedance into the MΩ range

Chapter 20

20-1 omitted

20-3 The gain is cut in half

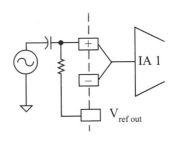

20-5

20-7 See Figure 20-8. The gain for IA1 = 8. The gain for IA2 = 2.

20-9

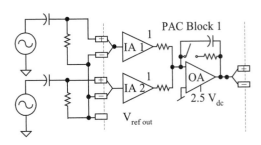

20-11 Figure 20-13. $R_f = 330$ kΩ, $R_i = 82$ kΩ

20-13 Figure 20-16. $R_f = 330$ kΩ, $R_i = 82$ kΩ, R = 620 kΩ, C = 100 pF

20-15 omitted

20-17 V+ = 5 V_{dc}, V− = −5 V_{dc}, $C_t = 2.2$ nF, $R_t = 2.2$ kΩ to pin 7, FSK pin to common or V+, $R_{pull-up} = 5.6$ kΩ

20-19 V+ = 5 V_{dc}, V− = −5V_{dc}, $C_t = 1$ µF, $R_t = 9.1$ kΩ, 180 Ω pin 13 to pin 14, $R_{mult} = 8.8$ kΩ

20-21 $R_{mult} = 18$ kΩ. Place an 820 Ω resistor between the control voltage and the AM pin (pin 1). Place a 3.3 kΩ resistor between the AM pin (pin 1) and common.

Appendix B

Systems of Units

The English system of units is given in Table 1. Note: length is also expressed in other units such as yard and mile.

Table 1 English system of units

Quantity	Unit	Abbreviation
length	foot	ft
mass	slug	
volume	gallon	gal
time	second	s
temperature	Fahrenheit	F
force	pound	lb
charge	coulomb	C
energy	foot-pound	ft·lb

The Metric CGS (centimeter-gram-second) system of units is given in Table 2.

Table 2 Metric CGS system of units

Quantity	Unit	Abbreviation
length	centimeter	cm
mass	gram	g
volume	milliliter (cm^3)	ml
time	second	s
temperature	Centigrade	C
force	dyne	dyn
charge	coulomb	C
energy	Newton-meter	N-m

The Metric MKS (meter-kilogram-second) system of units is given in Table 3.

Table 3 Metric MKS system of units

Quantity	Unit	Abbreviation
length	meter	m
mass	kilogram	kg
volume	liter	l
time	second	s
temperature	Centigrade	C
force	Newton	N
charge	coulomb	C
energy	Newton-meter	N-m

The International System (SI) system of units is given in Table 4. Note that the SI units are essentially the Metric MKS units with (1) temperature expressed as absolute temperature in degrees Kelvin ($K° = 273.15° + C°$) and energy expressed in Joules (J = N-m).

Table 4 International System (SI) of units

Quantity	Unit	Abbreviation
length	meter	m
mass	kilogram	kg
volume	liter	l
time	second	s
temperature	Kelvin	K
force	Newton	N
charge	coulomb	C
energy	Joule	J

Appendix C

Resistor Color Code

			Multiplier Band				
silver	gold	black	brown	red	orange	yellow	green
0.10	**1.0**	**10**	**100**	**1 000**	**10 000**	**100 000**	**1 000 000**
0.11	1.1	11	110	1 100	11 000	110 000	1 100 000
0.12	**1.2**	**12**	**120**	**1 200**	**12 000**	**120 000**	**1 200 000**
0.13	1.3	13	130	1 300	13 000	130 000	1 300 000
0.15	**1.5**	**15**	**150**	**1 500**	**15 000**	**150 000**	**1 500 000**
0.16	1.6	16	160	1 600	16 000	160 000	1 600 000
0.18	**1.8**	**18**	**180**	**1 800**	**18 000**	**180 000**	**1 800 000**
0.20	2.0	20	200	2 000	20 000	200 000	2 000 000
0.22	**2.2**	**22**	**220**	**2 200**	**22 000**	**220 000**	**2 200 000**
0.24	2.4	24	240	2 400	24 000	240 000	
0.27	**2.7**	**27**	**270**	**2 700**	**27 000**	**270 000**	
0.30	3.0	30	300	3 000	30 000	300 000	
0.33	**3.3**	**33**	**330**	**3 300**	**33 000**	**330 000**	
0.36	3.6	36	360	3600	36 000	360 000	
0.39	**3.9**	**39**	**390**	**3 900**	**39 000**	**390000**	
0.43	4.3	43	430	4 300	43 000	430 000	
0.47	**4.7**	**47**	**470**	**4 700**	**47 000**	**470 000**	
0.51	5.1	51	510	5 100	51 000	510 000	
0.56	**5.6**	**56**	**560**	**5 600**	**56 000**	**560 000**	
0.62	6.2	62	620	6 200	62 000	620 000	
0.68	**6.8**	**68**	**680**	**6 800**	**68 000**	**680 000**	
0.75	7.5	75	750	7 500	75 000	750 000	
0.82	**8.2**	**82**	**820**	**8200**	**82 000**	**820 000**	
0.91	9.1	91	910	9100	91 000	910 000	

American Wire Gauge (AWG) Table

Solid Copper Wire at 20°C

AWG Size	Diameter mm	Cross-section Area mm^2	Cross-section Area CM*	Resistance Ω/1000 feet	Safe Current Capacity Amps
0000 or 4/0	11.7	107	211 000	0.0490	230
000 or 3/0	10.4	85.0	168 000	0.0618	200
00 or 2/0	9.27	67.4	133 000	0.0780	175
0 or 1/0	8.25	53.5	105 000	0.0983	150
2	6.54	33.6	66 400	0.156	115
4	5.19	21.2	41 700	0.249	85
6	4.12	13.3	26 200	0.395	65
8	3.26	8.37	16 500	0.628	50
10	2.59	5.26	10 400	0.999	30
12	2.05	3.31	6 530	1.59	20
14	1.63	2.08	4 110	2.53	15
16	1.29	1.30	2 580	4.02	10
18	1.02	0.823	1 620	6.39	5
20	0.813	0.519	1 020	10.2	3
22	0.643	0.324	640	16.1	2
24	0.511	0.205	404	25.7	1.3
26	0.404	0.128	253	41.0	0.75
28	0.320	0.0804	159	65.3	
30	0.254	0.0507	100	104	
32	0.203	0.0324	64	162	
34	0.160	0.0201	40	261	
36	0.127	0.0127	25	415	
38	0.102	0.00811	16	648	
40	0.079	0.00487	9	1080	

*Circular mils (CM), CM = (diameter in mils)^2, 1 CM = (1 mil)^2, 1 mil = 0.001 inch

Index